T0297601

CELLULAR INTERNET OF THINGS

CELLULAR INTERNET OF THINGS

FROM MASSIVE DEPLOYMENTS TO CRITICAL 5G APPLICATIONS

SECOND EDITION

OLOF LIBERG
Ericsson Business Unit Networks

MÅRTEN SUNDBERG
Ericsson Business Unit Networks

Y.-P. ERIC WANG
Ericsson Research

JOHAN BERGMAN
Ericsson Business Unit Networks

JOACHIM SACHS
Ericsson Research

GUSTAV WIKSTRÖM
Ericsson Research

Academic Press is an imprint of Elsevier
125 London Wall, London EC2Y 5AS, United Kingdom
525 B Street, Suite 1650, San Diego, CA 92101, United States
50 Hampshire Street, 5th Floor, Cambridge, MA 02139, United States
The Boulevard, Langford Lane, Kidlington, Oxford OX5 1GB, United Kingdom

Copyright © 2020 Elsevier Ltd. All rights reserved.

No part of this publication may be reproduced or transmitted in any form or by any means, electronic or mechanical, including photocopying, recording, or any information storage and retrieval system, without permission in writing from the publisher. Details on how to seek permission, further information about the Publisher's permissions policies and our arrangements with organizations such as the Copyright Clearance Center and the Copyright Licensing Agency, can be found at our website: www.elsevier.com/permissions.

This book and the individual contributions contained in it are protected under copyright by the Publisher (other than as may be noted herein).

Notices
Knowledge and best practice in this field are constantly changing. As new research and experience broaden our understanding, changes in research methods, professional practices, or medical treatment may become necessary.

Practitioners and researchers must always rely on their own experience and knowledge in evaluating and using any information, methods, compounds, or experiments described herein. In using such information or methods they should be mindful of their own safety and the safety of others, including parties for whom they have a professional responsibility.

To the fullest extent of the law, neither the Publisher nor the authors, contributors, or editors, assume any liability for any injury and/or damage to persons or property as a matter of products liability, negligence or otherwise, or from any use or operation of any methods, products, instructions, or ideas contained in the material herein.

Library of Congress Cataloging-in-Publication Data
A catalog record for this book is available from the Library of Congress

British Library Cataloguing-in-Publication Data
A catalogue record for this book is available from the British Library

ISBN: 978-0-08-102902-2

For information on all Academic Press publications visit our website at https://www.elsevier.com/books-and-journals

Publisher: Mara Conner
Acquisition Editor: Tim Pitts
Editorial Project Manager: Peter Adamson
Production Project Manager: Nirmala Arumugam
Cover Designer: Greg Harris

Typeset by TNQ Technologies

Contents

Biography

Olof Liberg is a Master Researcher at Ericsson Business Unit Networks. After studies in Sweden, USA, Germany, and Switzerland, he received a bachelor's degree in Business and Economics and a master's degree in Engineering Physics, both from Uppsala University. He joined Ericsson in 2008 and has specialized in the design and standardization of cellular systems for machine-type communications and Internet of Things (IoT). He has, over the years, actively contributed to the work in several standardization bodies such as 3GPP, ETSI and the MulteFire Alliance. He was the chairman of 3GPP TSG GERAN and its Working Group 1, during the 3GPP study on new radio access technologies for IoT leading up to the specification of EC-GSM-IoT and NB-IoT.

Mårten Sundberg is a researcher at Ericsson Business Unit Networks, with a previous position as a Senior Specialist in GSM radio access technology. He joined Ericsson in 2005 after receiving his master's degree in Engineering Physics from Uppsala University. As Rapporteur of the 3GPP Work Item on EC-GSM-IoT he was leading the technical work to standardize the new GSM-based feature dedicated for IoT. In 2016, he started leading the work toward URLLC for LTE, being a Rapporteur for the Work Item introducing shortened TTI and shorter processing times. Apart from being active in the 3GPP standardization body, Mårten has also worked for many years in ETSI, harmonizing radio requirements in Europe.

Y.–P. Eric Wang is a Principal Researcher at Ericsson Research. He holds a PhD degree in electrical engineering from the University of Michigan, Ann Arbor. In 2001 and 2002, he was a member of the executive committee of the IEEE Vehicular Technology Society and served as the society's Secretary. Dr. Wang was an Associate Editor of the IEEE Transactions on Vehicular Technology from 2003 to 2007. He is a technical leader in Ericsson Research in the area of IoT connectivity. Dr. Wang was a corecipient of Ericsson's Inventors of the Year award in 2006. He has contributed to more than 150 US patents and more than 50 IEEE articles.

Johan Bergman is a Master Researcher at Ericsson Business Unit Networks. He received his master's degree in Engineering Physics from Chalmers University of Technology in Sweden. He joined Ericsson in 1997, initially working with baseband receiver algorithm design for 3G cellular systems. Since 2005, he has been working with 3G/4G physical layer standardization in 3GPP TSG RAN Working Group 1. As Rapporteur of the 3GPP TSG RAN Work Items on LTE for machine-type communications in Releases 13, 14, 15, and 16, he has led the technical work to standardize the new LTE-based features dedicated for IoT. He was a corecipient of Ericsson's Inventor of the Year award for 2017.

Joachim Sachs is a Principal Researcher at Ericsson Research. After studies in Germany, Norway, France and Scotland he received a diploma degree in electrical engineering from Aachen University (RWTH), Germany, and a PhD degree from the Technical University of Berlin. He joined Ericsson in 1997 and has worked on a variety of topics in the area of wireless communication systems, and has contributed to the standardization of 3G, 4G and 5G. Since 1995 he has been active in the IEEE and the German VDE Information Technology Society, where he is currently co-chair of the technical committee on communication networks. In 2009 he was a visiting scholar at Stanford University, USA.

Gustav Wikström is a Research Leader at Ericsson Research in the area of Radio Network Architecture and Protocols. He has a background in Experimental Particle Physics and received his PhD from Stockholm University in 2009, after Master studies in Engineering Physics in Lund, Uppsala, and Rennes. After Post-doc studies in Geneva, he joined Ericsson in 2011. He has been driving the evolution of network performance tools and studies and worked with WLAN enhancements toward IEEE. Until 2018 he was the driver of latency and reliability improvements (URLLC) in LTE and NR.

Preface

The Internet of Things is transforming the information and communications technology industry. It embodies the vision of connecting virtually anything with everything and builds on a global growth of the overall number of connected devices. To support, and further boost, this growth the *Third Generation Partnership Project* (3GPP) standards development organization has in its Releases 13, 14 and 15 developed the technologies *Extended Coverage GSM Internet of Things* (EC-GSM-IoT), *LTE for Machine-Type Communications* (LTE-M), *Narrowband Internet of Things* (NB-IoT) and *Ultra-Reliable and Low Latency Communications* (URLLC). These technologies provide cellular services to massive number of IoT devices with stringent requirements in terms of connection density, energy efficiency, reachability, reliability and latency.

This book sets out to introduce, characterize and in detail describe these new technologies that together are defining the concept known as the *Cellular Internet of Things.* After an introduction to the book in Chapter 1, Chapter 2 presents the 3GPP and the *MulteFire Alliance* (MFA) standardization organizations. Chapter 2 also gives an overview of the early work performed by 3GPP to support IoT on 2G, 3G and 4G, and introduces 3GPP's most recent work on the 5G *New Radio*. Chapters 3 to 8 focus on the work 3GPP successfully has performed on technologies supporting *massive machine-type communication* (mMTC). Chapters 3, 5 and 7 present descriptions of the physical layer design and the specified procedures for each of EC-GSM-IoT, LTE-M and NB-IoT. Chapters 4, 6 and 8 in detail evaluate the performance of each of these three technologies and, when relevant, compare it to the 5G mMTC performance requirements. Chapters 9 to 12 provide the design details and performance of LTE and 5G New Radio URLLC. The performance evaluations compare the achieved performance to the set of 5G performance requirements agreed for critical machine-type communication (cMTC) in terms of reliability and latency. Chapter 13 discusses the enhancements 3GPP Release 15 introduced on LTE for the support of *drone communication.*

Chapters 14 and 15 turn the attention from licensed spectrum operation which is commonly associated with the 3GPP technologies to wireless IoT systems operating in unlicensed frequency bands. Chapter 14 describes popular short- and long-range wireless technologies for providing IoT connectivity. Chapter 15 presents the work done by the MFA on adapting LTE-M and NB-IoT for operation in unlicensed frequency bands.

Chapter 16 summarizes the descriptions and performance evaluations provided in the earlier chapters and gives the reader guidance on how to best select an IoT system for meeting mMTC and cMTC market demands.

Chapter 17 provides an overall picture of the IoT technology. It is shown that the wireless connectivity is only one among many vital technical components in an IoT system. Internet technologies for IoT and the industrial IoT are discussed in this chapter. Chapter 18 wraps up the book with a look into the future and discusses where the cellular industry is turning its attention when continuing evolving 5G.

Acknowledgments

We would like to thank all our colleagues in Ericsson that have contributed to the development of the wireless IoT technologies described in this book. Without their efforts this book would never have been written. In particular we would like to express our gratitude to our friends Johan Sköld, John Diachina, Björn Hofström, Zhipeng Lin, Ulf Händel, Nicklas Johansson, Xingqin Lin, Uesaka Kazuyoshi, Sofia Ek, Andreas Höglund, Björn Nordström, Emre Yavuz, Håkan Palm, Mattias Frenne, Oskar Mauritz, Tuomas Tirronen, Santhan Thangarasa, Anders Wallén, Magnus Åström, Martin van der Zee, Bela Rathonyi, Anna Larmo, Johan Torsner, Erika Tejedor, Ansuman Adhikary, Yutao Sui, Jonas Kronander, Gerardo Agni Medina Acosta, Chenguang Lu, Henrik Rydén, Mai-Anh Phan, David Sugirtharaj, Emma Wittenmark, Laetitia Falconetti, Florent Munier, Niklas Andgart, Majid Gerami, Talha Khan, Vijaya Yajnanarayana, Helka-Liina Maattanen, Ari Keränen, Vlasios Tsiatsis, Viktor Berggren, Torsten Dudda, Piergiuseppe Di Marco, Janne Peisa, and Mohammad Mozaffari for their help in reviewing and improving the content of the book.

For their generous help with simulations and data analysis we would like to specially thank our skilled colleagues Kittipong Kittichokechai, Alexey Shapin, Osama Al-Saadeh, and Ikram Ashraf.

Further we would like to thank our colleagues in 3GPP for contributing to the successful standardization of EC-GSM-IoT, LTE-M, NB-IoT and URLLC. In particular we would like to thank Alberto Rico-Alvarino, Chao Luo, Gus Vos, Matthew Webb and Rapeepat Ratasuk for their help in reviewing the technical details of the book.

Finally, we would like to express our admiration for our families; Ellen, Hugo, and Flora; Matilda; Katharina, Benno, Antonia, and Josefine; Wan-Ling, David, Brian, Kuo-Hsiung, and Ching-Chih; Minka, Olof, and Dag. It is needless to say how important your inspiration and support have been during the long process of writing this book.

CHAPTER

1

The Internet of Things

OUTLINE

Abstract

This chapter introduces the overall content of the book. It contains an introduction to the *massive* and *critical machine-type communications* (mMTC, cMTC) categories of use cases, spanning a wide range of applications. When discussing these applications, consideration is given to the service requirements associated with mMTC and cMTC for example in terms of reachability and reliability. The chapter introduces the concept of the *cellular Internet of Things* which is defined by the *Third Generation Partnership Project* (3GPP) technologies: Extended Coverage Global System for Mobile Communications Internet of Things (EC-GSM-IoT), *Narrowband Internet of Things* (NB-IoT), *Long-Term Evolution for Machine-Type Communications* (LTE-M) and *ultra-reliable and low latency communications* (URLLC). The final part of the chapter looks beyond the 3GPP technologies and discusses a range of solutions that provides IoT connectivity in unlicensed spectrum.

1.1 Introduction

The *Internet of Things* (IoT) is part of a transformation that is affecting our entire society: industries, consumers and the public sector. It is an enabler in the broader digital transformation of the management of physical processes. It provides better insights and allow for

Cellular Internet of Things, Second Edition
https://doi.org/10.1016/B978-0-08-102902-2.00001-7

© 2020 Elsevier Ltd. All rights reserved.

more efficient operation. The IoT provides the capability to embed electronic devices into physical world objects and create smart objects that allow us to interact with the physical world by means of sensing or actuation. IoT enables networking among smart objects, applications and servers.

Fig. 1.1 depicts the instance of an IoT system. On the left-hand side there are physical assets – like machines, lights, meters; on the right-hand side there are applications interacting with the physical world. There can be a variety of different applications. If we assume as example, that the physical assets are sensors that monitor the vehicle flow on a street at different locations in a city, then the application could be to monitor traffic flows throughout the city in a traffic control center. In case that the physical assets include traffic lights, which can be activated via actuators, then the application could also steer the red-green periods of individual traffic lights, e.g. based on the observed traffic flow. This shows a simple example of digital transformation. A traffic infrastructure with traffic lights with fixed configuration is transformed into a smarter traffic infrastructure, where insights about the system states are collected and smart decisions are being taken and executed within the infrastructure. The applications themselves are running in the digital domain. A representation of the physical system (i.e. streets in the city) is created, based on a model (like a street map), and it is updated with information from the traffic sensors. The management and configuration of the traffic infrastructure (i.e. the traffic lights) is made in the traffic center and the execution is transferred back to the physical world, by means of switches in the traffic lights that steer the red-green phases.

The IoT system is the enabler for the service in the above example. IoT devices are connected to the physical assets and interact with the physical world via sensors and actuators. The IoT system connects the IoT devices to the specific application of the service and

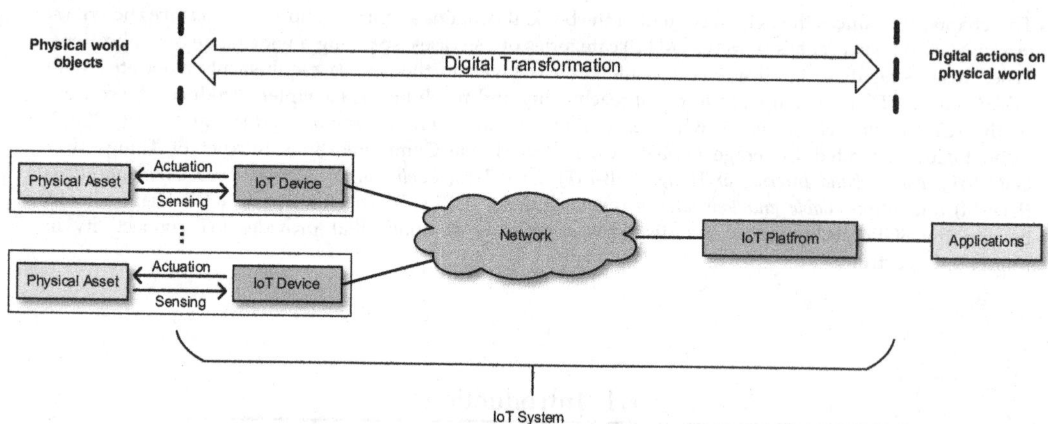

FIG. 1.1 IoT system providing connectivity, services and a digital representation of the physical world.

enables the application to control the physical assets via actuators connected to IoT devices. The IoT platform provides common functionality, which includes device and object identification and addressing, security functions, and management of IoT devices. The IoT connectivity, which is the focus of this book, provides a generic platform that can be used by many different services, as shown in Fig. 1.2.

1.2 IoT communication technologies

A significant number of communication technologies have been developed over the last two decades with significant impact on the IoT. In particular, *machine-to-machine* (M2M) communication solutions were developed to connect devices with applications. Most M2M communication solutions are purpose-build and designed to satisfy a very particular application and communication needs. Examples are connectivity for remote-controlled lighting, baby monitors, electric appliances, etc. For many of those systems the entire communication stack has been designed for a single purpose. Even if it enables, in a wider sense, an environment with a wide range of connected devices and objects, it is based on M2M technology silos, usually without end-to-end IP connectivity and instead via proprietary networking protocols. This is depicted on the left-hand side of Fig. 1.3. It is quite

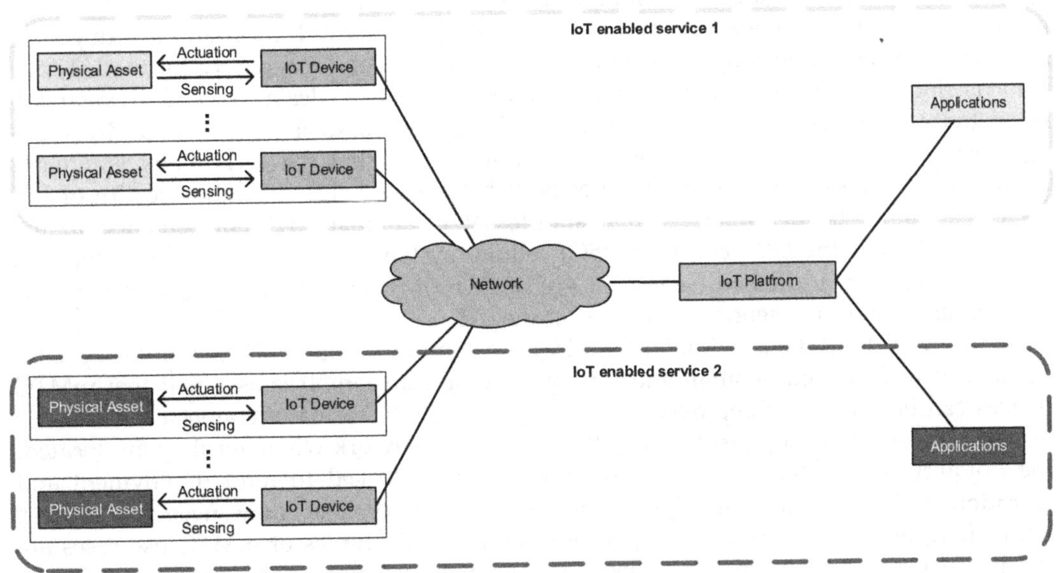

FIG. 1.2 IoT system as a platform to enable many services.

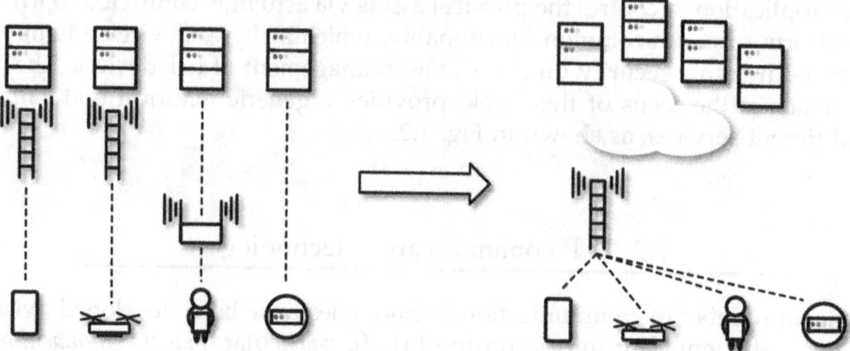

FIG. 1.3 From M2M silos to the IoT.

different from the vision of the IoT depicted on the right-hand side in Fig. 1.3, which is based on a common and interoperable IP-based connectivity framework for connecting devices and smart objects, which enables the IoT at full scale.

1.2.1 Cellular IoT

In recent years the *Third Generation Partnership Project* (3GPP) have evolved their cellular technologies to target a wide variety of IoT use cases. The second, third and fourth generations cellular communication systems provided since earlier connectivity for the IoT, but 3GPP is since its Release 13 developing technologies that by design provide *cellular IoT* connectivity. The 3GPP standardization of cellular networks is trying to address the requirements of novel IoT use case, in order to ensure that the technology standards evolution is addressing future market needs. It has become clear that the breadth of IoT use cases cannot be described with a simple set of cellular IoT requirements. In the standardization of the *fifth generations* (5G) cellular system, three requirements categories were defined to be addressed (see Fig. 1.4) [1]. Two of them are focused on *machine-type communication* (MTC), essentially addressing the IoT.

Massive MTC (mMTC) is defined for addressing communication of large volumes of simple devices with a need of small and infrequent data transfers. It is assumed that mMTC devices can be massively deployed, so that the scalability to many connected devices is needed, as well as the support to reach them with the network wherever they are located. The ubiquity of the deployment in combination with a need to limit deployment and operation cost motivates ultra-low complex IoT devices that may need to support non-rechargeable battery powered operation for years. Examples of mMTC use cases are utilities metering and monitoring, fleet management, telematics and sensor sharing in the automotive industry segment, or inventory management, asset tracking and logistics

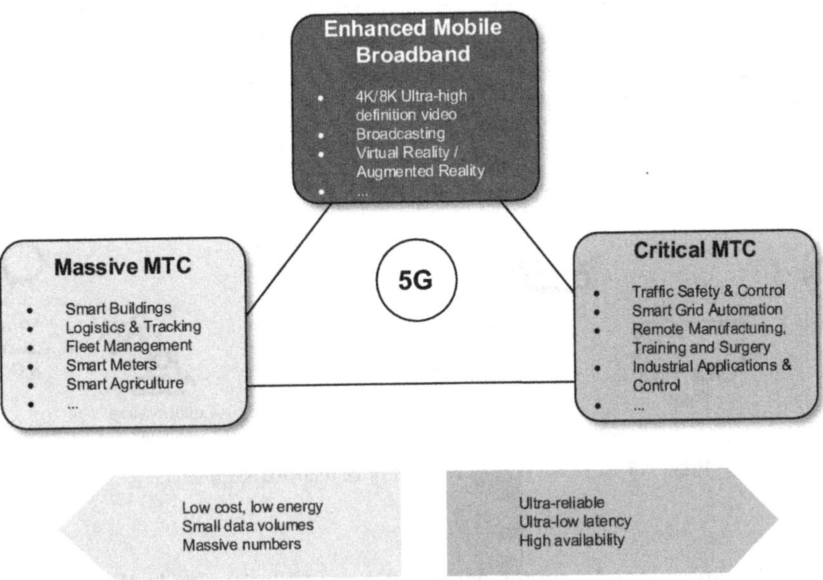

FIG. 1.4 Requirements on 5G.

in the manufacturing segment. 3GPP has specified the *Extended Coverage Global System for Mobile Communications Internet of Things* (EC-GSM-IoT), *Narrowband Internet of Things* (NB-IoT) and *Long-Term Evolution (LTE) for Machine-Type Communications* (LTE-M) technologies for support of mMTC. These solutions are thoroughly examined in Chapters 3–8.

Critical MTC (cMTC) is defined for addressing demanding IoT use cases with very high reliability and availability, in addition to very low latencies. Examples of cMTC use cases exist in various fields. In the automotive area, remote driving falls into this category, but also real-time sensor sharing, autonomous driving, cooperative maneuvers and cooperative safety. Other examples are teleprotection and distribution automation in a smart grid, automated guided vehicles in manufacturing or remote control of vehicles and equipment in smart mining. The cMTC requirement category is in the 3GPP standardization also referred to as *ultra-reliable and low latency communication* (URLLC). In this book we use the term URLLC for the technologies supporting the cMTC use cases. Chapters 9–12 provides the design details and performance of LTE URLLC and NR URLLC. The performance evaluations compare the achieved performance to the set of 5G performance requirements agreed for cMTC in terms of reliability and latency. It is shown that both LTE URLLC and NR URLLC meets all the minimum requirements, while the NR technology is shown to excels in terms of spectral efficiency, minimum achieved latency and deployment flexibility.

Even the categorization of mMTC and cMTC is rather coarse and does not address all IoT use cases. To better define the requirements that a certain use case puts on the devices and the supporting network, the information and communications technology (ICT) industry

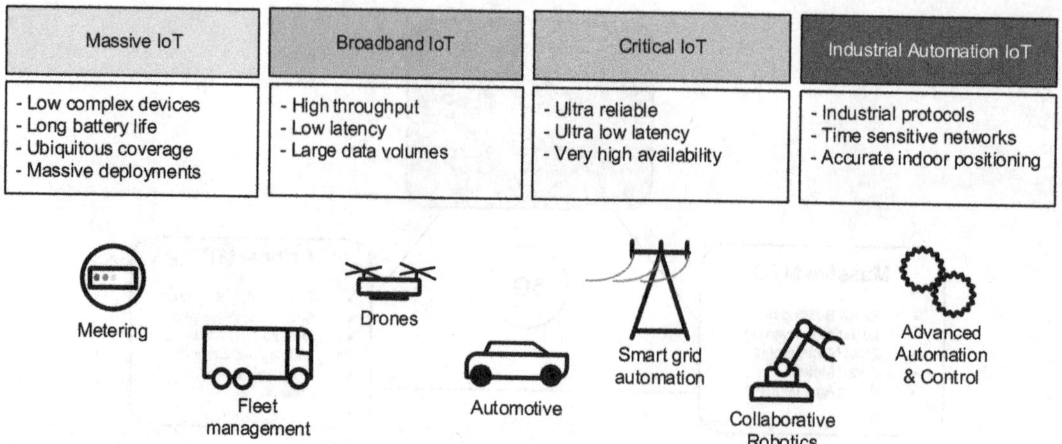

FIG. 1.5 Ericsson categorization of cellular IoT segments [2].

leader Ericsson has introduced a novel classification of Cellular IoT in the segments *massive IoT*, *broadband IoT*, *critical IoT* and *industrial automation IoT* as described in [2] and shown in Fig. 1.5.

Massive IoT and critical IoT are equivalent to mMTC and cMTC, respectively. Broadband IoT covers cellular IoT features, that are not explicitly addressed so far in standardization as a category. It applies to use cases with similar objectives as for massive IoT, in terms of battery efficient operation, device complexity and wide-area availability, but where in addition very high data rates are needed. To this end a certain comprise in terms of battery usage and device complexity is required to cater for high throughput. Examples are the transmission of high-definition maps for (semi-)autonomous vehicles, large software updates, computer vision systems, augmented or mixed reality systems, advanced wearables or aerial and ground vehicles. *Drone communication* is an example of an important broadband IoT use case that in recent years have grown quickly in importance and have a potential to bring significant social-economic benefits. Drones are increasingly used for aiding search, rescue, and recovery missions during or in the aftermath of natural disasters. Chapter 13 introduces the work 3GPP has done for the support for drone communication in LTE.

Industrial automation IoT covers cellular IoT features, that provide capabilities required for some industrial segments, in particular in the area of industrial automation. These are typically additional functional capabilities, rather than novel performance requirements. Often those features are needed in clearly localized solutions, like a cellular IoT system provided within a factory. An example for industrial automation IoT is the support for non-public network solutions, for native Ethernet transport with advanced LAN services and optimizations for *time sensitive networking*. Other examples are the support for precise time-synchronization that is provided from a time master clock over the cellular system to end-devices, or ultra-precise positioning. Chapter 16 is providing further insights into industrial automation IoT when discussing the concept of *industrial internet of things*.

1.2.2 Technologies for unlicensed spectrum

The 3GPP cellular technologies are not the only solutions competing for IoT traffic. Well-known technologies such as *Bluetooth* and *Wi-Fi* also serve as bearers for MTC traffic. A distinction between the group of cellular technologies and Bluetooth and Wi-Fi is that the former is traditionally intended for operation in licensed spectrum while the latter two belong to the group of systems operating in unlicensed spectrum, in so-called license exempt frequency bands.

Licensed spectrum corresponds to a part of the public frequency space that has been licensed by national or regional authorities to a private company, typically a mobile network operator, under the condition of providing a certain service to the public such as cellular connectivity. At its best, a licensed frequency band is globally available, which is of considerable importance for technologies aiming for worldwide presence. The huge success of GSM is, for example, to a significant extent built around the availability of the GSM 900 MHz band in large parts of the world. Licensed spectrum is, however, commonly associated with high costs, and the media frequently give reports of spectrum auctions bringing in significant incomes to national authorities across the world.

Unlicensed spectrum, on the other hand, corresponds to portions of the frequency space that can be said to remain public and therefore free of licensing costs. Equipment manufacturers using this public spectrum must meet a set of national or regional technical regulations for technologies deployed within that spectrum. Among of the most popular license exempt frequency bands are the so-called *Industrial, Scientific and Medical* (ISM) bands identified in article 5.150 of the ITU Radio Regulations [3]. Regional variations for some of these bands exist, for example, in the frequency range around 900 MHz while other bands such as the *2.4 GHz band* can be said to be globally available. In general, the regulations associated with license exempt bands aim at limiting harmful interference to other technologies operating within as well as outside of the unlicensed band.

Bluetooth and Wi-Fi, and thereto related technologies such as *Bluetooth Low Energy*, *ZigBee*, and *Wi-Fi Halow*, commonly use the ISM bands to provide relatively short-range communication, at least in relation to the cellular technologies. Bluetooth can be said to be part of a *wireless personal area network* while Wi-Fi provides connectivity in a *wireless local area network* (WLAN). In recent years, a new set of technologies have emerged in the category of *low power wide area networks (LPWAN)*. These are designed to meet the regulatory requirements associated with the ISM bands, but in contrast to WPAN and WLAN technologies they provide long-range connectivity, which is an enabler for supporting wireless devices in locations where WPAN and WLAN systems cannot provide sufficient coverage.

Chapter 14 reviews the most important license exempt spectrum regulations and introduces some of the most popular and promising IoT technologies for unlicensed spectrum operation. Chapter 15 describes the 3GPP based IoT systems specified by the *MulteFire Alliance* (MFA). The MFA is a standardization organization that develops wireless technologies for operation in unlicensed and shared spectrum. The MFA specifications are using the

3GPP technologies as baseline and add adaptations needed for operation in unlicensed spectrum.

1.3 Outline of the book

The content of the book is distributed over 18 chapters. Chapter 2 introduces the 3GPP and the MFA standardization forums. It presents an overview of the early work performed by 3GPP to support IoT on *2G*, *3G* and *4G*. The last part of the chapter provides an introduction to 3GPPs most recent work on 5G.

Chapters 3 to 8 focuses fully on the work 3GPP so successfully have carried out on technologies supporting mMTC. Chapters 3, 5 and 7 presents descriptions of the physical layer design and the higher and lower layer procedures for each of EC-GSM-IoT, LTE-M and NB-IoT. Chapters 4, 6 and 8 in detail evaluate the performance of each of the three technologies. For LTE-M and NB-IoT the performance evaluations show that the systems in all aspects meets the 5G performance requirements for mMTC services defined by 3GPP and the *International Telecommunications Union Radiocommunication sector (ITU-R)*.

Chapters 9–12 provides the design details and performance of LTE and NR URLLC. The performance evaluations compare the achieved performance to the set of 5G performance requirements agreed for cMTC in terms of reliability and latency. It is shown that both LTE and NR support the 5G requirements from ITU-R on cMTC in terms of reliability and latency. NR is shown to be more flexible in its design than LTE and offers a higher performance.

Chapter 13 discusses the enhancements 3GPP Release 15 introduced on LTE for the support of drone communication. It is described how LTE efficiently can support the reliable command-and-control communications required for drone operation.

Chapter 14 and 15 turns the attention from licensed spectrum operation associated with 3GPP and gives full attention to operation in unlicensed frequency bands. Chapter 14 describes the most popular short- and long-range wireless technologies for providing IoT connectivity in unlicensed spectrum. Chapter 15 presents the work of the MFA on adapting LTE-M and NB-IoT for operation in unlicensed spectrum bands. In this book we refer to these technologies as *LTE-M-U* and *NB-IoT-U*, where the 'U' stands for unlicensed.

Chapter 16 summarizes the descriptions and performance evaluations provided in the earlier chapters and gives the reader an insight in how to best select an IoT system for meeting mMTC and cMTC demands. This guidance is based on the technical capabilities and performance of each of the systems presented in the book.

Chapter 17 provides the reader with an overall picture of the IoT. It is shown that the wireless connectivity is only one among many vital technical components in an IoT system. IoT transfer protocols and the IoT application framework are discussed in this chapter.

Chapter 18 wraps up the book with a look into the future and discusses where the cellular industry is turning its attention when continuing evolving 5G (Fig. 1.6).

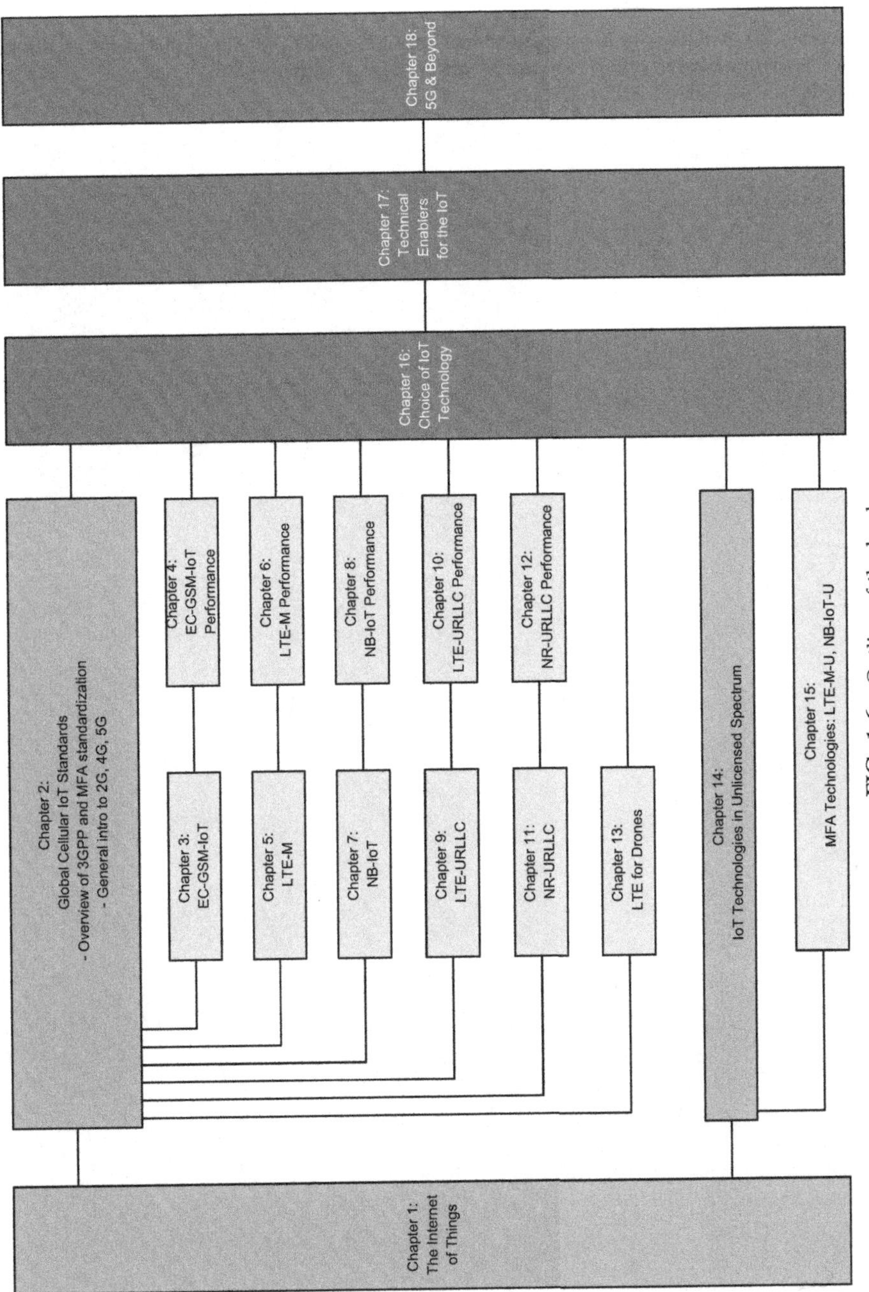

FIG. 1.6 Outline of the book.

References

[1] ITU-R. Report ITU-R M.2410, Minimum requirements related to technical performance for IMT-2020 radio interfaces(s), 2017.
[2] Ericsson. Cellular IoT evolution for industry digitalization, 2018. Website, [Online]. Available at: https://www.ericsson.com/en/white-papers/cellular-iot-evolution-for-industry-digitalization.
[3] ITU. Radio regulations, articles, 2012.

CHAPTER

2

Global cellular IoT standards

OUTLINE

Abstract

This chapter first presents the Third Generation Partnership Project (3GPP), including its ways of working, its organization, and its linkage to the world's largest regional standardization development organizations (SDOs).

Then, after providing a basic overview of the 3GPP cellular systems architecture, 3GPP's work on Cellular IoT is introduced. This introduction includes a summary of the early work performed by 3GPP in the area of *massive machine-type communications* (mMTC). The *Power Saving Mode (PSM)* and *extended Discontinuous Reception* (eDRX) features are discussed together with the feasibility studies of the technologies *Extended Coverage Global System for Mobile Communications Internet of Things* (EC-GSM-IoT), *Narrowband Internet of Things* (NB-IoT), and *Long-Term Evolution for Machine-Type Communications* (LTE-M).

Cellular Internet of Things, Second Edition
https://doi.org/10.1016/B978-0-08-102902-2.00002-9

© 2020 Elsevier Ltd. All rights reserved.

To introduce the work on *critical MTC* (cMTC) the 3GPP Release 14 feasibility *Study on Latency reduction techniques for Long Term Evolution* is presented. It triggered the specification of several features for reducing latency and increasing reliability in LTE.

Support for cMTC is a pillar in the design of the fifth generation (5G) *New Radio* (NR) system. To put the 5G cMTC work in a context an overview of NR is provided. This includes the Release 14 NR study items, the Release 15 normative work and the work on qualifying NR, and LTE, as *IMT-2020* systems.

Finally, an introduction to the *MulteFire Alliance* (MFA) and its work on mMTC radio systems operating in unlicensed spectrum is given. The MulteFire Alliance modifies 3GPP technologies to comply with regional regulations and requirements specified to support operation in unlicensed frequency bands.

2.1 3GPP

Third Generation Partnership Project (3GPP) is the global standardization forum behind the evolution and maintenance of the Global System for Mobile Communications (GSM), the *Universal Mobile Telecommunications System* (UMTS), the Long Term Evolution (LTE) and the fifth generation (5G) cellular radio access technology known as the *New Radio* (NR). The project is coordinated by seven regional standardization development organizations representing Europe, the United States, China, Korea, Japan, and India. 3GPP has since its start in 1998 organized its work in release cycles and has in 2019 reached Release 16.

A release contains a set of work items where each typically delivers a feature that is made available to the cellular industry at the end of the release cycle through a set of *technical specifications* (TSs). A feature is specified in four stages where stage 1 contains the service requirements, stage 2 a high-level feature description, and stage 3 the detailed description that is needed to implement the feature. The fourth and final stage contains the development of the performance requirements and conformance testing procedures for ensuring proper implementation of the feature. Each feature is implemented in a distinct version of the 3GPP TSs that maps to the release within which the feature is developed. At the end of a release cycle the version of the specifications used for feature development is frozen and published. In the next release a new version of each technical specification is created and edited as needed for new features associated with that release. Each release contains a wide range of features providing functionality spanning across GSM, UMTS, LTE, and NR as well as providing interworking between the four. In each release it is further ensured that GSM, UMTS, LTE, and NR can coexist in the same geographical area. That is, the introduction of, for example, NR into a frequency band should not have a negative impact on GSM, UMTS or LTE operation.

The technical work is distributed over a number of *technical specification groups* (TSGs), each supported by a set of *working groups* (WGs) with technical expertise representing different companies in the industry. The 3GPP organizational structure is built around three TSGs:

- *TSG Service and System Aspects* (SA),
- *TSG Core Network (CN) and Terminals* (CT), and,
- *TSG Radio Access Network* (RAN).

TSG SA is responsible for the SA and service requirements, i.e., the stage 1 requirements, and TSG CT for CN aspects and specifications. TSG RAN is responsible for the design and maintenance of the RANs. So, while TSG CT for example is working on the 4G *Evolved Packet Core* (EPC) and the *5G Core network* (5GC), TSG RAN is working on the corresponding radio interfaces know as LTE and NR, respectively.

The overall 3GPP project management is handled by the *Project Coordination Group* (PCG) that, for example, holds the final right to appoint TSG Chairmen, to adopt new work items and approve correspondence with external bodies of high importance, such as the International Telecommunications Union (ITU). Above the PCG are the seven SDOs: *ARIB* (Japan), *CCSA* (China), ETSI (Europe), *ATIS* (US), *TTA* (Korea), *TTC* (Japan), and *TSDSI* (India). Within 3GPP these standardization development organizations are known as the *Organizational Partners* that hold the ultimate authority to create or terminate TSGs and are responsible for the overall scope of 3GPP.

The Release 13 *massive MTC* (mMTC) specification work on EC-GSM-IoT, NB-IoT, and LTE-M was led by TSG GSM EGPRS RAN (GERAN) and TSG RAN. TSG GERAN was at the time responsible for the work on GSM/Enhanced Data Rates for GSM Evolution (EDGE) and initiated the work on EC-GSM-IoT and NB-IoT through a feasibility study resulting in *technical report (TR) 45.820 Cellular System Support for Ultra-Low Complexity and Low Throughput Internet of Things* [1]. It is common that 3GPP before going to normative specification work for a new feature performs a study of the feasibility of that feature and records the outcome of the work in a TR. In this specific case the report recommended to continue with normative work items on EC-GSM-IoT and NB-IoT. While TSG GERAN took on the responsibility of the EC-GSM-IoT work item, the work item on NB-IoT was transferred to TSG RAN. TSG RAN also took responsibility for the work item associated with LTE-M, which just as NB-IoT is part of the LTE series of specifications.

After 3GPP Release 13, i.e. after completion of the EC-GSM-IoT specification work, TSG GERAN and its WGs GERAN1, GERAN2, and GERAN3 were closed and their responsibilities were transferred to TSG RAN and its WGs RAN5 and RAN6. Consequently, TSG RAN is responsible for NB-IoT and GSM, including EC-GSM-IoT, in addition to being responsible for UMTS, LTE and the development of NR.

Fig. 2.1 gives an overview of the 3GPP organizational structure during Release 16, indicating the four levels: The Organizational Partners (OP) including the regional standards development organizations, the PCG, the three active TSGs, and the WGs of each TSG.

2.2 Cellular system architecture

2.2.1 Network architecture

The cellular systems specified by 3GPP have an architecture that is divided in a RAN part and a CN part. The RAN connects a device to the network via the radio interface, also known as the access stratum, while the CN connects the RAN to an external network. This can be a public network such as the Internet or a private enterprise network. The overall purpose of the radio and CNs is to provide an efficient data transfer between the external network and the devices served by the cellular system.

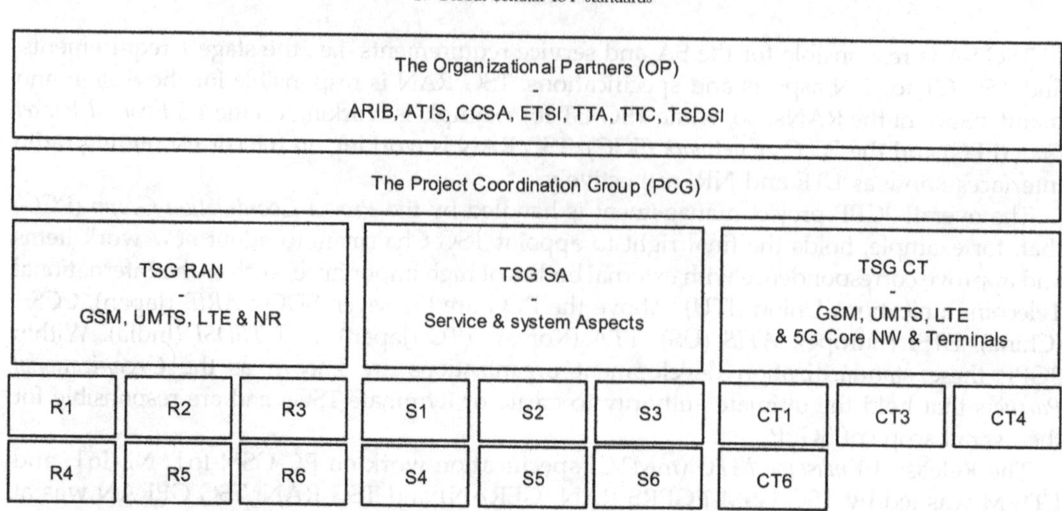

FIG. 2.1 Organisational structure of 3GPP.

Although the architecture has evolved over time similar components and functions can be recognized when comparing e.g. the GSM/EDGE and LTE architectures. Fig. 2.2 shows a set of important nodes and interfaces in an LTE and a GSM/EDGE core and radio network.

In case of LTE the radio network is known as E-UTRAN, or LTE, while the CN is named the EPC. Together E-UTRAN and the EPC define the EPS. In the EPC the Packet Data network Gateway (P-GW) provides the connection to an external packet data network. The Serving Gateway (S-GW) routes user data packets from the P-GW to an evolved Node B (eNB) that transmits them over the LTE radio interface (Uu) to an end user device. The connection between the P-GW and the device is established by means of a so-called EPS bearer, which is associated with certain Quality of Service (QoS) requirements. These correspond to for example data rate and latency requirements expected from the provided service.

Data and control signaling is separated by means of the user plane and control plane. The Mobility Management Entity (MME) which e.g. is responsible for idle mode tracking is connected to the eNB via the control plane. The MME also handles subscriber authentication and is connected to the Home Subscriber Service (HSS) data base for this purpose. It maps the EPS bearer to radio bearers that provides the needed QoS over the LTE radio interface.

In the GPRS core the GGSN acts as the link to the external packet data networks. The SGSN fills a role similar to the MME and handles idle mode functions as well as authentication toward the Home Location Register (HLR) which keeps track of the subscriber information. It also routes the user data to the radio network. In an LTE network the eNB is the single infrastructure node in the RAN. In case of GERAN the eNB functionality is distributed across a Base Station Controller and the Base Transceiver Station. One of the most fundamental difference between GSM/EDGE and the EPS architectures is that GSM/EDGE supports a circuit switched domain for the handling of voice calls, in addition to the packet switched domain. The EPS only operates in the packet switched domain. The Mobile Switching Center (MSC) is the GSM CN node that connects the classic Public Switched Telephone Network (PSTN) to GERAN. The focus of this book lies entirely in the packet switched domain.

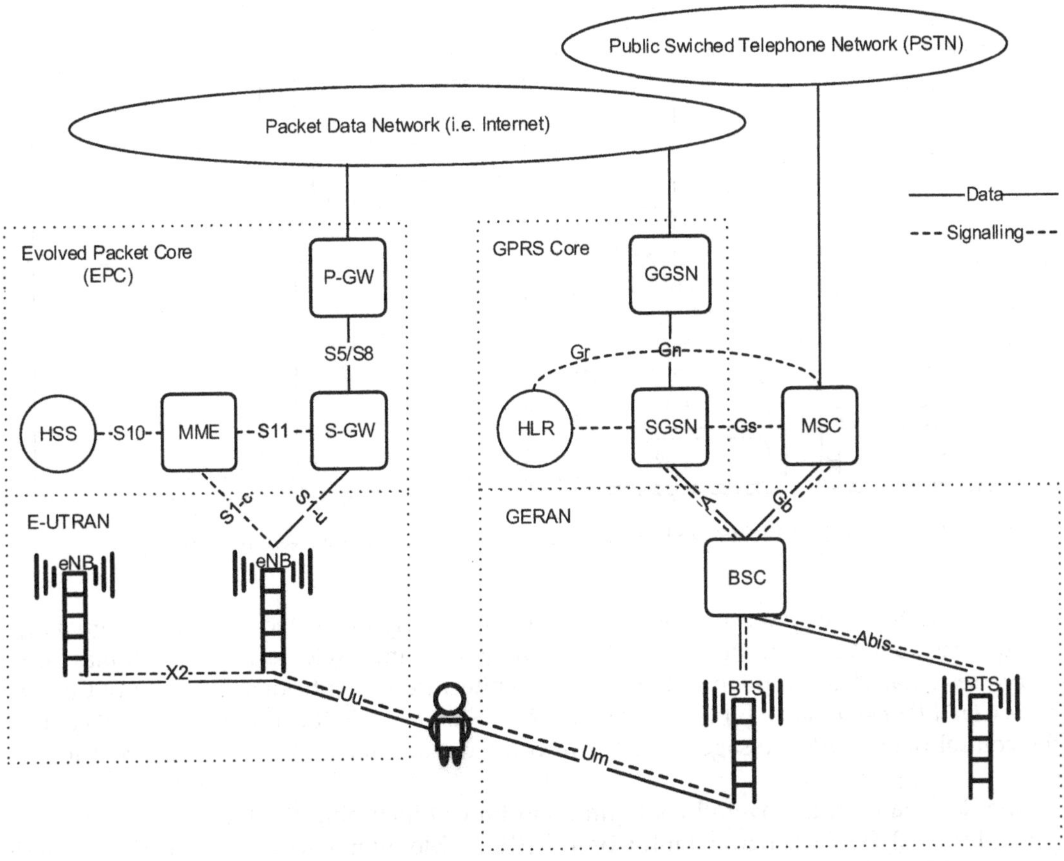

FIG. 2.2 Illustrating the Evolved Packet System (EPS) and GSM/EDGE radio and CNs.

Section 2.4 provides an architectural overview of NR and the 5G CN.

2.2.2 Radio protocol architecture

Understanding the 3GPP radio protocol stack and its applicability to the nodes and interfaces depicted in Fig. 2.2 is a good step towards understanding the overall system architecture. Fig. 2.3 depicts the LTE radio protocol stack including the control and user plane layers as seen from the device.

In the user plane protocol stack the highest layer is an IP layer, which carries application data and terminates in the P-GW. IP is obviously not a radio protocol, but still mentioned here to introduce the interface between the device and the P-GW. The IP packet is transported between the P-GW, the S-GW and the eNB using the GPRS Tunnel Protocol (GTP).

The Non-Access Stratum (NAS) and Radio Resource Control (RRC) layers are unique to the control plane. A message-based IP transport protocol known as the Stream Control Transmission Protocol (SCTP) is used between the eNB and MME for carrying the NAS messages. It provides a reliable message transfer between the eNB and MME.

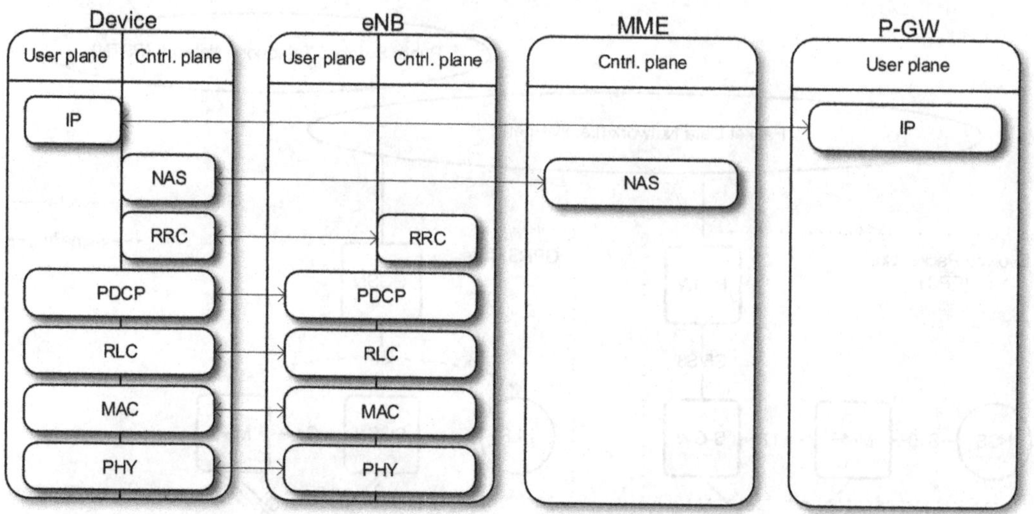

FIG. 2.3 The LTE control and user plane protocols and interfaces as seen from the device.

Originally the NAS protocol was intended to be entirely dedicated to support signaling, e.g. to attach and authenticate a device to the network. Since Release 13 the NAS may also carry user data. This exception to the general architecture was introduced as part of *Control plane CIoT EPS optimization* [9] feature discussed in Section 2.3.5. It is important to notice that the control plane NAS messages sent between the device and MME are transparent to the eNB.

The RRC handles the overall configuration of a cell including the Packet Data Convergence Protocol (PDCP), Radio Link Control (RLC), Medium Access Control (MAC) and physical (PHY) layers. It is responsible for the connection control, including connection setup, (re-)configuration, handover and release. The system information messages described in section 5.3.1.2 is a good example of RRC information.

The PDCP, the RLC, the MAC and the PHY layers are common to the control and user planes. The PDCP perform Robust Header Compression (RoHC) on incoming IP packets and manages integrity protection and ciphering of the control plane and ciphering of the user plane data sent over the access stratum. It acts as a mobility anchor for devices in RRC connected mode. It buffers, and in if needed retransmits, packets received during a handover between two cells. The PDCP packets are transferred to the RLC layer which handles a first level or retransmission in an established connection and makes sure that received RLC packets are delivered in sequence to the PDCP layers.

The RLC layer handles concatenation and segmentation of PDCP protocol data units (PDU) into RLC service data units (SDU). The RLC SDUs are mapped on RLC PDUs which are transferred to the MAC layer. Each RLC PDU is associated with a radio bearer and a logical channel. Two types of radio bearers are supported: signaling radio bearers (SRBs) and data radio bearers (DRBs). The SRBs are sent over the control plane and bears the logical channels known as the Broadcast, Common and Dedicated Control Channels (BCCH, DCCH, CCCH). The DRBs are sent over the user plane and are associated with the Dedicated Traffic Channel (DTCH). The distinction provided by the bearers and the logical channels allows a network to apply suitable access stratum configurations to provide a requested QoS for different types of signaling and data services.

MAC manages multiplexing of bearers and their logical channels with MAC control elements according to specified and configured priorities. The MAC control elements are used to convey information related to an ongoing connection such as the data buffer status report. MAC is also responsible for the random-access procedure and hybrid automatic repeat request (HARQ) retransmissions. The MAC PDUs are forwarded to the physical layer which is responsible for the physical layer functions and services such as encoding, decoding, modulation and demodulation.

Fig. 2.4 shows the data transfer through the protocol stack. At each layer a header (H) is appended to the SDU to form the PDU, and at the physical layer also a CRC is attached to the transport block.

The GPRS protocol stack also includes the RRC, PDCP, RLC, MAC and PHY layers. Although the same naming conventions are used in GPRS and LTE is should be understood that the functionality belonging to the different layers has evolved. GPRS non-access stratum signaling between the device and SGSN is defined by means of the Logical Link Control (LLC) and Sub-Network Dependent Convergence Protocol (SNDCP) protocols. LLC handles encryption and integrity protection, while SNDCP manages RoHC. This is functionality is similar to that provided by the LTE PDCP for providing compression and access stratum security. As a comparison remember that the PDCP terminates in the E-UTRAN while LLC and SNDCP terminates in the GPRS CN.

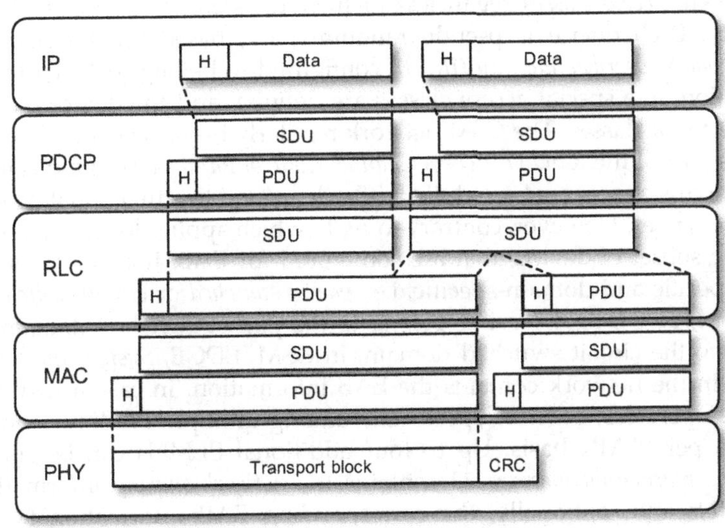

FIG. 2.4 LTE data flow.

2.3 From machine-type communications to the cellular internet of things

2.3.1 Access class and overload control

This section presents the early work done by 3GPP for GSM and LTE in the area of MTC from the very start of 3GPP in Release 99 until Release 14. UMTS is not within the scope of this overview, but the interested reader should note that many of the features presented for LTE are also supported by UMTS.

In 2007 and Release 8 *TSG SA WG1* working on the 3GPP system architecture published TR 22.868 *Study on Facilitating Machine to Machine Communication in 3GPP Systems* [2]. It highlights use cases such as metering and health, which are still of vital interest as 3GPP continues with the 5G specification effort. 3GPP TR 22.868 provides considerations in areas such as handling large numbers of devices, addressing of devices, and the level of security needed for machine-to-machine applications.

In 3GPP TSG SA typically initiates work for a given feature by first agreeing to a corresponding set of general service requirements and architectural considerations. In this case the SA WG1 work also triggered a series of Stage 1–3 activities in 3GPP Release 10 denoted *Network Improvement for Machine-Type Communications* [3]. The main focus of the work was to provide functionality to handle large numbers of devices, including the ability to protect existing networks from overload conditions that may appear in a network aiming to support a very large number of devices. For GSM/EDGE the overload control features *Extended Access Barring* (EAB) [4] and *Implicit Reject* [5] were specified as part of these Release 10 activities.

Already in the Release 99 specifications, i.e., the first 3GPP release covering GSM/EDGE, support for the *Access Class Barring* (ACB) feature is specified. It allows a network to bar devices of different *access classes* regardless of their registered *Public Land Mobile Network* (PLMN) identity. Each device is pseudo randomly, i.e., based on the last digit of their *International Mobile Subscriber Identity (IMSI)*, configured to belong to 1 of 10 normal access classes. In addition, five special access classes are defined, and the device may also belong to one of these special classes. The GSM network regularly broadcasts in its system information a bitmap as part of the *Random Access Channel Control Parameters* to indicate if devices in any of these 15 access classes are barred. EAB is built around this functionality and reuses the 10 normal access classes. However, contrary to ACB, which applies to all devices, EAB is only applicable to the subset of devices that are configured for EAB. It also allows a network to enable PLMN-specific and domain-specific, i.e., *packet switched* or *circuit switched*, barring of devices. For GSM/EDGE data services belong to the packet switched domain, while voice services belong to the circuit switched domain. In GSM/EDGE, *System Information message 21* broadcasted in the network contains the EAB information. In case a network is shared among multiple operators, or more specifically among multiple PLMNs, then EAB can be configured on a per PLMN basis. Up to four additional PLMNs can be supported by a network. *System Information message* 22 contains the *network sharing* information for these additional PLMNs and, optionally, the corresponding EAB information for each of the PLMNs [5].

ACB and EAB provide means to protect both the radio access and CN from congestion that may occur if a multitude of devices attempt to simultaneous access a network. The 10 normal access classes allow for ACB or EAB based barring of devices with a granularity of 10%. Because both ACB and EAB are controlled via the system information, these mechanisms have an inherent reaction time associated with the time it takes for a device to detect that the system information has been updated and the time required to obtain the latest barring information.

The GSM Implicit Reject feature introduces an Implicit Reject flag in a number of messages sent on the downlink (DL) *Common Control CHannel* (CCCH). Before accessing the network, a device configured for *Low Access Priority* [6] is required to decode a message on the DL CCCH and read the Implicit Reject flag therein. The support for low access priority is signaled by a device over the *Non-Access Stratum* (NAS) interface using the *Device Properties* information element [7] and over the *Access Stratum* in the *Packet Resource Request* message [6]. In case the Implicit Reject flag is set to "1" the device is not permitted to access the GSM network (NW) and is required to await the expiration of a timer before attempting a new access. Because it does not require the reading of the system information messages, Implicit Reject has the potential benefit of being a faster mechanism than either ACB or EAB type-based barring. When the Implicit Reject flag is set to "1" in a given downlink CCCH message then all devices that read that message when performing system access are barred from network access. By toggling the flag with a certain periodicity within each of the messages sent on the downlink CCCH, a partial barring of all devices can be achieved. Setting the flag to "1" within all downlink CCCH messages sent during the first second of every 10-s time interval will, for example, bar 10% of all devices. A device that supports the Implicit Reject feature may also be configured for EAB.

For LTE, ACB was included already in the first release of LTE, i.e. 3GPP Release 8, while the low-priority indicators were introduced in Release 10 [8]. An NAS low-priority indication was defined in the NAS signaling [15] and an *Establishment Cause* indicating delay tolerant access was introduced in the *Radio RRC Connection Request* message sent from the device to the base station [9]. These two indicators support congestion control of delay tolerant MTC devices. In case the RRC connection request message signals that the access was made by a delay tolerant device, the base station has the option to reject the connection in case of congestion and via the RRC *Connection Reject* message request the device to wait for the duration of a configured *extended wait timer* before making a new attempt.

In Release 11 the MTC work continued with the work item *System Improvements for MTC* [10]. In TSG RAN EAB was introduced in the LTE specifications. A new *System Information Block 14* (SIB14) was defined to convey the EAB-related information [9]. To allow for fast notification of updates of SIB14 the paging message was equipped with a status flag indicating an update of SIB14. As for TSG GERAN, barring of 10 different access classes is supported. In case of network sharing, a separate access class bitmap can, just as for GSM/EDGE, be signaled per PLMN sharing the network. A device with its low-priority indicator set needs to support EAB.

Table 2.1 summarizes the GSM/EDGE and LTE 3GPP features designed to provide overload control in different releases. It should be noted that ETSI was responsible for the GSM/EDGE specifications until 3GPP Release 99 when 3GPP took over the responsibility for the evolution and maintenance of GSM/EDGE. ACB was, for example, part of GSM/EDGE already before 3GPP Release 99. Note that after Release 99, the 3GPP release numbering was restarted from Release 4.

TABLE 2.1 3GPP features until Release 13 related to MTC overload control.

Release	GSM	LTE
99	Access class barring	–
8	–	Access class barring
10	Extended access class barring Implicit reject Low priority and access delay tolerant Indicators	Low priority and access delay tolerant Indicators
11	–	Extended access class barring

2.3.2 Small data transmission

In Release 12 the work item *Machine-Type Communications and other mobile data applications communications* [11] triggered a number of activities going beyond the scope of the earlier releases that to a considerable extent were focused on managing large numbers of devices. It resulted in TR 23.887 *Study on Machine-Type Communications (MTC) and other mobile data applications communications enhancements* [12] that introduces solutions to efficiently handle small data transmissions and solutions to optimize the energy consumptions for devices dependent on battery power.

MTC devices are to a large extent expected to transmit and receive small data packets, especially when viewed at the application layer. Consider, for example, street lighting controlled remotely where turning on and off the light bulb is the main activity. On top of the small application layer payload needed to provide the on/off indication, overhead from higher-layer protocols, for example, *User Datagram* and *Internet Protocols* and radio interface protocols need to be added thereby forming a complete protocol stack. For data packets ranging up to a few hundred bytes the protocol overhead from layers other than the application layer constitutes a significant part of the data transmitted over the radio interface. To optimize the power consumption of devices with a traffic profile characterized by small data transmissions it is of interest to reduce this overhead. In addition to the overhead accumulated over the different layers in the protocol stack, it is also vital to make sure various procedures are streamlined to avoid unnecessary control plane signaling that consumes radio resources and increases the device power consumption. Fig. 2.5 shows an overview of the message flow associated with an LTE *mobile originated* (MO) data transfer where a single uplink (UL) is sent between the user equipment (UE) and the eNB. It is clear from the depicted signaling flow that several signaling messages are transmitted before the uplink and downlink data packets are sent.

One of the most promising solutions for support of small data transmission is the *RRC Resume* procedure [9]. It aims to optimize, or reduce the number of signaling messages, that is needed to set up a connection in LTE. Fig. 2.5 indicates the part of the connection setup that becomes redundant in the RRC Resume procedure, including the Security mode command and the RRC connection reconfiguration messages. The key in this solution is to resume configurations established in a previous connection. Part of the possible optimizations is to suppress the RRC signaling associated with measurement configuration. This simplification

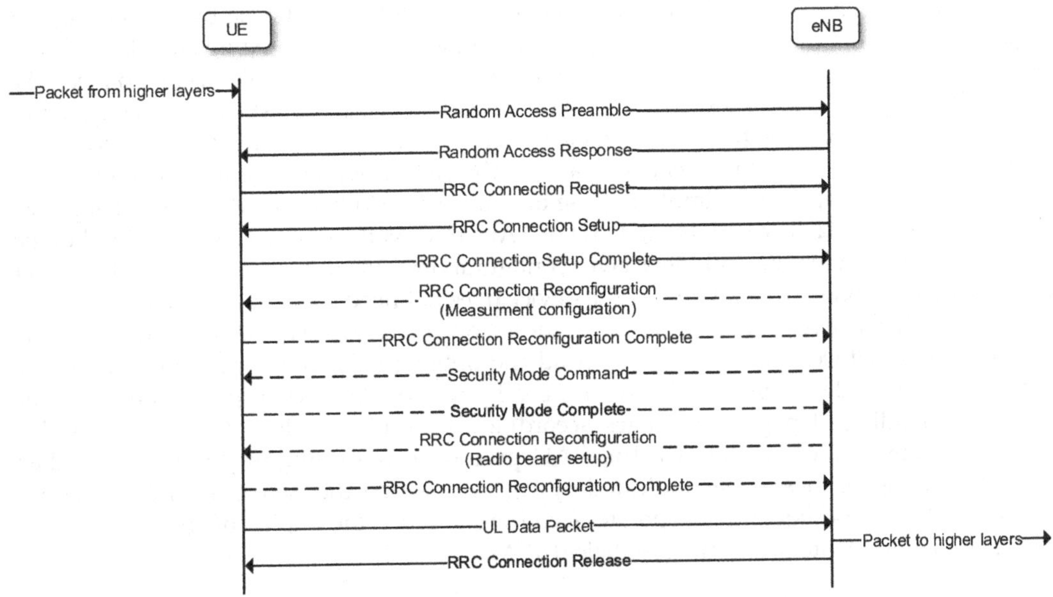

FIG. 2.5 LTE message transfer associated with the transmission of a single uplink and single downlink data package. Messages indicated with dashed arrows are eliminated in the *RRC Resume procedure* solution [9].

is justified by the short data transfers expected for MTC. For these devices measurement reporting is less relevant compared to when long transmissions of data are dominating the traffic profile. In 3GPP Release 13 this solution was specified together with the *Control plane CIoT EPS Optimization* [9] as two alternative solutions adopted for streamlining the LTE setup procedure to facilitate small and infrequent data transmission [13]. These two solutions are highly important to optimize latency and power consumption for LTE-M and NB-IoT evaluated in Chapters 6 and 8, respectively.

2.3.3 Device power savings

The 3GPP Release 12 study on MTC and other mobile data applications communications enhancements introduced two important solutions to optimize the device power consumption, namely Power Saving Mode (PSM) and extended Discontinuous Reception (eDRX). PSM was specified both for GSM/EDGE and LTE and is a solution where a device enters a power saving state in which it reduces its power consumption to a bare minimum [14]. While in the power saving state the mobile is not monitoring paging and consequently becomes unreachable for *mobile terminated* (MT) services. In terms of power efficiency this is a step beyond the typical *idle mode* behavior where a device still performs energy consuming tasks such as neighbor cell measurements and maintaining reachability by listening for paging messages. The device leaves PSM when higher layers in the device triggers a MO access, e.g., for an uplink data transfer or for a periodic Tracking Area Update/Routing Area Update (TAU/RAU). After the MO access and the corresponding data transfer have been completed, a device using

PSM starts an *Active Timer*. The device remains reachable for MT traffic by monitoring the paging channel until the Active timer expires. When the Active timer expires the device reenters the power saving state and is therefore unreachable until the next MO event. To meet MT reachability requirements of a service a GSM/EDGE device using PSM can be configured to perform a periodic RAU with a configurable periodicity in the range of seconds up to a year [7]. For an LTE device the same behavior can be achieved through configuration of a periodic TAU timer [15]. Compared to simply turning off a device, PSM has the advantage of supporting the mentioned MT reachability via RAU or TAU. In PSM the device stays registered in the network and may maintain its higher layer configurations. As such, when leaving the power saving state in response to a MO event the device does not need to first attach to the network, as it would otherwise need to do when being turned on after previously performing a complete power off. This reduces the signaling overhead and optimizes the device power consumption.

Fig. 2.6 depicts the operation of a device configured for PSM when performing periodic RAUs and reading of paging messages according to the idle mode DRX cycle applicable when the Active timer is running. The RAU procedure is significantly more costly than reading of a paging message as indicated by Fig. 2.6. During the Active timer the device is in idle mode and is required to operate accordingly. After the expiry of the Active timer the device again reenters the energy-efficient PSM.

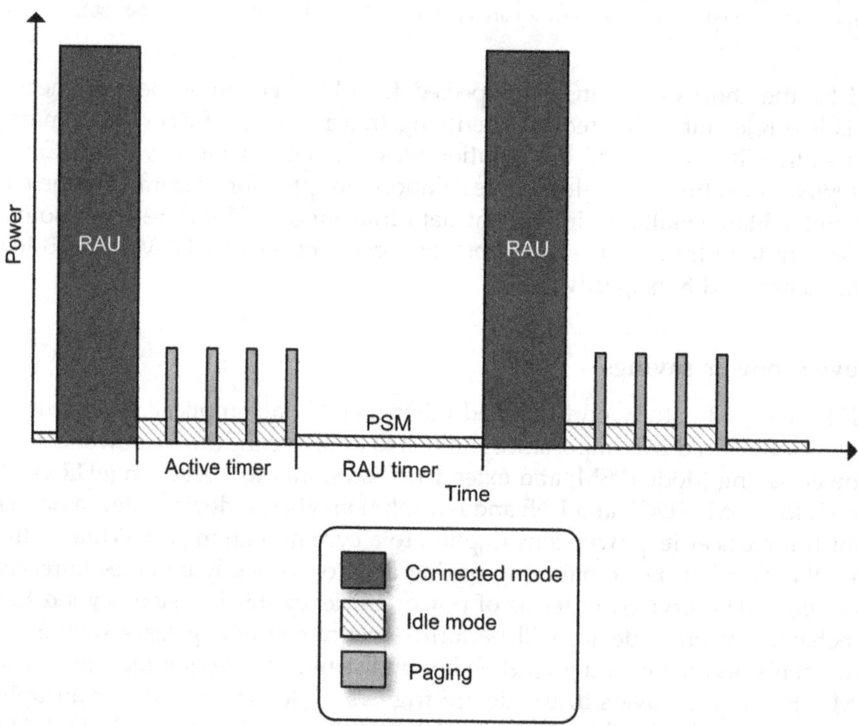

FIG. 2.6 Illustration of device transitions between connected state, idle state and PSM when waking up to perform periodic Routing Area Updates.

In Release 13 eDRX was specified for GSM and LTE. The general principle for eDRX is to extend the since earlier specified DRX cycles to allow a device to remain longer in a power saving state between *paging occasions* and thereby minimize its energy consumption. The advantage over PSM is that the device remains periodically available for MT services without the need to first perform e.g., a Routing or Tracking Area Update to trigger a limited period of downlink reachability. The *Study on power saving for Machine-Type Communication (MTC) devices* [16] considered, among other things, the energy consumption of devices using eDRX or PSM. The impacts of using PSM and eDRX on device battery life, assuming a 5 Watt-hour (Wh) battery, were characterized as part of the study. More specifically, the battery life for a device was predicted for a range of triggering intervals and reachability periods. A trigger may e.g., correspond to the start of a MT data transmission wherein an application server requests a device to transmit a report. After reception of the request the device is assumed to respond with the transmission of the requested report. The triggering interval is defined as the interval between two adjacent MT events. The reachability period is, on the other hand, defined as the period between opportunities for the network to reach the device using a paging channel. Let us, for example, consider an alarm condition that might only trigger on average once per year, but when it does occur there is near real-time requirement for an application server to know about it. For this example the ongoing operability of the device, capable of generating the alarm condition, can be verified by the network sending it a page request message and receiving a corresponding page response. Once the device operability is verified, the network can send the application layer message that serves to trigger the reporting of any alarm condition that may exist.

Fig. 2.7 presents the estimated battery life for a GSM/EDGE device in normal coverage when using PSM or eDRX. Reachability for PSM was achieved by the device by performing a periodic RAU, which initiates a period of network reachability that continues until the expiration of the Active timer. Both in case of eDRX and PSM it was here assumed that the device, before reading a page or performing a RAU, must confirm the serving cell identity and measure the signal strength of the serving cell. This is to verify that the serving cell remains the same and continues to be *suitable* from a signal strength perspective. When deriving the results depicted in Fig. 2.7 the energy costs of confirming the cell identity, estimating the serving cell signal strength, reading a page, performing a TAU, and finally transmitting the report were all taken from available results provided within the *Study on power saving for MTC devices* [16]. A dependency both on the reachability period and on the triggering interval is seen in Fig. 2.7. For eDRX a very strong dependency on the triggering interval is seen. The reason behind this is that the cost of sending the report is overshadowing the cost of being paged. Remember that the reachability period corresponds to the paging interval. For PSM the cost of performing a RAU is in this example of similar magnitude as sending the actual report so the reachability period becomes the dominating factor, while the dependency on triggering interval becomes less pronounced. For a given triggering interval Fig. 2.7 shows that eDRX is outperforming PSM when a shorter reachability is required, while PSM excels when the reachability requirement is in the same range as, or relaxed compared to, the actual triggering interval.

In the end GSM/EDGE eDRX cycles ranging up to 13,312 51-multiframes, or roughly 52 min, were specified. A motivation for not further extending the eDRX cycle is that devices with an expected reachability beyond 1 h may use PSM and still reach an impressive battery

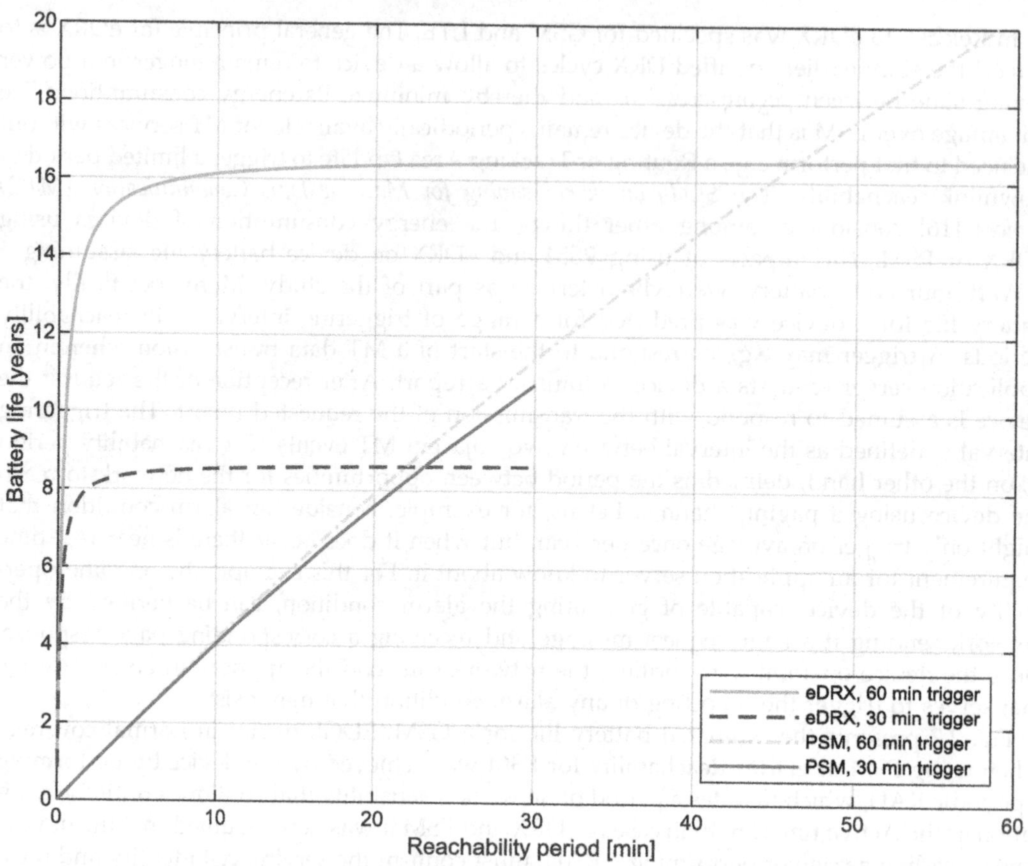

FIG. 2.7 Comparison between PSM and eDRX power consumption in a GSM network [16].

life, as seen in Fig. 2.7. The eDRX cycle can also be compared to the legacy max DRX cycle length of 2.1 s, which can be extended to 15.3 s if the feature *Split Paging Cycle* is supported [17].

For GSM/EDGE 3GPP went beyond PSM and eDRX and specified a new mode of operation denoted *Power Efficient Operation* (PEO) [18]. In PEO a device is required to support either PSM or eDRX, in combination with relaxed idle mode behavior. A PEO device is, for example, only required to verify the suitability of its serving cell shortly before its nominal paging occasions or just before a MO event. Measurements on a reduced set of neighbor cells is only triggered for a limited set of conditions such as when a device detects that the serving cell has changed or the signal strength of the serving cell has dropped significantly. PEO is mainly intended for devices relying on battery power where device power consumption is of higher priority than, e.g., *mobility* and latency, which may be negatively impacted by the reduced Idle Mode activities. Instead of camping on the best cell the aim of PEO is to assure that the device is served by a cell that is good enough to provide the required services.

For LTE, Release 13 specifies idle mode eDRX cycles ranging between 1 and 256 hyperframes. As one hyperframe corresponds to 1024 radio frames, or 10.24 s, 256 hyperframes correspond to roughly 43.5 min. As a comparison the maximum LTE idle mode DRX cycle length used before Release 13 equals 256 frames, or 2.56 s.

LTE does in addition to idle mode DRX support *connected mode* DRX to relax the requirement on reading the *Physical Downlink Control Channel* (PDCCH) for downlink assignments and uplink grants. The LTE connected mode DRX cycle was extended from 2.56 to 10.24 s in Release 13 [9].

Table 2.2 summarizes the highest configurable MT reachability periodicities for GSM and LTE when using idle mode eDRX, connected mode DRX, or PSM. For PSM the assumption is that the MT reachability periodicity is achieved through the configuration of periodic RAU or TAU.

In general, it is expected that the advantage of eDRX over PSM for frequent reachability periods is reduced for LTE compared with what can be expected in GSM/EDGE. The reason is that a typical GSM/EDGE device uses 33 dBm output power, while LTE devices typically use 23 dBm output power. This implies that the cost of transmission and a RAU or TAU in relation to receiving a page is much higher for GSM/EDGE than what is the case for LTE.

Table 2.3 summarizes the features discussed in this section and specified in Release 12 and 13 to optimize the mobile power consumption.

Chapters 3, 5, and 7 will further discuss how the concepts of relaxed monitoring of serving and neighbor cells, PSM, paging, idle, and connected mode DRX and eDRX have been designed for EC-GSM-IoT, LTE-M and NB-IoT. It will then be seen that the DRX cycles mentioned in Table 2.2 have been further extended for NB-IoT to support low power consumption and long device battery life.

2.3.4 Study on provision of low-cost MTC devices based on LTE

LTE uses the concept of *device categories* referred to as *user equipment* (UE) categories (Cat) to indicate the capability and performance of different types of equipment. As the LTE

TABLE 2.2 The maximum configurable mobile terminated reachability periodicities for GSM and LTE when using Idle mode eDRX, Connected mode eDRX or PSM with RAU/TAU based reachability.

	GSM	LTE
Idle mode eDRX	~52 min	~43 min
Connected mode eDRX	–	10.24 s
PSM	>1 year (RAU)	>1 year (TAU)

TABLE 2.3 3GPP Release 12 and 13 features related to device power savings.

Release	GSM	LTE
12	Power saving mode	Power saving mode
13	Extended DRX Power efficient operation	Extended DRX

specifications have evolved, the number of device categories has increased. Before Release 12, Category 1, developed in the first LTE Release, was considered the most rudimentary device category. Since it was originally designed to support mobile broadband services with data rates of 10 Mbps for the downlink and 5 Mbps for the uplink, it is still considered too complex to compete with GPRS in the low-end MTC segment. To change this, 3GPP TSG RAN initiated the Release 12 study item called *Study on Provision of low-cost MTC UEs based on LTE* [19], in this book referred to as the *LTE-M study item*, with the ambition to study solutions providing lower device complexity in combination with improved coverage.

A number of solutions for lowering the complexity and cost of the *radio frequency* (RF) and baseband parts of an LTE *modem* were proposed in the scope of the LTE-M study item. It was concluded that a reduction in transmission and reception bandwidths and peak data rates in combination with adopting a single RF receive chain and *half-duplex* operation would make the cost of an LTE device modem comparable to the cost of an EGPRS modem. A reduction in the maximum supported transmission and reception bandwidths and adopting a single RF receive chain reduces the complexity in both the RF and the baseband because of, e.g., reduced RF filtering cost, reduced sampling rate in the *analog-to-digital and digital-to-analog conversion* (ADC/DAC), and reduced number of baseband operations needed to be performed. The peak data rate reduction helps reduce the baseband complexity in both demodulation and decoding parts. Going from *full-duplex* operation as supported by Category 1 devices to half duplex allows the *duplex filter(s)* in the RF front end to be replaced with a less costly switch. A reduction in the transmission power can also be considered, which relaxes the requirements on the RF front-end *power amplifier* and may support integration of the power amplifier on the chip that is expected to reduce device complexity and manufacturing costs. Table 2.4 summarizes the findings recorded in the LTE-M study item for the individual cost reduction techniques and indicates the expected impact on coverage from each of the solutions. As the cost savings are not additive in all cases refer to Table 6.12 for cost estimates of combinations of multiple cost reduction techniques. The main impact on downlink coverage is caused by going to a single RF chain, i.e., one receive antenna instead of

TABLE 2.4 Overview of measures supporting an LTE modem cost reduction [19].

Objective	Modem cost reduction	Coverage impact
Limit full duplex operation to half duplex	7%–10%	None
Peak rate reduction through limiting the maximum transport block size (TBS) to 1000 bits	10.5%–21%	None
Reduce the transmission and reception bandwidth for both RF and baseband to 1.4 MHz	39%	1–3 dB DL coverage reduction due to loss in frequency diversity
Limit RF front end to support a single receive branch	24%–29%	4 dB DL coverage reduction due to loss in receive diversity
Transmit power reduction to support PA integration	10%–12%	UL coverage loss proportional to the reduction in transmit power

two. If a lower transmit power is used in uplink, this will cause a corresponding uplink coverage reduction. Reducing the maximum signal bandwidth to 1.4 MHz may cause coverage loss due to reduced frequency diversity. This can however be partly compensated for by use of frequency hopping.

Besides studying means to facilitate low device complexity the LTE-M study item [19] provided an analysis of the existing LTE coverage and presented means to improve it by up to 20 dB. Table 2.5 summarizes the frequency-division duplex LTE maximum coupling loss (MCL) calculated as:

$$MCL = P_{TX} - (SNR + 10log_{10}(k \cdot T \cdot BW) + NF) \tag{2.1}$$

P_{TX} equals the transmitted output power, SNR is the supported signal to noise ratio, BW is the signal bandwidth, NF the receiver Noise Figure, T equals an assumed ambient temperature of 290 K, and k is Boltzmann's constant.

It was assumed that the eNB supports two transmit and two receive antennas. The reference LTE device was assumed to be equipped with a single transmit and two receive antennas. The results were obtained through simulations assuming downlink *Transmission Mode* 2, i.e., downlink transmit diversity [20]. It is seen that the *Physical Uplink Shared Channel* (PUSCH) is limiting the LTE coverage to a MCL of 140.7 dB.

The initial target of the LTE-M study item [19] was to provide 20 dB extra coverage for low-cost MTC devices leading to a MCL of 160.7 dB. After investigating the feasibility of extending the coverage of each of the channels listed in Tables 2.5 to 160.7 dB through techniques such as *transmission time interval (TTI) bundling, Hybrid Automatic Repeat Request (HARQ) retransmissions* and repetitions, it was concluded that a coverage improvement of 15 dB leading to a MCL of 155.7 dB was an appropriate initial target for low-complexity MTC devices based on LTE.

The LTE-M study item triggered a 3GPP Release 12 work item [21] introducing an LTE device category (Cat-0) of low-complexity, and a Release 13 work item [22] introducing an LTE-M device category (Cat-M1) of even lower-complexity for low-end MTC applications

TABLE 2.5 Overview of LTE maximum coupling loss performance [19].

Performance/Parameters	**Downlink coverage**				**Uplink coverage**		
Physical channel	**PSS/SSS**	**PBCH**	**PDCCH Format 1A**	**PDSCH**	**PRACH**	**PUCCH Format 1A**	**PUSCH**
Data rate [kbps]	–	–	–	20	–	–	20
Bandwidth [kHz]	1080	1080	4320	360	1080	180	360
Power [dBm]	36.8	36.8	42.8	32	23	23	23
NF [dB]	9	9	9	9	5	5	5
#TX/#RX	2TX/2RX	2TX/2RX	2TX/2RX	2TX/2RX	1TX/2RX	1TX/2RX	1TX/2RX
SNR [dB]	-7.8	-7.5	-4.7	-4	-10	-7.8	-4.3
MCL [dB]	149.3	149	146.1	145.4	141.7	147.2	140.7

and the needed functionality to extend the coverage for LTE and LTE-M devices. Chapters 5 and 6 are in detail presenting the design and performance of LTE-M that is the result of these two work items and the two that followed in Release 14 and 15.

2.3.5 Study on cellular system support for ultra-low complexity and low throughput internet of things

In 3GPP Release 13 the study item on *Cellular System Support for Ultra-Low Complexity and Low Throughput Internet of Things* [1], here referred to as the *Cellular IoT study item*, was started in 3GPP TSG GERAN. It shared many commonalities with the *LTE-M study item* [19], but it went further both in terms of requirements and in that it was open to GSM backward compatible solutions as well as to non-backward compatible radio access technologies. The work attracted considerable interest, and 3GPP TR 45.820 Cellular system support for ultra-low complexity and low throughput IoT capturing the outcome of the work contains several solutions based on GSM/EDGE, on LTE and non-backwards compatible solutions, so-called *Clean Slate* solutions.

Just as in the LTE-M study item improved coverage was targeted, this time by 20 dB compared to GPRS. Table 2.6 presents the GPRS reference coverage calculated by 3GPP. It is based on the minimum GSM/EDGE *Block Error Rate* performance requirements specified in 3GPP TS 45.005 *Radio transmission and reception* [23]. For the downlink the specified device receiver *Sensitivity* of -102 dBm was assumed to be valid for a device noise figure (NF) of 9 dB. When adjusted to a NF of 5 dB, which was assumed suitable for IoT devices, the GPRS *Reference Sensitivity* ended up at -106 dBm. For the uplink 3GPP TS 45.005 specifies a GPRS single antenna base station sensitivity of -104 dBm that was assumed valid for a NF of 5 dB. Under the assumption that a modern base station supports a NF of 3 dB the uplink sensitivity

TABLE 2.6 Overview of GPRS maximum coupling loss performance [8].

#	Link direction	DL	UL
1	Power [dBm]	43	33
2	Thermal noise [dBm/Hz]	-174	-174
3	NF [dB]	5	3
4	Bandwidth [kHz]	180	180
5	Noise power [dBm] = (2)+(3)+$10\log_{10}$((4))	-116.4	-108.7
6	Single antenna receiver sensitivity according to 3GPP TS 45.005 [dBm]	-102 @ NF 9 dB	-104 @ NF 5 dB
7	Single antenna receiver sensitivity according to 3GPP TR 45.820 [dBm]	-106 @ NF 5 dB	-106 @ NF 3 dB
8	SINR [dB] = (7)−(5)	10.4	12.4
9	RX processing gain [dB]	0	5
10	MCL [dB] = (1)-((8)−(10))	149	144

reference also ended up at -106 dBm. To make the results applicable to a base station supporting receive diversity a 5-dB processing gain was added to the uplink reference performance.

The resulting GPRS MCL ended up at 144 dB because of limiting uplink performance. As the target of the Cellular IoT study item was to provide 20 dB coverage improvements on top of GPRS, this led to a stringent MCL requirement of 164 dB. The Cellular IoT study also specified stringent performance objectives in terms of supported data rate, latency, battery life, system capacity and device complexity.

After the Cellular IoT study item had concluded, normative work began in 3GPP Release 13 on EC-GSM-IoT [24] and NB-IoT [25]. Chapters 3, 4, 7, and 8 go into detail and present how EC-GSM-IoT and NB-IoT were designed to meet all the objectives of the Cellular IoT study item.

When comparing the initially targeted coverage for EC-GSM-IoT, NB-IoT, and LTE-M, it is worth to notice that Tables 2.5 and 2.6 are based on different assumptions, which complicates a direct comparison between the LTE-M target of 155.7 dBs MCL and the EC-GSM-IoT and NB-IoT target of 164 dB. Table 2.5 assumes, e.g., a base station NF of 5 dB, while Table 2.6 uses a NF of 3 dB. If those assumptions had been aligned, the LTE reference MCL had ended up at 142.7 dB and the LTE-M initial MCL target at 157.7 dB. If one takes into account that the LTE-M coverage target is assumed to be fulfilled for 20-dBm LTE-M devices, but that all the LTE-M coverage enhancement techniques are also available to 23-dBm LTE-M devices, the difference between the LTE-M coverage target and the 164-dB target shrinks to 3.3 dB. The actual coverage performance of EC-GSM-IoT, LTE-M, and NB-IoT is presented in Chapters 4, 6, and 8, respectively.

2.3.6 Study on Latency reduction techniques for LTE

In 3GPP Release 14 the *Study on Latency reduction techniques for LTE* [26] was carried out. It initiated the work to reduce the latency of LTE toward enabling support for *critical MTC* (cMTC). The attention of 3GPP in the area of IoT services had until Release 14 been on massive MTC, but from this point onwards 3GPP focused on two parallel cellular IoT streams: mMTC and cMTC.

The latency reduction study focused on optimizations of the connected mode procedures. As part of the study, reduced uplink transmission latency was investigated by means of shortening the *Semi Persistent Scheduling* uplink grant periodicity. Shortening of the minimum *transmission time interval* (TTI) to below 1 ms, and a reduction of device processing times were also considered candidate techniques for achieving latency improvements. This study led to the normative work on cMTC in LTE carried out in 3GPP Releases 14 and 15 which in detail are presented in Chapter 9. From Release 15 also NR supports cMTC, as presented in Chapters 11 and 12.

2.4 5G

2.4.1 IMT-2020

Approximately every 10 years a new generation cellular system is developed. 2G took cellular communication into the digital domain. 3G introduced mobile broadband services to the cellular market and 4G led the subscribers into the smartphone era. The pace of this evolution is set by the International Telecommunications Union (ITU). They define the requirements and qualities that needs to be met by a telecommunication system claiming to belong to one of the 'G's.

In 2017 the *International Telecommunications Union Radiocommunication sector* (ITU-R) defined the fundamental framework for the work on 5G by the publication of the report *Minimum requirements related to technical performance for IMT-2020 radio interfaces(s)* [27]. It presents the uses cases and requirements associated with the so called *International Mobile Telecommunication-2020* (IMT-2020) system. IMT-2020 is what in layman's terms is referred to as 5G. IMT-2020 is intended to support three major categories of use cases referred to as:

- mMTC,
- cMTC, and
- enhanced mobile broadband (eMBB).

The scope of this book includes mMTC and cMTC, which also may be referred to as *ultra-reliable and low latency communications* (URLLC). In this book we use the term cMTC for the set of critical use cases, and URLLC when discussing the technologies supporting the cMTC services and applications.

For each of the use cases, a set of requirements are defined that needs to be met by a 5G system. For mMTC, the 5G requirement to meet is known as *connection density*. It defines a set of deployment scenarios for which it needs to be shown that connectivity can be provided to 1,000,000 devices per square kilometer.

For cMTC, IMT-2020 requirements are defined in terms of latency and reliability. The latency requirements are divided into user plane and control plane latency. For user plane latency, a 1 ms timing budget is allowed for the successful transmission of a 32-byte data packet. The reliability requirement specifies that a data packet should be successfully delivered with a probability of 99.999%. For control plane latency a 20 ms timing budget is allowed for the transition from idle mode to the start of uplink data transmission with no associated reliability requirement. The set of IMT-2020 mMTC and cMTC requirements are summarized in Table 2.7.

2.4.2 3GPP 5G

2.4.2.1 5G feasibility studies

3GPP took on the challenge to develop its *5G System* (5GS) with start in Release 14. Following the regular procedures introduced in Section 2.1, the work began with a range of feasibility studies. The *Study on Scenarios and Requirements for Next Generation Access Technologies* [28] presents the 3GPP requirements on *NR* which is the 3GPP 5G radio interface.

TABLE 2.7 IMT-2020 mMTC and cMTC performance requirements [27].

mMTC connection density	cMTC latency	cMTC reliability
1,000,000 device/km^2	User plane: 1 ms Control plane: 20 ms	99.999%

TABLE 2.8 3GPP 5G mMTC performance requirements.

Connection density	Coverage	Latency	Device battery life
1000.0000 device/km^2	164 dB	10 s	10 years

It builds on and extends the set of IMT-2020 requirements. Table 2.8 presents the 3GPP 5G requirements in terms of connection density, coverage, latency and device battery life agreed for mMTC. The coverage, latency and battery life requirements are recognized from the Cellular IoT study introduced in Section 2.3.5. Chapters 6 and 8 discusses these requirements in detail.

For cMTC the requirement categories match those specified for IMT-2020. 3GPP did however go beyond ITU and tightened the latency requirements compared to the IMT-2020 requirements. Table 2.9 presents the 3GPP cMTC requirements. For latency a pair of user and control plane requirements with no associated reliability requirement is defined. The reliability requirement is associated with a 1 ms user plane latency. Chapters 10 and 12 discusses the interpretation of these requirements in further detail.

The Study on scenarios and requirements for next generation access technologies also specifies a set of operational requirements. Worth to notice is the requirement to provide support for a frequency range up to 100 GHz. Through this requirement 3GPP by far extends the in 4G supported range of frequencies to enable both increased capacity and enhanced mobile broadband services. In the *Study on channel model for frequencies from 0.5 to 100 GHz* [29] 3GPP defined new channel models applicable to this extended frequency range for evaluating the NR capabilities. Finally, in the *Study on NR Access Technology Physical layer aspects* [30] 3GPP considered the feasibility of various radio technologies for meeting the set of IMT-2020 and 3GPP 5G requirements.

In Release 15 the *Study on self-evaluation toward IMT-2020* [31] was initiated. It collected the 3GPP NR and LTE performance for the set of evaluation scenarios associated with the IMT-2020 requirements. The results from this work was used as basis for the formal 3GPP 5G submission to ITU. The evaluated performance included the mMTC connection density and the cMTC latency and reliability items. These evaluations are in detail presented in Sections 6, 8, 10 and 12.

TABLE 2.9 3GPP cMTC performance requirements.

Category	Latency	Reliability
User plane	0.5 ms	–
User plane	1 ms	99.999 %
Control plane	10 ms	–

2.4.2.2 5G network architecture

In 3GPP Release 15 the normative work on 5G started. The 3GPP 5G System is defined by the 5G Core network (5GC) and the Next-Generation RAN (NG-RAN). The NG-RAN includes both LTE Release 15 and the NR technologies, which means that both LTE and NR can connect to the 5GC. A base station connecting to the 5GC for providing NR services to the devices is known as a gNB. The base station providing LTE services is known as a ng-eNB. In Release 15 the 5GC does not support Cellular IoT optimizations, including e.g. PSM and eDRX, which are needed to make a connection to LTE-M or NB-IoT relevant. Release 16 have started to address the needed enhancements to prepare 5GC for IoT.

In the 5GC network the *Access and Mobility Management Function* (AMF) provides functionality such as authentication, tracking and reachability of devices in idle mode. The *User Plane Function* (UPF) connects the 5G system to an external packet data network and routes IP packets to the base stations [32]. Fig. 2.8 illustrates the 5G network architecture.

The 5G system supports several architectural options in its first release, i.e. 3GPP Release 15 [33]. In the *Standalone Architecture* NR operates on its own as a standalone system with the gNB responsible for both the control plane signaling and user plane data transmissions as depicted on the right side in Fig. 2.8.

An alternative and highly relevant setup is defined by the *Non-Standalone Architecture* option which is based on *E-UTRAN and NR Dual Connectivity* [34]. In this architecture, which is exemplified in Fig. 2.9, a primary LTE cell with a master eNB carries the EPC control plane signaling from the Mobility Management Entity (MME) over the S1 interface, and optionally also user plane data traffic. A secondary NR cell served by an en-gNB, is configured by the primary cell to carry user

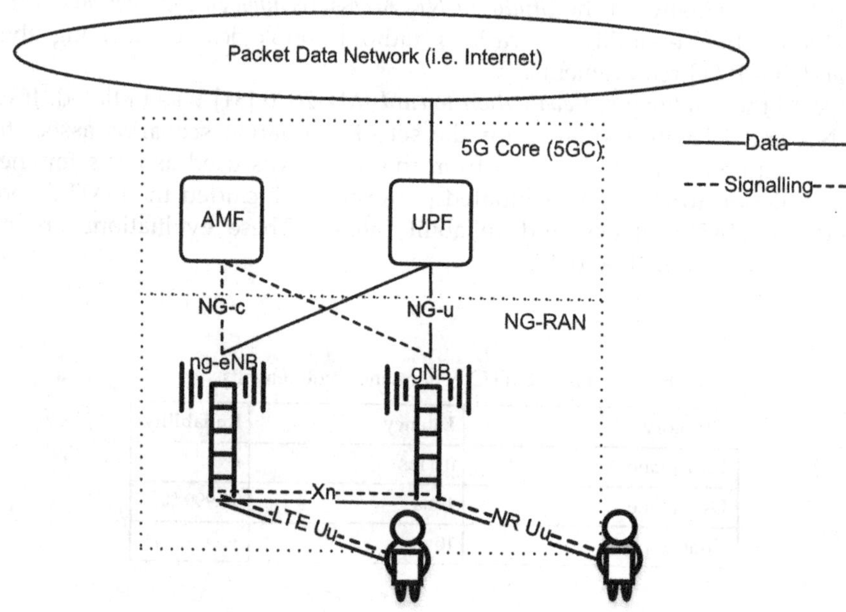

FIG. 2.8 5G Core and the NG-RAN architecture.

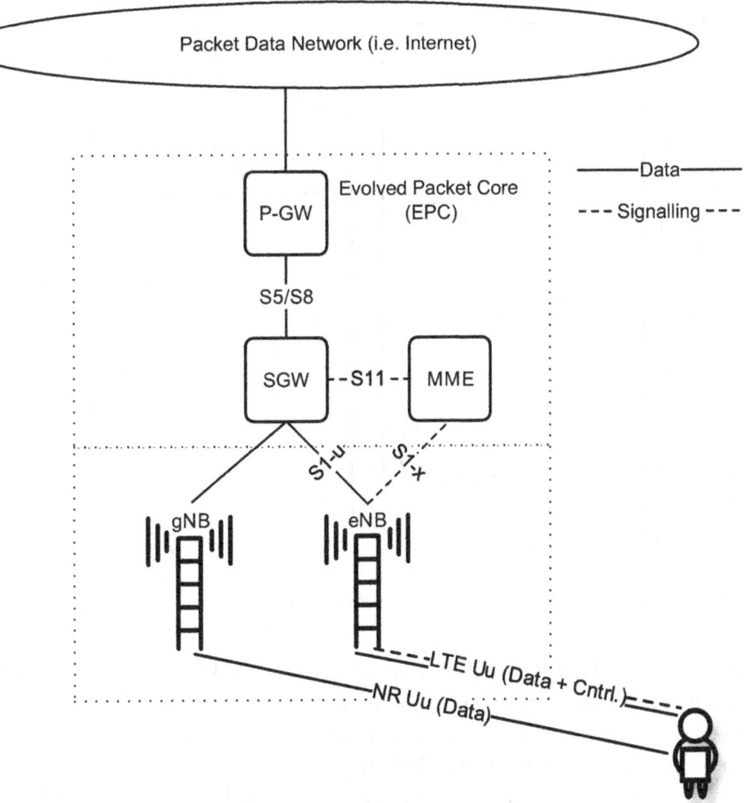

FIG. 2.9 Non-standalone architecture.

plane data from the Serving Gateway (S-GW) over the S1-u interface to add more bandwidth and capacity. This arrangement is intended to facilitate an initial deployment phase of NR during which the systems overall area coverage is expanding. In the E-UTRAN and NR Dual Connectivity solution LTE is intended to support continuous coverage both for the user and control plane. NR can be seen as a complement for boosting the user plane performance toward lower latencies, higher data rates and an overall higher capacity when the coverage permits.

2.4.2.3 *5G radio protocol architecture*

The NG-RAN radio protocol stack is divided in a control plane and a user plane. Fig. 2.10 depicts the radio protocol stack and the interfaces to the 5GC as seen from a device in case the device connects to a gNB over NR. Compared to the LTE radio protocol stack shown in Fig. 2.3 the Service Data Adaptation Protocol (SDAP) is added to the user plane stack. SDAP is responsible for the *QoS* handling, and maps data packets to radio bearers according to their *QoS flow*. The QoS flow is associated with attributes such as the required packet delay budget and packet error rate. A SDAP header is added to the IP packets which contains a *QoS Flow Identifier* (QFI) identifying the QoS flow.

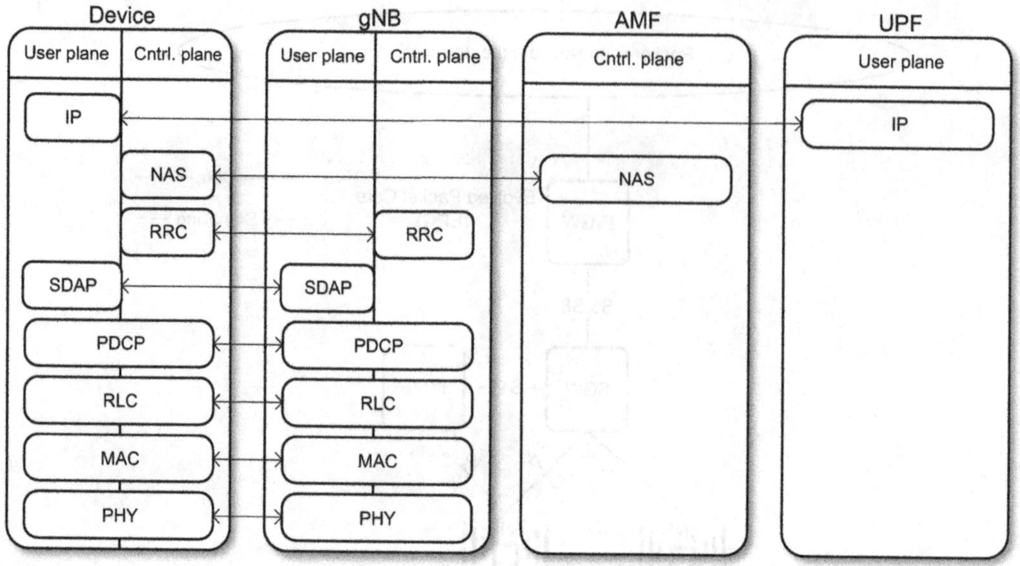

FIG. 2.10 NR control and user plane protocols and interfaces as seen from the device.

The overall functionality provided by the RRC, PDCP, RLC, and MAC layers are similar to that provided for LTE. But important changes have been made to for example facilitate the cMTC use cases with high requirements on low latency and high reliability. Chapter 11 discusses NR URLLC supporting the cMTC use cases.

Next Section 2.4.2.4 provides a high-level overview of the NR physical layer while Section 2.4.2.5 introduces specified mechanisms for supporting NR, LTE-M and NB-IoT coexistence.

2.4.2.4 *NR physical layer*

2.4.2.4.1 Modulation

The NR physical layer is similar to LTE in that the physical layer definitions are based on the *Orthogonal Frequency Division Multiplexing* (OFDM) modulation. LTE supports *cyclic prefix (CP)* based OFDM (CP-OFDM) in the downlink and *Single Carrier Frequency Division Multiple Access* (SC-FDMA), also known as *DFT-Spread-OFDM*, in the uplink. NR supports CP-OFDM in both link directions, and SC-FDMA in the uplink. In the uplink CP-OFDM is intended to facilitate high throughput, e.g. by the use of *multiple input multiple output (MIMO)* transmissions, while SC-FDMA with is reduced *peak to average power ratio* is intended for coverage limited scenarios.

The NR waveforms support modulation schemes PI/2-BPSK, QPSK, 16QAM, 64QAM and 256QAM.

2.4.2.4.2 Numerology

While LTE supports a basic subcarrier spacing of 15 kHz, NR support subcarrier spacings 15, 30, 60 and 120 kHz for its data and control channels. The synchronization signals and Physical Broadcast CHannel (PBCH) in addition supports the option of 240 kHz. This extension is motivated by the increased robustness offered by the larger subcarrier spacings toward higher levels of local oscillator phase noise and Doppler spread that is anticipated in the high frequency bands supported by NR. The larger subcarrier spacings also reduces the transmission time intervals to support lower latencies, which is a fundamental characteristic of cMTC.

For the 15 kHz subcarrier spacing the normal CP is identical to LTE with a length of 4.7 us. For the 30, 60, 120 and 240 kHz options, the NR normal CP length decreases in proportion to the increase in subcarrier spacing relative the basic 15 kHz subcarrier spacing option as seen in Fig. 2.11. The minimum required CP length is determined by the anticipated channel delay spread. The channel delay spread hence set a direct lower limit on the CP length, which for a given accepted CP overhead sets an upper limit on the acceptable subcarrier spacing. To support flexible use of the higher subcarrier spacing options the 60 kHz configuration in addition to the normal CP length of 1.2 us also supports an extended CP of length 4.2 us.

It's worth to mention that the delay spread is expected to be fairly independent of the carrier frequency. But it is typically larger for macro cells, in outdoor deployment scenarios, where low frequency bands are commonly used, than for indoor small cells where high frequency bands are a popular choice. The 15 and 30 kHz subcarrier spacing options with in absolute terms longer CP lengths are suitable for providing macro cell coverage, while the larger subcarrier spacings with shorter CP lengths are more suitable for small cell type of deployments with lower delay spreads. There are obviously exceptions to this rule, and the larger subcarrier spacings are useful as soon as the CP covers the delay spread anticipated in the targeted deployment.

2.4.2.4.3 Time and frequency resources

LTE supports a long list of frequency bands in the range 450 kHz to 5.9 GHz. For NR two ranges of frequency bands are specified, a first range spanning up to 6 GHz, and a second range for the *mm-wave* bands starting from 24 GHz and ending at 52.6 GHz. The 15, 30 and 60 kHz subcarrier spacings are intended for the lower frequency range, while the 60 and 120 kHz numerologies are for the higher frequency range.

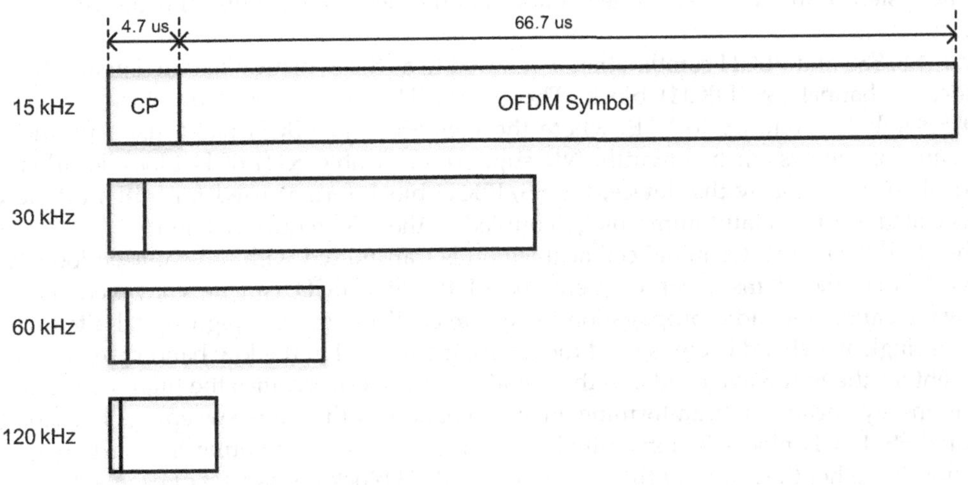

FIG. 2.11 OFDM symbol definition for 15–120 kHz subcarrier spacing.

NR supports frequency bands of width up to 100 MHz in the lower frequency range, and up to 400 MHz in the mm-wave region. The bands can be divided in multiple parts, each known as a *Bandwidth Part* (BWP) with individually configurable numerology. Due to the potentially very large system bandwidth a device does not need to support the full system bandwidth as in LTE, with the exception of LTE-M, and can instead operate on one of the BWPs. BWP adaptation supports adaptation of receive and transmit bandwidths and frequency location as well as the used numerology. This allows for reduced device complexity, device power savings and the use of service optimized numerologies.

The NR frequency grid is defined by *physical resource blocks* (PRB) which just as for LTE is defined by 12 subcarriers. The absolute frequency width of a PRB scales with the configured subcarrier numerology.

Also, the frame structure is dependent on the chosen numerology. The basic radio frame is of length 10 ms and contains 10 subframes each 1 ms long. The subframe contains 1 slot for the 15 kHz numerology. The slot is defined by 14 OFDM symbols for the case of normal CP, and 12 in case of extended CP. The slot length decreases, and the number of slots per subframe increases, as the numerology scales up.

In NR the smallest scheduling format is no longer a subframe. The concept of mini-slots allows 2, 4 or 7 OFDM symbols to be scheduled in the downlink to support low latency services including cMTC. In the uplink a mini-slot can be of any length.

NR do just as LTE, support both frequency and time division duplex modes. In contrary to LTE, in the time division duplex bands the scheduling of subframes can flexibly be configured for uplink or downlink traffic. This to accommodate dynamically changing traffic patterns.

2.4.2.4.4 Initial access and beam management

NR initial access builds on concepts established in LTE. NR does just as LTE make use of the *Primary Synchronization Signal* and *Secondary Synchronization Signal* for physical cell ID acquisition and synchronization to the downlink frame structure of a cell. The NR *PBCH* carries the master information block which contains the most critical system information such as system frame number, system information scheduling details, and the cell barring indication.

The PSS, SSS and PBCH combination is referred to as the Synchronization Signal/Physical Broadcast Channel (SS/PBCH) block. The SS/PBCH block spans 240 subcarriers and 4 OFDM symbols. Contrary to LTE, where the PSS, SSS and PBCH has a fixed position in the center of the system bandwidth, NR supports a flexible SS/PBCH block location. To avoid blind decoding by the device, the SS/PBCH block format used for initial cell access is associated with a default numerology coupled to the NR frequency band.

The SS/PBCH block for initial cell acquisition is transmitted with a default periodicity of 20 ms. Within the 20 ms several repetitions of the SS/PBCH can be conveyed. For the mm-wave bands, the radio propagation losses are challenging with high signal attenuation. While a single wideband beam is used for covering a full cell in the low band case, this is not sufficient for the mm-wave bands. In the initial access, NR overcomes the high band channel conditions by means of beamforming in combination with beam sweeping. Each of the repeated SS/PBCH blocks is transmitted using a narrow beam pointing in a certain spatial direction. Together the beams of the repeated SS/PBCH block forms a set of beams that spans the full spatial dimension of the served cell.

Also, the uplink time and frequency resources dedicated for the NR Physical Random-Access Channel (PRACH) are associated with a range of narrow spatial beams. After determining which of the SS/PBCH blocks, and transmit beams, that offers the best coverage a device select a PRACH time and frequency resource that is associated with a receive beam providing similar spatial coverage as the transmit beam of the selected SS/PBCH block.

Two PRACH sequences are defined. The first is just as the LTE PRACH based on an 839 length *Zadoff Chu* code, while the second is based on a length 139 Zadoff Chu code. For the long Zadoff Chu sequence four different PRACH formats are defined, supporting cell radiuses roughly in the range 15–120 km. The three first formats have inherited the LTE PRACH subcarrier spacing of 1.25 kHz. The fourth is based on 5 kHz subcarrier spacing catering for high speed scenarios. For the short sequence 9 different formats are defined for subcarrier spacings 15, 30, 60 and 120 kHz. These are mainly intended for the high frequency bands.

2.4.2.4.5 Control and data channels

In the downlink NR support the Physical Downlink Control CHannel (PDCCH) and the Physical Downlink Shared CHannel (PDSCH). In the uplink the Packet Uplink Control CHannel (PUCCH) and Packet Uplink Shared CHannel (PUSCH) are specified. The functionality of these channels is inspired by and closely related to the corresponding LTE physical channels. Notable is that the PDCCH bandwidth, in contradiction to LTE, does not need to span the full system bandwidth. The frequency location of the PUCCH is also flexible, and not restricted to the edges of the system bandwidth as is the case for LTE.

NR support *Low-Density Parity-Check* coding, *Polar coding* and *Reed-Muller* block codes. The Low-Density Parity-Check coding is used for the NR data channels and offers good performance for large transport block sizes (TBS). The Polar code gives good performance for short block sizes and is used for the NR control and broadcast channels, with the exception for the shortest control messages where Reed-Muller block codes are used.

NR has significant focus on MIMO technologies and supports eight downlink layers and four uplink layers. Single user MIMO, multi-user MIMO, and beamforming are supported. Beamforming is of particular importance for the high frequency bands to overcome the high attenuation associated with mm-waves.

2.4.2.5 NR and LTE coexistence

To support a gradual frequency refarming from LTE to NR, and to maintain NB-IoT and LTE-M deployments within a NR carrier, NR and LTE coexistence is important. NR support has therefore been specified for the most vital LTE bands. Reservation of NR time-frequency resources dedicated for LTE and NB-IoT transmissions within a NR carrier are supported. The NR PDSCH is rate matched around the reserved resources. The topic of NR and LTE coexistence is in greater detail covered in Chapters 5 and 7.

2.5 MFA

The *MulteFire Alliance* (MFA) is a standardization organization that was established in 2015 with the mission to enable the operation of the 3GPP technologies in unlicensed spectrum.

Its first suite of specifications in MFA Release 1.0 was published in 2017 and builds on 3GPP LTE Release 13 *License Assisted Access* and 3GPP LTE Release 14 *Enhanced License Assisted Access* features. The MFA specifications are based on the 3GPP specifications, with the addition of the needed modifications to support operation in unlicensed spectrum in various regions.

The main technical work in MFA is performed in a set of working groups (WGs) that are organized under a single technical specification group (TSG). The *Radio WG* is the largest WG. It focuses on the lower layers in the radio protocol stack. The *Minimum performance specification* WG defines the MFA radio requirements. The *End-to-end architecture* WG focuses on architectural aspects and higher layer specifications.

Besides the TSG, MFA contains the *Industry WG* and the *Certification group*. The Industry working group is working on the identification of services requirements with a focus on industrial applications. In general MFA aims to give significant attention to industrial use cases. The *Certification group* defines the test specification for the certification of MFA devices.

In Release 1.1, published in 2019, MFA defined versions of LTE-M and NB-IoT supporting operation in unlicensed frequency bands, which are referred to as *eMTC-U* and *NB-IoT-U* in the MFA specifications. Hereafter in this book eMTC-U will be referred to as *LTE-M-U*. Both these new technologies are in detailed described in Chapter 15.

References

[1] Third Generation Partnership Project, Technical Report 45.820, v13.0.0. Cellular System Support for Ultra-Low Complexity and Low Throughput Internet of Things, 2016.

[2] Third Generation Partnership Project, Technical Report 22.868, v8.0.0. Study on Facilitating Machine to Machine Communication in 3GPP Systems, 2016.

[3] Third Generation Partnership Project, TSG SA WG2. SP-100863. Update to Network Improvements for Machine Type Communication, 3GPP TSG SA, Meeting #50, 2010.

[4] Third Generation Partnership Project, Technical Specification 22.011, v14.0.0. Service Accessibility, 2016.

[5] Third Generation Partnership Project, Technical Specification 44.018, v14.00. Mobile Radio Interface Layer 3 Specification, Radio Resource Control (RRC) Protocol, 2016.

[6] Third Generation Partnership Project, Technical Specification 23.060, v14.0.0. General Packet Radio Service (GPRS), Service Description, Stage 2, 2016.

[7] Third Generation Partnership Project, Technical Specification 24.008, v14.0.0. Mobile Radio Interface Layer 3 Specification; Core Network Protocols, Stage 3, 2016.

[8] Third Generation Partnership Project, Technical Specification 23.401, v14.0.0. General Packet Radio Service (GPRS) Enhancements for Evolved Universal Terrestrial Radio Access Network (E-UTRAN) Access, 2016.

[9] Third Generation Partnership Project, Technical Specification 36.331, v14.0.0. Evolved Universal Terrestrial Radio Access (E-UTRA), Radio Resource Control (RRC), Protocol Specification, 2016.

[10] Third Generation Partnership Project, TSG SA WG3, SP-120848. System Improvements for Machine Type Communication, 3GPP TSG SA Meeting #58, 2012.

[11] Third Generation Partnership Project, TSG SA WG3, SP-130327. Work Item for Machine-Type and Other Mobile Data Applications Communications Enhancements, 3GPP TSG SA Meeting #60, 2013.

[12] Third Generation Partnership Project, Technical Report 23.887, v12.0.0. Study on Machine-Type Communications (MTC) and Other Mobile Data Applications Communications Enhancements, 2016.

[13] Third Generation Partnership Project, Technical Specification 36.300, v14.0.0. Evolved Universal Terrestrial Radio Access (E-UTRA) and Evolved Universal Terrestrial Radio Access Network (E-UTRAN), Overall Description, Stage 2, 2016.

[14] Third Generation Partnership Project, Technical Specification 23.682, v14.0.0. Architecture Enhancements to Facilitate Communications with Packet Data Networks and Applications, 2016.

[15] Third Generation Partnership Project, Technical Specification 24.301, v14.0.0. Non-Access-Stratum (NAS) Protocol for Evolved Packet System (EPS), Stage 3, 2016.
[16] Third Generation Partnership Project, Technical Report 43.869. GERAN Study on Power Saving for MTC Devices, 2016.
[17] Third Generation Partnership Project, Technical Specification 45.002, v14.0.0. Multiplexing and Multiple Access on the Radio Path, 2016.
[18] Third Generation Partnership Project, Technical Specification 43.064, v14.0.0. General Packet Radio Service (GPRS). Overall Description of the GPRS Radio Interface Stage 2, 2016.
[19] Third Generation Partnership Project, Technical Report 36.888, v12.0.0. Study on Provision of Low-Cost Machine-Type Communications (MTC) User Equipment's (UEs) Based on LTE, 2013.
[20] Third Generation Partnership Project, Technical Report 36.211, v15.0.0. Evolved Universal Terrestrial Radio Access (E-UTRA); Physical channels and modulation, 2018.
[21] Vodafone Group, et al. RP-140522 Revised Work Item on Low Cost & Enhanced Coverage MTC UE for LTE, 3GPP RAN Meeting #63, 2014.
[22] Ericsson, et al. RP-141660, Further LTE Physical Layer Enhancements for MTC, 3GPP RAN Meeting #65, 2014.
[23] Third Generation Partnership Project, Technical Specification 45.005, v14.0.0. Radio Transmission and Reception, 2016.
[24] Ericsson, et al. GP-151039, New Work Item on Extended Coverage GSM (EC-GSM) for Support of Cellular Internet of Things, 3GPP TSG GERAN, Meeting #67, 2015.
[25] Qualcomm Incorporated, et al. RP-151621, Narrowband IOT, 3GPP TSG RAN Meeting #60, 2015.
[26] Third Generation Partnership Project, Technical Report 36.881, v14.0.0. Study on latency reduction techniques for LTE, 2018.
[27] ITU-R. Report ITU-R M.2410, Minimum requirements related to technical performance for IMT-2020 radio interfaces(s), 2017.
[28] Third Generation Partnership Project, Technical Report 38.913, v15.0.0. Study on Scenarios and Requirements for Next Generation Access Technologies, 2018.
[29] Third Generation Partnership Project, Technical Report 38.901, v15.0.0. Study on channel model for frequencies from 0.5 to 100 GHz, 2018.
[30] Third Generation Partnership Project, Technical Report 38.802, v15.0.0. Study on New Radio Access Technology Physical layer aspects, 2018.
[31] Third Generation Partnership Project, Technical Report 37.910, v15.0.0. Study on self-evaluation towards IMT-2020, 2018.
[32] Third Generation Partnership Project, Technical Specification 38.300, v15.0.0. NR; NR and NG-RAN overall description, 2018.
[33] NTT DOCOMO. INC. RP-181378 WI Summary of New Radio Access Technology, 3GPP TSG RAN Meeting #80, 2017.
[34] Third Generation Partnership Project, Technical specification 37.340, v15.0.0. NR; Multi-connectivity, Overall description, Stage 2, 2018.

CHAPTER

3

EC-GSM-IoT

OUTLINE

Cellular Internet of Things, *Second Edition*
https://doi.org/10.1016/B978-0-08-102902-2.00003-0

© 2020 Elsevier Ltd. All rights reserved.

Abstract

This chapter presents the design of Extended Coverage Global System for Mobile Communications Internet of Things (EC-GSM-IoT). The initial section describes the background of the GSM radio access technology, highlighting the suitability of an evolved GSM design to support the Cellular IoT (CIoT) core requirements, which includes ubiquitous coverage, ultra-low-device cost, and energy efficient device operation. The following sections builds from the ground up, starting with the physical layer, going through fundamental design choices such as frame structure, modulation, and channel coding. After the physical layer is covered, the basic procedures for support of full system operation are covered, including, for example, system access, paging functionality, and improved security protocols. The reader will not only have a good knowledge of EC-GSM-IoT after reading the chapter but will also have a basic understanding of what characteristics a system developed for CIoT should possess. At the end, a look at the latest enhancements of the EC-GSM-IoT design is presented.

3.1 Background

This section provides the background as to why the standardization body 3GPP decided to undertake the extensive work of redesigning the since long existing GSM technology toward Cellular IoT (CIoT). The perspective is provided based on the market situation at the time when these improvements were first considered, i.e. around 2014–15.

3.1.1 The history of GSM

The GSM technology was first developed in Europe in 1980s, having its first commercial launch in 1991. In current writing (2019), the technology has already turned 28 years. Despite this it still maintains global coverage and is still among the most widely used cellular technologies. This is even true in the most mature markets, such as western Europe. Compared to previous analog cellular technologies, GSM is based on digital communication that allows for an encrypted and more spectrally efficient communication. The group developing the GSM technology was called Group Spécial Mobile (GSM) but with the global success of the technology it is now referred to as Global System for Mobile Communications (GSM).

The initial releases of the GSM standard were only defined for *circuit switched* (CS) services, including both voice and data calls. A call being CS implies that a set of radio resources are occupied for the duration of the call. Even if the transmitter is silent in a CS call, the resources are occupied from a network perspective and cannot be allocated to another user.

In 1996 work was started to introduce *packet switched* services (PS). A PS service no longer occupies the resources for the full duration of the call, but instead intends to only occupy resources when there are data to send. The first PS service was called General Packet Radio Service (GPRS) and was launched commercially in 2000. Following the success of GPRS, the PS domain was further enhanced by the introduction of Enhanced GPRS, also known as Enhanced Data Rates for GSM Evolution (EDGE), supporting higher end user data rates, primarily by the introduction of higher order modulation and improved protocol handling. In current GSM/EDGE networks, CS services are still used for speech calls, and the PS service is predominantly used for providing data services.

Since the deployment of the first GSM network in 1991, the technology has truly become the global cellular technology. It is estimated that it today reaches over 90% of the world's population [1] as it is deployed in close to all countries in the world. In all of these networks, voice services are supported and in the vast majority of networks there is also support for GPRS/EDGE.

3.1.2 Characteristics suitable for IoT

3.1.2.1 Global deployment

Considering that GSM has been deployed since 27 years, it is only natural that the characteristics of GSM/EDGE networks globally are vastly different. In countries where GSM/EDGE is deployed together with 3G and/or 4G, it is typically used as one of the main carriers of speech services, while serving as a fallback solution for data services. However, it is not always the case that devices are capable of 3G or 4G, even if the networks in many cases are. For example, in the Middle East and Africa around 75% of subscriptions 2015 were GSM/EDGE-only [1]. Furthermore, looking at the global subscription base of cellular technologies, GSM/EDGE-only subscriptions account for around 50% of the total subscription base [1]. For the remaining 50%, the vast majority of subscriptions supporting 3G and/or 4G technologies will also include GSM/EDGE capability as fallback when coverage is lost. This global presence is beneficial for all radio access technologies, including those providing Machine-Type Communication, e.g., not only to support roaming but also to reach a high level of economy of scale.

3.1.2.2 *Number of frequency bands*

An important aspect of the global success of GSM lies also in the number of frequency bands it is deployed in. Although the GSM specifications support a wider range of frequency bands, the deployment of GSM/EDGE is limited to four global frequency bands: 850, 900, 1800, and 1900 MHz. This global allocation of just four spectrum bands is much more aligned compared with other cellular technologies. The spectrum regulations used in different parts of the world typically pair the use of 900 MHz with 1800 MHz and 850 MHz with 1900 MHz. This means that in most regions only two of the frequency bands are allowed for GSM operation. For a device manufacturer, this aspect is important because to provide a single band or *dual band* device (one low band and one high band combination), which will already cover a large part of the world population, while a *quad band* device supporting all four bands truly provides global coverage. Having to support a lower number of frequency bands means less *radio frequency* (RF) components needed in the device, less optimization of components that are to be operable over multiple frequency bands, and in the end a lower *Bill of Material* and overall development cost.

Another important characteristic for cellular systems, especially for those targeting IoT, is to have extensive coverage to reach remote locations, such as deep indoor. Operation of GSM, as well as 3G and 4G technologies, is today defined from a few hundreds of MHz to a couple of GHz in the radio spectrum. It is a well known fact that the propagation loss is dependent on the carrier frequency, and that it is important to keep the carrier frequency low to minimize the path loss due to signal propagation and by that maximize the coverage provided. Because GSM is globally available either on the 900 MHz band or the 850 MHz band, it is well suited for a technology attempting to maximize coverage because the loss due to propagation is low.

Another important aspect of frequency band support in a device is roaming, which is also an important feature for CIoT. Take, for example, a tracking device where a module is integrated in a container, shipped around the world. If such a device would have to support a large variety of frequency bands just to be able to operate in the country it is shipped to, it will drive device complexity and cost.

In summary, the number of frequency bands deployed is an important factor when it comes to avoiding market fragmentation, reaching economies of scale, and allow for global roaming of devices.

3.1.2.3 *Small spectrum deployment*

Although GSM has a truly global footprint, 3G and 4G compete for the same scarce spectrum as GSM. With the advent of a fifth generation cellular system even more pressure will be put on the spectrum resources. Deployment of new technologies not only changes what technologies are available for the end consumer but will also impact how the traffic between the technologies is distributed. As the traffic shifts from GSM to 3G and/or 4G, it is possible to release parts of the spectrum used for GSM/EDGE and make room for the increased spectrum needs for improved end user experience and system capacity coming from the new technologies. This is referred to as *spectrum refarming*.

Refarming of GSM spectrum was started several years ago and will continue in the future. However, even in mature cellular markets there is one thing to reduce the spectrum operation of GSM and another to turn the GSM network completely off. Even if there are only, let's say, a few tenths of thousand devices in the network there might be contracts/subscriptions with the operator that are not possible to end prematurely. This is especially true for the *Machine-to-Machine* (M2M) market where, for example, devices can be placed in remote locations with contracts that can last for several decades. So, the spectrum allocated for GSM services is expected to shrink with time, in many markets the GSM networks will live for still a long time to come. To make future GSM deployments attractive for the operator, it is of interest to deploy the network as spectrally efficient as possible, in an as low spectrum allocation as possible.

3.1.2.4 Module price

The low price on GSM/EDGE devices is probably one of the more important reasons why GSM is the main M2M cellular carrier in networks today. In Fig. 3.1 [2], the M2M module price is estimated for a range of cellular technologies up to 2016. As can be seen, the GSM/EDGE module cost is considerably lower than other technologies reaching a global selling price of around USD 6–7. To this selling price there are of course regional variations, and, for example, the estimated average selling price in China for the same module is USD 4 [2]. There are several reasons why such low-average selling price can be achieved of which most have already been mentioned, such as a global deployment providing economies of scale, a mature and relatively low-complex technology that has been optimized in product implementations over the last 25 years, low number of frequency band-specific RF components due to the low number of frequency bands used for the technology.

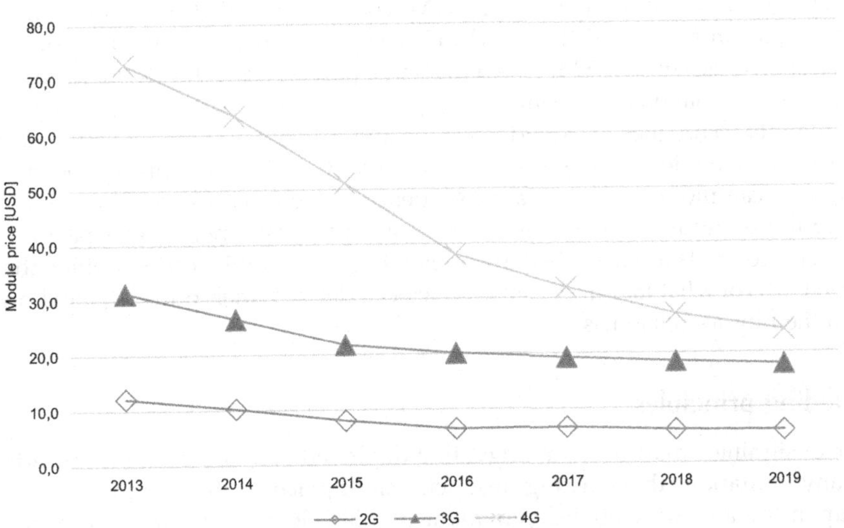

FIG. 3.1 M2M module price by technology. *From: IHS Markit, Technology Group, Cellular IoT Market Tracker – 1Q 2017.*

3.1.3 Enhancements undertaken by 3GPP

It was based on the knowledge above that Third Generation Partnership Project (3GPP) in its 13th release decided to further evolve GSM to cater for the requirements coming from the IoT. Beyond the requirements mentioned in Sections 1.2.3 and 2.5 to increase coverage, secure low-device cost, and high battery lifetime the work included the following:

- Provide a means to operate the technology in an as tight spectrum allocation as 600 kHz for an operator to deploy a GSM network completely removing, or at least minimizing, a conflict in spectrum usage with other technologies. In such small spectrum allocation, it can even be possible to deploy the GSM network in the guard band of wideband technologies such as 3G and 4G.
- Improve end user security to an (Long-Term Evolution) LTE/4G-grade security level to remove any security concerns that may exist in current GSM deployments.
- Ensure that all changes brought to the GSM standard by the introduction of EC-GSM-IoT ensure backward compatibility with already existing GSM deployments, to allow a seamless and gradual introduction of the technology sharing resources with existing devices.
- Ensure support of a massive number of CIoT devices in the network.

The remainder of this chapter will in detail outline how the above design guidelines have been followed by redesigning the GSM standard, and in Chapter 4 the performance evaluation of the technology will show how the performance objectives are fulfilled.

3.2 Physical layer

In this section the physical layer of EC-GSM-IoT is described. The section starts with the guiding principles in the overall EC-GSM-IoT design. The reader is then introduced to the ;basic physical layer design of GSM, which to a large part is reused for EC-GSM-IoT, including frame structure, modulation, and channel coding. The two main parts of this section are then in detail looking into techniques to extend coverage and increase system capacity. Extending the coverage is the new mode of operation introduced by EC-GSM-IoT, while the need to increase system capacity can more be seen as a consequence of the extended coverage where devices will take up more resources to operate in the system, having a negative impact on capacity. When the main concepts are described, the remaining parts of this section outline the network operation in terms of what logical channels have been defined, their purpose, and how they are mapped to the physical channels.

3.2.1 Guiding principles

When redesigning a mature technology, first deployed 27 years ago, it is important to understand any limitations that existing products might place on the design. For EC-GSM-IoT to build upon the already global deployment of GSM, it is of uttermost importance to be able to deploy it on already existing base stations in the field. However, it is also important for an existing GPRS/EGPRS device to be able to upgrade its implementation to either replace the GPRS/EGPRS implementation or to implement EC-GSM-IoT in addition to a GPRS/EGPRS implementation. In both cases, it is important for the design of EC-GSM-IoT to have a common design base as GPRS/EGPRS as possible.

A base station can be considered to consist of two main parts, a digital unit (or a baseband unit) and a radio unit. Supporting EC-GSM-IoT, in addition to GSM/EDGE, will mean that additional implementation effort is needed, at least on the digital unit to implement the new protocol stack and new physical layer designs. However, the support of the new feature cannot imply that hardware upgrades are needed, which would mean that products in the field need to be replaced. Assume, for example, that the base station's required processing power or sampling rate is increased to an extent where it is no longer possible to support it on existing digital units. This would imply a major investment for the operator to replace the base station hardware. In addition, for a technology that aims for extremely good coverage, it would result in a spotty network coverage until most/all of the base stations have been upgraded. Similarly, the radio unit has been designed with the current GSM signal characteristics in mind and, significantly, changing those such as increasing the signal dynamics could mean that the technology can no longer be supported by already deployed infrastructure.

A similar situation exists on the device side where a change to the physical layer that requires a new device platform to be developed would mean a significant investment in research, development, and verification. If in contrast, the basic principles of the physical layer are kept as close to GSM as possible, an already existing GPRS/EDGE platform would be able to be updated through a software upgrade to support EC-GSM-IoT, and if the additional complexity from EC-GSM-IoT is kept to a minimum, the same platform would be able to also support GPRS/EDGE operation. This, of course, does not prevent development of an EC-GSM-IoT-specific platform, more optimized in, for example, energy consumption and cost than the corresponding GPRS/EGPRS platform. However, if existing platforms can be used, a gradual introduction of the feature onto the market without huge investment costs is possible.

In addition to the aspects of product implementation mentioned above, also the network operation of EC-GSM-IoT needs to be considered. GSM networks have been operable for many years and are planned to be operable for many years to come as discussed in Section 3.1. The technology has hence been designed to be fully backward compatible with existing GSM/EDGE deployments and network configurations to seamlessly be able to multiplex GPRS/EDGE traffic with EC-GSM-IoT traffic on the same resources. By pooling the resources in this way, less overall resources will be consumed in the GSM network, and less spectrum resources will, in the end, be required for network operation.

Following these guiding principles will naturally mean that the physical layer to a large extent will be identical to already existing GSM/EDGE. At the same time, changes are required to meet the design objectives listed in Section 3.1.3.

The following section will, to some extent, be a repetition of the physical layer of GSM to serve as a basis for understanding EC-GSM-IoT. Focus will, however, be on new designs that have been added to the specifications by the introduction of EC-GSM-IoT, and details of the GSM physical layer that is not relevant for the understanding of EC-GSM-IoT have intentionally been left out.

3.2.2 Physical resources

3.2.2.1 *Channel raster*

GSM is based on a combination of frequency division multiple access and time division multiple access (TDMA).

The channels in frequency are each separated by 200 kHz, and their absolute placement in frequency, the so-called *channel raster*, is defined also in steps of 200 kHz. This means that the placement of the 200 kHz channels in frequency is not completely arbitrary, and that the center frequency of the channel needs to be divisible by 200 kHz. Because the placements in frequency, in a given frequency band, are limited by the channel raster, the channels can be numbered for easier reference and are referred to as *absolute RF carrier number* (*ARFCN*). That is, for a given frequency band and absolute carrier frequency the ARFCN value is fixed.

3.2.2.2 *Frame structure*

In time a frame structure is defined. Each, so-called *TDMA* frame, is divided into eight timeslots. To reference a specific point in time that exceeds the duration of a TDMA frame, the TDMA frames are grouped into hierarchical frame structure including *multiframes*, *superframes*, and *hyperframes*. The time reference in the overall frame structure is within a hyperframe accuracy, being roughly 3.5 h long.

To start with, the TDMA frame is grouped into one of two multiframes, either a 51 multiframe or a 52 multiframe.

The 51 multiframe carries channels that are used for time and frequency synchronization (Frequency Correction CHannel (FCCH), SCH, and EC-SCH), (common control channels and Extended Coverage Common Control CHannel (EC-CCCH), see Section 3.2.6 for more details), and (broadcast channels (BCCH) and Extended Coverage BroadCast CHannel (EC-BCCH), see Section 3.2.6 for more details) and are always mapped onto the broadcast carrier (the BCCH carrier).

The 52 multiframe is used by the packet traffic channels (Packet Data Traffic CHannel (PDTCH) and Extended Coverage Packet Data Traffic CHannel (EC-PDTCH), see Sections 3.2.6 and 3.2.7 for the downlink (DL) and uplink (UL), respectively) and their associated control channels (*Packet Associated Control CHannel* and Extended Coverage Packet Associated Control CHannel (EC-PACCH), see Sections 3.2.6 and 3.2.7 for the downlink and uplink, respectively).

The use of two different multiframes has its explanation in that traditionally a GSM device assigned resources on a packet traffic channel (52 multiframe) would still need to continuously monitor its surrounding environment by synchronizing to neighboring cells and to acquire cell-specific information (System Information (SI)). By ensuring that the two multiframes are drifting relative to each other over time, the channels of interest in the 51 multiframe will drift over time relative to a certain position in the 52 multiframe, and hence will not overlap over time, which would prevent acquisition of the traffic channel in the serving cell, and information from neighbor cells. For EC-GSM-IoT, there is no requirement to synchronize to or to acquire SI from neighboring cells, while being assigned packet traffic channel resources. Hence, there is no requirement from that perspective for a 51 multiframe to be used. However, since the existing FCCH, which is already mapped to a

51 multiframe, also has been chosen to be used for EC-GSM-IoT, the use of the 51 multiframe also for some of the EC-GSM-IoT logical channels is natural.

A set of 26 51 multiframes or 25.5 52 multiframes form a superframe. The superframes are in their turn grouped in sets of 2048, each set forming a hyperframe.

The overall frame structure is shown in Fig. 3.2.

Now that the frame structure is covered, let us turn our attention to the slot format. As stated above, and as shown in Fig. 3.2, each TDMA frame consists of eight TSs.

When using blind physical layer transmissions (see Section 3.2.8.2), which is a new transmission scheme introduced by EC-GSM-IoT, it is of importance for the receiver to be able to coherently combine the transmissions to be able to maximize the received signal-to-interference-plus-noise power ratio (SINR). Even a fractional symbol offset in the timing of the transmitter and receiver will cause suboptimum combination and, hence, loss in performance. Therefore, integral symbol length TS is defined for EC-GSM-IoT, as shown in Fig. 3.3. Considering that in GSM, it is allowed for each slot to have a duration of 156.25 symbols, the slot length need to alternate between 156 symbols and 157 symbols, to fit within the same TDMA frame duration.

3.2.2.3 *Burst types*

Each time slot is carrying a burst, which is the basic physical transmission unit in GSM. Different *burst types* are used depending on the logical channel and its use. For EC-GSM-IoT, five different burst types are used: *frequency correction bursts* (FB), *synchronization bursts* (SB), *access bursts* (AB), *dummy bursts* (DB), and *normal bursts* (NB).

Common to all burst types is that they per definition all occupy a full slot, even if active transmission is not required in the full burst. That is, different guard periods are used for the different burst types to extend the burst to the full slot. Common to the SB, AB, and NB is that they all contain tail bits, a training sequence/synchronization sequence, and encrypted bits (payload). Both the FB and the DB consist only of a field of fixed bits, apart from the tail bits. All burst types are shown in Fig. 3.4 with the number of bits per burst field in brackets.

As can be seen, all bursts occupy 156 or 157 symbols (depending on the TS they occupy, see Fig. 3.3). The burst types can be shortly described as:

- **Frequency correction burst**: The burst type is only used by the FCCH, see Section 3.2.6, and consists of 148 bits of state "0." Due to the properties of *Gaussian Minimum Shift Keying* (GMSK) modulation, the signal generated will be of sinusoidal waveform over the burst length, giving rise to a spike in the signal spectrum at 67.7 kHz $\left(\frac{f_s}{4}\right)$. The burst is used by the device to identify a GSM frequency, synchronize to the cell, and to perform a rough alignment in frequency and time with the base station reference structure. The channel is also, together with EC-SCH, used by the device in the cell (re)selection procedure, as well as in the *coverage class* (CC) selection, see Section 3.3.1.4.
- **Synchronization burst (SB)**: Similarly, to the FB, the SB is also only used by one type of logical channel; SCH and EC-SCH, see Section 3.2.6. After performing rough frequency and time synchronization using the FCCH, an EC-GSM-IoT device attempts to acquire the EC-SCH to more accurately get synchronized in time and frequency. This process also includes acquiring the frame number. Considering that the device will only be roughly synchronized to the base station reference after FCCH synchronization, the

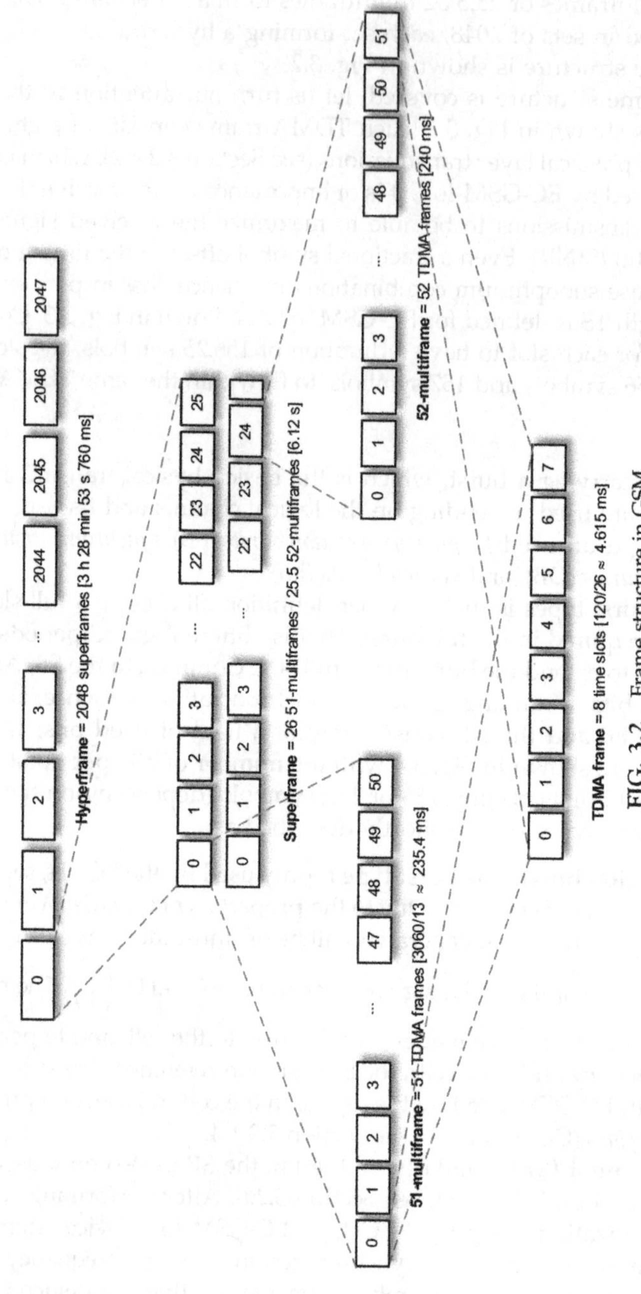

FIG. 3.2 Frame structure in GSM.

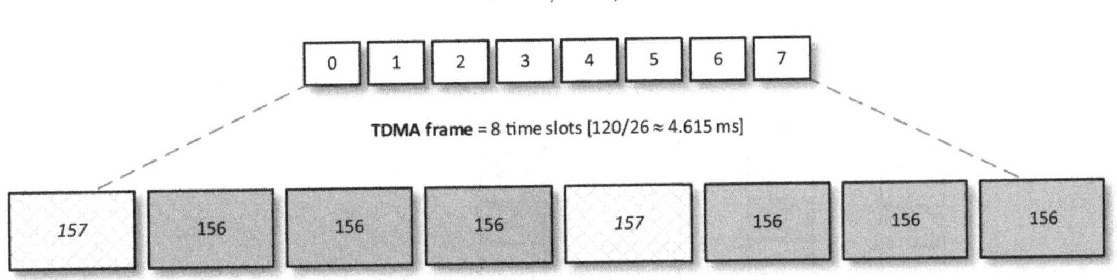

FIG. 3.3 Integral symbol length slot structure.

synchronization sequence of the EC-SCH has been designed longer than, for example, for the NB (which is a burst type only monitored by the device after fine synchronization) to provide a more reliable acquisition.

- **Access burst (AB)**: The AB is used when accessing the network on the Random Access CHannel/Extended Coverage Random Access CHannel (RACH/EC-RACH) after synchronization to the downlink frame structure has been obtained. To support synchronization to the uplink frame structure it has been designed with a longer guard period to be able to support a wide range of cell radii. The AB also contains a longer synchronization sequence because the base station does not know where to expect the AB due to the propagation delay (the burst will arrive at different positions within the slot depending on the distance between the device and the base station). Hence, a longer synchronization window in the base station receiver is needed compared to, for example, receiving a NB. The payload of the access burst typically contains an access cause and a random reference to avoid contention with other simultaneous accesses, see Section 3.3.1.7. The AB can also be used by the EC-PACCH channel when the network wants to estimate the *timing advance* (TA) of the device.
- **Normal burst**: The NB is the burst type used for most logical channels, including the EC-BCCH, EC-CCCH (DL), EC-PDTCH, and EC-PACCH. It consists of a 26-symbol-long training sequence code (TSC) and 58 payload symbols on each side of the TSC.
- **Dummy burst**: The DB could be seen as a special type of NB where the TSC and the payload part are replaced by a fixed bit pattern. It is only used on the Broadcast Channel (BCCH) when no transmission is scheduled for the network to always transmit a signal. The signal is used by devices in the network to measure on serving and neighboring cells.

3.2.3 Transmission schemes

3.2.3.1 Modulation

The symbol rate, f_s, for GSM and EC-GSM-IoT is $13 \times 10^6/48$ symbols/s or roughly 270.83 ksymbols/s, resulting in a symbol duration of roughly 3.69 μs. With a symbol rate exceeding the 200 kHz channel bandwidth there is a trade-off between the spectral response of the modulation used and by that the protection of the neighboring channels and the *inter-symbol interference* (ISI), caused by the modulation. In general, a contained duration in time

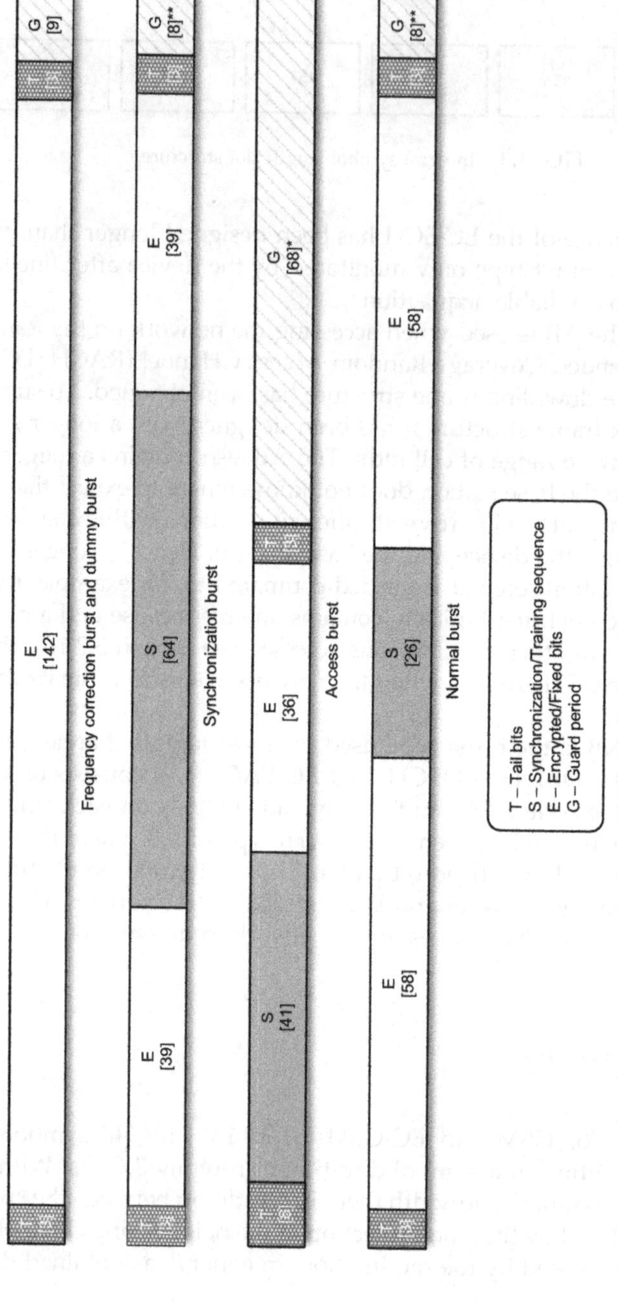

FIG. 3.4 Burst types.

of the modulation means less ISI, but results in a relatively wider frequency response, compared with a modulation with longer duration.

The basic modulation scheme used in GSM, and also by EC-GSM-IoT, is GMSK. GMSK modulation is characterized by being a constant envelope modulation. This means that there are no amplitude variations in the modulated signal, and the information is instead carried in the phase of the waveform. GMSK modulation is furthermore characterized by the *BT-product*, where B is the half-power, or −3 dB, Bandwidth and T is the symbol Time duration. The BT-product in GSM is 0.3 meaning that the double-sided −3 dB bandwidth is 162.5 kHz ($2 \times 0.3 \times 13 \times 10^6/48$). A BT-product of 0.3 also gives rise to an ISI of roughly 5 symbol periods (the duration of the modulation response where a nonnegligible contribution of other symbols can be observed). That is, already in the modulation process at the transmitter, distortions of interference from other symbols are added to the transmitted waveform. This result in a relatively spectrally efficient transmission with contained spectral characteristics, at the expense of increased complexity at the receiver that need to resolve the introduced ISI. The channel propagation may result in additional ISI and the filtering of the signal at the receiver. With a matched filter implementation in the receiver, i.e., the same filter response used as in the transmitter, the GMSK modulation provides roughly an 18 dB suppression of interference coming from adjacent channels. For more details on GMSK modulation and how to generate the modulated signal, see for example [3].

It is possible to make a linear decomposition of the nonlinear GMSK modulation, as shown by P. Laurent [4]. By using the main pulse response of the decomposition, a good approximation of GMSK can be obtained by the convolution of the main pulse with a $\pi/2$ rotated Binary Phase Shift Keying modulation. The same pulse shape is also the one defined for 8PSK modulation used by EGPRS and EC-GSM-IoT, and is referred to as a "Linearized GMSK pulse." Before pulse shaping, the 8PSK constellation is rotated by $3\pi/8$ radians to minimize the peaks in signal level caused by the ISI and also to avoid zero-crossings in the IQ-diagram. The knowledge of the modulation-specific rotation is also used by the receiver for the purpose of modulation detection.

In Fig. 3.5 the IQ-trace for both GMSK and 8PSK modulation is exemplified together with the amplitude response over the modulated bursts (mapped to a TS in the TDMA frame).

As can be seen, the GMSK modulation exhibits a constant envelope over the full burst, except for the start and stop of the burst, where the signal is ramping up and down in power, respectively. For 8PSK the situation is different with significant amplitude variations also within the burst. If one would look at the long-term characteristics of the 8PSK signal, one would see that peak-to-average-power ratio of the signal, i.e., how the absolute peak power relates to the average power of the signal, is 3.2 dB. That is, the peak signal power is roughly twice that of the average power. In addition, worth noting is that there are no zero-crossings of the signal in the IQ-diagram and this would also hold for the long-term characteristics, because of the symbol rotation angle used.

For EC-GSM-IoT, the modulation supported by the device can either be GMSK only or GMSK and 8PSK. The reason to allow only GMSK modulation to be supported is to allow for an ultra-low-cost device, for more information see Section 4.7. In terms of RF characteristics, the constant envelope modulation will allow for a power efficient implementation where the power amplifier (PA), in simplistic terms, can be optimized for a single operating point. In contrast, a PA used to support 8PSK modulation would have to be dimensioned for relatively

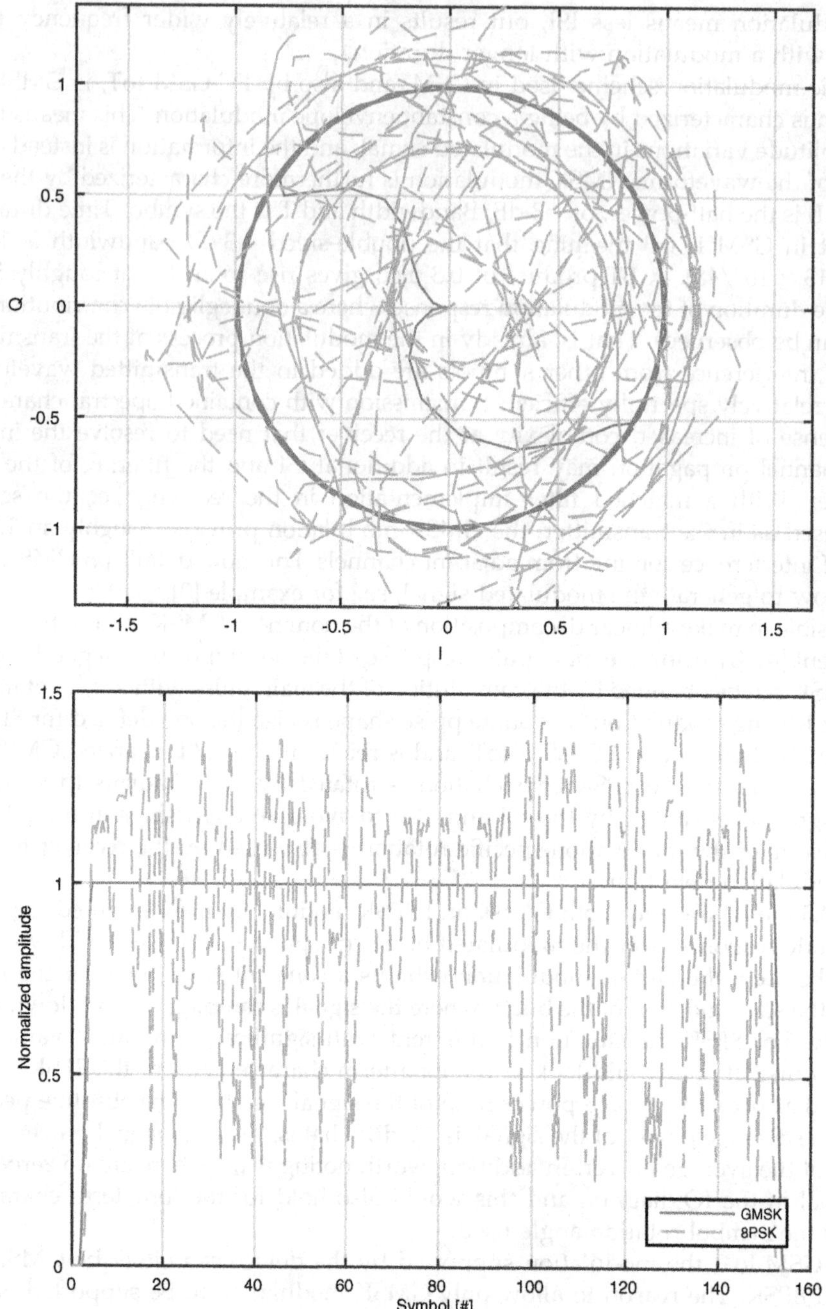

FIG. 3.5 Examples of IQ-trace (top) and amplitude response (bottom) of a GMSK and 8PSK modulated bursts.

rarely occurring peaks with twice the amplitude as the average signal. Although there are techniques that help bridge the gap in power efficiency, such as tracking the envelope of the signal, a difference is still there. In addition, considering the simple signal characteristics, only carrying information using the phase of the signal, a GMSK PA can, to a larger extent, distort the amplitude of the signal without impacting the performance of the radio link. Hence, apart from being more power efficient, a GMSK implementation can also be more cost and energy effective.

3.2.3.2 *Blind transmissions*

Blind transmissions, also referred to as *blind repetitions,* are the means by which an EC-GSM-IoT transmitter, instead of transmitting a block only once, transmits a predefined number of transmissions without any feedback from the receiving end indicating to the sending end that the block is erroneous. This is one of the main mechanisms used to extend the coverage of EC-GSM-IoT. Its usefulness is in detailed elaborate upon in Section 3.2.8.2.

A similar concept is used to extend the link coverage for NB-IoT and *LTE Machine-Type Communications* but is then simply referred to as *repetitions.*

3.2.3.3 *Coverage Classes*

To keep the EC-GSM-IoT feature implementation simple, at most four different numbers of blind repetitions are defined for any given logical channel. Each number is referred to as a CC and, hence, four different CCs are defined (CC1, CC2, CC3, and CC4). Logical channels using the concept of CCs are EC-CCCH, EC-PDTCH, and EC-PACCH. For synchronization and broadcast channels, EC-SCH and EC-BCCH, only a single set of blind transmissions are defined, dimensioned to reach the most extreme coverage conditions expected in the cell.

In Table 3.1 the blind transmissions defined and the associated CCs (where applicable) is shown.

3.2.4 Channel coding and interleaving

This section intends to give an introduction to the channel coding procedures for EC-GSM-IoT. The channel coding procedures here mirror the steps contained within the

TABLE 3.1 Blind transmissions and coverage classes for different logical channels.

Logical channel	Blind transmissions and coverage classes
EC-SCH	28
EC-BCCH	16
EC-CCCH/D	1 (CC1), 8 (CC2), 16 (CC3), 32 (CC4)
EC-CCCH/U	1 (CC1), 4 (CC2), 16 (CC3), 48 (CC4)
EC-PACCH	1 (CC1), 4 (CC2), 8 (CC3), 16 (CC4)
EC-PDTCH	1 (CC1), 4 (CC2), 8 (CC3), 16 (CC4)

channel coding specification, 3GPP TS 45.003 [5], which also includes interleaving and mapping of the encoded bits onto bursts. The interested reader is referred to the specification for more details, considering that the description of these procedures is kept relatively short.

The first step in the channel coding is typically to add a number of bits to the payload for detecting errors induced in the transmission. The error detecting code consists of a number of parity bits added before the forward error correction (FEC). The number of bits differs depending on the logical channel and/or the coding scheme. Generally speaking, the length has been chosen considering the implication of a block being erroneously interpreted by the receiving end.

After the error detecting capability has been added to the payload, a FEC code is applied. The FEC used in EC-GSM-IoT is based on convolutional codes that are fully reused from the EGPRS channel coding design. Two different mother codes are used, defined by either a 1/2 or a 1/3 code rate that are followed by optional puncturing/rate matching of encoded bits to reach the final code rate. The puncturing is typically defined by puncturing schemes (PSs), which basically is a list of bit positions identifying the ones to be removed from the encoded bit stream.

Both tail biting and zero padded convolutional codes are used. For coding schemes where the block length is limited, tail biting is typically used, which reduces the overhead from channel coding at the expense of an increased decoding complexity. Instead of a known starting and ending state of the decoder (as for the case of zero padding), the start and end state can only be assumed to be the same, but the state itself is not known.

The encoding process is typically followed by interleaving, which is simply a remapping of the bit order using a 1-to-1 mapping table.

After interleaving the bits are mapped onto the burst type associated with the logical channel, see Fig. 3.4. Different burst types are associated with different number of encrypted bits, and there can also be different number of bursts comprising the full block, depending on the logical channel, see Tables 3.2 and 3.3.

The general channel coding procedure is illustrated in Fig. 3.6.

The EC-PDTCH is different from the other logical channels in that the full block constitutes a number of separately encoded fields, see Table 3.3.

TABLE 3.2 Channel coding details for EC-SCH, EC-CCCH, EC-BCCH, EC-PACCH.

Logical channel	Uncoded bits	Parity bits	Mother code	Code rate	Tail biting	Interleaver	Burst type	Bursts per block
EC-SCH	30	10	1/2	0.56	No	No	SB	1
EC-CCCH/D	88	18	1/3	0.91[a]	Yes	No	NB	2
EC-CCCH/U	11	6	1/2	0.58	No	No	AB	1
EC-BCCH	184	40	1/2	0.50	No	Yes	NB	4
EC-PACCH/D	80	18	1/3	0.86[b]	Yes	No	NB	4
EC-PACCH/U	64	18	1/3	0.71[b]	Yes	No	NB	4

[a] *Code rate of a single block, which is then at least repeated once (for CC1).*
[b] *Code rate of a single block, which is then at least repeated three times (for CC1).*

TABLE 3.3 Channel coding details for EC-PDTCH.

Modulation and coding scheme	Modulation	SF code rate	USF code rate	RLC/MAC header			RLC data block		Burst type	Bursts per block
				Header type	Parity bits	Code rate[a]	Parity bits	Code rate		
MCS-1	GMSK	1/4	1/4	3	8	0.53 (DL) 0.49 (UL)	12	0.53	NB	4
MCS-2	GMSK			3				0.69	NB	4
MCS-3	GMSK			3				0.89	NB	4
MCS-4	GMSK			3				1.00	NB	4
MCS-5	8PSK	1/2	1/12	2		0.33 (DL) 0.33 (UL)		0.38	NB	4
MCS-6	8PSK			2				0.50	NB	4
MCS-7	8PSK	1/2		1				0.78	NB	4
MCS-8	8PSK			1				0.92	NB	4
MCS-9	8PSK			1				1.00	NB	4

[a]*Tail biting.*

Compared with EC-PDTCH, which follows the EGPRS design of PDTCH, the EC-PACCH is a new block format for EC-GSM-IoT. Because Uplink State Flag (USF)-based scheduling is not used in EC-GSM-IoT, as explained in see Section 3.3.2.1, the USF in the EC-PACCH is included in a rather unconventional way by simply removing some of the bits for the payload and replacing them by USF bits, effectively increasing the code rate of EC-PACCH control message content. This is referred to as bit stealing for USF and is only applicable to the EC-PACCH block on the downlink. The EC-GSM-IoT receiver is, however, not aware of if a USF is included in the EC-PACCH block or not and will treat the block in the same way in each reception. It can be noted that the inclusion USF is solely included for the purpose of scheduling other PS devices on the uplink.

The channel coding procedure for each logical channel is summarized in Tables 3.2 and 3.3.

3.2.5 Mapping of logical channels onto physical channels

In Sections 3.2.6 and 3.2.7 the downlink and uplink logical channels will be described. Each logical channel is mapped onto one or more basic physical channels in a predefined manner. A *basic physical channel* is defined by a set of resources using the one and the same TS in each TDMA frame over the full hyperframe. For packet logical channels the basic physical channel is also referred to as a *Packet Data CHannel* and is, in this case, always mapped to a 52 multiframe [6]. The logical channels that are mapped to PDCHs are EC-PDTCH and EC-PACCH.

To understand the mapping of the logical channels onto physical channels, it is good to understand the two dimensions in time used by the GSM frame structure (for more details on the frame structure, see Section 3.2.2), that is the TSs that a logical channel is mapped to, and the TDMA frames over which the logical channel spans.

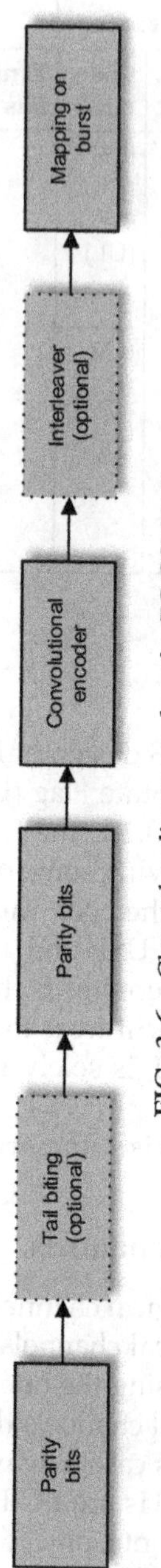

FIG. 3.6 Channel coding procedure for EC-GSM-IoT.

For illustration purposes, this book uses the principle of stacking consecutive TDMA frames horizontally, as shown in the bottom part of Fig. 3.7 (where the arrows show the time direction in each TDMA frame). This illustration will be used in the following sections when illustrating the mapping of the logical channels onto the physical frame structure.

3.2.6 Downlink logical channels

The set of logical channels used in the downlink by EC-GSM-IoT are shown in Fig. 3.8.

The purpose of each channel, how it is used by the device, and the mapping of the channel onto the physical resources will be shown in the following sections. The notation of "/D" for some of the channels, indicate that the channel is defined in both downlink and uplink, and that the downlink is here referred to.

3.2.6.1 FCCH

TS	0
TDMA frames	0, 10, 20, 30, 40 (see Fig. 3.9)
Mapping repetition period	51 TDMA frames
Multiframe	51
Burst type	Frequency correction burst
Block size	–
Carrier	BCCH

The Frequency Correction CHannel (FCCH) is the only channel that is the same as used by GSM devices not supporting EC-GSM-IoT. When the device is scanning the frequency band, the acquisition of FCCH will identify if a certain frequency belongs to a GSM network. The device also uses it to perform a rough frequency (to the base station reference) and time alignment.

As already mentioned in Section 3.2.2.3, the burst carrying the FCCH consists of 148 bits of state "0," which will generate a signal of sinusoidal character over the burst length, giving rise to a spike in the signal spectrum at 67.7 kHz ($f_s/4$).

The FCCH is mapped onto the 51 multiframe of TS0 of the BCCH carrier. This is the only basic physical channel, and the only RF channel, where the FCCH is allowed to be mapped.

As shown in Fig. 3.9 the FCCH is mapped onto TDMA frames 0, 10, 20, 30, and 40 in the 51 multiframe. This implies that the separation between the FB is 10, 10, 10, 10, 11 TDMA frame. Although the FCCH does not carry any information, the irregular structure of the FCCH mapping can be used by a device to determine the overall 51 multiframe structures.

How well the device is synchronized after the FCCH acquisition is very much dependent on the algorithm used and also the time spent to acquire the FCCH. The device oscillator accuracy is for typical oscillator components used in low-complexity devices equals 20 parts per million, i.e., for a device synchronizing to the 900 MHz frequency band, the frequency offset prior to FCCH acquisition can be up to 18 kHz ($900e6 \times 20 \times 10^{-6}$). After FCCH acquisition a

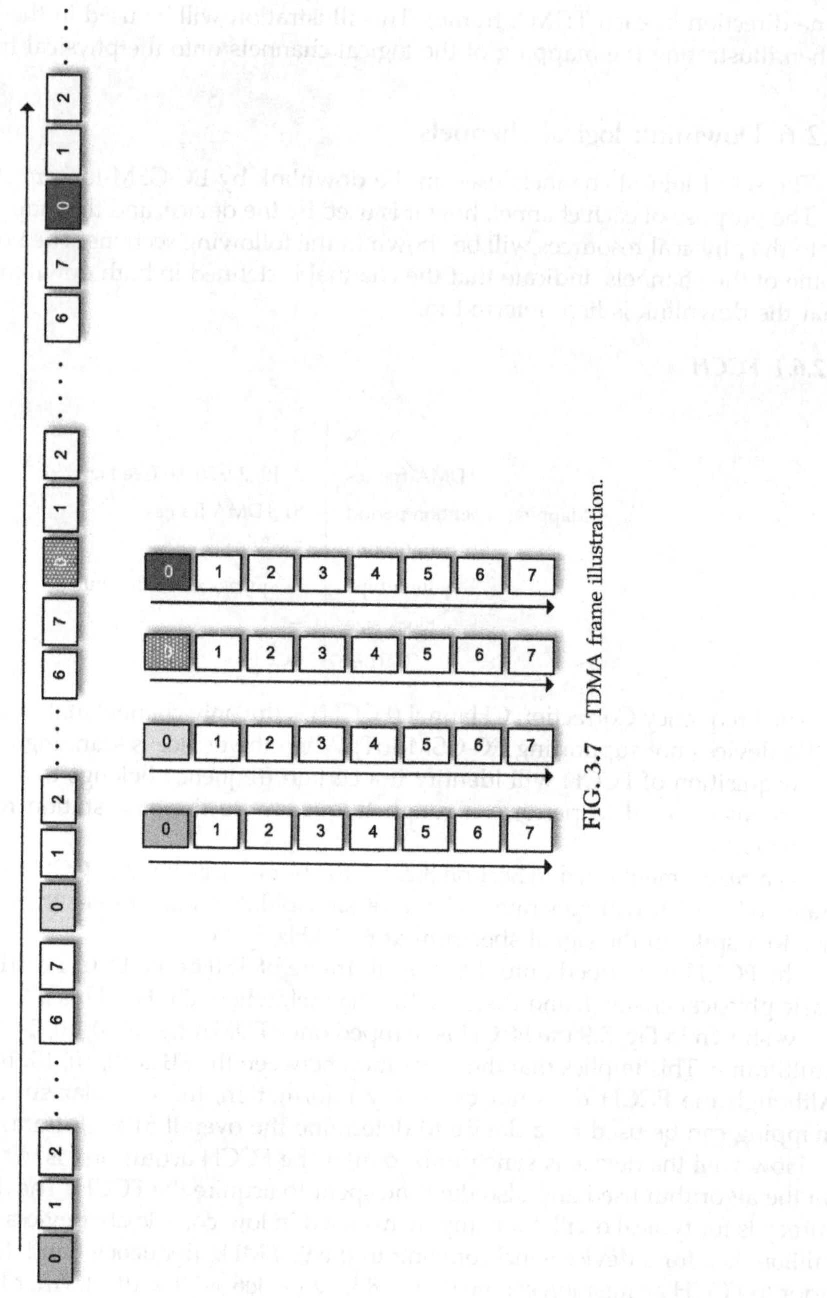

FIG. 3.7 TDMA frame illustration.

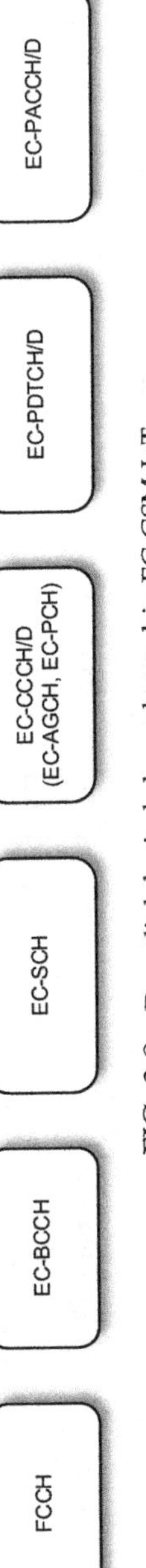

FIG. 3.8 Downlink logical channels used in EC-GSM-IoT.

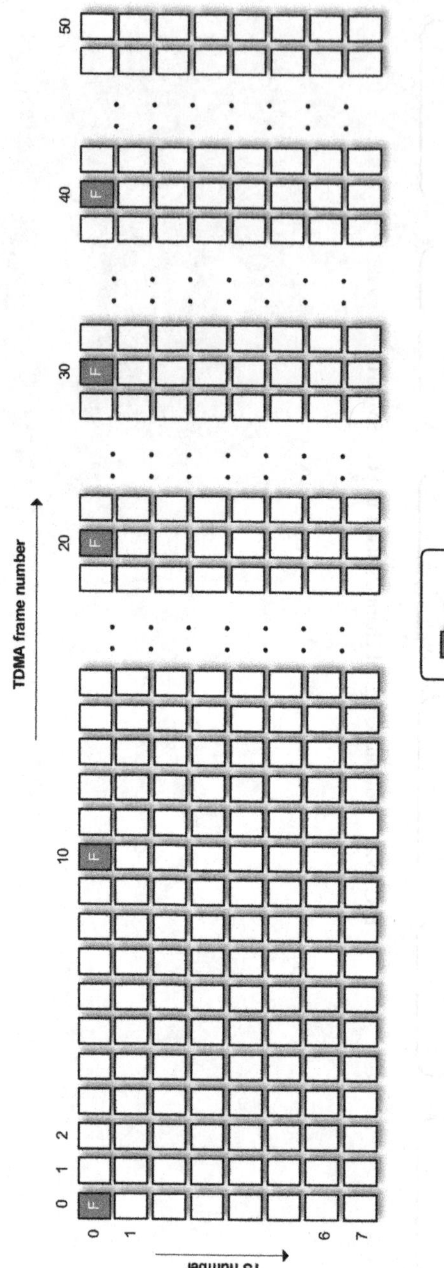

FIG. 3.9 FCCH mapping.

reasonable residual frequency and time offset is up to a few hundred Hz in frequency and a few symbol durations in time.

The next step for the device is to decode the EC-SCH channel.

3.2.6.2 EC-SCH

TS	1
TDMA frames	See Fig. 3.10
Mapping repetition period	204 TDMA frames
Blind physical layer transmission	28
Multiframe	51
Burst type	Synchronization burst
Block size	1 burst
Carrier	BCCH

After the rough frequency and time correction by the use of the FCCH, the device turns to the Extended Coverage Synchronization Channel (EC-SCH). As mentioned in Section 3.2.6.1 after acquiring the FCCH the device will have knowledge of the frame structure to a precision of the 51 multiframe but will not know the relation to the overall frame structure on a superframe and hyperframe level (see Section 3.2.2). The EC-SCH will assist the device to acquire the knowledge of the frame structure to a precision of a quarter hyperframe. There is no reason for the device at this point to know the frame structure more precisely. Instead, the missing piece of the puzzle is provided in the assignment message, where 2 bits will convey which quarter hyperframe the assignment message is received in.

The frame number on a quarter hyperframe level is communicated partly through information contained in the payload part of the EC-SCH. However, considering that the interleaving period of the EC-SCH is four 51 multiframes (i.e., the payload content will change after this period), see Fig. 3.10, and that the device will start its acquisition of the EC-SCH in any of these multiframes, a signaling, indicating which multiframe out of four is required. To solve this, a cyclic shift of the encoded bits is applied specific to each multiframe. The same cyclic shift is applied to all seven repetitions within the multiframe. This allows the receiver to accumulate, on IQ level, all bursts within each multiframe to maximize processing gain (following the same principle as illustrated in Fig. 3.21). The device would then need to decode the EC-SCH using up to four different hypotheses (one for each cyclic shift) before the block is decoded.

After EC-SCH acquisition the device should be synchronized in time and frequency, fulfilling the requirements of the specification on $^1/_2$ symbol synchronization accuracy and 0.1 ppm frequency accuracy (for the 900 MHz band this corresponds to residual time and frequency offset of at most 1.8 μs and 90 Hz, respectively).

The device is mandated to decode the EC-SCH each time, it is attempting to initiate a connection to receive on the downlink or transmit on the uplink. This makes the EC-SCH a powerful channel for indicating information to the device. It can be compared with the

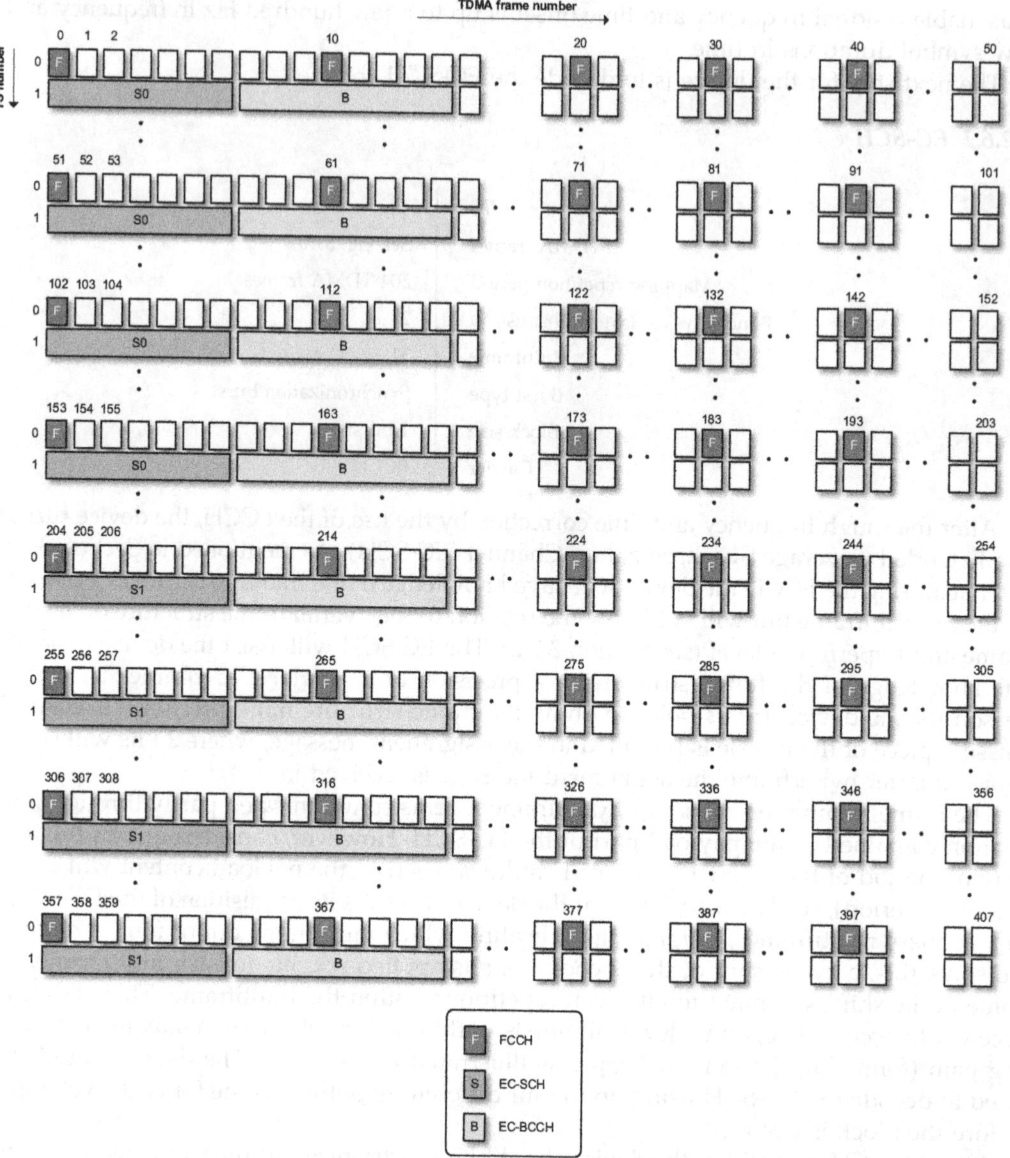

FIG. 3.10 Mapping of FCCH, EC-SCH, EC-BCCH.

NB-IoT and LTE Machine-Type Communications Master Information Block. Hence, apart from the remaining frame number that is communicated by the channel, the EC-SCH also communicates:

- The cell ID, i.e., the *Base Station Identity Code* (BSIC).
- System overload control (see Section 3.3.1.7).

- Load balancing on the random access channel (see Section 3.3.1.7).
- SI change (see Section 3.3.1.2)

The BSIC is a 9-bit field, which can address 512 different cell identities. In traditional GSM the BSIC is a 6-bit field but to accommodate future IoT deployments of GSM, 3GPP took the initiative to expand the space of possible cell identities from 64 to 512.

This was mainly motivated by three factors.

- **Tighter frequency reuse**: The frequency domain that provides orthogonality between cell IDs will be reduced.
- **Increased use of network sharing**: With IoT, operators are expected to more frequently share the networks that reduce the possibility to use the *Network Color Code*, which is part of the BSIC, as a means to separate cell IDs.
- **Reduced idle mode measurements**: With the reduced measurements by the device, it can wake up in a cell that uses the same BSIC as when previously was awake and with the reduced measurement activity, it can take up to 24 h before it will realize it.

The mapping of the EC-SCH channel is shown in Fig. 3.10 together with the FCCH and EC-BCCH, where TS 2–7 of each TDMA frame has been omitted to simplify the figure. It should be noted that the repetition period of the EC-SCH channel only spans four 51 multi-frames, using 28 blind transmissions, and hence two EC-SCH blocks are (0 and 1) are depicted in the figure.

3.2.6.3 EC-BCCH

TS	1
TDMA frames	See Fig. 3.10
Mapping repetition period	408 TDMA frames
Blind physical layer transmissions	16
Multiframe	51
Burst type	Normal burst
Block size	4 bursts
Carrier	BCCH

After EC-SCH acquisition the device may potentially continue with acquiring the EC SI, which is transmitted on the Extended Coverage Broadcast Control CHannel (EC-BCCH). As the name implies the EC SI is system-specific information to EC-GSM-IoT and will convey not only information related to the specific cell it is transmitted in, but can also provide information on other cells in the system.

The coding scheme used by the EC-BCCH channel is CS-1, which is also the modulation and coding scheme (MCS) used by GPRS/EGPRS for control signaling purposes. After channel coding and interleaving, the block consists of four unique bursts that are mapped onto one TS over four consecutive TDMA frames. Each *EC SI message instance* (where each *EC SI message* can consist of multiple instances) is mapped onto in total 16 block periods, using two consecutive block periods in each 51 multiframe, mapped over in total eight 51 multiframes. The EC SI messages, and the multiple EC SI message instances (if any), are mapped onto the physical frame in ascending order. The sequence of EC SI messages and their associated EC SI message instances can be seen as an EC SI cycle that repeats itself over time.

By defining the EC SI placement in this static manner, the device will always know where to look for EC SI message instances, once the frame number has been acquired from the EC-SCH. In addition, the network will not know which devices are in a specific cell, or which conditions the devices are deployed in, it will have to always repeat each message the same and maximum number of times to be certain to reach all devices.

Fig. 3.10 shows TS0 and TS1 of the TDMA frame with TS2−TS7 omitted (showed by a dashed line), illustrating the mapping of FCCH, EC-SCH, and EC-BCCH. It can be noted that the FCCH bursts are all identical (there is no distinction between them in the figure), while EC-SCH is transmitted in two unique blocks (S0 and S1) over the eight 51 multiframes because the repetition length is four 51 multiframes as explained in Section 3.2.6.2. Only one EC-BCCH block is transmitted (B) because the repetition period is eight 51 multiframes (see above).

3.2.6.4 *EC-CCCH/D (EC-AGCH, EC-PCH)*

TS	1, and optionally 3, 5, 7
TDMA frames	See Fig. 3.11
Mapping repetition period	51 (CC1), 102 (CC2), 102 (CC3), 204 (CC4)
Blind physical layer transmissions	1 (CC1), 8 (CC2), 16 (CC3), 32 (CC4)
Multiframe	51
Burst type	Normal burst
Block size	2 bursts
Carrier	BCCH

The Extended Coverage Common Control CHannel (EC-CCCH) is used by the network to reach one or more devices on the downlink using either Extended Coverage Access Grant CHannel (EC-AGCH) or Extended Coverage Paging CHannel (EC-PCH). From a physical layer point of view, the two logical channel types are identical and they differ only in which TDMA frames the channels can be mapped onto. In some TDMA frames only EC-AGCH can be mapped, whereas in other TDMA frames both EC-AGCH and EC-PCH are allowed. The reason for having a restriction on how to map the EC-PCH is related to how paging groups are derived, and how cells in the network are synchronized (see Section 3.3.1.5 for more information).

Whether the message sent is mapped onto EC-AGCH or EC-PCH is conveyed through a message type field in the message itself. That is, it is only after decoding the block that the device will know whether the message sent was carried by EC-AGCH or EC-PCH. In case of EC-AGCH, only one device can be addressed by the message sent, whereas for EC-PCH up to two devices can be addressed by the same message.

Compared with FCCH, EC-SCH, and EC-BCCH that have been described in Section 3.2.6, the EC-CCCH/D channel makes use of CCs introduced in Section 3.2.8, to be able to reach users in different coverage conditions effectively.

Each block for each CC is mapped onto predefined frames in the overall frame structure. In Fig. 3.11 the mapping of the EC-CCCH/D blocks are shown.

As can be seen, the CC1 blocks are mapped to two TDMA frames, whereas in case of CC4 32 blind transmissions are used, spread over four 51 multiframes, to reach devices in extreme

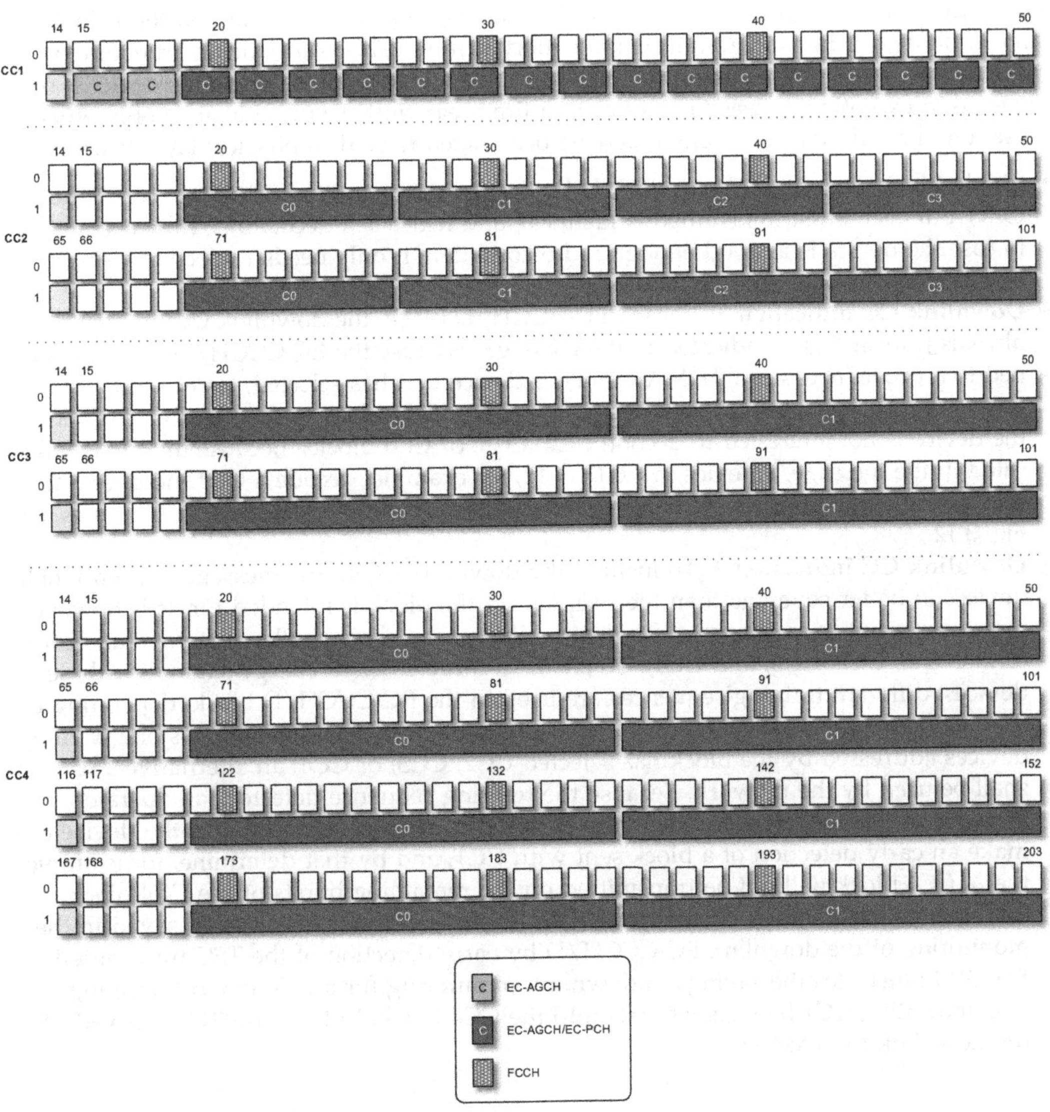

FIG. 3.11 Mapping of EC-CCCH/D for different CCes.

coverage conditions. To spread the transmissions over several multiframes instead of transmitting them consecutively in time, will provide time diversity, improving the reception of the block. It can be noted that this is also the case for completely stationary devices, as long as the surrounding environment provides time variations in the radio propagation (for example, cars driving by, leaves in trees caught by wind etc.).

One of the main objectives with serving the IoT segment is not only to support challenging coverage conditions but also to provide an energy efficient operation reaching up to 10 years of battery lifetime as evaluated in Section 4.5. Here, the EC-CCCH plays an important role. Each time a device accesses the system it will have to monitor the EC-CCCH to provide the device with dedicated resources. Hence, an energy efficient operation on the EC-CCCH can contribute considerably to the overall battery lifetime. This is specifically true for devices that may only monitor the paging channel during its whole lifetime, where the device is only triggered in exceptional cases (for example, a fire break-out). To provide an energy efficient operation, three design considerations have been taken related to physical layer processing:

- **Single block decoding**: The EC-CCCH/D is at least transmitted using two bursts (for CC1), but each burst transmitted is identical, and hence self-decodable. This means that in case the device is in good enough radio condition, it only needs to decode a single burst before the block can be decoded.
- **Downlink CC indication 1**: In each EC-CCCH/D block, the downlink CC by which the block is transmitted is indicated in the message. Because the EC-CCCH/D block is mapped to a predefined set of TDMA frames, a device that has selected, for example, CC1 knows that the network is mandated to transmit a CC1 block to that device, and hence the device is not interested to decode CC2, CC3, or CC4 blocks. Because the CC is indicated in the message, a device in CC1 could, for example, decode a CC4 message after only a single burst (see single block decoding) and sleep for the remaining 63 bursts, see Fig. 3.12.
- **Downlink CC indication 2**: To include the downlink CC in the message will only help devices in better coverage than the recipient of the block is intended for. It is however more important to save energy for devices that are in more challenging coverage conditions (i.e., CC2, CC3, and CC4). To enable a more efficient energy saving for these devices, different training sequences are used in the EC-CCCH/D block, depending on if all devices addressed by the block are in CC1 or not. That is, as long as one or more devices addressed by the block has selected CC2, CC3, or CC4, an alternative TSC shall be used by the network. Because the training sequence detector can operate correctly at lower SNR than where the EC-CCCH/D block is decodable, the device can make an early detection of a block sent with CC1, and by that determine, for example, that a CC4 block will not be transmitted on the remaining bursts of the CC4 block. Simulations have shown [7] that roughly 80%–85% of the energy can be saved in the monitoring of the downlink EC-CCCH/D by early detection of the TSC transmitted. Fig. 3.13 illustrates the savings in downlink monitoring for a CC4 device detecting the TSC indicating CC1 block, six bursts into the CC4 block. In this case 91% is saved of the downlink monitoring.

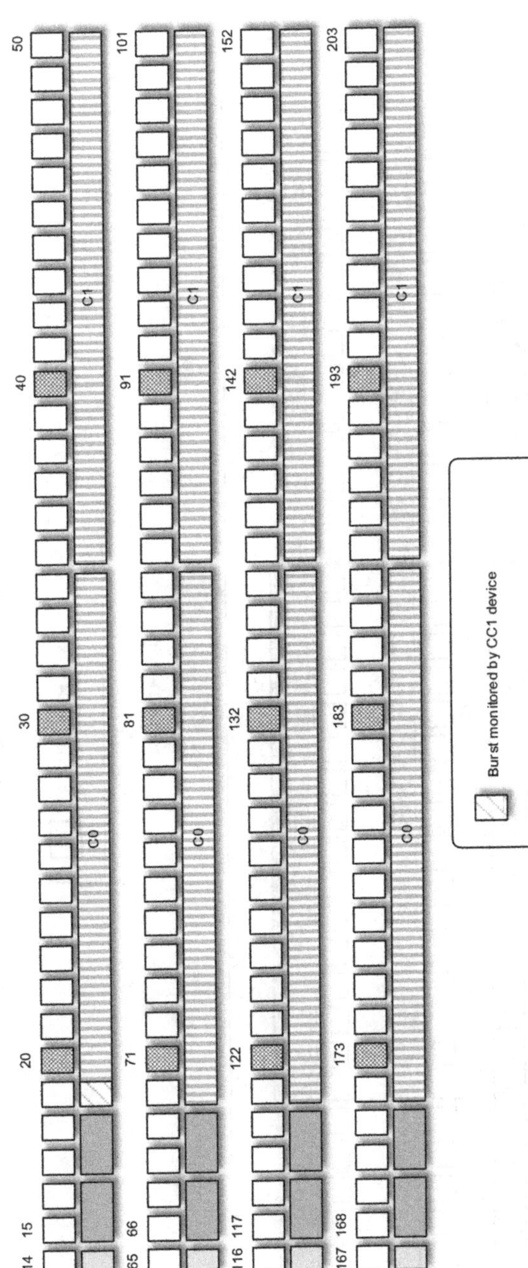

FIG. 3.12 Downlink CC indicator – in message content.

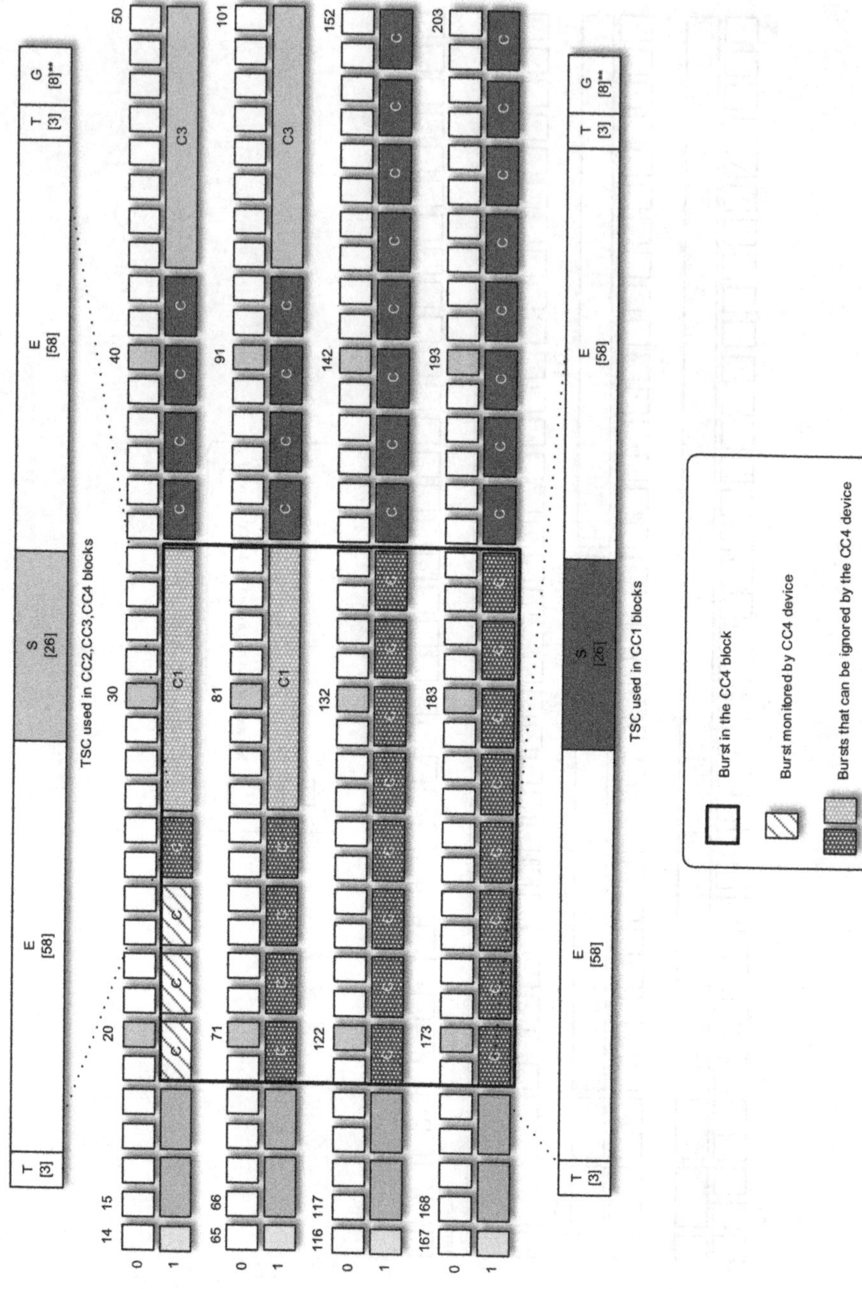

FIG. 3.13 Downlink CC indicator – by training sequence.

3.2.6.5 EC-PDTCH/D

TS	Any
TDMA frames	See Fig. 3.15
Mapping repetition period	52
Blind physical layer transmissions	1 (CC1), 4 (CC2), 8 (CC3), 16 (CC4)
Multiframe	52
Burst type	Normal
Block size	4 bursts
Carrier	Any

The Extended Coverage Packet Data Traffic CHannel (EC-PDTCH) is one of the logical channels for EC-GSM-IoT, which is almost identical to the corresponding logical channel used in EGPRS. It is used for carrying payload from the network to the device or, in case of uplink transmission, from the device to the network. The EC-PDTCH block consists of four different bursts mapped to the same TS over four consecutive TDMA frames. The block is referred to as a *radio block* and is transmitted over a *Basic Transmission Time Interval* of 20 ms. The EC-PDTCH is mapped onto one or more physical channels referred to as *PDCH* (*Packet Data CHannel*).

To adapt to more extreme coverage conditions, CCs are also used for EC-PDTCH, see Table 3.1. When monitoring the downlink EC-PDTCH (EC-PDTCH/D), it is up to the device to blindly detect the modulation scheme used in the block. This is done by detecting the rotation angle of the predefined training sequence. After synchronization, channel estimation and modulation detection, the device equalizes the symbols of the burst to provide a set of estimated probabilities that a certain bit has been transmitted. This probability is denoted a "soft bit." The higher the magnitude of the soft bit, the more certain the receiver is that a specific bit value was transmitted. Because different *Radio Link Control* (RLC)/*Medium Access Control* (MAC) headers can be used for the same modulation scheme the next step is to read the Stealing Flags (see Section 3.2.4), to determine the header type used in the block. After the RLC/MAC header has been decoded, the device will have knowledge of the MCS used in the block and also the redundancy version or PS, used in the *Hybrid Automatic Repeat Request* (HARQ) process, see Section 3.3.2. The procedure is illustrated in Fig. 3.14.

As with the EC-CCCH channel, the EC-PDTCH also includes an indication of the downlink CC in each block to allow energy efficient operation. Fig. 3.12 illustrates this mechanism for EC-CCCH.

In case of CC1, the PDTCH block is mapped onto one PDCH, over four consecutive TDMA frames. For CC2, CC3, and CC4 the blind repetitions of the block are mapped onto four consecutive TSs and over 4, 8, or 16 TDMA frames, respectively.

Increasing the mapping for CC1 from one to four PDCHs was decided based on multiple reasons. The more TSs that are used in the TDMA frame, the better it is for the receiver that can rely on coherent transmissions within the TDMA frame, see Section 3.2.8, improving the processing gain compared to mapping the blind repetitions over a smaller number of PDCHs

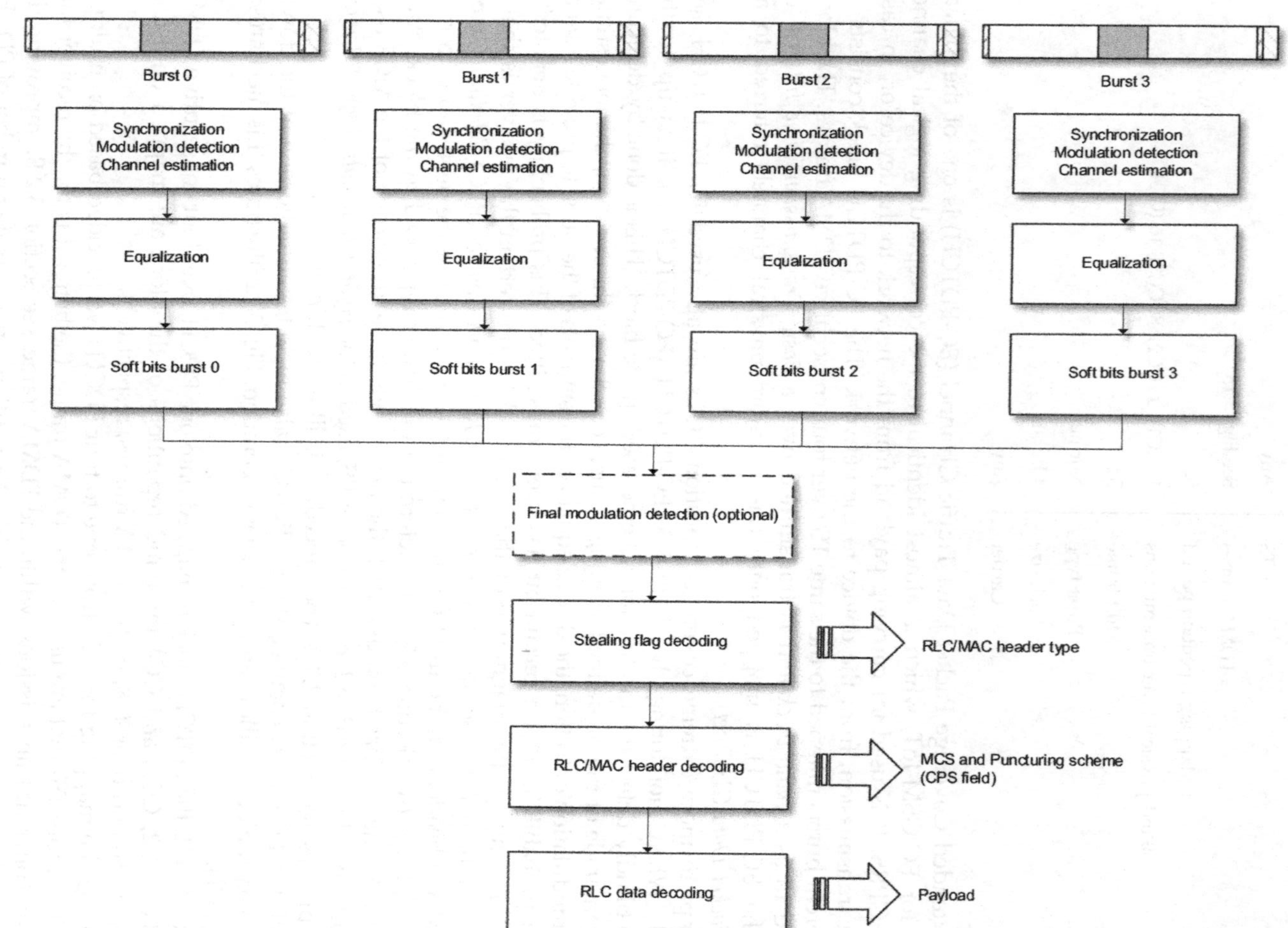

FIG. 3.14 Illutration of procedure to equalize an EC-GSM-IoT radio block.

(and instead more TDMA frames). Hence, from a performance point of view, it would be best to map the repetitions over all eight TSs in the TDMA frame. This poses potential problems of PA heat dissipation and resource management restrictions. If the EC-PDTCH would take up all 8 TSs, the only channel possible to transmit in addition on that carrier would be the associated control channel, the EC-PACCH, as well as the traffic channel and associated control channel for non-EC-GSM-IoT traffic (the PDTCH and PACCH). Hence, from a network deployment point of view, a shorter allocation over TSs in a TDMA frame is required. As a trade-off between network deployment flexibility, radio link level performance, and device implementation, four consecutive TSs have been chosen for all higher CCs, although a second option is introduced in Release 14 introduced in Section 3.4.

The mapping of the EC-PDTCH is shown in Fig. 3.15. Although CC1 is only mapped to one PDCH, four consecutive PDCHs are shown to illustrate the same amount of resources,

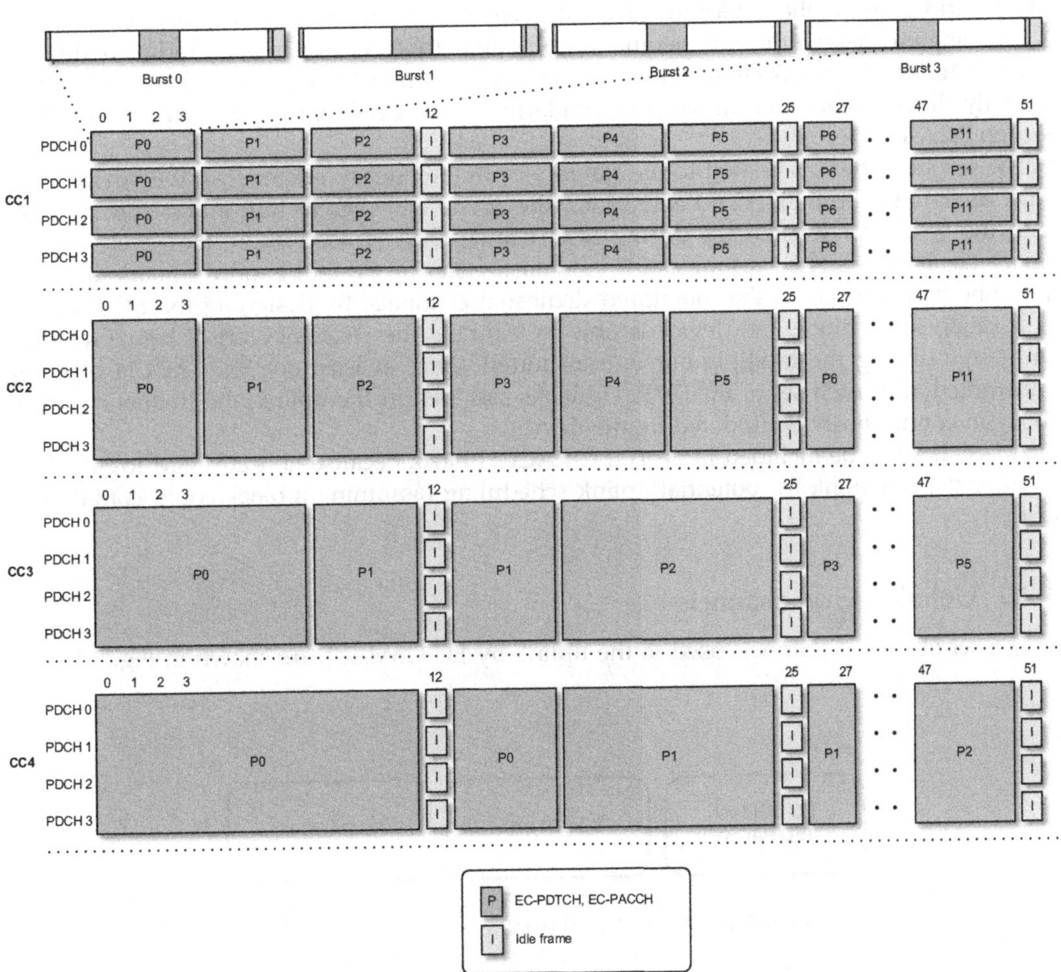

FIG. 3.15 Mapping of EC-PDTCH and EC-PACCH for different CCes.

irrespective of the CC. It can be noted that there are no restrictions on the mapping over the four consecutive PDCHs as long as the block is kept within the TDMA frame, i.e., "PDCH 0" in Fig. 3.15 can be mapped to TS index 0, 1, 2, 3, or 4 (in which, for the latter case, the block is mapped to TS4, TS5, TS6, TS7).

3.2.6.6 EC-PACCH/D

The Extended Coverage Packet Associated Control CHannel (EC-PACCH) is, as the name implies, an associated control channel to the EC-PDTCH. This means that it carries control information related to the EC-PDTCH operation, e.g., output power level to be used (power control), resource (re)assignment, and Ack/Nack bitmap (for the uplink HARQ process).

The EC-PACCH/D information is encoded onto a single normal burst, carrying 80 information bits, which is always repeated four times to construct an EC-PACCH/D block. The reason for this is twofold to ensure an energy efficient operation for users in good radio conditions, and to allow the device to correct its frequency to the base station reference.

Ensuring an energy efficient operation implies here that a device in good radio conditions can monitor the downlink channel for a single burst, attempts to decode the block, and if successfully decoded, go back to sleep in similarity to the EC-CCCH operation presented in Section 3.2.6.4.

The second benefit is related to the stability of the frequency reference in the device over time. As stated in Sections 3.2.6.1 and 3.2.6.2, the device will align its time and frequency base with the base station reference during FCCH and EC-SCH acquisition. However, once a device has been assigned resources for its data transfer, it would be more suitable if it can fine tune its reference on the monitored dedicated channels. By designing the EC-PACCH as a single burst block, the device is able to estimate the frequency offset by correlating and accumulating the multiple bursts transmitted. Since at least one EC-PACCH block is transmitted after each set of EC-PDTCH blocks assigned in the uplink, the frequency reference can continuously be updated by the device.

A USF need to be included in every EC-PACCH block because non-EC-GSM-IoT devices monitor the downlink for potential uplink scheduling (assuming a block could contain the USF).

3.2.7 Uplink logical channels

The set of logical channels used in the uplink by EC-GSM-IoT are shown in Fig. 3.16.

FIG. 3.16 Uplink logical channels used in EC-GSM-IoT.

3.2.7.1 EC-CCCH/U (EC-RACH)

TS	0, 1, 3, 5, 7 and [0,1] [2,3], [4,5], [6,7]
TDMA frames	See Figs. 3.17 and 3.18
Mapping repetition period	51 (CC1), 51 (CC2), 51 (CC3), 102 (CC4)
Blind physical layer transmissions	1 (CC1), 4 (CC2), 16 (CC3), 48 (CC4)
Multiframe	51
Burst type	Access burst
Block size	1 burst
Carrier	BCCH

Before a connection between the device and the network can be setup, the device needs to initiate an access on the random access channel. The initiation of the system access request can either be triggered by the device referred to as *Mobile Originated* traffic or by the network referred to as *Mobile Terminated* traffic, initiated, for example, when paging the device. Before using the random access channel, the device needs to be synchronized to the network in time and frequency and not being barred for access (see Sections 3.2.6 and 3.3.1).

The device sends an AB to initiate the communication, which contains 11 bits, containing, for example, an indication of downlink CC. In addition, the uplink CC is implicitly indicated to the network by the choice of the TSC used.

The device will transmit the Packet Channel Request using either one or two TSs per TDMA frame. This is referred to as a 1 or 2 TS mapping of the Extended Coverage Random Access Channel (EC-RACH). At first glance, using two different mappings for transmitting the same information seems unnecessary complex. There are, however, reasons for this design. The 2 TS mapping extends coverage further than the 1 TS mapping. The reason is that the device needs to keep the coherency of the transmission over the two TSs in the TDMA frame and hence, the processing gain at the base station is improved. However, it comes at a cost of having to multiplex system accesses from EC-GSM-IoT devices with the access of non-EC-GSM-IoT devices accessing on TS 0. Hence, an operator would, depending on the load in the network from non-EC-GSM-IoT devices, and the need for extending the coverage, allow the use of 2 TS mapping by the EC-GSM-IoT devices. Only one of the mapping options will be used at a specific point in time in the cell.

The performance gain of the 2 TS EC-RACH mapping has been evaluated [8] where it was seen that up to 1.5 dB performance gain is achieved for the users in the worst coverage.

Since the device accessing the network only knows the timing of the downlink frame structure (from the synchronization procedure, see Sections 3.2.6.1 and 3.2.6.2), it will not know the distance from the device position to the base station receiver. To accommodate for different base stations to device distances (propagation delays) in the cell, the AB has been designed to be shorter than a regular NB (see Section 3.2.2.2 for the burst structures). The shorter burst duration provides a 68-symbol-long guard period, which allows for roughly a 35 km cell radius ($3e8 \times (T_{symb} \times 68)/2 \approx 35$ km). Because the downlink synchronization will also be affected by the propagation delay, the uplink transmission will have to be shifted in time to take into account a propagation delay twice the distance between the device and the base station. The propagation delay is estimated by the base station (comparing its reference timing with the timing of the received AB from the device) and the amount of TA to be used by the device is communicated to it in the message sent on the access grant channel (see Section 3.2.6.4).

The mapping of the logical channel onto the physical resources is shown in Figs. 3.17 and 3.18.

3.2.7.2 *EC-PDTCH/U*

As for the downlink traffic channel, the EC-PDTCH/U is almost identical to the corresponding channel used in EGPRS. Hence, the description of EC-PDTCH/D applies equally well to the EC-PDTCH/U with some exceptions. The differences lie in, for example, the exclusion of the USF in the overall block structure (as for EC-PACCH/U compared with EC-PACCH/D, see Section 3.2.7.3). In addition, the RLC/MAC header format will not be the same comparing the uplink and downlink EC-PDTCH blocks.

The EC-PDTCH/U follows the mapping of EC-PDTCH/D, see Fig. 3.15.

3.2.7.3 *EC-PACCH/U*

From a physical layer point of view, the design of the EC-PACCH in the uplink is very similar to the corresponding downlink channel presented in Section 3.2.6.6.

The EC-PACCH/U block is constructed of a single burst repeated at least four times, using the same number of blind block repetitions as for the downlink for the different CCs.

One difference lies in not transmitting the USF in the uplink. Because the USF's sole purpose is for the network to schedule devices in the uplink, there is no reason to include this field in the uplink blocks.

Also the payload size of the EC-PACCH in the uplink differs from the downlink format, carrying a payload size of at most 64 bits.

The EC-PACCH/U contains, apart from Ack/Nack information relating to downlink data transmissions, also optionally a channel quality report where the average bit quality and the variation in bit quality over the received bursts are reported, together with an estimation of the signal level measurements.

The EC-PACCH/U follows the mapping of EC-PACCH/D, see Fig. 3.15.

3.2.8 Extending coverage

3.2.8.1 *Defining maximum coupling loss*

Now that the main principles of the physical layer design are covered, we turn our attention to perhaps the most important part of the EC-GSM-IoT feature—the extension of the coverage limit of the system.

How do we define a coverage limit of a system? A mobile phone user would probably consider a loss of coverage when it is no longer possible to make a phone call, when an ongoing call is abruptly stopped, or when, for example, no web page is loading when opening a web browser.

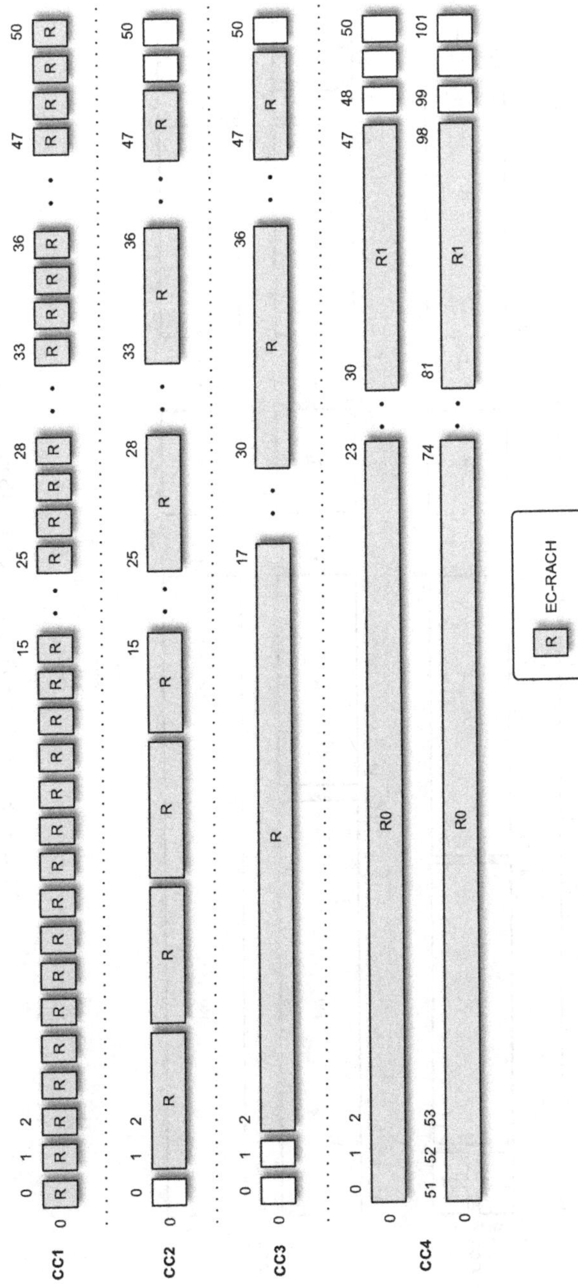

FIG. 3.17 EC-CCCH/U (EC-RACH), 1 TS mapping.

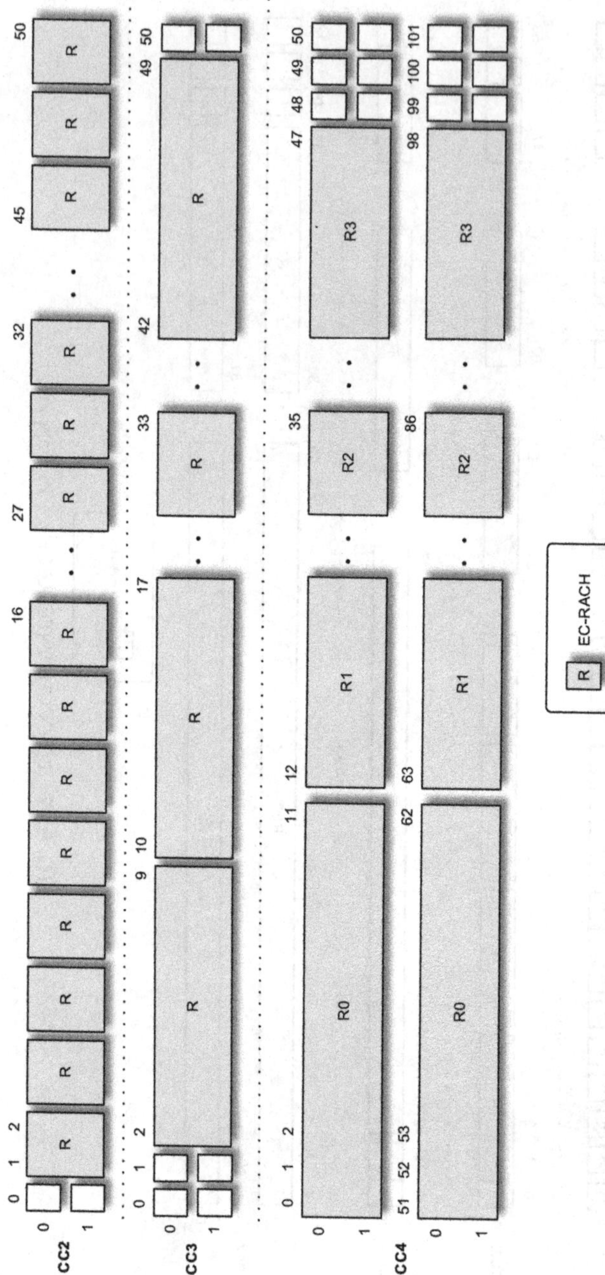

FIG. 3.18 EC-CCCH/U (EC-RACH), 2 TS mapping.

For a data service, a suitable definition of coverage is at a specified throughput target. That is, at what lowest signal level can still X bps of throughput be achieved. For EC-GSM-IoT, this target was set to 160 bps. In addition to the traffic channel reaching a certain minimum throughput, also the signaling/control channels need to be operable at that point for the device to synchronize to the network, acquire the necessary associated control channels etc.

When designing a new system, it is important that the coverage is balanced. When comparing the coverage of different logical channels, it is easily done in the same direction (uplink or downlink) using the same assumptions of transmit power and thermal noise levels (see below) at the receiver. When we are to balance the uplink and downlink, assumption need to be made on device and base station transmit levels, and also what level of thermal noise that can typically be expected to limit the performance.

Multiple factors impact the thermal noise level at the receiver, but they can be separated into a dependency of:

- **Temperature**: The higher the temperature, the higher the noise level will be.
- **Bandwidth**: Because the thermal noise is spectrally white (i.e., the same spectral density irrespective of frequency) the larger the bandwidth of the signal, the higher the absolute noise power level will be.
- **Noise Figure (NF)**: The overall NF of a receiver stem from multiple sources, which we will not cover in this book, but one can see the NF as an increased noise level after the receiver chain, compared to the ideal thermal noise without any imperfections. The NF is expressed in decibel and is typically defined at room temperature.

Lowering the NF would improve the coverage in direct proportion to the reduction in NF, but a lower NF also implies a more complex and costly implementation, which is directly in contrast to one of the main enablers of a CIoT.

Another simple means to improve coverage is to increase the transmit power in the system, but this would imply an increase in implementation complexity and cost. Also the bandwidth needs to be kept as in GSM to minimize implementation and impact to existing deployments.

An alternative way to express coverage is in terms of its *Maximum Coupling Loss* (MCL). The MCL defines the maximum loss in the radio link that the system can cope with between a transmitter and a receiver, before losing coverage.

The MCL can be defined as,

$$\mathrm{MCL(dB)} = P_{\mathrm{Tx}} - (\mathrm{NF} + \mathrm{SNR} - 174 + 10\log_{10}(\mathrm{B})), \quad (3.1)$$

where P_{Tx} is the output power [dBm], B is the bandwidth [Hz], 174 is the ideal thermal noise level expressed as [dBm/Hz] at temperature 300 K, NF is the Noise Figure [dB], SNR is the signal-to-noise ratio [dB].

We have already concluded that P_{Tx} (the output power), B (the signal bandwidth), and NF (the noise figure) are not quantities suitable to use in improving coverage for EC-GSM-IoT. Hence, only the SNR is the quantity left.

A simple means for improving the experienced SNR at the receiver and at the same time reusing existing system design and ensuring backward compatibility are to make use of blind repetitions (see Section 3.2.8.2).

The feature also relies on an improved channel coding for control channels (see Section 3.2.8.3), a more efficient HARQ retransmissions for the data channels (see Sections 3.2.8.4 and 3.3.2.2), and an increased allowed acquisition time (see Section 3.2.8.5).

3.2.8.2 Maximizing the receiver processing gain

In case of extended coverage, the receiver will, in worst case, not even be able to detect that a signal has been received. This problem is solved by the introduction of blind repetitions where the receiver first can accumulate the IQ representation of blindly transmitted bursts to a single burst with increased SNR before synchronizing to the burst and performing channel estimation. After the SNR is increased the rest of the receiver chain can be kept identical to that of GSM. This can be seen in Fig. 3.19.

To maximize SNR, the signals should be coherently, i.e., aligned in amplitude and phase, combined. Because the white noise is uncorrelated it will always be noncoherently combined, resulting in a lower rise in the noise power than in the wanted signal. Taking this to a more generic reasoning, when combining N number of transmissions, the SNR compared to a single transmission can, in the ideal case, be simply expressed as in Eq. (3.2).

$$SNR_N(dB) = 10\ \log_{10}(N) + SNR_1 \tag{3.2}$$

However, in reality a combination of two signals is rarely perfect, and instead signal impairments will result in an overall processing gain lower than expressed in Eq. (3.2).

To assist the receiver in the coherent combination, the blind repetitions should also be coherently transmitted. This means that multiple signals from the transmitting antenna will follow the same phase trajectory. As long as no other impairments are added to the signal, and as long as the propagation channel is stationary, the receiver could simply blindly accumulate the signal and achieve the expected processing gain. As we will see later, reality is not as forgiving as the simplest form of theory. Furthermore, to ensure coherent transmissions the phase reference in the transmitter need to be maintained between all blind repetitions. For the TDMA structure in GSM, where blocks are typically transmitted in one or a few TSs, the components in the transmitter would have to be active to keep the phase reference over the full TDMA frame. This not only consumes more energy, specifically of interest in the device, but might not be easily supported by existing GSM devices because before the introduction of EC-GSM-IoT there is no requirement on coherent transmissions and the phase is allowed to be random at the start of every burst.

Fig. 3.20 illustrates three different ways in the EC-GSM-IoT specification to transmit blocks, using one, two, or four bursts in each TDMA frame.

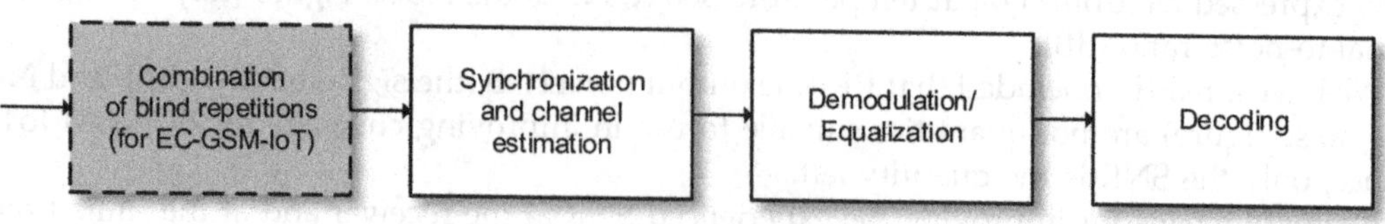

FIG. 3.19 EC-GSM-IoT receiver chain.

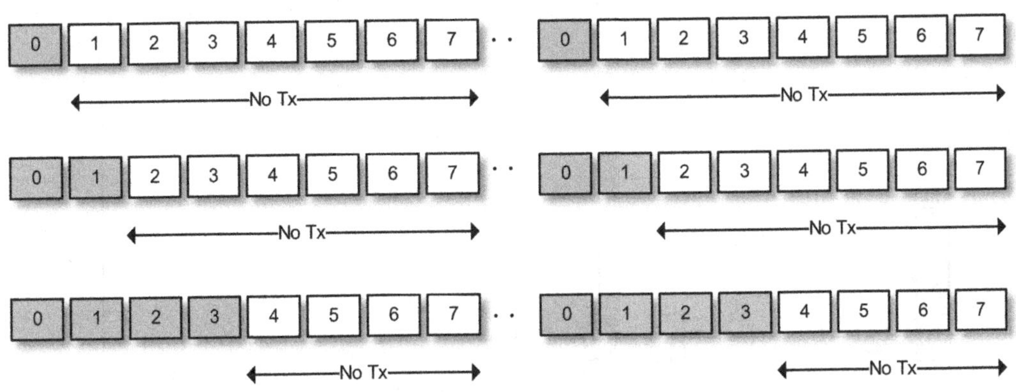

FIG. 3.20 Different ways to blindly perform repetitions over multiple TDMA frames.

In all cases there will be a substantial time where no transmission occurs, and when the device can turn off several components and basically only keep the reference clock running. What can also be seen from Fig. 3.20 is that when transmitting on consecutive TSs there is no interruption in the transmission, and hence coherency can more easily be kept. This is also the approach that the 3GPP specifications have taken, i.e., those transmissions need only be coherent in case bursts are transmitted over consecutive TSs. Between TDMA frames any phase difference could hence be expected.

To maximize the processing gain, and hence the experienced SNR by the receiver, preferably all blind repetitions are to be combined before calling the demodulator including the channel estimator. This can be achieved by first blindly combining the transmissions within a TDMA frame. Blind in this regard means that the bursts are not compensated in any way before the combination. This can be done because the receiver knows that the bursts have been transmitted coherently.

To combine the blind repetitions across TDMA frames, the receiver would have to estimate the random phase between the transmissions (because no coherency is ensured). This combination will result in a potential error in that case the processing gain from the combination will not follow Eq. (3.2).

An example of a receiver implementation (an expansion of the dashed block in Fig. 3.19) over four consecutive TSs and four consecutive TDMA frames, following the procedure described above, is shown in Fig. 3.21.

The combinations are sequentially numbered in time starting with the blind accumulation over the four blind transmissions in the first TDMA frame (**1**–**3**), followed by the blind accumulation in the second TDMA frame (**4**–**6**). The phase shift between the two accumulated bursts is then estimated (**7**) and compensated for (**8**) before accumulation (**9**). The two last TDMA frames of the block follow the same sequence and procedure.

To illustrate the nonideal processing gain from a nonideal combination of transmissions, simulations have been carried out with blind transmissions of 2^N with N ranging from 1 to 10. For each increment of N (doubling of the number of transmissions), a 3 dB gain is expected from the coherent combination. However, as can be seen in Fig. 3.22, the ideal gain (solid line) according to Eq. (3.2) can only be reached without any visible degradation

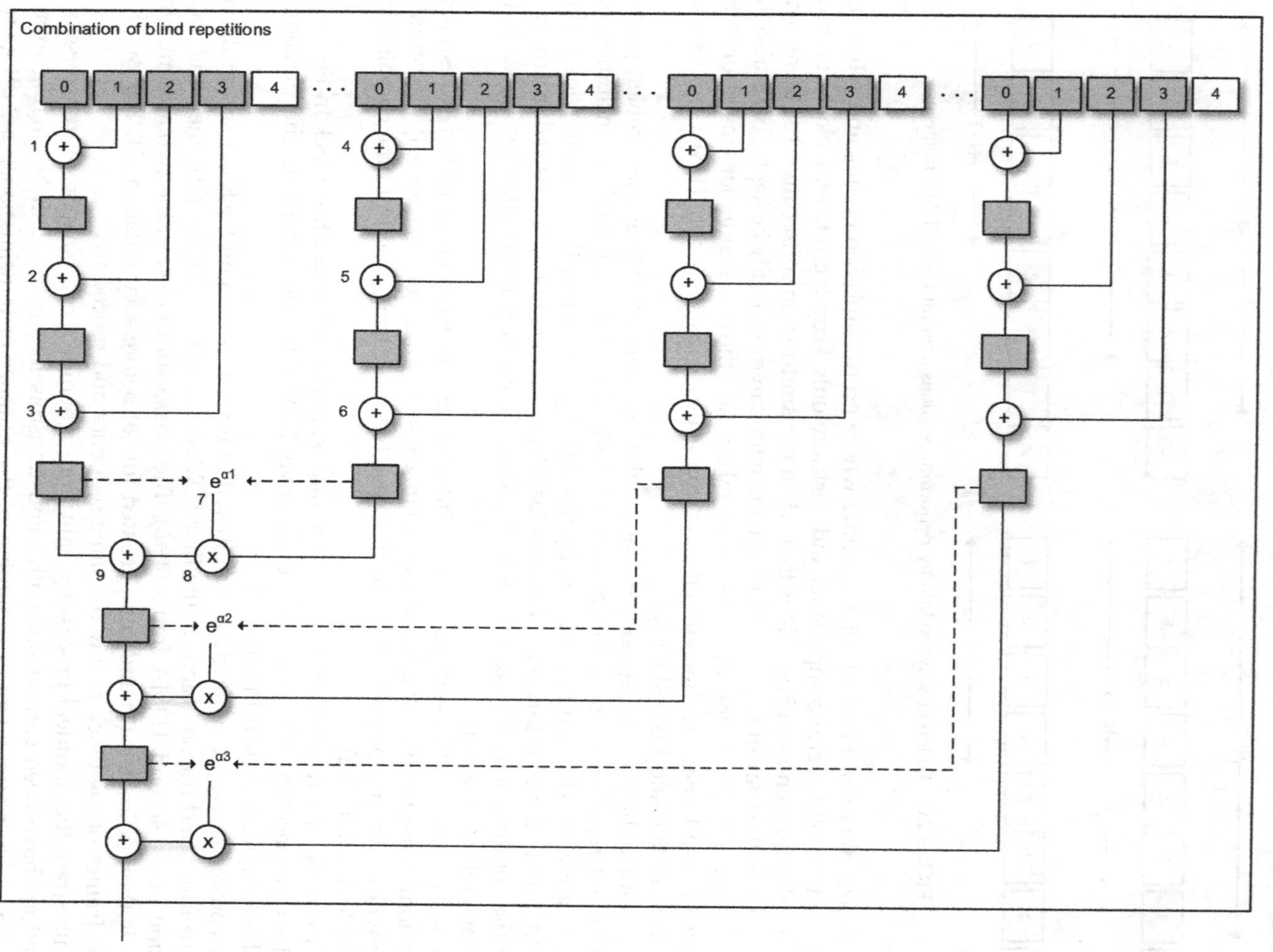

FIG. 3.21 Combination of blind repetitions.

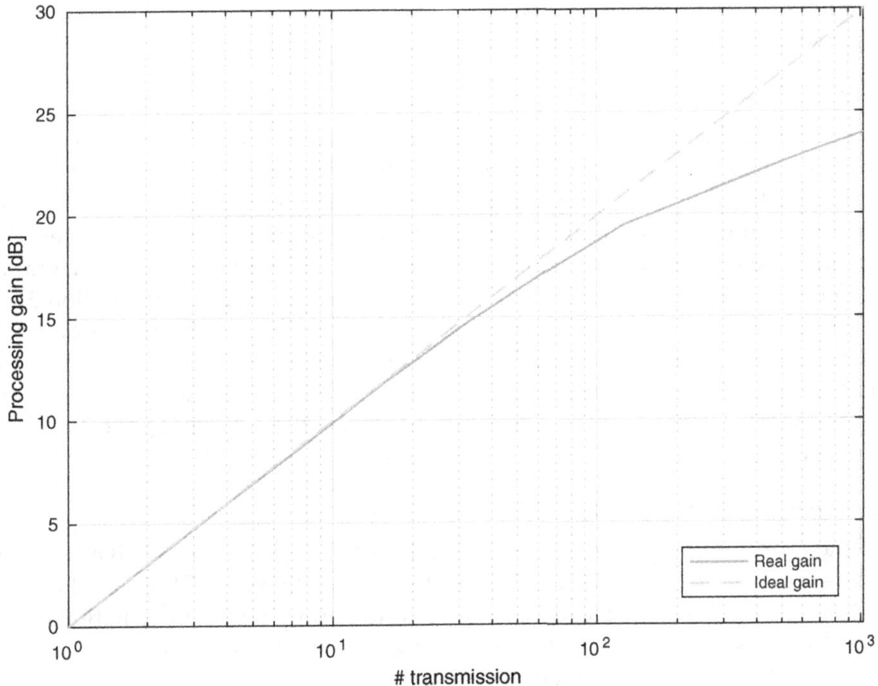

FIG. 3.22 Real versus ideal processing gain at different number of blind repetions.

between 2 and 16 blind repetitions. At higher number of repetitions, the experienced gain is lower than the ideal gain. Four times more transmissions are, for example, required to reach 24 dB gain (1024 in real gain vs. 256 with ideal gain), leading to a waste of radio resources.

It can be noted that in the 3GPP specifications, blind repetitions are referred to as *blind physical layer transmissions*. In this book when discussing EC-GSM-IoT, blind transmissions or blind repetitions are simply referred to.

3.2.8.3 Improved channel coding

The compact protocol implementation of EC-GSM-IoT compared with GPRS/EGPRS opens up for a reduced message size of the control channels. The overall message size of the control channels have, for example, been reduced from 23 octets as defined for GPRS/EGPRS to 8, 10, and 11 octets for EC-PACCH/U, EC-PACCH/D, and EC-CCCH/D, respectively. In addition to, the already mentioned blind transmissions, the improved channel coding from the reduced payload space also contributes to the coverage extensions on these logical channels.

3.2.8.4 More efficient HARQ

HARQ type II was introduced with EGPRS and is also used for EC-GSM-IoT. More details are provided in Section 3.3.2.2 on the HARQ operation for EC-GSM-IoT, but one can note here that for uplink operation, the use of *Fixed Uplink Allocation* (FUA) for allocation/assignment of resources will (which is more elaborated upon in Section 3.3.2.1) allow the receiver to operate at a higher BLock Error Rate (BLER) on the uplink, because no detection of the sequence number is required. A higher BLER of the traffic channel will effectively increase coverage (as long as the targeted minimum throughput is reached).

3.2.8.5 *Increased acquisition time*

To see an increased acquisition time as an extension of coverage might seem strange at an initial thought but this could actually be a means that can improve coverage for all type of channels. For EC-GSM-IoT, the only channel that is extended in coverage by an increased acquisition time is the channel used for initial synchronization to the network/cell. At initial detection of the network the device will have to scan the RFs to find a suitable cell to camp on (for more details on the cell selection and cell reselection procedures, see Section 3.3.1.2). In EC-GSM-IoT the channel used for this purpose is the FCCH, see Section 3.2.6.1 for more details. Having a device synchronize to a cell during a longer period will not have a negative impact on the network performance but will have an effect on the latency in the system. For the IoT applications targeted by EC-GSM-IoT, however, the service/application is usually referred to as delay tolerant.

3.2.8.6 *Increasing system capacity*

One of the more profound impacts to the system by serving an extensive set of users in extreme coverage situations is the impact on the system capacity. That is, when users are positioned in extreme coverage locations, they need to blindly repeat information (see Section 3.2.8.2) and by this eat up capacity that otherwise could serve other users. To combat this negative effect, EC-GSM-IoT has been designed using what is referred to as overlaid Code Division Multiple Access. The code is applied across the blind physical layer transmissions (i.e., on a burst-per-burst basis), and, consequently, has no impact to the actual spectral properties of the signal.

The basic principle is to assign multiple users different codes that allow them to transmit on the uplink, simultaneously. The codes applied are orthogonal to allow the base station to receive the superpositioned signal from the (up to four) users and still separate the channel from each user. The code consists of shifting each burst with a predefined phase shift, 7according to the assigned code. The codes are picked from the rows of a 4 × 4 *Hadamard* matrix, see Eq. (3.3), where "0" implies that no phase shift is performed, whereas a "1" means a 180° (π radians) phase shift.

$$H = \begin{bmatrix} 0 & 0 & 0 & 0 \\ 0 & 1 & 0 & 1 \\ 0 & 0 & 1 & 1 \\ 0 & 1 & 1 & 0 \end{bmatrix} \tag{3.3}$$

A 180° phase difference is the same as having the signal completely out of phase, and hence the total signal will be canceled out if adding a 180° phase shifted signal to the original signal.

To receive the signal from the multiple users the base station simply applies the code of the user it will try to decode. In this way all paired users are canceled out since the codes are

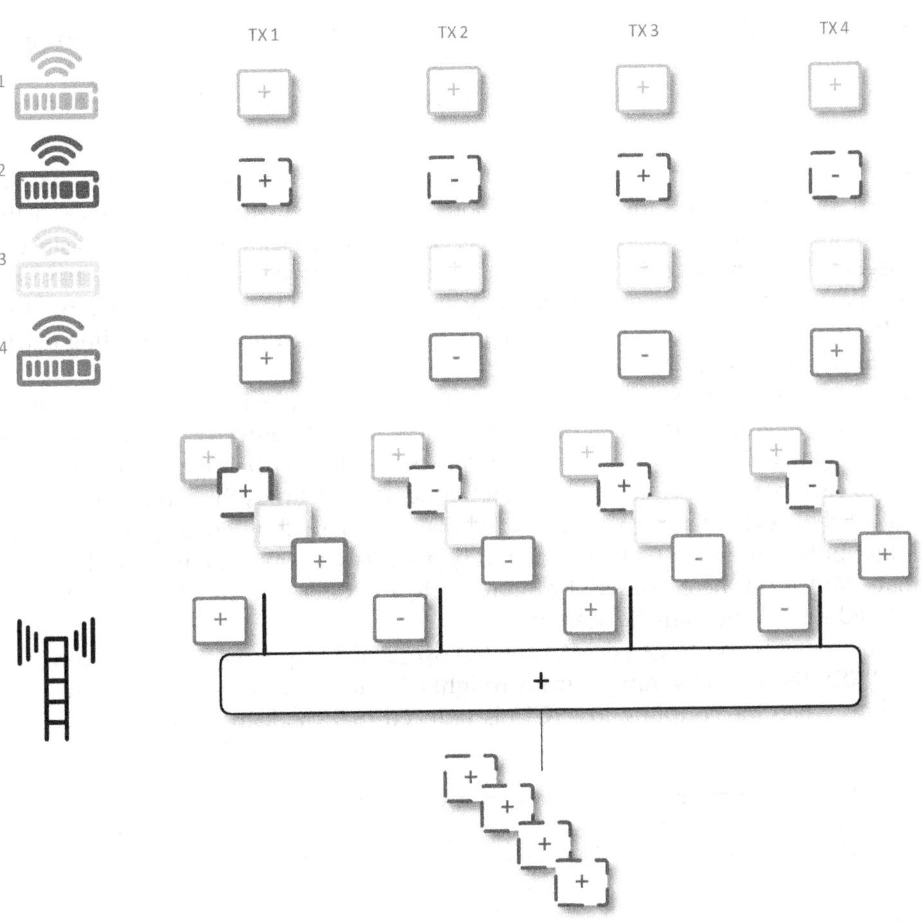

FIG. 3.23 Illustration of overlaid CDMA.

orthogonal to the applied code, whereas the signal for the user of interest still will benefit from the processing gain of adding the repeated bursts together.

The procedure is illustrated in Fig. 3.23 where four devices are, simultaneously, transmitting on the uplink, each device using a different code. The codes are illustrated as "+" (no change of phase) and "−" (180° phase shift) on a per burst basis.

Compared to having a single user transmitting on the channel, the same processing gain is achieved, i.e., the SNR is increased by 6 dB (10 $\log_{10}(4)$). Hence, the channel capacity has increased by a factor of four!

In reality the ideal capacity increase is not always reached, and this mainly depends on the following:

1. Not being able to pair users that, simultaneously, want to use the channel using blind repetitions.
2. Impairments in the signal, e.g., frequency offsets, that partly destroys the orthogonality.

3. Large power imbalances between the received signal that makes it sensitive to signal imperfections.
4. An increased power used on the uplink channel that can cause interference to other devices.

Considering (1) this is not really a problem for the technique itself because the main target is to increase capacity, and if there are not enough users to pair to increase the capacity, the need for the extra capacity is not that evident, (4) is also not considered a problem since it will only be users in bad coverage, i.e., with a low signal level received at the serving base station, that will make use of the technique. Hence, any contribution of increased interference to other cells is expected to be small. It is thus (2) and (3) that are of main concern regarding reaching a suboptimum channel capacity.

In Fig. 3.24 [9] the increase in channel capacity from multiplexing four users on the same channel, compared with allocating a single user to the same resources is shown. For a specific realization of power imbalance, the device can be up to 20 dB lower in received power compared to each paired device. The residual frequency offset (i.e., the remaining error in frequency after the device has synchronized to the base station reference) is picked from a normal distribution of mean 0 Hz and a standard deviation of either 10 or 45 Hz. It can be noted that N(0,10) can be considered a more typical frequency offset, whereas N(0,45) can be considered a more pessimistic scenario.

As can be seen, at the targeted MCL of 164 dB (20 dB improved coverage compared to GPRS/EGPRS), the capacity ranges from roughly 3.5 to 4.0 in capacity increase, reaching higher values when the coupling loss (CL) is reduced (SNR is increased).

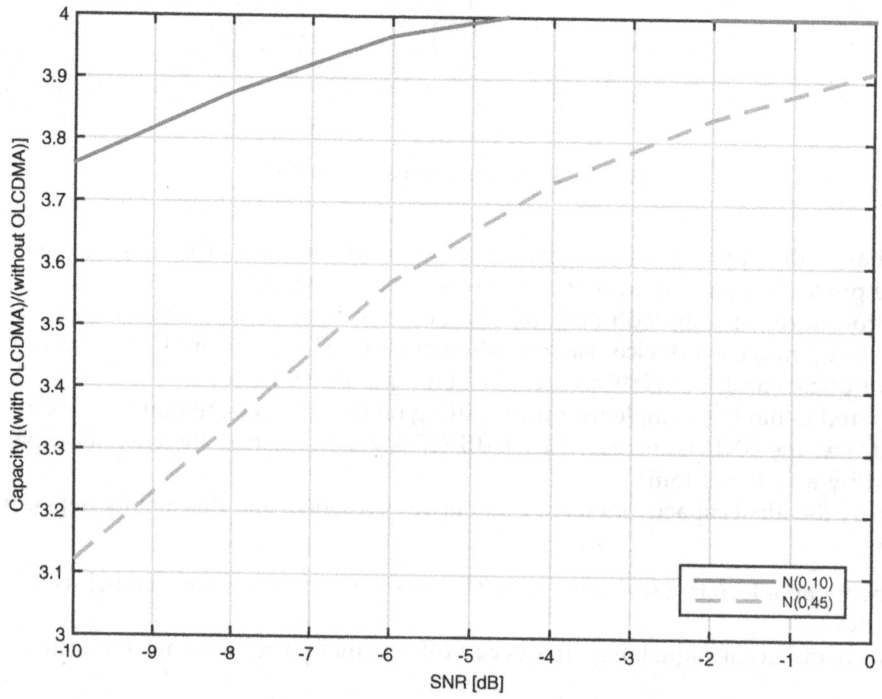

FIG. 3.24 Capacity increase from overlaid CDMA (OLCDMA).

The codes used for the technique are always of length four and are only applied over blind repetitions sent with a coherent phase, i.e., over the four blind repetitions in a TDMA frame introduced in Section 3.2.8.2. This implies that it can be used for EC-PACCH and EC-PDTCH transmissions for CC2, CC3, and CC4. For more details of the resource mapping and different uplink channels, see Section 3.2.7.

Overlaid Code Division Multiple Access (OLCDMA) is only applied on the uplink.

3.3 Idle and connected mode procedures

In Section 3.2 the physical layer of EC-GSM-IoT was described. In this section, the procedures related to the physical layer operation will be outlined. That is, apart from knowing how information is transferred over the physical layer, this section will describe the related behavior of the device and the network in different modes of operation. This section reflects the sequential order, of events that typically occur when a device attempts a system access to initiate a data transfer. First the idle mode operation is described, with all its related activities from the device, including descriptions of how the network reaches the device when in idle mode, by, for example, paging the device. The idle mode description is followed by the system access procedure where the device initiates a connection with the network. After this follows the assignment of resources and how resources can be managed during the data transfer. The section will also look into how to protect the network from overload and how the security of the network has been improved to protect devices from different type of attacks.

The description of the higher layers is limited to the layers directly above the physical layer. To understand the full GSM protocol stack, and how different protocols maps to different nodes in the architecture, the interested reader is referred to Ref. [10].

3.3.1 Idle mode procedures

Idle mode operation, or simply *idle mode*, is the state of the device when a device is not connected to the network. For example, an electricity meter that is not providing a report to the server will, to save power, not constantly stay connected to the network and instead operate in idle mode. For a device in idle mode there is still be some level of activity, e.g., to support mobility and paging-based reachability, but substantially less than when connected to the network, in that case the device is in *packet transfer mode*.

There are six main idle mode procedures for which the behavior of an EC-GSM-IoT device differs from that of a GSM/EDGE device:

- Cell selection
- Cell reselection
- SI acquisition
- CC selection
- Paging
- Power Saving Mode (PSM)

3.3.1.1 *Cell selection*

For a legacy GSM device to always be connected to the most suitable cell, near continuous measurements are typically performed of the surrounding radio environment. This means measuring its currently selected cell as well as the surrounding neighboring cells, or if no cell has been selected yet, a scan of the supported frequency band(s) is required. Two important modes of cell selection are defined: *cell selection* and *cell reselection*.

When performing cell selection, the device will, in worst case, not have knowledge of the surrounding radio environment and hence need to acquire knowledge of what cells are around it and which cell is most suitable to connect to. This activity is hereafter referred to as *normal cell selection*. If a device would not perform this task and, for example, simply connect to the first cell detected, it would typically mean a less than optimum cell would be used, thereby resulting in high interference levels in the network because the device would have to use a higher power level to reach the base station (and the base station in its turn a higher power level to reach the device). Apart from causing interference to other devices in the network, this would also mean an increased level of packet resource usage (e.g., due to data retransmission), as well as energy consumption for the device.

Cell selection is, for example, performed at power on of the device, or when the device is roaming and the radio environment is not previously known. The task in normal cell selection is to perform a full scan of the frequency band supported by the device and connect to what is referred to as a *suitable cell*. A suitable cell shall fulfill certain criteria [11] among which the most important ones are that the cell is allowed to be accessed (for example, that it is not barred), and that the path loss experienced by the device is sufficiently low (this criterion is referred to as *C1* in the GSM/EDGE specifications). After finding a suitable cell, the device is allowed to *camp* on that cell.

An alternative to the normal cell selection described above is the use of a *stored list cell selection*; in that case the device uses stored information to limit the scan of the frequency band to a stored set of frequencies. These set of frequencies can, for example, be based on EC SI, see Section 3.3.1.3, acquired from previous connections to the network and stored at power off of the devices. Using a stored list cell selection will speed up the process of finding a suitable cell as well as help reduce energy consumption.

The scan of the frequency band is, to a larger extent, left up to implementation, but there are certain rules a device needs to follow. One important difference between a legacy GSM device and an EC-GSM-IoT device is that the latter can be in extended coverage when performing the search, which means that scanning a certain ARFCN will take longer time (see Section 3.2.8.4) and by that consume more energy. If a device is not in extended coverage, a GSM frequency can be detected by simply scanning the frequency for energy and acquiring a *received signal strength indication (RSSI)*, referred to in the specification as *RLA_C*. Cells are then ranked according to their RSSI. For legacy GSM, the cell scan stops here and a cell selection is made. For EC-GSM-IoT however, the RSSI is not enough. Because only the total energy level of the signal is considered, the measurement will also include contributions from interference and thermal noise at the receiver. Hence, to get a reasonably correct RSSI estimate, the SINR should be sufficiently high. In GSM the cell edge has traditionally been assumed to have an SINR of roughly 9 dB. This means that noise and interference at most contribute to an error in the signal estimate of 0.5 dB.

Considering the refarming of GSM frequencies (see Section 3.1) and that EC-GSM-IoT is designed to be able to operate in rather extreme coverage together with a system frequency allocation down to 600 kHz, both interference and high levels of thermal noise (compared to the wanted signal level) become more of a problem for EC-GSM-IoT than in traditional GSM networks. The EC-GSM-IoT system is, instead of designed to operate at +9 dB SINR, roughly aiming at an operation at −6 dB SINR (see Chapter 4). If only measuring RSSI the signal level estimation will be off by around 7 dB (because the total power will be dominated by the thermal noise level, 6 dB higher than the wanted signal).

Hence, after the RSSI scan, an EC-GSM-IoT device is required to perform a second scan wherein more accurate measurements are made on the strongest set of cells identified by the RSSI scan. The number of cells to include in the second scan is determined by the difference in signal strength to the strongest cell. The maximum difference is set by the network and indicated within the information devices acquire from the EC SI. This refined measurement intends to only measure the signal of a single cell, and exclude the contribution from interference and noise, and is referred to as *RLA_EC*. The device is only allowed to estimate RLA_EC using FCCH and/or EC-SCH. This is in contrast to the RSSI scan, which could measure any of the physical channels on the BCCH carrier. Requiring only FCCH and EC-SCH to be measured will, for example, allow the network to down-regulate the other channels on the BCCH carrier (except the EC-BCCH) and by this lower the overall network interference. Especially for a tight frequency reuse deployment, see Chapter 4, this will have a large impact on the overall network operation since only 5% of all bursts on the BCCH carrier are required be transmitted with full power. Fig. 3.25 shows that when EC-GSM-IoT-only is supported in a given cell, resources in a 51 multiframe except FCCH, EC-SCH, and EC-BCCH can be down-regulated. It can be noted that in case also non-EC-GSM-IoT devices are supported in the cell, also the SCH, CCCH, and BCCH are transmitted with full power, in that case 16% of the resources cannot be down-regulated.

In case the device itself is in an extreme coverage situation, it could be that the RSSI scan does not provide a list of frequencies for a second RLA_EC scan (because the cell signals are all below the noise floor in the receiver). In this case, a second scan over the band using the RLA_EC measurement is been performed. Each ARFCN will at most be scanned for 2 s before a device is required to select another ARFCN to scan.

After the second scan is completed, the device attempts to find a suitable cell (from the set of cells for which RLA_EC measurements were performed) by acquiring the EC-SCH (identifying the BSIC, frame number, and additional access information regarding the cell) and the EC SI. In both EC-SCH and EC SI information regarding barring of devices can be found. In EC SI further details related to camping of the cell can be found, such as, minimum allowed signal level measured to access the cell, see Section 3.3.1.3. If a suitable cell is found, the device can camp on it.

3.3.1.2 Cell reselection

After a cell has been selected, a legacy GSM the device will continuously monitor the cell and its surrounding neighbor cells to always camp on the most suitable cell. The procedure when a device changes the cell it is camping on is referred to as cell reselection. Compared to legacy GSM devices that are almost continuously monitoring the surrounding environment, an EC-GSM-IoT device need only perform measurements of the neighboring cells under

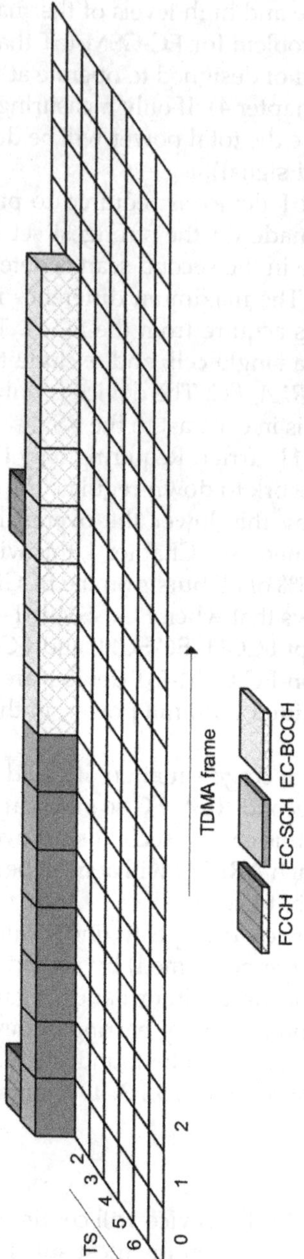

FIG. 3.25 Power down-regulation on the BCCH carrier.

certain conditions. The reason for performing continuous measurement is to ensure that the best cell is camped on and thereby guarantee e.g., good call quality should call establishment be triggered at any point in time, as well as quick mobility (handover or cell reselection) between cells. In case of CIoT the mobility of the devices is expected to be limited, the frequency of data transmission is expected to be low, and many of the applications are expected to be delay tolerant. This allows the requirements to be relaxed. In addition, considering that continuous measurements have a large impact on the battery lifetime, a minimization of these activities should be targeted.

To allow a reasonable trade-off between camping always on the most suitable cell, and saving battery, the measurements for cell reselection are only triggered by certain events, of which the most important ones are as follows:

- **Signaling failure**: If the device fails to decode the EC-SCH within 2.5 s.
- **A change in BSIC is detected**: For example, if the device wakes up to reconfirm the previously camped on cell and detects when acquiring EC-SCH that the BSIC is different from previously recorded. This implies that it has moved during its period of sleep and that it needs to reacquire the surrounding radio environment.
- **Failed path loss criterion**: The camped on cell is no longer allowed to camp on (no longer considered suitable) because the abovementioned path loss criterion (C1) fails (i.e., the measured signal level of the cell is below the minimum allowed level)
- **24 h**: More than 24 h have passed since the last measurement for cell reselection was performed
- **Change in signal level**: If the measured signal level of the camped on cell drops more than a certain threshold value compared to the best signal level measured on that cell. The threshold value, C1_DELTA, is set to the difference between the C1 value of the selected cell and the strongest neighbor determined at the most recent instance of performing the cell reselection measurement procedure.

An illustration of the fifth criterion is provided in Fig. 3.26. At time N the device has selected cell A as its serving cell after performing measurements for cell reselection. The strongest neighbor cell recorded, at this point in time, is cell B, and the difference in signal strength between the two cells, *C1_DELTA*, is recorded. At a later point in time (M) the device wakes up to reconfirm cell A as a suitable cell to remain camping on. It measures a signal strength (A″) therein and determines it has degraded by more than C1_DELTA compared with the previously measured signal level (A′). This triggers measurements for cell reselection and based on the measurements, the device detects cell B to be the one with the strongest signal level and hence reselects to it. It should be noted that the above criteria only trigger *measurements* for cell reselection. It might well be so that after the measurements are performed, the device remains camping on the same serving cell.

The metric used when performing measurements for cell reselection is the same as used for cell selection, i.e., RLA_EC. To get an RLA_EC estimate for a specific cell, at least 10 samples of the signal strength of the cell, taken over at least 5 s, need to be collected. This is to ensure that momentary variations of the signal level are averaged out. It should be noted that the RLA_EC measurements performed for a set of cells need not be performed in serial but could done in parallel to minimize the overall measurement period.

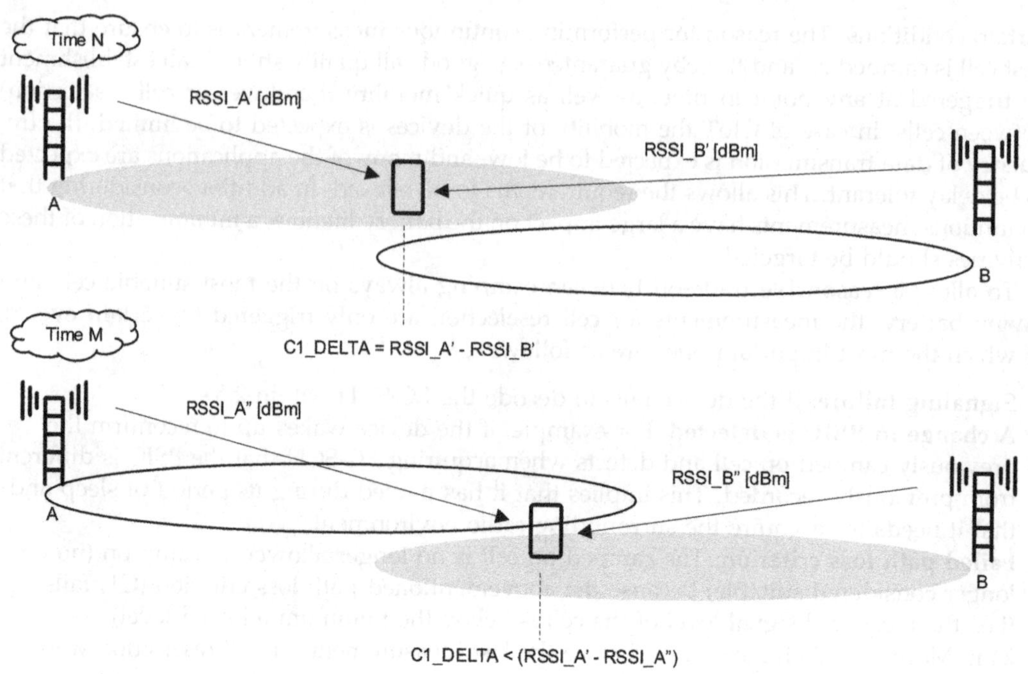

FIG. 3.26 Triggering of measurements for cell reselection.

Other means to save the battery lifetime consist of only reacquiring EC SI if the EC-SCH indicates that the EC SI content has changed. Because the EC SI is not expected to frequently change, this is an easy way to keep SI reading to a minimum. The network will toggle the states of a change indicator information in the EC-SCH at the time of EC SI change, which informs the device that EC SI needs to be reacquired. However, the change indicator information indicates which EC SI message that has been changed, and hence it need not acquire the full EC SI. A complete reacquisition of the EC SI is required if more than 24 h has passed since the last reading.

3.3.1.3 *Extended coverage system information (EC SI)*

The EC SI contains, as the name indicates, information needed to operate in the system. The SI messages are sent on the BCCH with a certain period, see the EC-BCCH description in Section 3.2.6.3.

There are currently four messages defined, EC SI 1, EC SI 2, EC SI 3, and EC SI 4. Each message can vary in size depending on the parameters transmitted in the cell, and hence the number of EC SI message instances per EC SI message can also vary. Each EC SI message instance is mapped to an EC-BCCH block mapped to the physical resources as described in Section 3.2.6.3.

For the device to save energy and to avoid acquiring EC SI too frequently, the synchronization channel (EC-SCH) will assist the device by changing the EC-BCCH CHANGE MARK field, thereby indicating that one or more EC SI messages has changed. The device then acquires any EC SI message and reads the EC SI_CHANGE_MARK field included therein. This field includes 1 bit for each of four EC SI messages where a change of state for any of these 4 bits (relative to the last time EC SI_CHANGE_MARK field was acquired) indicates to the device that the corresponding EC SI message must be reacquired. The EC SI_CHANGE_MARK field also includes one overflow bit, which can be toggled to indicate to the device that all EC SI messages need to be reacquired. There are rules associated with how often bit states are allowed to be changed by the network, see Ref. [12], and the specification recommends that changes to the EC SI information do not occur more frequently than 7 times over a 24-h period. In addition, if a device has not detected any changes to EC-BCCH CHANGE MARK over a period of 24 h, the full EC SI message set is reacquired.

The following principles apply for the EC SI messages regardless of how many message instances are required to transmit any given EC SI message:

- The EC SI messages comprising a cycle of EC SI information are sent in sequence in ascending order.
- In case any given EC SI message contains multiple message instances (up to four) they are also sent in ascending order.
- The sequence of EC SI messages and associated message instances are repeated for each cycle of EC SI information.

The transmission of the first three EC SI messages is mandatory. The fourth message is related to network sharing and is only required if network sharing operation is activated in the network.

The content of each EC SI message is shortly described in Fig. 3.27.

3.3.1.4 Coverage Class selection

Cell selection and cell reselection are activities performed both for EC-GSM-IoT and in legacy GSM operation. CC selection, however, is something specific to EC-GSM-IoT.

The most important improvement to the system operation introduced with EC-GSM-IoT is the ability to operate devices in more challenging coverage conditions (i.e., extended coverage) than what is supported by GPRS/EGPRS. To accommodate the extended coverage and still provide a relatively efficient network operation, CCs have been introduced, as described in Section 3.2.3.3. In case a logical channel makes use of CCs, a CC is defined by a certain number of blind repetitions used, each repetition being mapped onto the physical resources in a predetermined manner, see Sections 3.2.7 and 3.2.8 for more details for downlink and uplink channels, respectively.

When a device is in idle mode the network will have no knowledge of the whereabouts of the device or in what coverage condition it is in. Hence, in EC-GSM-IoT it is under device control to select the CC, or more correctly put, it is under device control to perform measurements that form the basis for the CC selection. The network provides, through EC SI (broadcasting to all devices in the cell), information used by a device to determine which CC to select, given a certain measured signal.

The procedure is illustrated in Fig. 3.28.

In Fig. 3.28 the device measures a signal level of −110 dBm. This in itself will not allow the device to select the appropriate CC but it needs information from the network to determine the CC to be selected. In this case, −110 dBm implies that CC3 should be selected.

Two different CC selection procedures are supported by the specification in the downlink, either a signal level-based selection (RLA_EC) or an SINR-based selection (*SLA*).

EC-SI 1

Provides cell allocation information to mobile stations that have enabled EC operation

EC Mobile Allocation list : Possible frequency lists that can be used by the device

EC-SI 2

Provides EC-RACH/RACH control information and cell selection information

EC Cell Selection Parameters : Cell/Routing Area/Local Area identity and minimum signal levels to access the cell, and maximum UL power allowed on EC-CCCH

Coverage Class Selection Parameters : Defies if SINR/RSSI based selection is used and the thresholds that define each CC

(EC-)RACH control parameters : How a device can access the (EC-)RACH, including for example resources to use and CC adaptation

EC Cell Options : Power control (alpha) and timer related settings

EC-SI 3

Provides cell reselection parameters for the serving and neighbour cells and EC-BCCH allocation information for the neighbour cells

EC Cell Reselection Parameters : Information required for reselection for both serving and neighbor cells, including for example trigger to perform cell reselection measurements.

EC Neighbour Cell Reselection Parameters : BSIC of neighbor cells, and access related information, such as minimum signal level to access the cell.

EC-SI 4 [optional]

Provides information related to network sharing. For example if a common PLMN is used, and the number of additional PLMNs defined. For each additional PLMN, a list is provided on the NCC permitted. Also the access barring information for additional PLMNs are provided together with information if an exception report is allowed to be sent in each PLMN.

FIG. 3.27 EC SI message content (non-exhaustive).

Which measurement method to be used by the device is broadcasted by the base station in the EC SI. Using a SINR-based selection can greatly improve the accuracy by which a device determines the applicable CC. In contrast, for the case where the CC selection is based on signal level, any interference would be ignored by the device and essentially the network would have to compensate for a high interference level by setting the CC thresholds in the network more conservatively, meaning that a device would start using blind repetitions earlier (at higher signal levels). This in turn will increase the interference in the network, and consume more energy in the device (longer transmission time).

The two procedures have been evaluated by system level simulations where interference from other devices and the measurement performed by the device is in detail modeled [13]. As can be seen in Fig. 3.29, the distributions of CC over SINR are much more concentrated in the bottom plot than in the upper one. Ideally only a single CC should be selected at a given SINR, but due to different imperfections, this will not happen. The device will,

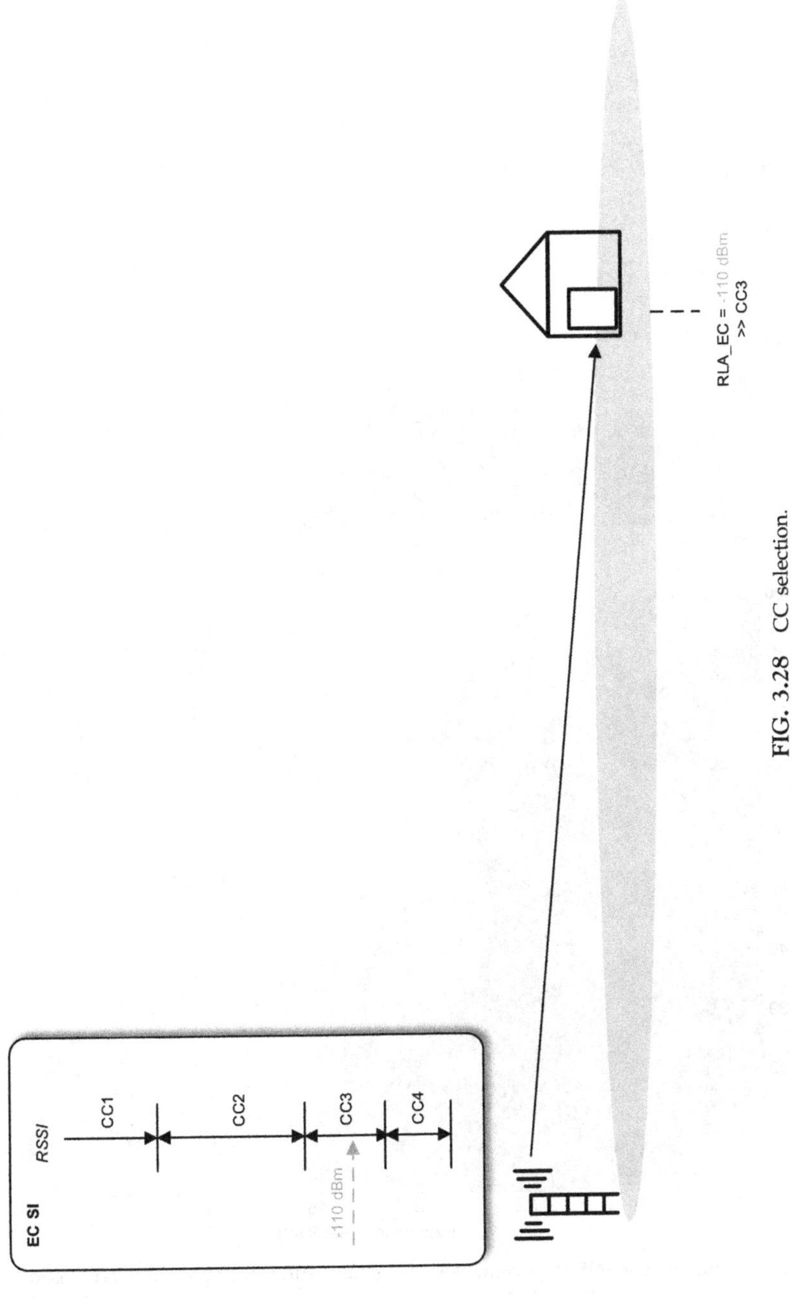

FIG. 3.28 CC selection.

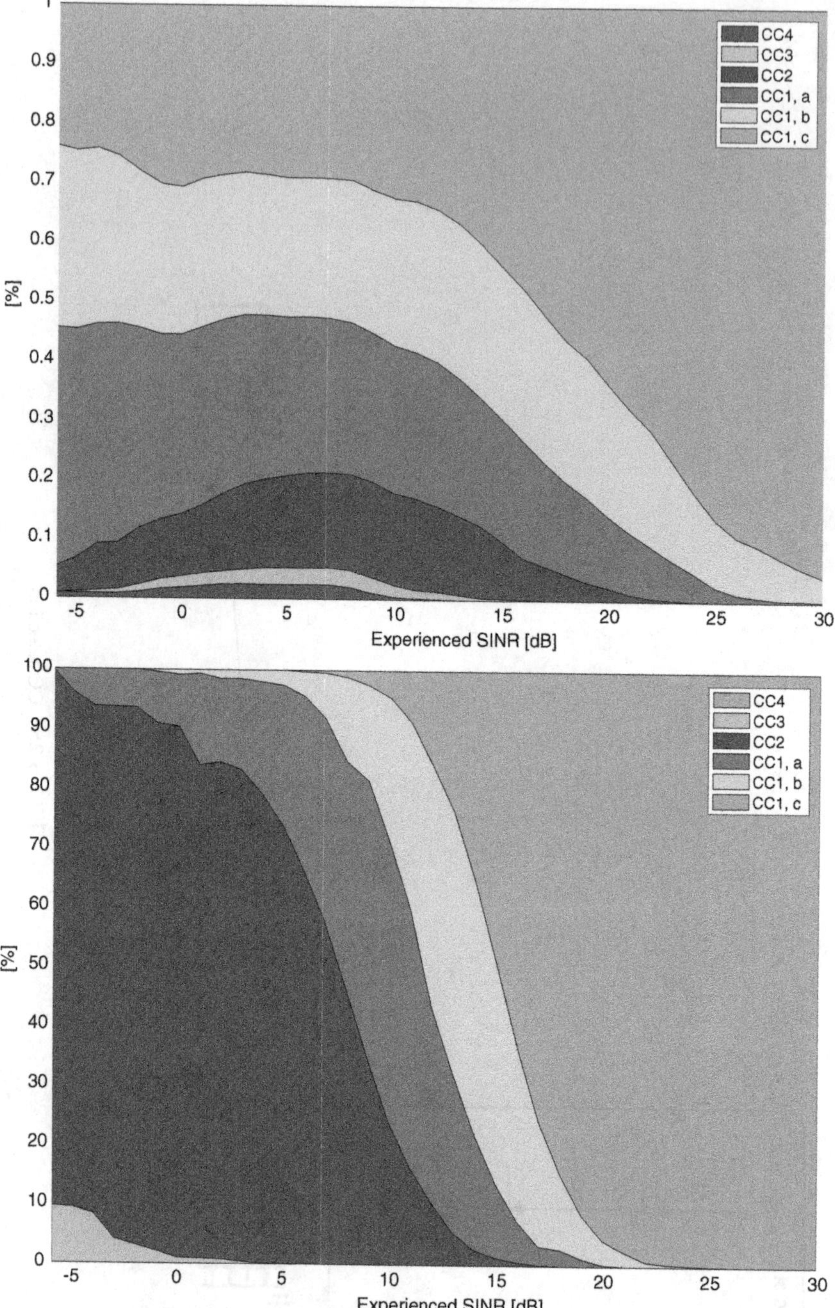

FIG. 3.29 CC distribution over SNR (top: signal level based measurements; bottom: SINR based measurements).

for example, not perfectly measure the SINR during the measurement period and the SINR in the system will not be the same during the measurements as when the actual transmission takes place (although the thermal noise can be considered stable over time, the interference is difficult to predict). One can note that more than four curves are seen in the figure, which is related to the possibility for a device to report the signal level above the signal level where the device switches from CC1 to CC2, which is explained in more detailed in Section 3.3.2 (here this is only referred to as CC1, a, b, and c respectively). It can further be noted that a specific reading of the percentage value for a specific CC should not be the focus when interpreting the figures. For the purpose of understanding the benefits with SINR-based CC selection, the overall shape of the curves is what matters.

Although the device is under control of the CC selection, it is important that the network knows the CC selected when the device is in idle mode, in case the network wants to reach the device using paging (see Section 3.3.1.2). There are a set of specific conditions specified where the device need to inform the network about a change in CC. One of them is, for example, if the selected CC is increased, the device experiences worse coverage than before. For the opposite case (lowering the selected CC) the device needs to only inform the network when experiencing an extreme coverage improvement, switching from CC4 to CC1. Based on this, there could, for example, be situations where the selected CC last time communicated to the network is CC3, while the device currently selected CC is CC2. In this case the device is not required to, but could, report the updated selection of CC2. Section 3.3.1.2 provides more details on why this type of operation is allowed.

Although all measurements for CC selection are performed on the downlink, the device also needs to estimate its uplink CC. This is done based on the estimated CL in the downlink, which can be assumed to be the same in both directions. To help devices estimate their CL the base station sends in EC SI the output power level (P_{out}) used by the base station on FCCH and EC-SCH, and hence by knowing the received signal power (P_{rx}), the CL can be derived (see Eq. 3.4)

$$CL(dB) = P_{out}(dBm) - P_{rx}(dBm) \tag{3.4}$$

Because the device is aware of its own output power capability, the received power at the base station can be derived using Eq. (3.4) with CL assumed the same in uplink and downlink, and P_{out} being the device output power.

It can be noted that only signal level-based CC selection is supported on the uplink since the reciprocity of the CL does not apply to the interference (i.e., knowledge of the downlink interference level does not provide accurate information about the uplink interference level).

3.3.1.5 *Paging*

For an EC-GSM-IoT device, reachability from the network (i.e., means by which the network initiates communication with the device) is supported either by paging when Discontinuous Reception (DRX) operation is used or in conjunction with Routing Area Updates in the case of PSM (see Section 3.3.1.3).

In case of paging-based reachability, the functionality is significantly different than what is traditionally used in legacy GSM, where all devices in a cell make use of the same DRX cycle configured using the SI. In this case, the DRX setting from the network is a trade-off between having a fast reachability and enabling energy savings in the device.

The same principle applies for EC-GSM-IoT, but in this case the used DRX cycle is negotiated on a per device basis. In case a DRX cycle is successfully negotiated (during Non Access

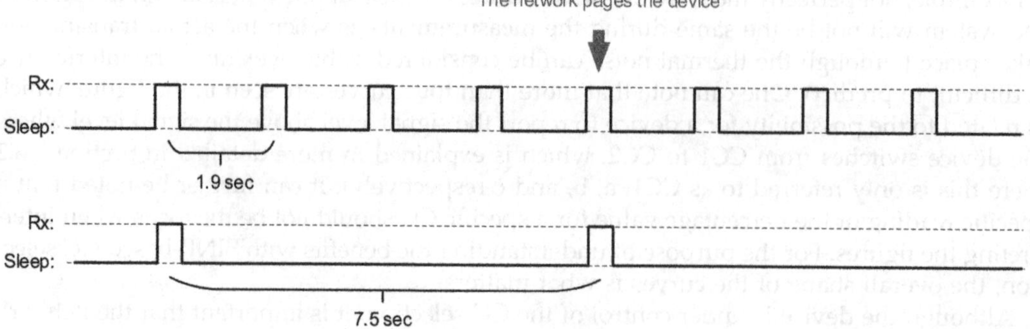

FIG. 3.30 Discontinuous reception (DRX).

TABLE 3.4 DRX cycles.

Approximate DRX cycle length	Number of 51-MF per DRX cycle
2 s	8
4 s	16
8 s	32
12 s	52
25 s	104
49 s	208
1.6 min	416
3.3 min	832
6.5 min	1664
13 min	3328
26 min	6656
52 min	13312

Stratum signaling), the device can consider that DRX cycle to be supported in all cells of the Routing Area (a large set of cells defined by the network operator), and that the same negotiated DRX cycle therefore applies upon reselection to other cells in the same Routing Area.

A DRX cycle is the time period between two possible paging opportunities when the network might want to (but is not required to) reach the device. In-between these two opportunities the device will not be actively receiving and can be considered to be in a level of sleep, hence the reception is discontinuous over time. The functionality is illustrated in Fig. 3.30 by the use of two possible DRX cycle configurations. It can be noted that if two devices share the same DRX cycle, they will still most probably not wake up at the same time given that they will typically use different paging opportunities within that DRX cycle. In other words, it is only the periodicity with which the devices wake up that is the same between them.

The full set of DRX cycles supported for EC-GSM-IoT is shown in Table 3.4.

As can be seen, there is a large range of reachability periods supported, from around 2 s up to 52 min. This can be compared to the regular DRX functionality in legacy GSM that only allows DRX cycles to span a few seconds at most.

The negotiation of the DRX cycle is performed between the core network and device using Non Access Stratum signaling.

The paging group to monitor for a specific DRX cycle is determined using the mobile subscription ID, International Mobile Subscriber Identity (*IMSI*). This provides a simple means to spread out the load on the paging channel over time because the International Mobile Subscriber Identity can be considered more or less a random number between different devices. The shorter the DRX cycle, the fewer are the number of different nominal paging groups available within that DRX cycle. For example, for CC1 using the lowest DRX cycle of 2 s, 128 different nominal paging groups are supported therein, while for the DRX cycle of 52 min, around 213,000 different nominal paging groups are supported. The higher the CC the smaller number of nominal paging groups supported for each DRX cycle because each nominal paging group consists of a set of physical resources used to send a single paging message for a given CC. For example, using a DRX cycle of 2 s, the number of paging groups for CC4 is 4 (compared to 128 for CC1). To increase paging capacity, the network can configure up to four physical channels in the cell to be used for access grant and paging. In addition, up to two pages can be accommodated in the same paging group, in that case the paging message needs to be sent according to the highest CC of the two devices multiplexed in the same paging message, see Section 3.2.6.4.

The determination of the nominal paging group selected by a device for a given DRX cycle has been designed so that the physical resources used by that device for a lower CC are always a subset of the physical resources used by the same device for a higher CC (i.e., the resources overlap). This is to ensure that if the network assumes a device to be in a higher CC than the device has selected, the device will still be able to receive the paging message. The network will in this case send additional blind transmission that will not be monitored by the device, which can be seen as a waste of network resources. The gain with this approach is that instead the device need not continuously inform the network when going from a higher CC to a lower CC, which saves energy (device transmission is much more costly compared to reception, which is more elaborated upon in Chapter 4). The principle of the overlapping paging groups for different CCs is illustrated in Fig. 3.31. In this case, the paging groups for CC1 occur over TDMA frame 27 and 28 in 51 multiframe N. If the negotiated DRX cycle would be eight 51 multiframes (see Table 3.4), the same group would occur in 51 multiframe N + 8, N + 16, etc. As can be seen, TDMA frame 27 and 28 are also contained in the paging group for CC2 that spans TDMA frames 27 to 34 over 51-multiframe N and N + 1. The same principle follows for higher CCs as well, e.g., for a given device and DRX cycle, the physical resources of the nominal paging group used for CC3 are fully contained within the physical resources used for the nominal paging group for CC4 for that device.

Before each time the device wakes up to monitor its nominal paging group, it needs to perform a number of tasks:

- Synchronize to the cell, reconfirm the BSIC, and possibly read EC SI (see Sections 3.2.7.1 and 3.2.7.2)
- Evaluate the path loss criterion to ensure the cell is still suitable (see Section 3.3.1.2)
- Perform CC selection (see Section 3.3.1.4)

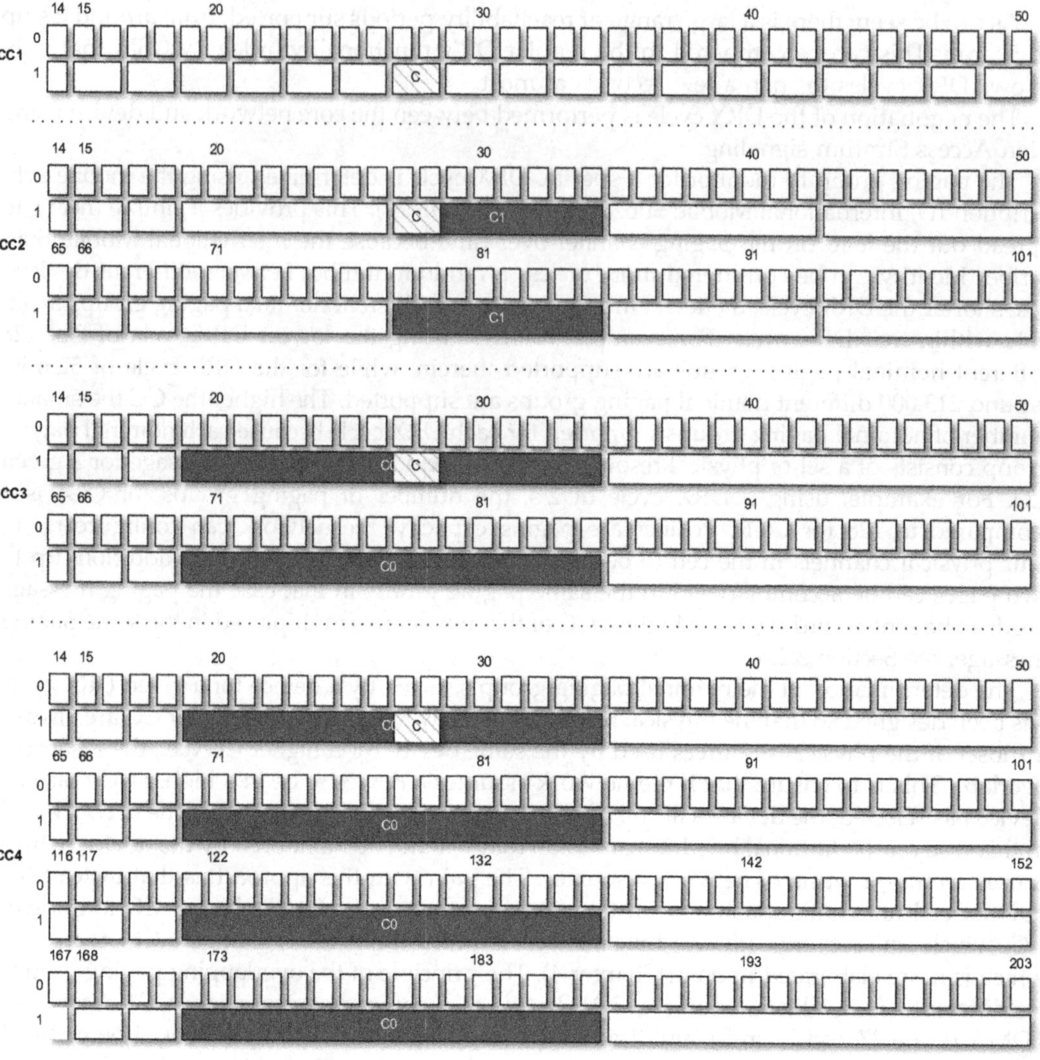

FIG. 3.31 Overlapping paging blocks between CCs.

The device needs to wake up sufficiently in advance of its nominal paging group because the above tasks require some time to perform. However, if the tasks above have been performed within the last 30 s, they are still considered valid, and need not be performed.

On network level, the cells within a given Routing Area need to be synchronized within a tolerance of 4 s regarding the start of 51 multiframes. This is to ensure that if a device moves around in the network, the time of its nominal paging group will occur roughly at the same time, irrespective of the actual serving cell. If this requirement would not be in place, then a device with a long paging cycle, say 26 min, would, in worst case, have to wait 52 min to be reached if the synchronization of cells differ enough for the device to miss its nominal paging group upon reselection to a new cell in the same Routing Area.

Although using long DRX cycles is critical for energy saving purposes in the device, it is not required by an EC-GSM-IoT device to support the full range of DRX cycles in Table 3.4

(see Section 3.3.1.5). Also PSM (see Section 3.3.1.3), is an optional functionality to support. If neither the full set of DRX cycles nor PSM is supported, the device at least needs to support reachability using the lowest DRX cycle in the set, i.e., 2 s.

It can be noted that in the 3GPP specifications, the DRX cycles and the related functionalities are referred to as extended DRX, eDRX (extended DRX).

3.3.1.6 *PSM*

Generally speaking, using DRX is more beneficial for devices that prefer a lower periodicity of reachability, whereas PSM becomes an attractive alternative for devices that can accept longer periods of being unreachable. When considering devices that can accept being reachable about once every 30 min or more, both DRX and PSM can be considered to be equal from the perspective of battery lifetime targets associated with EC-GSM-IoT devices. More details of the PSM functionality is provided in Section 2.2.3. In Chapter 2 there is also some evaluation shown between DRX and PSM depending on the requirement on reachability.

3.3.1.7 *System access procedure*

Before being able to send information across the network, the device needs to identify itself and establish a connection. This is done using the random access (system access) procedure that allows a device to make the transition from idle mode to packet transfer mode (connected mode).

The random access by EC-GSM-IoT devices can either be triggered using RACH or the EC-RACH channel. Which of the two channels to use is indicated to the device through EC-SCH (using the parameter *RACH Access Control*). Because the device is always required to acquire EC-SCH before performing a random access, it will always read the RACH Access Control parameter and thereby acquire the most up-to-date control information applicable to its serving cell. It is only applicable if the device is in CC1 that the RACH is allowed to be used, and hence devices that have selected CC2, CC3, and CC4 will always initiate system access using EC-RACH (irrespective of the RACH Access Control bit). The use of the RACH by CC1 devices depends on the traffic situation in the cell and because the control thereof is put in EC-SCH, the RACH usage can dynamically vary over time (and can also be toggled by the network to for example offload traffic in 40% of the CC1 access attempts from EC-RACH to RACH). If allowing CC1 users to access on RACH mapped on TS0 it will share the random access resources with non-EC-GSM-IoT devices. Hence, in case of an already loaded RACH channel (high traffic load) it is advisable not to offload CC1 users to the RACH. On the other hand, if they stay on the EC-RACH they are more likely to interfere with CC2, CC3, and CC4 users, which, because of operating in extended coverage, are received with a low signal level at the base station. This trade-off needs to be considered in the network implementation of the feature.

3.3.1.7.1 EC packet channel request

The message sent on the (EC-)RACH contains the following:

- A random reference (3 bits)
- The number of blocks in the transmit buffer (3 bits)
- Selected downlink CC/Signal level indication (3 bits)
- Priority bit (1 bit)

The random reference helps the network to distinguish a given device from other devices that happen to select the same physical resource to initiate access on. As the name implies the field is set randomly, and hence there is a risk of 1/8 that two users accessing the same (EC-)RACH resource will select the same random reference (see Section 3.3.1.7 for details on how this is resolved).

The number of blocks to transmit provides information to the network on how many blocks it should allocate in the FUA, see Section 3.3.2.1. In this context, a block is based on the smallest MCS size, MCS-1, carrying 22 bytes of payload.

Since the device has performed idle mode measurements, see Section 3.3.1, before initiating system access, it will not only indicate the selected downlink CC, but, in case CC1 is selected, it will also indicate the region in which its measured downlink signal level falls (using a granularity set by the network) thereby allowing the network to know the extent to which the signal level is above the minimum CC1 level. This then leads to the possibility of the network down regulating the power level used for sending subsequent messages to the device in the interest of interference reduction.

The final bit is related to priority, and can be set by the device to indicate that the access is to be prioritized, for example, in an alarm situation (also referred to as sending an Exception report). However, allowing the priority bit to be arbitrarily set would open up for a misuse of the functionality. That is, why would a device manufacturer not set priority bit if he/she knows that the associated data transfer would be prioritized in the network. To avoid this abusive behavior, the priority bit is forwarded to the core network that would, for example, support a different charging mechanism for exception reports.

3.3.1.7.2 Coverage Class adaptation

Before initiating system access, the device need to perform measurements to select an appropriate CC, see Section 3.3.1.4. However, measurements are always associated with a level of uncertainty, and there is always a risk that the CC thresholds set by the network (indicated within EC SI) are not conservative enough to provide a successful system access. In contrast, setting too conservative threshold to ensure successful system access would imply a waste of resources (devices using a higher CC than required). To combat these drawbacks, a CC adaptation can be used in the system access procedure wherein a device can send multiple access requests during a single access attempt.

This means that the device can increase its initially selected CC in case a failure is experienced during a given access attempt (i.e., an access request is transmitted but the device fails to receive an expected response from the network within a certain time frame). The number of failures experienced by a device before increasing CC during a given access attempt is controlled by the network (in EC SI). At most two increments of the CC are allowed during an access attempt. That is, if CC1 is selected initially, the device will at most send an access request using CC3 as a result of repeated failures being experienced during that access attempt (i.e., if access requests sent using CC1 and CC2 have both failed).

It can be noted that a device will not know if it is the downlink or the uplink CC that is the reason for a failed access request sent during a given access attempt. In addition, the failure could also be the result of an unfortunate collision with an access request sent in the same or partially overlapping TDMA frames by another device. As such, to remove as much ambiguity as possible following an access request failure both the downlink and uplink CC are incremented at the same time.

Simulations have shown [14] that, at the same access attempt success rate, the resource usage can be reduced between 25% and 40% by allowing CC adaptation. This does not only benefit the energy consumption of the device (less transmissions) but will also help reduce the overall interference level in the network.

3.3.1.7.3 Contention resolution

The network will not know to which device the channel request belongs because the device only includes a "Random Reference" of 3 bits in the EC Packet Channel Request (see Section 3.3.1.7.1). This problem is resolved during the contention resolution phase.

For EC-GSM-IoT, there are two procedures that can be used for contention resolution (which one to be used is controlled by the EC SI).

The first procedure is very similar to the contention resolution used in an EGPRS access. In this case, the device will include its unique ID, the *Temporary Logical Link Identity* (*TLLI*), 4 bytes long, in each uplink RLC data block until the network acknowledges the reception of the device ID. Considering the use of FUA (see Section 3.3.2.1), this becomes a relatively inefficient way of communicating the device ID to the network because essentially 4 bytes need to be reserved for TLLI in each uplink RLC data block transmitted by the device during the FUA, as illustrated in Fig. 3.32.

As can be seen from Fig. 3.32, if a low MCS is used, the overhead from 4 bytes of TLLI in each RLC data block can be significant (18% for MCS-1, with each block carrying 22 bytes). Assume, for example, that the payload to be transferred is 62 bytes. This would fit into 3 MCS-1 blocks. However, if a TLLI of 4 bytes need to be included in each block, the device need instead to transmit 4 RLC data blocks, which is an increase of 33% in transmitted resources and energy.

To minimize overhead an alternative contention resolution phase can be used, referred to as enhanced AB procedure. In this case the TLLI is only included in the first uplink RLC data block, and, in addition, a reduced TLLI (rTLLI) is included in the RLC/MAC header of the subsequent RLC data blocks. The rTLLI comprises the 4 least significant bits of the full TLLI. The procedure is illustrated in Fig. 3.33.

The inclusion of the rTLLI in the RLC/MAC header will not consume RLC data block payload space and hence, given the example above only 3 MCS-1 blocks would be required to transmit the 62 bytes of payload (18 + 22 + 22).

Comparing the two options for contention resolution procedure, the use of the enhanced AB procedure will result in a somewhat higher risk that two devices accessing at the same time will stay on the channel for a longer time simultaneously transmitting. This can occur only if the first RLC block has not been received by the network at the point when it sends the device an acknowledgment to the FUA. In this case only the rTLLI received by the network within the RLC/MAC header can be used to distinguish the users (i.e., there is a risk that the same rTLLI is used by the devices accessing), and this inability to complete contention resolution continues until the RLC data block with the full TLLI has been received. Hence, both implementation options are kept in the specification.

3.3.1.7.4 Access Control

To protect cellular networks from overload, there are typically multiple mechanisms in place. Most such mechanisms are expected to be used in exceptional situations, when, for example,

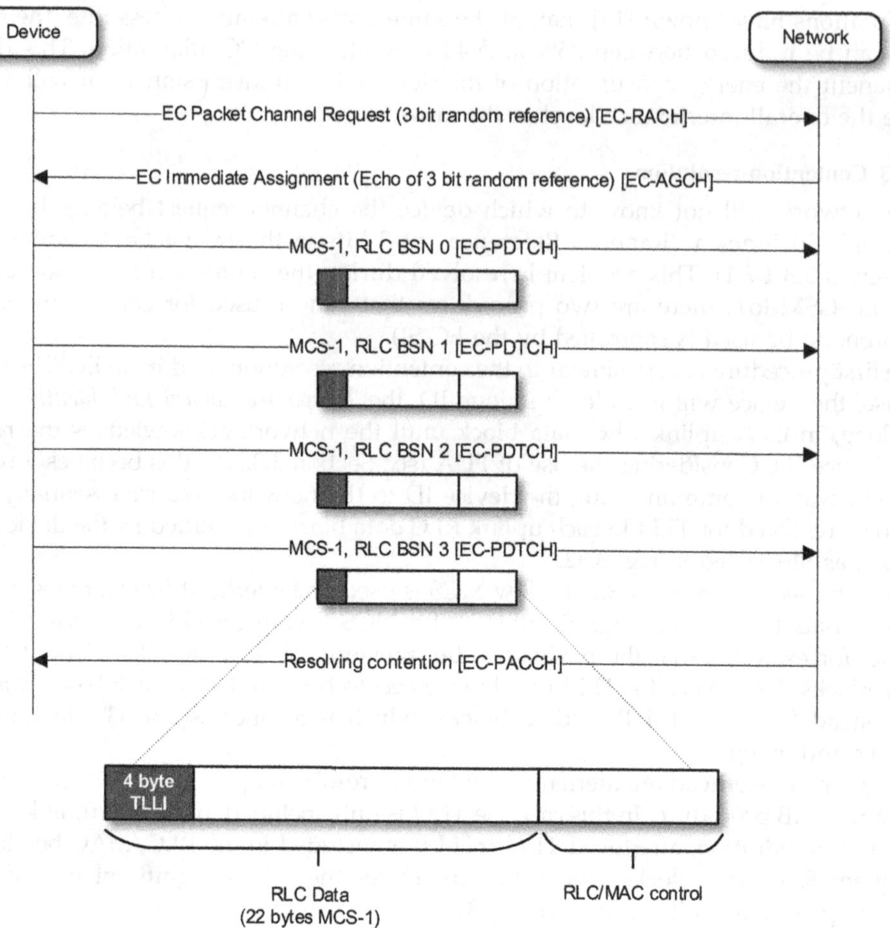

FIG. 3.32 Contention resolution, not using enhanced access burst procedure.

turning on a network/parts of a network after a power outage. It also serves as a protection between operators, if, for example, operator A's overlapping network is shut down; in that case the devices can start camping on, and trying to access, operator B's network.

As described in Chapter 2, when 3GPP started to adapt the GSM network toward IoT by the work on MTC, the networks were enhanced to protect them from a flood of new devices. Two levels of overload control were specified: Extended Access Barring (EAB) and Implicit Reject (IR), see Chapter 2 for more details. With EC-GSM-IoT the same functionality is used but it is implemented in a somewhat different fashion.

The most important difference to legacy functionality is to make the devices aware of any potential barring of system accesses in an energy efficient manner. To achieve this, the first level of barring is placed in the EC-SCH, which is anyway required to be read by the device before any communication with the network, see Section 3.3.1.7. A two-bit *implicit reject status* flag is defined that allows for indicating either no barring, barring of all devices or two possible barring levels of roaming devices for the serving cell.

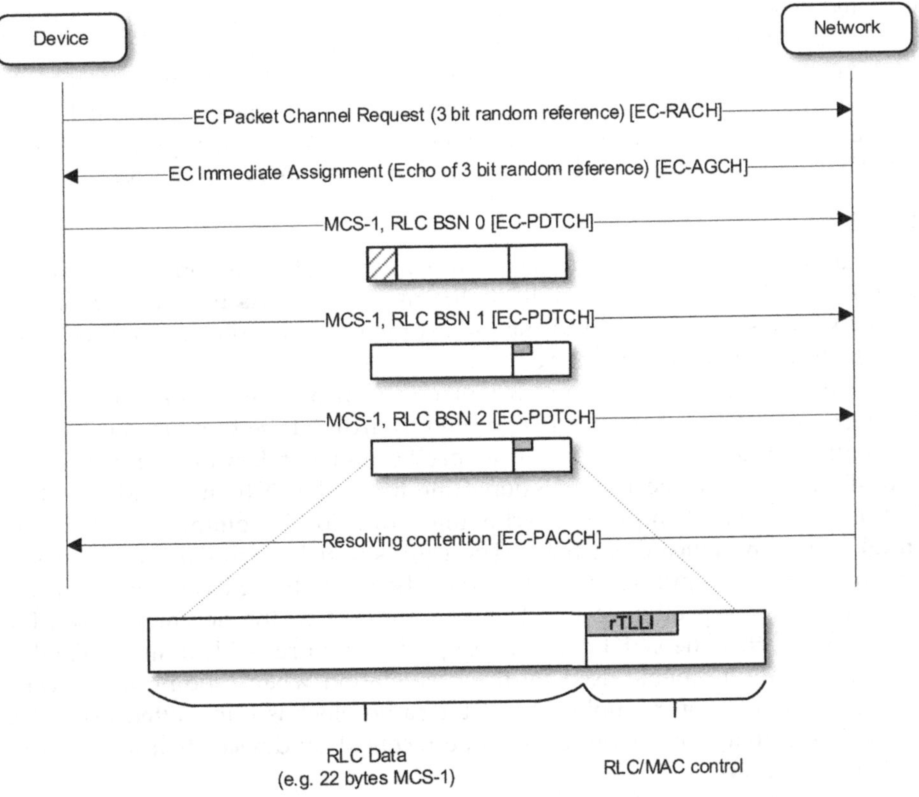

FIG. 3.33 Contention resolution, using enhanced access burst procedure.

As mentioned in Section 3.1, roaming is an important functionality for CIoT devices, and for many of them, they will be deployed in networks where they are roaming for their entire life-cycle. A device using EC-GSM-IoT is not subject to the legacy EAB-based barring and instead supports a second level of barring wherein information sent in EC SI is used to determine if PLMN-specific barring is in effect for its serving cell. The PLMN-specific barring information allows for indicating whether or not barring applies to exception reports, a subset of Access Classes 0–9 or a subset of Access Classes 10–15. In addition, EC SI may indicate when a cell is barred for all access attempts (i.e., regardless of the registered PLMN used by a device).

3.3.2 Connected mode procedures

3.3.2.1 *Assignment and allocation of resources*

3.3.2.1.1 Downlink

The downlink resource allocation for EC-GSM-IoT is very similar to the procedure used for legacy GPRS/EGPRS.

The data connection, or the *extended coverage temporary block flow (EC TBF)*, is established by the network (see Section 3.3.1.7) wherein the downlink resources assigned to the device are communicated in the *EC Immediate Assignment* sent on EC-AGCH. The device is also assigned a *Temporary Flow Identity (TFI)* associated with the EC TBF.

The device monitors the assigned resources looking for an (RLC)/MAC header containing the TFI assigned to it. If the assigned TFI is found, the device considers the block to be allocated to it, and decodes the remainder of the block. The physical layer process of the downlink EC-PDTCH block reception is described more in detail in Section 3.2.8.2.

3.3.2.1.2 Uplink

One of the main differences in the resource management comparing EC-GSM-IoT with previous PS allocations for GPRS/EGPRS is the use of FUA. As the name implies, it is a means to allocate resources in a fixed manner, compared with previously used allocation schemes, which were dynamic in nature.

In legacy GPRS/EGPRS a device gets resources assigned in an assignment message, and after that, resources can be dynamically allocated (scheduled) on those resources. The decision to dynamically allocate resources to a specific user is taken on a transmission time interval basis (20 ms for Basic Transmission Time Interval and 10 ms for Reduced TTI) by the scheduler in the base station. The scheduling decision is communicated by the USF that can take up to 8 values (3-bit field). The USF is sent in each dedicated block on the downlink. In order not to put restrictions on the scheduler, the USF can address any device on the uplink, i.e., irrespective of which device the rest of the downlink radio block is intended to. Due to this, the USF has been designed to perform with similar reliability irrespective of modulation scheme used by the base station when sending any given radio block on the downlink. For example, even if the radio block is transmitted using 32QAM, the USF carried in that block is designed to be received by devices that are scheduled to use GMSK modulation.

Hence, from a perspective of increasing coverage compared to legacy GPRS/EGPRS the robust baseline performance of the USF can be viewed as a good starting point. However, looking at it more closely, improving the performance further will have rather large implications to the rest of the system if the dynamic scheduling aspects of USF are to be maintained. Considering the means to improve coverage in Section 3.2.8, any attempt to evolve the legacy USF to support extended coverage using either blind repetitions or allowing an increased acquisition time would have severe implications to the resource efficiency on the uplink. Improving the channel coding performance by increasing redundancy would lead to less users possible to multiplex in the uplink (using a 1-bit or 2-bit field instead of 3-bits). If anything, considering the massive number of devices expected for IoT, the addressing space currently offered by USF should be increased, not decreased.

The above realizations lead 3GPP to adopt instead the FUA approach to uplink traffic channel resource management, which eliminates the need for the legacy USF-based uplink scheduling. With FUA, the network allocates multiple blocks in the uplink (contiguous or noncontiguous in time) based on device request. The basic procedure is as follows:

1. The device requests an uplink allocation (on EC-RACH) and informs the network on the size of the uplink transfer (by an integer number of MCS-1 blocks required to transmit the data in its buffer).
2. The network allocates, by an assignment message (on EC-AGCH), one or more blocks in the uplink to the device.

3. After the transmission of the allocated blocks (using EC-PDTCH), the network sends an Ack/Nack report and a new assignment message (using EC-PACCH) for any remaining blocks and/or for block that are to be retransmitted. This step is repeated until all blocks have been correctly received.
4. If the device has more data in its buffer than indicated to the network in step (1), it can indicate the need to continue its uplink connection by a *follow-on indicator* informing the network on the number of blocks it still has available in its buffer.

The procedure is illustrated in Fig. 3.34, which illustrates the full packet flow of an uplink transfer.

3.3.2.2 *Hybrid ARQ*

HARQ Type II using *incremental redundancy* is deployed on the EC-PDTCH. The functionality is similar as what is used for EGPRS although there are some important differences that will be outlined in this section.

3.3.2.2.1 EGPRS

The basic HARQ functionality in EGPRS is implemented using the following principles:

Each MCS is designed with a separately coded RLC/MAC header in addition to the RLC data block(s) transmitted in the radio block (see Section 3.2.6.5 for more details). The header is

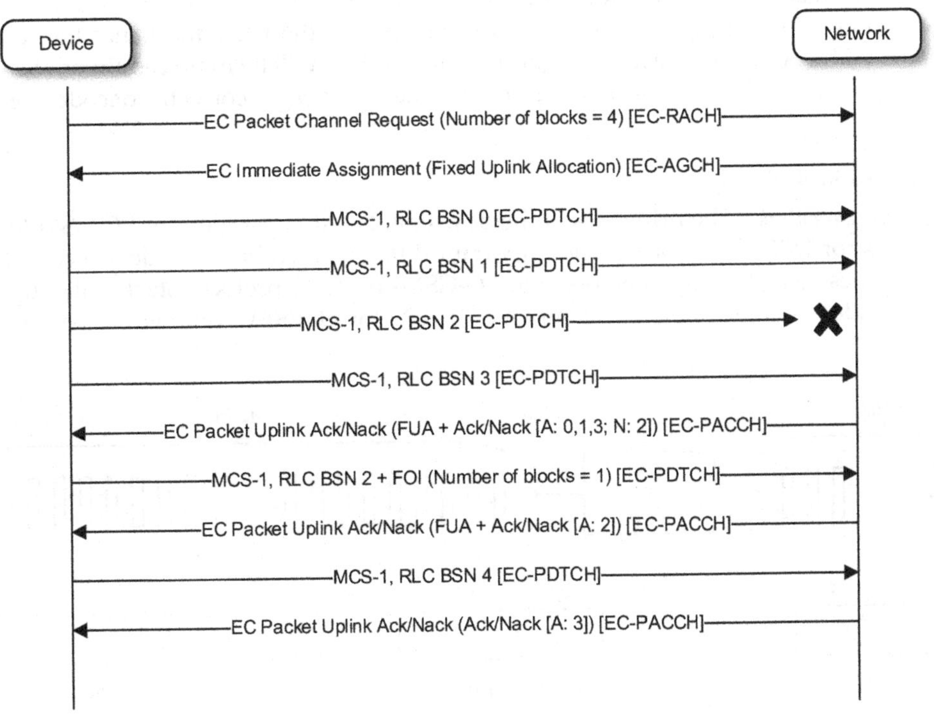

FIG. 3.34 FUA.

designed to be more robust than the corresponding associated data block, and hence during reception the header can be correctly received even when the data part is erroneous. In case this happens, the receiver will know the *Block Sequence Number* of the transmitted block(s) and can store the estimated bits (soft bits) of the data block in its soft bit buffer. The receiver also needs to know the MCS and the redundancy version, or the *PS*, used to store the soft bits in the correct position. This is communicated with the *Coding and Puncturing Scheme Indicator Field* in the RLC/MAC header. Different numbers of PSs (2 or 3) are used depending on the code rate of the MCS. The number of PSs used is determined by how many redundancy versions that are required to cover all bits from the encoder. This is illustrated by a simple example in Fig. 3.35.

In this case, after receiving all three PSs the code rate of the MCS is R = 1, and hence three PSs are needed to cover all bits from the encoder, and every third bit of the encoded bit stream is added to the respective PS.

At the receiver, the bits transmitted can be estimated and put in the soft bit buffer as long as the RLC/MAC header is received. In case no soft bit information is available for some of the bits, the decoding of the block is still possible, but those bit positions will not add any information to the overall decoding procedure.

The receiver behavior is illustrated in Fig. 3.36 where the header is correctly received in both the first and second transmission (while the data was not possible to decode). In the third transmission (using PS2 at the transmitter) neither the header nor the data was decodable, and hence no additional information can be provided for further decoding attempts (since no information on BSN, MCS, and/or PS was obtained). In the fourth transmission PS0 is used again (the transmitter will cyclically loop over the PSs), and hence the soft bits associated with PS0 will increase in magnitude (the receiver will be more certain on their correct value). This additional information enables the receiver to correctly decode the data block.

3.3.2.2.2 EC-GSM-IoT

3.3.2.2.2.1 Downlink The principles of the downlink HARQ operation for EC-GSM-IoT are the same as for EGPRS except for one important difference. To limit the device complexity when it comes to soft buffer memory, the EC-GSM-IoT RLC protocol stack only supports an RLC window size of 16 blocks (compared to 1024 for EGPRS). Although a device is not

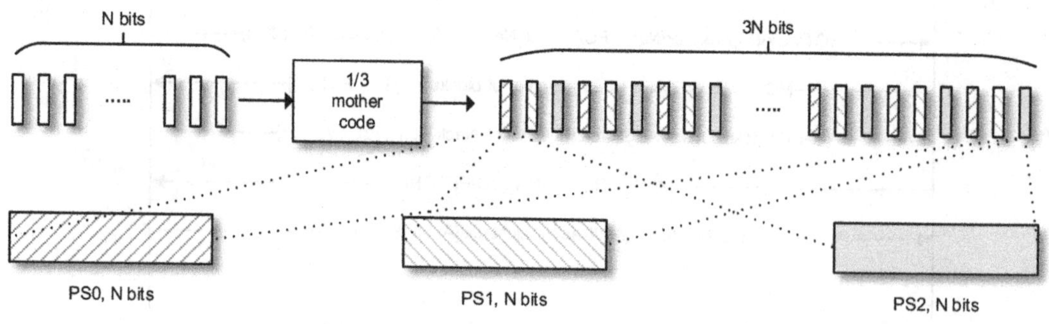

FIG. 3.35 Puncturing schemes for HARQ.

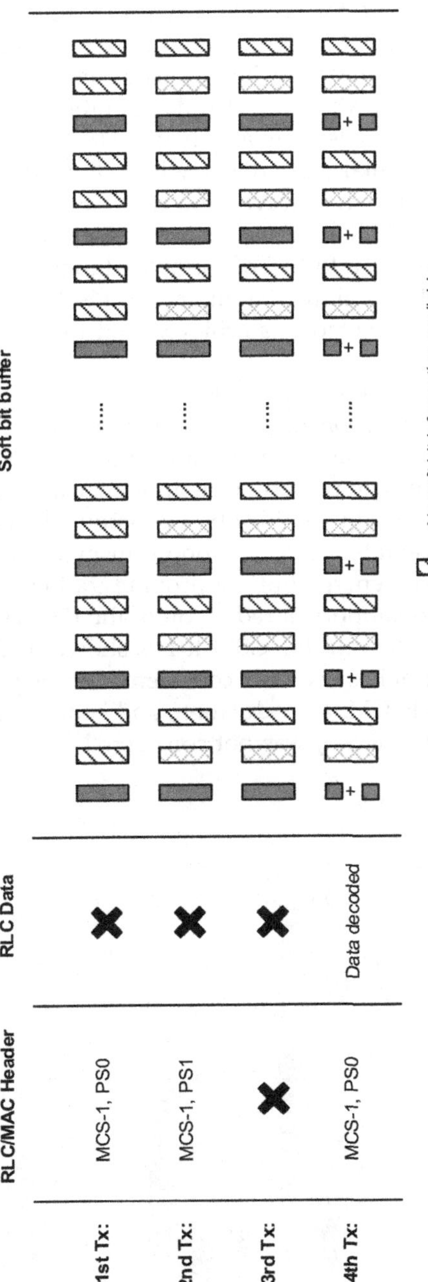

FIG. 3.36 HARQ soft buffer handling in the receiver.

required to be able to store the soft bits for every RLC block in the RLC window, it still implies a significantly reduced requirement on soft buffer capacity in the device.

A further reduction in complexity for downlink HARQ is the reduction of the number of PSs used for the GMSK-modulated MCSs. Simulations have shown that reducing the number of PS for MCS-1 and MCS-2 from two to one has no visible impact on performance [15]. This can be explained by the fact that the MCSs are already operating at a relatively low code rate, and furthermore, the RLC/MAC header is relatively close to the RLC data performance (i.e., the achievable gains with incremental redundancy will be limited by the RLC/MAC header performance). For MCS-3 and MCS-4, however, the code rate is relatively high, see Section 3.2.4, and only using chase combining (every retransmission contains the same bits) will result in a link performance loss of up to 4–5 dB compared to incremental redundancy using three redundancy versions [15]. If limiting the number of PSs to two instead of three, a 33% lower requirement on soft buffer memory is achieved and the performance degradation is limited.

Fig. 3.37 shows the expected performance difference between the approach of using two or three redundancy versions when considering a non frequency hopping channel based on simulations [15]. As can be seen, for the first two transmissions (where PS0 and PS1 are used respectively in both cases) there is no performance difference (as expected). The largest performance difference (0.6 dB) is seen in the third transmission when using PS2 (if supporting three PSs) will decrease the code rate, while a second transmission of PS0 (if using two PSs) will only increase the received bit energy but not provide coding gain. All in all, the performance difference is limited and supports a reduction of the PSs. The number of PSs for the first four MCSs comparing EGPRS and EC-GSM-IoT is shown in Table 3.5.

The reduction in the number of PSs was only considered for the purpose of targeting ultra-low-cost devices supporting only GMSK modulation, and hence a similar complexity reduction for higher MCSs (MCS-5 to MCS-9) was not considered.

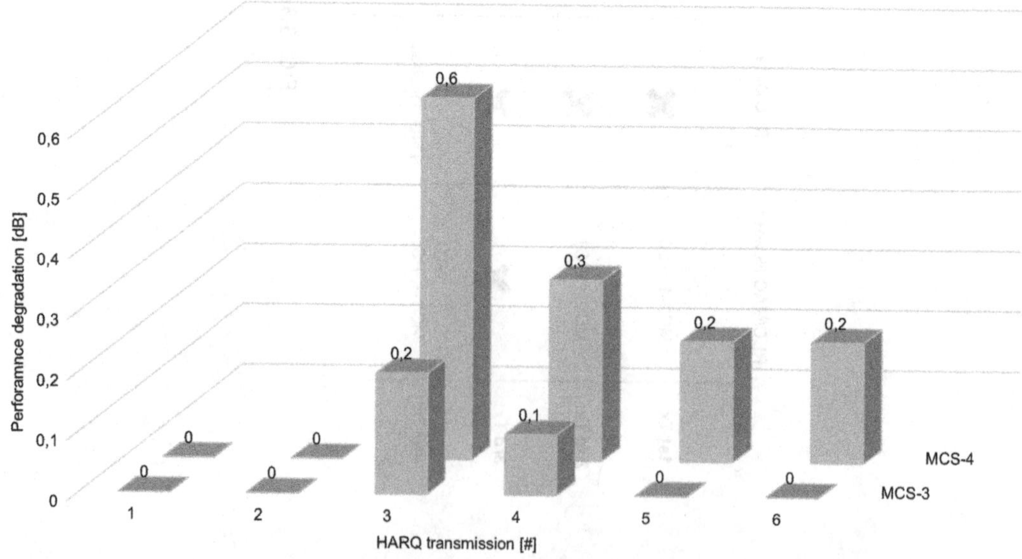

FIG. 3.37 Performance difference comparing three or two PSs.

TABLE 3.5 Number of PSs.

	EGPRS	EC-GSM-IoT
MCS-1	2	1
MCS-2	2	1
MCS-3	3	2
MCS-4	3	2

The same number of PSs for MCS-1 to MCS-4 applies both on the downlink and uplink.

3.3.2.2.2.2 Uplink

In the uplink there is a more profound difference in the HARQ design comparing EGPRS and EC-GSM-IoT.

In typical HARQ operation the receiver would be required to decode the RLC/MAC header for the HARQ to work efficiently (see Section 3.3.2.2). For EC-GSM-IoT, due to the use of FUA (see Section 3.3.2.1), this requirement does not apply. Instead of the base station acquiring the information on BSN, MCS, and PS from the RLC/MAC header, the information can instead be provided by the Base Station Controller (BSC). Since the BSC sends the Ack/Nack information to the device, and also the allocation of the resources to be used for any given FUA, it will know how many blocks will be transmitted and over which physical resources. Furthermore, the device is required to follow a certain order in the transmission of the allocated blocks. For example, retransmissions (Nacked blocks) need to be prioritized before new transmissions, and new transmissions need to be transmission in order. Hence, for the purpose of soft buffering at the base station there is no longer a requirement to receive and correctly decode the RLC/MAC header over the air interface. The HARQ shown in Fig. 3.36 instead becomes more optimized for EC-GSM-IoT as shown in Fig. 3.38.

However, before the decoded data block is delivered to upper layers, also the RLC/MAC header needs to be decoded, to verify the information received by the BSC and to acquire the remaining information (not related to the HARQ operation) in the header. Considering that the RLC/MAC header will not change over multiple transmissions for MCS-1 and MCS-2 (there is only one PS supported, see Table 3.5) the receiver can also apply soft combining on the RLC/MAC header to improve its performance.

The principle of the uplink HARQ operation between BTS (Base Transceiver Station, or base station) and BSC is shown in Fig. 3.39. In case of EC-GSM-IoT, the information on BSN, MCS, and PS over the air interface can be ignored by the BTS in the uplink HARQ operation since the same information is acquired from the BSC.

The more robust HARQ operation specific to EC-GSM-IoT will allow the receiver to operate at a higher BLER level than for HARQ operation specific to EGPRS, implying an improved resource efficiency, which will partly compensate for the increased use of resources for CC2, CC3, and CC4 when using blind repetitions.

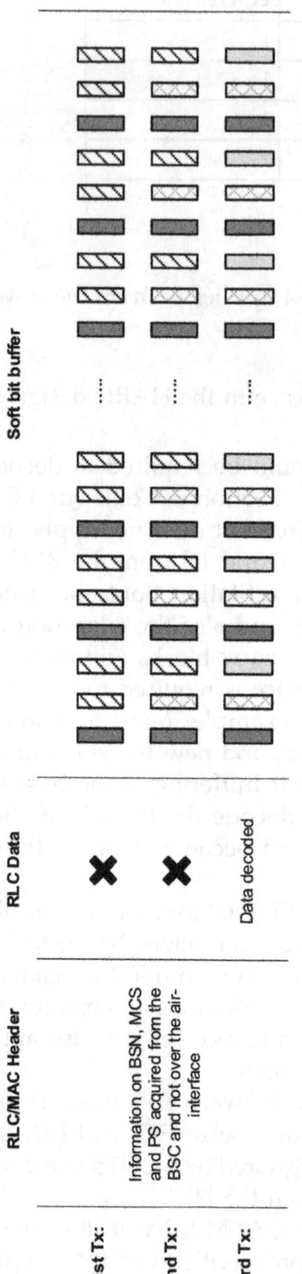

FIG. 3.38 EC-GSM-IoT uplink HARQ.

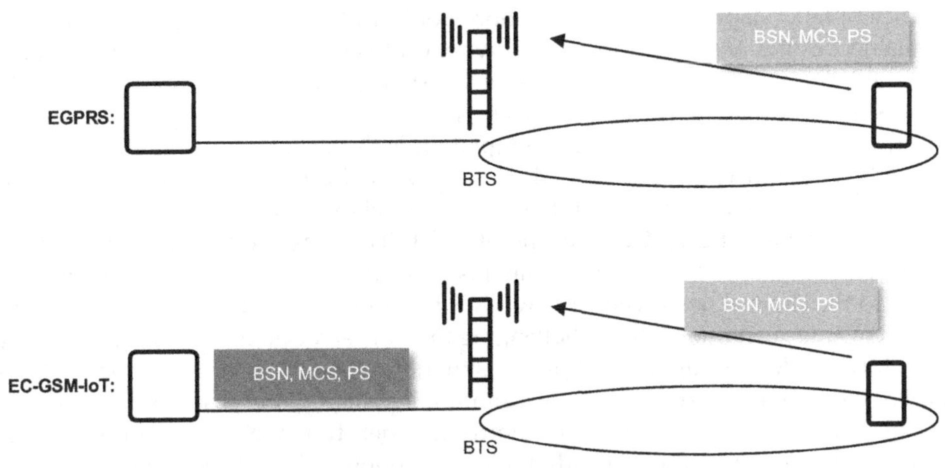

FIG. 3.39 Uplink HARQ operation.

3.3.2.3 *Link adaptation*

Link adaptation is the means where network adapts the MCS transmitted (downlink), or instructs the device to adapt its MCSs (uplink). For EC-GSM-IoT link adaptation also involves the adaptation of the CC used.

For the network to take a decision on the appropriate downlink MCS, the device is required to continuously measure (and potentially report) what is referred to as a *C-value* on EC-PDTCH and EC-PACCH blocks addressed to it (where it finds a TFI value matching the assigned value). The C-value is filtered over time and can either be signal level-based (RLA_EC, see Section 3.3.1.1) or SINR-based (SLA, see Section 3.3.1.4), as indicated by the network in EC SI. From the C-value reported the network will hence get an understanding of the signal level or SINR level experienced at the receiving device.

In addition to the signal level estimation or SINR estimation the device also reports an estimation of the mean (*MEAN_BEP*) and standard deviation (*CV_BEP*) of the raw bit error rate probability of the demodulated bursts over the radio blocks transmitted on all channels (TSs) assigned to it in the downlink. In case blind repetitions are used the MEAN_BEP and CV_BEP are calculated after the combination of any blind repetitions. The reason to, in addition to the mean, also estimate the standard deviation is to get an estimate of the variations of the channel. A channel with a large diversity will benefit more from using MCSs with a lower code rate, while if a low standard deviation is experienced, MCSs with higher code rate will be more suitable.

A *channel quality report* for the purpose of updating the downlink link adaptation is requested from the network on a perneed basis. That is, the network can decide when, and how often, a report should be requested. On the device side, the channel quality report can be multiplexed with the Ack/Nack reporting sent on the uplink EC-PACCH.

The corresponding uplink link adaptation in the base station is left to implementation, and there are no requirements on measurements by the specifications. It is however reasonable to believe that a similar estimation as provided by the device, also exist in the base station.

An efficient link adaptation implementation would furthermore make use of all information available, i.e., not only the channel quality report from the device. Such additional information could, for example, be estimations of interference levels in the network and BLER estimations based on the reported Ack/Nack bitmap.

A further functionality related to link adaptation is the CC adaptation in Packet Transfer Mode (in idle mode there is also a CC adaptation on the EC-CCCH, see Section 3.3.2.2). For the more traditional MCS link adaptation described above, there is always an associated packet control channel (EC-PACCH) to the EC-PDTCH where the link adaptation information is carried. The associated control channel is used to transmit the channel quality report, and it is also the channel used when a new MCS is commanded. However, when starting to operate in coverage conditions more challenging than the associated control channel has been designed for, the device might end up in a situation where it is in a less robust CC than needed for the current conditions, but it cannot adopt its CC because the network cannot assign a more robust one due to a failing control channel. To make the system robust against this type of event, a timer-based CC adaptation is (optionally) deployed. In short the device will increment its assigned downlink CC if no downlink EC-PACCH block has been received within a (configurable) time from the previous FUA. The same CC adaptation is of course needed at the base station and is also based on the most recent FUA assigned to the device. The downlink CC will only be incremented once, and if still no EC-PACCH is received from the network, the device will leave the EC TBF and return to idle mode.

3.3.2.4 *Power control*

The uplink power control for EC-GSM-IoT devices is the same as used for GPRS/EGPRS operation.

The network can choose to completely control the power level used by the device by assigning a power level reduction (Γ_{CH}). In this case the network needs to base its decision on previous feedback from the device to avoid guessing an appropriate power level to be used. It is hence a closed-loop power control. Alternatively, the network can allow the device to decide its output power based on the measured C-value (see Section 3.3.5) representing more of an open-loop power control. Still, the network is in control regarding to what extent the C-value should be weighted in the power loop equation, as determined by a parameter *alpha* (α). Alpha is broadcasted in EC SI, and hence the same value applies to all devices in the cell. However, whether to apply alpha in the power control equation is controlled by the uplink assignment message sent to each device.

The resulting output power level based on different settings of α and Γ_{CH} is illustrated in Fig. 3.40 (for the 900 MHz band and a device with a maximum output power of 33 dBm). It can be noted that in all possible configurations of alpha, the maximum output power of 33 dBm will always be used if Γ_{CH} (indicated by Γ in the figure legend) is set to 0 (not shown in the figure). However, by setting Γ_{CH} to a higher value (at most 62) a power reduction can be achieved. As stated above, if setting $\alpha = 0$ the downlink signal strength (C-value) will not have an impact to the output power level, which can be seen by the horizontal line in the figure, while if $\alpha > 0$ there is a C-value dependency on the used output power.

The output power equation is shown in Eq. (3.5).

$$P_{CH} = \min(\Gamma_0 - \Gamma_{CH} - \alpha \times (C + 48), PMAX)(dBm) \quad (3.5)$$

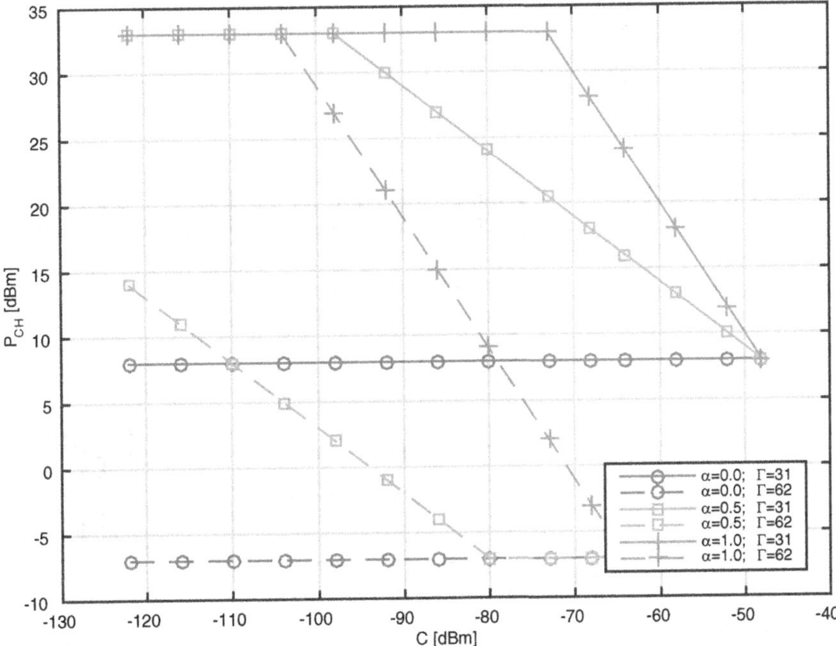

FIG. 3.40 EC-GSM-IoT uplink Power Control.

where Γ_0 is a constant of either 39 or 36 (depending on frequency band), and PMAX is the maximum output power supported by the device. The lowest power level defined in the 900 MHz frequency band is −7 dBm, which can also be seen in Fig. 3.40.

3.3.3 Backward compatibility

As stated already in Section 3.1, backward compatibility is an important part of the EC-GSM-IoT feature. The backward compatibility applies not only to the deployment of the system but also to the implementation of it in existing device and networks. That is, for a device manufacturer and a network vendor, the feature has been designed with as much commonality as possible with legacy devices and networks, thereby allowing it to be implemented on existing device and network platforms along with existing GSM features.

For a network operator the deployment of the EC-GSM-IoT feature should consist of a software upgrade of the existing installed base. No further replanning of the network or installment of additional hardware is expected. Hence a smooth transition from a network without EC-GSM-IoT devices to a more IoT-centric network can easily be supported in a seamless fashion. The only dedicated resources required for EC-GSM-IoT operation is the new synchronization, BCCH and common control channels on TS1 of the BCCH carrier.

Fig. 3.41 illustrates a cell where EC-GSM-IoT has been deployed where apart from the broadcast carrier, also two traffic (TCH) carriers are used to increase cell capacity. As can be seen, it is only TS1 of the BCCH carrier that need to be exclusively allocated to

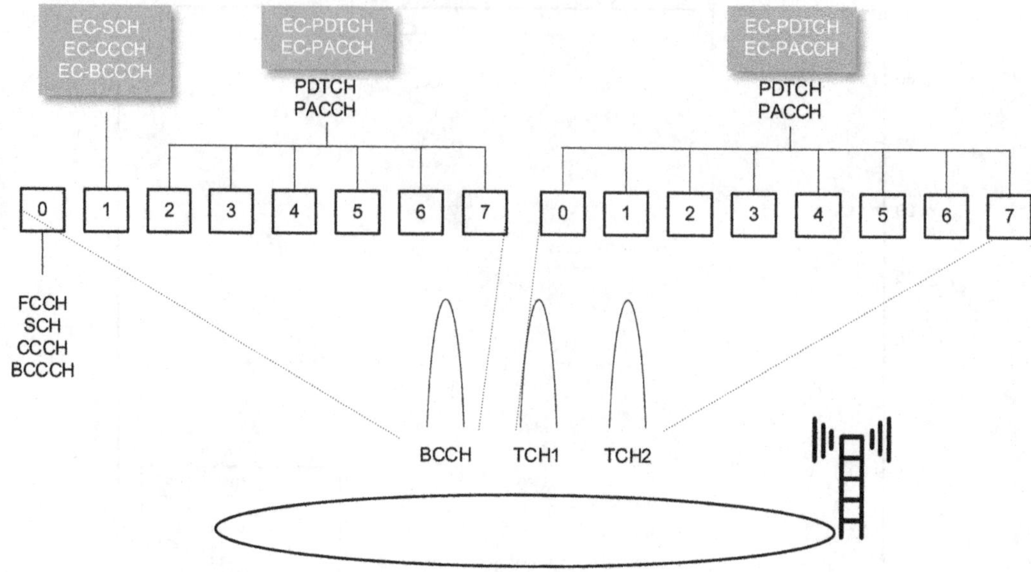

FIG. 3.41 Channel configuration.

EC-GSM-IoT-related channels. In all other TSs (apart from TS0 on the BCCH carrier that need to be allocated for non-EC-GSM-IoT traffic) the PS traffic of GPRS/EGPRS can be fully multiplexed, without any restrictions, with the corresponding EC-GSM-IoT traffic.

It can be noted that in the figure it is assumed that no resources in the cell is allocated for speech services, which would not be able to share dedicated traffic resources with neither GPRS/EGPRS nor EC-GSM-IoT packet services, and hence are not of relevance for backward compatibility in this context. It should, however, be noted that there support of CS and PS services are not impacted by the introduction of EC-GSM-IoT. That is, any limitations, or possibilities, of resource multiplexing are not changed with the introduction of EC-GSM-IoT.

3.3.4 Improved security

CIoT devices are designed to be of ultra-low cost, and by that, to have limited system complexity. This does, however, not imply that the security of the system is not of major concern - rather the opposite. Consider, for example, a household wherein 40 devices are connected to different applications. Having a person with malicious intent being able to control of all these devices could have severe implications to personal integrity, security, and perhaps even health. Hence, a decision was taken in 3GPP to further enhance the security of GSM to state-of-the-art grade security, targeting IoT. Although the driving force behind the improvements was IoT, the enhanced security features can also be implemented in GSM/EDGE devices not supporting EC-GSM-IoT.

GSM was designed more than 20 years ago, considering, at that time, known security threats and computational power available in those days. Although the security in GSM has evolved over the years, there are still basic design principles of the system that pose threats to security, such as bidding-down attacks (an attacker acting as fake base station, requesting the device to downgrade its level of security).

Apart from providing state-of-the-art grade security, the characteristics of CIoT compared to regular cellular operation need also to be considered. These include, for example, longer inactivity periods between connections to the network and attention to minimizing the impact to battery lifetime from security-related procedures. Minimizing the impact to battery lifetime includes simply minimizing the messages being transmitted and received over the radio interface related to security. In particular, it is of interest to minimize transmissions, which consumes roughly 40 times the energy of reception (see Section 4.2). The longer inactivity periods stem, for example, from the use of DRX cycles up to 52 min (see Section 3.3.1.5) or from the use of Power Saving Mode (see Section 2.2). In general, the less frequent the security context is updated the less impact on battery lifetime but the more vulnerable the security context becomes. It could, however, be argued that a less frequency reauthentication for devices supporting CIoT applications will also provide less opportunities for a potential attacker to decipher the device communication. It can also be noted that the frequency with which the authentication procedure is performed is up to the operator deployment and is not specified in 3GPP.

The improvements to the GSM security-related procedures for EC-GSM-IoT include the following:

- **Mutual authentication**: In GSM, only the device is authenticated from the network perspective, whereas the network is never authenticated at the device. This increases the risk of having what is usually referred to as a fake-BTS (an attacker acting as a network node) that the device can connect to. With EC-GSM-IoT the protocol for mutual authentication used in 3G networks has been adopted. Mutual authentication here refers to both the device and the network being authenticated in the connection establishment.
- **Improved encryption/ciphering**: The use of various 64-bit long encryption algorithms in GSM has since a few years back been shown to be vulnerable to attacks. Although 128-bit ciphering has been supported in GPRS (GEA4) for some time, a second 128-bit algorithm (GEA5, also making the use of 128-bit encryption keys mandatory) has been defined, and its use is mandatory for EC-GSM-IoT devices and networks. The same length of encryption keys is also used in 3G and 4G networks.
- **Rejection of incompatible networks**: One of the issues already mentioned with GSM is the bidding-down attack, where the attacker acts as a network node instructing the device to lower its encryption level (to something that can be broken by the attacker) or to turn the encryption completely off. With EC-GSM-IoT there is a minimum security context that the network needs to support, and in case the network is found not to support these, the device shall reject the network connection.
- **Integrity protection**: Integrity protection of control plane has been added which was earlier not supported for GPRS/EGPRS. The protection profile has been updated to match protection profiles used in 4G. Integrity protection can also optionally be added to the user plane data, which especially is of interest in countries where encryption cannot be used.

The changes introduced to EC-GSM-IoT puts the security on the same level as what is used by 3G and 4G networks today.

For more details on the security-related work, the interested reader is referred to Refs. [16,17].

3.3.5 Device and network capabilities

To support EC-GSM-IoT for a device implies the support of all functionality described in this chapter, with the following exceptions:

- The device has two options of support when it comes to downlink and uplink modulation schemes, either to only support GMSK modulation scheme or to support both GMSK and 8PSK modulation (see Section 3.2.2 for more details).
- The device has also two options of supporting different output power levels. Almost all GSM devices are capable of 33 dBm maximum output power. With EC-GSM-IoT also an additional power class of 23 dBm has been added to the specifications. Since both 33 and 23 dBm devices follow the same specification (i.e., no additional blind physical layer transmissions are, for example, used by 23 dBm devices), the result of lowering the output power is a reduction in maximum uplink coverage by 10 dB. At a first glance, this might seem counterproductive for a feature mainly aiming at coverage improvement. The reason to support a second output power class is the ability to support a wider range of battery technologies and to be able to integrate the PA onto the chip, which decreases implementation cost, and hence will allow an even lower price point for 23 dBm devices compared to 33 dBm devices. It should also be noted that the percentage of devices outside of a given coverage level rapidly decreases with increasing coverage. Hence, the majority of the devices outside of GPRS/EGPRS coverage will be covered by a 23 dBm implementation for which a 10 dB uplink coverage improvement can be realized.
- For DRX-related functionality the device needs to only support the shortest DRX cycle and not the full range as shown in Table 3.4. Although energy efficiency is of utmost importance for many of the CIoT modules and related applications, it does not apply to all of them, in that case a DRX cycle of 2 s can be acceptable. In addition, Power Saving Mode can be implemented as an alternative to longer DRX cycles, if reduced reachability is acceptable, reaching years of battery life (see Chapter 2 for more details).
- The number of TSs that can be simultaneously allocated in uplink and downlink respectively is related to the *multislot class* of the device (the class to support being up to device implementation). The multislot class that EC-GSM-IoT is implemented on must, however, align with the multislot class for EGPRS (in case EGPRS is also supported by the device). The minimum supported capability is 1 TS uplink and 1 TS downlink.

In contrast to the device, a network implementation of EC-GSM-IoT can leave many aspects of the feature for later implementation. For example, the capacity enhancing feature described in Section 3.2.5 is not of use in a network where only a small amount of EC-GSM-IoT device is accessing the system simultaneously. Other aspects of network implementation could, for example, be to only support one of the two contention resolution options in Section 3.3.2.3 or one of the two EC-RACH mapping options described in Section 3.2.8.1. Since the main intention of the feature is to support improved coverage it is, however, also mandatory for the network to support at least CC1 and CC4, leaving CC2 and CC3 optional. This ensures that any network implementation of EC-GSM-IoT can support the intended 164 dB in MCL, exceeding the coverage limit of GPRS/EGPRS by roughly 20 dB.

3.4 Other features

In this section, features that have been specified in addition to the baseline EC-GSM-IoT functionality, is shortly described. As with other features, the support for these features are optional. The areas of improvement are positioning, coverage and flexibility in network deployment of EC-GSM-IoT.

3.4.1 Improved positioning of devices

In smartphones, today the use of GPS to estimate the position of the device is commonly used in, for example, map applications or other location-based services. In low-cost cellular devices targeted for IoT, adding a GPS to the IoT module will increase the price point for it to no longer be competitive in the CIoT market. At the same time, positioning is of great importance for certain IoT applications, such as the tracking of goods.

In GSM there are multiple positioning methods specified. Several of them require, however, dedicated hardware and/or synchronized networks to perform the positioning estimation and have therefore not be deployed to a wide extent in current networks. The task of the Release 14 work on positioning enhancements [18] is, hence, to provide an improvement to device positioning without requiring network synchronization or dedicated hardware used specifically for positioning.

The reference positioning method, which does not require dedicated hardware or network synchronization, is a method that makes use of the cell ID and TA estimation (Fig. 3.42). When a device accesses a cell, the position of that base station is known to the network operator. Furthermore, the access will include an estimation of the TA. The TA is basically an estimate of the radial distance between the device and the base station. That is, the device will be placed in a "donut"-like figure defined by two radii R_1 and R_2 from the base station position. The width of the donut, i.e., R_2-R_1, is defined by how accurate the base station and device can synchronize to the overall frame structure, and also how accurately the synchronization is reported (for example, if the base station uses symbol resolution on the measured TA to the network node that determines the location, the Serving Mobile Location Center).

The improvements looked at by 3GPP in Release 14 is to use the known method of multilateration to more precisely determine the device position. This means that the device will synchronize to multiple base stations to acquire the associated TA (or estimated distance), which after reported to a location server can be used to more accurately determine the device position. The angle of arrival from the base stations used for multilateration should from a device perspective be as evenly spread as possible to provide a good positioning estimate.

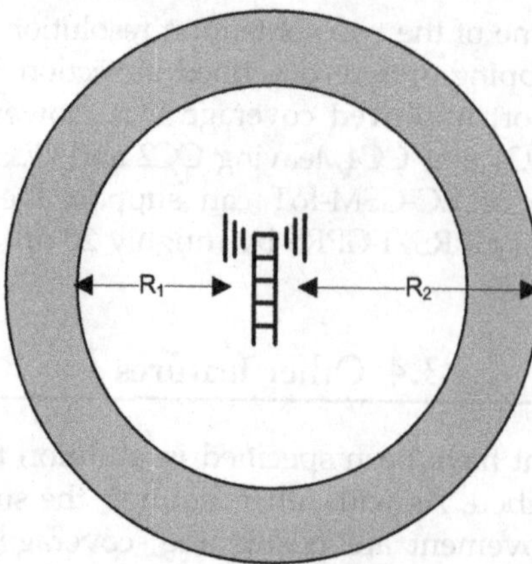

FIG. 3.42 Cell ID together with estimated TA for positioning.

For example, if using three base stations for the positioning attempt, they should preferably be separated with an angle of 120° (360/3).

Fig. 3.43 illustrates this procedure where three base stations are used for positioning. After the TA values to the three base stations are known, the network can calculate the device position as the intersection of the three "donut" (the black circle in the figure). The more base stations that are used for positioning, the smaller will the area of intersection be, and hence the better the estimate.

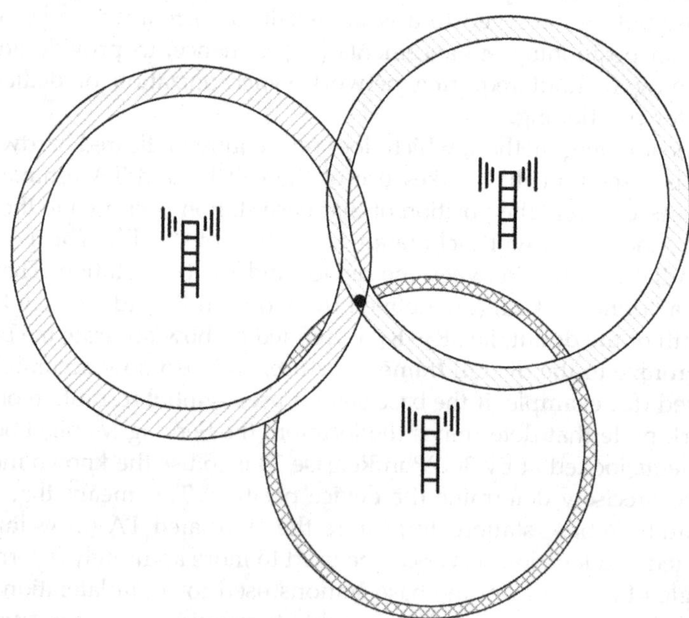

FIG. 3.43 Multilateration.

The accuracy of the distance to each of the base stations need not be the same (i.e., the width of the donut-like shape) because the estimation error is dependent on the synchronization performance, which in turn will be SINR dependent, and hence could be quite different between different base stations.

The selection of cells can be either under network or device control. In case the network selects the base stations, it would first need to know the possible base stations to select from (i.e., which base stations are reachable by the device), and in a second step, select the subset the device is to acquire TA from. If the network would have information about SINR and/or RSSI to each base station, it could consider both the geometry of the cells and the expected device TA synchronization accuracy in its selection of cells. In case the device selects the base stations, it would instead select the base stations with the highest signal level (indicating also a higher SINR), because it will have no perception of the angular location of the base stations. It could however be informed by the network which base stations are colocated, in that case it can avoid including more than one base station from the same site in the selection process.

The multilateration positioning method has been evaluated in Chapter 4.

A second positioning method is also considered for the Release 14 specification work where, in addition to the multilateration method, a method not relying on acquiring TA values from multiple base stations can be used under certain conditions. To understand this approach, we introduce the notion of Real Time Difference (RTD) and Observed Time Difference (OTD). RTD is the actual time difference of the frame structure between cells. OTD is the time difference observed at the device between the frame structures of different cells. The difference between them is, hence, the time it has taken for the signal to propagate from each base station to the receiving device. The propagation delay is also what the TA estimates. In simple terms, if two out of OTD, RTD, and the TAs are known, the third can be calculated.

A two-step approach is taken, where the first step involves one or more TA-based multilateration positioning attempts by one or more devices. In these attempts, not only the TA is reported from the device to the network but also the OTD. Hence, based on the previous reasoning, the RTD can be derived. The second step occurs at a later point in time where another positioning attempt is to be performed (not necessarily by the same device(s) as in the first step). If the new positioning attempts involves the same set of cells as in the first step, and if the RTD is still considered valid for these cells (the time base at the network will slowly drift over time, and hence depending on the accuracy of the network timing, the previously acquired RTD can only be considered valid for a limited time), then only the OTD need to be acquired, and the TA can instead be derived. Because the acquisition of the OTD only involves downlink monitoring (i.e., reception), the method is more energy efficient than the multilateration based on TA acquisition at the cost of a more complex network implementation and lower positioning accuracy.

3.4.2 Improved coverage for 23 dBm devices

A new power class for the device was introduced in Release 13 (see Section 3.3.5), which lowers the output power level by 10 dB compared with previously used levels. Allowing a lower output power class allows a more cost efficient implementation at the expense of a 10 dB loss in uplink coverage. In Release 14, work has been initiated to minimize this loss. Although the full 10 dB gap is not expected to be bridged, an improvement of roughly 5 dB could still be achieved without significant impact to network operation. What is considered is to introduce a new CC, CC5, and making use of a more extensive number of blind repetitions. It is also considered to further minimize protocol overhead to allow a higher level of redundancy in the transmission. One logical channel that needs special consideration is the EC-RACH where a simple increase in the number of repetitions also would mean an increase of interference and collision-risk. Hence, means to improve the EC-RACH is also considered where, for example, a longer burst, using a longer known synchronization sequence is considered to improve the link budget.

3.4.3 New TS mapping in extended coverage

One of the implications of using CC2, CC3, and CC4 on dedicated traffic channels (EC-PDTCH and EC-PACCH) is that they all need four consecutive TSs on the same carrier to be operable. For a network operator, this results in restrictions in how the feature can be deployed. Consider, for example, a cell with only a single BCCH carrier. As seen from earlier chapters (see Section 3.2.6) TS0 and TS1 will be occupied by synchronization channels, control channels, and BCCH for GSM and EC-GSM-IoT, respectively. In addition to this, the cell might serve voice calls on CS speech channels that are not possible to multiplex with PS channels to which GPRS, EGPRS, and EC-GSM-IoT belongs. Serving voice calls are often prioritized over serving PS calls; in that case EC-GSM-IoT can only be deployed in the cell if the speech channels are not taking up more than 2 TS, see Fig. 3.44.

To improve deployment flexibility, it is possible to allow CC2, CC3, and CC4 to be mapped onto, for example, 2 TS instead of 4 TS. Considering that the 4 TS mapping was chosen to maximize coverage and increase coherent reception (see Section 3.2.8) there will be a negative impact on coverage by the new mapping, but this is certainly better than not being able to operate EC-GSM-IoT using CC2, CC3, or CC4 in the cell at all. The difference of the two mapping options is illustrated in Fig. 3.45 for CC2.

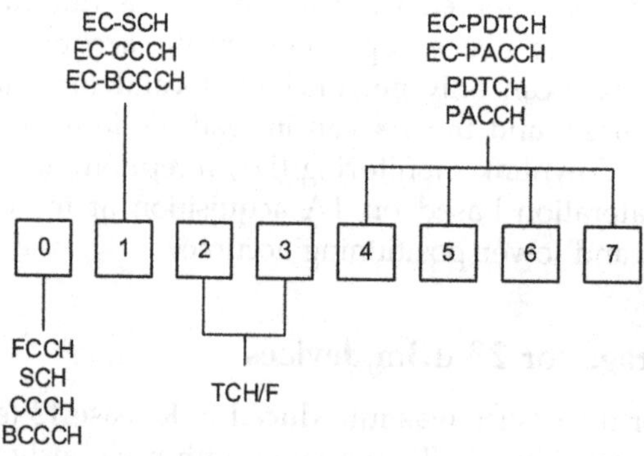

FIG. 3.44 BCCH channel configuration including speech channels.

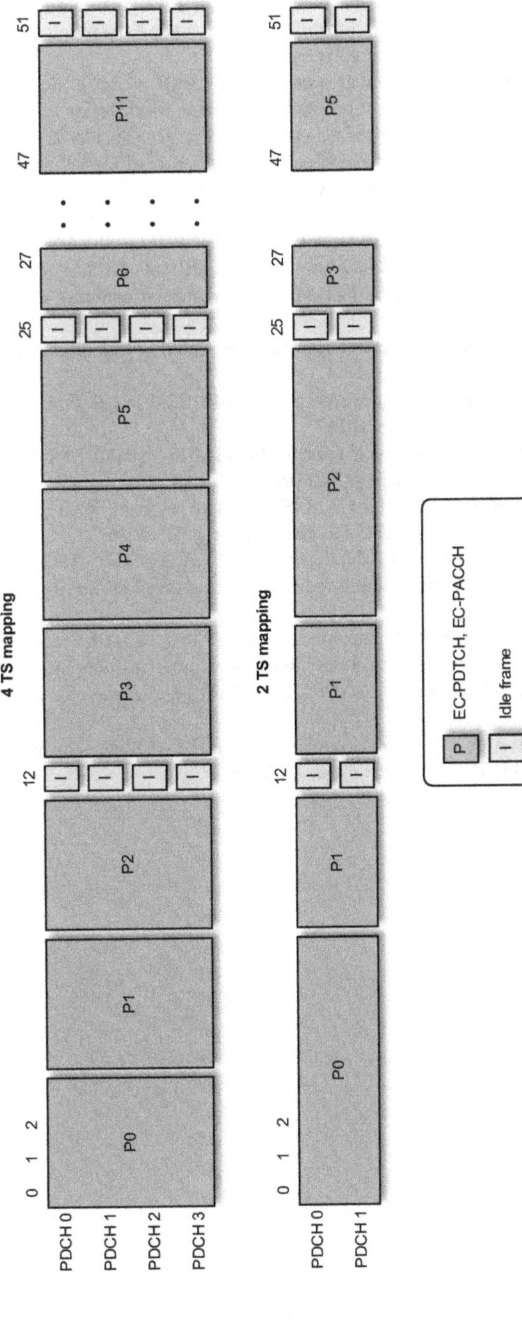

FIG. 3.45 4 TS versus 2 TS mapping for CC2.

References

[1] Ericsson. Ericsson mobility report, May 2016.

[2] IHS Markit, Technology Group. Cellular IoT market tracker, Q1, 2017. Results based on IHS Markit, technology group, cellular IoT market tracker – 1Q 2017. Information is not an endorsement. Any reliance on these results is at the third party's own risk, 2017. technology.ihs.com.

[3] Third Generation Partnership Project, Technical specification 45.004, v15.0.0. Modulation, 2016.

[4] P. Laurent. Exact and approximate construction of digital phase modulations by superposition of amplitude modulated pulses (AMP). IEEE Trans. Commun, February 1986, Vol. 34, No. 2, 150–60.

[5] Third Generation Partnership Project, Technical specification 45.003, v15.0.0. Channel coding, 2016.

[6] Third Generation Partnership Project, Technical specification 45.002, v15.0.0. Multiplexing and multiple access on the radio path, 2016.

[7] Nokia. GP-160448 EC-PCH enhancements for EC-GSM-IoT, 3GPP TSG GERAN meeting 70, 2016.

[8] L. M. Ericsson. GP-160032 Additional EC-RACH mapping, 3GPP TSG GERAN meeting 69, 2016.

[9] L. M. Ericsson. GP-160096 EC-EGPRS, overlaid CDMA design and performance evaluation, 3GPP TSG GERAN meeting 69, 2016.

[10] Third Generation Partnership Project, Technical specification 43.051, v15.0.0. GSM/EDGE Overall description; stage 2, 2016.

[11] Third Generation Partnership Project, Technical specification 43.022, v15.0.0. Functions related to mobile station (MS) in idle mode and group receive mode, 2016.

[12] Third Generation Partnership Project, Technical specification 44.018, v15.0.0. Mobile radio interface layer 3 specification; GSM/EDGE radio resource control (RRC) protocol, 2016.

[13] L. M. Ericsson. GP-160266. Impact on PDCH for EC-GSM-IoT in a reduced BCCH spectrum allocation (including carrier SS vs SINR measurements), 3GPP TSG GERAN meeting 70, 2016.

[14] L. M. Ericsson. GP-160035. Coverage class Adaptation on EC-RACH, 3GPP TSG GERAN meeting 69, 2016.

[15] L. M. Ericsson. GP-150139. EC-GSM, Incremental redundancy and chase combining, 3GPP GERAN TSG meeting 65, 2016.

[16] Third Generation Partnership Project, Technical specification 33.860, v13.0.0. Study on enhanced general packet radio service (EGPRS) access security enhancements with relation to cellular Internet of things (IoT), 2016.

[17] Vodafone, Ericsson, Orange, TeliaSonera, SP-160203. New GPRS algorithms for EASE, 3GPP TSG SA meeting 71, 2016.

[18] L. M. Ericsson. Orange, Mediatek inc., Sierra wireless, Nokia, RP-161260 New work item on positioning enhancements for GERAN, 3GPP TSG RAN meeting 72, 2016.

CHAPTER

4

EC-GSM-IoT performance

OUTLINE

Abstract

This chapter presents the performance of EC-GSM-IoT in terms of coverage, throughput, latency, power consumption, system capacity, and device complexity. It is shown that EC-GSM-IoT meets commonly accepted targets for these metrics in 3GPP, even in a system bandwidth as low as 600 kHz. In addition to the presented performance, an overview of the methods and assumptions used when deriving the presented

Cellular Internet of Things, Second Edition
https://doi.org/10.1016/B978-0-08-102902-2.00004-2

© 2020 Elsevier Ltd. All rights reserved.

results is given. This allows for a deeper understanding of both the scenarios in which EC-GSM-IoT is expected to operate and the applications EC-GSM-IoT is intended to serve.

4.1 Performance objectives

The design of the EC-GSM-IoT physical layer and the idle and connected mode procedures presented in Chapter 3 were shaped by the objectives agreed during the study on *Cellular System Support for Ultra-low Complexity and Low Throughput Internet of Things* [1], in this book referred to as the Cellular IoT study item, i.e.:

- Maximum coupling loss (MCL) of 164 dB
- Data rate of at least 160 bits per second
- Service latency of 10 s
- Device battery life of up to 10 years
- System capacity of 60,000 devices/km^2
- Ultra-low device complexity

During the subsequent standardization of EC-GSM-IoT [2] it was, in addition, required to introduce necessary improvements to enable EC-GSM-IoT operation in a spectrum deployment as narrow as 600 kHz.

While presenting the EC-GSM-IoT performance for each of these objectives, this chapter introduces the methodologies used in the performance evaluations. Similar methodologies are used in the LTE-M and NB-IoT performance evaluations presented in Chapters 6 and 8, respectively.

The results and methodologies presented in this chapter are mainly collected from 3GPP TR 45.820 Cellular System Support for Ultra-low Complexity and Low Throughput Internet of Things [1] and TS 45.050 *Background for Radio Frequency (RF) requirements* [3]. In some cases, the presented results are deviating from the performance agreed by 3GPP, for example, because of EC-GSM-IoT design changes implemented during the normative specification phase subsequent to the Cellular IoT study item and the publishing of 3GPP TR 45.820. These deviations are minor and the results presented in this chapter have to a large extent been discussed and agreed by the 3GPP community.

4.2 Coverage

One of the key reasons for starting the Cellular IoT study item was to improve the coverage beyond that supported by GPRS. The target was to exceed the GPRS MCL, defined in Section 2.2.5, by 20 dB to reach a MCL of 164 dB. Coverage is meaningful first when coupled with a quality of service target and in the Cellular IoT study items it was required that the MCL was fulfilled at a minimum guaranteed data rate of 160 bits per second.

To determine the EC-GSM-IoT MCL, the coverage performance of each of the EC-GSM-IoT logical channels is presented in this section. The section also discusses the criteria's used to define adequate performance for each of the channels at the MCL and the radio-related assumptions used in the evaluation of their performance. In the end of the section the achieved MCL and data rates are presented.

4.2.1 Evaluation assumptions

4.2.1.1 Requirements on logical channels

4.2.1.1.1 Synchronization channels

The EC-GSM-IoT synchronization channels, i.e., the Frequency Correction CHannel (FCCH) and Extended Coverage Synchronization CHannel (EC-SCH), performance is characterized by the time required by a device to synchronize to a cell. During the design of EC-GSM-IoT, no explicit synchronization requirement was defined, but a short synchronization time is important in most idle and connected mode procedures to provide good latency and power-efficient operation.

While the FCCH is defined by a sinusoidal waveform, the EC-SCH is a modulated signal that contains cell-specific broadcast information that needs to be acquired by a device before initiating a connection. Beyond the time to synchronize to a cell also the achieved Block Error Rate (BLER) is a good indicator of the EC-SCH performance. A 10% BLER is a relevant target for the EC-SCH to provide adequate system access performance in terms of latency and reliability.

4.2.1.1.2 Control and broadcast channels

The performances of the control and broadcast channels are typically characterized in terms of their BLER. A BLER level of 10% has traditionally been targeted in the design of GSM systems [4] and is well proven. At this BLER level, the Extended Coverage Packet Associated Control Channel (EC-PACCH), Extended Coverage Access Grant Channel (EC-AGCH), Extended Coverage Paging Channel (EC-PCH), and the Extended Coverage Broad-Cast CHannel (EC-BCCH) are considered to achieve sufficiently robust performance to support efficient network operation.

For the Extended Coverage Random Access CHannel (EC-RACH), a 20% BLER is a reasonable target. Aiming for a higher BLER level on the EC-RACH than for the downlink Extended Coverage Common Control CHannels (EC-CCCH/D), which delivers the associated assignment of resources, or pages a device, reflects that this is a best effort channel where it is not critical if a device needs to perform multiple attempts before successfully accessing the system. Furthermore, because the EC-RACH is a collision-based channel, using a slightly higher BLER level operating point will support an efficient use of the spectrum and will increase the utilization of the channel capacity.

4.2.1.1.3 Traffic channels

For the Extended Coverage Packet Data Traffic CHannel (EC-PDTCH) data rate is a suitable design criterion. EC-GSM-IoT uses, besides blind transmissions, hybrid automatic repeat request (HARQ) to achieve high coupling loss, as explained in Section 3.3.2.2. To use HARQ efficiently, an average EC-PDTCH BLER significantly higher than 10% is typically targeted. To derive the achievable EC-PDTCH coupling loss and data rate, the HARQ procedure needs to be evaluated. The high-level HARQ process flow, in terms of EC-PDTCH data packets

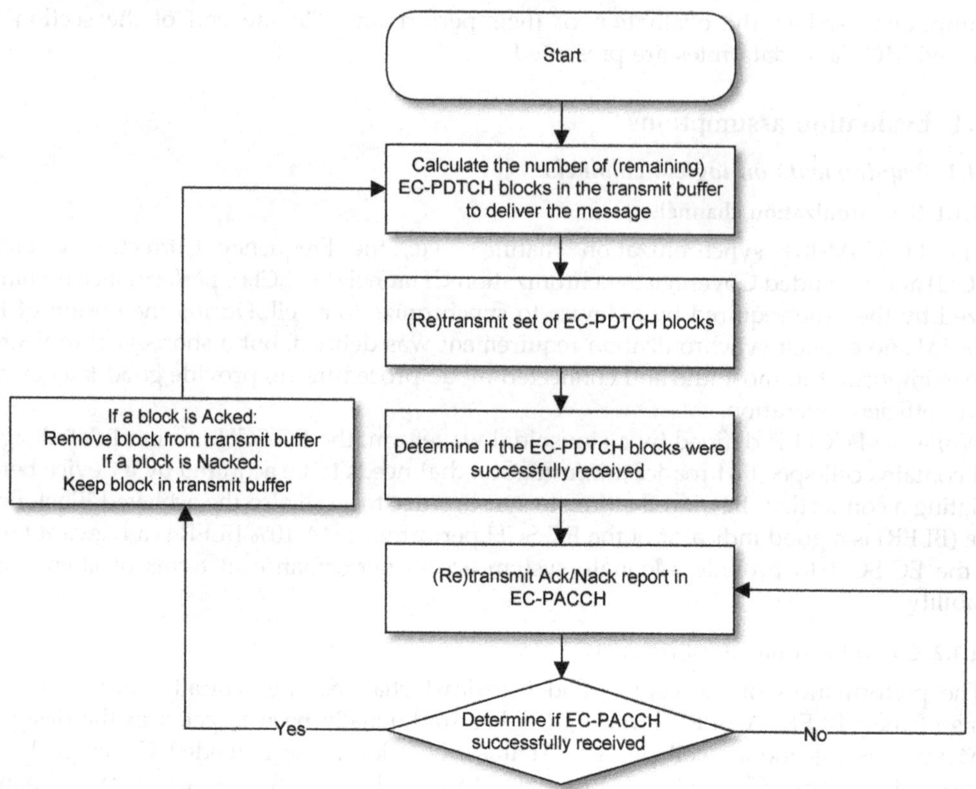

FIG. 4.1 HARQ packet flow used in the modeling of the EC-PDTCH performance.

transmitted in one direction and EC-PACCH HARQ feedback transmitted in the opposite direction, is depicted in Fig. 4.1.

As a response to each set of transmitted EC-PDTCH blocks, an EC-PACCH control block is transmitted to positively acknowledge (Ack) or negatively acknowledge (Nack) the reception of the EC-PDTCH block. Failed blocks are retransmitted. At each stage shown in Fig. 4.1, both total processing delays and transmission times are incremented to derive the total latency associated with the HARQ transmission of an uplink or downlink report. To generate reliable performance results, and to construct a cumulative distribution function (CDF) of the EC-PDTCH HARQ latency, a large number of instances of the HARQ packet flow was simulated. The latency achieved at the 90th percentile (i.e., the time below which 90% of the simulated reports are delivered) is used in the EC-PDTCH performance evaluations presented later in this chapter.

4.2.1.2 Radio-related parameters

For the EC-GSM-IoT performance evaluations presented herein, Typical Urban (TU) propagation conditions with a Rayleigh fading of 1 Hz maximum Doppler spread is assumed. This is intended to model a stationary device with slowly time varying fading induced by mobility of items in its spatial proximity. The *root mean square delay spread* of the TU channel is around 1 μs, which is fairly challenging and not as typical as the name indicates; see Ref. [6] for more details.

To derive the noise level in the receiver, the ideal thermal noise density at 300 K, or 27 °C (−174 dBm/Hz; see Section 3.2.8.1), in combination with a device noise figure (NF) of 5 dB and a base station NF of 3 dB are used. These NFs are assumed to correspond to realistic device and base station implementations although it needs to be stressed that the NFs between different implementations can vary substantially, even between units using the same platform design.

An initial frequency offset in the device of 20 ppm when synchronizing to the FCCH and EC-SCH is assumed. The source of this initial frequency error is described in Section 4.7.1. In addition to the initial frequency error, a continuous frequency drift of 22.5 Hz/s is assumed. This is intended to model the frequency drift expected in a temperature-controlled crystal oscillator-based frequency reference.

In terms of output power, it is assumed that the base station is configured with 43 dBm output power per 200 kHz channel. This corresponds to a typical GSM macro deployment scenario. For the device side a 33 dBm power class is assumed, which again corresponds to a common GSM implementation. Also, a new lower power class of 23 dBm is investigated. This power class is of particular interest because it is commonly understood to facilitate reduced device complexity, as elaborated on in Section 4.7.4.

For the evaluation of the EC-PDTCH coverage the MCS-1 (modulation and coding scheme 1) is used. MCS-1 uses Gaussian Minimum Shift Keying (GMSK) modulation and a code rate of roughly 0.5, which makes it the most robust EC-GSM-IoT MCS. Each MSC-1 block carries 22 bytes of payload.

An overview of the radio related simulations assumptions used in the EC-GSM-IoT performance evaluations is given in Table 4.1.

TABLE 4.1 Simulation assumptions for EC-GSM-IoT coverage performance.

Parameter	Value
Frequency band	900 MHz
Propagation condition	Typical Urban (TU)
Fading	Rayleigh, 1 Hz
Device initial oscillator inaccuracy	20 ppm (applied in FCCH/EC-SCH evaluations)
Device frequency drift	22.5 Hz/second
Device NF	5 dB
Base station NF	3 dB
Device power class	33 dBm or 23 dBm
Base station power class	43 dBm
Modulation and coding scheme	MCS-1

4.2.1.3 Coverage performance

Table 4.2 presents the downlink coverage performance for each of the logical channels according to the evaluation's assumptions introduced in Section 4.2.1. The coverage is defined in terms of the MCL as:

$$MCL = P_{TX} - (SNR + 10log_{10}(k \cdot T \cdot BW) + NF) \quad (4.1)$$

The power (P_{TX}), the operating SNR, the signal bandwidth (BW) and the receiver NF are all given in Table 4.2. T equals an assumed ambient temperature of 290 K, and k is Boltzmann's constant.

The FCCH and EC-SCH synchronization performance is presented in terms of the time required to synchronize to a cell at the MCL of 164 dB, taken at the 90th percentile of the synchronization acquisition delay CDF. The control and broadcast channels performances are presented as the MCL at which 10% BLER is achieved.

For the EC-PDTCH/D Table 4.2 presents a *MAC-layer data rate* of 0.5 kbps achieved at the 90th percentile of the throughput CDF generated from a simulation modeling the HARQ procedure depicted in Fig. 4.1. In this case a 33 dBm device is assumed to feedback the EC-PACCH/U control information. The EC-PDTCH/D performance is also presented at 154 dB's coupling loss under the assumption of a 23 dBm device sending the EC-PACCH/U Ack/Nack feedback. In this case a MAC-layer data rate of 2.3 kbps is achievable. The 10-dB reduction in coverage is motivated by the 10 dB lower output power of the 23 dBm device. The presented data rates correspond to MAC-layer throughput over the access stratum where no special consideration is given to the SNDCP, LLC, RLC, and MAC overheads accumulated across the higher layers. Simplicity motivates the use of this metric, which is also used in the LTE-M and NB-IoT performance evaluations in Chapters 6 and 8.

The uplink performance is summarized in Table 4.3. Two device power classes are evaluated, i.e., 33 and 23 dBm. At 33 dBm output power, 164 dB MCL is achievable while for 23 dBm, support for 154 dB coupling loss is accomplished. Although the 164 dB MCL target is not within reach for the 23 dBm case, it is still of interest as the lower power class reduces

TABLE 4.2 EC-GSM-IoT downlink Maximum Coupling Loss performance [1,7].

Channel/Parameter	FCCH/ EC-SCH	EC-SCH	EC-BCCH	EC-CCCH/D	EC-PACCH/D	EC-PDTCH/D	
Bandwidth [kHz]	271	271	271	271	271	271	271
Power [dBm]	43	43	43	43	43	43	43
NF [dB]	5	5	5	5	5	5	5
Performance	1.15 s	10% BLER	10% BLER	10% BLER	10% BLER	0.5 kbps[1)]	2.3 kbps[2)]
SINR [dB]	-6.3	-8.8	-6.5	-8.8	-6.4	-6.3	3.7
MCL [dB]	164	166.5	164.2	166.5	164.1	164	154

[1)] *Assuming a 33 dBm device feedbacks the EC-PACCH/U.*
[2)] *Assuming a 23 dBm device feedbacks the EC-PACCH/U.*

TABLE 4.3 EC-GSM-IoT uplink maximum coupling loss performance [1,7,8].

Channel/Parameter	EC-RACH		EC-PACCH/U		EC-PDTCH/U	
Bandwidth [kHz]	271	271	271	271	271	271
Power [dBm]	33	23	33	23	33	23
NF [dB]	3	3	3	3	3	3
Performance	20% BLER	20% BLER	10% BLER	10% BLER	0.5 kbps	0.6 kbps
SINR [dB]	-15	-15	-14.3	-14.3	-14.3	-14.3
MCL [dB]	164.7	154.7	164.0	154.0	164.0	154.0

device complexity. A low output power is also beneficial because it lowers the requirement in terms of supported power amplifier drain current that needs to be provided by the battery feeding the device with power. The presented EC-PDTCH/U MAC-layer data rates were derived based on the HARQ model depicted in Fig. 4.1.

The results in Tables 4.2 and 4.3 show that EC-GSM-IoT meets the targeted MCL requirement of 164 dB for a MAC-layer data rate of 0.5 kbps.

4.3 Data rate

Section 4.2 presents EC-PDTCH MAC-layer data rates in the range of 0.5–0.6 kbps and 0.5–2.3 kbps in the uplink and downlink, respectively. These data rates are applicable under extreme coverage conditions. To ensure a spectrally efficient network operation and a high end-user throughput, it is equally relevant to consider the throughput achievable for radio conditions sufficiently good to guarantee no or a limited level of block errors. Under such conditions, the network can configure the use of the highest supported modulation and coding scheme on the maximum number of supported time slots. Up to eight time slots can be supported by EC-GSM-IoT according to the 3GPP specifications, although it is expected that support for four or five time slots in practice will be a popular design choice.

In the downlink, a device is dynamically scheduled on its assigned resources and a base station will in best case transmit eight MCS-9 blocks on the eight assigned time slots during four consecutive TDMA frames. Each MCS-9 block contains an RLC/MAC header of 5 bytes and two RLC blocks, each of 74 bytes. The maximum supported EC-GSM-IoT RLC window size of 16 limits the number of RLC blocks that at any given time can be outstanding with a pending acknowledgment status. The base station uses the *RRBP* field in the RLC header of the EC-PDTCH/D block to poll the device for a Packet Downlink Ack/Nack (PDAN) report. The device responds earliest 40 ms after the end of the EC-PDTCH/D transmission time interval (TTI) as illustrated in Fig. 4.2. Assuming that the base station needs 20 ms to process the PDAN report before resuming the EC-PDTCH/D transmission implies that eight MCS-9 blocks each of 153 bytes size can be transmitted every 100 ms. This limits the peak downlink MAC-layer data rate of EC-GSM-IoT to 97.9 kbps.

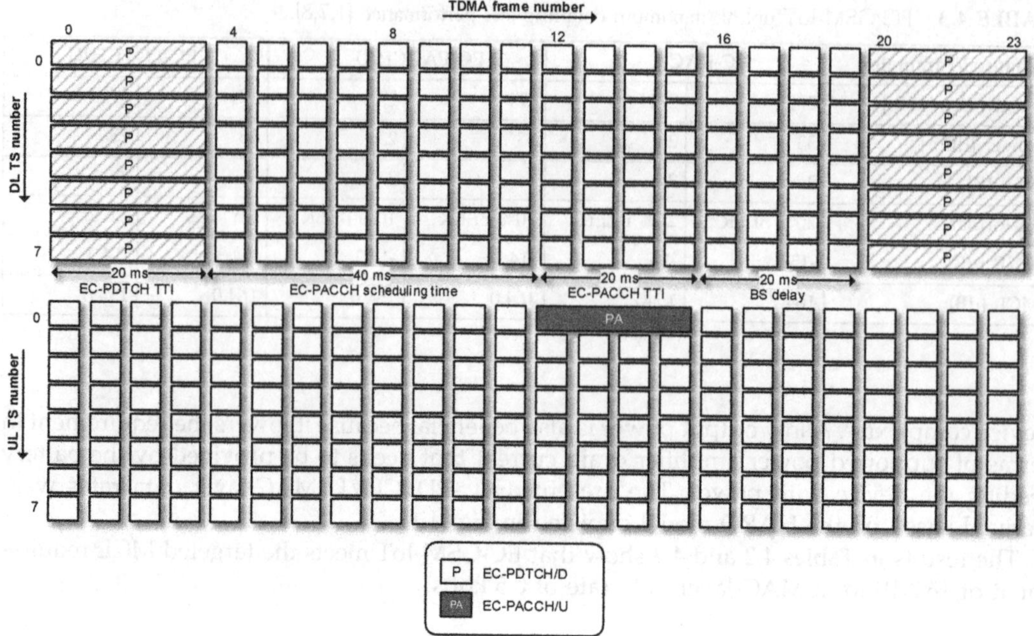

FIG. 4.2 EC-GSM-IoT downlink scheduling cycle.

This data rate can be compared with the often referred to *physical layer data rate* of 489.6 kbps that can be reached across the EC-PDTCH/D 20 ms TTI. High data rates on link level can be translated into a high spectral efficiency, which is of importance for the system as a whole in terms of system capacity. For the individual device the support of a flexible range of data rates in combination with a proper link adaptation equates to improved latency and battery life when radio conditions improve.

In the uplink, EC-GSM-IoT uses the concept of Fixed Uplink Allocations (FUA) (see Section 3.3.2.1.2) to schedule traffic. For devices supporting 8PSK the best performance is achieved when eight MCS-9 blocks are scheduled on eight time slots. Again, the RLC window size of 16 sets a limitation on the number of scheduled blocks. After the EC-PACCH/D carrying the FUA information element has been transmitted, a minimum scheduling gap of 40 ms delays the EC-PDTCH/U transmission of the MCS-9 blocks as illustrated in Fig. 4.3. After the end of the EC-PDTCH/U transmission the timer *T3326* [9] needs to expire before the network can send the next EC-PACCH/D containing an Ack/Nack report as well as a new FUA. Just as for the downlink, this implies that eight MCS-9 blocks can be transmitted every 100 ms. This limits the uplink peak MAC-layer data rate of EC-GSM-IoT to 97.9 kbps.

The EC-PDTCH/U peak physical layer data rate matches the EC-PDTCH/D 489.6 kbps across the 20 ms TTI. For devices only supporting GMSK modulation on the transmitter side, the highest modulation and coding scheme is MCS-4, which contains a RLC/MAC header of 4 octets and a single RLC block of 44 octets. In this case 16 MCS-4 RLC blocks can be scheduled during 40 ms every 120 ms leading to an uplink peak MAC-layer data rate of 51.2 kbps.

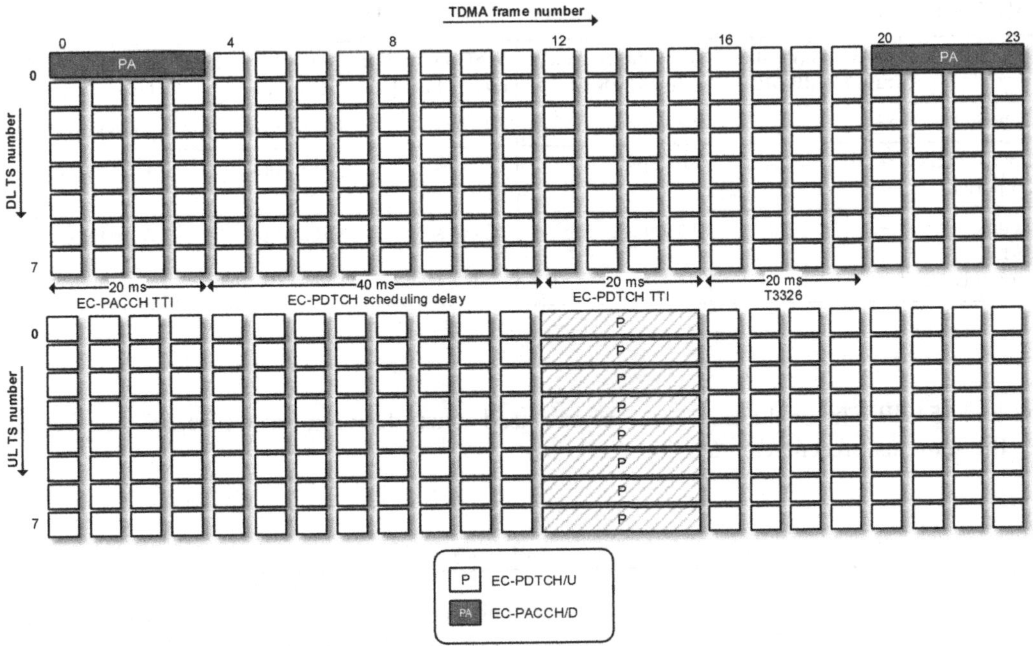

FIG. 4.3 EC-GSM-IoT uplink scheduling cycle.

The EC-PDTCH/U peak physical layer data rate for a GMSK only device is limited to 153.6 kbps over the 20 ms TTI.

Tables 4.4 and 4.5 summarizes the findings of this and the previous section in terms of MAC-layer data rates supported at 164 dB MCL and the peak physical layer data rates experienced under error-free conditions. In addition, it presents the MAC-layer data rates simulated at coupling losses of 154 and 144 dB. For the 33 dBm device MCS-1 is providing the best performance at 164 and 154 dB coupling loss. At 144 dB coupling loss MCS-3 is the best choice in the uplink even when 8PSK is supported, while MCS-4 provides the highest data rate for the downlink. For the 23 dBm device MCS-1 is giving best performance at

TABLE 4.4 EC-GSM-IoT data rates for 33 dBm device power class [5].

	MAC-layer data rate 164 dB MCL	MAC-layer data rate 154 dB CL	MAC-layer data rate 144 dB CL	Peak MAC-layer data rate	Peak physical layer data rate
Downlink	0.5 kbps	3.7 kbps	45.6 kbps	97.9 kbps	489.6 kbps
Uplink, 8PSK supported	0.5 kbps	2.7 kbps	39.8 kbps	97.9 kbps	489.6 kbps
Uplink, GMSK supported	0.5 kbps	2.7 kbps	39.8 kbps	51.2 kbps	153.6 kbps

TABLE 4.5 EC-GSM-IoT data rates for 23 dBm device power class [5].

	MAC-layer data rate 164 dB MCL	MAC-layer data rate 154 dB CL	MAC-layer data rate 144 dB CL	Peak MAC-layer data rate	Peak physical layer data rate
Downlink	–	2.3 kbps	7.4 kbps	97.9 kbps	489.6 kbps
Uplink, 8PSK supported	–	0.6 kbps	2.7 kbps	97.9 kbps	489.6 kbps
Uplink, GMSK supported	–	0.6 kbps	2.7 kbps	51.2 kbps	153.6 kbps

144 and 154 dB. The evaluation assumptions used when deriving these performance figures are the same as presented in Section 4.2.1.

4.4 Latency

For large data transmissions the data rate is decisive for the user experience. For short data transfers expected in IoT networks the latency, including the time to establish a connection and transmitting the data, is a more relevant metric for characterizing the experienced quality of service. Hence, to guarantee a minimum level of service quality also under the most extreme conditions, EC-GSM-IoT should be capable of delivering a so-called *Exception report* within 10 s after waking up from it most energy-efficient state.

4.4.1 Evaluation assumptions

The Exception report is a concept specified in 3GPP Release 13 and corresponds to a message of high urgency that is prioritized by the radio and core network (CN). The EC-RACH channel request message contains a code point that indicates the transmission of an Exception report, which allows the network to prioritize the scheduling of the same. The 96-byte Exception report here studied for EC-GSM-IoT includes 20 bytes of application layer data, 65 bytes of protocol overhead from the *COAP* application, *DTLS* security, *UDP* transport, and *IP* protocols, and 11 bytes of overhead from *SNDCP* and *LLC* layers, i.e., the GPRS CN protocols [5].

Table 4.6 summarizes the packet size definitions assumed in the evaluation of the EC-GSM-IoT latency performance. It should be noted that the 40-byte IP overhead can optionally be reduced to 4 bytes if robust header compression is successfully applied in the CN. This would significantly reduce the message size and improve the latency performance.

Besides the time to transmit the 96-byte Exception report once a connection has been established, using the EC-PDTCH/U and EC-PACCH/D, the latency calculations include the time to synchronize to the network over the FCCH and EC-SCH and the time to perform the random access procedure using the EC-RACH and the EC-AGCH. In addition to the actual transmission times for the various channels, also processing delays in the device and base

TABLE 4.6 EC-GSM-IoT packet definitions including application, security, transport, IP and GPRS CN protocol overhead [1].

Type	Exception report
Application data	20
COAP	4
DTLS	13
UDP	8
IP	40
SNDCP	4
LLC	7
Total	96

station as well as scheduling delays are here accounted for when evaluating the EC-GSM-IoT service latency.

The acquisition of system information is not included in the latency calculations because its content can be assumed to be known by the device due to its semi-static characteristics. EC-GSM-IoT actually mandates reading of the system information not more often than once in every 24 h. It should also be remembered that the EC-SCH is demodulated as part of the synchronization procedure, and it contains the most crucial information concerning frame synchronization, access barring, and modification indication of the system information.

Fig. 4.4 illustrates the signaling and packet transfers considered in the latency evaluation [10]. Three specific parts are identified, namely the time to acquire synchronization T_{SYNC}, the time to perform the Random Access procedure to access the system T_{RA}, and the time to transmit the data T_{DATA}. In the depicted example it is assumed that a first EC-PDTCH/U transmission is followed by three HARQ retransmissions.

4.4.2 Latency performance

Fig. 4.5 shows the time to detect a cell and perform time, frequency, and frame synchronization using the FCCH and EC-SCH under the assumption that a device wakes up from deep sleep with frequency error as large as 20 ppm corresponding to a frequency offset of 18 kHz in the 900 MHz band. This is a reasonable requirement on the frequency accuracy of the real time clock (RTC) responsible for keeping track of time and scheduled events during periods of deep sleep, as discussed in Section 4.7.1. The synchronization time T_{SYNC} used in the latency calculations is derived from the 90th percentile in the synchronization time CDF depicted in Fig. 4.5.

T_{RA} corresponding to the time needed to perform random access and receive the Fixed Uplink Allocation is dependent on the assumed coverage class, i.e., the number of blind transmissions, of the EC-RACH and EC-AGCH that guarantees 20% and 10% BLER respectively, for the applicable coupling loss of the studied scenario.

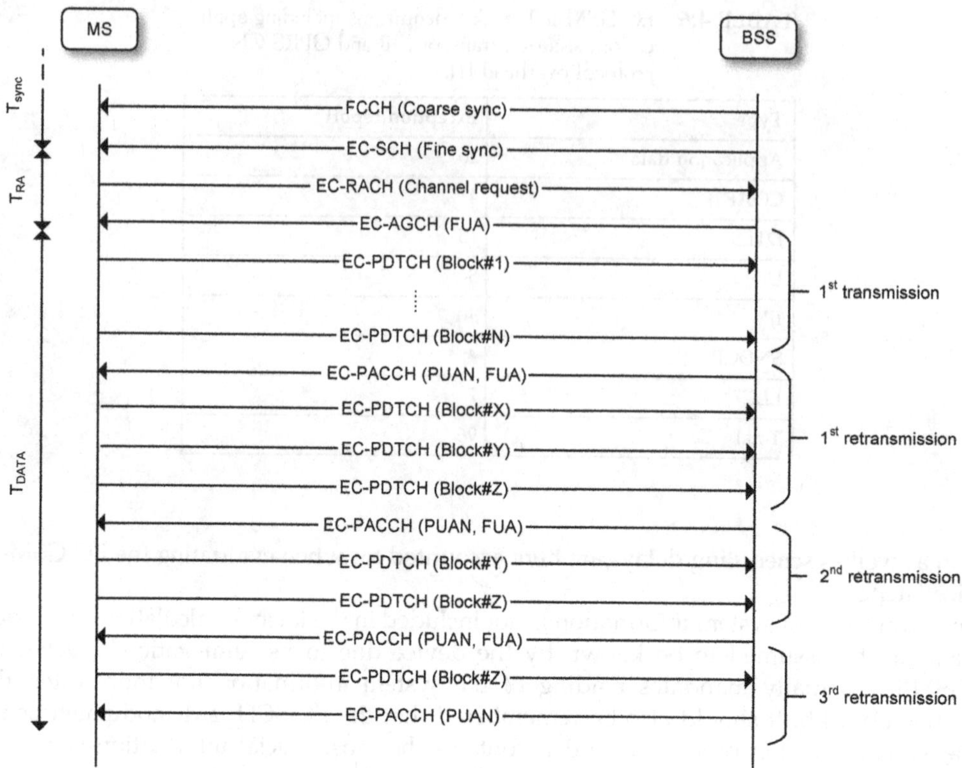

FIG. 4.4 EC-GSM-IoT message transfer.

T_{DATA} is based on the HARQ transmission of the six EC-MCS-1 blocks needed to deliver the 96-byte Exception report, following the procedure illustrated in Fig. 4.1. The packet transfer delay associated with the 90th percentile of the EC-PDTCH latency CDF described in Section 4.2.1.1.3 is used for determining T_{DATA}.

Table 4.7 summarizes the total latency associated with Exception report transmission at coupling losses of 144, 154 and 164 dB.

4.5 Battery life

To support large-scale deployments of IoT systems with minimal maintenance requirements, it was required that an EC-GSM-IoT device for a traffic scenario characterized as small infrequent data transmission supports operation over at least 10 years on a pair of AA batteries providing 5 Wh.

4.5.1 Evaluation assumptions

The EC-GSM-IoT battery life is evaluated for a scenario where a device after waking up from it most energy-efficient state transmits an uplink report and receives a downlink

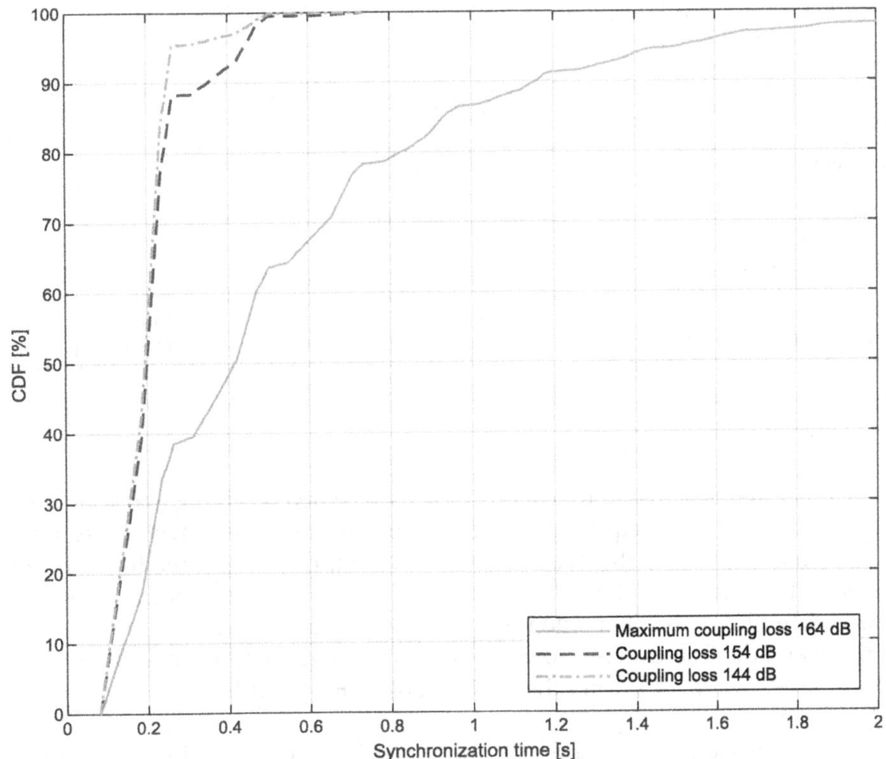

FIG. 4.5 EC-GSM-IoT time to FCCH and EC-SCH synchronization.

application acknowledgment. The uplink report size is in these evaluations set to 50 or 200 bytes, while the downlink application layer acknowledgment packet size is assumed to equal 65 bytes at the entry of the GPRS CN. These packet sizes are assumed to contain the protocol overheads reported in Table 4.6. Two different report triggering intervals, of 2 and 24 h, are assumed. Table 4.8 summarizes these assumptions.

Before the higher layers in a device triggers the transmission of a report, the device is assumed to be in idle mode in which it may enter Power Save Mode to suspend all its idle mode tasks and optimize energy consumption. Ideally it is only the real time clock

TABLE 4.7 EC-GSM-IoT exception report latencies for devices using 23 or 33 dBm output power [5,10,11].

Coupling loss	23 dBm device	33 dBm device
144 dB	1.2 s	0.6 s
154 dB	3.5 s	1.8 s
164 dB	–	5.1 s

TABLE 4.8 EC-GSM-IoT packet sizes at the entry of the GPRS CN for evaluation of battery life [1].

Message type	UL report		DL application acknowledgment
Size	200 bytes	50 bytes	65 bytes
Triggering interval	Once every 2 h or once every day		

TABLE 4.9 EC-GSM-IoT power consumption [1].

TX, 33 dBm	TX, 23 dBm	RX	Idle mode, light sleep	Idle mode, Deep sleep
4.051 W	0.503 W	99 mW	3 mW	15 uW

(see Section 4.7.1) that is active in this deep sleep state. When receiving and transmitting, the device baseband and radio frequency (RF) front end increase the power consumption. The transmit operation dominates the overall power consumption due to the high output power and the moderate power amplifier (PA) efficiency. This is especially the case for the 33 dBm power class where the transmitter side is expected to consume roughly 40 times more power than the receiver side. The different modes of device operation and their associated power consumption levels assumed during the evaluations of EC-GSM-IoT battery life are summarized in Table 4.9.

Fig. 4.6 illustrates the uplink and downlink packet flows modeled in the battery life evaluation. Not illustrated is a 1 s period of light sleep between the end of the uplink report

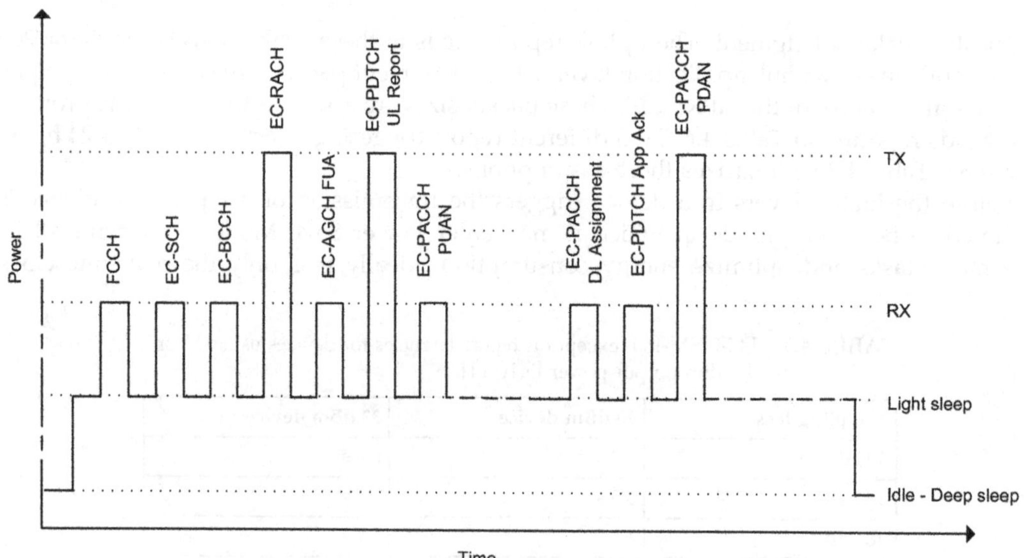

FIG. 4.6 EC-GSM-IoT packet flow used in the evaluation of battery life.

transmission and the start of the downlink application acknowledgment message. A period of light sleep after the end of the final EC-PACCH/U transmission is also modeled. This period is assumed to be configured by the *Ready Timer* to 20 s, during which the device uses a discontinous reception (DRX) cycle that allows for two paging opportunities to enable downlink reachability. An important difference compared with the latency evaluation described in Section 4.4 is that the number of EC-PDTCH blocks modeled in this evaluation corresponds to the average number of blocks needed to be sent, including retransmissions, to secure that the uplink report and downlink application acknowledgment are successfully received. For the synchronization time, the average FCCH and EC-SCH acquisition time was used and not the 90th percentile value used in the latency evaluations. This is a reasonable approach for the modeling of device power consumption for over more than 10 years of operation.

4.5.2 Battery life performance

The resulting battery life for the investigated scenarios are presented in Tables 4.10 and 4.11. It is seen that a 10-year battery life is feasible for the reporting interval of 24 hours. It is also clear that the 2 hours reporting interval is a too aggressive target when the devices are at the MCL of 164 dB. Under these assumptions a battery life of a couple of years is achievable for the assumed pair of AA batteries. For devices with requirements on longer battery life than presented in Table 4.10, this can obviously be achieved by adding battery capacity.

TABLE 4.10 EC-GSM-IoT, 33 dBm device, battery life time [11].

			Battery life		
Reporting interval	**DL packet size**	**UL packet size**	**144 dB CL**	**154 dB CL**	**164 dB CL**
2 h	65 bytes	50 bytes	22.6	13.7	2.8
		200 bytes	18.4	8.5	1.2
24 h		50 bytes	36.0	33.2	18.8
		200 bytes	35.0	29.5	11.0

TABLE 4.11 EC-GSM-IoT, 23 dBm device, battery life time [11].

			Battery life	
Reporting interval	**DL packet size**	**UL packet size**	**144 dB CL**	**154 dB CL**
2 h	65 bytes	50 bytes	26.1	12.5
		200 bytes	22.7	7.4
24 h		50 bytes	36.6	32.5
		200 bytes	36.0	28.3

In these evaluations, an ideal battery power source was assumed. It delivers 5 Wh without any losses or imperfections that typically can be associated with most battery types. EC-GSM-IoT requires a drain current in the order of 1 A, which may require extra consideration when selecting the battery technology to support an EC-GSM-IoT device. A highly optimized RF front end is also assumed in these investigations with a PA efficiency of 50%. A lower efficiency will deteriorate the reported battery life.

4.6 Capacity

As a carrier of IoT services, it is required of EC-GSM-IoT to support a large volume of devices. More specifically, a supported system capacity of at least 60,680 devices per square kilometer (km^2) is expected to be supported by EC-GSM-IoT [1]. This objective is based on an assumed deployment in downtown London with a household density of 1517 homes/km^2 and 40 devices active per household. For a hexagonal cell deployment with an inter-site distance of 1732 m, this results in 52,547 devices per cell. This is clearly an aggressive assumption that includes the underlying assumption that EC-GSM-IoT will serve all IoT devices in every household, while in real life we use a multitude of different solutions to connect our devices. Table 4.12 summarizes the assumptions behind the targeted system capacity.

4.6.1 Evaluation assumptions

For the cellular IoT solutions, coverage is a key criterion, which makes the use of the sub-GHz frequency bands very attractive. Therefore, in the evaluation of EC-GSM-IoT capacity it is assumed that the studied system is deployed in the 900 MHz frequency band, which since long is supported by GSM. To model coverage, a distance-dependent path loss model is assumed in combination with a large-scale shadow fading with a standard deviation of 8 dB and correlation distance of 110 meters. This is intended to model an urban environment where buildings and infrastructure influence the received signal characteristics. All devices modeled in the network are assumed to be stationary and indoor. In addition to the distance-dependent path loss, a very aggressive outdoor to indoor penetration loss model with losses ranging up to 60 dB is assumed to achieve an overall MCL of 164 dB in the studied system.

Just as for the coverage evaluations, a Rayleigh fading Typical Urban channel model with 1 Hz Doppler spread is assumed to model the expected small scale signal variations due to items moving in the proximity of the assumed stationary EC-GSM-IoT devices.

TABLE 4.12 Assumption on required system capacity [1].

Household density [homes/km^2]	Devices per home	Devices per km^2	Inter-site distance [m]	Devices per hexagonal cell
1517	40	60,680	1732	52,547

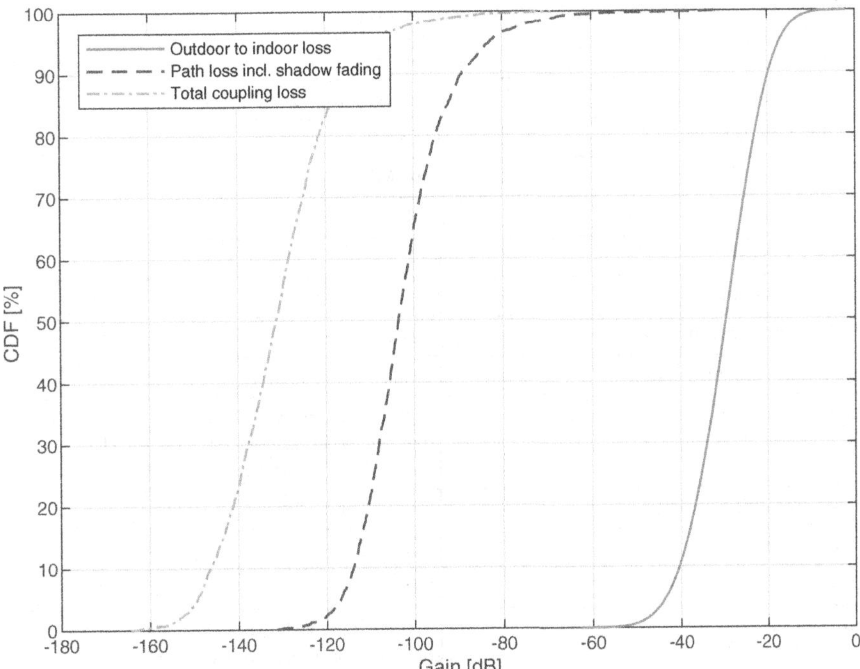

FIG. 4.7 Distance dependent path loss, outdoor to indoor penetration loss, and overall coupling loss distributions recorded from an EC-GSM-IoT system simulator.

The overall coupling loss distribution is presented in Fig. 4.7 together with its path loss and outdoor to indoor loss components. The coupling loss is defined as the loss in signal power calculated as the difference in power measured at the transmitting and receiving nodes antenna ports. Besides the path loss, outdoor to indoor loss, and shadow fading the coupling loss captures base station and device antenna gains. For the base station, a single transmit and two receive antennas with 18 dBi directional gain is assumed. This corresponds to a macro deployment with over-the-rooftop antennas. For the device side a −4 dBi antenna gain is assumed. This antenna loss is supposed to model an antenna integrated in a device where form factor is prioritized over antenna efficiency.

The simulated system assumes that a single 200 kHz channel is configured per cell. The first time slot in every TDMA frame is configured with GSM synchronization, broadcast, and control channels, while the second time slot is used for the EC-GSM-IoT version of the same logical channels. The remaining six time slots are reserved for the EC-PDTCH and EC-PACCH transmissions. The EC-GSM-IoT channel mapping is in detailed covered in Section 3.2.5.

Table 4.13 captures a set of the most relevant EC-GSM-IoT system simulation parameter settings.

The system capacity is determined for two types of traffic scenarios as described in the next two sections.

TABLE 4.13 System level simulation assumptions [1].

Parameter	Model
Cell structure	Hexagonal grid with 3 sectors per size
Cell inter site distance	1732 m
Frequency band	900 MHz
System bandwidth	2.4 MHz
Frequency reuse	12
Frequency channels per cell	1
Base station transmit power	43 dBm
Base station antenna gain	18 dBi
Channel mapping	TS0: FCCH, SCH, BCCH, CCCH TS1: EC-SCH, EC-BCCH, EC-CCCH TS2-7: EC-PACCH, EC-PDTCH
Device transmit power	33 dBm or 23 dBm
Device antenna gain	-4 dBi
Device mobility	0 km/h
Path loss model	$120.9 + 37.6 \cdot \mathrm{LOG}_{10}(d)$, with d being the base station to device distance in km
Shadow fading standard deviation	8 dB
Shadow fading correlation distance	110 m

4.6.1.1 *Autonomous reporting and network command*

In the first traffic scenario, corresponding to a deployment of smart utility meters, it is assumed that 80% of all devices autonomously triggers an uplink report with a Pareto distributed payload size ranging between 20 and 200 bytes as illustrated in Fig. 4.8.

For the part of the meters sending the autonomous report, a set of different triggering intervals ranging from twice per hour to once per day, as captured in Table 4.14, is investigated. In 50% of the cases the device report is assumed to trigger an application-level acknowledgment resulting in a downlink transmission following the uplink report. The payload of the application level acknowledgment is for simplicity assumed to be zero bytes which means that the content of the downlink transmission is defined by the 76 bytes protocol overhead defined in Table 4.6.

The network is assumed to send a 20-byte downlink command to the 20% of the devices not transmitting an autonomous uplink report. The network command follows the distribution and periodicity captured in Table 4.14. Every second device is expected to respond to the network command with an uplink report. The packet size of this report follows the Pareto distribution depicted in Fig. 4.8 with a range between 20 and 200 bytes. Given the assumptions presented in Table 4.14, it can be concluded that a device on average makes the transition from idle to connected mode once in every 128.5 min.

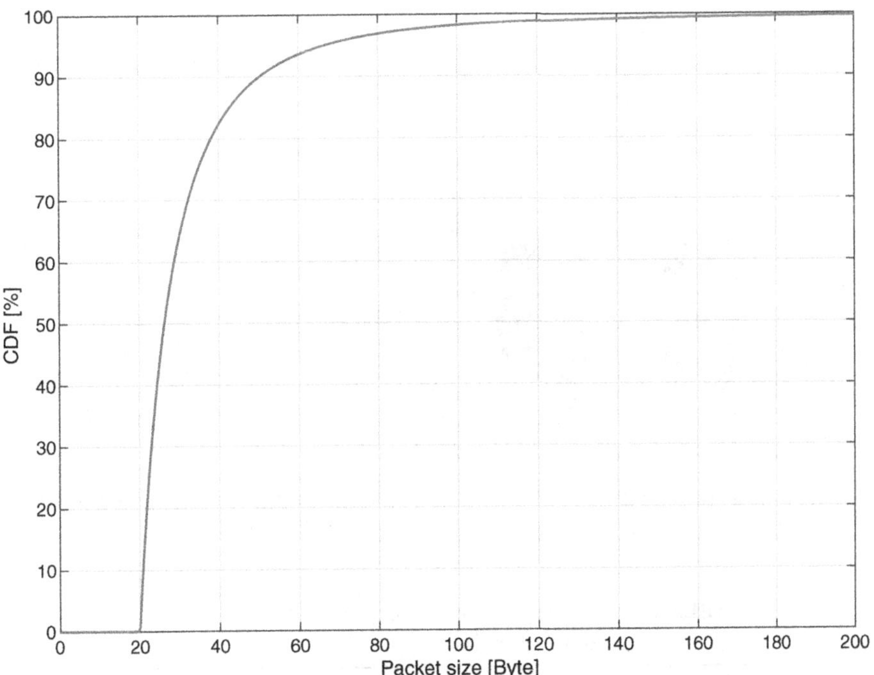

FIG. 4.8 Pareto distributed payload ranging between 20 and 200 bytes and with a mean size of 33.6 bytes.

Fig. 4.9 summarizes the overall uplink and downlink message sizes and periodicities taking the details of the device autonomous reporting and network command assumptions into account. The presented packet sizes do not account for the protocol overheads of Table 4.6, which should be added to get a complete picture of the data volumes transferred over the access stratum.

Fig. 4.9 indicates that the traffic model is uplink heavy, which is a typical aspect of IoT traffic scenarios. At the same time, it can be seen that downlink traffic constitutes a substantial amount of the overall traffic generated.

TABLE 4.14 Device autonomous reporting and network command periodicity and distribution [1].

Device report and network command periodicity [hours]	Device distribution [%]
24	40
2	40
1	15
0.5	5

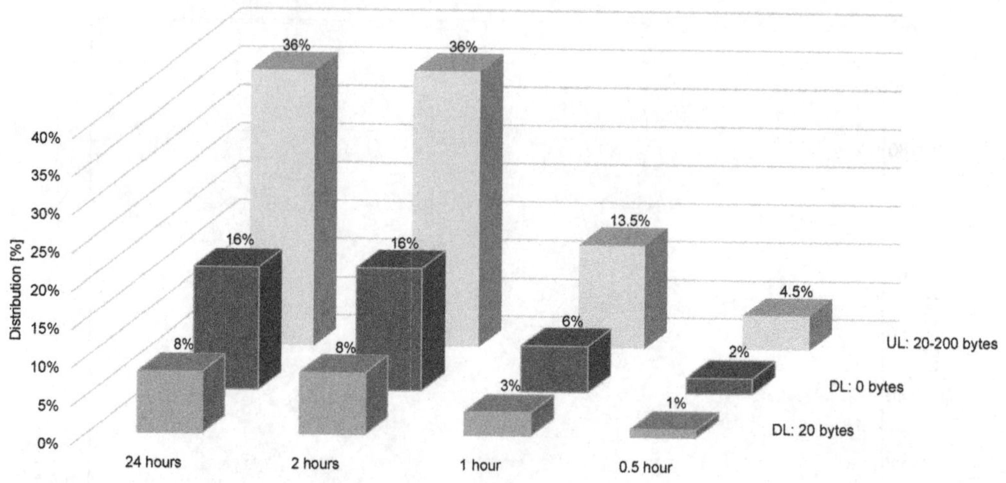

FIG. 4.9 Mobile autonomous reporting and network command periodicity and packet size distribution [1].

TABLE 4.15 Software download periodicity and distribution [1].

Periodicity [days]	Device distribution [%]
180	100

4.6.1.2 *Software download*

In the second traffic scenario, the system's ability to handle large downlink transmission to all devices in a network is investigated. It is in general foreseen that uplink traffic will dominate the load in LPWANs, but that one of the important exceptions is the download of software e.g., to handle firmware upgrades. A Pareto distributed download packet size between 200 and 2000 bytes is here considered to model a firmware upgrade targeting all devices in the network. Again, the protocol overheads of Table 4.6 are added on top of the payload packet size. The devices are expected to receive a software upgrade on average once in every 180 days. These assumptions are presented in Table 4.15.

4.6.2 Capacity performance

In the first scenario, the periodicity of the device autonomous reports and network commands in combination with the targeted load of 52,547 users per cell results in an overall 6.8 users per second and cell making the transition from idle to connected mode to access the system.

The EC-GSM-IoT radio resource utilization in a system configured according to table 4.13 at an access load of 6.8 users per second and cell is presented in Fig. 4.10. The EC-CCCH, EC-PDTCH, and EC-PACCH usage is presented in terms of average fraction

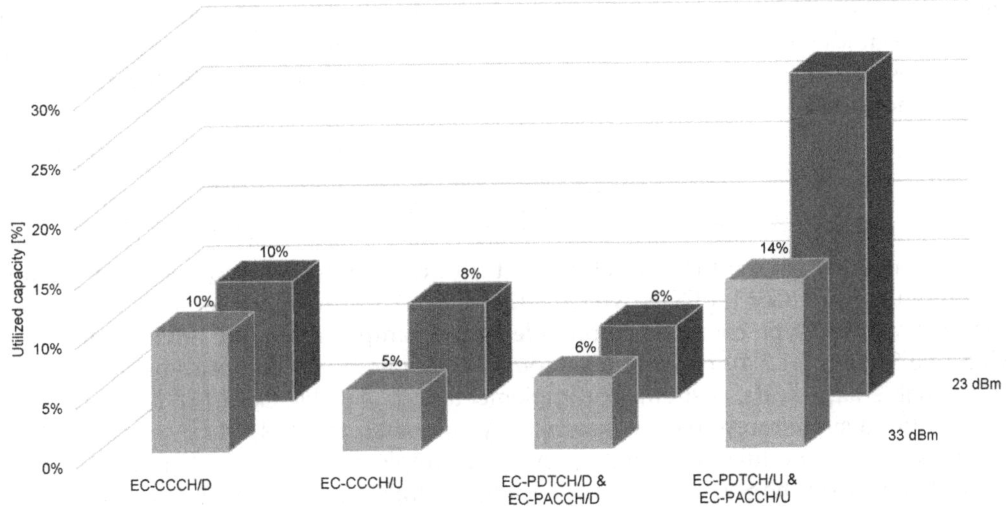

FIG. 4.10 Average percentage of radio resources consumed to serve 52,547 users per cell [3].

of the available resources used. The downlink common control signaling required to set up 6.8 connections per second on average occupies roughly 10% of the available EC-CCCH/D resources on TS1. For the EC-CCCH/U the load is even lower. For the case of the 33 dBm device power class the EC-PDTCH/U and EC-PACCH/U consume on average around 14% of the six configured time slots to deliver the uplink data and Ack/Nack reports. For the 23 dBm power class roughly 27% of the six uplink time slots are on average occupied. The downlink resource consumption is significantly lower than the uplink resource consumption because of the uplink heavy traffic model. At this level of system load the percentage of failed connection attempts was kept below 0.1%. These results indicate that a significantly higher load than the targeted load of 52,547 users per cell can be supported by an EC-GSM-IoT system.

It should be noted that no paging load is considered on the EC-CCCH/D. In reality the load on the paging channel will be dependent not only on the number of devices that the network needs to page, but also on the paging strategy taken by the network. That is, when the network tries to reach a device, it will not exactly know where the device is located and needs to send the paging message to multiple cells to increase the chance of getting a response. With a device that has negotiated a long eDRX cycle, it can take a very long time to reach the device. Hence, there is a clear trade-off between paging load and paging strategy that will have an impact on the overall mobile-terminated reachability performance. Any load caused by paging should be added to the resource usage presented in Fig. 4.10, specifically to the EC-CCCH/D load. In case of a too significant paging load increase, the network can allocate up to four time slots, i.e., TS 1, 3, 5, and 7, for EC-CCCH/D, and by that increase the EC-CCCH capacity by well over 400%.

In the second scenario, where a software download is studied, it is assumed that the load can be evenly spread over time. The download periodicity of 180 days in combination with the load of 52,547 users per cell resulted in a single user making the transition from idle to connected mode on average once in every 5 min. With this assumption the consumed number of resources is close to being negligible despite the assumption of larger packet sizes.

4.7 Device complexity

To be competitive in the IoT market place, it is of high importance to offer a competitive device module price. GSM/EDGE, which is the still in 2018 the most popular cellular technology for machine-type communication, offers, for example, a module price in the area of USD 5 (see Section 3.1.2.4). However, for some IoT applications, this price-point is still too high to enable large-scale, cost-efficient implementations. For EC-GSM-IoT it is therefore a target to offer a significantly reduced complexity compared to GSM/EDGE.

An EC-GSM-IoT module can, to a large extent, be implemented as a system on chip (SoC). The functionality on the SoC can be divided into the following five major components:

- Peripherals
- RTC
- Central processing unit (CPU)
- Digital signal processor (DSP) and hardware accelerators
- Radio transceiver (TRX)

In addition to these blocks, a number of parts may be located outside the SoC, as discrete components on a printed circuit board. The power amplifier (PA) defining the device power class and crystal oscillators (XO) providing the module frequency references are such components.

4.7.1 Peripherals and real time clock

The peripherals block provides the module with external interfaces to, e.g., support a SIM, serial communication, graphics, and general-purpose input and output. This is a generic block that can be found in most communication modules. The range of supported interfaces is more related to the envisioned usage of the device than to the radio access technology providing its connectivity.

The RTC is a low power block keeping track of time and scheduled tasks during periods of deep sleep. Ideally, it is the only component consuming power when a device is in deep sleep, e.g., during periods of eDRX or Power Save Mode. The RTC is typically fed by a low power XO running at 32 kHz, which may be located outside of the SoC as illustrated in Fig. 4.11. For EC-GSM-IoT the accuracy of the XO during periods of sleep is assumed to equal 20 ppm.

Both the Peripherals block and the RTC are generic components that can be expected to be found in all cellular devices regardless of the supported access technology. It is important to understand that the cost associated with functionality related to the radio access technology is only a part of the total price on a communications module.

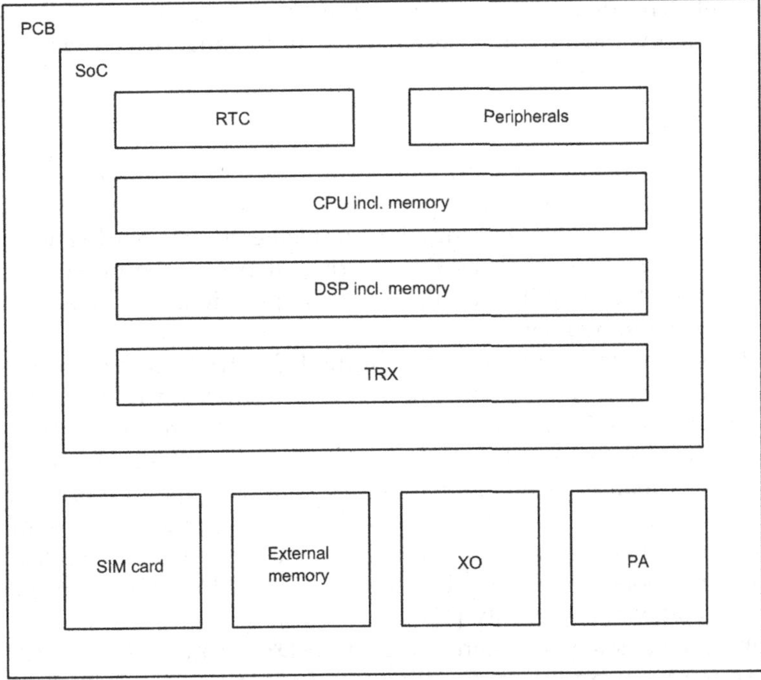

FIG. 4.11 EC-GSM-IoT module architecture.

4.7.2 CPU

The CPU is responsible for generic functions such as booting, running drivers, and applications. It also contains the supported protocol stacks including the GSM protocol stack, i.e., SNDCP, LLC, RLC, and MAC. It contains a controller as well as a memory. A reduction in the protocol stack reduces the CPU memory requirements. But a reduction in and simplifications of the applications supported by the module will also allow reduced computational load and memory requirements to facilitate a less advanced implementation.

The protocol stack in an EC-GSM-IoT device is favorably impacted by the following facts:

- Circuit switched voice is not supported.
- The only mandated modulation and coding schemes are MCS-1 to MCS-4.
- The RLC window size is only 16 (compared to 64 for GPRS or 1024 for EGPRS).
- There is a significant reduction in the number of supported RLC/MAC messages and procedures compared with GPRS.
- Concurrent uplink and downlink data transfers are not supported.

A reduction in RLC buffer memory down to 2 kB and a reduction in SNDCP/LLC buffer memory down to 43 kB are explicitly mentioned in TR 45.820 [1]. Compared with a GPRS implementation, TR 45.820 also concludes that an overall 35%–40% reduction in protocol stack memory requirements is feasible for EC-GSM-IoT. The significance of these reductions is to a large extent dependent on how large the memory consumed by the 3GPP protocol stack is in relation to the overall controller memory for booting, drivers, and applications.

4.7.3 DSP and transceiver

The DSP feeds, and is fed by, the CPU with RLC/MAC headers, data, and control blocks. It handles the modem baseband parts and performs tasks such as symbol mapping, encoding, decoding, and equalization. The DSP may be complemented by hardware accelerators to optimize special purpose tasks such as FIR filtering. It passes the bit stream to the TRX that performs tasks such as GMSK modulation, analog to digital conversion, filtering, and mixing the signal to radio frequency.

For the DSP baseband tasks, the reception of the EC-PDTCH is the most computational demanding task consuming an estimated 88 × 103 DSP cycles per TDMA frame, i.e., per 4.6 ms. For coverage class 2, 3, and 4, four repeated bursts are mapped on consecutive time slots. Assuming that the four bursts can be combined on IQ-level (see Section 3.2.8.2) allows the device to equalize a single burst and not four as in the case of GPRS. Therefore, although EC-PDTCH reception is the most demanding operation, it is significantly less demanding than GSM/EDGE PDTCH reception. Compared with a GPRS reference supporting four receive time slots the 88×10^3 DSP cycles per TDMA frame correspond to a 66% reduction in computational complexity [1].

The IQ-combination poses new requirements on the DSP memory. Four bursts, each of 160 samples, stored using 2 × 16 bit IQ representation will, e.g., consume $4 \times 160 \times 2 \times 16 = 2.56$ kB. This is, however, more than compensated for by the reduced requirements on soft buffer size stemming from the reduced RLC window and a reduced number of redundancy versions supported for EC-GSM-IoT, as explained in Section 3.3.2.2.

Based on the these observations, the overall reduction in the DSP memory size compared to an EGPRS reference results in an estimated saving in ROM and RAM memory of 160 kB. This corresponds to a ROM memory savings of 48% and RAM memory savings in the range of 19%–33% [1].

For the TRX RF components, it is positive that EC-GSM-IoT supports only four global frequency bands. This minimizes the need to support frequency-specific variants of the RF circuitry. Also, the fact the EC-GSM-IoT operates in half duplex has a positive impact on the RF front end as it facilitates the use a RX-TX antenna switch instead of a duplexer.

4.7.4 Overall impact on device complexity

Based on the findings presented in sections 4.7.1–4.7.3, in terms of reduction in higher and lower layers' memory requirements, procedures, and computational complexity, it has been concluded that a 20% reduction in the EC-GSM-IoT SoC size compared to GPRS is within reach [1].

In addition to the components on the chip, it is mainly the PA that is of interest to consider for further complexity reduction. For EC-GSM-IoT, the in GSM commonly used 33 dBm power class is supported by the specification. However, because of its high power and drain current, it needs to be located outside of the chip. At 50% PA efficiency the PA would, e.g., generate 4 W power, of which 2 W will be dissipated as heat. The 23 dBm power class was therefore specified to allow the PA to be integrated on the chip. At 3.3 V supply voltage and an on-chip PA efficiency of 45%, the heat dissipation is reduced to 250 mW and the drain current is down at 135 mA, which is believed to facilitate a SoC including the PA. This will further reduce the overall module size and complexity. The potential cost/complexity benefit from integrating the PA onto the chip has not been quantified for EC-GSM-IoT but is more in detailed investigated for LTE-M (see Chapter 6), which can at least give an indication of the potential complexity reduction also for other technologies.

4.8 Operation in a narrow frequency deployment

GSM is traditionally operating the BCCH frequency layer over at least 2.4 MHz by using a 12-frequency reuse. This is also the assumption used when evaluating EC-GSM-IoT capacity in Section 4.6. For a LPWAN network it is clearly an advantage to support operation in a smaller frequency allocation. For EC-GSM-IoT operation over 9 or 3 frequencies, i.e., using 1.8 MHz or 600 kHz, is investigated in this section. More specifically, the operation is evaluated in the areas of idle mode procedures, common control channels, and data traffic and dedicated control channel performance.

The results presented in Sections 4.8.1 and 4.8.2 clearly show that EC-GSM-IoT can be deployed in a frequency allocation as tight as 600 kHz, with limited impact on system performance.

4.8.1 Idle mode procedures

Reducing the frequency reuse in the BCCH layer may impact tasks such as synchronization to, identification of, and signal measurements on a cell via the FCCH and EC-SCH. Especially, the FCCH detection is vulnerable because the FCCH signal definition (see Section 3.2.6.1) is the same in all cells. A suboptimal acquisition of the FCCH may negatively influence tasks such as Public Land Mobile Network (PLMN) selection, cell selection, and cell reselection.

4.8.1.1 PLMN and cell selection

For initial PLMN or cell selection, a device may need to scan the full range of supported bands and absolute radio frequency channel numbers (ARFCNs) in the search for an EC-GSM-IoT deployment. A quad band device supporting GSM 850, 900, 1850, and 1900 frequency bands needs to search in total 971 ARFCNs. In worst case, a device requires 2 seconds to identify an EC-GSM-IoT cell when at 164 dB MCL, as depicted in Fig. 4.5. This was proven to be the case regardless of the frequency reuse, since thermal noise dominates over external interference in deep coverage locations even in a tight frequency deployment. In a scenario where only a single base station is within coverage, and where this base station is configured using the last frequency being searched for by a device, a sequential scan over the 971 ARFCNs would demand 971 × 2 s = 32 min of active RF reception and baseband processing. By means of an interleaved search method the search time can be reduced to 10 min as

TABLE 4.16 Worst-case of full band search time for a quad band device at 164 dB coupling loss from the serving cell [3].

System bandwidth	600 kHz	1.8 MHz	2.4 MHz
Time of PLMN selection	10 min		

TABLE 4.17 The probability for a EC-GSM-IoT device to select the optimal serving cell [3].

System bandwidth	600 kHz	1.8 MHz	2.4 MHz
Probability of selecting strongest cell as serving cell	89.3 %	89.7 %	90.1 %

TABLE 4.18 The probability and time required for an EC-GSM-IoT device to successfully reconfirm the serving cell after a period of deep sleep [3].

System bandwidth	600 kHz	1.8 MHz	2.4 MHz
Probability of reconfirming serving cell	98.7 %	99.9 %	99.9 %
Synchronization time, 99th percentile	0.32 s	0.12 s	0.09 s

presented in Table 4.16 [3]. In practice it is also expected that it is sufficient for an EC-GSM-IoT device to support the two sub-GHz bands for global coverage, which has the potential to further reduce the worst-case full band search time.

After the initial full band scan, a serving cell needs to be selected. To improve performance in an interference limited network the signal strengths of a set of highly ranked cells are measured over the FCCH and EC-SCH (see Section 3.3.1.1) with the measurements ideally excluding contributions from interference and noise. With this new approach of measuring for cell selection, the likelihood of selecting the strongest cell as serving cell is, as summarized in Table 4.17, close to being independent of the frequency reuse.

4.8.1.2 *Cell reselection*

After the initial cell selection, EC-GSM-IoT mobility relies on the idle mode cell reselection procedure where significant importance is put on accurate evaluation of the quality of the serving cell (see Section 3.3.1.2). One important scenario for EC-GSM-IoT is that a device waking up after a long period of deep sleep can successfully synchronize to and reconfirm the identity of the serving cell. The reconfirmation of the serving cell is, as seen in Table 4.18, only slightly impacted by the tighter frequency reuse.

4.8.2 Data and control channel performance

The data and control channel capacity are evaluated under the same assumptions as elaborated on in Section 4.6.1. In addition to the 12-frequency reuse, consuming 2.4 MHz

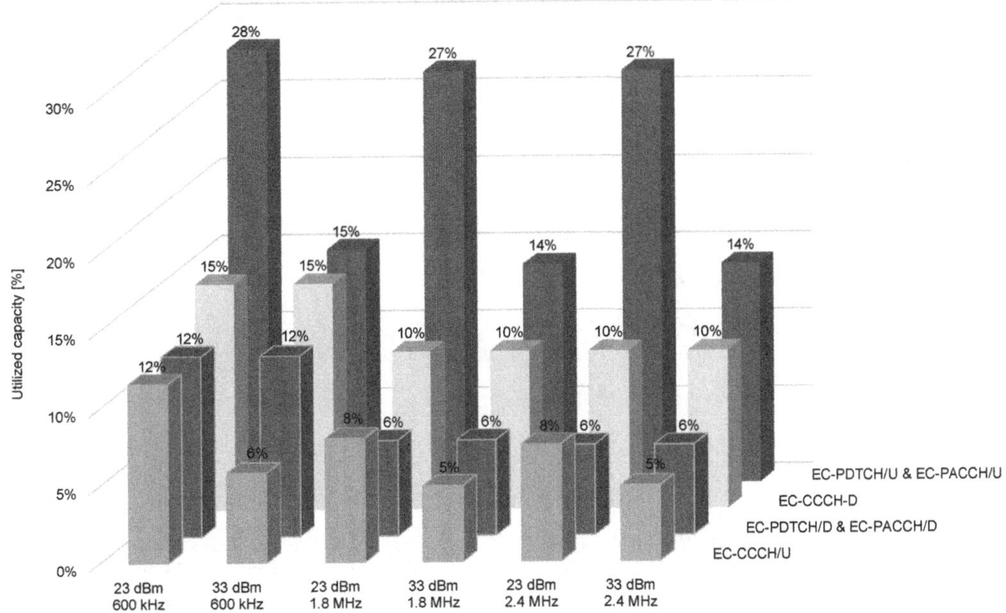

FIG. 4.12 Radio resource utilization at a load of 52547 users per cell [3].

system bandwidth, also 9- and 3-frequency reuse patterns are investigated. The relative radio resource consumption is summarized in Fig. 4.12 in terms of average fraction of the available radio resources consumed. In all three deployment scenarios the percentage of failed connection attempts are kept below 0.1% at the investigated system load.

The impact on resource utilization is seen as negligible when going from 2.4 to 1.8 MHz, while it becomes noticeable when going down to a 600 kHz deployment. This is especially noticeable for the downlink traffic channels that are more severely hit by the increased interference levels stemming from the tightened frequency reuse.

The targeted load of 52,547 devices is comfortably met under all scenarios. In relation to the available radio resources on the BCCH carrier, the presented figures are relatively modest indicating that a load well beyond 52,547 users per cell may be supported even when only using a 600 kHz frequency deployment.

Besides the increased resource consumption presented in Fig. 4.12, the reduced frequency reuse also results in increased service delays mainly because of more retransmissions caused by increased interference levels, and users selecting higher coverage classes. Fig. 4.13 and 4.14 illustrate the impact on the time to successfully transmit a device autonomous report and on the time to transmit a downlink application acknowledgment once a connection has been established, including EC-PACCH and EC-PDTCH transmission times and thereto associated delays. Here a 33 dBm device is studied. The uplink and downlink packet sizes follows the characteristics specified in Section 4.6.1. The impact when going from 2.4 to 1.8 MHz is negligible for both cases. When taking a further step to 600 kHz the impact becomes more accentuated but is still acceptable for the type of services EC-GSM-IoT targets.

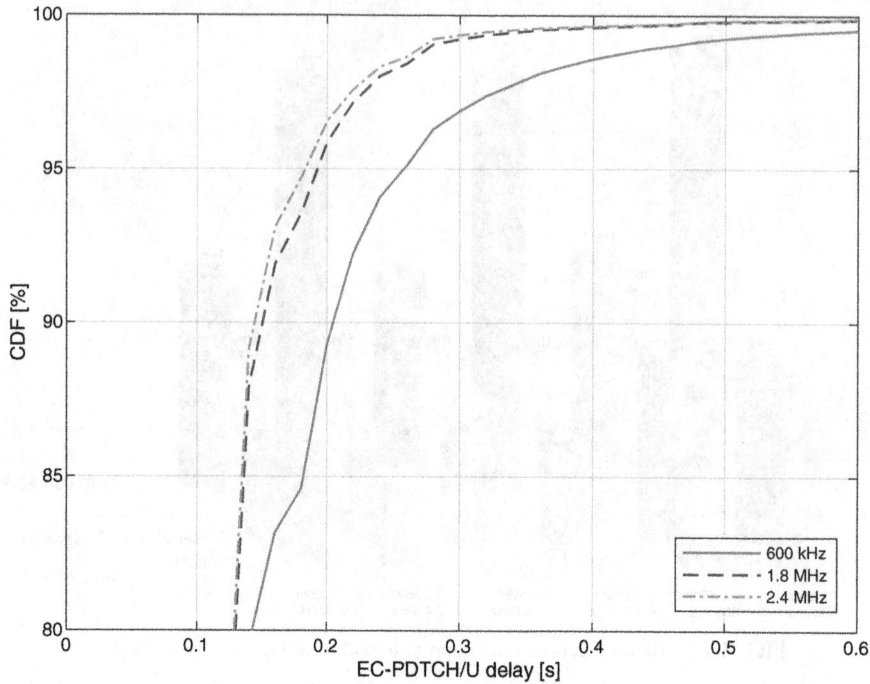

FIG. 4.13 EC-PDTCH/U transmission delay during the delivery of a device autonomous report.

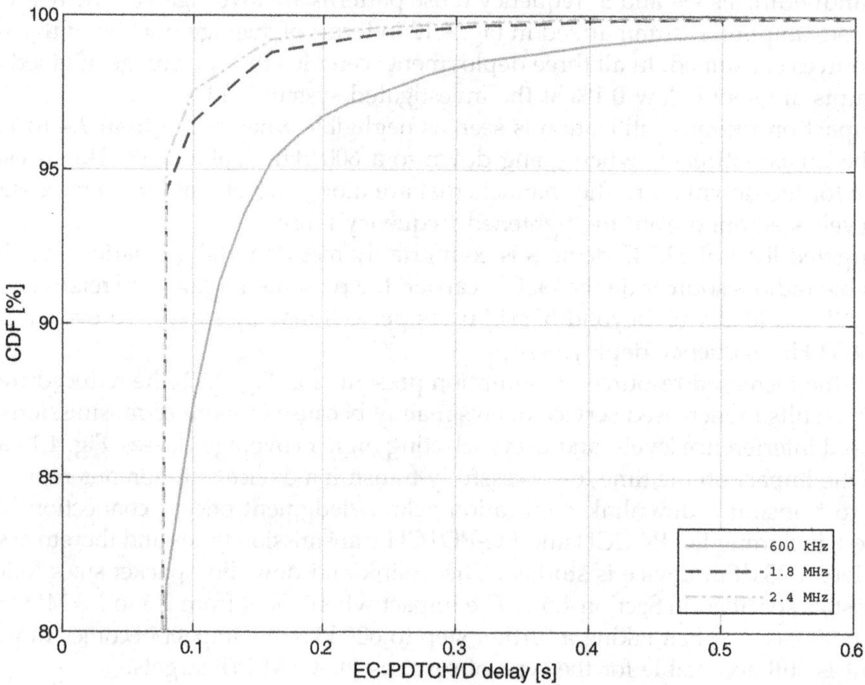

FIG. 4.14 EC-PDTCH/D transmission delay during the delivery of an application acknowledgment.

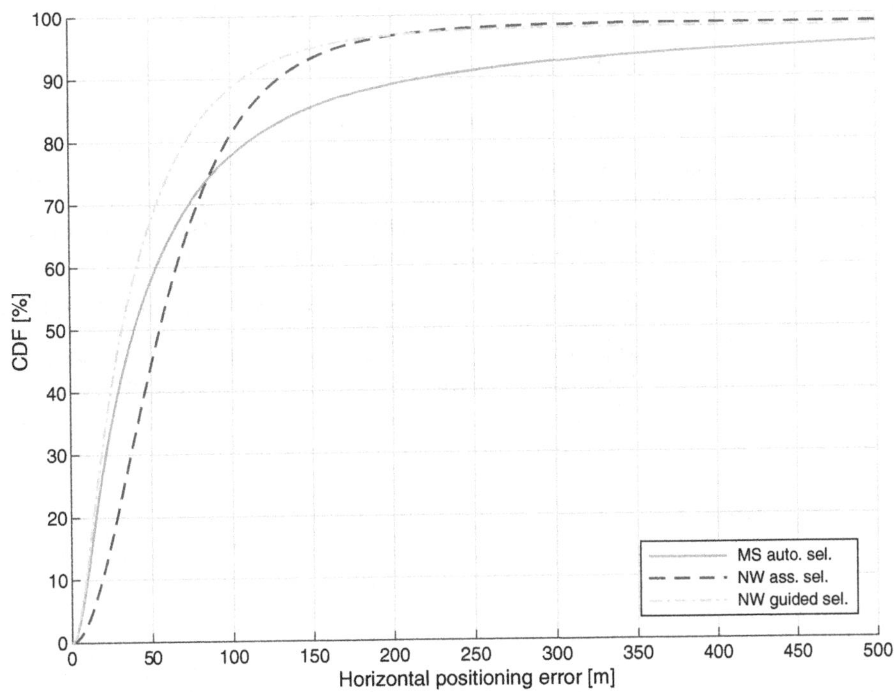

FIG. 4.15 Positioning accuracy of multilateration.

4.9 Positioning

Multiple methods for positioning of devices in the network has been described in Section 3.4.1. In this section, the multilateration method is evaluated.

The positioning accuracy has been evaluated [12] by system simulations, modeling, for example, the SINR dependent synchronization accuracy. The results are presented in Fig. 4.15. For the multilateration case, three base stations are used in the positioning attempt. It can be noted that the more base stations that are used, the better the accuracy, but also the more energy will be consumed by the device. From the figure it is seen that the most suitable method is where the device selects the base stations (based on descending SINR), excluding cells that are co-sited (legend: "NW guided sel."). It should be noted that another method could be the optimal one for another number of base stations used in the positioning attempt [12].

References

[1] Third Generation Partnership Project, Technical Report 45.820, v13.0.0. Cellular System Support for Ultra-Low Complexity and Low Throughput Internet of Things, 2016.

[2] Ericsson, et al. GP-151039. New Work Item on Extended Coverage GSM (EC-GSM) for Support of Cellular Internet of Things, 3GPP TSG GERAN Meeting #67, 2015.

[3] Third Generation Partnership Project, Technical Specification 45.050, v13.1.0. Background for Radio Frequency (RF) Requirements, 2016.

[4] Third Generation Partnership Project, Technical Specification 45.005, v14.0.0. Radio Transmission and Reception, 2016.
[5] Ericsson, et al. GP-150419 EC-GSM, Throughput, Delay and Resource Analysis, 3GPP TSG GERAN Meeting #66, 2015.
[6] Ericsson, et al. GP-140608 Radio Propagation Channel for the Cellular IoT Study, 3GPP TSG GERAN Meeting #63, 2014.
[7] Ericsson, et al. GP-160292 Introduction of Radio Frequency Colour Code, 3GPP TSG GERAN Meeting #70, 2016.
[8] Ericsson, et al. GP-160032 Additional EC-RACH Mapping, 3GPP TSG GERAN Meeting #69, 2016.
[9] Third Generation Partnership Project, Technical Specification 44.060, v14.0.0. General Packet Radio Service (GPRS); Mobile Station (MS) - Base Station System (BSS) Interface; Radio Link Control/Medium Access Control (RLC/MAC) Protocol, 2016.
[10] Ericsson, et al. GP-150449 EC-GSM – Exception Report Latency Performance Evaluation, 3GPP TSG GERAN Meeting #66, 2015.
[11] Ericsson, et al. GP-160351 EC-GSM-IoT Performance Summary, 3GPP TSG GERAN Meeting #70, 2015.
[12] Ericsson. R6–160235 Enhanced Positioning – Positioning Performance Evaluation 3GPP TSG RAN Working Group 6 Meeting #2, 2016.

CHAPTER

5

LTE-M

OUTLINE

Cellular Internet of Things, Second Edition
https://doi.org/10.1016/B978-0-08-102902-2.00005-4

© 2020 Elsevier Ltd. All rights reserved.

Abstract

In this chapter, we describe the *Long-Term Evolution for Machine-Type Communications* (LTE-M) technology with an emphasis on how it is designed to fulfill the objectives that LTE-M targets, namely achieving low device cost, deep coverage and long battery lifetime while maintaining capacity for a large number of devices per cell, with performance and functionality suitable for both low-end and mid-range applications for the Internet of Things.

Section 5.1 describes the background behind the introduction of LTE-M in the Third Generation Partnership Project (3GPP) specifications and the design principles of the technology. Section 5.2 focuses on the physical channels with an emphasis on how these channels were designed to fulfill the objectives that LTE-M was intended to achieve. Section 5.3 covers LTE-M procedures in idle and connected mode, including all activities from initial cell selection to completing a data transfer. The idle mode procedures include the initial cell selection, which is the procedure that a device has to go through when it is first switched on or is attempting to select a new cell to camp on. Idle mode activities also include acquisition of system information, paging, random access, and multicast. Descriptions of some fundamental connected mode procedures include scheduling, power control, mobility, and positioning. Both the fundamental functionality introduced in 3GPP Release 13 and the improvements introduced in Release 14 and Release 15 are covered. Finally, coexistence between LTE-M and *Fifth Generation* (5G) *New Radio* (NR) is presented in Section 5.4. The performance of LTE-M including its fulfillment of 5G mMTC requirements is covered in Chapter 6.

5.1 Background

In this section we describe the introduction of *Long-Term Evolution for Machine-Type Communications* (LTE-M) into the 3GPP specifications and the design principles. The design principles were adopted in order to achieve the required low device complexity and cost, coverage enhancement, long device battery lifetime and support of massive number of devices for the massive Internet of Things (IoT) segment. In addition, the design was required to support adequate peak rates and mobility support to be able to address more demanding applications such as voice services, and ensuring high deployment flexibility and coexistence with ordinary LTE.

5.1.1 3GPP standardization

LTE-M extends LTE with features for improved support for *Machine-Type Communications* (MTC) and IoT. These extensions have their origin in the 3GPP study item *Study on provision of low-cost MTC User Equipments based on LTE* [1], in this book referred to as the *LTE-M study item*.

Since the conclusion of the LTE-M study item, a number of related 3GPP work items have been completed, starting with an initial Release 12 work item that can be seen as a precursor for the more ambitious Release 13 work item:

- Release 12 work item *Low cost and enhanced coverage MTC UE for LTE* [2], sometimes referred to as the *MTC work item*, which introduced LTE device category 0 (Cat-0)
- Release 13 work item *Further LTE Physical Layer Enhancements for MTC* [3], sometimes referred to as the *eMTC work item*, which introduced the Coverage Enhancement (CE) modes A and B as well as LTE device category M1 (Cat-M1)

In this book we use the term LTE-M when we refer to the Cat-M device category series, the CE modes, and all functionality that can be supported by the Cat-M devices or the CE modes, such as the power consumption reduction techniques *Power Saving Mode* (PSM) and *Extended Discontinuous Reception* (eDRX). According to this definition, all LTE devices (including Cat-0 devices) that implement CE mode support are considered LTE-M devices, but if they do not implement CE mode support then they are not considered LTE-M devices. Cat-M devices have mandatory support for CE mode A and are always considered LTE-M devices.

Already, many LTE-M networks have been deployed and a device ecosystem has been established. The GSM Association (GSMA), which is an organization that represents the interests of mobile network operators worldwide, tracks the status of commercial launches of LTE-M. Since the completion of Release 13 LTE-M in March 2016, there had been more than 30 LTE-M launches in over 25 markets as of June 2019, according to GSM Association [4]. On the device side, the Global Mobile Suppliers Association (GSA) published a research report in 2018 [5] stating that as of August 2018, there were 101 modules supporting LTE-M, 48 of which also support NB-IoT.

Release 13 laid the foundation for LTE-M in the form of low-cost devices and coverage enhancements, but the LTE-M standard has been further evolved in Releases 14 and 15:

- Release 14 work item *Further Enhanced MTC for LTE* [6], sometimes referred to as the *feMTC work item*, which introduced various improvements for support of higher data rates, improved VoLTE support, improved positioning, multicast support, as well as the new LTE device category M2 (Cat-M2)
- Release 15 work item *Even Further Enhanced MTC for LTE* [7], sometimes referred to as the *efeMTC work item*, which introduced further improvements for reduced latency and power consumption, improved spectral efficiency, new use cases, and more

TABLE 5.1 New LTE-M features introduced in 3GPP Releases 14 and 15.

Release 14 (2017)	Section	Release 15 (2018)	Section
Support for higher data rates		**Support for new use cases**	
• New device category M2	5.2.3	• Support for higher device velocity	5.3.2.5
• Higher uplink peak rate for Cat-M1	5.2.3	• Lower device power class	5.1.2.1
• Wider bandwidth in CE mode	5.2.2.3		
• More downlink HARQ processes in FDD	5.3.2.1.1	**Reduced latency**	
• ACK/NACK bundling in HD-FDD	5.3.2.1.1	• Resynchronization signal	5.2.4.2.2
• Faster frequency retuning	5.2.4.1	• Improved MIB/SIB performance	5.3.1.2.1
		• System info update indication	5.3.1.2.3
VoLTE enhancements			
• New PUSCH repetition factors	5.2.5.4	**Reduced power consumption**	
• Modulation scheme restriction	5.2.5.4	• Wake-up signals	5.2.4.5
• Dynamic ACK/NACK delays	5.3.2.1.1	• Early data transmission	5.3.1.7.3
		• ACK/NACK feedback for uplink data	5.3.2.1.2
Coverage improvements		• Relaxed monitoring for cell reselection	5.3.2.5
• SRS coverage enhancement	5.2.5.3.2		
• Larger PUCCH repetition factors	5.2.5.5	**Increased spectral efficiency**	
• Uplink transmit antenna selection	5.2.5.4	• Downlink 64QAM support	5.2.4.7
		• CQI table with large range	5.3.2.2
Multicast support	5.3.1.9	• Uplink sub-PRB allocation	5.2.5.4
Improved positioning	5.3.2.6	• Flexible starting PRB	5.3.2.1
Mobility enhancements	5.3.2.5	• CRS muting	5.2.4.3.1
		Improved access control	5.3.1.8

Table 5.1 provides a summary of the LTE-M enhancements introduced in 3GPP Releases 14 and 15, with references to the relevant book sections. All the Releases 14 and 15 features can be enabled through a software upgrade of the existing LTE network equipment. In many cases it may also be possible to upgrade the software/firmware in existing devices to support the new features.

In Release 15, 3GPP evaluated LTE-M against a set of agreed *Fifth Generation* (5G) performance requirements defined for the *massive machine-type communications* (mMTC) use case [8]. As shown in Chapter 6, LTE-M meets these requirements with margins and does in all relevant aspects qualify as a 5G mMTC technology. As we will see in Section 5.4, LTE-M is also able to coexist efficiently with the 5G New Radio (NR) air interface introduced in Release 15.

5.1.2 Radio Access Design Principles

5.1.2.1 Low device complexity and cost

During the LTE-M study item [1], various device cost reduction techniques were studied, with the objective to bring down the LTE device cost substantially to make LTE attractive for low-end MTC applications that have so far been adequately handled by GSM/GPRS. It was estimated that this would correspond to a device modem manufacturing cost in the order of 1/3 of that of the simplest LTE modem, which at that time was a single-band LTE device Cat-1 modem.

The study identified the following cost reduction techniques as most promising:

- Reduced peak rate
- Single receive antenna
- Half-duplex operation
- Reduced bandwidth
- Reduced maximum transmit power.

A first step was taken in Release 12 with the introduction of LTE device Cat-0 that supported a reduced peak rate for user data of 1 Mbps in downlink and uplink (instead of at least 10 Mbps in downlink and 5 Mbps in uplink for Cat-1 and higher categories), a single receive antenna (instead of at least two), and optionally *half-duplex frequency-division duplex* (HD-FDD) operation.

The next step was taken in Release 13 with LTE device Cat-M1 that includes all the cost reduction techniques of Cat-0, plus a reduced bandwidth of 1.4 MHz (instead of 20 MHz), and optionally a lower device power class with a maximum transmit power of 20 dBm (instead of 23 dBm). Release 15 introduces an even lower 14-dBm power class to enable implementation of devices with low power consumption and small form factor, see Section 5.2.3.

With the cost reduction techniques introduced in Release 13, the *Bill of Material* cost for the Cat-M1 modem was estimated to reach that of an Enhanced GPRS modem. For further information on LTE-M cost estimates, refer to Section 6.7.

5.1.2.2 Coverage enhancement

The LTE-M study item [1] also studied coverage enhancement (CE) techniques, with the objective to improve coverage of LTE networks at that time by 20 dB to provide coverage for devices with challenging coverage conditions, for example stationary utility metering devices located in basements.

The study identified various forms of prolonged transmission time as the most promising coverage enhancement techniques. The fact that many of the IoT applications of interest have very relaxed requirements on data rates and latency can be exploited to enhance the coverage through repetition or retransmission techniques. The study concluded that 20 dB coverage enhancement can be achieved using the identified techniques.

Release 13 standardized two *CE modes*: CE mode A, supporting up to 32 repetitions for the data channels, and CE mode B, supporting up to 2048 repetitions. Recent evaluations show that the initial coverage target of 20 dB can be reached using the repetitions available in CE mode B. For further information on LTE-M coverage and data rate estimates, refer to Sections 6.2 and 6.3.

In this book we refer to LTE devices with CE mode support as *LTE-M devices*. These devices may be low-cost *Cat-M devices*, or they may be higher LTE device categories configured in a CE mode. For more information on the CE modes refer to Section 5.2.2.3.

5.1.2.3 *Long device battery lifetime*

Support for a device battery lifetime of many years, potentially decades, has been introduced in a first step in the form of the PSM in Release 12 and in a second step in the form of the eDRX in Release 13. These features are supported for LTE-M devices and also for other 3GPP radio access technologies.

These techniques reduce the power consumption primarily by minimizing any unnecessary "on" time for the receiver and the transmitter in the device. Compared to ordinary LTE devices, LTE-M devices can have a further reduced power consumption during their "on" time mainly thanks to the reduced transmit and receive bandwidths.

PSM and eDRX are described in Sections 2.2.3, 5.3.1.4, and 5.3.1.5, and the battery lifetime for LTE-M is evaluated in Section 6.5.

5.1.2.4 *Support of massive number of devices*

The handling of massive numbers of devices in LTE was improved already in Releases 10 and 11, for example in the form of access class barring (ACB) and overload control, as discussed in Section 2.2.1. Further improvements have been introduced later on, for example in the form of the *Radio Resource Control (RRC) Suspend/Resume* mechanism described in Section 2.2.2, which helps reduce the required signaling when resuming an RRC connection after a period of inactivity as long as the device has not left the cell in the meanwhile.

For more information on LTE-M capacity estimates, refer to Section 6.6.

5.1.2.5 *Deployment flexibility*

LTE-M can be deployed in a wide range of frequency bands, as can be seen from the list of supported bands in Table 5.2. Both paired bands for frequency-division duplex (FDD) operation and unpaired bands for time-division duplex (TDD) operation are supported, and new bands have been added in every release. Even though the simplest LTE-M devices (i.e. the Cat-M devices) only support a reduced bandwidth, LTE-M supports the same system bandwidths at the network side as LTE (1.4, 3, 5, 10, 15, and 20 MHz).

5.1.2.6 *Coexistence with LTE*

LTE-M extends the LTE physical layer with features for improved support for MTC. The LTE-M design therefore builds on the solutions already available in LTE.

The fundamental downlink and uplink transmission schemes are the same as in LTE, meaning *Orthogonal Frequency-Division Multiplexing* (OFDM) in downlink and *Single-Carrier Frequency-Division Multiple-Access* (SC-FDMA) in uplink, with the same numerologies (channel raster, subcarrier spacing, cyclic prefix (CP) lengths, resource grid, frame structure, etc.). This means that LTE-M transmissions and LTE transmissions related to, for example, smartphones and mobile broadband modems can coexist in the same LTE cell on the same LTE carrier and the resources can be shared dynamically between LTE-M users and ordinary LTE users.

If an operator has a large spectrum allocation for LTE, then there is also a large bandwidth available for LTE-M traffic. The downlink and uplink resources on an LTE carrier can serve as a resource pool that can be fully dynamically shared between LTE traffic and LTE-M traffic. It may furthermore be possible to schedule delay-tolerant LTE-M traffic during periods when the ordinary LTE users are less active, thereby minimizing the performance impact from the LTE-M traffic on the LTE traffic.

TABLE 5.2 Frequency bands defined for Cat-M1/M2 as of Release 15 [9].

Band	Duplex mode	Uplink [MHz]	Downlink [MHz]
1	FDD	1920–1980	2110–2170
2	FDD	1850–1910	1930–1990
3	FDD	1710–1785	1805–1880
4	FDD	1710–1755	2110–2155
5	FDD	824–849	869–894
7	FDD	2500–2570	2620–2690
8	FDD	880–915	925–960
11	FDD	1427.9–1447.9	1475.9–1495.9
12	FDD	699–716	729–746
13	FDD	777–787	746–756
14	FDD	788–798	758–768
18	FDD	815–830	860–875
19	FDD	830–845	875–890
20	FDD	832–862	791–821
21	FDD	1447.9–1462.9	1495.9–1510.9
25	FDD	1850–1915	1930–1995
26	FDD	814–849	859–894
27	FDD	807–824	852–869
28	FDD	703–748	758–803
31	FDD	452.5–457.5	462.5–467.5
39	TDD	1880–1920	1880–1920
40	TDD	2300–2400	2300–2400
41	TDD	2496–2690	2496–2690
66	FDD	1710–1780	2110–2200
71	FDD	636–698	617–652
72	FDD	451–456	461–466
73	FDD	450–455	460–465
74	FDD	1427–1470	1475–1518
85	FDD	698–716	728–746

5.2 Physical layer

In this section, we describe the LTE-M physical layer design with an emphasis on how these physical signals and channels are designed to fulfill the objectives that LTE-M targets, namely low device cost, deep coverage, and long battery lifetime, while maintaining capacity for a large number of devices per cell, with performance and functionality suitable for both low-end and mid-range applications for the IoT.

5.2.1 Physical resources

5.2.1.1 *Channel raster*

LTE-M supports many frequency bands (see Table 5.2 for a list of supported bands) and the same system bandwidths as ordinary LTE (1.4, 3, 5, 10, 15, and 20 MHz). A *channel raster* defines the allowed carrier frequencies.

LTE-M is reusing LTE's *Primary Synchronization Signal* (PSS), *Secondary Synchronization Signal* (SSS), and the core part of the *Physical Broadcast Channel* (PBCH) carrying the *Master Information Block* (MIB). These physical signals are located in the center of the LTE system bandwidth and this center frequency is aligned with a channel raster of 100 kHz. The absolute frequency that the LTE and LTE-M system is centered around can be deduced from the *E-UTRA Absolute Radio Frequency Channel Number* (EARFCN). For more details on synchronization signals and procedures, see Sections 5.2.4.2 and 5.3.1.1.

When an LTE-M device is operating in a so-called *narrowband* or *wideband*, its center frequency is not necessarily aligned with the 100-kHz channel raster of LTE. For more information on narrowband and wideband operation, see Section 5.2.2.2.

5.2.1.2 *Frame structure*

The overall time frame structure on the access stratum for LTE and LTE-M is illustrated in Fig. 5.1. On the highest level one hyperframe cycle has 1024 hyperframes that each consists of 1024 frames. One frame consists of 10 subframes, each dividable into two slots of 0.5 ms as shown in the figure. Each slot is divided into 7 OFDM symbols in case of *normal CP* length and 6 OFDM symbols in case of *extended CP* length. The normal CP length is designed to support propagation conditions with a delay spread up to 4.7 μs, while the extended CP is intended to support deployments where the delay spread is up to 16.7 μs. All illustrations in this book assume the normal CP length because it is much more commonly used than the extended CP length.

Each subframe can be uniquely identified by a *hyper system frame number* (H-SFN), a *system frame number* (SFN), and *subframe number*. The ranges of H-SFN, SFN, and subframe number are 0–1024, 0–1024, and 0–9, respectively.

5.2.1.3 *Resource grid*

One *physical resource block* (PRB) spans 12 subcarriers, which with the 15-kHz subcarrier spacing correspond to 180 kHz. When full-PRB transmission is used, the smallest time-frequency resource that can be scheduled to a device is one PRB pair mapped over two slots,

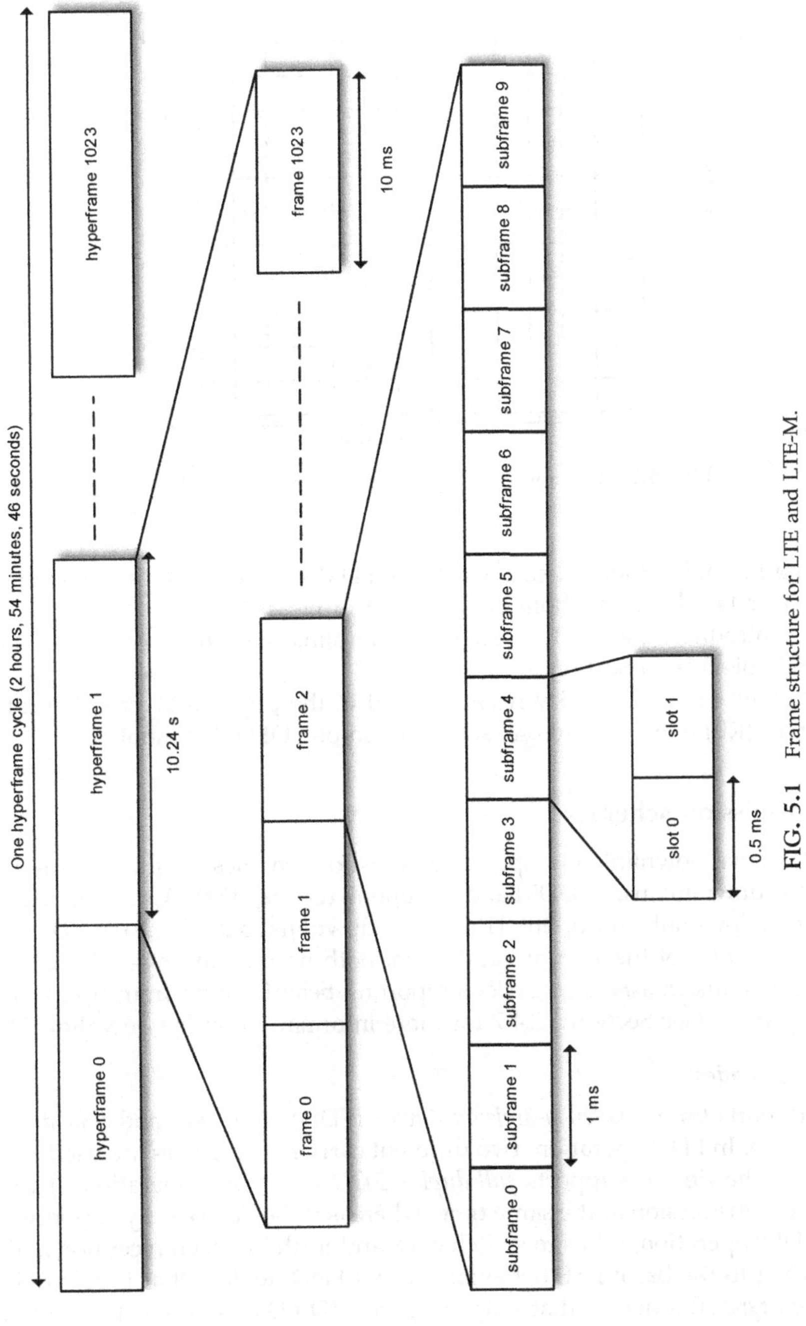

FIG. 5.1 Frame structure for LTE and LTE-M.

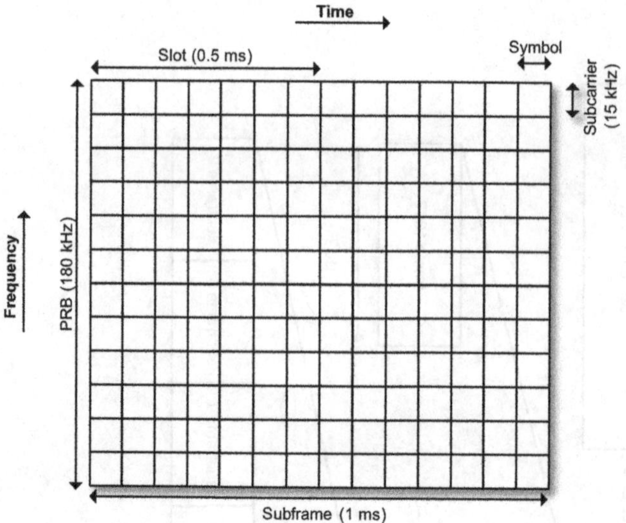

FIG. 5.2 Physical resource block (PRB) pair in LTE and LTE-M.

which for the normal CP length case (with 7 OFDM symbols per slot) corresponds to 12 subcarriers over 14 OFDM symbols as illustrated in Fig. 5.2.

Release 15 introduces sub-PRB transmission in uplink, with *resource unit* (RU) definitions according to Table 5.16 in Section 5.2.5.4.

An even smaller time-frequency resource used in the physical layer specifications is the *resource element* (RE) that refers to one subcarrier in one OFDM symbol.

5.2.2 Transmission schemes

The fundamental downlink and uplink transmission schemes are the same as in LTE. This means that the downlink uses OFDM and the uplink uses SC-FDMA, with 15 kHz subcarrier spacing in both downlink and uplink [10]. In the downlink, a *direct current (DC) subcarrier* is reserved at the center of the system bandwidth. Both normal and extended CP lengths are supported. Downlink *transmission modes* supporting beamforming from up to four antenna ports are supported (see Section 5.2.4.7 for more information on the downlink TM).

5.2.2.1 Duplex modes

LTE-M supports both *frequency-division duplex* (FDD) operation and *time-division duplex* (TDD) operation. In FDD operation, two different carrier frequencies are used for downlink and uplink. If the device supports *full-duplex FDD* (FD-FDD) operation, it can perform reception and transmission at the same time, whereas if the device only supports *half-duplex FDD* (HD-FDD) operation, it has to switch back and forth between reception and transmission. According to the basic LTE behavior for HD-FDD devices that is referred to as *HD-FDD operation type A*, a device that only supports HD-FDD is only expected to be able to do downlink reception in subframes where it does not perform uplink transmission. In

HD-FDD operation type A, the switching back and forth between reception and transmission is fast but relies on the existence of two separate local oscillators for downlink and uplink carrier frequency generation. However, to facilitate implementation of low-cost devices employing just a single local oscillator for carrier frequency generation for both downlink and uplink, *HD-FDD operation type B* was introduced [5,6], and it is HD-FDD operation type B that is used for LTE-M devices (and also for LTE device Cat-0). In HD-FDD operation type B, a guard subframe is inserted at every switch from downlink to uplink and from uplink to downlink, giving the device time to retune its carrier frequency.

In TDD operation, where the same carrier frequency is used for downlink and uplink transmission, the division of so-called *normal subframes* within a frame into downlink and uplink subframes depends on the cell-specific *UL–DL configuration* as indicated in Table 5.3. The switching from downlink to uplink takes place during a guard period within a so-called *special subframe,* indicated by "S" in the table. The symbols before the guard period are used for downlink transmission and the symbols after the guard period are used for uplink transmission. The location and length of the guard period within the special subframe is given by a cell-specific *special subframe configuration*. Interested readers can refer to Ref. [10] for more details.

LTE-M devices can be implemented with support for FD-FDD, HD-FDD operation type B, TDD, or any combination of these duplex modes. This means that LTE-M can be deployed both in paired FDD bands and unpaired TDD bands (see Table 5.2 for a list of supported bands), and that both full-duplex and half-duplex device implementations are possible, allowing for trade-off between device complexity and performance.

5.2.2.2 *Narrowband and wideband operation*

The supported LTE system bandwidths are {1.4, 3, 5, 10, 15, 20} MHz including guard bands. Discounting the guard bands, the maximum bandwidth that can be scheduled in the largest system bandwidth (20 MHz) is 100 PRBs or 18 MHz. Ordinary LTE devices support transmission and reception spanning the full system bandwidth.

TABLE 5.3 UL–DL configurations for TDD operation in LTE and LTE-M.

	Subframe number									
UL–DL configuration	**0**	**1**	**2**	**3**	**4**	**5**	**6**	**7**	**8**	**9**
0	DL	S	UL	UL	UL	DL	S	UL	UL	UL
1	DL	S	UL	UL	DL	DL	S	UL	UL	DL
2	DL	S	UL	DL	DL	DL	S	UL	DL	DL
3	DL	S	UL	UL	UL	DL	DL	DL	DL	DL
4	DL	S	UL	UL	DL	DL	DL	DL	DL	DL
5	DL	S	UL	DL	DL	DL	DL	DL	DL	DL
6	DL	S	UL	UL	UL	DL	S	UL	UL	DL

LTE-M introduces low-cost devices that are only required to support a reduced bandwidth for transmission and reception. These low-cost devices are sometimes referred to as *Bandwidth-reduced Low-complexity* (BL) devices in the standard specifications. The simplest LTE-M device was introduced in Release 13 and it supports a maximum channel bandwidth of 6 PRBs [11]. In many cases, the transmissions to or from LTE-M devices are restricted to take place within one out of a number of nonoverlapping *narrowbands* of size 6 PRBs as illustrated in Fig. 5.3 for the 15-MHz system bandwidth case.

For all system bandwidths except for the smallest one, the system bandwidth cannot be evenly divided into narrowbands which means that there are some PRBs that are not part of any narrowband. For the system bandwidths which have an odd total number of PRBs, the PRB at the center is not included in any narrowband, and if there are any remaining PRBs not included in any narrowband, they are evenly distributed at the edges of the system bandwidth, i.e., with the lowest and highest PRB indices, respectively [10]. The number of narrowbands and the PRBs not belonging to any narrowband are listed in Table 5.4.

In Release 13, the PRBs not belonging to any narrowband cannot be used for LTE-M related transmissions on the physical channels *MTC Physical Downlink Control Channel* (MPDCCH), *Physical Downlink Shared Channel* (PDSCH), and *Physical Uplink Shared Channel* (PUSCH) but can be used for LTE-M related transmissions on other physical channels/signals and for any ordinary LTE transmissions in the cell.

In Release 14, support for *larger data channel bandwidths* than 6 PRBs is introduced in order to allow transmission of larger *transport block size* (TBS) for the data channels PDSCH and PUSCH (see Section 5.2.2.3), which motivates the definition of nonoverlapping *widebands*, each one composed of up to 4 adjacent nonoverlapping narrowbands. For small system bandwidths (1.4, 3, 5 MHz) the wideband contains the whole system bandwidth. In odd system bandwidths (3, 5, 15 MHz) the central wideband contains the center PRB as well. The number of widebands for each system bandwidth are listed in the rightmost column in Table 5.4.

In Release 15, the scheduling flexibility in the frequency domain is improved also for devices configured with a maximum channel bandwidth of 6 PRBs through the introduction of a more *flexible starting PRB* for the data channels PDSCH and PUSCH. The main benefit with this increased flexibility is that it allows LTE-M data transmissions to be more

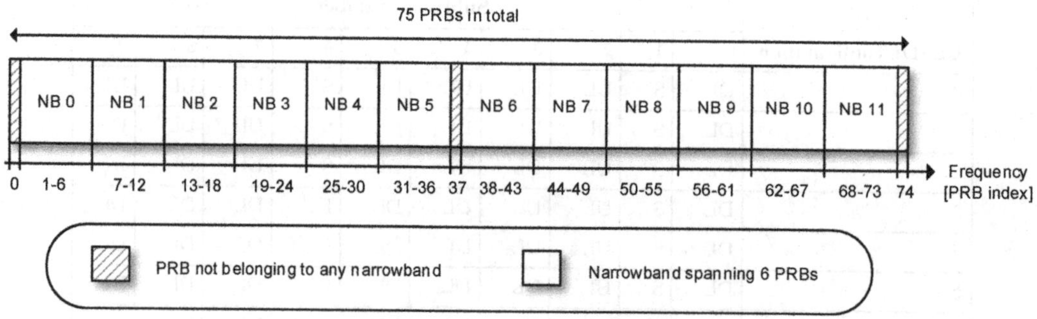

FIG. 5.3 LTE-M narrowbands in 15 MHz LTE system bandwidth.

TABLE 5.4 LTE-M narrowbands and widebands.

LTE system bandwidth including guard bands (MHz)	Total number of PRBs in system bandwidth	Number of narrowbands	PRBs not belonging to any narrowband	Number of widebands (introduced in release 14)
1.4	6	1	None	1
3	15	2	3 (1 on each edge + 1 at the center)	1
5	25	4	1 (at the center)	1
10	50	8	2 (1 on each edge)	2
15	75	12	3 (1 on each edge + 1 at the center)	3
20	100	16	4 (2 on each edge)	4

efficiently multiplexed with other LTE and LTE-M transmissions (see Section 5.3.2.1 for more detailed information). This means that resource allocations can contain almost any contiguous 6 PRBs rather than being confined within a single narrowband. So, while the control channel MPDCCH is still confined within a narrowband, the data channels are now less tied to the narrowbands than they used to be in Release 13.

The center frequency of a narrowband or wideband is not necessarily aligned with the 100-kHz channel raster of LTE (and as already mentioned, the center frequency of an LTE-M device is not necessarily aligned with the center frequency of a narrowband or wideband). However, as explained in Section 5.2.1, the signals and channels essential for cell search and basic system information (SI) acquisition, i.e., PSS, SSS, and PBCH (see Section 5.2.4), are common with LTE, and therefore still located at the center of the LTE system bandwidth (around the DC subcarrier) and aligned with the 100-kHz channel raster [9].

To ensure good frequency diversity even for devices with reduced bandwidth, frequency hopping is supported for many of the physical signals and channels (see Section 5.3.3.2 for more information on frequency hopping).

5.2.2.3 *Coverage enhancement modes*

LTE-M implements a number of coverage enhancement (CE) techniques, the most significant one being the support of repetition of most physical signals and channels. The motivation for the coverage enhancement is twofold.

First, low-cost LTE-M devices may implement various simplifications to drive down the device complexity, for example, a single-antenna receiver and a lower maximum transmission power. These simplifications are associated with some performance degradation that would result in a coverage loss compared to LTE unless it is compensated for through some coverage enhancement techniques.

Second, it is expected that some LTE-M devices will experience very challenging coverage conditions. Stationary utility metering devices mounted in basements serve as an illustrative example. This means that it may not be sufficient that LTE-M provides the same coverage as LTE, but in fact the LTE-M coverage needs to be substantially improved compared to LTE.

To address these aspects, LTE-M introduces two CE modes. *CE mode A* provides sufficient coverage enhancement to compensate for all the simplifications that can be implemented by low-cost LTE-M devices and then some additional coverage enhancement beyond normal LTE coverage. *CE mode B* goes a step further and provides the deep coverage that may be needed in more challenging coverage conditions. CE mode A is optimized for moderate coverage enhancement achieved through a small amount of repetition, whereas CE mode B is optimized for substantial coverage enhancement achieved through a large amount of repetition. If a device supports CE mode B, then it also supports CE mode A.

The low-cost LTE-M device categories (Cat-M1 and Cat-M2) have mandatory support for CE mode A and can optionally also support CE mode B. These low-cost devices always operate in one of the two CE modes. The CE modes support efficient operation of low-cost LTE-M devices, which, for example, means that resource allocation in CE mode is based on the narrowbands and widebands introduced in Section 5.2.2.2.

Higher LTE device categories (Cat-0, Cat-1, and so on) can optionally support the CE modes—either just CE mode A or both CE mode A and B. These more capable devices will typically only operate in CE mode if this is needed in order to stay in coverage, i.e., when they are outside the normal LTE coverage. When these devices are in normal LTE coverage, they will typically use normal LTE operation rather than CE mode and enjoy the higher performance available in normal LTE operation in terms of, e.g., data rates and latency.

The maximum data channel bandwidths supported by the CE modes have been increased above 6 PRBs in Release 14 as shown in Table 5.5. This helps to increase the achievable data rates which enables LTE-M to address use cases with more demanding throughput requirements. As can be seen from the table, the new LTE-M device category M2 has a maximum data channel bandwidth of 5 MHz (see Section 5.2.3), and ordinary LTE devices that support CE mode can be configured with a maximum data channel bandwidth of up to 20 MHz. In CE mode B, the maximum uplink channel bandwidth is limited to 1.4 MHz, as shown in

TABLE 5.5 Maximum channel bandwidths for LTE-M in Release 14.

CE mode data channel bandwidth capability (MHz)	Introduced in	Associated Cat-M device	CE mode A		CE mode B	
			Downlink (MHz)	Uplink (MHz)	Downlink (MHz)	Uplink (MHz)
1.4	Release 13	Cat-M1	1.4	1.4	1.4	1.4
5	Release 14	Cat-M2	5	5	5	1.4
20	Release 14	—	20	5	20	1.4

Table 5.5, since devices configured with CE mode B are expected to be so power limited that they cannot exploit a larger uplink bandwidth.

As already mentioned, an ordinary LTE device can indicate support for CE mode A, or A and B. Furthermore, the device can since Release 14 indicate support for a maximum data channel bandwidth (1.4, 5, 20 MHz) in CE mode. A device indicating support for a particular bandwidth must also support bandwidths smaller than the indicated bandwidth.

The base station decides what maximum data channel bandwidth to configure for a device. Typically, an ordinary LTE device would only be configured in CE mode if it needs the coverage enhancement provided by the CE modes. However, even for a device in good coverage, it may be beneficial to be configured in CE mode with a relatively small bandwidth to save power. Therefore, Release 14 introduces assistance signaling that allows the device to indicate to the base station that the device would prefer to be configured in CE mode with a particular maximum bandwidth, and then the base station may choose to take this information into account when configuring the device.

5.2.3 Device categories and capabilities

LTE-M defines two low-cost device categories, *Category M1* (Cat-M1) and *M2* (Cat-M2). Cat-M1 was introduced in Release 13 and Cat-M2 was introduced in Release 14. They are differentiated by the parameters listed in Table 5.6 [11]. Furthermore, note that the whole range of ordinary LTE device categories can implement support for CE mode A or B if desired (see Section 5.2.2.3).

LTE-M device category Cat-M1 is suitable for MTC applications with low data rate requirements. Many utility metering applications would fall into this category. For these applications, the data rates supported by GSM/GPRS are fully adequate, and the design goal for Cat-M1 was to achieve similar device complexity and cost as an Enhanced GPRS (EGPRS) device. Cat-M1 devices have instantaneous physical layer peak rates in downlink

TABLE 5.6 Cat-M1 and Cat-M2 physical layer parameters.

Device category	Cat-M1	Cat-M1 with extra-large uplink TBS	Cat-M2
Introduced in	Release 13	Release 14	Release 14
Maximum channel bandwidth [MHz]	1.4	1.4	5
Maximum uplink transport block size [bits]	1000	2984	6968
Maximum downlink transport block size [bits]	1000	1000	4008
Total number of soft channel bits for decoding	25344	25344	73152
Total layer 2 buffer sizes [bits]	20000	40000	100000
Half-duplex FDD operation type (see Section 5.2.2.1)	Type B	Type B	Type B

and uplink of 1 Mbps. In Release 13, taking MAC-layer scheduling delays into account, Cat-M1 devices supporting FD-FDD have MAC-layer peak rates of 800 kbps in downlink and 1 Mbps in uplink, and Cat-M1 devices supporting HD-FDD have MAC-layer peak rates of 300 kbps in downlink and 375 kbps in uplink. In Release 14, various data rate improvements are introduced, including scheduling improvements that can increase the downlink MAC-layer peak rate for Cat-M1 to 1 Mbps in FD-FDD and to 588 kbps in HD-FDD (see Section 5.3.2.1.1).

Table 5.6 also contains a column for *Cat-M1 with extra-large uplink TBS*. In downlink-heavy TDD configurations, support of a larger TBS in uplink than in downlink will help balance the downlink and uplink peak rates, and it may be possible to do this without increasing the device complexity significantly. For this reason, Release 14 introduces the possibility to support a larger maximum uplink TBS of 2984 bits instead of 1000 bits for Cat-M1. The larger uplink TBS is an optional device capability that can be supported in any duplex mode. If the new maximum TBS is used in FD-FDD, it gives Cat-M1 an uplink MAC-layer peak rate of around 3 Mbps instead of 1 Mbps.

There is a range of IoT applications with requirements on low device cost and long battery lifetime where LTE-M would be an attractive solution if the supported data rates would be even higher, closer to that of 3G devices or LTE Cat-1 devices. An important class of such applications is wearables such as smart watches. For this reason, a new LTE-M device category (Cat-M2) was introduced in Release 14, with 5 MHz transmit and receive bandwidths instead of the 1.4 MHz supported by Cat-M1. The larger bandwidth allows data transmission in downlink (on PDSCH) and uplink (on PUSCH) with a maximum channel bandwidth of 24 PRBs (a wideband) instead of just 6 PRBs (a narrowband).

A maximum TBS of 4008 bits in downlink and 6968 bits in uplink gives Cat-M2 instantaneous physical layer peak rates of ~4 Mbps in downlink and ~7 Mbps in uplink. Again, the reason for the larger maximum TBS in uplink compared to downlink is that it helps balance the downlink and uplink peak rates in some downlink-heavy TDD configurations (see e.g., UL–DL configuration #2 in Table 5.3). As mentioned earlier, increasing the uplink TBS typically has a relatively small impact on the device complexity compared to increasing the downlink TBS. Furthermore, the decoder complexity increase (in terms of number of *soft channel bits* that need to be stored) caused by the larger downlink TBS is rather moderate thanks to the use of *Limited Buffer Rate Matching* [12].

The maximum channel bandwidth for control channels (MPDCCH, SIBs, etc.) is still 6 PRBs, because there is no strong need to increase the data rates for the control channels. This means that the implementation efforts required to upgrade existing LTE-M networks to support the higher data rates of Cat-M2 will be small because most physical channels and procedures are the same as in Release 13. Note that a Cat-M2 device can operate as a Cat-M1 device in an LTE-M network that has not been upgraded because Cat-M2 is fully backward compatible with Cat-M1, and a Cat-M2 device only activates the advanced features when configured to do so by a base station.

An LTE-M device (Cat-M1, Cat-M2, or an ordinary LTE device supporting CE mode) can indicate the support of various *device capabilities* [11]. A selection of these device capabilities is listed in Table 5.7 (note that this is a simplified view, i.e. the actual RRC signaling does

TABLE 5.7 Some important LTE-M device capabilities (simplified view).

Release	Device capability	Parameter value	Section
13/14	Device power class [dBm]	14, 20, or 23	5.1.2.1
13	Coverage enhancement mode A support (mandatory for device categories M1 and M2)	YES or NO	5.2.2.3
13	Coverage enhancement mode B support	YES or NO	5.2.2.3
13	Support for downlink transmission mode 6 in CE mode A	YES or NO	5.2.4.7
13	Support for downlink transmission mode 9 in CE mode A	YES or NO	5.2.4.7
13	Support for downlink transmission mode 9 in CE mode B	YES or NO	5.2.4.7
13	Support for frequency hopping for unicast transmission	YES or NO	5.3.3.2
13	Support for basic OTDOA positioning	YES or NO	5.3.2.6
14	Support for enhanced OTDOA positioning	YES or NO	5.3.2.6
14	Support for multicast transmission of 1.4 MHz or 5 MHz	Not indicated	5.3.1.9
14	Support for extra-large uplink TBS for Cat-M1	YES or NO	5.2.3
14	Max data channel bandwidth in CE mode [MHz]	1.4, 5, or 20	5.2.2.3
14	Support for 10 downlink HARQ processes in FDD	YES or NO	5.3.2.1.1
14	Support for HARQ-ACK bundling in HD-FDD	YES or NO	5.3.2.1.1
14	Support for faster frequency retuning (number of symbols)	0, 1, or 2	5.2.4.1
14	Support for additional PUSCH repetition factors and PUSCH/PDSCH modulation scheme restriction	YES or NO	5.2.5.4
14	Support for dynamic HARQ-ACK delays	YES or NO	5.3.2.1.1
14	Support for SRS repetition	YES or NO	5.2.5.3.2
14	Support for additional PUCCH repetition factors	YES or NO	5.2.5.5
14	Support for closed-loop transmit antenna selection	YES or NO	5.2.5.4
15	Support for resynchronization signal	Not indicated	5.2.4.2.2
15	Support for wake-up signal	YES or NO	5.2.4.5
15	Minimum gap for wake-up signal in eDRX [ms]	40, 240, 1000, or 2000	5.2.4.5
15	Support for mobile-originated early data transmission	YES or NO	5.3.1.7.3
15	Support for relaxed cell reselection monitoring	YES or NO	5.3.2.5
15	Support for CRS muting	YES or NO	5.2.4.3.1
15	Support for downlink 64QAM transmission	YES or NO	5.2.4.7
15	Support for alternative CQI table	YES or NO	5.3.2.2
15	Support for HARQ feedback for uplink data	YES or NO	5.3.2.1.2
15	Support for PUSCH sub-PRB allocation	YES or NO	5.2.5.4
15	Support for flexible starting PDSCH PRB in CE mode A	YES or NO	5.3.2.1.1
15	Support for flexible starting PDSCH PRB in CE mode B	YES or NO	5.3.2.1.1
15	Support for flexible starting PUSCH PRB in CE mode A	YES or NO	5.3.2.1.2
15	Support for flexible starting PUSCH PRB in CE mode B	YES or NO	5.3.2.1.2

not necessarily look like in the table). In most cases it is optional for LTE-M devices and LTE-M networks to support these new features. If a device supports a feature, it indicates its capability to the network, and then it is up to the network whether and how to take the capability into account when configuring the device. Most of the Release 14 and 15 features listed in Table 5.1 have signaling support for capability indication (from the device to the network) as well as configuration parameters (from the network to the device).

5.2.4 Downlink physical channels and signals

LTE-M supports the set of downlink channels and signals depicted in Fig. 5.4. The physical layer provides data transport services to higher layers through the use of *transport channels* via the *Medium Access Control* (MAC) layer [13]. The *Downlink Control Information* (DCI) is not a transport channel, which is indicated by the dashed line. The MAC layer in turn provides data transport services through the use of *logical channels* that are also shown in the figure for completeness [14]. For more information on the higher layers, refer to Section 5.3.

In this section we focus on the downlink physical channels and signals. PSS, SSS, and PBCH are transmitted periodically in the center of the LTE carrier. MPDCCH and PDSCH are transmitted in a narrowband (see Section 5.2.2.2). Downlink *Reference Signals* (RS) are transmitted in all PRBs.

5.2.4.1 *Downlink subframes*

A cell-specific *subframe bitmap* can be broadcasted in the SI (see Section 5.3.1.2) to indicate which downlink subframes are valid for LTE-M transmission. The bitmap length is 10 or 40 bits corresponding to the subframes within 1 or 4 frames. A network can, for example, choose to indicate subframes that are used as *Positioning Reference Signal* (PRS) or *Multimedia Broadcast Multicast Service (MBMS) Single Frequency Network* (MBSFN) subframes as invalid for LTE-M, but this is up to the network implementation.

Fig. 5.5 shows an example with a 10-bit LTE-M subframe bitmap indicating that subframes #5 and #7 are invalid. Assume that the downlink (MPDCCH or PDSCH)

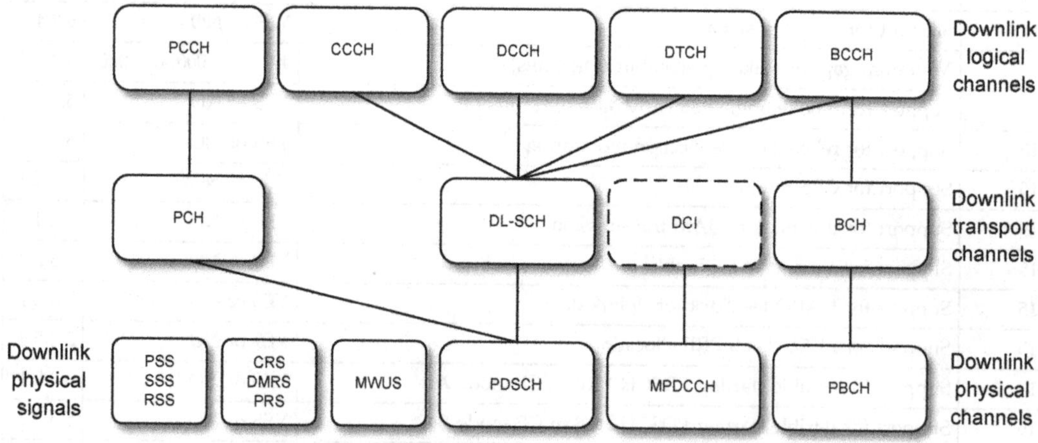

FIG. 5.4 Downlink channels and signals used in LTE-M.

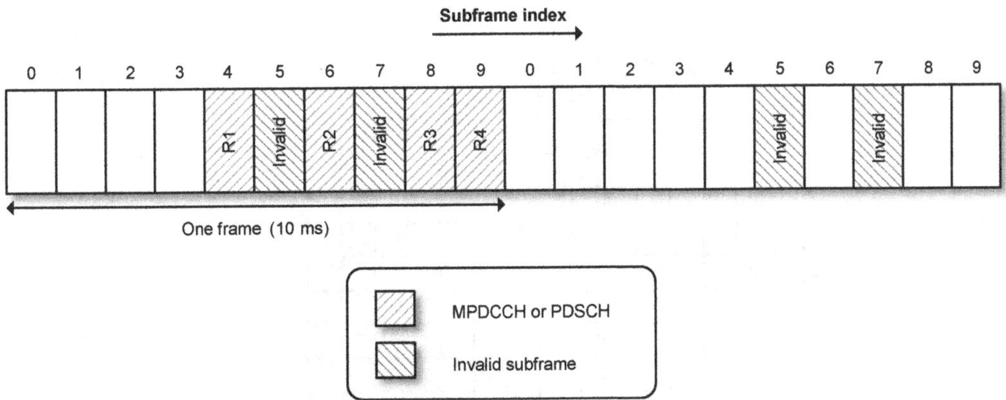

FIG. 5.5 LTE-M subframe bitmap example.

transmission that starts in subframe #4 in the first frame should use subframe repetition factor 4. If all subframes were valid, the repetitions denoted R1, R2, R3, and R4 in the figure would be mapped to subframes #4, #5, #6, and #7, respectively, but due to the invalid subframes, the repetitions are instead mapped to valid subframes #4, #6, #8, and #9, respectively.

The downlink subframe structure in LTE-M only uses a part of the downlink subframe REs in LTE. As shown in Fig. 5.6, the downlink subframe structure in LTE consists of an *LTE control region* and an *LTE data region*. The LTE control region consists of one or more OFDM symbols in the beginning of the subframe and the LTE data region consists of the remaining OFDM symbols in the subframe. In LTE, data transmissions on PDSCH are mapped to the LTE data region, whereas a number of control channels (*Physical Control Format Indicator Channel, Physical Downlink Control Channel* (PDCCH), and *Physical Hybrid Automatic Repeat Request (HARQ) Indicator Channel* (PHICH)) are mapped to the LTE control region. These control channels are all wideband channels spanning almost the whole LTE system bandwidth, which can be up to 20 MHz.

Because LTE-M devices can be implemented with a reception bandwidth as small as one narrowband, the mentioned wideband LTE control channels are not used for LTE-M. Instead, a new narrowband control channel (MPDCCH) is used for LTE-M devices and it is mapped to the LTE data region rather than the LTE control region to avoid collisions between the LTE control channels and the new LTE-M control channel. This means that in LTE-M, both the control channel (MPDCCH) and the data channel (PDSCH) are mapped to the LTE data region. (The MPDCCH shares this property with the *Enhanced Physical Downlink Control Channel* (EPDCCH) channel that was introduced in LTE Release 11, and as we will see in Section 5.2.4.6, the MPDCCH design is in fact based on the EPDCCH design).

The LTE-M *starting symbol* for MPDCCH/PDSCH transmissions is cell-specific and broadcasted in the SI (see Section 5.3.1.2). An early LTE-M starting symbol can be configured if the LTE control channel load is not expected to require an LTE control region longer than one symbol. If a larger LTE control region is deemed necessary, then a later LTE-M starting symbol should be configured to avoid collisions between LTE and LTE-M transmissions. The possible LTE-M starting symbols are the second, third, and fourth symbol in the subframe,

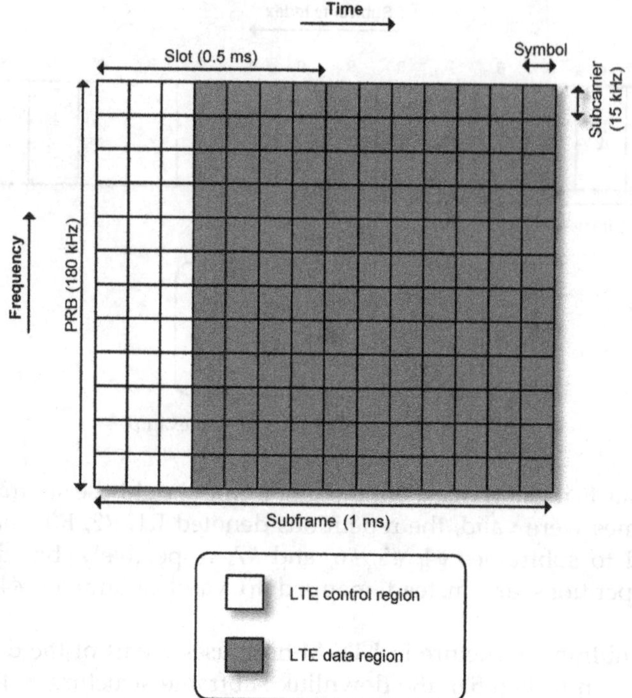

FIG. 5.6 Downlink subframe structure in LTE.

except for the smallest system bandwidth (1.4 MHz) where the possible LTE-M starting symbols are the third, fourth, and fifth symbol [15]. In the example in Fig. 5.6, the starting symbol is the fourth symbol. In the TDD case, in subframes #1 and #6, the LTE-M starting symbol is no later than the third symbol because of the position of PSS/SSS (see Section 5.2.4.2).

When an LTE-M device needs to retune from one downlink narrowband in a first subframe to another downlink narrowband in a second subframe (or from an uplink narrowband to a downlink narrowband with a different center frequency in case of TDD), the device is allowed to create a *guard period for narrowband retuning* by not receiving the first two OFDM symbols in the second subframe [10]. This means that the guard period falls partly or completely within the LTE control region and that the impact on the LTE-M transmission can be expected to be minimal. A similar retuning gap is inserted in uplink (see Section 5.2.5.1).

In Release 14, it is possible for the device to indicate that it can do faster frequency retuning (in downlink and uplink) so that the guard period can be smaller than two symbols. The device can thus indicate that it needs one symbol or even zero symbols—the latter value is mainly intended for ordinary LTE devices with CE mode capabilities, which may have no need to do retuning to move between different narrowbands when operating in CE mode (because ordinary LTE devices can receive and transmit the full LTE system bandwidth rather than just a narrowband or wideband). Faster retuning means somewhat less truncation of the transmitted signal and therefore somewhat better link performance.

5.2.4.2 *Synchronization signals*

5.2.4.2.1 PSS and SSS

Subframes in FDD	#0 and #5 for both PSS and SSS
Subframes in TDD	#1 and #6 for PSS, #0 and #5 for SSS
Subframe periodicity	5 ms for both PSS and SSS
Sequence pattern periodicity	5 ms for PSS, 10 ms for SSS
Subcarrier spacing	15 kHz
Bandwidth	62 subcarriers (not counting the DC subcarrier)
Frequency location	At the center of the LTE system bandwidth

LTE-M devices rely on LTE's *Primary Syncronization Signal* (PSS) and *Secondary Synchronization Signal* (SSS) for acquisition of a cell's carrier frequency, frame timing, CP length, duplex mode, and *Physical Cell Identity* (PCID). For more details on the cell selection procedure including the time and frequency synchronization and cell identification, see Section 5.3.1.1.

The LTE signals can be used without modification even by LTE-M devices in challenging coverage conditions. Because PSS and SSS are transmitted periodically, the device can accumulate the received signal over multiple frames to achieve sufficient acquisition performance, without the need to introduce additional repetitions on the transmit side (at the cost of increased acquisition delay).

LTE supports 504 PCIDs divided into 168 groups where each group contains 3 identities. In many cases the 3 identities correspond to 3 adjacent cells in the form of sectors served by the same base station. The 3 identities are mapped to 3 PSS sequences and one of these PSS sequences is transmitted every 5 ms in the cell, which enables the device to acquire the "half-frame" timing of the cell. For each PSS sequence there are 168 SSS sequences indicative of the PCID group. Like PSS, SSS is transmitted every 5 ms, but the 2 SSSs within every 10 ms are different. This enables the device to acquire the PCID as well as the frame timing. The same SSS sequence pattern repeats itself every 10 ms. The same PCID can be used in two or more cells as long as they are far apart enough to avoid ambiguity due to overhearing, so the number of PCIDs does not impose a limit on the total number of cells in a network.

Fig. 5.7 and 5.8 illustrate the PSS/SSS resource mapping for FDD cells and TDD cells, respectively. In the FDD case, PSS is mapped to the last OFDM symbol in slots #0 and #10 and SSS is mapped to the symbol before PSS. In the TDD case, PSS is mapped to the third OFDM symbol in subframes #1 and #6 and SSS is mapped to the symbol three symbols before PSS [10]. This means that the duplex mode (FDD or TDD) can be detected from the synchronization signals, although this is normally not needed because a given frequency band typically only supports one of the duplex modes [9]. The exact PSS/SSS symbol positions vary slightly depending on the CP length, which means that the device can also detect whether it should assume normal or extended CP length based on the detection of the synchronization signals.

As shown in the figures, PSS and SSS are mapped to the center 62 subcarriers (around the DC subcarrier) of the LTE carrier. This means that the signal fits within the smallest LTE-M

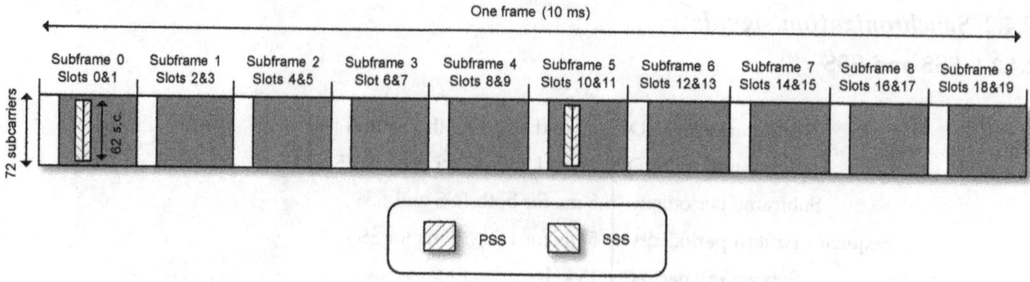

FIG. 5.7 Primary and secondary synchronization signals in LTE FDD.

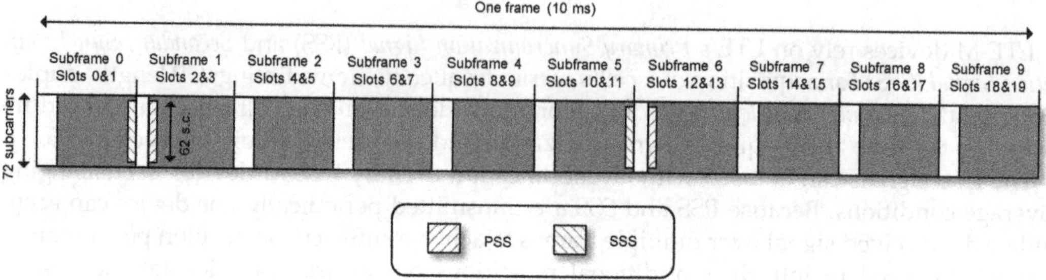

FIG. 5.8 Primary and secondary synchronization signals in LTE TDD.

device bandwidth that corresponds to 72 subcarriers. The PSS/SSS region is not aligned with any of the narrowbands (see Section 5.2.2.2) except when the smallest system bandwidth (1.4 MHz) is used, which means that the LTE-M device may need to do frequency retuning (see Section 5.2.4.1) whenever it needs to receive PSS/SSS.

5.2.4.2.2 RSS

Subframe	Configurable starting frame
Basic transmission time interval (TTI)	8, 16, 32, 40 ms
Periodicity	160, 320, 640, 1280 ms
Subcarrier spacing	15 kHz
Bandwidth	2 PRBs
Frequency location	Any 2 adjacent PRBs

The *Resynchronization Signal* (RSS) was introduced in 3GPP Release 15 for enhancing energy efficiency when a device needs to re-acquire time and frequency synchronization toward a cell. It is optional for the base station to transmit RSS and optional for the device to use it. If RSS is transmitted, it is transmitted less often than PSS/SSS, but each RSS transmission contains more energy, which can substantially help reduce the device power consumption related to reacquisition of time and frequency synchronization toward a cell. For low-mobility devices experiencing very challenging coverage conditions, it may be possible to reduce the resynchronization time from over a second using PSS/SSS to the duration of an RSS, for example 40 ms. The device still needs to rely on PSS/SSS for initial synchronization to a cell, but once the device has received SI containing an RSS configuration, the device can use RSS. The sparse nature of RSS may also allow the device to skip SFN acquisition through MIB decoding.

The RSS sequence depends on the PCID and on a *System Info Unchanged* flag. The flag allows the device to detect from RSS whether the SI has been updated in the cell, something that would otherwise require the device to monitor MIB or SIB1, which would typically be a more energy consuming operation for the device compared to receiving RSS (for more details on the flag see Section 5.3.1.2.3).

In frequency domain, RSS is mapped to 2 adjacent PRB pairs, and there is no frequency hopping. In time domain, RSS is mapped to the last 11 OFDM symbols in each subframe in the configured RSS duration. The RSS base sequence is generated using a pseudo-random sequence based on PCID that has a length of one subframe, and extended to multiple subframes by following a binary code that maps the base sequence or its conjugate to as many subframes as the RSS length. Furthermore, RSS is punctured by CRS (described in Section 5.2.4.3.1). The resulting RSS resource mapping within a subframe looks the same as the MWUS (described in Section 5.2.4.5) resource mapping illustrated in Fig. 5.14.

RSS can be used alone or in tandem with MWUS (see Section 5.2.4.5) in a wake-up receiver, something that will be discussed further in Section 5.3.1.4. The RSS bandwidth of 2 PRBs (360 kHz) is small enough to facilitate efficient wake-up receiver implementations using a relatively low sampling rate. The low complexity for RSS detection is particularly welcome in case multiple hypotheses regarding the time/frequency error need to be tested, which may be the case if the device experiences a large frequency error, e.g. as a result of a long sleep time. Additionally, the repeated structure of RSS allows for low-complexity reception due to the possibility of reusing previous outputs of a time-domain correlator.

The RSS configuration indicates a frequency location, periodicity, time offset, duration, and potential power boosting. The network can choose the RSS frequency location to be any 2 adjacent PRBs within the LTE system bandwidth. The RSS periodicity is configured to 160, 320, 640, or 1280 ms. The RSS starting frame is given by a time offset relative to SFN #0. For the lowest periodicities 160 and 320 ms, any frame can be configured as the starting frame, while for the higher periodicities 640 and 1280 ms, the allowed time offsets are restricted to every second and every fourth frame, respectively. The duration is configured to 8, 16, 32, or 40 ms. Finally, RSS can be configured with a power boosting of 0, 3, 4.8, or 6 dB. Power boosting allows RSS to be configured with a similar power level as it would have had without boosting if it had been designed to be transmitted over the full 6-PRB narrowband rather than over just 2 PRBs.

In case an RSS PRB pair overlaps in a subframe with any PRB pair carrying PSS, SSS, PBCH or a PDSCH carrying system information, then that RSS subframe is dropped (not transmitted). This behavior is needed for backwards compatibility reasons since not all devices may be aware of the presence of RSS.

5.2.4.3 *Downlink reference signals*

5.2.4.3.1 CRS

Subframes	Any
Subcarrier spacing	15 kHz
Bandwidth	Full system bandwidth (Release 15 supports CRS muting)
Frequency location	According to Fig. 5.9 in affected PRBs

Downlink reference signals (RS) are predefined signals transmitted by the base station to allow the device to estimate the downlink propagation channel to be able to demodulate the downlink physical channels [10] and perform downlink reference signal strength or quality measurements [16]. The physical layer allows demodulation and measurements to be performed even at relatively high device velocities, as discussed in Section 5.3.2.5.

The *Cell-specific Reference Signal* (CRS) can be used for demodulation of PBCH or PDSCH and is transmitted from one, two, or four logical antenna ports numbered 0–3, where in the typical case each logical antenna port corresponds to a physical antenna. The CRS for different antenna ports is mapped to REs in every PRB and in every (non-MBSFN) subframe in the cell as shown in Fig. 5.9, unless *CRS muting* is used. The CRS mapping shown in Fig. 5.9 is one example and may be frequency shifted up by one or more subcarriers depending on the PCID value.

The CRS muting feature is introduced in Release 15, and it enables the base station to turn off some of the CRS transmissions when they are not needed by Cat-M devices for demodulation or measurements. Somewhat simplified, Cat-M devices that indicate support of CRS muting can assume that the base station transmits CRS within the channel bandwidth of the device during transmission to the device and can furthermore assume that CRS is always transmitted in the central region of the system bandwidth and also over the full system bandwidth every 10th or 20th subframe (for more details, see Refs. [9,17]). This feature helps reduce the inter-cell interference experienced in neighbor cells, thereby improving the downlink throughput in neighbor cells, and may also be useful in NR coexistence scenarios (see Section 5.4).

5.2.4.3.2 DMRS

Subframe	Any
Subcarrier spacing	15 kHz
Bandwidth	Same as associated MPDCCH/PDSCH
Frequency location	According to Fig. 5.10 in affected PRBs

The downlink *Demodulation Reference Signal* (DMRS) can be used for demodulation of PDSCH or MPDCCH and is configured per device and is not PCID dependent, except in case of *MPDCCH Common Search Space* (see Section 5.3.3.1) where the *DMRS sequence initialization* is based on PCID. The DMRS is transmitted on the same logical antenna port as the associated PDSCH or MPDCCH. If the logical antenna port is mapped to multiple

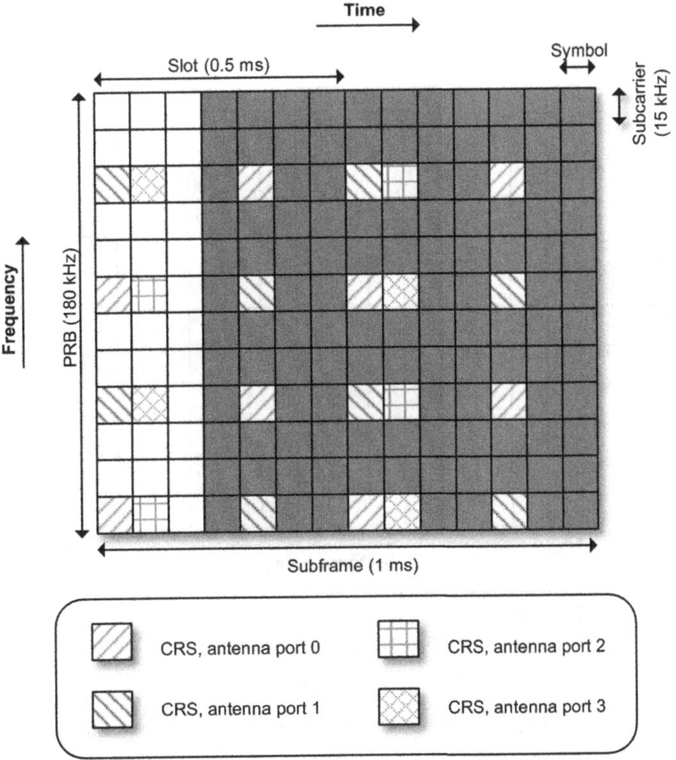

FIG. 5.9 Cell-specific CRS in LTE and LTE-M.

physical antennas, the coverage and capacity can be improved through antenna techniques such as beamforming. DMRS can be transmitted to different devices from up to 4 logical antenna ports, numbered 7–10 for PDSCH and 107–110 for MPDCCH, mapped to REs as shown in Fig. 5.10. CRS is also transmitted but not shown in Fig. 5.10. As can be seen from the figure, DMRS for antenna ports 7 and 8 is mapped to the same set of REs but separated by an *orthogonal cover code*, and the same holds for DMRS for antenna ports 9 and 10. The DMRS for the four different antenna ports can thus be distinguished by the device.

5.2.4.3.3 PRS

Subframe	Configurable by LPP signaling
Basic transmission time interval (TTI)	1 ms
Repetitions	1, 2, 4, 6 ms (Release 14 also supports 10, 20, 40, 80, 160 ms)
Periodicity	160, 320, 640, 1280 ms (Release 14 also supports 10, 20, 40, 80 ms)
Subcarrier spacing	15 kHz
Bandwidth	1.4, 3, 5, 10, 15, 20 MHz
Frequency location	At the center of the LTE system bandwidth
Frequency hopping	Between 2 or 4 locations if configured (the location in the center plus 1 or 3 narrowbands)

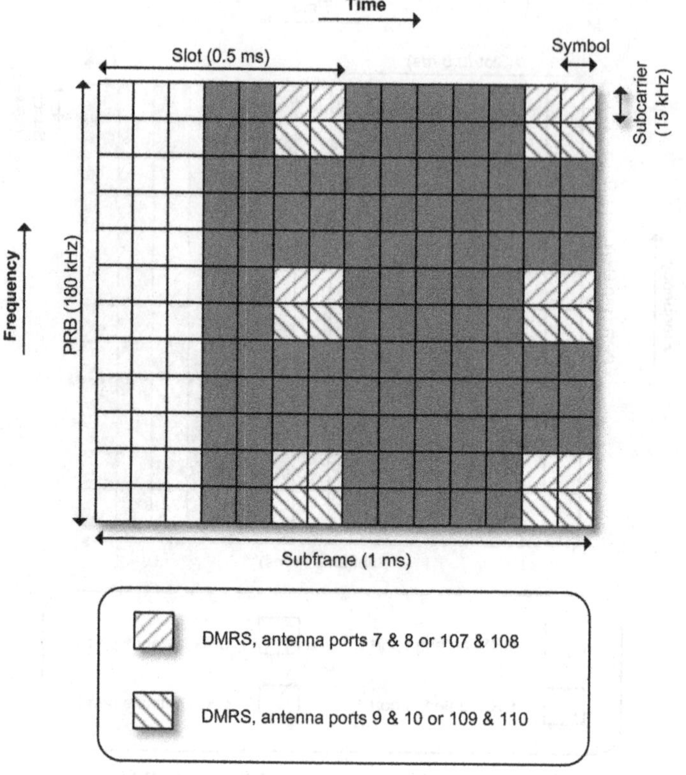

FIG. 5.10 User-specific DMRS in LTE and LTE-M.

The *Positioning Reference Signal* (PRS) is used for the *Observed Time Difference of Arrival* (OTDOA) multilateration positioning method, where the position of a receiving device is determined based on differences in time of arrival between PRS signals from different, time-synchronized base stations (see Section 5.3.2.5 for general information about the positioning methods in LTE-M).

PRS is a broadcast signal. The device receives the PRS configuration through the *LTE Positioning Protocol* (LPP) [18] from a positioning server known as the *Evolved Serving Mobile Location Center* (E-SMLC). The E-SMLC negotiates the PRS configuration with the base stations via the *LPPa* protocol [19].

Resource mapping in a PRS subframe is illustrated in Fig. 5.11. A PRS symbol sequence in a PRS subframe is a pseudo-random sequence, and each symbol is quadrature phase shift keying (QPSK) modulated. The pseudo-random sequence is cell dependent and the sequence changes in different PRS subframes. Observe that the resource mapping patterns can be shifted up or down in the frequency dimension to create 6 different,

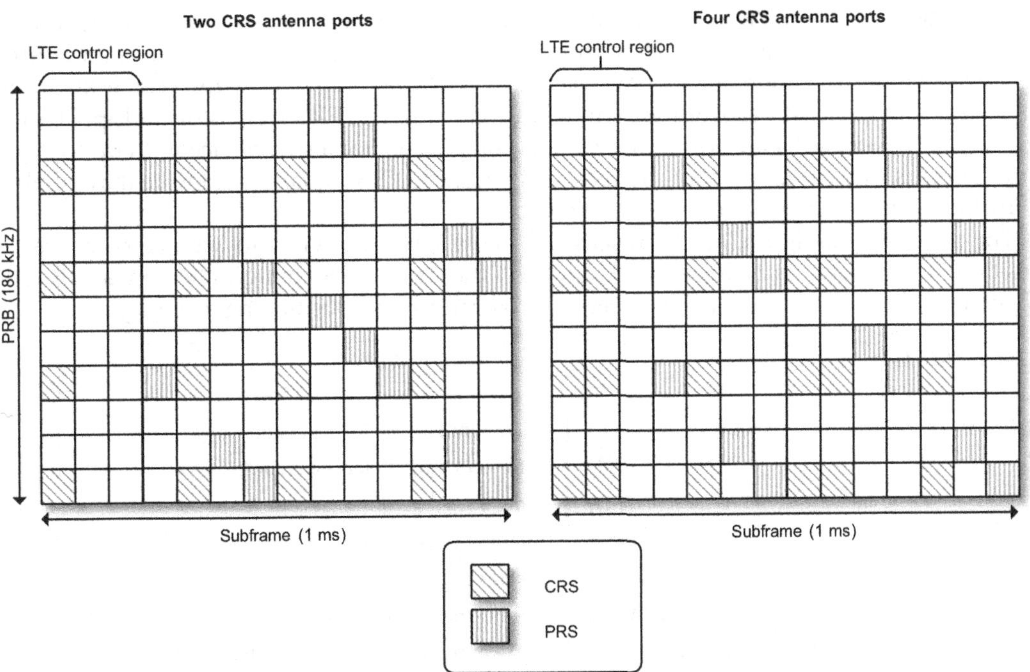

FIG. 5.11 Positioning reference signal in LTE and LTE-M.

orthogonal mapping patterns. These orthogonal mapping patterns can be used in neighboring cells in a synchronized network to avoid inter-cell interference. The mapping pattern used by a cell is determined by a configurable PRS identity, with a default value equal to the *Cell Identity* (CID).

Release 14 introduces *OTDOA enhancements* with respect to the configuration of the associated PRS the in time and frequency domains. Because the low-cost LTE-M devices have limited receive bandwidth (1.4 MHz for Cat-M1, and 5 MHz for Cat-M2), they will benefit from a PRS that is mapped over a longer duration in the time domain rather than over a wide bandwidth in the frequency domain. Therefore, Release 14 introduces the possibility to configure PRS that are transmitted with a longer duration at every PRS occasion and/or at more frequent PRS occasions (the available parameter values are listed in the table in the beginning of this section).

Also, optional PRS frequency hopping is introduced in Release 14 to provide frequency diversity gains also to narrowband LTE-M devices (configuration of PRS frequency hopping is described in Table 5.35).

Furthermore, Release 14 allows configuring a cell with multiple PRS configurations, for example a first PRS configuration with 20 MHz bandwidth but short duration for ordinary

LTE devices, a second PRS configuration with 5 MHz bandwidth and somewhat longer duration for Cat-M2 devices, and a third PRS configuration with 1.4 MHz bandwidth and even longer duration for Cat-M1 devices. A device that is able to receive all these PRS signals (or at least parts of the signals) can make use of PRS signals of multiple bandwidths, since Release 14 allows configuring not only a cell but also a device with up to three simultaneous PRS configurations. By receiving multiple PRS signals, the device may be able to further improve its positioning performance.

The new PRS configurations in Release 14 allow LTE-M devices to achieve comparable positioning accuracy as ordinary LTE devices. The exact PRS configuration in a cell will depend on the desired trade-off between PRS overhead and positioning accuracy.

5.2.4.4 *PBCH*

Subframes in FDD	#0 for core part, #9 for repetitions
Subframes in TDD	#0 for core part, #5 for repetitions
Basic transmission time interval (TTI)	40 ms
Repetitions	Core part plus 0 or 4 repetitions
Subcarrier spacing	15 kHz
Bandwidth	72 subcarriers (not counting the DC subcarrier)
Frequency location	At the center of the LTE system bandwidth

The *Physical Broadcast Channel* (PBCH) is used to deliver the MIB that provides essential information for the device to operate in the network (see Section 5.3.1.1.3 for more details on MIB).

The PBCH is mapped to the center 72 subcarriers in the LTE system bandwidth. The PBCH of LTE serves as the *PBCH core part* in LTE-M and the LTE-M specification adds additional *PBCH repetitions* for improved coverage. The repetitions can be used in all system bandwidths except the smallest one (1.4 MHz). It is up to the network whether to enable the PBCH repetitions in a cell. Enabling the repetitions is only motivated in cells that need to support deep coverage.

The TTI for PBCH is 40 ms and the *transport block size* (TBS) is 24 bits. The MIB content changes from TTI to TTI but typically in a predictable way, which makes it possible to improve the reception performance by accumulating the PBCH transmissions from two consecutive 40-ms periods (see Section 5.3.1.1.3).

A 16-bit *cyclic redundancy check* (CRC) is attached to the transport block. The CRC is masked with a bit sequence that depends on the number of CRS transmit antenna ports on the base station (see Section 5.2.4.3), which means that the device learns the number of CRS transmit antenna ports as a by-product in the process of decoding PBCH [20].

Together, the 40 bits from the 24-bit transport block and the 16-bit CRC are encoded using the *LTE tail-biting convolutional code* (TBCC), and rate matched to generate 1920 encoded bits.

The encoded bits are scrambled with a cell-specific sequence (for randomization of inter-cell interference) and segmented into four segments distributed to four consecutive frames. Each segment is 480 bits long and mapped to 240 QPSK symbols distributed over the 72 subcarriers. Transmit diversity is applied for PBCH based on *Space-Frequency Block Coding* (SFBC) in case of two antenna ports and on a combination of SFBC and *Frequency-Switched Transmit Diversity* (FSTD) in case of four antenna ports [10].

The PBCH core part is always transmitted as four OFDM symbols in subframe #0 in every frame. When the PBCH repetitions are enabled, they are transmitted in subframes #0 and #9 in the FDD case and in subframes #0 and #5 in the TDD case, as illustrated in Fig. 5.12 and 5.13. Note that the zoomed-in parts only show the first 12 of 72 subcarriers. The PBCH repetitions part contains four copies of each one of the four OFDM symbols in the PBCH core part, resulting in a repetition factor of five for each OFDM symbol. If the copied OFDM symbol contains CRS, the CRS is also copied. In the FDD case, the fact that subframes 0 and 9 are adjacent facilitates coherent combination across PBCH repetitions, for example for frequency estimation purposes. In the TDD case, the fact that subframes 0 and 5 have been selected means that PBCH repetition can be supported in all UL–DL configurations because these subframes are downlink subframes in all UL–DL configurations (see Table 5.3).

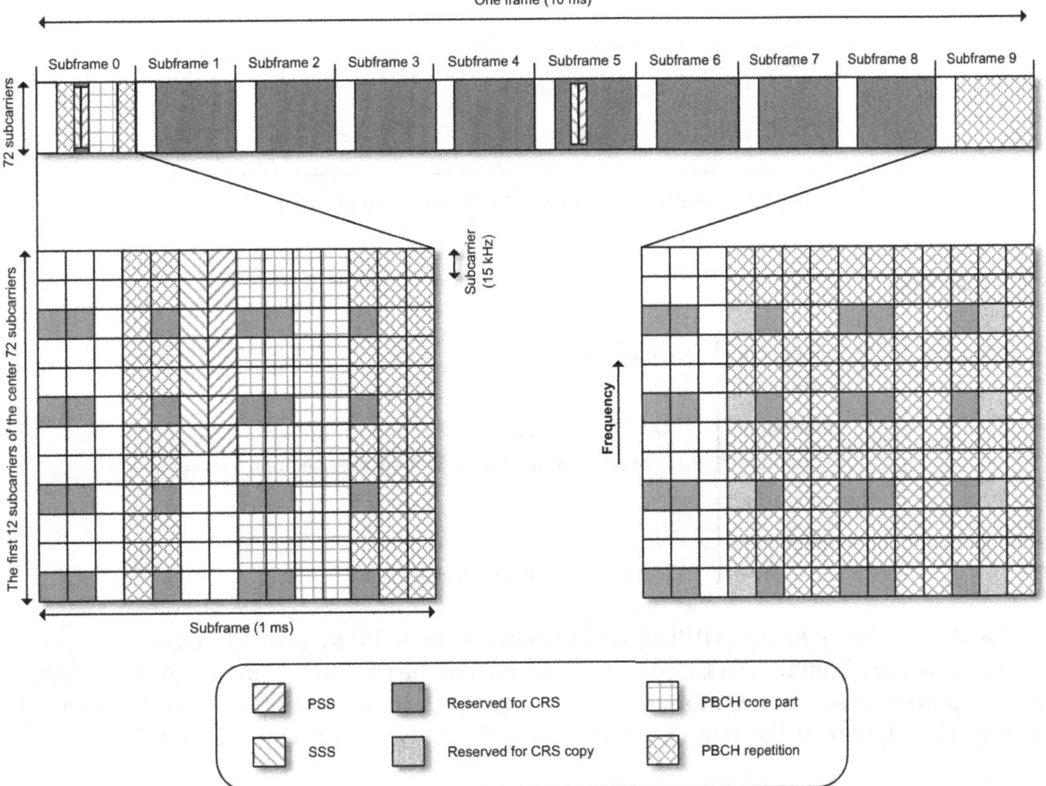

FIG. 5.12 PBCH core part and PBCH repetition in LTE FDD.

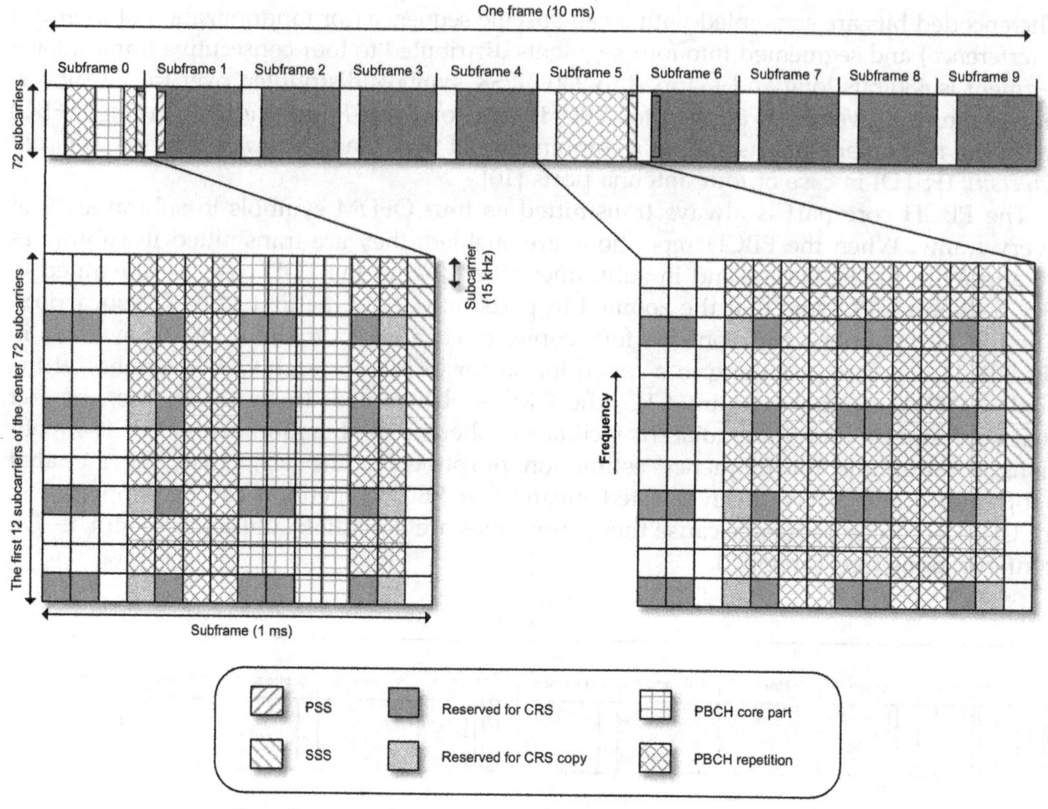

FIG. 5.13 PBCH core part and PBCH repetition in LTE TDD.

5.2.4.5 *MWUS*

Subframe	Configurable
Basic TTI	1 ms
Repetitions	$R_{max}/4$, $R_{max}/8$, $R_{max}/16$, $R_{max}/32$ R_{max}: maximum number of MPDCCH repetitions for paging
Subcarrier spacing	15 kHz
Bandwidth	2 PRBs
Frequency location	2 adjacent PRBs within a narrowband

The *MTC Wake-Up Signal* (MWUS) was introduced in 3GPP Release 15 to further improve the device battery lifetime. An LTE-M device is most of the time in idle mode, during which it wakes up periodically to monitor a *paging occasion* (PO) to determine whether it is *paged*. A detailed description of the paging procedure will be given in Section 5.3.1.4. For now, it suffices to note that a paging indicator is transmitted in MPDCCH by using DCI format 6-2, which has 10-13 information bits (see Table 5.23). As in most paging occasions no paging indicator is sent, a device usually ends up waking up to look for a paging signal only to find that no paging indicator is sent. By only providing a 1-bit wake-up indication, a much shorter MWUS can be transmitted before a device would need to wake up to look for paging to indicate whether the device needs to monitor the subsequent paging occasion(s) or can go back to sleep immediately. In the following, we focus on the MWUS physical layer aspects, but more information about the paging procedure including the use of wake-up signaling can be found in Section 5.3.1.4.

MWUS resource mapping within a subframe is shown in Fig. 5.14. In frequency domain, MWUS is mapped to 2 PRB pairs within a 6-PRB narrowband (either the 2 lowest PRB pairs, or the 2 center PRB pairs, or the 2 highest PRB pairs), and there is no frequency hopping.

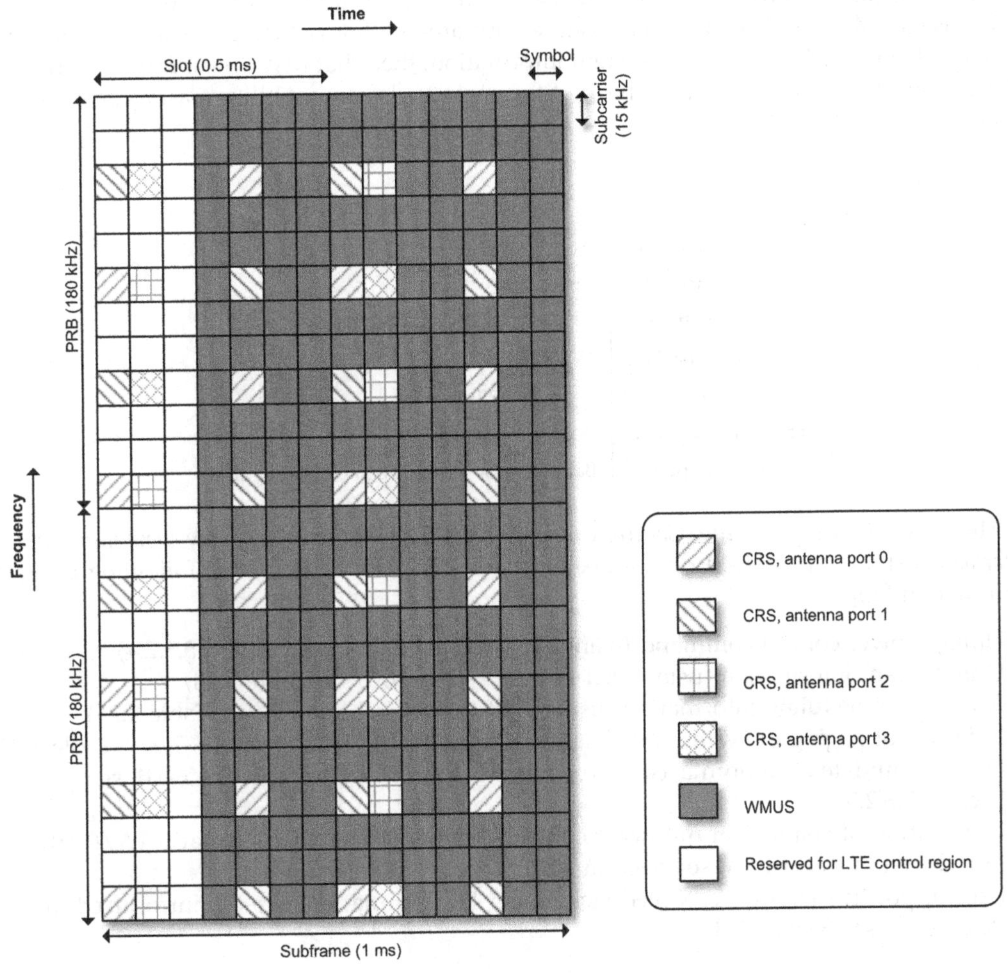

FIG. 5.14 MWUS resource mapping.

In time domain, MWUS is mapped to the last 11 OFDM symbols in a subframe, and subframe repetition is applied to reach enough coverage. The number of MWUS repetitions R_{MWUS} is configured to be 1/4, 1/8, 1/16, or 1/32 of the maximum number of MPDCCH repetitions configured for paging, $R_{\max}$ (see Sections 5.2.4.6, 5.3.1.4, and 5.3.3.1).

The same length-132 MWUS sequence is mapped to each one of the two PRB pairs. The sequence is generated based on an extended length-131 *Zadoff–Chu* (ZC) sequence, which is scrambled using a pseudo-random scrambling mask that extends over the whole MWUS transmission and varies from subframe to subframe when R_{MWUS} exceeds 1. As illustrated in Fig. 5.14, the MWUS sequence is then *punctured* by CRS.

The MWUS sequence design is the same as for the corresponding NB-IoT signal NWUS (see Section 7.2.4.8) except that the MWUS has the sequence mapped to 2 PRB pairs instead of just 1. The MWUS bandwidth of 2 PRBs (360 kHz) is small enough to facilitate efficient wake-up receiver implementations using a relatively low sampling rate compared to the sampling rate required to receive the full MPDCCH narrowband of 6 PRBs (1.08 MHz).

In case an MWUS PRB pair overlaps in a subframe with any PRB pair carrying PSS, SSS, RSS, PBCH or a PDSCH carrying system information, then that MWUS subframe is dropped (not transmitted). This behavior is needed for backwards compatibility reasons since not all devices support MWUS.

5.2.4.6 MPDCCH

Subframe	Any
Basic TTI	1 ms
Repetitions	1, 2, 4, 8, 16, 32, 64, 128, 256
Subcarrier spacing	15 kHz
Bandwidth	2, 4 or 6 PRBs
Frequency location	Within a narrowband
Frequency hopping	Between 2 or 4 narrowbands if configured

The *MTC Physical Downlink Control Channel* (MPDCCH) is used to carry *Downlink Control Information* (DCI). An LTE-M device needs to monitor MPDCCH for the following types of information [20].

- Uplink power control command (using DCI Format 3/3A, see Section 5.3.2.4)
- Uplink grant information (using DCI Format 6-0A/6-0B, see Table 5.28)
- Downlink scheduling information (using DCI Format 6-1A/6-1B, see Table 5.25)
- Indicator of paging or system information update (using DCI Format 6-2, see Table 5.23)
- Order to initiate a random access procedure (using DCI Format 6-1A/6-1B, see Section 5.3.2.3)
- Notification of changes in multicast control channel (using DCI Format 6-1A/6-1B/6-2, introduced in Release 14, see Section 5.3.1.9)
- Explicit positive HARQ-ACK feedback (using DCI Format 6-0A/6-0B, introduced in Release 15, see Section 5.3.2.1.2)

The MPDCCH design is based on the EPDCCH of LTE, which was introduced in LTE in Release 11. This means that the REs in one PRB pair are divided into 16 *enhanced resource-element groups* (EREGs) with each EREG containing 9 REs, as illustrated in Fig. 5.15, where EREG #0 is highlighted. Furthermore, EREGs can be further combined into *enhanced control channel elements* (ECCEs). In normal subframes with normal CP length, each ECCE is composed of 4 EREGs and thus 36 REs.

An MPDCCH can span 2, 4, or 6 PRB pairs, and within these PRB pairs the transmission can either be *localized* or *distributed*. Localized transmission means that each ECCE is composed of EREGs from the same PRB pair, and distributed transmission means that each ECCE is composed of EREGs from different PRB pairs [10]. Distributed transmission provides frequency diversity, but localized transmission is more suitable for beam forming or when it is desired to frequency multiplex MPDCCH with other transmissions (e.g. PDSCH) within a narrowband.

To achieve sufficient coverage, multiple ECCEs can furthermore be aggregated in an MPDCCH, according to the *ECCE aggregation level* of the MPDCCH. In normal subframes with normal CP length, aggregation of 2, 4, 8, 16, or 24 ECCEs is supported, where the highest

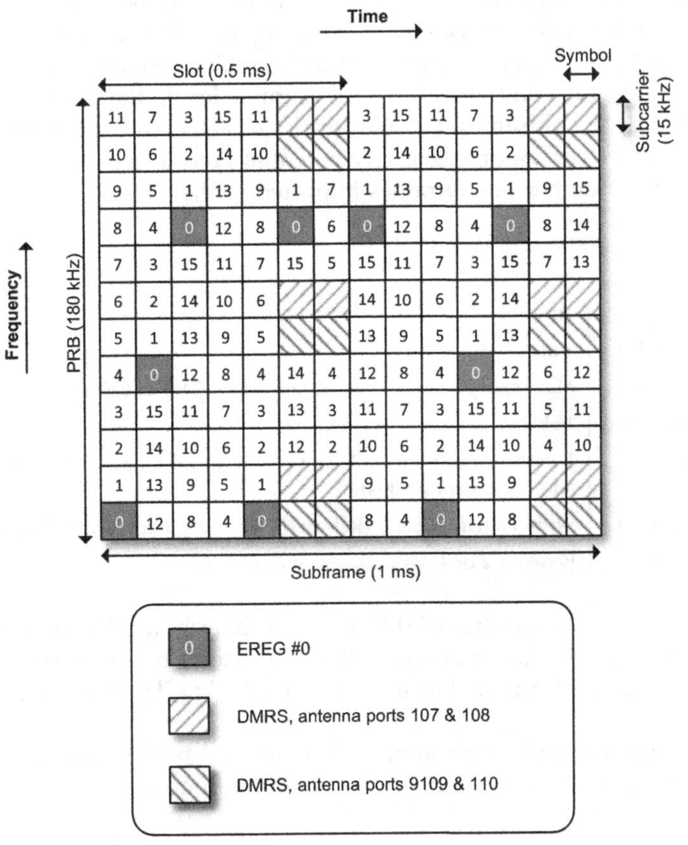

FIG. 5.15 Enhanced resource-element groups (EREGs) for MPDCCH.

aggregation level corresponds to aggregation of all REs in 6 PRB pairs. The device attempts to decode multiple MPDCCH candidates according to the *MPDCCH search space* as discussed in Section 5.3.3.1.

The MPDCCH carries the DCI and a 16-bit CRC is attached to the DCI. The CRC is masked with a sequence determined by the *Radio Network Temporary Identifier* (RNTI). The RNTI is an identifier used for addressing one or more devices and the RNTIs that can be monitored by a device are listed in Table 5.32 in Section 5.3.3.1. After the CRC attachment and RNTI masking, TBCC encoding and rate matching is used to generate a code word with a length matched to the number of encoded bits available for MPDCCH transmission. The determination of the number of available bits takes into account the MPDCCH aggregation level, modulation scheme (QPSK) and the REs not available for MPDCCH, i.e., the REs before the LTE-M starting symbol (see Section 5.2.4.1) in the subframe and the REs occupied by CRS and DMRS (see Section 5.2.4.3). The MPDCCH transmission and its associated DMRS are masked with a scrambling sequence which is cell- or user-specific depending on whether it addresses a common or dedicated RNTI (see mapping between MPDCCH search spaces, RNTIs, DCI formats, and their uses in idle and connected mode in Table 5.32 in Section 5.3.3.1).

Further MPDCCH coverage enhancement beyond what can be achieved with the highest ECCE aggregation level can be provided by repeating the subframe up to 256 times. For CE mode B, to simplify frequency error estimation and combination of the repetitions in receiver implementations that use combining on I/Q sample level, the scrambling sequence is repeated over multiple subframes (4 subframes in FDD and 10 subframes in TDD). Furthermore, the device can assume that any potential precoding matrix (for beamforming) stays the same over a number of subframes indicated in the SI (see Section 5.3.3.2).

5.2.4.7 PDSCH

Subframe	Any
Basic TTI	1 ms
Repetitions	Maximum 32 in CE mode A, maximum 2048 in CE mode B
Subcarrier spacing	15 kHz
Bandwidth	1–6 PRBs in CE mode A, 4 or 6 PRBs in CE mode B (Release 14 supports additional bandwidths)
Frequency location	Within a narrowband (Release 15 supports more flexible location)
Frequency hopping	Between 2 or 4 narrowbands if configured

The *Physical Downlink Shared Channel* (PDSCH) is primarily used to transmit unicast data. The data packet from higher layers is segmented into one or more *transport blocks* (TB), and PDSCH transmits one TB at a time. For information on PDSCH scheduling, refer to Section 5.3.2.1.1.

PDSCH is also used to broadcast information such as system information (see Section 5.3.1.2), paging messages (see Section 5.3.1.4), and random access related messages (see Section 5.3.1.6).

In this section we first go through the Release 13 functionality before we turn to the Release 14 and 15 enhancements for PDSCH. The Release 14 and 15 enhancements mainly concern unicast transmission in connected mode, while broadcast transmissions are in general still according to the Release 13 specification.

Table 5.8 shows the *modulation and coding schemes* (MCS) and TBS for PDSCH in CE mode A and B in Release 13. However, the low-cost LTE-M device Cat-M1 is restricted to a maximum TBS of 1000 bits, so the TBS values larger than 1000 bits do not apply to Cat-M1, only to higher device categories configured with CE mode A (see Section 5.2.2.3 for information on the CE modes and Section 5.2.3 for information about device categories).

TABLE 5.8 PDSCH modulation and coding schemes and transport block sizes in LTE-M (configured with max 1.4 MHz channel bandwidth and not configured with Release 14 modulation scheme restriction or Release 15 feature for 64QAM support).

MCS index	Modulation scheme	TBS index	Transport block sizes in CE mode A						Transport block sizes in CE mode B	
			# PRB pairs						# PRB pairs	
			1	2	3	4	5	6	4	6
0	QPSK	0	16	32	56	88	120	152	88	152
1	QPSK	1	24	56	88	144	176	208	144	208
2	QPSK	2	32	72	144	176	208	256	176	256
3	QPSK	3	40	104	176	208	256	328	208	328
4	QPSK	4	56	120	208	256	328	408	256	408
5	QPSK	5	72	144	224	328	424	504	328	504
6	QPSK	6	328	176	256	392	504	600	392	600
7	QPSK	7	104	224	328	472	584	712	472	712
8	QPSK	8	120	256	392	536	680	808	536	808
9	QPSK	9	136	296	456	616	776	936	616	936
10	16QAM	9	136	296	456	616	776	936	Unused	
11	16QAM	10	144	328	504	680	872	1032		
12	16QAM	11	176	376	584	776	1000	1192		
13	16QAM	12	208	440	680	904	1128	1352		
14	16QAM	13	224	488	744	1000	1256	1544		
15	16QAM	14	256	552	840	1128	1416	1736		

Further restrictions apply when PDSCH is used for broadcast. The modulation scheme is then restricted to QPSK and special TBS may apply (see Section 5.3.1).

A 24-bit CRC is attached to the TB. The channel coding is the standard *LTE turbo coding* with 1/3 code rate, 4 redundancy versions (RV), rate matching, and interleaving 12]. PDSCH is not mapped to REs before the LTE-M starting symbol (see Section 5.2.4.1) and not to REs occupied by RS (see Section 5.2.4.3).

In CE mode A in Release 13, the PDSCH is modulated with QPSK or 16QAM and mapped to between 1 and 6 PRB pairs anywhere within a narrowband. In CE mode B in Release 13, the PDSCH is modulated with QPSK and mapped to 4 or 6 PRB pairs within a narrowband. The modulation scheme restrictions facilitate low-cost LTE-M device receiver implementations with relaxed requirements on demodulation accuracy compared to ordinary LTE devices which support at least up to 64QAM.

Coverage enhancement can be provided by repeating the subframe up to 2048 times. The maximum numbers of repetitions in CE modes A and B, respectively, are configurable per cell according to Tables 5.10 and 5.11.

LTE-M supports the following PDSCH transmission modes inherited from LTE:

- **TM1**: Single-antenna transmission (supported in both CE mode A and B)
- **TM2**: Transmit diversity (supported in both CE mode A and B)
- **TM6**: Closed-loop codebook-based precoding (supported in CE mode A only)
- **TM9**: Non-codebook-based precoding (supported in both CE mode A and B)

TM2 is based on *Space-Frequency Block Coding* (SFBC) in case of two antenna ports and on a combination of SFBC and *Frequency-Switched Transmit Diversity* (FSTD) in case of four antenna ports [10]. Feedback of precoding matrix recommendations for TM6 and TM9 and other feedback (downlink channel quality indicator and HARQ feedback) are discussed in Sections 5.2.5.5 and 5.3.2.2. Because most LTE-M devices are expected to be low-cost devices with a single receive antenna, *multiple-input multiple-output* (MIMO) operation is not supported.

For PDSCH demodulation, the device uses CRS for TM1/TM2/TM6 and DMRS for TM9 (see Section 5.2.4.3). The PDSCH is masked with a scrambling sequence generated based on the PCID and the RNTI. For CE mode B, to simplify frequency error estimation and combination of the repetitions in receiver implementations that use combining on I/Q sample level, the same scrambling sequence and the same RV are repeated over multiple subframes (4 subframes in FDD and 10 subframes in TDD). Furthermore, the device can assume that any potential precoding matrix (for beamforming) stays the same over a number of subframes indicated in the SI (see Section 5.3.3.2).

Release 14 introduces the possibility to *restrict the PDSCH modulation scheme to QPSK* in connected mode. It has been observed that the combinations of TBS and modulation scheme standardized in Release 13 (shown in Table 5.8) are not always optimal when repetition is applied. The link performance can sometimes be better if the modulation scheme is restricted to QPSK when 16QAM should be selected according to the tables. When a device is configured with this feature, PDSCH transmission will be limited to QPSK whenever PDSCH is scheduled with repetition. Support of PDSCH modulation scheme restriction is bundled into a single capability indication and configuration parameter with two other related Release 14 features (PUSCH modulation scheme restriction, and new PUSCH repetition factors) described in Section 5.2.5.4.

Release 14 also introduces the possibility to use *larger maximum channel bandwidth* than 6 PRBs for PDSCH in CE mode A and B (and for PUSCH in CE mode A), as shown in Table 5.5. A device configured with 5 MHz maximum PDSCH channel bandwidth supports a maximum downlink TBS of 4008 bits (which matches the maximum downlink TBS of device category Cat-M2 and can also be configured for higher-category devices that support CE mode operation), and a device configured with 20 MHz maximum PDSCH channel bandwidth supports a maximum downlink TBS of 27376 bits. The associated resource allocation methods are summarized in Table 5.26.

Release 15 introduces *downlink 64QAM support* for PDSCH unicast transmission without repetition in CE mode A to increase the spectral efficiency. When this feature is configured, MCS and TBS are according to Table 5.9, where the MCS field has been extended from 4 bits to 5 bits (see Section 5.3.2.1.1). However, there is no intention to increase the device peak rate, so the TBS is capped by the maximum TBS for the device category. See Section 5.3.2.2 for information on *Channel Quality Information* (CQI) reporting for 64QAM support.

5.2.5 Uplink physical channels and signals

LTE-M supports the set of uplink channels depicted in Fig. 5.16. The physical layer provides data transport services to higher layers through the use of transport channels via the MAC layer [13]. The *Uplink Control Information* (UCI) is not a transport channel, which is indicated by the dashed line. The MAC layer in turn provides data transport services through the use of logical channels, which are also shown in the figure for completeness [14]. For more information on the higher layers, refer to Section 5.3.

In this section we focus on the uplink physical channels. Due to the adopted uplink transmission scheme in LTE (i.e., SC-FDMA), the transmission from a device needs to be contiguous in the frequency domain. To maximize the chances that large contiguous allocations are available for uplink data transmission on the PUSCH for LTE and LTE-M users, it is often considered beneficial to allocate the resources for *Physical Random Access Channel* (PRACH) and *Physical Uplink Control Channel* (PUCCH) near the edges of the system bandwidth. The uplink RS are not shown in Fig. 5.16 but are transmitted together with PUSCH or PUCCH or separately for sounding of the radio channel.

5.2.5.1 Uplink subframes

A cell-specific *subframe bitmap* can be broadcasted in the SI (see Section 5.3.1.2) to indicate which subframes are valid for LTE-M transmission. For FDD uplink, the bitmap length is 10 bits corresponding to the uplink subframes within 1 frame. For TDD, the bitmap length is 10 or 40 bits corresponding to the subframes within 1 or 4 frames. This bitmap could, for example, facilitate so-called dynamic TDD operation within the LTE cell. Typically, all uplink subframes are configured as valid.

When an LTE-M device needs to retune from one uplink narrowband in a first subframe to another uplink narrowband in a second subframe, the device creates a *guard period for narrowband retuning* by not transmitting two of the SC-FDMA symbols [10]. If the two

TABLE 5.9 PDSCH modulation and coding schemes and transport block sizes in LTE-M when configured with Release 15 feature for 64QAM support.

MCS index	Modulation scheme	TBS index	CE mode A # PRB pairs 1	2	3	4	5	6
0	QPSK	0	16	32	56	88	120	152
1	QPSK	1	24	56	88	144	176	208
2	QPSK	2	32	72	144	176	208	256
3	QPSK	3	40	104	176	208	256	328
4	QPSK	4	56	120	208	256	328	408
5	QPSK	5	72	144	224	328	424	504
6	QPSK	6	328	176	256	392	504	600
7	QPSK	7	104	224	328	472	584	712
8	QPSK	8	120	256	392	536	680	808
9	QPSK	9	136	296	456	616	776	936
10	16QAM	9	136	296	456	616	776	936
11	16QAM	10	144	328	504	680	872	1032
12	16QAM	11	176	376	584	776	1000	1192
13	16QAM	12	208	440	680	904	1128	1352
14	16QAM	13	224	488	744	1000	1256	1544
15	16QAM	14	256	552	840	1128	1416	1736
16	16QAM	15	280	600	904	1224	1544	1800
17	64QAM	15	280	600	904	1224	1544	1800
18	64QAM	16	328	632	968	1288	1608	1928
19	64QAM	17	336	696	1064	1416	1800	2152
20	64QAM	18	376	776	1160	1544	1992	2344
21	64QAM	19	408	840	1288	1736	2152	2600
22	64QAM	20	440	904	1384	1864	2344	2792
23	64QAM	21	488	1000	1480	1992	2472	2984
24	64QAM	22	520	1064	1608	2152	2664	3240
25	64QAM	23	552	1128	1736	2280	2856	3496
26	64QAM	24	584	1192	1800	2408	2984	3624
27	64QAM	25	616	1256	1864	2536	3112	3752
28	64QAM	26	712	1480	2216	2984	3752	4392
29	QPSK	Reserved	TBS from previous DCI for TB					
30	16QAM	Reserved	TBS from previous DCI for TB					
31	64QAM	Reserved	TBS from previous DCI for TB					

TABLE 5.10 PDSCH/PUSCH repetition factors in CE mode A.

Broadcasted maximum number repetitions for CE mode A (separately configured for PDSCH and PUSCH)	PDSCH repetition factors that can be selected from the DCI on MPDCCH	PUSCH repetition factors that can be selected from the DCI on MPDCCH	
		When Release 14 feature for new PUSCH repetition factors is NOT configured	When Release 14 feature for new PUSCH repetition factors is configured
No broadcasted value (default)	1, 2, 4, 8	1, 2, 4, 8	1, 2, 4, 8, 12, 16, 24, 32
16	1, 4, 8, 16	1, 4, 8, 16	1, 2, 4, 8, 12, 16, 24, 32
32	1, 4, 16, 32	1, 4, 16, 32	1, 2, 4, 8, 12, 16, 24, 32

TABLE 5.11 PDSCH/PUSCH repetition factors in CE mode B.

Broadcasted maximum number repetitions for CE mode B (separately configured for PDSCH and PUSCH)	PDSCH/PUSCH repetition factors that can be selected from the DCI on MPDCCH
No broadcasted value (default)	4, 8, 16, 32, 64, 128, 256, 512
192	1, 4, 8, 16, 32, 64, 128, 192
256	4, 8, 16, 32, 64, 128, 192, 256
384	4, 16, 32, 64, 128, 192, 256, 384
512	4, 16, 64, 128, 192, 256, 384, 512
768	8, 32, 128, 192, 256, 384, 512, 768
1024	4, 8, 16, 64, 128, 256, 512, 1024
1536	4, 16, 64, 256, 512, 768, 1024, 1536
2048	4, 16, 64, 128, 256, 512, 1024, 2048

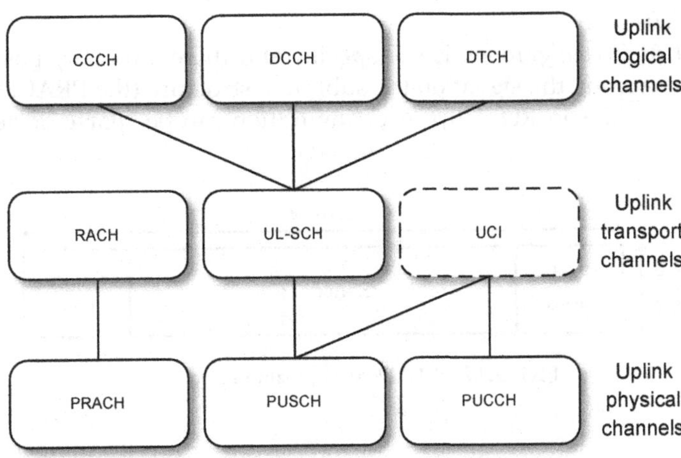

FIG. 5.16 Uplink channels used in LTE-M.

subframes both carry PUSCH or they both carry PUCCH, the guard period is created by truncating the last symbol in the first subframe and the first symbol in the second subframe. If one of the subframes carries PUSCH and the other one carries PUCCH, truncation of PUCCH is avoided by instead truncating up to two symbols of PUSCH. The rationale for this rule is that PUSCH is protected by more robust channel coding and retransmission scheme compared to PUCCH. A similar retuning gap is inserted in downlink (see Section 5.2.4.1). In Release 14, it is possible for the device to indicate that it can do faster frequency retuning (in uplink and downlink) so that the guard period can be smaller than two symbols (for details see Section 5.2.4.1).

A *shortened format* may be used for PUSCH/PUCCH to make room for *Sounding Reference Signal* (SRS) transmission in the last SC-FDMA symbol in an uplink subframe (see Section 5.2.5.3).

5.2.5.2 *PRACH*

Subframe	Any
Basic TTI	1, 2, or 3 ms
Repetitions	1, 2, 4, 8, 16, 32, 64, 128
Subcarrier spacing	1.25 kHz
Bandwidth	839 subcarriers (ca. 1.05 MHz)
Frequency location	Any
Frequency hopping	Between two frequency locations if configured

The *Physical Random Access Channel* (PRACH) is used by the device to initialize connection and allows the serving base station to estimate the time of arrival of uplink transmission. For more information on the random access procedure in idle and connected mode, see Sections 5.3.1.6 and 5.3.2.3.

The time of arrival of the received PRACH signal reflects the round-trip propagation delay between the base station and device. Fig. 5.17 shows the structure of the LTE PRACH preamble.

LTE-M reuses the LTE PRACH formats listed in Table 5.12, where T_s is the *basic LTE time unit* $1/(15{,}000 \times 2{,}048)$ s.

In LTE, the *PRACH configuration* is cell-specific and there are many possible configurations in terms of mapping the signal on the subframe structure (the PRACH configurations are listed in Section 5.7.1 in Ref. [10]). A configuration can be sparse or dense in time, as

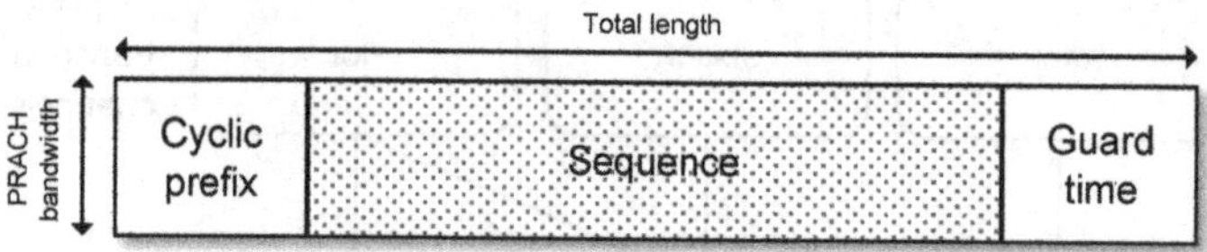

FIG. 5.17 LTE PRACH preamble structure.

TABLE 5.12 PRACH formats in LTE-M.

PRACH format	CP length	Sequence length (ms)	Total length	Cell range from guard time	FDD PRACH configurations	TDD PRACH configurations
0	3,168 $T_s \approx$ 0.10 ms	0.8	1 ms	15 km	0–15	0–19
1	21,024 $T_s \approx$ 0.68 ms	0.8	2 ms	78 km	16–31	20–29
2	6,240 $T_s \approx$ 0.20 ms	1.6	2 ms	30 km	32–47	30–39
3	21,024 $T_s \approx$ 0.68 ms	1.6	3 ms	108 km	48–63	40–47

illustrated by the examples for PRACH Format 0 in Fig. 5.18, where FDD PRACH configuration 2 uses every 20th subframe and FDD PRACH configuration 14 uses every subframe. A device can make a PRACH attempt in any PRACH opportunity using one out of the (max 64) configured PRACH preamble sequences.

LTE-M introduces PRACH coverage enhancement (CE) through up to 128 times repetition of the basic PRACH preamble structure in Fig. 5.17. The repetitions are mapped onto the PRACH subframes that are included in the PRACH configuration.

In a cell supporting CE mode B, up to 4 *PRACH CE levels* can be defined. If the cell only supports CE mode A, up to 2 PRACH CE levels can be defined. The network has several options for separating the PRACH resources that correspond to different PRACH CE levels.

- **Frequency domain**: The network can separate the PRACH resources of the PRACH CE levels by configuring different *PRACH frequencies* for different PRACH CE levels.

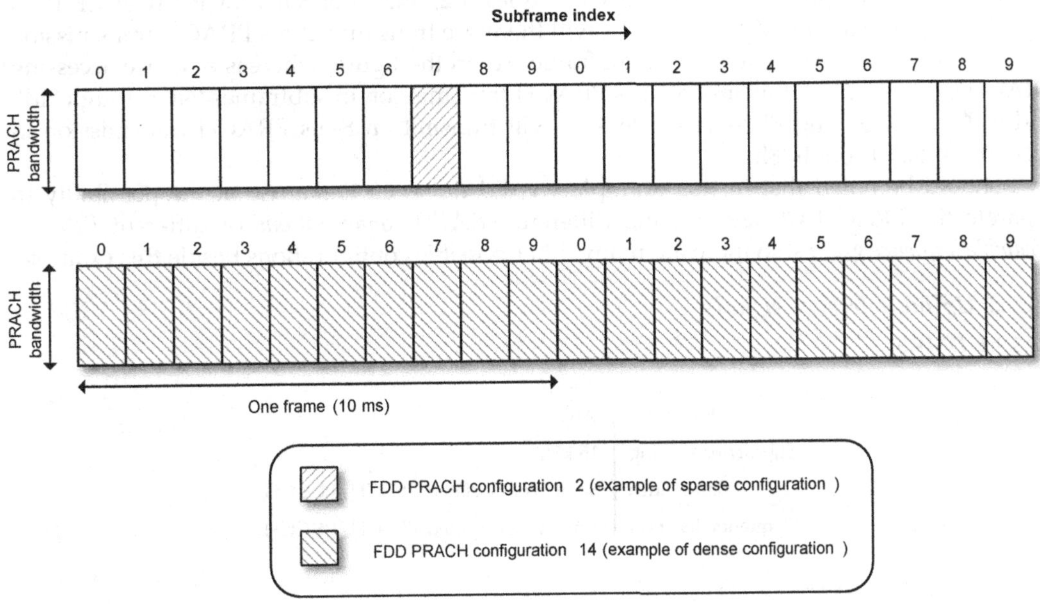

FIG. 5.18 Example PRACH configurations for FDD PRACH Format 0 in LTE and LTE-M.

- **Time domain**: The network can separate the PRACH resources of the PRACH CE levels by configuring different *PRACH configurations* and *PRACH starting subframe periodicities* for different PRACH CE levels.
- **Sequence domain**: The network can separate the PRACH resources of the PRACH CE levels by configuring nonoverlapping *PRACH preamble sequence groups* for different PRACH CE levels.

Fig. 5.19 shows a rather elaborate example of how 4 PRACH CE levels can be multiplexed in the frequency and time domains. For simplicity, this example uses relatively small repetition factors (1, 2, 4, and 8) for the 4 PRACH CE levels, and all PRACH CE levels are configured with the dense PRACH configuration 14 (see Fig. 5.18). PRACH CE level 0 is configured with its own PRACH frequency so that devices accessing PRACH CE level 0 can transmit a 1-ms PRACH transmission in any subframe. PRACH CE levels 1, 2, and 3 are configured to time share a second PRACH frequency.

The reason that the all 4 PRACH CE levels in this example seem to use both PRACH frequencies is that all PRACH CE levels have been configured to use frequency hopping, and the PRACH frequency hopping offset has been carefully selected to be 25 PRBs which is half the system bandwidth of 50 PRBs, which results in the two PRACH frequencies swapping places with each other at every frequency hop (i.e. every 2 ms, since this is the configured uplink frequency hopping interval). For more information on frequency hopping, see Section 5.3.3.2.

The time sharing between PRACH CE levels 1, 2, and 3 in this example is achieved by configuring a PRACH starting subframe periodicity of 16 ms. When a PRACH starting subframe periodicity is configured for a PRACH CE level, the PRACH opportunity in each period for that level starts after one repetition factor, meaning that is starts after 2 ms for PRACH CE level 1, after 4 ms for PRACH CE level 2, and after 8 ms for PRACH CE level 3. A device accessing PRACH CE level 1 will therefore transmit a 2-ms PRACH transmission in a subframe '2a' and the following subframe '2b' in the figure, whereas a device accessing PRACH CE level 2 will transmit a 4-ms PRACH transmission in subframes '4a' through '4d', and a device accessing PRACH CE level 3 will transmit an 8-ms PRACH transmission in subframes '8a' through '8h'.

It should be noted that in this example we did not even make use of the possibility to separate the PRACH CE levels using different *PRACH configurations* or different *PRACH preamble sequence groups*, so there are many other possible configurations beside this example.

5.2.5.3 *Uplink reference signals*

5.2.5.3.1 DMRS

Subframe	Any
Subcarrier spacing	15 kHz
Bandwidth	Same as associated PUSCH/PUCCH
Frequency location	Same as associated PUSCH/PUCCH

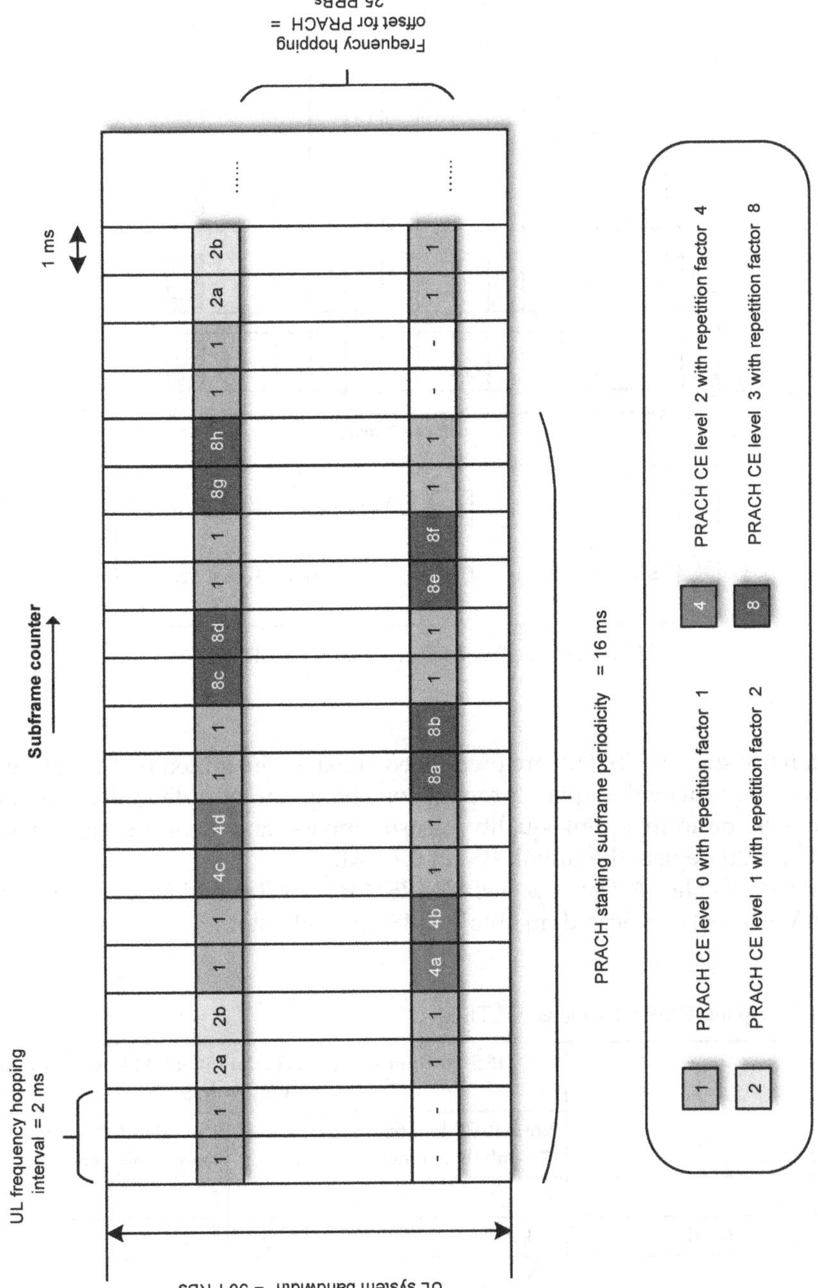

FIG. 5.19 Example of PRACH CE level multiplexing in LTE-M.

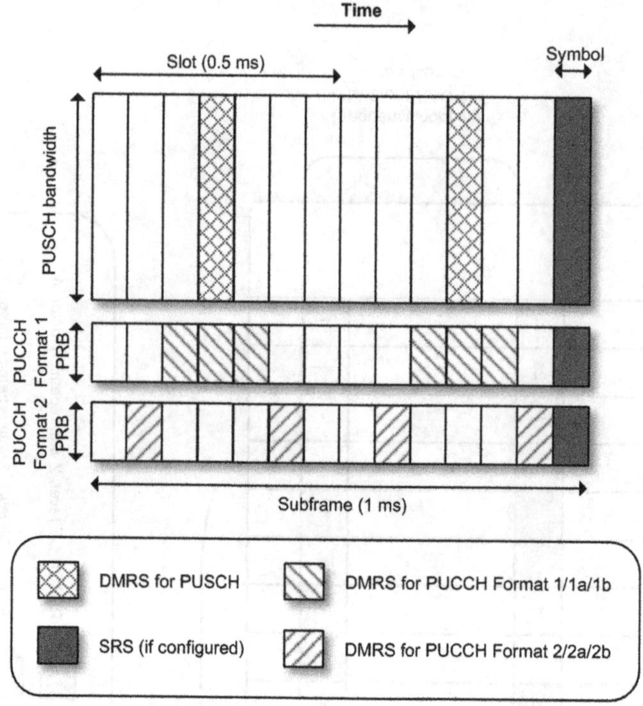

FIG. 5.20 Uplink reference signals for LTE-M.

Uplink reference signals (RS) [10] are predefined signals transmitted by the device to allow the base station to estimate the uplink propagation channel to be able to demodulate uplink physical channels, perform uplink quality measurements, and issue timing advance (TA) commands. Fig. 5.20 depicts the uplink RS for LTE-M.

The uplink *Demodulation Reference Signal* (DMRS) for PUSCH and PUCCH is transmitted in the SC-FDMA symbols indicated in Table 5.13 in each slot in the transmitted uplink

TABLE 5.13 Uplink DMRS locations in LTE-M.

	DMRS position within each slot (SC-FDMA symbol indices starting with 0)	
Physical channel	**Normal CP length (7 symbols per slot)**	**Extended CP length (6 symbols per slot)**
PUSCH	3	2
PUCCH Format 1/1a/1b	2, 3, 4	2, 3
PUCCH Format 2	1, 5	3
PUCCH Format 2a/2b	1, 5	N/A

subframe. The DMRS bandwidth is equal to the channel bandwidth of the associated PUSCH or PUCCH transmission. For PUCCH the channel bandwidth is always 1 PRB, but for PUSCH the channel bandwidth is variable (see Section 5.2.5.4 for details).

Multiplexing of multiple PUCCH in the same time-frequency resource is enabled by the possibility to generate multiple orthogonal DMRS sequences by applying a *cyclic time shift*, and in case of PUCCH Format 1/1a/1b also by applying an *orthogonal cover code* on top of the cyclic time shift.

5.2.5.3.2 SRS

Subframe	Any
Subcarrier spacing	15 kHz
Bandwidth	4 PRBs
Frequency location	Configurable

The network can reserve the last SC-FDMA symbol of some uplink subframes in a cell for *Sounding Reference Signal* (SRS) transmission for sounding of the radio channel. The device will then use shortened format for PUSCH and PUCCH in the affected subframes to make room for potential SRS transmission from itself or another device. Periodic SRS transmission can be configured through RRC configuration, whereas aperiodic SRS transmission can be triggered by setting the SRS request bit in DCI (see Tables 5.25 and 5.28). CE mode A supports both periodic and aperiodic SRS transmission. CE mode B does not support SRS transmission but will use shortened formats for PUSCH and PUCCH according to the SRS configuration in the cell to avoid collision with SRS transmissions from other devices.

All LTE-M physical channels support coverage enhancement through repetition in Release 13 but the SRS does not. Release 14 introduces the possibility to use the uplink part of the special subframe in TDD (described in Section 5.2.2.1) for transmission of SRS symbol repetitions. This can be used to improve the link adaptation in both uplink and downlink by exploiting the uplink–downlink channel reciprocity in TDD.

5.2.5.4 *PUSCH*

Subframe	Any
Basic TTI	1 ms (for full-PRB transmission) or 2/4/8 ms (for Release 15 sub-PRB transmission)
Repetitions	Maximum 32 in CE mode A, maximum 2048 in CE mode
Subcarrier spacing	15 kHz
Bandwidth	1–6 PRBs in CE mode A, 1 or 2 PRBs in CE mode B (Release 14 and 15 support additional bandwidths)
Frequency location	Within a narrowband (Release 15 supports more flexible location)
Frequency hopping	Between 2 narrowbands if configured

The *Physical Uplink Shared Channel* (PUSCH) is primarily used to transmit unicast data. The data packet from higher layers is segmented into one or more TB, and PUSCH transmits one TB at a time. For information on PUSCH scheduling, refer to Section 5.3.2.1.2.

PUSCH is also used for transmission of UCI when aperiodic *Channel State Information* (CSI) transmission (see Table 5.30) is triggered by setting the CSI request bit in DCI (see Table 5.28) or in case of collision between PUSCH and PUCCH (see Section 5.2.5.5).

In this section we first go through the Release 13 functionality before we turn to the Release 14 and 15 enhancements for PUSCH.

Table 5.14 shows the MCS and TBS for PUSCH in CE mode A and B in Release 13. However, in Release 13 the low-cost LTE-M device Cat-M1 is restricted to a maximum TBS of 1000 bits, so the TBS values larger than 1000 bits do not apply to Cat-M1, only to higher

TABLE 5.14 PUSCH modulation and coding schemes and transport block sizes in LTE-M (when not configured with Release 14 feature for larger uplink TBS and not using sub-PRB allocation introduced in Release 15).

MCS index	Modulation scheme	TBS index	Transport block sizes in CE mode A # PRB pairs: 1	2	3	4	5	6	Transport block sizes in CE mode B # PRB pairs: 1	2
0	QPSK	0	16	32	56	88	120	152	56	152
1	QPSK	1	24	56	88	144	176	208	88	208
2	QPSK	2	32	72	144	176	208	256	144	256
3	QPSK	3	40	104	176	208	256	328	176	328
4	QPSK	4	56	120	208	256	328	408	208	408
5	QPSK	5	72	144	224	328	424	504	224	504
6	QPSK	6	328	176	256	392	504	600	256	600
7	QPSK	7	104	224	328	472	584	712	328	712
8	QPSK	8	120	256	392	536	680	808	392	808
9	QPSK	9	136	296	456	616	776	936	456	936
10	QPSK	10	144	328	504	680	872	1032	504	1032
11	16QAM	10	144	328	504	680	872	1032	Unused	
12	16QAM	11	176	376	584	776	1000	1192		
13	16QAM	12	208	440	680	904	1128	1352		
14	16QAM	13	224	488	744	1000	1256	1544		
15	16QAM	14	256	552	840	1128	1416	1736		

device categories configured with CE mode A (see Section 5.2.2.3 for more information on the CE modes and Section 5.2.3 for information about device categories).

A 24-bit CRC is attached to the TB. The channel coding is the standard LTE turbo coding with 1/3 coding rate, 4 redundancy versions (RV), rate matching, and interleaving [20]. PUSCH is mapped to the SC-FDMA symbols that are not used by DMRS (see Section 5.2.5.3).

In CE mode A in Release 13, the PUSCH is modulated with QPSK or 16QAM and mapped to between 1 and 6 PRB pairs anywhere within a narrowband. In CE mode B in Release 13, the PUSCH is modulated with QPSK and mapped to 1 or 2 PRB pairs within a narrowband.

Coverage enhancement can be provided by repeating the subframe up to 2048 times. The maximum numbers of repetitions in CE modes A and B, respectively, are configurable per cell according to Tables 5.10 and 5.11. HD-FDD devices supporting CE mode B can indicate to the network that they need to insert periodic *uplink transmission gaps* in case of long PUSCH transmissions in which case the device will insert a 40-ms gap every 256 ms [10]. This gap is used by the device to correct frequency and time errors by measuring the downlink RS.

PUSCH is masked with a scrambling sequence generated based on the PCID and the RNTI. For CE mode B, to simplify frequency error estimation and combination of the repetitions in receiver implementations that use combining on I/Q sample level, the same scrambling sequence and the RV are repeated over multiple subframes (4 subframes in FDD, 5 subframes in TDD).

Release 14 introduces the possibility for a device to support a *new range of PUSCH repetition factors* in order to facilitate efficient, delay-sensitive scheduling. The set of available PUSCH repetition factors for CE mode A in Release 13 listed in Table 5.10 are powers of two, which is not considered optimal in order to support delay-sensitive services such as VoLTE in CE mode A. During a VoLTE call, a *speech frame* is produced every 20 ms. If no *speech frame bundling* is applied, it needs to be possible to transmit a transport block each direction (uplink and downlink) every 20 ms. If speech frame bundling is applied, each transport block can carry multiple (e.g., 2 or 3) speech frames, meaning that the TBs are larger but less frequent (e.g., every 40 or 60 ms). The new range is {1, 2, 4, 8, 12, 16, 24, 32}, where 12 and 24 are new values compared to Release 13. Furthermore, the whole range is immediately available through a 3-bit field in the DCI, i.e., the interpretation of the DCI field for number of PUSCH repetitions no longer depends on any higher layer parameter (the one referred to as *Broadcasted maximum number of repetitions for CE mode A* in Table 5.10).

Release 14 introduces the possibility to *restrict the PUSCH modulation scheme to QPSK*. It has been observed that the standardized combinations of TBS and modulation scheme (shown in Table 5.14) are not always optimal when repetition as applied. The link performance can sometimes be better if the modulation scheme is restricted to QPSK when 16QAM should be selected according to the tables. In uplink, QPSK may also have an additional benefit in terms of lower *Peak-to-Average Power Ratio* (PAPR) compared to 16QAM, allowing for higher transmit power. When a device is configured with this feature, PUSCH transmission will be limited to QPSK when a new 1-bit field in the DCI indicates that QPSK should be used instead of the default modulation scheme for the indicated MCS. Support of PUSCH modulation scheme restriction and the new PUSCH repetition factors mentioned above are bundled into a single capability indication and configuration parameter with another related Release 14 feature (PDSCH modulation scheme restriction) described in Section 5.2.4.7.

TABLE 5.15 PUSCH modulation and coding schemes and transport block sizes in LTE-M when configured with Release 14 feature for larger uplink TBS.

MCS index	Modulation scheme	TBS index	Transport block sizes in CE mode A # PRB pairs 1	2	3	4	5	6
0	QPSK	0	16	32	56	88	120	152
1	QPSK	2	32	72	144	176	208	256
2	QPSK	4	56	120	208	256	328	408
3	QPSK	5	72	144	224	328	424	504
4	QPSK	6	328	176	256	392	504	600
5	QPSK	8	120	256	392	536	680	808
6	QPSK	10	144	328	504	680	872	1032
7	16QAM	10	144	328	504	680	872	1032
8	16QAM	12	208	440	680	904	1128	1352
9	16QAM	14	256	552	840	1128	1416	1736
10	16QAM	16	328	632	968	1288	1608	1928
11	16QAM	17	336	696	1064	1416	1800	2152
12	16QAM	18	376	776	1160	1544	1992	2344
13	16QAM	19	408	840	1288	1736	2152	2600
14	16QAM	20	440	904	1384	1864	2344	2792
15	16QAM	21	488	1000	1480	1992	2472	2984

Release 14 also introduces support for *larger uplink TBS for Cat-M1* in CE mode A. The background is described in Section 5.2.3, and the MCS and TBS values that apply when the feature is configured are listed in Table 5.15.

Release 14 also introduces the possibility to use *larger maximum channel bandwidth* than 6 PRBs for PUSCH in CE mode A (and for PDSCH in CE mode A and B), as shown in Table 5.5. A device configured with 5 MHz maximum PUSCH channel bandwidth supports a maximum uplink TBS of 6968 bits (which matches the maximum uplink TBS of device category Cat-M2). The associated resource allocation method is briefly described in Section 5.3.2.1.2.

Release 14 also introduces support for *device transmit antenna selection* in CE mode A. For (higher-category LTE) devices that support it, the base station can indicate what antenna the device should transmit from by masking the CRC bits in DCI Format 6-0A. This can provide substantial gains in case the device is physically oriented in such a way that the two antennas experience very different radio conditions.

TABLE 5.16 PUSCH resource unit (RU) lengths in Release 15.

Resource allocation size	Modulation scheme	Resource unit (RU) length
1 or more PRBs (as in Release 13/14)	QPSK or 16QAM	1 subframe
6 subcarriers	QPSK	2 subframes
3 subcarriers	QPSK	4 subframes
2 out of 3 subcarriers	$\pi/2$-BPSK	8 subframes

Release 15 introduces *PUSCH sub-PRB allocation* in CE mode A and B, allowing a smaller PUSCH resource allocation than 1 PRB, increasing the multiplexing capacity significantly for users that are coverage limited rather than bandwidth limited. New allocation sizes are $^1/_2$ PRB (6 subcarriers) and $^1/_4$ PRB (3 subcarriers), which are used with QPSK modulation. In the latter case, a new *π/2-BPSK modulation* can also be used to produce a single subcarrier commuting between two adjacent subcarrier positions leaving the third subcarrier unused in the cell (the exact subcarrier positions also depend on whether the PCID is odd or even, to achieve some inter-cell interference randomization). The DMRS alternates between the same two subcarriers. This new modulation can achieve near 0 dB peak-to-average power ratio (PAPR), which is beneficial for uplink data coverage and for device power consumption. Rate-matching is performed over time-domain *resource units* (RU) whose lengths are defined according to Table 5.16 (the case with a resource allocation length of 1 subframe is not explicitly referred to as a resource unit in the specifications but is nevertheless included in the table for completeness). The maximum number of repetitions of the RUs is limited by the maximum total transmission time of 32 subframes in CE mode A and 2048 subframes in CE mode B. The allowed TBS values for sub-PRB allocation are listed in Table 5.17.

5.2.5.5 *PUCCH*

Subframe	Any
Basic TTI	1 ms
Repetitions in CE mode A	1, 2, 4, 8
Repetitions in CE mode B	4, 8, 16, 32 (Release 14 also supports 64 and 128)
Subcarrier spacing	15 kHz
Bandwidth	1 PRB
Frequency location	Any PRB
Frequency hopping	Between 2 PRB locations

The *Physical Uplink Control Channel* (PUCCH) is used to carry the following types of *Uplink Control Information* (UCI).

TABLE 5.17 PUSCH modulation and coding schemes and transport block sizes in LTE-M when using sub-PRB allocation introduced in Release 15.

MCS index	Modulation scheme	Transport block sizes				
		CE mode A			CE mode B	
		1 RU	2 RUs	4 RUs	2 RUs	4 RUs
0	QPSK or $\pi/2$-BPSK	32	88	328	88	328
1	QPSK or $\pi/2$-BPSK	56	144	408	144	408
2	QPSK or $\pi/2$-BPSK	72	176	504	176	504
3	QPSK or $\pi/2$-BPSK	104	208	600	208	600
4	QPSK or $\pi/2$-BPSK	120	224	712	224	712
5	QPSK or $\pi/2$-BPSK	144	256	808	256	808
6	QPSK or $\pi/2$-BPSK	176	328	936	328	936
7	QPSK or $\pi/2$-BPSK	224	392	1000	392	1000

- Uplink scheduling request (SR)
- Downlink HARQ feedback (ACK or NACK)
- Downlink channel state information (CSI)

A PUCCH transmission is mapped to a configurable *PUCCH region* that consists of two PRB locations with equal distance to the center frequency of the LTE system bandwidth, typically chosen to be close to the edges of the system bandwidth. Inter-subframe (not intra-subframe) frequency hopping takes place between the two PRB locations in the PUCCH region (see Fig. 5.25 and 5.26 and Section 5.3.3.2). PUCCH is mapped to the SC-FDMA symbols that are not used by DMRS (see Section 5.2.5.3). Coverage enhancement can be provided by repeating the subframe up to 32 times in Release 13 (and up to 128 times in Release 14).

The PUCCH formats supported by LTE-M are listed in Table 5.18. If a device in connected mode has a valid periodic PUCCH resource for *Scheduling Request* (SR), it can use it to request an uplink grant when needed, otherwise it has to rely on the random access procedure for this. A PUCCH resource for *HARQ feedback (ACK or NACK)* is allocated when a downlink data (PDSCH) transmission is scheduled, and since Release 14 the HARQ feedback for up to 4 downlink TB can be bundled into a single ACK or NACK using the *HARQ-ACK bundling* feature (see Section 5.3.2.1.1). *Channel State Information* (CSI) reporting is described in Section 5.3.2.2.

Typically, PUCCH Format 1/1a/1b and PUCCH Format 2/2a/2b would be mapped to different PUCCH regions, meaning at least 2 PUCCH regions corresponding to at least $2+2=4$ PRB locations. However, if the system bandwidth is small, one possibility is to map PUCCH Format 1/1a/1b and PUCCH Format 2/2a/2b so that they are both allocated to a PRB at a system bandwidth edge, both of them frequency hopping between the two

TABLE 5.18 PUCCH formats in LTE-M.

PUCCH format	Description	Modulation scheme	Comment
1	Scheduling request	On-off keying (OOK)	Supported in CE mode A and B
1a	1-bit HARQ feedback	BPSK	Supported in CE mode A and B
1b	2-bit HARQ feedback for TDD	QPSK	Only supported in CE mode A
2	10-bit CSI report	QPSK	Only supported in CE mode A
2a	10-bit CSI report + 1-bit HARQ feedback	QPSK + BPSK	Only supported in CE mode A
2b	10-bit CSI report + 2-bit HARQ feedback in TDD	QPSK + QPSK	Only supported in CE mode A

edges at the same time, but never mapped to the same edge at the same time. In this way, fewer PRB location are consumed by PUCCH and more PRB locations are available for uplink data (PUSCH) transmission.

It is in principle possible to multiplex up to 36 PUCCH Format 1/1a/1b in the same time-frequency resource by applying different *cyclic time shifts* and *orthogonal cover codes*, and up to 12 PUCCH Format 2/2a/2b in the same time-frequency resource by applying different cyclic time shifts (see Section 5.2.5.3). However, usually only half of these cyclic time shifts are considered useable in practical radio propagation conditions, i.e. 18 for PUCCH Format 1/1a/1b and 6 for PUCCH Format 2/2a/2b.

The PUCCH resources for SR and CSI are semi-statically configured via user-specific RRC signaling. The allocation of a PUCCH resource for HARQ feedback is more complicated since it would be too costly to allocate it semi-statically per user. The PUCCH resource for HARQ feedback is determined based on both semi-static and dynamic information, including a semi-static PUCCH resource starting offset configured per MPDCCH PRB set, a dynamic offset within the MPDCCH PRB set derived from the ECCE index for the first ECCE used by the MPDCCH (see Section 5.2.4.6) carrying the DCI scheduling the downlink data (PDSCH) transmission, another dynamic offset (0, -1, -2, or +2) indicated by a 2-bit field in the DCI (see Section 5.3.2.1.1) allowing the base station to avoid collision between PUCCH transmissions from different users, and additional offsets in case of localized MPDCCH transmission or TDD (for details see Section 10.1 in Ref. [15]).

The number of PUCCH repetitions to use in connected mode is configured per device. In CE mode A, the possible PUCCH repetition numbers are {1, 2, 4, 8} and different repetition numbers can be configured for PUCCH Format 1/1a/1b and PUCCH Format 2/2a/2b. In CE mode B, the possible repetition numbers for PUCCH Format 1/1a/1b are {4, 8, 16, 32} in Release 13 and {4, 8, 16, 32, 64, 128} in Release 14. Similar ranges apply for the PUCCH carrying HARQ feedback for the PDSCH carrying Message 4 during the random access procedure described in Section 5.3.1.6, but the values are broadcasted in System Information Block 2 (SIB2).

Simultaneous transmission of more than one PUSCH or PUCCH transmission from the same device is not supported. If a device is scheduled to transmit both PUSCH and UCI in a subframe, and both PUSCH and PUCCH are without repetition, then the UCI is multiplexed into the PUSCH according to ordinary LTE behavior, but if PUSCH or PUCCH is with repetition then PUSCH is dropped in that subframe. If a device is scheduled to transmit two or more of HARQ feedback, SR, and periodic CSI in a subframe, and PUCCH is without repetition, then ordinary LTE behavior applies, but if PUCCH is with repetition then only the highest priority UCI is transmitted, where HARQ feedback has the highest priority and periodic CSI has the lowest priority.

5.3 Idle and connected mode procedures

In this section, we describe LTE-M physical layer procedures and higher layer protocols, including all activities from initial cell selection to setting up and controlling a connection. This section uses physical layer related terminology introduced in Section 5.2.

The idle mode procedures include the initial cell selection, system information (SI) acquisition, cell reselection, paging, and multicast procedures. The transition from idle to connected mode involves the procedures for random access, and access control. The connected mode operation includes procedures for scheduling, retransmission, power control, mobility support, and positioning. Idle mode procedures and connected mode procedures are treated in Sections 5.3.1 and 5.3.2, respectively. Additional physical layer procedures common for idle and connected mode are treated in Section 5.3.3.

The LTE-M radio protocol stack is inherited from LTE [13] and is described in Section 2.2. The main changes are in the physical layer, but there are also changes to the higher layers. Changes to the control plane are mainly covered in Section 5.3.1 and changes to the user plane are mainly covered in Section 5.3.2. The mappings between physical channels, transport channels, and logical channels are illustrated in Fig. 5.4 and 5.16.

5.3.1 Idle mode procedures

The first idle mode procedure that the device needs to carry out is cell selection. Once a cell has been selected, most of the interaction between the device and the base station relies on transmissions addressed by the base station to the device using a 16-bit *Radio Network Temporary Identifier* (RNTI) [14]. The RNTIs monitored by LTE-M devices in idle mode are listed in Table 5.32 in Section 5.3.3.1 together with references to the relevant book sections.

5.3.1.1 Cell selection

The main purpose of cell selection is to identify, synchronize to, and determine the suitability of a cell. The general steps in the LTE-M cell selection procedure (which to a large extent follows the LTE cell selection procedure) are as follows:

1. Search for the PSS to identify the presence of an LTE cell and to synchronize in time and frequency to the LTE carrier frequency and half-frame timing.

2. Synchronize to the SSS to identify the frame timing, PCID, CP length, and duplex mode (FDD or TDD).
3. Acquire the MIB to identify the SFN, the downlink system bandwidth, the scheduling information for the LTE-M-specific SIB1, and (since Release 15) a flag bit for unchanged SI.
4. Acquire the SIB1 to identify, for example, the H-SFN, public land mobile network (PLMN) identity, tracking area, unique cell identity, UL–DL configuration (in case of TDD), and scheduling information for other SI messages and to prepare for verification of the cell suitability.

These procedures are in detail described in the next few sections.

5.3.1.1.1 Time and frequency synchronization

The *initial cell selection* procedure aims to time-synchronize to PSS and to obtain a *carrier frequency error* estimation. As shown in Fig. 5.7 and 5.8, PSS is transmitted every 5 ms at the center 62 subcarriers of the downlink carrier. The device can assume that the allowed carrier frequencies are aligned with a 100-kHz channel raster [9], i.e., the carrier frequencies to search for are multiples of 100 kHz. The initial oscillator inaccuracy for a low-cost device may be as high as 20 ppm (parts per million), corresponding to, for example, 18 kHz initial carrier frequency error for a 900-MHz band. This means that there is relatively large uncertainty both regarding time and frequency during initial cell selection, which means that time and frequency synchronization at initial cell selection can take significantly longer than at *non-initial cell selection* or *cell reselection* (cell reselection is described in Section 5.3.1.3).

By time synchronizing to PSS the device detects the 5-ms (half-frame) timing. PSS synchronization can be achieved by correlating the received signal with the three predefined PSS sequences. Time and frequency synchronization can be performed in a joint step using subsequent PSS transmissions. For further details on the synchronization signals, refer to Section 5.2.4.2.

Release 15 introduces a possibility for *EARFCN pre-provisioning,* where the device is provided with an EARFCN list (see Section 5.2.1.1) and the geographical area where each EARFCN applies, which can help speed up the initial cell acquisition [21]. The information can be stored in SIM [22]. It is up to the device implementation whether and how to use this information. This feature is introduced for both LTE-M and NB-IoT.

Release 15 furthermore introduces the *Resynchronization Signal* (RSS) which cannot be used for initial synchronization to a cell but for maintaining or re-acquiring time and frequency synchronization toward the cell if needed (for details, see Section 5.2.4.2.2).

5.3.1.1.2 Cell identification and initial frame synchronization

Like the PSS, the SSS is transmitted every 5 ms at the center 62 subcarriers of the downlink carrier. As discussed in Section 5.2.4.2, the SSS can be used to acquire the frame timing, the PSS and SSS sequences together can be used to identify the cell's physical cell identity (PCID), and the relative positions of the PSS and SSS transmissions within a frame can be used to detect the duplex mode (FDD or TDD) and the CP length (normal or extended).

5.3.1.1.3 MIB acquisition

After acquiring the PCID, the device knows the CRS placement within a resource block as the subcarriers that CRS REs are mapped to are determined by PCID. It can thus demodulate and decode PBCH, which carries the *Master Information Block* (MIB). For further details on PBCH, refer to Section 5.2.4.4. One of the information elements carried in the MIB is the 8 most significant bits of the SFN. Because the SFN is 10 bits long, the 8 MSBs of SFN change every 4 frames, i.e., every 40 ms. As a result, the TTI of PBCH is 40 ms. A MIB is encoded to a PBCH code block, consisting of 4 code subblocks. PBCH is transmitted in subframe 0 in every frame, and each PBCH subframe carries a code subblock. A code subblock can be repeated as explained in Section 5.2.4.4 for enhanced coverage. Initially the device does not know which subblock is transmitted in a specific frame. Therefore, the device needs to form four hypotheses to decode a MIB during the initial cell selection process. This is referred to as blind decoding. In addition, to correctly decode the MIB CRC the device needs to hypothesize whether 1, 2, or 4 antenna ports are used for CRS transmission at the base station. A successful MIB decoding is indicated by having a correct CRC. At that point, the device has acquired the information listed below:

- Number of antenna ports for CRS transmission
- System frame number (SFN)
- Downlink system bandwidth
- Scheduling information for the LTE-M-specific SIB1
- Flag bit for unchanged system information (introduced in Release 15, see Section 5.3.1.2.3)

Typically, the system bandwidth is the same in downlink and uplink but in principle they can be different. The downlink system bandwidth is indicated in MIB, whereas the uplink system bandwidth is indicated in SIB2, which is described in Section 5.3.1.2.2. Table 5.4 lists the supported system bandwidths.

The presence of the scheduling information for the LTE-M-specific SIB1 is an indication that LTE-M is supported by the cell. Hereafter the LTE-M-specific SIB1 will be referred to as SIB1 for short. (However, in the standard specifications it is referred to as *SIB1-BR*, where BR stands for *bandwidth-reduced*.)

The SFN field in MIB changes every 40 ms but most other MIB fields do not change very often, if ever. Therefore, the MIB content changes from TTI to TTI, but typically in a predictable way. This knowledge can be exploited to improve the reception performance by accumulating the PBCH transmissions from two or more consecutive 40-ms periods, and therefore Release 15 introduces more stringent reading delay requirements for the *Cell Global Identity* (CGI) reading based on PBCH accumulation across two 40-ms periods (see Section 5.3.1.2.1).

5.3.1.1.4 CID and H-SFN acquisition

After acquiring the MIB including the scheduling information about SIB1 the device is able to locate and decode SIB1. We will describe more about how device acquires SIB1 in Section 5.3.1.2.1. From a cell search perspective, it is important to know that the SIB1 carries the hyper

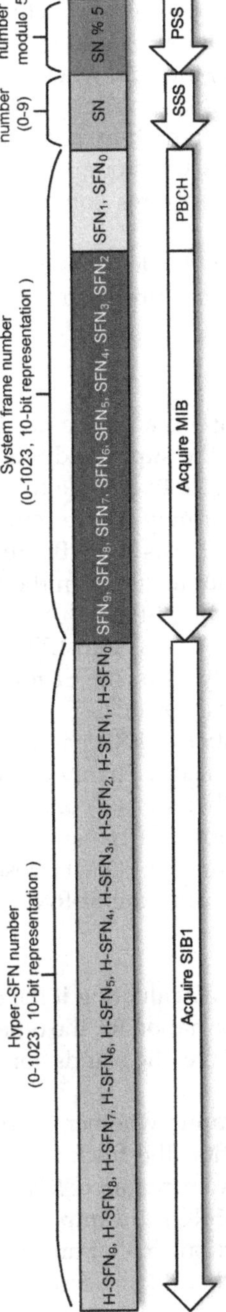

FIG. 5.21 Illustration of how the LTE-M device acquires complete timing information during the initial cell search.

system frame number (H-SFN), the public land mobile network (PLMN) identity, tracking area, and cell identity (CID). Unlike the PCID, the CID is unique within the PLMN. After acquiring SIB1, the device has achieved complete synchronization to the framing structure shown in Fig. 5.1. Based on the information provided in SIB1, the device will be able to determine whether the cell is suitable for camping, and whether the device can attempt to attach to the network. Somewhat simplified, a cell is considered suitable if the PLMN is available, the cell is not barred, and the cell's signal strength exceeds a minimum requirement. There are separate signal strength requirements for CE mode A and CE mode B.

Fig. 5.21 illustrates how the device acquires complete framing information during the initial cell search procedure.

After completing the initial cell search, the device is expected to have a time and frequency accuracy that offers robust operation in subsequent transmission and reception during connected and idle mode operations.

5.3.1.2 *System Information acquisition*

After selecting a suitable cell to camp on a device needs to acquire the full set of *system information* (SI). Table 5.19 summarizes the supported SIB types [13]. SIB1 and SIB2 contain the most critical SI, required by the device to be able to select the cell for camping and for accessing the cell. The other SIBs may or may not be transmitted in a given cell depending on the network configuration. LTE-M inherits the SIBs from LTE. LTE SIBs related to functionality not supported by LTE-M are not included in the table. Interested readers can refer to Ref. [23] for additional details regarding SIBs.

Although LTE-M reuses the SIB definitions from LTE, it should be noted that the SI for LTE-M devices is transmitted separately from the SI for ordinary LTE devices. The main reason for this is that the SI transmissions in LTE are scheduled using LTE's PDCCH, and may be scheduled to use any number of PRB pairs for PDSCH, which together means SI transmission occupies in general a too large channel bandwidth to be received by LTE-M devices with reduced receive bandwidth, and as a consequence LTE-M devices are not able to receive LTE's ordinary SI transmissions. SI transmissions for LTE-M devices are transmitted without an associated PDCCH on the PDSCH described in Section 5.2.4.7, masked with a scrambling sequence generated using the SI-RNTI defined in the standard [14].

5.3.1.2.1 System Information Block 1

SIB1 carries information relevant when evaluating if the device is allowed to camp on and access the cell, as well as scheduling information for the other SIBs. SIB1 for LTE-M is referred to as *SIB1-BR* in the specifications, where BR stands for *bandwidth-reduced,* but it is here referred to as SIB1 for brevity.

The information relevant when evaluating whether the device is allowed to access the cell includes, for example, the PLMN identity, the tracking area identity, cell barring and cell reservation information, and the minimum required *Reference Signal Received Power* (RSRP) and *Reference Signal Received Quality* (RSRQ) to camp on and access the cell. Cell selection and cell reselection are described in Sections 5.3.1.1 and 5.3.1.3.

The scheduling information for the other SIBs is described in Section 5.3.1.2.2. In addition, SIB1 contains information critical for scheduling of downlink transmissions in general.

TABLE 5.19 SIB types in LTE-M.

SIB	Content
SIB1	Information relevant when evaluating if a device is allowed to access a cell and scheduling information for other SIBs (see Section 5.3.1.2.1)
SIB2	Configuration information for common and shared channels
SIB3	Cell reselection information, mainly related to the serving cell
SIB4	Information about the serving frequency and intrafrequency neighboring cells relevant for cell reselection
SIB5	Information about other frequencies and interfrequency neighboring cells relevant for cell reselection
SIB6	Information about UMTS 3G frequencies and neighboring cells relevant for cell reselection
SIB7	Information about GSM frequencies cells relevant for cell reselection
SIB8	Information about CDMA2000 3G frequencies and neighboring cells relevant for cell reselection
SIB9	Name of the base station in case it is a small so-called home base station (femto base station)
SIB10	Earthquake and Tsunami Warning System (ETWS) primary notification
SIB11	ETWS secondary notification
SIB12	Commercial Mobile Alert System warning notification
SIB14	Information about Extended Access Barring for access control (see Section 5.3.1.8)
SIB15	Information related to multicast (introduced in Release 14, see Section 5.3.1.9)
SIB16	Information related to GPS time and Coordinated Universal Time
SIB17	Information relevant for traffic steering between E-UTRAN and WLAN
SIB20	Information related to multicast (introduced in Release 14, see Section 5.3.1.9)

This information includes the H-SFN, the UL−DL configuration (in case of TDD), the LTE-M downlink subframe bitmap, and the LTE-M starting symbol. For further information on these timing aspects, refer to Sections 5.2.1.2, 5.2.2.1 and 5.2.4.1.

The scheduling information for SIB1 itself is signaled using 5 bits in MIB [15]. If the value is zero then the cell is an LTE cell that does not support LTE-M devices, otherwise the TBS and number of repetitions for SIB1 are derived according to Table 5.20.

A SIB1 transport block is transmitted on PDSCH according to an 80 ms long pattern that starts in frames with SFN exactly divisible by 8 [10]. As indicated in Table 5.20, the PDSCH is repeated in 4, 8, or 16 subframes during this 80-ms period (except for system bandwidths smaller than 5 MHz, which only support 4 repetitions). Exactly what subframes are used for SIB1 transmission depends on the PCID and duplex mode according to Table 5.21. The PCID helps randomize the interference between SIB1 transmissions from different cells.

The LTE-M downlink subframe bitmap signaled in SIB1 has no impact on the SIB1 transmission itself. Similarly, the LTE-M starting symbol signaled in SIB1 does not apply to the

TABLE 5.20 Transport block size and number of PDSCH repetitions for SIB1 in LTE-M.

SIB1 scheduling information signaled in MIB	SIB1 transport block size in bits	Number of PDSCH repetitions in an 80-ms period
0	LTE-M not supported in the cell	
1	208	4
2		8
3		16
4	256	4
5		8
6		16
7	328	4
8		8
9		16
10	504	4
11		8
12		16
13	712	4
14		8
15		16
16	936	4
17		8
18		16
19–31	Reserved values	

TABLE 5.21 Subframes for SIB1 transmission.

System bandwidth (MHz)	Number of PDSCH repetitions in an 80-ms period	PCID	Subframes with SIB1 transmission in FDD case	Subframes with SIB1 transmission in TDD case
<5	4	Even	Subframe #4 in even frames	Subframe #5 in odd frames
		Odd	Subframe #4 in odd frames	Subframe #5 in odd frames
≥5	4	Even	Subframe #4 in even frames	Subframe #5 in odd frames
		Odd	Subframe #4 in odd frames	Subframe #0 in odd frames
	8	Even	Subframe #4 in every frame	Subframe #5 in every frame
		Odd	Subframe #9 in every frame	Subframe #0 in every frame
	16	Even	Subframes #4 and #9 in every frame	Subframes #0 and #5 in every frame
		Odd	Subframes #0 and #9 every frame	Subframes #0 and #5 in every frame

SIB1 transmission. Instead, the starting symbol for the PDSCH carrying SIB1 is always the fourth OFDM symbol in the subframe except for the smallest system bandwidth (1.4 MHz) where it is always the fifth OFDM symbol in the subframe [15].

SIB1 is transmitted on PDSCH with QPSK modulation using all 6 PRB pairs in a narrowband and RV cycling across the repetitions. The frequency locations and frequency hopping for SIB1 are fixed in the standard as described in Section 5.3.3.2. In case the scheduling causes collision between SIB1 and other MPDCCH/PDSCH transmission in a narrowband in a subframe, the SIB1 transmission takes precedence and the other transmission is dropped in that narrowband in that subframe.

The SIB1 information is unchanged at least during a *modification period* of 5.12 s, except in the rare cases of *Earthquake and Tsunami Warning System* (ETWS) or *Commercial Mobile Alert System* (CMAS) notifications, or when the ACB information changes (see Section 5.3.1.8 for more information). In practice the time between SI updates is typically much longer than 5.12 s. Release 15 introduces more stringent reading delay requirements for the *Cell Global Identity* (CGI) reading based on PBCH accumulation across two 40-ms periods (see Section 5.3.1.1.3) and SIB1 accumulation within one modification period assuming MIB and SIB1 do not change frequently. This enables reduction of the allowed CGI identification time for Cat-M1 from 5120 ms to 3200 ms at a low SNR level of -15 dB [24].

5.3.1.2.2 System Information Blocks 2-20

SIB1 contains scheduling information for the other SIBs listed in Table 5.19. SIBs other than SIB1 are carried in *SI messages*, where each SI message can contain one or more SIBs [23]. Each SI message is configured with an *SI periodicity*, a TBS and a starting narrowband. The possible periodicities are {8, 16, 32, 64, 128, 256, 512} frames, and the possible TBS are {152, 208, 256, 328, 408, 504, 600, 712, 808, 936} bits. Each SI message can also be configured with its own *SI value tag* as described in Section 5.3.1.2.3.

The SI messages are periodically broadcasted during specific, periodic, and nonoverlapping time domain windows known as *SI windows* of configurable length. Possible SI window lengths are {1, 2, 5, 10, 15, 20, 40, 60, 80, 120, 160, 200} ms, where the smallest values are inherited from LTE and probably not that relevant for LTE-M. If the periodicity for the nth SI message is T_n frames, then that SI message is transmitted in the nth SI window after every frame that has an SFN evenly divisible by T_n. The intention with the specified behavior is to map different SI messages to different SI windows even if they have the same periodicity.

Furthermore, to support operation in extended coverage, SI messages can be repeated within their respective SI windows. Possible repetition patterns are {every frame, every second frame, every fourth frame, and every eighth frame} throughout the SI window. All SI messages have the same repetition pattern.

Each SI message is transmitted on PDSCH with QPSK modulation using all 6 PRB pairs in its narrowband and RV cycling across the repetitions. Frequency hopping is supported as described in Section 5.3.3.2. In case the scheduling causes collision between SI messages and other MPDCCH/PDSCH transmission than SIB1 in a narrowband in a subframe, the SI message transmission takes precedence and the other transmission is dropped in that narrowband in that subframe [15].

The SI message content is unchanged during a configurable modification period where the possible values are {2, 4, 8, 16, 64} times the cell paging cycle. In practice the time between SI updates is typically much longer than this.

5.3.1.2.3 System Information update

When the network modifies the SI in a cell, it can indicate this to the devices through the SI value tag [23]. The SI value tag is a 5-bit field in SIB1 that is changed every time the SI content has changed. It is also possible for the network to signal an SI value tag per SI message, which enables the device to just reacquire the SI messages that have actually changed instead of having to reacquire them all, which can be power consuming for the device. These SI value tags are 2-bit fields.

When the SI has changed, the network can also explicitly notify the devices through so-called *direct indication* in the DCI used for paging [20]. The meaning of the 8-bit direct indication field is described in Table 5.22 [23]. Paging and eDRX are discussed in Section 5.3.1.4.

Note that some SI updates, for example regularly changing parameters such as time information and access barring information, may not result in any SI value tag changes or explicit SI update notifications from the network. Also, when a device is configured with an eDRX cycle that is longer than the SI modification period, the device verifies that the stored SI is valid before trying to establish a connection or receive paging.

An LTE-M device considers its stored SI to be valid for several hours from the moment it was acquired, where the number of hours is either 3 or 24 h depending on network configuration [23]. When the SI has become invalid, the device should reacquire the SI before it accesses the network. An LTE-M device also verifies the validity of the SI every time it is released from connected mode to idle mode.

Release 15 introduces a flag bit in MIB indicating unchanged SI, and this flag bit helps reduce the need for the device to reacquire SIB1 just to read the SI value tag (see Section 5.3.1.1.3). The flag bit is set to true when no change has occurred in SIB1 or SI during the last 3 or 24 h (the configured SI validity time). Furthermore, if the resynchronization signal

TABLE 5.22 SI update notification through direct indication in DCI format for paging in LTE-M.

Bit	Meaning
1	General SI update notification to devices not configured with eDRX
2	SIB10/11 update notification
3	SIB12 update notification
4	SIB14 update notification
5	General SI update notification to devices configured with eDRX
6	Reserved
7	Reserved
8	Reserved

(RSS, also introduced in Release 15) is transmitted in the cell, then the device may even be able to skip MIB reacquisition, since the mentioned flag bit is also replicated in RSS (see Section 5.2.4.2.2).

5.3.1.3 Cell reselection

After selecting a cell, a device is required to evaluate the serving cell using a cell selection criterion, *S*, in terms of RSRP and RSRQ. If the serving cell can no longer fulfill the criterion, the device initiates the measurements of neighbor cells. In simple words, in case the device detects that a neighbor cell has become stronger in terms of the RSRP and RSRQ than the currently serving cell, then the cell reselection procedure is triggered. A hysteresis value provided in SIB3 helps prevent ping-pong reselection [23].

If both the cell and the device support CE mode B, and the device has not been restricted by the network from using CE mode B, then separate signal strength conditions apply for CE mode A and B, so that a cell supporting CE mode B can be selected at lower RSRP or RSRQ levels than a cell that only supports CE mode A.

Devices that are in good coverage, i.e., experience a sufficiently high RSRP and RSRQ levels in the serving cell, can choose to not perform measurements for cell reselection, provided that the current cell fulfills the cell selection/suitability criterion, *S*. The thresholds for RSRP and RSRQ are configured in SIB3 separately for the intra-frequency and inter-frequency measurement cases. This helps improve the battery life of these devices. Besides securing that a device camps on the best cell, the cell reselection procedure is the main mechanism for idle mode mobility.

Release 15 introduces a possibility for relaxed monitoring of neighbor cells also for devices that are experiencing weak coverage provided that certain relaxation criteria are met. If the feature is enabled in the cell, after 5 minutes with normal neighbor cell measurements, the device can suspend neighbor cell measurements unless it detects that the serving-cell RSRP degrades and the degradation exceeds a configured hysteresis value, in which case the device again performs normal neighbor cell measurements as usual for 5 more minutes. If measurements do not lead to a cell reselection within a 5-minute window, the device can stop neighbor cell monitoring for up to 24 hours, after which the process repeats. This will allow a stationary device in challenging coverage conditions to avoid wasting power on trying to find a stronger cell when it has already found the strongest cell in the vicinity.

5.3.1.4 Paging, DRX and eDRX

In idle mode, the device monitors periodic *paging occasions* (PO) in a *paging narrowband* in the downlink for potential attempts from the network to contact the device [25]. The paging transmission can contain multiple *paging records* intended for different devices in the cell. When a device receives a paging transmission at its paging occasion in its paging narrowband, it checks whether any of the paging records matches its device identity, and if there is a match, the device responds by initiating a connection to the cell using the random access procedure described in Section 5.3.1.6. The device can be identified either using a *System Architecture Evolution Temporary Mobile Subscriber Identity* (S-TMSI) or the more rarely used global *International Mobile Subscriber Identity* (IMSI).

The monitoring of paging has implications on device battery lifetime and the latency of downlink data delivery to the device. A compromise is achieved by configuring a *discontinuous reception* (DRX) cycle and/or an *extended DRX* (eDRX) cycle. The maximum DRX cycle is 256 frames (2.56 s) in both idle and connected mode, and the maximum eDRX cycle is 256 hyperframes (about 44 min) in idle mode and 1024 frames (10.24 s) in connected mode. After each eDRX cycle, a *paging time window* (PTW) occurs, configurable in multiples of 128 frames up to 2048 frames (20.48 s) during which downlink reachability is achieved through the configured DRX cycle. Fig. 7.49 can serve as an illustration of these concepts, although it should be noted that the maximum values of the parameters are different in LTE-M and NB-IoT. The paging occasions for a device are determined by the DRX/eDRX configuration and the device identity (IMSI for DRX, S-TMSI for eDRX).

The total number of narrowbands that are used for paging is configurable per cell, and among these narrowbands the device determines its paging narrowband based on its device identity (IMSI). Frequency hopping is supported as described in Section 5.3.3.2.

The monitored physical channel is MPDCCH. MPDCCH repetition is supported using the *Type-1 MPDCCH Common Search Space* (CSS) described in Section 5.3.3.1. The MPDCCH carries DCI using DCI Format 6-2 [20]. This DCI format can either be used for carrying a direct indication field according to Table 5.22 or for scheduling a PDSCH carrying paging record(s) according to Table 5.23. The DCI CRC is masked with the *Paging RNTI* (P-RNTI), which is defined in the standard [14].

TABLE 5.23 DCI Format 6-2 for paging and direct indication for LTE-M.

Information	Size [bits]	Possible settings
Flag for paging/ direct indication	1	Paging or direct indication (If this flag bit indicates direct indication then the remaining DCI content is according to Table 5.22)
PDSCH narrowband	1–4	Any narrowband in the system bandwidth
PDSCH TBS	3	{40, 56, 72, 120, 136, 144, 176, 208} bits
Number of PDSCH repetitions	3	One of the following ranges, depending on the setting of the DCI field "Number of MPDCCH repetitions": 00: {1, 2, 4, 8, 16, 32, 64, 128} 01: {4, 8, 16, 32, 64, 128, 192, 256} 10: {32, 64, 128, 192, 256, 384, 512, 768} 11: {192, 256, 384, 512, 768, 1024, 1536, 2048}
Number of MPDCCH repetitions	2	One of the following ranges, depending on the setting of the SIB2 parameter for max number of repetitions R_{max}: $R_{max} = 1$: {1} $R_{max} = 2$: {1, 2} $R_{max} = 4$: {1, 2, 4} $R_{max} = 8$: {1, 2, 4, 8} $R_{max} = 16$: {1, 4, 8, 16} $R_{max} = 32$: {1, 4, 16, 32} $R_{max} = 64$: {2, 8, 32, 64} $R_{max} = 128$: {2, 16, 64, 128} $R_{max} = 256$: {2, 16, 64, 256}

When the DCI schedules a PDSCH, the PDSCH is transmitted with QPSK modulation using all 6 PRB pairs in the narrowband [15]. PDSCH repetition is supported in a similar way as for downlink unicast transmission as described in Section 5.3.2.1.1. The number of paging records that can be carried in one transport block depends on the size of each device identity. The size of each paging record can vary between 25 and 61 bits, meaning that the largest TBS (208 bits) can carry between 3 and 8 paging records.

When the *Mobility Management Entity* (MME) in the *core network* needs to page an LTE-M device, it will inform the involved base station(s) that the device is an LTE-M device so that the paging can be transmitted using the right format (with DCI Format 6-2, MPDCCH, and so on). The MME can also optionally provide a *Paging Coverage Enhancement Level*, which is an estimate between 1 and 256 of the required number of repetitions for MPDCCH [26]. In that case, the MME has been keeping the value as device history information since an earlier session in the same cell [23]. If the device is, for example, a stationary metering device that needs large coverage enhancement because it is in a basement, it might be a good *paging strategy* to page the device right away with many MPDCCH repetitions in the cell where the device last accessed the network. However, if the device moves around when in idle mode there may not be any adequate history information in the MME because an LTE-M device in idle mode does not inform the network when the coverage situation changes [13]. For potential IoT use cases where the device is mobile but anyway frequently experiences bad coverage conditions, it may be difficult to find a suitable paging strategy, because it may not be acceptable from overhead point of view to page the device with many repetitions in multiple cells. In this case, some level of *Mobile Originated* (MO) traffic may be used to assist the network in keeping track of the device and thereby improve the downlink reachability for the device. One example of this is device-triggered tracking area updates.

To further improve device battery lifetime, Release 15 introduced the *MTC Wake-Up Signal* (MWUS) described in Section 5.2.4.5. MWUS facilitates device energy saving by indicating whether a paging indicator will be sent in an associated paging occasion. As this in essence is a single bit information (i.e., paging indicator present or not present), the total transmission time required is much shorter than that required for MPDCCH with DCI format 6-2. The number of repetitions for MWUS, R_{MWUS}, is a fraction (1/4, 1/8, 1/16, or 1/32) of the maximum number of repetitions configured for MPDCCH for paging, $R_{\max}$ (with the restriction that R_{MWUS} is never less than 1). Thus, MWUS enables the device to go back to sleep sooner, thereby achieving energy saving.

The time gap between the MWUS and the paging occasion (where the MPDCCH starts) is configurable. When MWUS is used with either DRX or eDRX, a "short" gap length of 40, 80, 160 or 240 ms can be configured, but in the eDRX case it is also possible to configure a "long" gap length of 1 or 2 s. The longer gaps are intended to facilitate device implementations with *wake-up receivers*, where a less complex detector is used for MWUS and the ordinary receiver is only started up if an MWUS was indeed detected.

When MWUS is used with eDRX, it is furthermore possible to configure MWUS to be associated with more than one paging occasion, achieving even bigger energy saving. In this case, when the device detects an MWUS, it will stay awake for the configured number of paging occasions (allowed values are 1, 2, and 4). This allows an MWUS to apply to (a part of) a paging time window (as illustrated in Fig. 7.50 for the corresponding mechanism in NB-IoT).

MWUS can be used in tandem with the *Resynchronization Signal* (RSS), which was also introduced in Release 15. Since MWUS is only transmitted when there is an actual paging, the device may need to receive some other signal than MWUS in order to maintain a sufficient level of time and frequency synchronization for reliable MWUS detection performance. RSS is a relatively dense synchronization signal transmitted according to a known pattern (see Section 5.2.4.2.2) which is particularly well suited for this purpose in a low-complexity wake-up receiver.

5.3.1.5 Power Saving Mode

The DRX and eDRX mechanisms described in the previous section provide minimum device power consumption for IoT applications where it needs to be possible to reach the device for *Mobile Terminated* (MT) traffic through paging within seconds or minutes. For applications where it is not required to be able to reach the device through paging faster than within half an hour or more, the PSM may be able to provide further power saving. The device will still be able to perform MO transmission in uplink without delay. PSM is a standalone feature applicable to all 3GPP radio access technologies. For information about PSM, refer to Sections 2.2.3 and 7.3.1.5.

5.3.1.6 Random access in idle mode

The random access procedure in LTE-M follows the same steps as in LTE [14]. After synchronizing to the network, confirming that access is not barred (as described in Section 5.3.1.8) and reading the PRACH configuration information in SIB2, the device can send a PRACH preamble to access the network (as described in Section 5.2.5.2). The random access procedure is illustrated in Fig. 5.22, and it is also used when the device responds to a paging message. Use of random access in connected mode is described in Section 5.3.2.3.

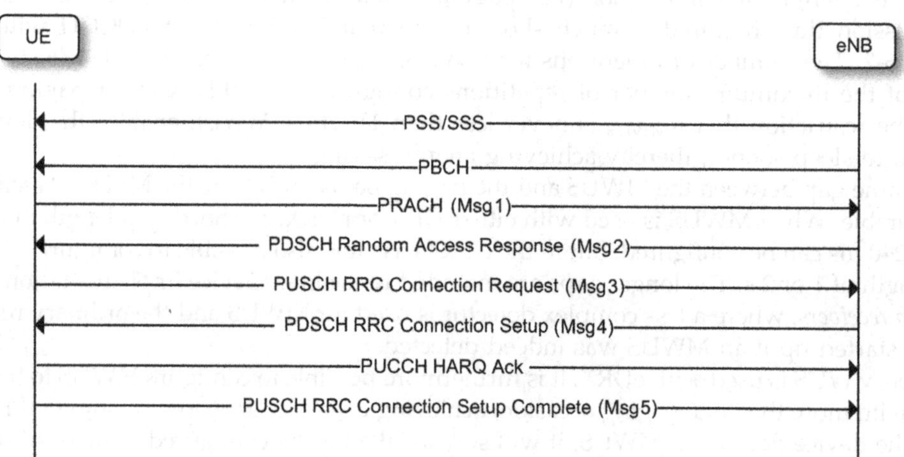

FIG. 5.22 Random access procedure in LTE and LTE-M.

If the base station detects a PRACH preamble, it sends back a *Random Access Response* (RAR), also known as *Message* 2. The RAR contains an uplink *timing advance* (TA) command. The RAR further contains scheduling information pointing to the radio resources that the device can use to transmit a request to connect, also known as *Message* 3. In Message 3, the device will include its identity as part of an RRC message. The device can in some cases include its buffer status in Message 3 to facilitate the scheduling for subsequent uplink transmissions. In *Message 4*, the network transmits a connection setup/resume message and contention resolution data that resolves any contention due to multiple devices transmitting the same preamble in the first step. The device finally replies with a connection setup/resume complete message to terminate the random access procedure and complete the transition to connected mode. The device may also append uplink data in the MAC layer of this message to optimize the latency of the data transfer.

LTE-M supports the ordinary RRC connection setup procedure described in Section 2.3.2 with the message transfer illustrated in Fig. 2.5 and the latency evaluated in Section 6.4. For devices for which small infrequent data transmission is the dominating use case, Release 13 and 15 furthermore introduce three optimizations reducing the connection setup signaling (*RRC Suspend/Resume*, *Data over NAS*, and *Early Data Transmission*), and these are described in Section 5.3.1.7.

Before the initial PRACH transmission, the device needs to determine an appropriate PRACH resource configuration according to its coverage level estimation. The cell can configure up to three RSRP thresholds that are used by the device to select the PRACH resource configuration appropriate for its level of coverage. The PRACH resource configurations are signaled in SIB2. An example is given in Fig. 5.23, in which three RSRP thresholds are configured and therefore there are four PRACH resources for four *PRACH CE levels*, respectively. The device performs a CRS-based RSRP measurement and selects a PRACH CE level in line with the measurement result. The higher the PRACH CE level, the larger the number of PRACH repetitions. If the device does not receive a RAR message in response to a PRACH attempt, it will make further attempts until it receives RAR or until it has reached the maximum allowed number of attempts. As in ordinary LTE operation, PRACH preamble power ramping is applied, but in LTE-M the device will also be able to do PRACH CE level ramping, meaning that the device moves to the next PRACH CE

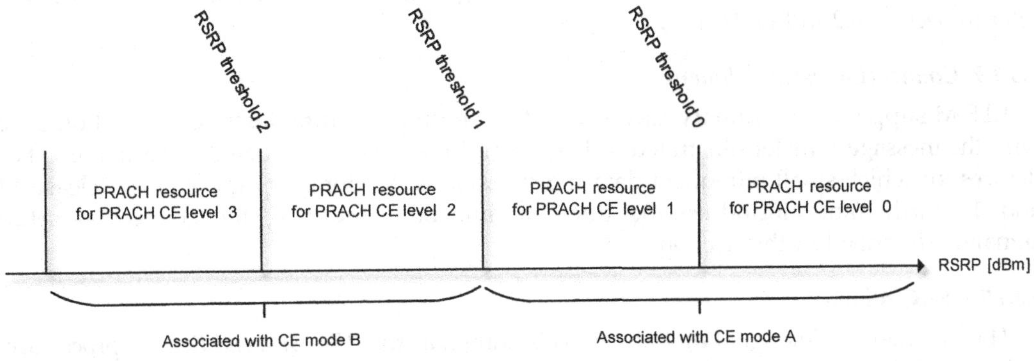

FIG. 5.23 PRACH configurations and RSRP thresholds for LTE-M.

level (increasing the number of PRACH repetitions per attempt) after a few unsuccessful attempts on one PRACH CE level.

After the PRACH preamble transmission, the device monitors a *RAR window* in the downlink for a potential MPDCCH transmission that schedules a PDSCH transport block containing 56-bit RAR messages for one or more devices. The MPDCCH uses the *Type-2 MPDCCH CSS* described in Section 5.3.3.1. The DCI CRC is masked with the *Random Access RNTI* (RA-RNTI), which is determined from the PRACH transmission time according to a predefined rule in the standard [14]. The RAR message contains a *Temporary Cell-RNTI* (TC-RNTI) and a *RAR grant* and these are used to schedule the initial Message 3 transmission on PUSCH. Potential PUSCH HARQ retransmissions for Message 3 and all PDSCH HARQ (re) transmissions for Message 4 are scheduled using MPDCCH in Type-2 MPDCCH CSS with TC-RNTI. The HARQ-ACK feedback for Message 4 is transmitted on PUCCH, so the random access procedure makes use of five physical channels (PRACH for Message 1, MPDCCH + PDSCH for Message 2, PUSCH + MPDCCH for Message 3, and MPDCCH + PDSCH + PUCCH for Message 4).

The device is strictly speaking not configured in CE mode A or B (as described in Section 5.2.2.3) until it has entered connected mode. However, the PRACH CE levels are associated with the CE modes as illustrated in Fig. 5.23. This means that if a PRACH preamble is successfully received on PRACH CE level 0 or 1 then the following messages (RAR, Message 3, and Message 4) will use DCI formats and various parameters signaled in SIB (e.g., maximum numbers of repetition and frequency hopping intervals) intended for CE mode A, and similarly if the PRACH preamble is successfully received on PRACH CE level 2 or 3 then the following messages will use DCI formats and SIB parameters intended for CE mode B. This means that the description of downlink scheduling in connected mode in Section 5.3.2.1.1 is to a large extent valid for RAR and Message 4, and the description of uplink scheduling in Section 5.3.2.1.2 is to a large extent valid for Message 3.

However, there are some differences compared to ordinary unicast in connected mode. RAR is restricted to QPSK modulation and does not support HARQ retransmission, and the initial Message 3 transmission is not scheduled from MPDCCH but from a grant field in the RAR message. The RAR grant contains a PUSCH grant, which includes narrowband index, resource allocation within the narrowband and TBS (typically 56, 72, or 88 bits), and an MPDCCH narrowband for scheduling of both potential HARQ retransmission(s) of Message 3 and HARQ (re)transmission(s) of Message 4. For further details on the RAR grant, refer to Section 6.2 in Ref. [15].

5.3.1.7 Connection establishment

LTE-M supports the ordinary RRC connection setup procedure described in Section 2.3.2 with the message transfer illustrated in Fig. 2.5 and the latency evaluated in Section 6.4. For devices for which small infrequent data transmission is the dominating use case, Release 13 and 15 furthermore introduce the three optimizations reducing the connection setup signaling described in this section.

5.3.1.7.1 RRC resume

The first method for signaling reductions is known as the *RRC Suspend/Resume* procedure, or just the *RRC Resume* procedure. It is part of the *User Plane CIoT EPS optimizations*. It allows a device to resume a connection previously suspended including the PDCP state, the access stratum security and RRC configurations. This eliminates the need to negotiate access stratum security as well as configuring the radio interface, including the *data radio bearers* carrying the data over the air interface, at connection setup. It also supports the PDCP to make efficient use of its *Robust Header Compression* (RoHC) already from the first data transmission in a resumed connection.

This functionality is based on a resume identity which identifies a suspended connection. It is signaled in the RRC Connection release message from the network to a device when a connection is suspended. The device signals the resume identity back to the network when it wants to resume a connection using Message 3 including the RRC Connection resume request message. Fig. 5.24 illustrates the complete procedure.

The RRC resume procedure allows uplink data to be multiplexed with the RRC signaling already in Message 5. This multiplexing between RLC packet data units containing user data and control signaling is achieved in the MAC layer.

5.3.1.7.2 Data over Non-access Stratum

The second method for signaling reductions is known as the *Data over NAS* (DoNAS) procedure and it is part of the *Control Plane CIoT EPS optimization*. In Message 5, the RRC Connection setup complete message, a *Non-Access Stratum* (NAS) container is used for transmitting uplink user data over the control plane. This method was primarily designed with NB-IoT use cases in mind but is optionally supported also in LTE-M. For details on this method, refer to Section 7.3.1.7.2.

5.3.1.7.3 Early Data Transmission

As described above uplink data and downlink data can at the earliest be delivered in *Message 5* and *Message 6*, respectively. This leaves room for further enhancements. Release 15 introduces a feature called *Early Data Transmission* (EDT) that allows a device to transmit

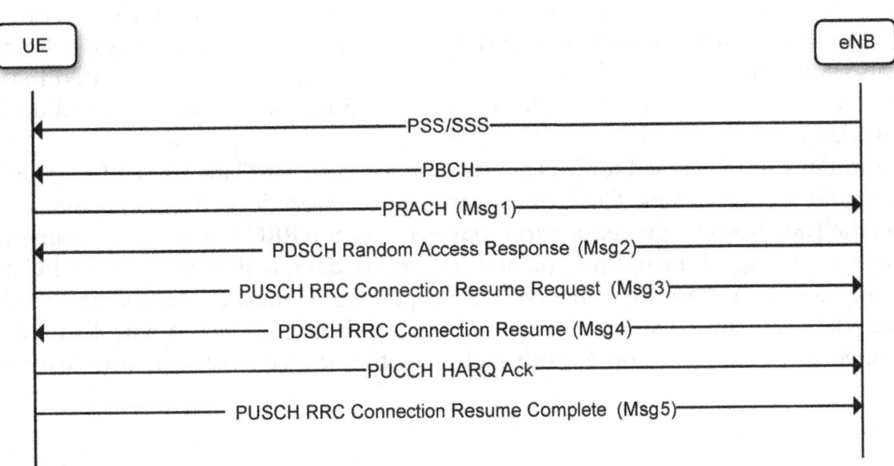

FIG. 5.24 RRC Resume procedure.

its uplink data in *Message 3*. In this case, the device can complete its data transmission in idle mode without a need to transition to connected mode.

EDT is however limited to small data payloads. Uplink EDT supports the TBS range {328, 408, 504, 600, 712, 808, 936, 1000} bits in CE mode A and {328, 408, 456, 504, 600, 712, 808, 936} bits in CE mode B. While these TBS values are not very large, they are significantly larger than an ordinary Message 3 which is typically in the range 56–88 bits (corresponding to the size of an RRC Connection Request or RRC Connection Resume Request message). The SI indicates a maximum TBS per PRACH CE level. Thus, a device can use the EDT procedure to transmit its uplink data only when the number of payload bits is less than the maximum TBS permitted. A device indicates its intent to use EDT by initiating a random access procedure with an PRACH preamble randomly selected from a set of preambles configured per PRACH CE level for the EDT procedure. In this case, upon detecting the PRACH preamble, the base station knows that the device attempts to transmit its uplink data through EDT, and therefore can include an EDT uplink grant for *Message 3* in *Message 2*. The base station can optionally allow devices to use a smaller TBS than the configured maximum TBS, and in this case the device can select a TBS for Message 3 based on its data buffer status. Message 2 indicates how many subframe repetitions the device should use for Message 3 transmission assuming that the maximum TBS is used, but if the device selects a smaller TBS then it scales down the number of repetitions proportionally, which helps reduce power consumption.

Since the device-selected TBS requires the base station to do blind decoding of a number of possible TBS hypotheses, it is up to the base station to indicate (per PRACH CE level) whether it is allowed or not, and the maximum number of TBS values that the device can select from (1, 2, or 4). The frequency domain (PRB) resource allocation is according to the UL grant indicated in Message 2 regardless of whether the device uses the maximum TBS or chooses a smaller TBS, and the modulation scheme is fixed to QPSK, so the base station does not need to use multiple hypotheses in the demodulation process, only in the decoding process.

Furthermore, since a Message 3 carrying user data is more resource-demanding than an ordinary Message 3, the base station always has the possibility to deny an EDT request by indicating in Message 2 that the device should fall back to ordinary random access procedure, in which case the device transmits an ordinary Message 3 instead of a Message 3 for EDT.

EDT also supports downlink data transmission in Message 4. This may be used to provide an application layer acknowledgment to the Message 3 uplink transmission and additional data if available.

EDT is in Release 15 enabled for MO access both for the User Plane CIoT EPS optimization procedure and Control Plane CIoT EPS optimization procedure. The user plane version builds on the RRC Resume procedure and makes use of the RRC Connection resume request in Message 3. Message 4, including potential downlink data, is defined by the RRC connection release message in case the connection is suspended or released immediately following the uplink data transmission. For the control plane solution, a pair of new RRC messages were defined for Message 3 and 4, both including the needed NAS container carrying the data.

5.3.1.8 Access control

LTE-M supports *access class barring* (ACB) and *extended access barring* (EAB) as described in Section 2.2.1. SIB1 contains the scheduling information for SIB14 that carries the extended access barring information. Absence of SIB14 scheduling information in SIB1 implies that barring is not activated, whereas presence of SIB14 scheduling information informs the devices that a barring is activated. A change of the barring parameters can occur at any time, triggering the base station to also change the SIB14 scheduling information.

A device in bad coverage locations requires a high repetition factor to be configured for its dedicated physical channels. During high network loads, it may be desirable to bar devices in bad coverage locations and use the available resources to serve many more devices in good coverage locations. Release 15 introduces PRACH CE level specific barring to prevent devices at certain coverage level, or worse, from accessing the network. This is enabled by providing an RSRP threshold in SIB14. If a device has measured RSRP below this threshold, it is barred from accessing the network. It should back off and then reattempt access at a later point in time.

5.3.1.9 Multicast

Support for multicast transmission is seen as beneficial for some IoT applications where there is, for example, a need for efficient distribution of software/firmware upgrades to a large number of devices. However, due to the inherent narrowband property of LTE-M, it does not support LTE's Multimedia Broadcast Multicast Service Single Frequency Network functionality since that is based on a channel bandwidth equal to the full LTE system bandwidth.

Release 14 introduces support for LTE-M multicast transmission based on the *Multimedia Broadcast Multicast Service* (MBMS) framework in the form of *Single-Cell Point-to-Multipoint* (SC-PTM) transmission. The solution is based on the SC-PTM feature introduced for ordinary LTE devices in Release 13, which follows LTE's MBMS architecture shown in Fig. 5.25, where the *Broadcast/Multicast Service Center* (BM-SC) and the *MBMS Gateway* are the MBMS-specific nodes.

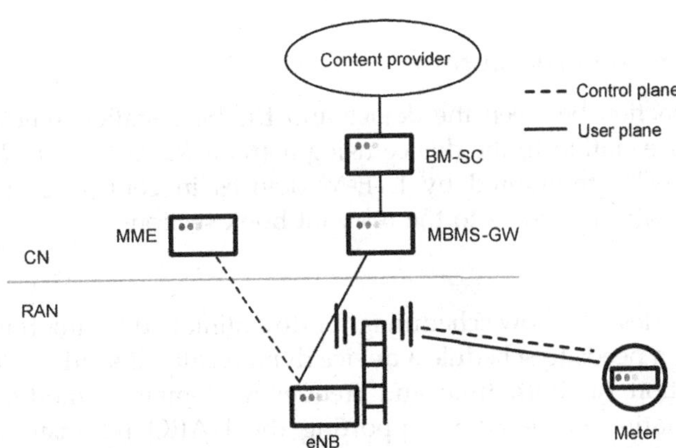

FIG. 5.25 MBMS architecture.

TABLE 5.24 MPDCCH search spaces, RNTIs, and DCI formats for SC-PTM for LTE-M.

Mode	MPDCCH search space	RNTI	Usage	DCI format
Idle	–	SI-RNTI	Broadcast of system information	–
	Type-1A common	SC-RNTI	Scheduling of SC-MCCH	6–2
	Type-2A common	G-RNTI	Scheduling of SC-MTCH	6-1A, 6-1B

SC-PTM support was introduced in Release 14 for both LTE-M and NB-IoT, and the standardization work was partly carried out as a joint effort because the objectives were the same (to extend Release 13 SC-PTM to support narrowband operation and coverage enhancement), so the SC-PTM solutions are very similar, although they use different physical channels in the physical layer. NB-IoT SC-PTM is described in Section 7.3.1.11, where the general procedure illustrated in Fig. 7.55 is applicable also for LTE-M SC-PTM.

SC-PTM for LTE-M (and NB-IoT) is only supported in idle mode. The new SIB20 can contain scheduling information for one *Single Cell Multicast Control Channel* (SC-MCCH) per cell, and SC-MCCH can contain scheduling information for one *Single Cell Multicast Traffic Channel* (SC-MTCH) per multicast service. The maximum channel bandwidth of an SC-MTCH can be either 1.4 or 5 MHz, with a maximum TBS of 1000 and 4008 bits, respectively (matching the maximum capabilities of Cat-M1 and Cat-M2). Both SC-MCCH and SC-MTCH are transmitted on PDSCH transmissions scheduled by MPDCCH. There is no retransmission mechanism, but both MPDCCH and PDSCH can be repeated to achieve the required coverage.

Table 5.24 lists the MPDCCH search spaces, RNTIs, and DCI formats used for SC-PTM. When the listed DCI formats are used for scheduling SC-MCCH or SC-MTCH, they include a DCI field for indication of SC-MCCH update notification, which means that the device does not need to waste power on re-acquiring SC-MCCH unless it has actually changed. Configuration of MPDCCH search spaces is further discussed in Section 5.3.3.1, and configuration of frequency hopping is discussed in Section 5.3.3.2.

5.3.2 Connected mode procedures

Most of the interaction between the device and the base station relies on transmissions addressed by the base station to the device using a 16-bit *Radio Network Temporary Identifier* (RNTI) [14]. The RNTIs monitored by LTE-M devices in connected mode are listed in Table 5.32 together with references to the relevant book sections.

5.3.2.1 Scheduling

In this section, we describe how scheduling for downlink and uplink transmissions works. When the base station needs to schedule a device dynamically, it sends a DCI which includes the resource allocation (in both time and frequency domains), modulation and coding scheme, and information needed for supporting the HARQ retransmission scheme. The DCI is carried on an MPDCCH, which is transmitted in an MPDCCH search space that the device is known to be monitoring (see Section 5.3.3.1) and the DCI has a CRC attached, which is masked with a user-specific *Cell RNTI* (C-RNTI) so that only the device for which the DCI is intended will decode it successfully, whereas other devices monitoring the same MPDCCH search space will discard the DCI because the CRC does not pass for them when they try to unmask the CRC using their C-RNTIs.

This section describes this *dynamic scheduling* of downlink and uplink as well as *semipersistent scheduling (SPS).*

5.3.2.1.1 Dynamic downlink scheduling

To allow low-complexity device implementation, LTE-M adopts the following scheduling principles:

- Cross-subframe scheduling (i.e., DCI and the scheduled data transmission do not occur in the same subframe) with relaxed processing time requirements.
- Optionally, HD-FDD operation at the device (i.e., no simultaneous transmission and reception at the device) allows time for the device to switch between transmission and reception modes.

Fig. 5.26 shows an LTE-M downlink scheduling example without repetition, with MPDCCH and PDSCH transmissions scheduled in parallel to a device in the same downlink narrowband, and with the HARQ-ACK feedback transmitted on PUCCH in the uplink. In ordinary LTE, the PDCCH or EPDCCH carrying the DCI and the PDSCH carrying the data are transmitted in the same downlink subframe, but LTE-M has a cross-subframe scheduling delay of 2 ms that shows in the figure as a 1-ms gap between the end of the MPDCCH transmission carrying the DCI and the start of the scheduled PDSCH transmission. Other than that, the timing relationships are similar to ordinary LTE, with a 3-ms gap between the end of the PDSCH transmission and the start of the PUCCH transmission carrying the associated HARQ-ACK feedback, and another 3-ms gap before a potential HARQ retransmission of the same data is scheduled. Due to the extra 2-ms scheduling delay, the downlink HARQ *round-trip time* (RTT) is increased from $4 + 4 = 8$ to $2 + 4 + 4 = 10$ ms. The maximum number of downlink HARQ processes depends on the duplex mode and CE mode as indicated in Table 5.25. HD-FDD devices cannot be scheduled more frequently than in this example because they cannot transmit and receive simultaneously and furthermore need a guard subframe at every switching between downlink and uplink (as discussed in Section 5.2.2.1). For a discussion on the impact of the 10-ms RTT on the downlink peak rate in (half-duplex and full-duplex) FDD, see Section 6.3 and later in this section.

The base station schedules downlink transmission on PDSCH dynamically using DCI Format 6-1A and 6-1B in CE mode A and B, respectively. Table 5.25 shows the information carried in these DCI formats [20]. Some of the fields are specific to LTE-M and some of them are basically inherited from the ordinary LTE DCI formats. An effort has been made to make the DCI format for CE mode B as compact as possible because it is intended for situations where the device experiences a weak downlink signal implying large numbers of repetitions, and every extra bit is expensive in terms of coverage and/or resource consumption. The most important fields are present in both DCI formats, such as the PDSCH MCS, PDSCH resource block assignment, number of PDSCH repetitions, and number of repetitions of the MPDCCH carrying the DCI itself. The information about the number of MPDCCH repetitions is needed

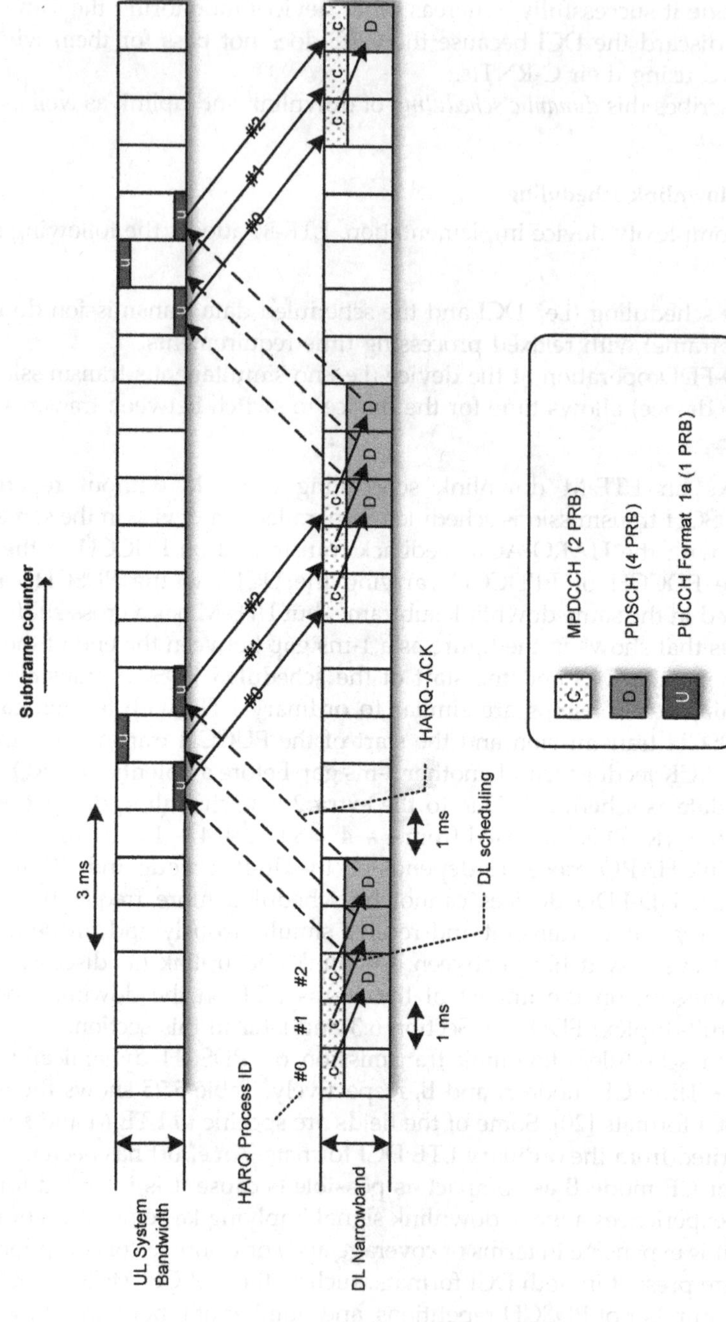

FIG. 5.26 LTE-M downlink scheduling example with MPDCCH and PDSCH transmitted without repetition in the same downlink narrowband.

TABLE 5.25 DCI Formats 6-1A and 6-1B used for scheduling PDSCH in CE modes A and B in Release 13.

Information	DCI format 6-1A		DCI format 6-1B	
	Size [bits]	**Possible settings**	**Size [bits]**	**Possible settings**
Format 6-0/6-1 differentiation	1	1	1	1
Frequency hopping flag	1	See Section 5.3.3.2	–	–
MCS	–	–	4	See Table 5.8
Resource block assignment	0–4	Narrowband index	0–4	Narrowband index
	5	0–20: Allocation of 1–6 PRB pairs	1	0: 4 PRB pairs (#0, …, #3)
		21–31: Unused in Release 13		1: 6 PRB pairs (#0, …, #5)
MCS	4	See Table 5.8	–	–
Number of PDSCH repetitions	2	See Table 5.10	3	See Table 5.11
HARQ process number	3–4	FDD: 0–7, TDD: 0–15	1	0–1
New data indicator	1	Toggle bit for new data	1	Toggle bit for new data
Redundancy version	2	0–3	–	–
PUCCH power control	2	See Section 5.3.2.4	–	–
Downlink assignment index	2	TDD-specific field	–	–
Antenna port and scrambling ID	2	TM9-specific field	–	–
SRS request	1	See Section 5.2.5.3	–	–
Precoding information	2 or 4	TM6-specific field	–	–
PMI confirmation	1	TM6-specific field	–	–
HARQ-ACK resource offset	2	See Section 5.2.5.5	2	See Section 5.2.5.5
Number of MPDCCH repetitions	2	See Table 5.33	2	See Table 5.33

when calculating the starting subframe for the PDSCH transmission. When these DCI formats are used to schedule RAR as described in Section 5.3.1.6, some fields are reserved or repurposed (refer to Reference [20] for the detailed definition).

Fig. 5.27 shows an LTE-M downlink scheduling example where some repetition has been applied to enhance the coverage of the transmissions. The MPDCCH carrying the DCI is repeated in four subframes, the PDSCH carrying the data in eight subframes, and the PUCCH carrying the HARQ-ACK feedback in four subframes. All subframes are assumed to be configured as valid for transmission (see Sections 5.2.4.1 and 5.2.5.1). Note that the

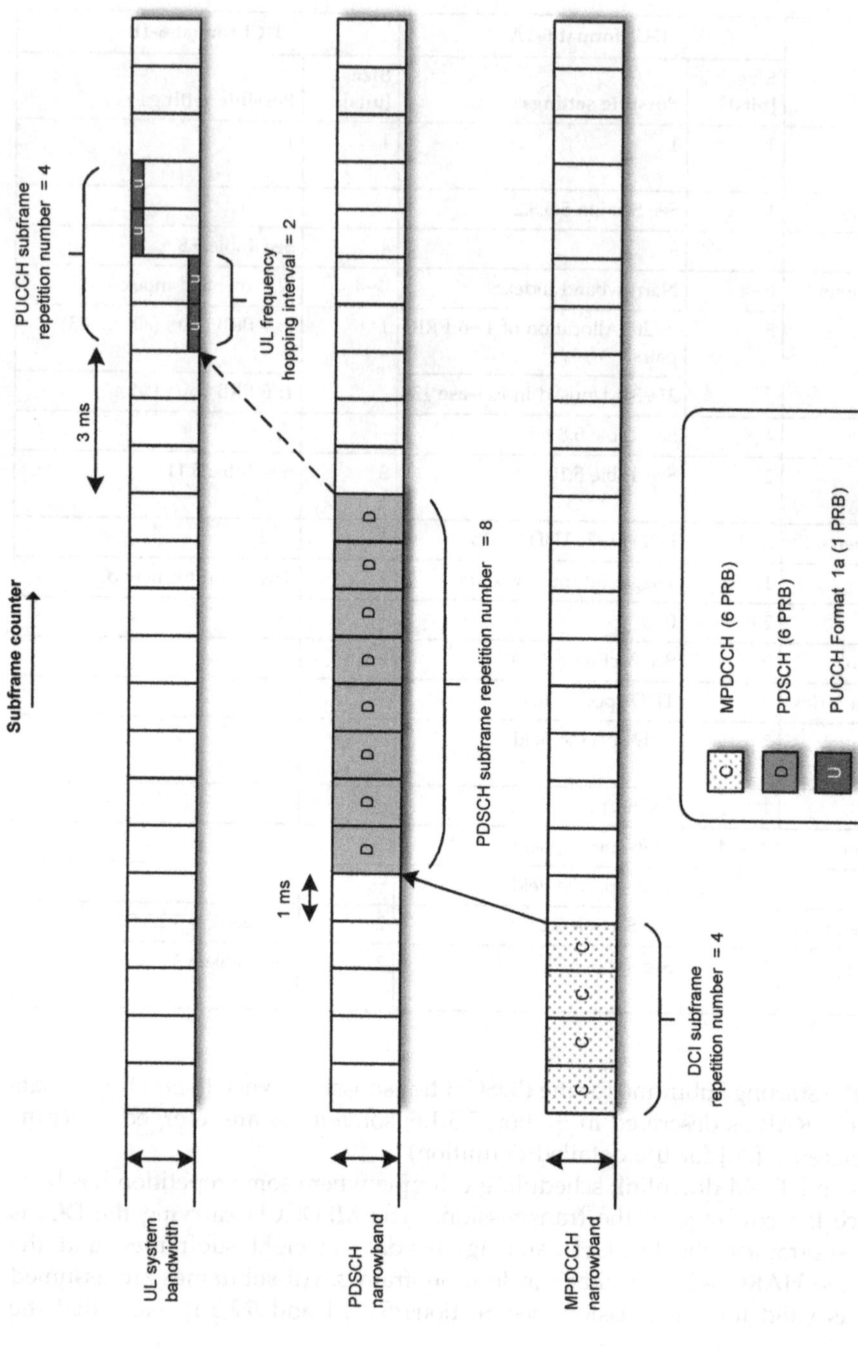

FIG. 5.27 LTE-M downlink scheduling example with MPDCCH and PDSCH transmitted with repetition in different downlink narrowbands.

example makes use of the possibility to schedule PDSCH in a different narrowband than the MPDCCH narrowband. Furthermore, both MPDCCH and PDSCH are transmitted with the maximum channel bandwidth supported in the CE modes in Release 13, which is 1 narrowband (6 PRBs). This is in general beneficial for the device from coverage point of view. The reason for this is that because the total downlink transmit power on the cell-carrier is typically (more or less) evenly distributed over all the PRBs in the system bandwidth, a large downlink channel bandwidth usually also means a large chunk of the available transmit power.

The DCI field for *Number of PDSCH repetitions* contains a 2- or 3-bit value which is interpreted according to Tables 5.10 and 5.11. The DCI field for *Number of MPDCCH repetitions* is interpreted according to Table 5.33. Configuration of repetitions for PUCCH is discussed in Section 5.2.5.5. In the example in Fig. 5.27, a 2-ms uplink frequency hopping interval is assumed. Frequency hopping is discussed in Section 5.3.3.2. In CE mode A, the base station's scheduling decisions for PDSCH can be assisted by periodic or aperiodic CSI reports from the device, which is described in Section 5.3.2.2.

Release 14 introduces several options for *larger maximum channel bandwidths* for PDSCH and PUSCH as described in Sections 5.2.2.3 and 5.2.4.7. In Release 13 LTE-M, no physical channels have a larger channel bandwidth than 6 PRBs. Table 5.26 shows a summary of the associated PDSCH resource allocation methods (refer to Section 7.1.6 in Ref. [15] for details). Each resource allocation method has been designed to provide enough scheduling flexibility to exploit the wider bandwidth without increasing the DCI size more than necessary.

Release 14 introduces the possibility to support up to *10 downlink HARQ processes in FDD*. As discussed in Sections 5.3.2.1.1 and 6.3, there is a relationship between the HARQ RTT and number of HARQ processes. LTE-M has inherited the number of downlink HARQ processes for FDD (eight processes) from LTE even though LTE-M has a somewhat larger RTT than LTE (10 ms for LTE-M vs. 8 ms for LTE). This means that in CE mode A in FD-FDD in Release 13, downlink data can be scheduled in eight consecutive subframes, but then there will be a 2-ms gap before the downlink data transmission can continue. Therefore, FD-FDD devices supporting 10 downlink processes will be able to receive the maximum downlink TBS in every subframe. This can be done without the need to increase the number of soft channel bits that need to be stored in the decoder in the device because the maximum number of processes is only expected to be utilized under good channel conditions when relatively high code rate is used (see description of *Limited Buffer Rate Matching* in Ref. [12]). This increases the downlink MAC-layer peak rate for Cat-M1 from 800 kbps to 1 Mbps in FD-FDD. The HARQ process number field in DCI format 6-1A is increased from 3 to 4 bits to be able to indicate 10 processes.

Release 14 also introduces the possibility to support *HARQ-ACK bundling in HD-FDD*. In Release 13, every downlink data transmission is associated with a HARQ-ACK feedback transmission in uplink, as described in Section 5.2.5.5. Because the LTE-M uplink in Release 13 does not support transmission of more than one HARQ-ACK feedback per subframe, a device in HD-FDD operation will (in the non-repetition case) spend as long transmitting HARQ-ACK feedback in uplink as it spends on the actual downlink data reception. As can be seen in Fig. 5.26, an HD-FDD device will not be able to spend more than 3 of 10 subframes on actual downlink data reception. Instead of just receiving three consecutive downlink data subframes with the corresponding HARQ-ACK, a device supporting HARQ-ACK bundling

TABLE 5.26 PDSCH resource allocation methods in CE mode A and B.

Configured maximum PDSCH channel bandwidth	CE mode A	CE mode B
1.4 MHz	A narrowband index is indicated in the DCI and 1-6 contiguous PRBs can be allocated anywhere within that narrowband.	A narrowband index is indicated in the DCI and the first 4 or all 6 PRBs can be allocated anywhere within that narrowband.
5 MHz	The narrowband index in the DCI is used to indicate a starting narrowband. Anywhere within the starting narrowband, 1-6 contiguous PRBs can be allocated. A 3-bit bitmap indicates whether the allocation within the starting narrowband also applies to one or more of the 3 next narrowbands (i.e., the allocation used in the starting narrowband is reused in the other enabled narrowbands).	Up to 2 bits in the DCI are used to indicate a wideband index (see Section 5.2.2.2). For the indicated wideband, a 4-bit bitmap indicates which up to 4 narrowbands to allocate. All PRBs within the indicated narrowbands are allocated.
20 MHz	A 1-bit field in the DCI indicates whether to interpret the remainder of the DCI as in the 5-MHz case described above or as a resource block group (RBG) bitmap, where the RBG size is 6 PRBs (in 10–15 MHz system bandwidth) or 9 PRBs (in 20 MHz system bandwidth).	Like the 5-MHz case described above except that the 2-bit wideband index is extended to form a 3-bit wideband combination index where the first 4 values indicate the 4 widebands (respectively) and the next 4 values indicate the 2 lowest widebands, the 2 highest widebands, the 3 lowest widebands, and all 4 widebands (respectively)

will be able to transmit up to three consecutive so-called *HARQ-ACK bundles* where each bundle contains acknowledgements for up to 4 downlink data subframes, as illustrated in Fig. 5.28. If the device also supports 10 downlink HARQ processes as described above, then the 3 bundles can acknowledge up to 10 consecutive downlink data subframes. In the illustrated example, the bundles are sent back-to-back and then the corresponding HARQ-ACK feedbacks are sent back-to-back. There will only be a single HARQ-ACK bit per bundle—the device sends ACK if all downlink TBs within a bundle were successfully decoded, otherwise it sends NACK (Negative Acknowledgment) which means that the whole bundle will be retransmitted by the base station. This means that use of HARQ-ACK bundling is mainly useful for device that experience rather good channel conditions (which is often the case for many devices in a typical cell). The result is that a Cat-M1 device supporting HARQ-ACK bundling and 10 downlink processes can receive data in 10 out of 17 subframes in HD-FDD and thereby almost double its downlink MAC-layer peak rate from 300 to 588 kbps.

Release 14 furthermore introduces a *dynamic HARQ-ACK delay* allowing the base station to control the delay between the end of a PDSCH transmission and the beginning of the associated HARQ-ACK feedback on PUCCH or PUSCH dynamically through a new 3-bit field in the DCI. In LTE-M in Release 13, for FDD, this delay is fixed to 4 ms as shown in Fig. 5.21 and 5.22. With a less rigid timing relationship, efficient scheduling of VoLTE

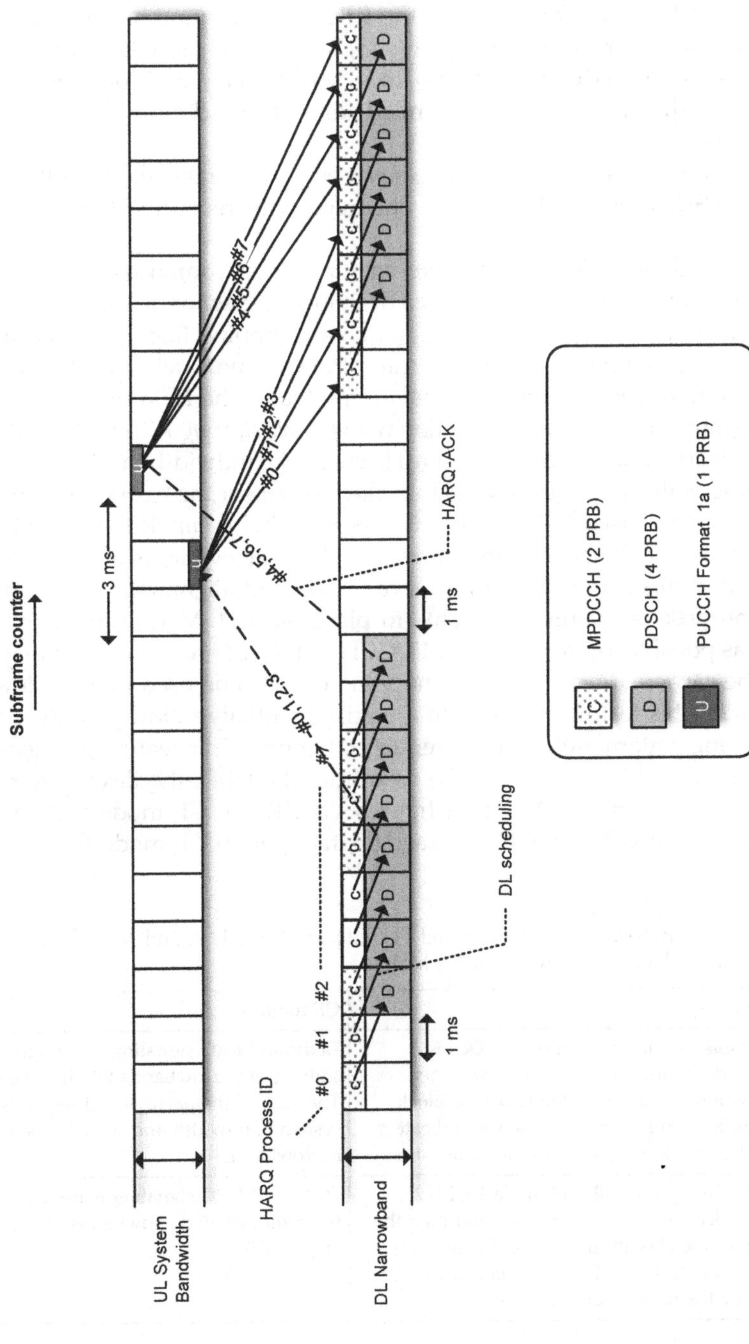

FIG. 5.28 HARQ-ACK bundling in CE mode A.

transmissions is facilitated, especially in HD-FDD. This feature can be seen as a subset of the functionality introduced by the HARQ-ACK bundling feature described above. The interpretation of the 3-bit DCI field depends on the RRC configuration and it is either {4, 5, 6, 7, 8, 9, 10, 11} subframes or {4, 5, 7, 9, 11, 13, 15, 17} subframes, where the former range is intended for VoLTE scheduling with up to 4 MPDCCH repetitions (and for HARQ-ACK bundling) and the latter range is intended for VoLTE scheduling with more than 4 MPDCCH repetitions.

Release 14 also introduces a *PDSCH modulation scheme restriction*, allowing the modulation to be limited to QPSK whenever PDSCH is scheduled with repetition (for more details see Section 5.2.4.7).

Release 15 introduces *downlink 64QAM support* for devices configured with max 1.4 MHz PDSCH channel bandwidth in CE mode A, as described in Section 5.2.4.7. The MCS field is extended from 4 to 5 bits by repurposing the frequency hopping flag in DCI Format 6-1A. If the device is configured with 64QAM enabled and the DCI indicates that PDSCH is without repetition, then the frequency hopping flag is interpreted as the MSB of the 5-bit MCS.

Release 15 also introduces resource allocation with *flexible starting PRB* for both PDSCH (and PUSCH) for devices configured with max 1.4 MHz channel bandwidth in CE mode A and B. In case of PDSCH, one of the main purposes is to make it easier for the base station to pack transmissions for narrowband-based LTE-M users and *resource block group* (RBG) based LTE users as efficiently as possible within the system bandwidth. The narrowbands and the RBGs do not typically line up, and this feature tries to achieve the desired alignment in downlink. In case of PUSCH, the purpose is mainly to be able to place the LTE-M transmissions as close to the carrier edges as possible, right beside any PUCCH or PRACH resources that may be configured closest to the carrier edges. Frequency hopping can also be used both in downlink and uplink together with this feature but note that it still essentially follows the Release 13 LTE-M frequency hopping pattern, not the LTE frequency hopping. The feature for flexible starting PRB is summarized in Table 5.27. As can be seen from the table, the flexible starting PRB is signaled in the DCI in CE mode A but configured via RRC in CE mode B (to avoid a DCI size increase and associated MPDCCH coverage degradation in CE mode B).

TABLE 5.27 Flexible starting PRB for PDSCH and PUSCH in CE mode A and B (with configured maximum channel bandwidth of 1.4 MHz).

	CE mode A	CE mode B
PDSCH	Use of flexible resource allocation via DCI is enabled by dedicated RRC signaling, and then previously unused values in the Resource block assignment field in the DCI are used to indicate a number of more RGB-aligned resource allocations.	Dedicated RRC signaling is used to enable shifting of narrowbands to align them with RBG. The shift of the narrowband depends on the system bandwidth and the allocated narrowband.
PUSCH	Use of flexible resource allocation via DCI is enabled by dedicated RRC signaling, and then the DCI can indicate allocation of up to 6 consecutive PRBs anywhere in the LTE system bandwidth (not restricted by the narrowband borders).	Dedicated RRC signaling is used to configure a fixed shift of all narrowbands by -1, +1, +2, or +3 PRBs.

5.3.2.1.2 Dynamic uplink scheduling

If a device in connected mode has data to transmit but no PUSCH resource, it can request a PUSCH resource by transmitting a *scheduling request* (SR) on PUCCH as described in Section 5.2.5.5. If the device has no valid PUCCH resource either, it will use the random access procedure instead as described in Section 5.3.2.3.

Fig. 5.29 shows an LTE-M uplink scheduling example without repetition. Similar to LTE, a DCI carried on MPDCCH schedules a PUSCH transmission with a 3-ms gap between the end of the MPDCCH transmission and the start of the PUSCH transmission. A difference compared to LTE is that the uplink HARQ scheme is asynchronous in LTE-M, whereas it is synchronous in LTE with a *Physical HARQ Indicator Channel* (PHICH) for HARQ feedback. This means that HARQ retransmissions in LTE-M are always explicitly scheduled using a DCI on an MPDCCH, i.e., there is no PHICH in LTE-M. Other than that, the HARQ operation in LTE-M is similar to the HARQ operation in LTE (this can be said for both uplink HARQ and downlink HARQ). The maximum number of uplink HARQ processes depends on the CE mode as indicated in Table 5.28.

The base station schedules uplink transmission on PUSCH dynamically using DCI Format 6-0A and 6-0B in CE mode A and B, respectively. Table 5.28 shows the information carried in these DCI formats [20]. Many aspects are similar to the PDSCH case described in the previous section. For example, there are fields indicating the number of repetitions for the PUSCH and for the MPDCCH itself, respectively, and a field for indicating the PUSCH narrowband. One notable difference compared to the downlink case described in the previous section is that PUSCH transmission in CE mode B is always scheduled on very few PRB pairs (1 or 2 PRB pairs), whereas PDSCH transmission in CE mode B is always scheduled on a large portion of the allocated narrowband (4 or 6 PRB pairs). Unlike in downlink, in uplink an increase in channel bandwidth may not enable the transmitter to allocate higher power to the transmission—the device is probably already transmitting with maximum power and a larger bandwidth may simply be a waste of bandwidth.

The following uplink resource allocation types are defined for LTE-M in Section 8.1 in Ref. [15]:

- **Uplink resource allocation type 0** (introduced in Release 13) is used for allocating 1-6 contiguous PRBs in CE mode A when the device is configured with 1.4 MHz max PUSCH channel bandwidth.
- **Uplink resource allocation type 2** (introduced in Release 13) is used for allocating 1-2 contiguous PRBs in CE mode B.
- **Uplink resource allocation type 4** (introduced in Release 14) is used for allocating 9-24 contiguous PRBs (with a granularity of 3 PRBs) in CE mode A when the device is configured with 5 MHz max PUSCH channel bandwidth.
- **Uplink resource allocation type 5** (introduced in Release 15) is used for allocating 2-6 contiguous subcarriers (i.e. sub-PRB allocation) in CE mode A and B.

Release 14 introduces support for *larger uplink TBS for Cat-M1* in CE mode A. The background is described in Sections 5.2.3, and when the feature is configured, the MCS field in DCI Format 6-0A is interpreted according to Table 5.15.

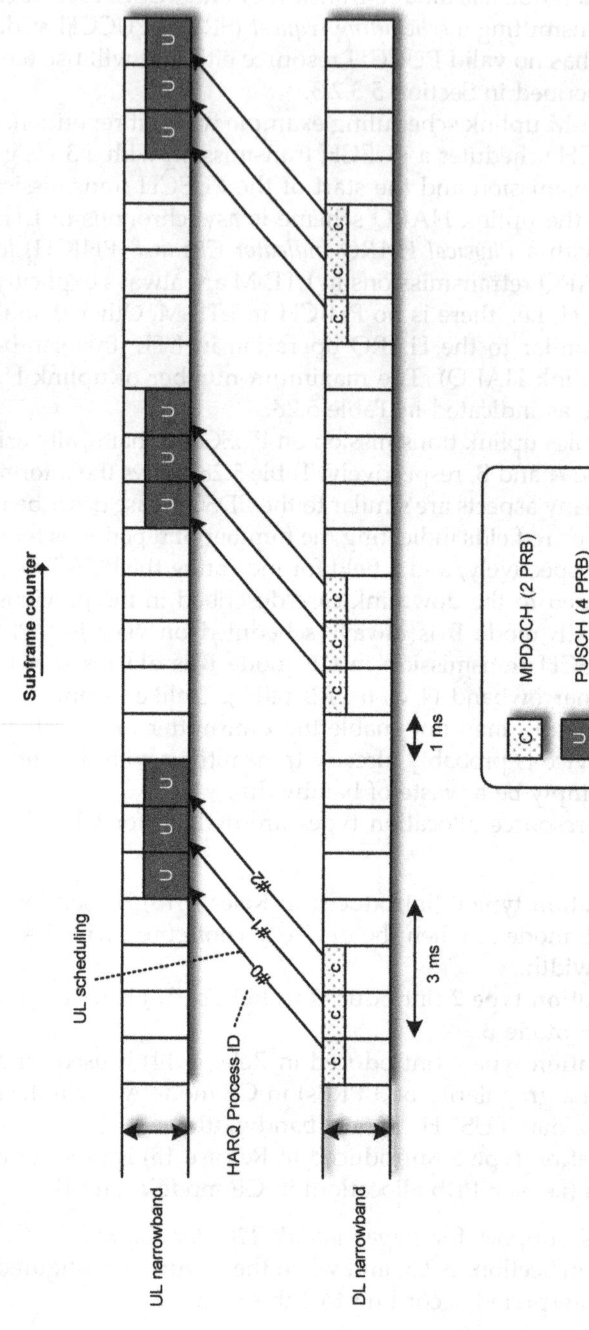

FIG. 5.29 LTE-M uplink scheduling example without repetition.

TABLE 5.28 DCI Formats 6-0A and 6-0B used for scheduling PUSCH in CE modes A and B in Release 13.

Information	DCI format 6-0A		DCI format 6-0B	
	Size [bits]	**Possible settings**	**Size [bits]**	**Possible settings**
Format 6-0/6-1 differentiation	1	0	1	0
Frequency hopping flag	1	See Section 5.3.3.2	–	–
Resource block assignment	0–4	Narrowband index	0–4	Narrowband index
	5	0–20: Allocation of 1–6 PRB pairs	3	0–5: PRB index for 1 PRB pair
				6: 2 PRB pairs (#0 and #1)
		21–31: Unused in Release 13		7: 2 PRB pairs (#2 and #3)
MCS	4	See Table 5.14	4	See Table 5.14
Number of PUSCH repetitions	2	See Table 5.10	3	See Table 5.11
HARQ process number	3	0–7	1	0–1
New data indicator	1	Toggle bit for new data	1	Toggle bit for new data
Redundancy version	2	0–3	–	–
PUSCH power control	2	See Section 5.3.2.4	–	–
Uplink index	2	TDD-specific field	–	–
Downlink assignment index	2	TDD-specific field	–	–
CSI request	1	See Section 5.3.2.2	–	–
SRS request	1	See Section 5.2.5.3	–	–
Number of MPDCCH repetitions	2	See Section 5.3.3.1	2	See Section 5.3.3.1

Release 14 introduces several options for *larger maximum channel bandwidths* for PUSCH and PDSCH as described in Sections 5.2.2.3 and 5.2.5.4. In Release 13 LTE-M, no physical channels have a larger channel bandwidth than 6 PRBs. When the device is configured with max 5 MHz PUSCH channel bandwidth in CE mode A, the ordinary DCI Format 6-0A is still used for allocation of 1–6 PRBs, but in addition some previously reserved values in the Resource block assignment field in the DCI are now used for allocation of 9–24 PRBs with a granularity of 3 PRBs.

Release 14 introduces a *PUSCH modulation scheme restriction*, allowing the modulation to be limited to QPSK when a new 1-bit field in the DCI indicates that QPSK should be used instead of the default modulation scheme for the indicated MCS. This feature is bundled with *new PUSCH repetition factors* which can be selected more dynamically with a finer granularity using another new 1-bit field in the DCI. Both sub-features are described in Section 5.2.5.4.

Also introduced in Release 14 is the support for *device transmit antenna selection* in CE mode A as described in Section 5.2.5.4. The antenna selection is controlled by the base station through masking of the CRC in DCI format 6-0A.

Release 15 introduces resource allocation with *flexible starting PRB* in CE mode A and B for both PDSCH and PUSCH. Both cases are described in Section 5.3.2.1.1 and Table 5.27.

Release 15 also introduces *PUSCH sub-PRB allocation* in CE mode A and B, as described in Section 5.2.5.4. In CE mode A, the DCI is extended with a 2-bit field for number of *resource units* (RU), and if its value is '00' then the rest of the DCI is interpreted as in earlier releases, otherwise the DCI is a grant for sub-PRB allocation with the indicated number of RUs (1, 2, or 4). In CE mode B, the DCI is extended with a 1-bit flag indicating whether the DCI should be interpreted as in earlier releases or as a grant for sub-PRB allocation, in which case a 1-bit field is used to indicate the number of RUs (2 or 4). In CE mode A, the DCI can indicate a narrowband and a PRB position (0, 1, 2, 3, 4, 5) within that narrowband and a set of allocated subcarriers within the PRB, but in CE mode B, the DCI only indicates a narrowband and a set of allocated subcarriers within a PRB where the PRB position (0, 1, 2, 3, 4, or 5) within the narrowband is configured by RRC (in order to keep the DCI size small). In both CE mode A and B, the MCS index field is reduced from 4 bits to 3 bits (see Table 5.17) and the set of allocated subcarriers within the PRB is indicated using 10 values according to Table 5.29.

Release 15 introduces a possibility to send a *positive HARQ-ACK feedback* to the device via DCI Format 6-0A/6-0B. For FD-FDD and TDD devices, this enables the base station to order *early termination of PUSCH transmission* in CE mode A or B. If the base station has been able to decode the uplink data before the scheduled end of the PUSCH transmission, there is no

TABLE 5.29 PUSCH sub-PRB allocation options in CE mode A and B in Release 15.

Resource allocation field	Resource allocation size	Modulation scheme	Set of allocated subcarriers
0	2 out of 3 subcarriers	π/2-BPSK	{0, 1} in cell with even PCID {1, 2} in cell with odd PCID
1	2 out of 3 subcarriers	π/2-BPSK	{3, 4} in cell with even PCID {4, 5} in cell with odd PCID
2	2 out of 3 subcarriers	π/2-BPSK	{6, 7} in cell with even PCID {7, 8} in cell with odd PCID
3	2 out of 3 subcarriers	π/2-BPSK	{9, 10} in cell with even PCID {10, 11} in cell with odd PCID
4	3 subcarriers	QPSK	{0, 1, 2}
5	3 subcarriers	QPSK	{3, 4, 5}
6	3 subcarriers	QPSK	{6, 7, 8}
7	3 subcarriers	QPSK	{9, 10, 11}
8	6 subcarriers	QPSK	{0, 1, 2, 3, 4, 5,}
9	6 subcarriers	QPSK	{6, 7, 8, 9, 10, 11}

reason for the device to waste time and energy on continuing to transmit, so the base station can then either send an *explicit HARQ-ACK* in DCI (this is done by setting the *'Resource block assignment'* field in DCI Format 6-0A or the *'MCS'* field in DCI Format 6-0B to all '1's) which will make the device stop its ongoing transmission, or an *implicit HARQ-ACK* by sending an ordinary uplink grant in DCI which will make the device stop its ongoing transmission and start the new transmission defined in the uplink grant. Early termination of PUSCH transmission cannot be supported by HD-FDD devices since they cannot monitor the downlink while performing uplink transmission. However, the HARQ-ACK also enables an *early termination of MPDCCH monitoring* which saves device energy. The HARQ-ACK can be used when a connection is being released, to inform the device that the final RLC ACK from the device has been successfully received and that the device does not need to monitor the MPDCCH for potential retransmission requests (which it would otherwise do until a retransmission timer expires) and can then enter idle mode. This *early termination of MPDCCH monitoring* can be supported by HD-FDD, FD-FDD and TDD devices.

5.3.2.1.3 Semipersistent scheduling

Beside the dynamic scheduling described in the previous sections, LTE-M supports *semi-persistent scheduling* (SPS) in CE mode A (but not in CE mode B) for downlink and uplink in a similar manner as LTE [14]. In LTE, SPS is mainly motivated by *Voice over Internet Protocol* (VoIP) services where periodic speech frames need to be scheduled and it is desired to avoid the physical control channel overhead associated with dynamic scheduling. For LTE-M devices, periodic sensor reporting could be a potential use case for SPS beside Voice over Internet Protocol.

When SPS is configured, the device is configured by higher layers with a SPS-C-RNTI and a time interval. The SPS operation can then be activated or deactivated by a DCI addressed to the SPS-C-RNTI of the device. The activation DCI indicates what frequency resources, MCS, number of repetitions, etc. that should be used at the periodic persistent resources. The SPS-C-RNTI is also used for scheduling potential HARQ retransmissions. Note that SPS can be overridden by dynamic scheduling at any time if needed.

5.3.2.2 Channel quality reporting

CE mode A supports downlink *Channel State Information* (CSI) reporting from the device in order to assist the base station's scheduling decisions. Both RRC-configured periodic reporting on PUCCH (see Section 5.2.5.5) and DCI-triggered aperiodic reporting on PUSCH (see Section 5.2.5.4 and Table 5.28) are supported. The CSI modes supported by LTE-M are listed in Table 5.30 [15].

The *Channel Quality Information* (CQI) report reflects the device's recommendation regarding what PDSCH MCS to use when targeting 10% BLock Error Rate for the first HARQ transmission. The *Precoding Matrix Indicator* (PMI) report is the device's recommendation (a 2-bit or 4-bit value depending on the number of antenna ports) regarding what precoding matrix to use in PDSCH TM6 and TM9 (see Section 5.2.4.7). For PDSCH TM9, either closed-loop or open-loop beamforming may be used, and in the latter case no PMI reporting is needed. In LTE-M, the CQI and PMI reports are based on CRS measurements in the narrowbands monitored by the device for MPDCCH monitoring. The wideband CQI report reflects the quality when all monitored narrowbands are used for a transmission

TABLE 5.30 Downlink CSI modes in LTE-M.

CSI mode	Description	Triggering	Physical channel	Comment
1–0	Wideband CQI in TM1/TM2/TM9	Periodic	PUCCH Format 2/2a/2b	Only supported in CE mode A
1–1	Wideband CQI and PMI in TM6/TM9	Periodic	PUCCH Format 2/2a/2b	Only supported in CE mode A
2–0	Subband CQI in TM1/TM2/TM6/TM9	Aperiodic	PUSCH	Only supported in CE mode A

TABLE 5.31 Downlink CQI tables in LTE-M.

	Release 13 CQI table			Release 15 CQI table with 64QAM			Release 15 alternative CQI table		
CQI index	Modulation	Code rate x 1024 x R^{CSI}	Efficiency x R^{CSI}	Modulation	Code rate x 1024	Efficiency	Modulation	Code rate x 1024	Repetition
0	Out of range			Out of range			Out of range		
1	QPSK	40	0.0781	QPSK	40	0.0781	QPSK	56	32
2	QPSK	78	0.1523	QPSK	78	0.1523	QPSK	207	16
3	QPSK	120	0.2344	QPSK	120	0.2344	QPSK	266	4
4	QPSK	193	0.3770	QPSK	193	0.3770	QPSK	195	2
5	QPSK	308	0.6016	QPSK	308	0.6016	QPSK	142	1
6	QPSK	449	0.8770	QPSK	449	0.8770	QPSK	266	1
7	QPSK	602	1.1758	QPSK	602	1.1758	QPSK	453	1
8	16QAM	378	1.4766	16QAM	378	1.4766	QPSK	637	1
9	16QAM	490	1.9141	16QAM	490	1.9141	16QAM	423	1
10	16QAM	616	2.4063	16QAM	616	2.4063	16QAM	557	1
11	Reserved			64QAM	466	2.7305	16QAM	696	1
12	Reserved			64QAM	567	3.3223	16QAM	845	1
13	Reserved			64QAM	666	3.9023	64QAM	651	1
14	Reserved			64QAM	772	4.5234	64QAM	780	1
15	Reserved			64QAM	873	5.1152	64QAM	888	1

(averaged over these narrowbands), whereas the subband CQI report additionally contains one separate CQI report for the best one of the monitored narrowbands, information that can help guide the base station's scheduling decisions.

Table 5.31 shows the three CQI tables supported by LTE-M in Release 15. The reported 4-bit CQI index corresponds to a recommended modulation, code rate, and number of useful bits per symbol (efficiency or number of repetitions). The leftmost column shows the original Release 13 CQI table which is calibrated according to the coverage situation for the device using the parameter for the *number of subframes for the CSI reference resource* (R^{CSI}) in the range {1, 2, 4, 8, 16, 32} subframes, where R^{CSI} is an RRC configuration parameter. The base station is expected to configure the device with an R^{CSI} that roughly matches the foreseen required number of repetitions for reliable PDSCH transmission. Release 15 introduces two new CQI tables to support the downlink 64QAM feature (see Section 5.2.4.7). The CQI table in the center column is optimized for good coverage and assumes an R^{CSI} of 1 subframe, whereas the CQI table in the rightmost column is an *alternative CQI table* which spans a larger SNR range. The CQI range of the alternative CQI table can also be supported by LTE-M devices that do not support 64QAM, but in that case the reported CQI index is up to 12, which is the highest CQI value for 16QAM in the alternative CQI table. The alternative CQI table has a coarser granularity but a larger range than the other CQI tables, which means that it may be suitable for devices experiencing varying channel conditions which might otherwise require impractically frequent RRC reconfiguration of the R^{CSI} parameter in order to adjust the CQI reporting range.

5.3.2.3 Random access in connected mode

The device can initiate the random access procedure in connected mode when it needs to request an uplink TA command and/or an uplink grant. A *contention-based random access* is then performed, with similar RAR and random access message 3 transmissions as in the idle mode random access procedure described in Section 5.3.1.6. However, unlike in idle mode, message 3 does not include an RRC message, and because the device has already been assigned with a C-RNTI, which the device includes in message 3, the contention resolution in the fourth step is in this case performed using C-RNTI rather than TC-RNTI.

The base station can also order a device in CE mode A or B to initiate a random access procedure by sending a so-called *MPDCCH order* to the device. This is useful during handover to another cell or when downlink data transmission is resumed after a period of inactivity and the base station wants the device to reacquire uplink time alignment for the expected uplink responses to the downlink data transmission. A modified version of DCI Format 6-1A or 6-1B is used to transmit the order. A starting PRACH CE level can be indicated in the order. As in LTE, a dedicated PRACH preamble index can be indicated already in the MPDCCH order to allow *contention-free random access*, and in this case no explicit contention resolution phase is needed and the random access procedure ends already with the reception of the RAR. If no PRACH preamble index is indicated in the order, a contention-based random access is performed in the same way as for device-triggered connected mode random access.

5.3.2.4 Power control

Uplink closed-loop *transmit power control* (TPC) commands for PUSCH/PUCCH can be sent to LTE-M devices in CE mode A using the transmit power control field in the DCI Format 6-0A/6-1A described in Sections 5.3.2.1.1 and 5.3.2.1.2, or using DCI Format 3/3A addressed with TPC-PUSCH-RNTI or TPC-PUCCH-RNTI in *Type-0 MPDCCH CSS* (see Section 5.3.3.1). A single DCI with DCI Format 3/3A can carry power control commands to multiple devices. The standard supports several power control command mappings but the most commonly used one uses 2 bits to command a transmit power change of {-1, 0, +1, +3} dB. This is similar to the power control behavior in LTE.

However, LTE-M devices in CE mode B are expected to be in bad coverage and will therefore always transmit using the configured maximum transmission power. Similarly, in the random access procedure, when a device reaches the highest PRACH CE level (PRACH CE level 3), it will always transmit at maximum power during PRACH transmission.

5.3.2.5 Mobility support

Beside the cell selection and cell reselection mechanisms in idle mode described in Sections 5.3.1.1 and 5.3.1.3, LTE-M devices support connected mode mobility mechanisms such as handover, RRC connection release with redirection, RRC reestablishment, measurement reporting, etc., similarly as LTE devices [13].

Unlike ordinary LTE devices, low-cost LTE-M devices with reduced bandwidth support need *measurement gaps* not only for *interfrequency* measurements in connected mode but also for *intrafrequency* measurements in connected mode because they may need to retune their narrowband receiver to the center of the system bandwidth to receive the center 72 subcarriers where the PSS/SSS signals (used for acquisition and maintenance of downlink time and frequency synchronization and for identification of cells) are transmitted. The device may also perform *Radio Resource Management* (RRM) measurements such as RSRP while it has its receiver retuned to the center.

LTE-M supports intrafrequency RSRP measurements in connected mode in Release 13, and Release 14 introduces complete mobility support in idle and connected mode in the form of intrafrequency RSRP/RSRQ measurements and interfrequency RSRP/RSRQ measurements. These measurements are important for mobile and real-time use cases, such as wearables and voice services.

The LTE-M standard was initially developed with applications characterized by relatively low mobility in mind. However, the specified LTE-M physical layer is robust enough to be able to maintain good link quality also at somewhat higher device velocities. Since this is an attractive property for many IoT application, *enhanced performance requirements for higher velocity* were introduced for CE mode A in Release 15. These requirements are defined assuming that devices fulfilling the requirements should be able to support at least 240 km/h at 1 GHz or 120 km/h at 2 GHz carrier frequency.

A device in connected mode performs *Radio Link Monitoring* to determine whether it is *in sync* or *out of sync* with respect to the serving cell by comparing CRS-based measurements with thresholds known as *Qin* and *Qout* that correspond to 2% and 10% BLock Error Rate of hypothetical MPDCCH transmissions, respectively [17]. The comparison is done over a period known as the *evaluation period*. If the evaluation results in out of sync for more than a certain number of times (which is a configurable parameter *N310*) over a certain period of time (upon expiry of the *T310* timer), the device declares *Radio Link Failure* and turns off its transmitter to avoid causing unwanted interference. The device may be able to find a better cell through cell selection and reestablish the connection. The radio link monitoring procedure was further improved in Release 14 where two new reporting events were introduced, known as *earlyQin* and *earlyQout*, which are triggered earlier in time than the *Qin* and *Qout* events to give the network more time adapt the radio sources accordingly.

5.3.2.6 *Positioning*

Beside basic *Cell Identity* (CID) based positioning, LTE-M supports the LTE positioning techniques *Enhanced Cell Identity* (E-CID) and *Observed Time Difference of Arrival* (OTDOA). From protocol point of view, E-CID and OTDOA were supported already in Release 13, but complete measurement performance requirements for E-CID and OTDOA were not introduced for LTE-M devices until in Release 14. For a description of the general principles behind the E-CID and OTDOA positioning techniques, see Section 7.3.2.6 (the section concerns NB-IoT, but the principles are equally applicable to LTE-M).

Release 14 furthermore introduces *OTDOA enhancements* with respect to the configuration of the associated PRS the in time and frequency domains (see Section 5.2.4.3.3). The new PRS configurations in Release 14 allow LTE-M devices to achieve comparable positioning accuracy as ordinary LTE devices. The exact PRS configuration in a cell will depend on the desired trade-off between PRS overhead and positioning accuracy.

Release 15 introduced *new PRS measurement gap patterns* that enable devices to do OTDOA positioning measurements during measurement gaps in connected mode for PRS durations longer than 6 subframes, which facilitates positioning in challenging coverage conditions.

5.3.3 Procedures common for idle and connected mode

5.3.3.1 *MPDCCH search spaces*

The transmission opportunities for MPDCCH (described in Section 5.2.4.6) are defined in the form of *search spaces*. Each device monitors an MPDCCH search space for potential DCI transmissions addressed to one of the RNTIs monitored by the device. An MPDCCH search space typically contains *blind decoding candidates* with different numbers of MPDCCH repetitions.

Release 13 supports the following MPDCCH search spaces [15]:

1. *Type-1 common search space* (Type1-CSS) is monitored by the device at its paging occasions in idle mode.
2. *Type-2 common search space* (Type2-CSS) is monitored by the device during the random access procedure in idle and connected mode.
3. *UE-specific search space* (USS) is a user-specific search space monitored by the device in connected mode. This is where scheduling of downlink and uplink data transmissions usually take place.
4. *Type-0 common search space* (Type0-CSS) is monitored by the device when it is configured with CE mode A in connected mode. It can be used for power control commands and as fallback if the user-specific search space fails.

Release 14 introduces the following additional MPDCCH search spaces for multicast (see Section 5.3.1.9):

1. *Type-1A common search space* (Type1A-CSS) is monitored by the device at the transmission opportunities for the multicast control channel (SC-MCCH) in idle mode. It is largely based on the Type1-CSS design.

2. *Type-2A common search space* (Type2A-CSS) is monitored by the device at the transmission opportunities for the multicast traffic channel (SC-MTCH) in idle mode. It is largely based on the Type2-CSS design.

An MPDCCH search space is defined by the following key parameters [23]:

1. *The MPDCCH narrowband index* indicates one of the narrowbands within the system bandwidth. The total number of narrowbands for each system bandwidth is shown in Table 5.4.
2. *The number of MPDCCH PRB pairs* is in the range {2, 4, 6} PRB pairs. For Type1-CSS and Type2-CSS, the number of MPDCCH PRB pairs is fixed to 6 PRB pairs.
3. *The MPDCCH resource block assignment* indicates the positions of the PRB pairs. If the number of PRB pairs is six (as is always the case for Type1-CSS and Type2-CSS) then this parameter is not needed because all PRB pairs within the narrowband are then included in the resource block assignment.
4. *The maximum MPDCCH repetition factor* (R_{max}) indicates the repetition factor for the candidate with the largest repetition factor in the search space. The range for this parameter is {1, 2, 4, 8, 16, 32, 64, 128, 256}.
5. *The relative MPDCCH starting subframe periodicity* (G) is used to determine the starting subframe periodicity for the search space. The range for this parameter is {1, 1.5, 2, 2.5, 4, 5, 8, 10} in FDD and {1, 2, 4, 5, 8, 10, 20} in TDD. The *absolute MPDCCH starting subframe periodicity* (*T*) in terms of subframes is calculated as $T = R_{max}G$.

The configuration parameters for Type1-CSS and Type2-CSS are signaled in SIB2, whereas the configuration parameters for USS and Type0-CSS are signaled in user-specific RRC signaling. A device in CE mode A monitors both USS and Type0-CSS, but this is facilitated in the device by the fact that these two search spaces share the same properties listed above. LTE-M devices in CE mode B or in idle mode never need to monitor more than a single search space at a time.

Further details about the search spaces are listed in Table 5.32 [15].

Within an MPDCCH search space, different candidates can have different ECCE aggregation levels and different repetition factors (*R*). As explained in Section 5.2.4.6, in normal subframes with normal CP length, aggregation of 2, 4, 8, 16 or 24 ECCEs is supported. With the smallest ECCE aggregation level (i.e., 2), half of the ECCEs available in a PRB pair are aggregated, and with the highest ECCE aggregation level (i.e., 24), all the ECCEs available in a narrowband are aggregated. The available repetition levels R_1, R_2, R_3, and R_4 depend on R_{max} according to Table 5.33. For more details on search space definitions, see Section 9.1.5 in Reference [15].

Fig. 5.30 shows an MPDCCH search space example with $R_{max} = 4$ and $G = 1.5$. In this search space, an MPDCCH can be scheduled without repetition ($R = 1$) in the subframes marked *A*, *B*, *C*, and *D*, or with two times repetition ($R = 2$) in the subframes marked *AB* and *CD*, or with four times repetition ($R = 4$) in the subframes marked *ABCD*. If an MPDCCH is transmitted according to candidate *A* then candidates *AB* and *ABCD* are blocked in that *T* period but candidates *B*, *C*, *D*, and *CD* can still be used in the same *T* period. The $T - R_{max} = 2$ subframes between consecutive search spaces are not included in the search

TABLE 5.32 MPDCCH search spaces, RNTIs, and DCI formats monitored by LTE-M devices.

Mode	MPDCCH search space	RNTI	Usage	DCI format	Section
Idle	–	SI-RNTI	Broadcast of system information	–	5.3.1.2
	Type-1 common	P-RNTI	Paging and SI update notification	6–2	5.3.1.4
	Type-2 common	RA-RNTI	Random access response	6-1A, 6-1B	5.3.1.6
		TC-RNTI	HARQ retransmission of random access message 3	6-0A, 6-0B	5.3.1.6
		TC-RNTI	Random access contention resolution with message 4	6-1A, 6-1B	5.3.1.6
	Type-1A common	SC-RNTI	Scheduling of SC-MCCH	6–2	5.3.1.9
	Type-2A common	G-RNTI	Scheduling of SC-MTCH	6-1A, 6-1B	5.3.1.9
Connected	UE-specific	C-RNTI	Random access order	6-1A, 6-1B	5.3.2.3
		C-RNTI	Dynamic downlink scheduling	6-1A, 6-1B	5.3.2.1.1
		C-RNTI	Dynamic uplink scheduling	6-0A, 6-0B	5.3.2.1.2
		SPS-C-RNTI	Semipersistent downlink scheduling	6-1A	5.3.2.1.3
		SPS-C-RNTI	Semipersistent uplink scheduling	6-0A	5.3.2.1.3
	Type-0 common (only in CE mode A)	C-RNTI	Random access order	6-1A	5.3.2.3
		C-RNTI	Dynamic downlink scheduling	6-1A	5.3.2.1.1
		C-RNTI	Dynamic uplink scheduling	6-0A	5.3.2.1.2
		SPS-C-RNTI	Semipersistent uplink scheduling	6-0A	5.3.2.1.3
		TPC-PUCCH-RNTI	PUCCH power control	3, 3A	5.3.2.4
		TPC-PUSCH-RNTI	PUSCH power control	3, 3A	5.3.2.4
	Type-2 common	RA-RNTI	Random access response	6-1A, 6-1B	5.3.2.3
		TC-RNTI	HARQ retransmission of random access message 3	6-0A, 6-0B	5.3.2.3
		C-RNTI	Random access contention resolution	6-0A, 6-0B, 6-1A, 6-1B	5.3.2.3

space (which means that the device can go to sleep during these subframes unless it has some other reason to stay awake during these subframes).

Fig. 5.31 shows an MPDCCH search space example for Type1-CSS, the CSS used for paging. As can be seen, all candidates in the search space start in the same subframe.

TABLE 5.33 Repetition levels for MPDCCH search spaces.

	Repetition levels for Type1-CSS				Repetition levels for USS, Type0-CSS and Type2-CSS			
R_{max}	R_1	R_2	R_3	R_4	R_1	R_2	R_3	R_4
256	2	16	64	256	32	64	128	256
128	2	16	64	128	16	32	64	128
64	2	8	32	64	8	16	32	64
32	1	4	16	32	4	8	16	32
16	1	4	8	16	2	4	8	16
8	1	2	4	8	1	2	4	8
4	1	2	4	–	1	2	4	–
2	1	2	–	–	1	2	–	–
1	1	–	–	–	1	–	–	–

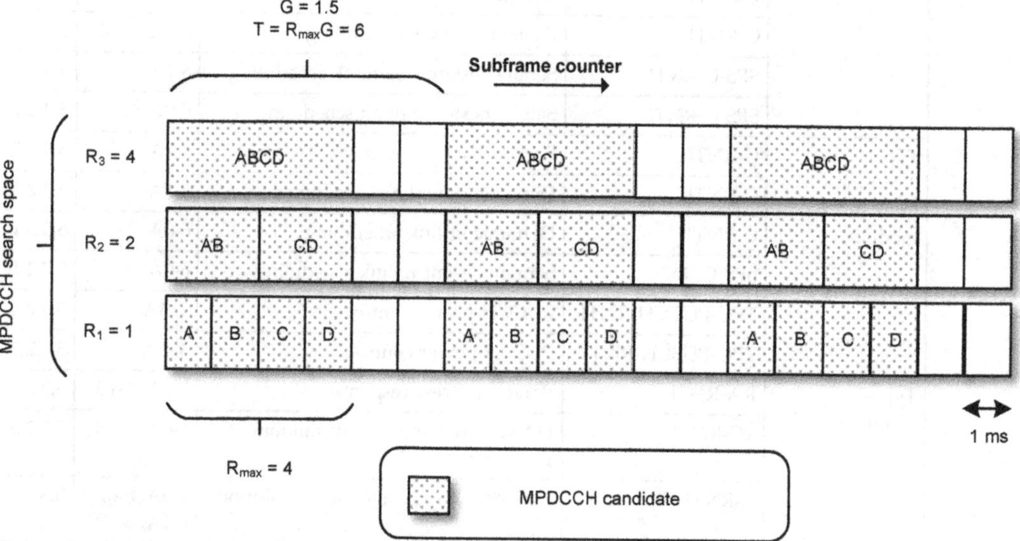

FIG. 5.30 MPDCCH search space example for USS, Type0-CSS and Type2-CSS.

This allows devices in good coverage to go to sleep after they have detected that there is no transmission intended for them in the first subframe of the search space (i.e., in candidate *A*). If a search space such as the one in Fig. 5.30 was used also for paging, the device would have to stay awake longer since it would have to be prepared for the eventuality that the base station choses to page the device in candidates *B*, *C* or *D*.

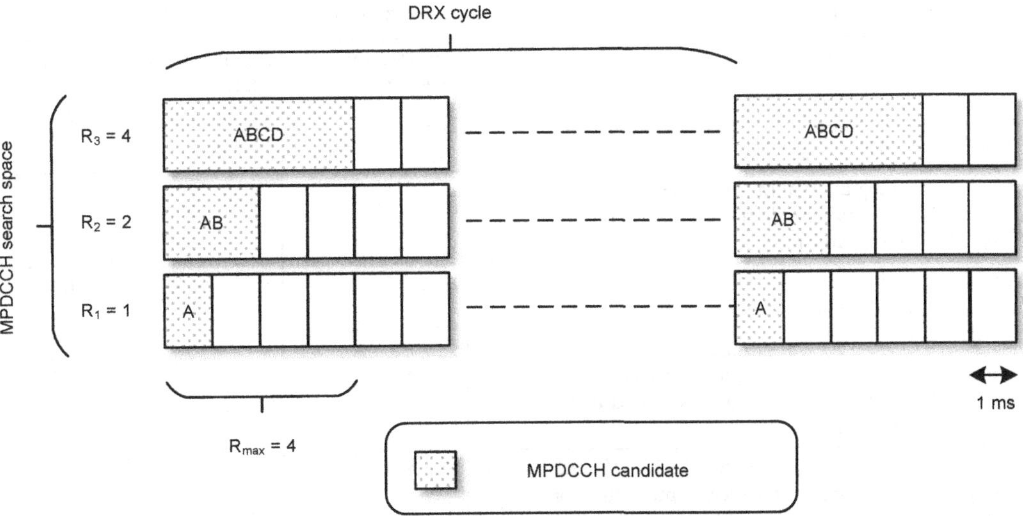

FIG. 5.31 MPDCCH search space example for Type1-CSS.

5.3.3.2 Frequency hopping

LTE-M transmissions are restricted to a narrowband, but LTE-M provides means for frequency diversity through frequency hopping between different narrowbands for all physical channels and signals except PSS/SSS, PBCH, RSS, MWUS, and SRS. As described in Sections 5.2.4.2 and 5.2.4.4, PSS/SSS and PBCH are always located at the center of the system bandwidth, similarly as in ordinary LTE. If RSS or MWUS is supported in the cell, the configuration information in SIB2 includes their frequency locations (see Sections 5.2.4.2.2 and 5.2.4.5).

Table 5.34 lists cell-specific configuration parameters for the time intervals and frequency offsets for frequency hopping in LTE-M. The time intervals indicate when the hops should take place in the time domain and the offsets indicate how large the hops should be in the frequency domain. The time intervals are synchronized so that the frequency hops for transmissions to/from different devices can take place at the same time. The frequency hops take place during the guard periods for frequency retuning described in Sections 5.2.4.1 and 5.2.5.1. The parameters for frequency hopping intervals for MPDCCH/PDSCH serve a double purpose since they also indicate the interval during which the device can assume that the MPDCCH/PDSCH precoding remains the same (as mentioned in Sections 5.2.4.6 and 5.2.4.7).

For SIB1 and PUCCH, the frequency hopping is fixed in the LTE-M standard but for all other types of transmission it is up to the network whether to use frequency hopping or not. Table 5.35 lists the frequency hopping activation methods for different types of LTE-M transmission.

When frequency hopping is used in downlink, a parameter in SIB1 (listed in Table 5.34) controls whether the hopping is between 2 and 4 narrowbands. The number of narrowbands

TABLE 5.34 Cell-specific time intervals and frequency offsets for frequency hopping in LTE-M.

Parameter	Possible values in FDD	Possible values in TDD	Signaled in
Number of narrowbands for frequency hopping for MPDCCH/PDSCH	2, 4	2, 4	SIB1
Frequency hopping interval [ms] for MPDCCH/PDSCH in CE mode A and during random access procedure for PRACH CE levels 0 and 1	1, 2, 4, 8	1, 5, 10, 20	SIB1
Frequency hopping interval [ms] for MPDCCH/PDSCH in CE mode B and during random access procedure for PRACH CE levels 2 and 3	2, 4, 8, 16	5, 10, 20, 40	SIB1
Frequency hopping interval [ms] for PUCCH/PUSCH in CE mode A and during random access procedure for PRACH CE levels 0 and 1	1, 2, 4, 8	1, 5, 10, 20	SIB2
Frequency hopping interval [ms] for PUCCH/PUSCH in CE mode B and during random access procedure for PRACH CE levels 2 and 3	2, 4, 8, 16	5, 10, 20, 40	SIB2
Frequency hopping offset for MPDCCH/PDSCH [narrowbands]	1–16	1–16	SIB1
Frequency hopping offset for PUSCH [narrowbands]	1–16	1–16	SIB2
Frequency hopping offset for PRACH [PRBs]	0–94	0–94	SIB2

used for frequency hopping for the PDSCH transmission that carries the SIB1 itself is fixed in the standard (as described in Table 5.35). The hopping pattern for the SIB1 transmission starts in a frame with an SFN evenly divisible by 8, in a narrowband that depends on the PCID, and hops between 2 and 4 narrowbands that have been selected so that they avoid the two center narrowbands in the system bandwidth to avoid collision with the center 72 subcarriers (the PSS/SSS/PBCH region).

When frequency hopping is used in uplink, the hopping is always between two frequency locations. For PUCCH, frequency hopping is always active, and the hopping occurs between two PRB locations that are symmetric with respect to the center frequency of the LTE system bandwidth. It should be noted that typically it is only PUCCH transmissions that are longer than the uplink frequency hopping interval that enjoy a frequency diversity gain from one or more frequency hops during the PUCCH transmission, whereas shorter PUCCH transmissions are short enough to be transmitted in their entirety in one of the two PRB locations before the time comes for a frequency hop. Fig. 5.32 shows an example of uplink frequency hopping with an uplink frequency hopping interval of 2 ms, a PUCCH transmission with repetition factor 4, a PUSCH transmission with repetition factor 8, and a PUSCH frequency hopping offset of two narrowbands.

TABLE 5.35 Frequency hopping activation methods available in an LTE-M cell.

Type of transmission	Physical channel(s)	Frequency hopping activation method(s)
SIB1	PDSCH	The number of narrowbands for SIB1 transmission depends strictly on the downlink system bandwidth signaled in MIB: For 1.4 MHz: no frequency hopping. For 3, 5, or 10 MHz: hopping between 2 narrowbands. For 15 or 20 MHz: hopping between 4 narrowbands.
SI message	PDSCH	Frequency hopping for SI messages and paging messages is activated by a common activation bit in SIB1.
Paging	MPDCCH, PDSCH	
Random access preamble	PRACH	Frequency hopping for PRACH is activated by an activation bit per PRACH CE level in SIB2.
Random access response and random access messages 3 and 4	MPDCCH, PDSCH, PUSCH, PUCCH	Frequency hopping for RAR, message 3, and message 4 is activated by an activation bit per PRACH CE level in SIB2.
Unicast downlink data transmission	MPDCCH, PDSCH	Frequency hopping for unicast downlink data transmission is activated by an activation bit in user-specific RRC signaling.
		In CE mode A, the frequency hopping can furthermore be deactivated by the frequency hopping bit in DCI Format 6-1A.
Unicast uplink data transmission	MPDCCH, PUSCH	Frequency hopping for unicast uplink data transmission is activated by an activation bit in user-specific RRC signaling.
		In CE mode A, the frequency hopping can furthermore be deactivated by the frequency hopping bit in DCI Format 6-0A.
HARQ-ACK, SR, CSI	PUCCH	Frequency hopping for PUCCH is always activated.
Multicast control channel	MPDCCH, PDSCH	Frequency hopping for the multicast control channel (SC-MCCH) is activated by a parameter in SIB20, and the parameter also indicates whether the hopping should follow the SIB2 frequency hopping parameters for CE mode A or B.
Multicast traffic channel	MPDCCH, PDSCH	Frequency hopping for a multicast traffic channel (SC-MTCH) is activated by a parameter in SC-MCCH, and another parameter indicates whether the hopping should follow the SIB2 frequency hopping parameters for CE mode A or B. If the hopping follows CE mode A, it can furthermore be deactivated by the frequency hopping bit in DCI Format 6-1A.
Positioning reference signal	PRS	Frequency hopping for PRS can be configured in LTE Positioning Protocol (LPP) between 2 or 4 frequency locations if the PRS bandwidth is 6 PRBs. The first frequency location is in the center of the LTE system bandwidth and the other frequency locations(s) are in configurable narrowbands.

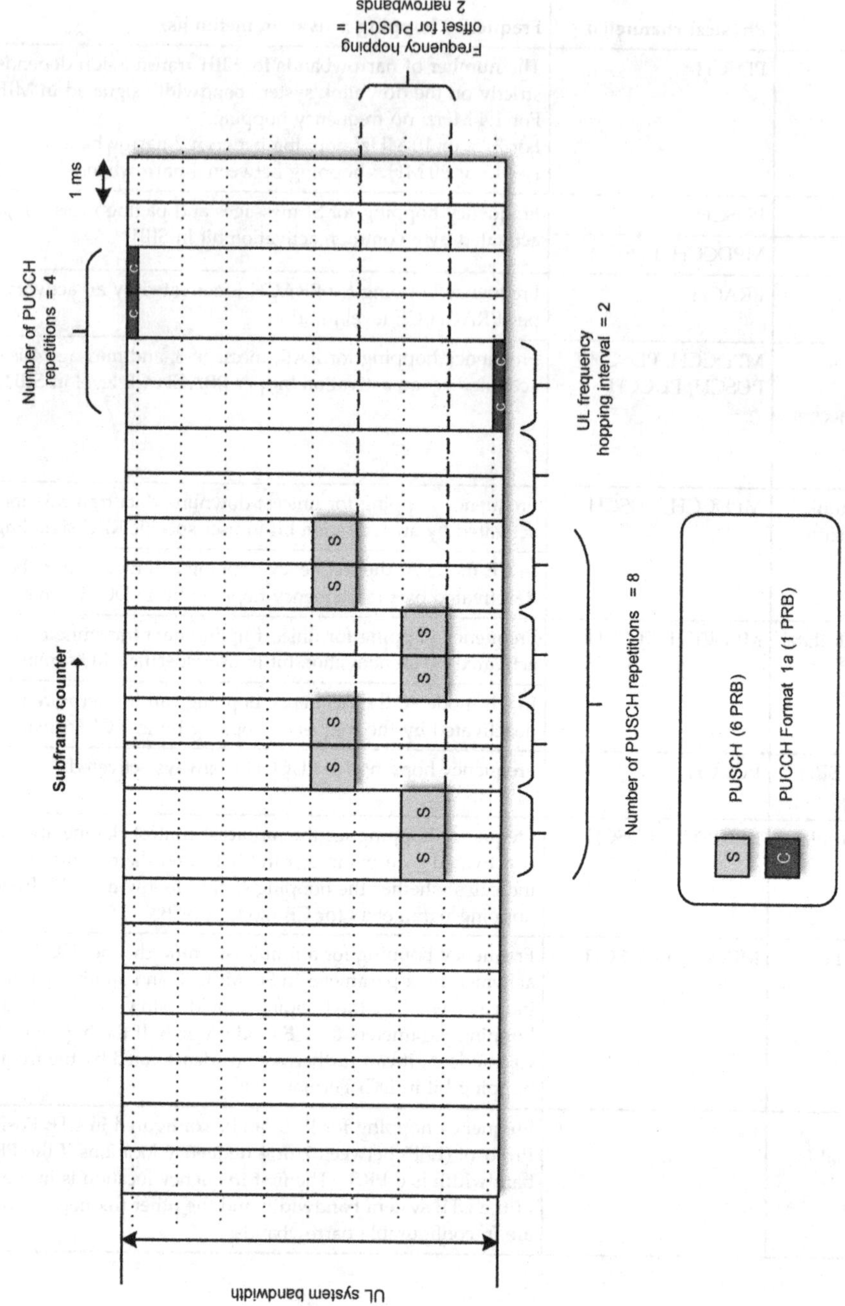

FIG. 5.32 Uplink frequency hopping in LTE-M.

5.4 NR and LTE-M coexistence

3GPP Release 15 introduces a NR-access technology known as *New Radio* (NR), which offers significant performance advantages over LTE in terms of data rates, latency, deployment flexibility and energy efficiency (see Chapter 2.4 for an introduction of NR and Reference [27] for detailed description of NR). NR is designed for enhancing the performance of mobile broadband and *ultra-reliable low-latency communications* (URLLC) services (described in Chapter 11), whereas *low-power wide-area* (LPWA) IoT use cases are expected to already be adequately addressed by existing 3GPP technologies such as LTE-M and NB-IoT. It has been shown that LTE-M and NB-IoT fulfill the 5G massive MTC requirements (see performance evaluations in Chapters 6 and 8, respectively) and thus qualify as 5G component technologies. Therefore, it is desired that LTE-M and NB-IoT can coexist efficiently with NR also after a potential migration from LTE to NR. NR coexistence with LTE-M is covered in this section, and NR coexistence with NB-IoT is covered in Section 7.4.

LTE-M is defined in LTE bands 1, 2, 3, 4, 5, 7, 8, 11, 12, 13, 14, 18, 19, 20, 21, 25, 26, 27, 28, 31, 39, 40, 41, 66, 71, 72, 73, 74 and 85 (see Table 5.3 for the frequency ranges of these bands). Many of these bands are also defined for NR [28]. Table 5.36 lists all the bands that are defined for both NR and LTE-M. These bands can thus be used to deploy both NR and LTE-M.

LTE-M is an extension of LTE and can therefore operate seamlessly within an LTE carrier, as described in Section 5.1.2.6. As we will see in this section, LTE-M and NR support several mechanisms that can enable almost as seamless operation of LTE-M within an NR carrier. It should be noted that in an NR coexistence scenario, the LTE-M system bandwidth can be smaller than the NR system bandwidth (this is different from LTE coexistence scenarios, where the LTE-M system bandwidth is always the same as the LTE system bandwidth).

It is up to the base station implementation whether the time-frequency resource split between LTE-M and NR is completely static or more dynamic. Fig. 5.33 illustrates static resource sharing, where a small slice of the available spectrum is allocated to LTE-M and not used for NR, and the rest of the spectrum is used for NR. The slice corresponds to the number of PRBs in the configured LTE-M system bandwidth (see Table 5.4), i.e. it can be as narrow as 6 PRBs and only contain a single LTE-M narrowband if it is desired to minimize the impact on the amount of resources available for NR. If the foreseen amount of LTE-M traffic is relatively small, a static resource split between NR and LTE-M with a small LTE-M system bandwidth may be a suitable configuration.

A more flexible resource sharing is illustrated in Fig. 5.34, where a large portion of the available resources serves as a common resource pool for NR and LTE-M, and it is up to the base station to allocate resources dynamically from this resource pool to NR and LTE-M devices. The shared part corresponds to the number of PRBs in the configured LTE-M system bandwidth. As shown in Table 5.4, the LTE-M system bandwidth can range from 6 to 100 PRBs and contain as few as a single narrowband or as many as 16 narrowbands. A more dynamic resource sharing has the potential to provide more efficient resource utilization through resource pooling but may require a more complex base station implementation. The standardized mechanisms for supporting static or dynamic resource sharing will be described in the remainder of this section.

TABLE 5.36 Frequency bands that are defined for both NR and LTE-M.

Band	Duplex mode	Uplink [MHz]	Downlink [MHz]	NR channel bandwidth for 15 kHz subcarrier spacing [MHz]	NR channel raster [kHz]
1	FDD	1920–1980	2110–2170	5, 10, 15, 20	100
2	FDD	1850–1910	1930–1990	5, 10, 15, 20	100
3	FDD	1710–1785	1805–1880	5, 10, 15, 20, 25, 30	100
5	FDD	824–849	869–894	5, 10, 15, 20	100
7	FDD	2500–2570	2620–2690	5, 10, 15, 20	100
8	FDD	880–915	925–960	5, 10, 15, 20	100
12	FDD	699–716	729–746	5, 10, 15	100
20	FDD	832–862	791–821	5, 10, 15, 20	100
25	FDD	1850–1915	1930–1995	5, 10, 15, 20	100
28	FDD	703–748	758–803	5, 10, 15, 20	100
39	TDD	1880–1920	1880–1920	5, 10, 15, 20, 25, 30, 40	100
40	TDD	2300–2400	2300–2400	5, 10, 15, 20, 25, 30, 40, 50	100
41	TDD	2496–2690	2496–2690	10, 15, 20, 40, 50	15 or 30
66	FDD	1710–1780	2110–2200	5, 10, 15, 20, 40	100
71	FDD	636–698	617–652	5, 10, 15, 20	100
74	FDD	1427–1470	1475–1518	5, 10, 15, 20	100

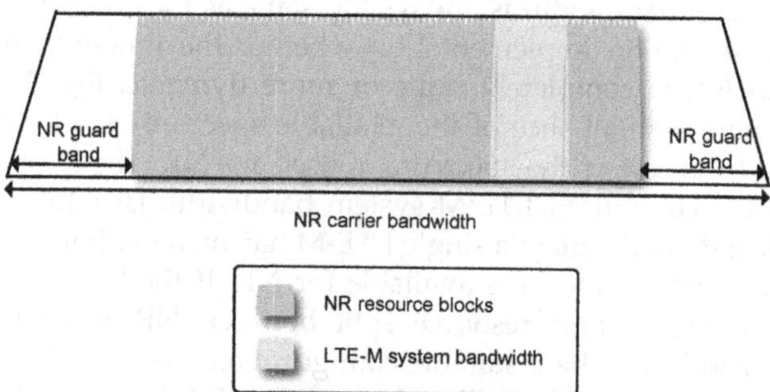

FIG. 5.33 Static resource sharing between LTE-M and NR.

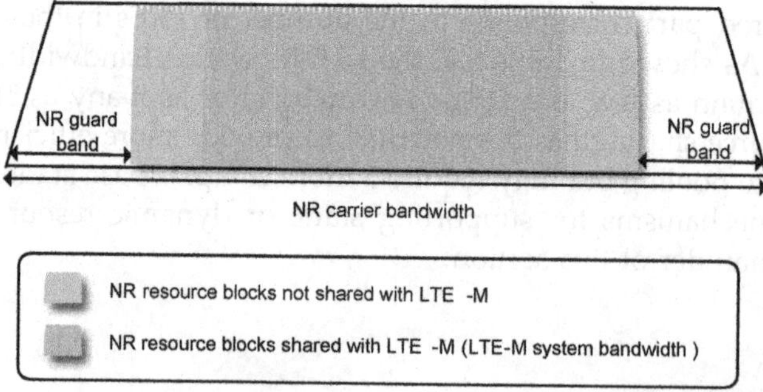

FIG. 5.34 Dynamic resource sharing between LTE-M and NR.

To ensure that LTE-M legacy devices can operate without knowing any NR-specific operation, one fundamental aspect is that an LTE-M device can identify an LTE-M cell during the initial cell selection process (see Section 5.3.1.1). To achieve this, the LTE-M carrier center frequency needs to be placed according to a 100-kHz channel raster (see Section 5.2.1.1). Furthermore, as PSS, SSS and PBCH are the signals used by the LTE-M device during cell search, they need to be preserved.

Similarly, to ensure that NR devices can operate regardless of whether there is an LTE-M carrier deployed in the same band or not, the NR carrier needs to be placed according to NR channel raster shown in Table 5.36. In most cases, the NR channel raster is 100 kHz. Furthermore, the NR *synchronization signal block* (SSB) needs to be preserved as it is used by the NR device during cell search.

Meanwhile, the interference between NR and LTE-M should be minimized so that the impact on NR and LTE-M performance is negligible when they are deployed in the same band. Unlike LTE and LTE-M, NR supports multiple subcarrier configurations. If the NR carrier is configured with the same subcarrier spacing as LTE-M, i.e., 15 kHz subcarrier spacing, it is desirable to align the NR and LTE-M subcarriers on the same subcarrier grids, with the frequencies of an NR subcarrier and an LTE-M subcarrier differing by an integer multiple of 15 kHz. With this, if the NR and LTE-M networks are synchronized, NR subcarriers and LTE-M subcarriers are mutually orthogonal. If the NR carrier is configured with subcarrier spacing other than 15 kHz, a guard band is needed between NR and LTE-M to ensure minimal inter-subcarrier interference.

It can be shown that for 15 kHz NR subcarrier spacing and 100 kHz NR carrier raster, every 20th NR subcarrier (i.e., every 300 kHz) relative to the NR channel raster will coincide with an LTE-M raster position [29]. One out of three possible LTE-M channel raster positions fulfills this condition. If the carrier positions are selected according to this relation, subcarrier grid alignment between NR and LTE-M is achieved in downlink (but not necessarily in uplink, which will be discussed next).

LTE-M, like LTE but unlike NR, has an unused *DC subcarrier* at the center of the downlink system bandwidth (see Section 5.2.2). However, no DC subcarrier is inserted in uplink. As a result, the LTE-M downlink system bandwidth is one subcarrier wider than the LTE-M uplink system bandwidth and one subcarrier wider than the NR system bandwidth. This is illustrated in Fig. 5.35. In order to achieve subcarrier grid alignment also in uplink, the DC subcarrier needs to be considered. In LTE-M, the channel raster points to the unused DC subcarrier in the center of the LTE downlink system bandwidth, whereas the NR channel raster points to an ordinary NR subcarrier near the middle of the NR carrier. For an NR carrier with N resource blocks, both in downlink and uplink there are $12\,N$ subcarriers indexed from 0 to $12N\text{-}1$, and the channel raster is mapped to subcarrier $6\,N$ [27]. For an LTE-M carrier with M resource blocks, there are $12\,M+1$ downlink subcarriers (including the DC subcarrier) indexed from 0 to $12\,M$ and the channel raster is mapped to subcarrier $6\,M$, but there are only $12\,M$ uplink subcarriers (no DC subcarrier). In order to enable simultaneous subcarrier grid alignment between NR and LTE-M in both downlink and uplink, NR supports a *configurable uplink half-subcarrier shift*. When the base station has enabled this shift in the cell, a half subcarrier (+7.5 kHz) shift is applied to the NR uplink. Release 15 supports this shift in NR FDD but not in NR TDD, so coexistence between NR and LTE-M in TDD

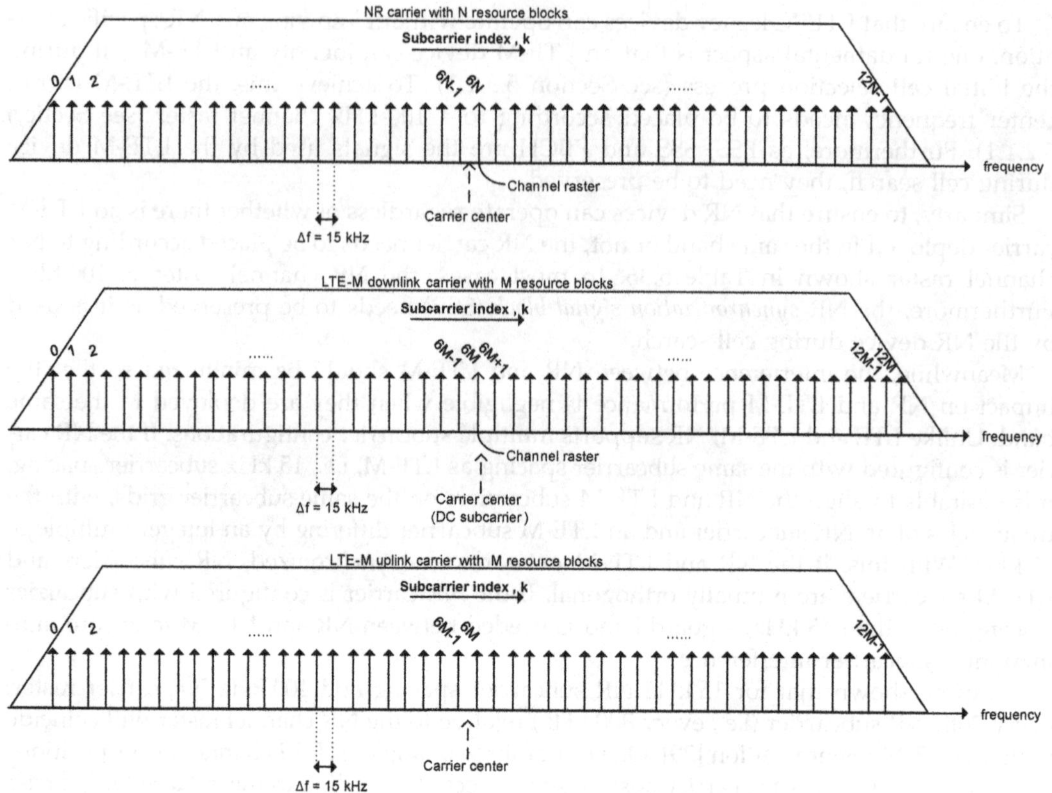

FIG. 5.35 Locations of NR and LTE-M channel raster (assuming the NR carrier is configured with 15 kHz subcarrier spacing).

bands cannot use this shift and instead need to rely on a sufficiently large inter-carrier guard band between NR and LTE-M.

It should be noted that due to the mentioned DC subcarrier, perfect alignment between NR resource blocks and LTE-M resource blocks is not possible. When alignment cannot be achieved, it may be necessary to reserve *M +1* NR downlink resource blocks in order to fit *M* LTE-M resource blocks.

Collision between NR and LTE-M transmissions can be avoided by exploiting the mechanisms for resource reservation that are available in LTE-M and NR.

On the LTE-M side, the base station can use the *LTE-M downlink subframe bitmap* and the *LTE-M starting symbol* (both described in Section 5.2.4.1) to restrict the LTE-M transmissions within the cell. The downlink subframe bitmap can be used to protect periodic NR transmissions such as NR SSB, and the LTE-M starting symbol should be configured large enough to avoid collision with NR PDCCH transmissions in the beginning of the downlink subframe. Similarly, the *LTE-M uplink subframe bitmap* (described in Section 5.2.5.1) can be used to protect potential periodic NR uplink resources if needed.

On the NR side, the base station can use *NR downlink resource reservation bitmaps* to indicate to NR devices that some particular resource blocks and/or OFDM symbols are restricted from NR PDSCH transmission. The NR devices will take these restrictions into account in the NR PDSCH rate-matching. These reservation mechanisms can be used for protecting periodic LTE-M transmissions such as PSS, SSS, PBCH, SIB1 and SI messages.

NR also supports resource element (RE) level *reservation of REs occupied by LTE CRS*. If the base station indicates that an LTE or LTE-M carrier is transmitted in some part of the NR carrier, the NR devices will take the LTE CRS into account in the NR PDSCH rate-matching. This mechanism is an important enabler for the more dynamic resource sharing approach illustrated in Fig. 5.34, where a portion of the NR system bandwidth can be used as a resource pool shared between NR and LTE-M, achieving the same LTE-M scheduling flexibility within an NR carrier as within an LTE carrier (as described in Section 5.1.2.6). The *CRS muting* feature introduced in LTE-M in Release 15 (see Section 5.2.4.3.1) can potentially be useful in this context for reducing the amount of LTE CRS transmission that needs to take place within the NR carrier.

When deciding how to use these tools for coexistence between NR and LTE-M, the foreseen traffic mix should be considered. If only a small LTE-M traffic load is expected, it may be enough to configure a small LTE-M system bandwidth, whereas LTE-M traffic with high average load or large traffic peaks may require configuration of a larger LTE-M system bandwidth. If a large LTE-M system bandwidth is configured, dynamic resource sharing (as depicted in Fig. 5.33) between LTE-M and NR may be needed in order to be able to utilize the resource efficiently, whereas static resource sharing (as depicted in Fig. 5.33) may be considered sufficiently efficient if the LTE-M system bandwidth is small enough compared to the NR system bandwidth.

References

[1] Third Generation Partnership Project, Technical Report 36.888, v12.0.0. Study on Provision of Low-Cost Machine-Type Communications (MTC) User Equipments (UEs) based on LTE, 2013.

[2] Vodafone Group, et al. RP-140522, Revised Work Item on Low Cost & Enhanced Coverage MTC UE for LTE, 3GPP RAN Meeting #63, 2014.

[3] Ericsson, et al. RP-141660, Further LTE Physical Layer Enhancements for MTC, 3GPP RAN Meeting #65, 2014.

[4] GSM Association. LTE-M Deployment Guide to Basic Feature Set Requirements, June 2019. Available from: https://www.gsma.com/iot/wp-content/uploads/2019/07/201906-GSMA-LTE-M-Deployment-Guide-v3.pdf.

[5] Global Mobile Suppliers Association. NB-IoT and LTE-M: Global Market Status, August 2018.

[6] Ericsson, et al. RP-170532, Revised WID Proposal on Further Enhanced MTC for LTE, 3GPP RAN Meeting #75, 2017.

[7] Ericsson, et al. RP-172811, Revised WID on Even further enhanced MTC for LTE, 3GPP RAN Meeting #77, 2017.

[8] Third Generation Partnership Project, Technical Report 37.910, v1.1.0. Study on Self Evaluation towards IMT-2020 Submission, 2018.

[9] Third Generation Partnership Project, Technical Specification 36.101, v15.6.0. Evolved Universal Terrestrial Radio Access (E-UTRA); User Equipment (UE) Radio Transmission and Reception, 2019.

[10] Third Generation Partnership Project, Technical Specification 36.211, v15.5.0. Evolved Universal Terrestrial Radio Access (E-UTRA); Physical Channels and Modulation, 2019.

[11] Third Generation Partnership Project, Technical Specification 36.306, v15.4.0. Evolved Universal Terrestrial Radio Access (E-UTRA); User Equipment (UE) Radio Access Capabilities, 2019.

[12] Intel Corporation, R1-1611936, Soft Buffer Requirement for feMTC UEs with Larger Max TBS, 3GPP RAN1 Meeting #87, 2016.

[13] Third Generation Partnership Project, Technical Specification 36.300, v15.4.0. Evolved Universal Terrestrial Radio Access (E-UTRA) and Evolved Universal Terrestrial Radio Access Network (E-UTRAN); Overall Description; Stage 2, 2019.
[14] Third Generation Partnership Project, Technical Specification 36.321, v15.5.0. Evolved Universal Terrestrial Radio Access (E-UTRA); Medium Access Control (MAC) Protocol Specification, 2019.
[15] Third Generation Partnership Project, Technical Specification 36.213, v15.5.0. Evolved Universal Terrestrial Radio Access (E-UTRA); Physical Layer Procedures, 2019.
[16] Third Generation Partnership Project, Technical Specification 36.214, v15.3.0. Evolved Universal Terrestrial Radio Access (E-UTRA); Physical Layer; Measurements, 2019.
[17] Third Generation Partnership Project, Technical Specification 36.133, v15.6.0. Evolved Universal Terrestrial Radio Access (E-UTRA); Requirements for Support of Radio Resource Management, 2019.
[18] Third Generation Partnership Project, Technical Specification 36.355, v15.2.0. Evolved Universal Terrestrial Radio Access (E-UTRA), LTE Positioning Protocol (LPP), 2019.
[19] Third Generation Partnership Project, Technical Specification 36.455, v15.2.0. Evolved Universal Terrestrial Radio Access (E-UTRA), LTE Positioning Protocol A (LPPa), 2019.
[20] Third Generation Partnership Project, Technical Specification 36.212, v15.5.0. Evolved Universal Terrestrial Radio Access (E-UTRA), Multiplexing and Channel Coding, 2019.
[21] Third Generation Partnership Project, Technical Specification 24.368, v15.1.0. Non-Access Stratum (NAS) configuration; Management Object (MO), 2018.
[22] Third Generation Partnership Project, Technical Specification 31.102, v15.5.0. Characteristics of the Universal Subscriber Identity Module (USIM) application, 2019.
[23] Third Generation Partnership Project, Technical Specification 36.331, v15.5.0. Evolved Universal Terrestrial Radio Access (E-UTRA); Radio Resource Control (RRC), Protocol Specification, 2019.
[24] Ericsson. R4-1806756, Simulation results of CGI reading for eFeMTC UE, in: 3GPP RAN4 Meeting #87, 2018.
[25] Third Generation Partnership Project, Technical Specification 36.304, v15.3.0. Evolved Universal Terrestrial Radio Access (E-UTRA); User Equipment (UE) Procedures in Idle Mode, 2019.
[26] Third Generation Partnership Project, Technical Specification 36.413, v15.5.0. Evolved Universal Terrestrial Radio Access Network (E-UTRAN); S1 Application Protocol (S1AP), 2019.
[27] E. Dahlman, S. Parkvall, J. Sköld. 5G NR: The Next Generation Wireless Access Technology, Academic Press, Oxford, 2018.
[28] Third Generation Partnership Project, Technical Specification 38.104, v15.5.0. NR; Base Station (BS) radio transmission and reception, 2019.
[29] M. Mozaffari, Y.-P. E. Wang, O. Liberg, J. Bergman. Flexible and Efficient Deployment of NB-IoT and LTE-MTC in Coexistence with 5G New Radio. Proc. IEEE int. Conf. on Computer Commun., Paris, France, 29 April–2 May, 2019.

CHAPTER

6

LTE-M performance

OUTLINE

Abstract

This chapter presents LTE-M performance in terms of coverage, data rate, latency, and system capacity based on the functionality described in Chapter 5. The presented performance evaluations are largely following the International Mobile Telecommunication 2020 (IMT-2020) and 5G evaluation frame work defined by the International Telecommunications Union Radiocommunication sector (ITU-R) and 3GPP, respectively. It is shown that LTE-M in all aspects meets the massive machine-type communications (mMTC) part of the requirements defined by ITU-R and 3GPP. The reduction in device complexity achieved by LTE-M compared to higher LTE device categories is also presented. While LTE-M has been specified for half-duplex frequency-division duplexing (HD-FDD), full-duplex FDD (FD-FDD) and time-division duplexing (TDD) operation, this chapter focuses on the performance achievable for LTE-M HD-FDD.

6.1 Performance objectives

The work on LTE-M started in the 3GPP with the *Study on provision of low-cost machine-type communications (MTC) user equipment's based on LTE* [1], in this book referred to as the *LTE-M study item*. The initial objective of the LTE-M study item was to identify solutions providing reduced device complexity. Later a second objective was added to define solutions providing

Cellular Internet of Things, Second Edition
https://doi.org/10.1016/B978-0-08-102902-2.00006-6

© 2020 Elsevier Ltd. All rights reserved.

substantial coverage enhancement (CE). Some of the identified complexity reduction techniques were specified in 3GPP Release 12 in the form of a new LTE device category (Cat-0). Release 13 enabled further device cost reduction in the form of a second low complexity device category (Cat-M1) and specifies two CE modes originally targeting at least 15 dB improved coverage compared to normal LTE coverage. This work is briefly introduced in Section 2.2.4, where the LTE reference coverage is shown to correspond to 140.7 dB maximum coupling loss (MCL).

In Release 15 the International Telecommunications Union Radiocommunication sector (ITU-R) defined the set of International Mobile Telecommunication 2020 (IMT-2020) requirements for enhanced mobile broad band, critical MTC and massive MTC (mMTC). The mMTC objective on connection density required the support of 1,000,0000 devices per km^2 [2]. 3GPP reused this objective in its work on 5G and did in addition to this define four more requirements for mMTC [3]:

- A coverage corresponding to a MCL of 164 dB should be supported.
- A sustainable data rate of at least 160 bits per second should be supported at the 164 dB MCL.
- A small data transmission latency of no more than 10 seconds should be supported at the 164 dB MCL.
- A battery-powered device should support small infrequent data transmission during at least 10 years at the 164 dB MCL.

These requirements are recognizable from the initial work on EC-GSM-IoT and NB-IoT carried out in the 3GPP Release 13 study item *Cellular system support for ultra-low complexity and low throughput Internet of Things* [4], in this book referred to as the *Cellular IoT study item.* While the 5G performance objectives match the requirements defined for EC-GSM-IoT and NB-IoT, the evaluation assumptions defined for 5G and those used in the Cellular IoT study item differ somewhat. Sections 2.3 and 2.4 in further detail discuss the Cellular IoT study, IMT-2020 and 5G.

This chapter presents the expected LTE-M half-duplex frequency-division duplexing (HD-FDD) performance for each of the 5G performance objectives. It is shown that LTE-M in all aspects meets the massive MTC part of the requirements defined by ITU-R and 3GPP.

6.2 Coverage

3GPP defines coverage in terms of MCL, which between two communicating nodes specifies the maximum tolerable signal attenuation between the transmitting and the receiving node's antenna ports. The MCL is a function of the transmitted output power (P_{TX}), the supported signal to noise ratio (SNR), the signal bandwidth (BW) and the receiver noise figure (NF):

$$MCL = P_{TX} - (SNR + 10\log_{10}(\mathrm{k}\cdot T\cdot BW) + NF) \tag{6.1}$$

T equals an assumed ambient temperature of 290 K, and k is Boltzmann's constant ($1.38\cdot 10^{-23}$ J/K). 3GPP requires that a 5G system supports 164 dB MCL which is intended to cater for deep indoor deployments of devices served by a macro network.

To evaluate the coverage of LTE-M the performance of all supported physical signals and channels was evaluated according to the simulation assumptions presented in Table 6.1. The tapped delay line (TDL) channel model is based on Rayleigh fading taps with a 2-Hz Doppler spread and a root mean square delay spread of 363 ns. This is short compared to the LTE-M cyclic prefix and does not challenge the orthogonality of the modulated OFDM subcarriers. The base station is assumed to map the modulated signal to 2 transmit antenna ports and make use of transmission mode 2 for the PBCH, MPDCCH and PDSCH. When evaluating the PSS and SSS synchronization performance the signal is mapped over 4 antenna ports. This extra space diversity has proven to be beneficial for the synchronization performance. The evaluated LTE-M narrowband (NB) is assumed to be transmitted within a 10-MHz LTE carrier configured with a total output power of 46 dBm. This results in 29 dBm per physical resource blocks (PRB) or 36.8 dBm per narrowband. To improve the initial cell acquisition time a 3-dB power boosting is applied to the PSS, SSS and PBCH transmissions.

The base station receiver is associated with 4-way receive diversity and a NF of 5 dB. The device uses 23 dBm output power and a single transmit and single receive branch. It supports a NF of 7 dB. Both the base station and device models the use of realistic receiver implementations.

Table 6.2 presents the LTE-M coverage for a block error rate (BLER) of at most 1% on the physical control channels, i.e. the PRACH, PUCCH and MPDCCH, 10% BLER for the initial hybrid automatic repeat request (HARQ) transmission on the physical data channels, i.e. the PDSCH and PUSCH, and 10% BLER on PSS/SSS and PBCH which are used for synchronization and system information acquisition.

TABLE 6.1 Assumptions made in the evaluations of LTE-M MCL [5].

Parameter	Value
Physical channels and signals	DL: PSS/SSS, PBCH, MPDCCH, PDSCH UL: PUCCH Format 1a, PRACH Format 0, PUSCH
Frequency band	700 MHz
TDL channel model	TDL-iii
Fading	Rayleigh
Doppler spread	2 Hz
Device NF	7 dB
Device antenna configuration	1 TX and 1 RX
Device power class	23 dBm
Base station NF	5 dB
Base station antenna configuration	2 or 4 TX and 4 RX
Base station power level	29 dBm per PRB 3 dB power boosting on PSS, SSS and PBCH.

TABLE 6.2 LTE-M coverage.

Performance/Parameters	**Downlink coverage**				**Uplink coverage**		
Physical channel	PSS/SSS	PBCH	MPDCCH	PDSCH	PRACH	PUCCH	PUSCH
TBS [bits]	–	24	18	328	–	1	712
Bandwidth [kHz]	945	945	1080	1080	1048.75	180	30
Power [dBm]	39.2	39.2	36.8	36.8	23	23	23
NF	7	7	7	7	5	5	5
#TX/#RX	4TX/1RX	2TX/1RX	2TX/1RX	2TX/1RX	1TX/4RX	1TX/4RX	1TX/4RX
Transmission, Acquisition time [ms]	1500 ms	800 ms	256 ms	768 ms	64 ms	64 ms	1536 ms
BLER	10%	10%	1%	2%	1%	1%	2%
SNR [dB]	-17.5	-17.5	-20.8	-20.5	-32.9	-26	-16.8
MCL	164	164	164.2	164	164.7	165.5	164

The downlink transmissions are configured to span the full narrowband to optimize the downlink power allocation and frequency diversity. Remember that a base station generally offers a flat and constant power spectral density, which means that the total output power of a transmission scales linearly with its frequency allocation. The MPDCCH make use of the highest available aggregation level 24 to optimize the code rate for achieving lowest possible transmission time. In the uplink the smallest possible resource allocation is used, including the 2-out-of-3 tone PUSCH transmission scheme presented in Section 5.2.5.4. This optimizes the uplink power spectral density and received SINR which is important when operating at an extreme coupling loss. For all signals and channels an increased acquisition time in combination with time repetitions are used to maximize the coverage.

The 164 dB MCL is met with the downlink synchronization signals and the PUSCH being the limiting channels with the longest acquisition times. The MPDCCH needs to be configured with 256 repetitions to achieve the 1% BLER target set for the control channel transmission. This is the maximum configurable repetition number, so the MPDCCH coverage is also a limiting factor unless a higher control channel BLER than 1% is acceptable. A certain supported MCL is only meaningful when associated with requirements on the link quality. In the next sections we will see that LTE-M given the performance in Table 6.2 meets the 5G performance requirements for data rate, latency and battery life defined at the 164 dB MCL. If an application can tolerate relaxed performance requirements and can e.g. live with a higher latency than 10 seconds or a shorter battery life than 10 years, then the MCL can be pushed beyond 164 dB.

6.3 Data rate

In this section the physical-layer data rates and the MAC-layer throughputs achievable for device categories Cat-M1 and Cat-M2 are examined. It is shown that LTE-M meets the 5G requirement of 160 bits per second at the 164 dB MCL.

TABLE 6.3 LTE-M HD-FDD Cat-M1 and Cat-M2 max TBS.

Device	Rel-13 PDSCH	Rel-13 PUSCH	Rel-14 PDSCH	Rel-14 PUSCH
Cat-M1	1000 bits	1000 bits	1000 bits	2984 bits
Cat-M2	–	–	4008 bits	6968 bits

The physical-layer data rate is estimated over the PDSCH and PUSCH transmission time intervals and serves, when normalized by the channel bandwidth, as an indication of the spectral efficiency that can be achieved for LTE-M. The maximum physical-layer data rates are determined by the 1-ms transmission time intervals and the maximum TBSs supported by Cat-M1 and Cat-M2 in Release 13 and 14, as summarized in Table 6.3.

The MAC-layer data rate is used to estimate the sustainable data rate offered by Cat-M1 and Cat-M2 devices. It is defined as the data rate at which MAC *protocol data units* are delivered to the physical layer. This is a powerful and yet simple metric that corresponds to the data rate offered by the physical layer to the higher layers in the radio protocol stack. It takes all relevant scheduling and processing delays at the access stratum into account and considers all data mapped to a transport block as useful data. To convert this to a sustainable data rate offered to an application provider both the delays and the overhead introduced at PDCP, RLC and MAC must be accounted for. This requires, for example, a detailed model of the RLC *service data units* segmentation and concatenation into RLC PDUs which is beyond the scope of this book. A good rule of thumb is that the radio protocol stack overhead per transport block sent over the user plane corresponds to roughly 1 byte from PDCP, 2 bytes from RLC and 2 bytes from MAC. Fig. 6.5 presents the data flow through the LTE protocol stack.

6.3.1 Downlink data rate

For Cat-M1 and Cat-M2, downlink physical-layer data rates of 1 Mbps and 4 Mbps, respectively, are achievable.

The maximum Cat-M1 downlink MAC-layer data rate, according to the Release 13 design baseline, is achieved when three HARQ processes are scheduled back-to-back as shown in Fig. 6.1. Although LTE-M supports up to eight HARQ processes in FDD in Release 13 the timing restrictions of the technique are such that for half-duplex FDD three HARQ processes gives the maximum PDSCH data rate. In this example, the MPDCCH carrying the downlink control information is mapped to 2 PRBs and schedules the PDSCH, containing the maximum 1000-bit transport block for Cat-M1, over 4 PRBs. Fig. 6.1 illustrates an MPDCCH-to-PDSCH scheduling gap of 1 ms, a downlink-to-uplink switching gap of 1 ms, and a PDSCH-to-PUCCH gap of 3 ms. The PUCCH is transmitted on a single PRB location that is frequency hopping across the system bandwidth. This configuration gives us a scheduling cycle of 10 ms, which leads to a peak MAC-layer throughput of 300 kbps.

In Release 14 the possibility to bundle feedback from four HARQ processes in one PUCCH Format 1a ACK/NACK transmission is supported. This reduces the PUCCH transmission overhead and allows 8 PDSCH transport blocks to be sent over 15 subframes as shown

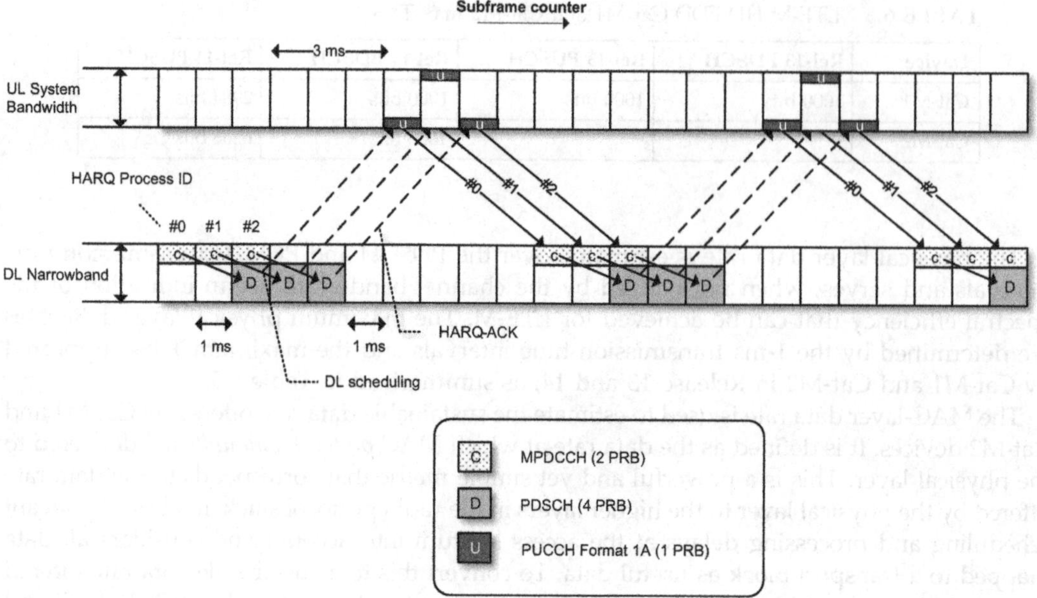

FIG. 6.1 Cat-M1 Release 13 PDSCH scheduling cycle.

in Fig. 6.2. This configuration gives Cat-M1 a peak MAC-layer throughput of 533 kbps. Release 14 also introduces support for up to 10 HARQ processes in downlink in FDD, and if this feature is used together with the HARQ bundling, the peak MAC-layer throughput is increased to 588 kbps.

The Cat-M2 maximum MAC-layer data rates are achieved for the same scheduling strategies as used for Cat-M1 in Fig. 6.1 and 6.2. As Cat-M2 supports up to 5 MHz PDSCH bandwidth the system can e.g. assign a PDSCH spanning 15 PRBs, evenly distributed across 3 narrowbands, to carry the maximum transport block of 4008 bits. This gives us a Cat-M2 peak MAC-layer throughput of 1.202 Mbps when not using HARQ bundling, 2.137 Mbps when HARQ bundling is configured, and 2.357 Mbps when both HARQ bundling and 10 HARQ processes are configured.

For Cat-M1 the results presented in Section 6.2 suggest that the following allocations should be considered when estimating the MAC-layer data rate at the MCL:

- MPDCCH using aggregation level 24 and a transmission time of 256 ms
- PDSCH carrying a 328 bits transport block using a transmission time of 768 ms
- PUCCH Format 1a using a transmission time of 64 ms

By configuring the MPDCCH user-specific search space (described in Section 5.3.3.1) with $R_{max} = 256$ and $G = 1.5$, it is possible to schedule a PDSCH transmission once every third scheduling cycle, meaning once every 1,152 ms. Given the PDSCH TBS of 328 bits and a 2% PDSCH BLER (see Table 6.2) this gives us a MAC-layer data rate of 279 bps:

$$THP = \frac{(1 - BLER) \cdot TBS}{MPDCCH\ Period} = \frac{0.98 \cdot 328}{1.152} = 279 \text{ bps} \tag{6.2}$$

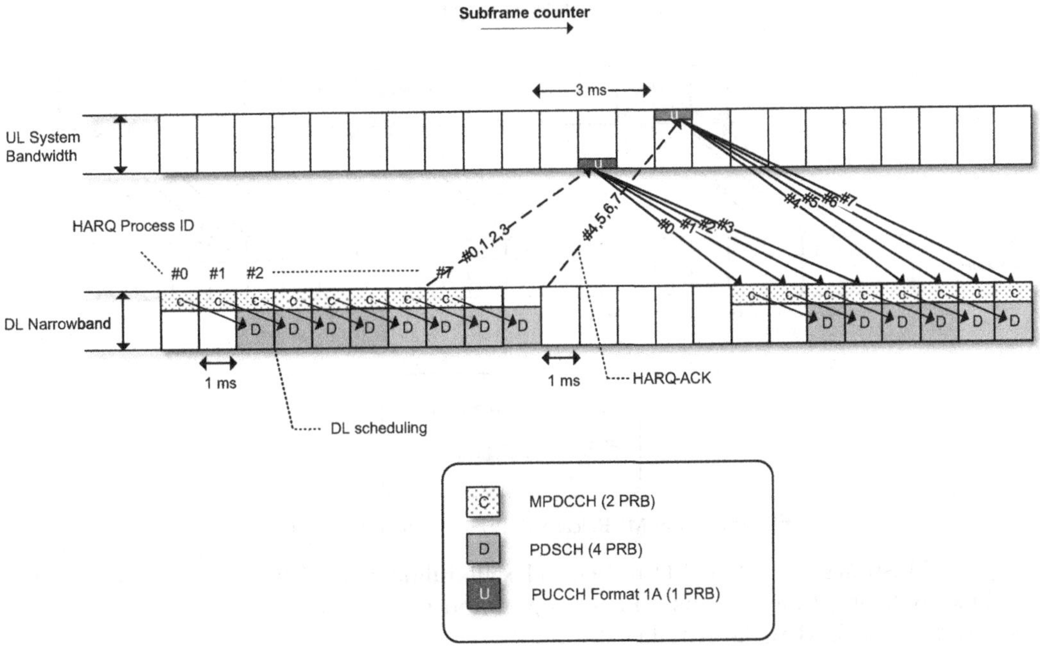

FIG. 6.2 Cat-M1 Release 14 PDSCH scheduling cycle using HARQ bundling.

TABLE 6.4 LTE-M HD-FDD PDSCH data rates for Cat-M1.

Scenario	MAC-layer at 164 dB MCL	MAC-layer peak	MAC-layer peak for Rel-14 HARQ bundling	MAC-layer peak for Rel-14 HARQ bundling and 10 HARQ processes	PHY-layer peak
Cat-M1	279 bps	300 kbps	533 kbps	588 kbps	1 Mbps
Cat-M2	> 279 bps	1.202 Mbps	2.137 Mbps	2.357 Mbps	4.008 Mbps

Table 6.4 summarizes the Cat-M1 and Cat-M2 PDSCH data rates. Note that the Cat-M2 data rate at the MCL will be at least as good as the Cat-M1 data rate.

6.3.2 Uplink data rate

The maximum uplink physical-layer data rates are 1 Mbps and 7 Mbps for Cat-M1 and Cat-M2, respectively.

The peak MAC-layer throughputs are reached when three of the eight available HARQ processes are scheduled as shown for Cat-M1 in Fig. 6.3. In the illustrated example, the MPDCCH schedules the PUSCH over 4 PRBs containing the largest Release 13 Cat-M1 transport block of 1000 bits. In Release 14, the Cat-M1 maximum PUSCH TBS was increased to 2984 bits, which is supported for a PUSCH allocation of 6 PRBs. The Cat-M2 maximum TBS is 6968 bits which is available for a 24 PRB PUSCH allocation.

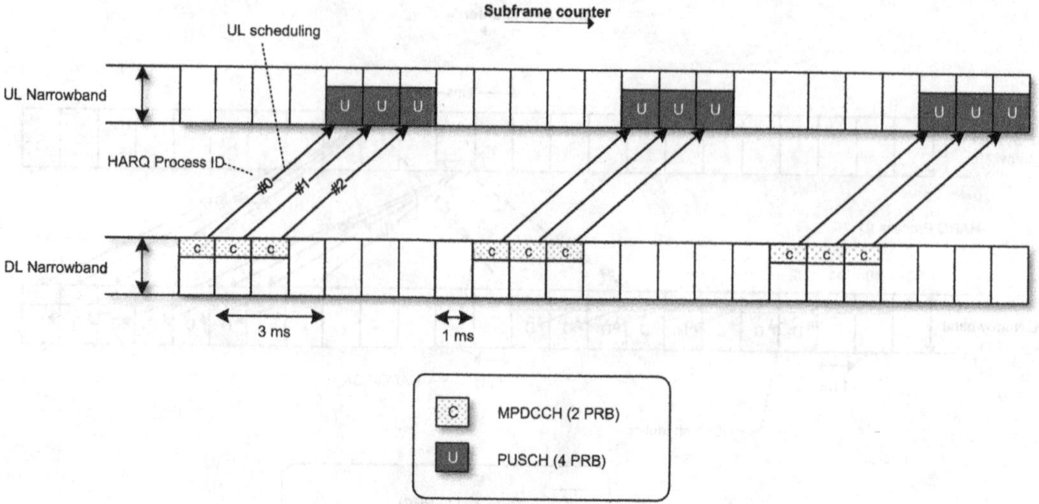

FIG. 6.3 Cat-M1 Release 13 PUSCH scheduling cycle.

Fig. 6.3 illustrates an MPDCCH-to-PUSCH scheduling gap of 3 ms, and an uplink-to-downlink switching time of 1 ms. This configuration gives us a scheduling cycle of 8 ms, which leads to peak MAC-layer data rates of:

- 375 kbps for Cat-M1 using the Release 13 max TBS.
- 1.119 Mbps for Cat-M1 using the Release 14 max TBS.
- 2.613 Mbps for Cat-M2

Cat-M1 and Cat-M2 offers the same PUSCH performance at the MCL. The results presented in Section 6.2 suggests that the following allocations should be considered when estimating the 164 dB MCL data rate:

- MPDCCH using aggregation level 24 and a transmission time of 256 ms
- PUSCH carrying a 728 bits transport block using a transmission time of 1536 ms

By configuring the MPDCCH user-specific search space (described in Section 5.3.3.1) with $R_{max} = 256$ and $G = 1.5$, it is possible to schedule a PUSCH transmission once every fifth scheduling cycle, meaning once every 1,920 ms. Given the PUSCH TBS of 728 bits and a 2% PUSCH BLER (see Table 6.2) this gives us a MAC-layer data rate of 363 bps:

$$THP = \frac{(1 - BLER) \cdot TBS}{MPDCCH\ Period} = \frac{0.98 \cdot 728}{1.920} = 363 \text{ bps} \tag{6.3}$$

Table 6.5 summarizes the Cat-M1 and Cat-M2 PUSCH data rates.

TABLE 6.5 LTE-M HD-FDD PUSCH data rates.

Scenario	MAC-layer at 164 dB MCL	MAC-layer peak for Rel-13 TBS	MAC-layer peak for Rel-14 TBS	PHY-layer peak for Rel-13 TBS	PHY-layer peak for Rel-14 TBS
Cat-M1	363 bps	375 kbps	1.119 Mbps	1 Mbps	2.984 Mbps
Cat-M2	363 bps		2.609 Mbps		6.968 Mbps

6.4 Latency

LTE-M is designed to support a wide range of mMTC use cases. For those characterized by small data transmission the importance of the data rates presented in the previous section is overshadowed by the latency required to set up a connection and perform a single data transmission. In this section we focus on the latency to deliver a small uplink packet. We consider both the lowest latency achievable under error-free conditions and the worst-case latency calculated for devices located at the 164 dB MCL. It is shown that the 5G requirement of 10 seconds latency at the 164 dB MCL is met.

For applications requiring a consistent and short latency it is recommended to keep a device in RRC connected mode and configure it for semi-persistent scheduling (SPS). This supports transmission opportunities occurring e.g. with a periodicity of 10 ms. The latency for SPS is determined by the wait time for a SPS resource, the MPDCCH and PUSCH transmission times (t_{MPDCCH}, t_{PUSCH}) and the MPDCCH to PUSCH scheduling time. Assuming a worst case wait time, the lowest latency offered by SPS equals:

$$t_{wait} + t_{MPDCCH} + t_{sched} + t_{PDSCH} = 10 + 1 + 1 + 1 = 13 \text{ ms} \tag{6.4}$$

SPS supports operation in CE mode A but not in CE mode B. Keeping a device in RRC connected mode is also not a long-term energy efficient strategy. Next, we therefore look at the latency achievable for a device that triggers a mobile originated data transmission in RRC idle mode. Fig. 6.4 presents the Release 13 RRC Resume connection establishment procedure for which a device can resume a previously suspended connection including resuming access stratum security and an earlier configured *data radio bearer*. The figure also indicates the assumed latency definition.

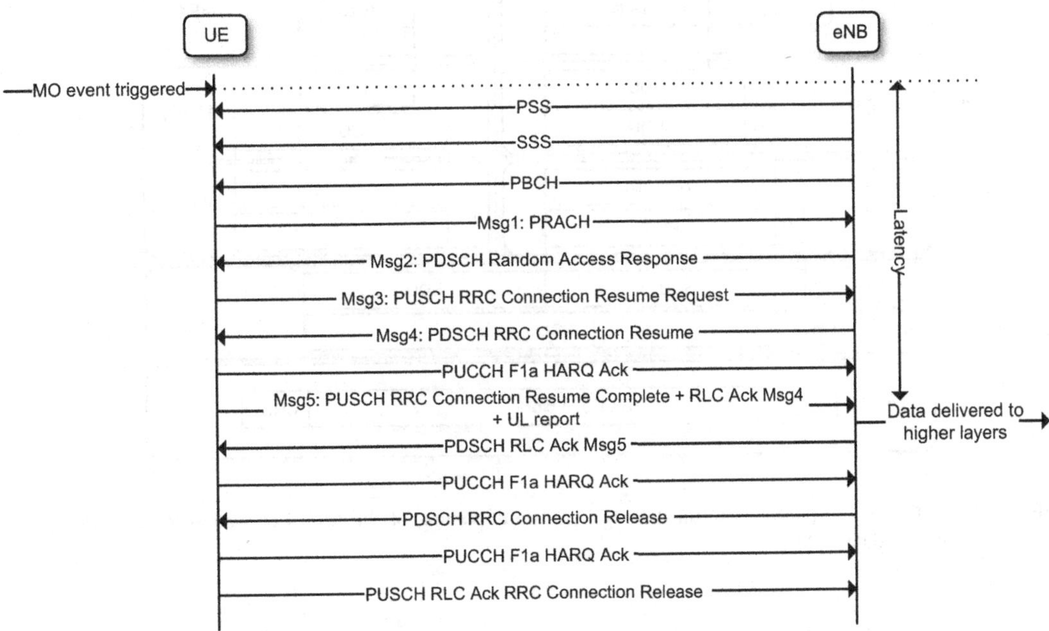

FIG. 6.4 LTE-M RRC Resume procedure for latency evaluations.

Before establishing a connection, the device performs cell acquisition during which it acquires synchronization to the downlink frame structure and confirm the status of the system information. Since Release 15, the system information validity check is supported by a single bit flag in the master information block. The second step is to go through the random access procedure which provides synchronization to the uplink frame structure and contention resolution. Finally, the device can complete the connection establishment procedure and convey the uplink data transmission in Message 5 by configuring the MAC to multiplex the RRC Connection Resume Complete message with the data packet. In MAC, the RRC Connection Resume Complete message is sent on the *signaling radio bearer 1* on the logical *Dedicated Control Channel (DCCH)*, while the data packet is sent over a resumed data radio bearer on the logical *Dedicated Traffic Channel (DTCH)* as shown in Fig. 6.5. The figure also illustrates the:

- PDCP layers *robust header compression* of the IP layer headers
- PDCP layers ciphering of the SRB SDUs by means of the 4-byte *Message Authentication Code – Integrity (MAC-I)* field
- Headers appended by the PDCP, RLC and MAC-layers
- 3-byte CRC added by the PHY layer

The 3GPP 5G evaluation of small data transmission latency requires that the latency should be studied for the delivery of a 105-byte MAC protocol data unit. Based on these assumptions and the performance presented at the 164 dB MCL in Table 6.1 a latency of 7.7 seconds can be achieved for LTE-M using the RRC Resume procedure [5].

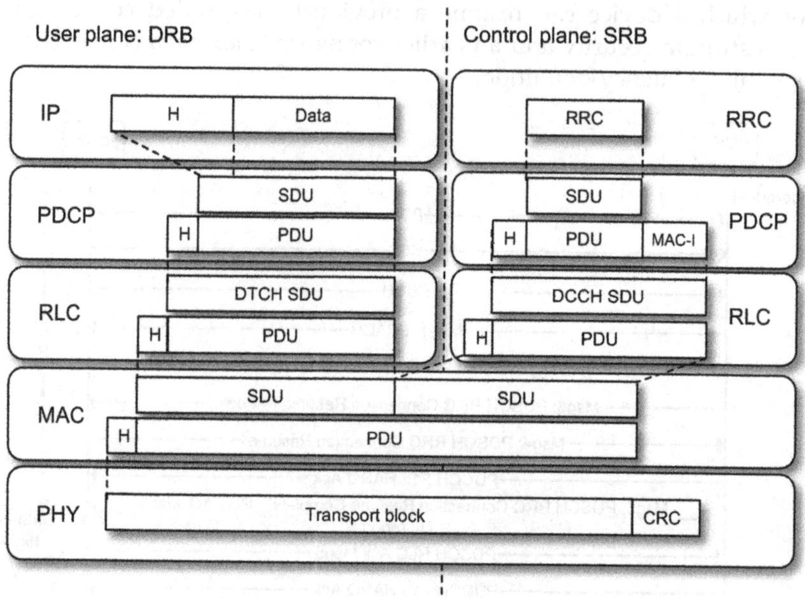

FIG. 6.5 RRC Resume packet flow illustrating MAC multiplexing of the RRC Connection Resume Complete message with user data in Msg5.

TABLE 6.6 LTE-M latency.

Method	Latency
SPS under error free conditions	13 ms
EDT under error free conditions	33 ms
EDT at 164 dB MCL	5.0 s
RRC resume at 164 dB MCL	7.7 s

3GPP Release 15 went beyond the RRC Resume procedure and specified the Early Data Transmission (EDT) procedure. With this procedure, user data on the dedicated traffic channel can be MAC multiplexed with the RRC Connection Resume Request message already in Message 3. In an error-free case a device may deliver an uplink report according to the following EDT timing based on the LTE-M specifications:

- t_{SSPB}: The PSS/SSS synchronization signals and the PBCH master information block can be acquired in a single radio frame, i.e. within 10 ms.
- t_{PRACH}: The LTE-M PRACH is highly configurable, and a realistic assumption is that a PRACH resource is available at least once every 10 ms.
- $t_{RAR,\ wait}$: The random access response window starts 3 ms after a PRACH transmission.
- t_{RAR}: The random access response transmission including MPDCCH and PDSCH transmission times and the cross-subframe scheduling delays requires 3 ms.
- T_{Msg3}: The Message 3 transmission may start 6 ms after the RAR, and requires a 1-ms transmission time, i.e. in total 7 ms.

Summarizing the above timings gives us a best-case latency for EDT of 33 ms. At the 164 dB MCL a latency of 5 seconds can be achieved for LTE-M using the EDT procedure [5]. From the results summarized in Table 6.6 it is clear the LTE-M not only meets the 5G requirement with margin, but it is also capable of serving applications requiring short response times.

6.5 Battery life

The massive MTC use cases should be supported in ubiquitous deployments of massive number of devices. To limit the deployment and operation cost, the deployed devices may need to support non-rechargeable battery powered operation for years.

The battery life of LTE-M Cat-M1 is evaluated in this section, and it is shown that the 5G requirement of 10 years battery life is within reach. The traffic model assumed for this investigation is based on the reporting intervals and packet sizes defined by 3GPP for the 5G evaluations. The packet sizes presented in Table 6.7 are defined to apply on top of the PDCP layer depicted in Fig. 6.5.

The device power consumption levels used in the evaluations are presented in Table 6.8. They distinguish between power levels at transmission (TX), reception (RX), in inactive state (e.g. in-between transmit and receive operations) and in the RRC idle Power Saving Mode

TABLE 6.7 Packet sizes on top of the PDCP layer for evaluation of battery life [3].

Message type	UL report	DL application acknowledgment
Size	200 bytes	20 bytes
Arrival rate	Once every 24 h	

TABLE 6.8 LTE-M power consumption [4].

TX (23-dBm power class)	RX	Inactive	PSM
500 mW	80 mW	3 mW	0.015 mW

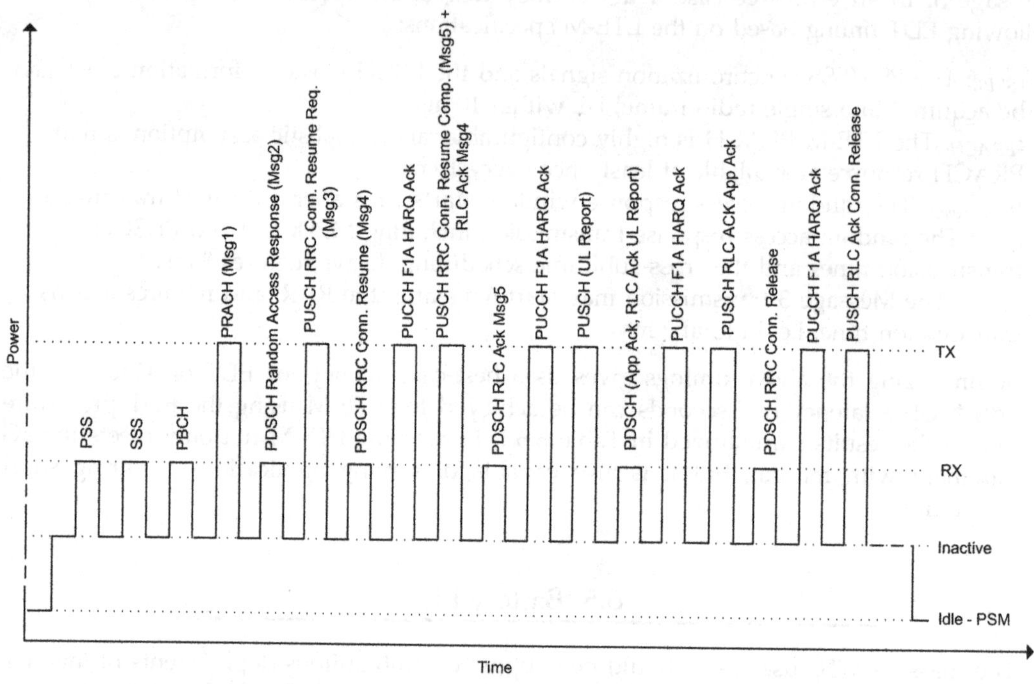

FIG. 6.6 LTE-M RRC Resume procedure for battery life evaluations.

(PSM). These power levels are reused from the Cellular IoT study item [4]. It should however be noted that recent publications have shown that these power level assumptions are optimistic.

The RRC Resume procedure is assumed for the Connection Establishment procedure. The complete packet flow used in these evaluations is shown in Fig. 6.6. Not depicted are the MPDCCH transmissions scheduling each transmission. Between the mobile originated events triggered once every 24 h, the device is assumed to use PSM to optimize its power consumption. The EDT procedure could in principle have been used also in this evaluation if it had not been for the uplink packet size of 200 bytes plus overheads, which exceeds the by EDT maximum supported uplink packet size of 1000 bits.

An ideal 5-Wh battery power source was assumed in the evaluations with no power leakage or other imperfections. Under these assumptions, a battery life of 11.9 years is achievable at the 164 dB MCL which comfortably meets the 5G requirement [5].

6.6 Capacity

The only IMT-2020 requirement defined by ITU-R on mMTC is connection density. It requires that a mMTC technology can support 1,000,000 devices per square kilometer for a traffic model when each device accesses the system once every 2 hours and transmits a 32-byte message. Per square kilometer the system hence needs to facilitate 1,000,000 connections over 2 hours, or ~280 connection establishments per second. Each connection needs to provide a latency of at most 10 seconds, within which the 32-byte message should be successfully delivered to the network with 99% reliability.

IMT-2020 requires that the connection density target is fulfilled for four different urban macro (UMA) scenarios defined by:

- Base station inter-site distances of 500 and 1732 meters.
- Two different channel models named *Urban Macro A* (UMA A) and *Urban Macro B* (UMA B).

The two channel models provide different path loss and outdoor-to-indoor signal loss models. In combination with the two inter-site distances the channel models define the coupling gain statistics for each of the four IMT-2020 scenarios as presented in Fig. 6.7.

Table 6.9 summarizes the most important assumptions used when evaluating the LTE-M system capacity for IMT-2020. A detailed description is found in Ref. [6]. It is assumed that the base station is configured with 46 dBm transmit power, which is equally divided across the 50 PRBs in the 10-MHz LTE system bandwidth. This gives us 29 dBm/PRB or 36.8 dBm over the simulated LTE-M narrowband. The studied LTE-M narrowband is assumed to be located outside of the center 72 subcarriers, meaning that the narrowband does not carry any load from mandatory PSS, SSS, and PBCH transmissions in the downlink. To cope with the high anticipated access load, the simulated narrowband reserves 10% of all the uplink resources for random access preamble transmissions.

Fig. 6.8 shows the supported connection density per narrowband versus the latency required to successfully deliver the 32-byte packet. LTE-M supports a very high capacity especially in the deployment corresponding to a 500-meter inter-site distance. In this case a single narrowband, not taking PSS, SSS and PBCH transmissions into consideration, can handle more than 5 million connections. For the 1732-m inter-site distance we face a 12 times larger cell size. This explains the reductions in supported connection density observed for the 1732-meter inter-site distance scenarios, which are in the same order of magnitude as the increase in cell size.

Table 6.10 summarizes the achieved connection density per simulated narrowband, and the needed system resource to cater for the required 1,000,000 connections per km^2. Note that LTE-M PUCCH transmissions are configured at the edges of the LTE system bandwidth and are typically not part of an LTE-M narrowband. This is indicated by the addition of

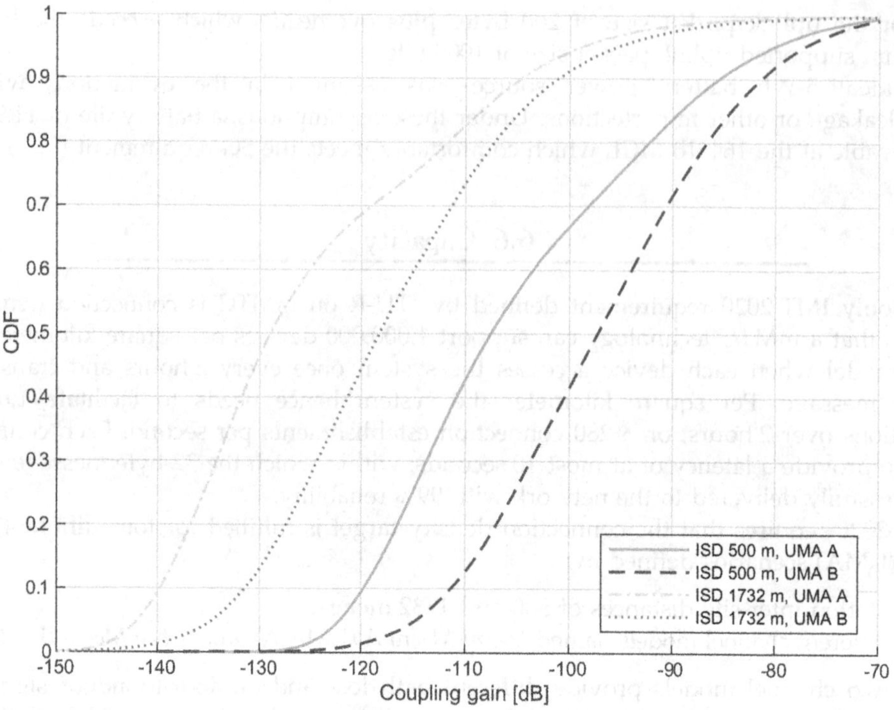

FIG. 6.7 The cumulative distribution function of the coupling gain according to the ITU-R defined IMT-2020 mMTC evaluation scenarios.

TABLE 6.9 System level simulation assumptions.

Parameter	Model
Cell structure	Hexagonal grid with 3 sectors per size
Cell inter site distance	500 and 1732 m
Frequency band	700 MHz
LTE system bandwidth	10 MHz
Frequency reuse	1
Base station transmit power	46 dBm
Power boosting	0 dB
Base station antenna configuration	2 TX, 2 RX
Base station antenna gain	17 dBi
Device transmit power	23 dBm
Device antenna gain	0 dBi
Device mobility	0 km/h
Pathloss model	UMA A, UMA B

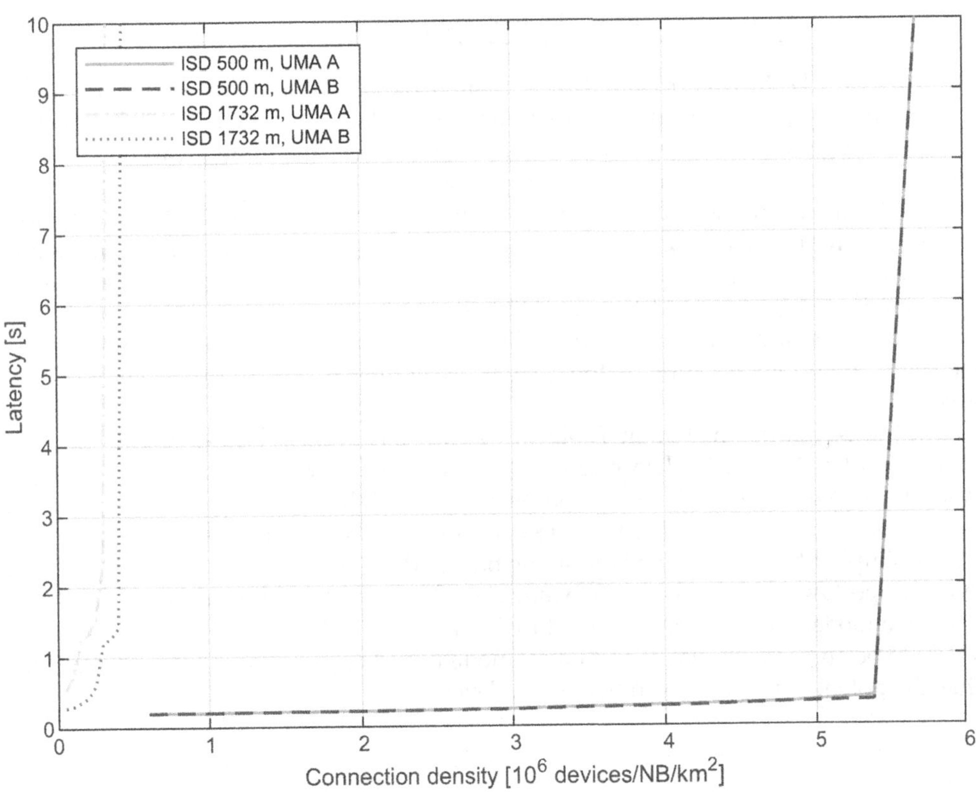

FIG. 6.8 LTE-M connection density per narrowband.

TABLE 6.10 LTE-M connection density [6].

Scenario	Connection density	Resources to support 1,000,000 connections per km^2
ISD 500 m, UMA A	5,680,000 devices/NB	1 NB + 2 PRBs
ISD 500 m, UMA B	5,680,000 devices/NB	1 NB + 2 PRBs
ISD 1732 m, UMA A	342,000 devices/NB	3 NBs + 2 PRBs
ISD 1732 m, UMA B	445,000 devices/NB	3 NBs + 2 PRBs

2 PRBs in the third column of Table 6.10. The load due to PSS, SSS, PBCH and SI transmissions were also not accounted for in these simulations. A coarse estimation is that these transmissions make use of around 40%–50% of the available downlink resources in a single narrowband within the LTE system bandwidth. It can be noted that the LTE-M capacity can be further improved by means of the PUSCH sub-PRB feature introduced in Release 15 (see Section 5.2.5.4), which was not used in this evaluation.

6.7 Device complexity

The work on LTE-M was triggered by a desired reduction in device cost, and the target was to go down significantly in complexity and cost relative earlier LTE device categories. This enables large-scale deployments of IoT devices, where the system can be competitive in the IoT landscape, competing, for example, with low-power wide-area network alternatives in the unlicensed spectrum domain. LTE-M intends at the same time to address a large range of mMTC use cases, involving high-throughput and low-latency applications. This motivates higher requirements on computational complexity and memory requirements than those adopted for EC-GSM-IoT and NB-IoT.

To get a better understanding of the LTE-M complexity, Table 6.11 summarizes some of the more important features of the LTE-M basic device Cat-M1 that was specified in Release 13.

To put the design parameters in Table 6.11 in a context, Table 6.12 estimates the modem cost reduction for the LTE-M device categories introduced in Release 12 (Cat-0) and Release 13 (Cat-M1) based on the cost reduction estimates in Table 7.1 in the LTE-M study item technical report [1]. The cost reductions are expressed in terms of modem cost reduction relative to the simplest LTE device available at the time of the LTE-M study item, which was a Cat-1 device supporting a single frequency band. The LTE-M study item concluded that the *bill of material* for a modem would need to be reduced to about 1/3 of that for a single-band LTE Cat-1 modem to be on par with that of an EGPRS modem, and as can be seen from Table 6.12, Cat-M1 has the potential to reach even below this level.

TABLE 6.11 Overview of Release 13 LTE-M device category M1.

Parameter	Value
Duplex modes	HD-FDD, FD-FDD, TDD
Half-duplex operation	Type B
Number of receive antennas	1
Transmit power class	14, 20, 23 dBm
Maximum DL/UL bandwidth	6 PRB (1.080 MHz)
Highest DL/UL modulation order	16QAM
Maximum number of supported DL/UL spatial layers	1
Maximum DL/UL transport block size	1000 bits
Peak DL/UL physical layer data rate	1 Mbps
DL/UL channel coding type	Turbo code
DL physical layer memory requirement	25,344 soft channel bits
Layer 2 memory requirement	20,000 bytes

TABLE 6.12 Overview of measures supporting an LTE-M modem cost reduction [1].

Combination of modem cost reduction techniques	Modem cost reduction
Single-band 23-dBm FD-FDD LTE Category 1 modem - Reference modem in the LTE-M study item	0%
Single-band 23-dBm FD-FDD LTE Category 1bis modem - Reduced number of receive antennas from 2 to 1	24%–29%
Single-band 23-dBm FD-FDD LTE Category 0 modem - Reduced peak rate from 10 to 1 Mbps - Reduced number of receive antennas from 2 to 1	42%
Single-band 23-dBm HD-FDD LTE Category 0 modem - Reduced peak rate from 10 to 1 Mbps - Reduced number of receive antennas from 2 to 1 - Half-duplex operation instead of full-duplex operation	49%–52%
Single-band 23-dBm FD-FDD LTE Category M1 modem - Reduced peak rate from 10 to 1 Mbps - Reduced number of receive antennas from 2 to 1 - Reduced bandwidth from 20 to 1.4 MHz	59%
Single-band 23-dBm HD-FDD LTE Category M1 modem - Reduced peak rate from 10 to 1 Mbps - Reduced number of receive antennas from 2 to 1 - Reduced bandwidth from 20 to 1.4 MHz - Half-duplex operation instead of full-duplex operation	66%–69%
Single-band 20-dBm FD-FDD LTE Category M1 modem - Reduced peak rate from 10 to 1 Mbps - Reduced number of receive antennas from 2 to 1 - Reduced bandwidth from 20 to 1.4 MHz - Reduced transmit power from 23 to 20 dBm	69%–71%
Single-band 20-dBm HD-FDD LTE Category M1 modem - Reduced peak rate from 10 to 1 Mbps - Reduced number of receive antennas from 2 to 1 - Reduced bandwidth from 20 to 1.4 MHz - Half-duplex operation instead of full-duplex operation - Reduced transmit power from 23 to 20 dBm	76%–81%

It should be emphasized that the modem baseband and radio frequency cost is only one part of the total device cost. As pointed out for EC-GSM-IoT in Section 4.7, also components to support peripherals, real time clock, central processing unit, and power supply need to be taken into consideration to derive the total cost of a device. The potential of mass production is also highly important. LTE-M, as all LTE-based technologies, has significant benefits in this area due to the widespread use of the technology.

References

[1] Third Generation Partnership Project, Technical Report 36.888, v12.0.0. Study on Provision of Low-cost Machine-Type Communications (MTC) User Equipment's (UEs) Based on LTE, 2013.
[2] ITU-R, Report ITU-R M.2410, Minimum requirements related to technical performance for IMT-2020 radio interfaces(s), 2017.
[3] Third Generation Partnership Project, Technical Report 38.913, v15.0.0. Study on Scenarios and Requirements for Next Generation Access Technologies, 2018.
[4] Third Generation Partnership Project, Technical Report 45.820, v13.0.0. Cellular System Support for Ultra-Low Complexity and Low Throughput, Internet of Things, 2016.
[5] Ericsson. Sierra Wireless, R1-1903119, IMT-2020 self-evaluation: mMTC coverage, data rate, latency & battery life, 3GPP RAN1 Meeting #96, 2019.
[6] Ericsson. R1-1903120, IMT-2020 self-evaluation: mMTC non-full buffer connection density, 3GPP RAN1 Meeting #96, 2019.

CHAPTER

7

NB-IoT

OUTLINE

Cellular Internet of Things, Second Edition
https://doi.org/10.1016/B978-0-08-102902-2.00007-8

© 2020 Elsevier Ltd. All rights reserved.

Abstract

This chapter presents the design of Narrowband Internet of Things (NB-IoT). The first part of this chapter describes the background behind the introduction of NB-IoT in the Third Generation Partnership Project (3GPP) specifications and the design principles of the technology.

The second part of this chapter focuses on the physical channels with an emphasis on how these channels were designed to fulfill the objectives that NB-IoT was intended to achieve, namely, deployment flexibility, ubiquitous coverage, ultra-low device cost, long battery lifetime, and capacity sufficient for supporting a massive number of devices in a cell. Detailed descriptions are provided regarding both downlink and uplink transmission schemes and how each of the NB-IoT physical channels is mapped to radio resources in both the frequency and time dimensions.

The third part of this chapter covers NB-IoT idle and connected mode procedures and the transition between these modes, including all activities from initial cell selection to completing a data transfer.

NB-IoT since its first introduction has been further enhanced both in terms of performance and features. This chapter highlights the improvements accomplished in Release 14 and Release 15 of the 3GPP specifications. As NB-IoT devices are expected to have long life cycles, how a network operator migrates from *Long-Term Evolution* (LTE) to *Fifth Generation* (5G) *New Radio* (NR) while continuing to honor its contracts toward NB-IoT service providers becomes an important aspect. At the end of this chapter, we describe NB-IoT and NR coexistence.

7.1 Background

7.1.1 3GPP standardization

In early 2015, the market for *low-power wide area networks* (LPWAN) was rapidly developing. Sigfox was building out their *Ultra Narrowband Modulation* networks in France, Spain, the Netherlands, and the United Kingdom. The *LoRa Alliance*, founded with a clear ambition to provide IoT connectivity with wide-area coverage, released LoRaWAN R1.0 specification in June 2015 [1]. The alliance at that point quickly gathered significant industry interest and strong membership growth. Up until then, *Global System for Mobile Communications/General Packet Radio Service* (GSM/GPRS) had been the main cellular technology of choice for serving wide-area IoT use cases, thanks to it being a mature technology with low modem cost. This position was challenged by the emerging LPWAN technologies that presented an alternative technology choice to many of the IoT verticals served by GSM/GPRS.

Anticipating the new competition, 3GPP (see Section 2.1) started a feasibility study on *Cellular System Support for Ultra-low Complexity and Low Throughput Internet of Things* [2], referred to as the *Cellular IoT study* for short in the following sections. As explained in Section 2.2.5, ambitious objectives on coverage, capacity, and battery lifetime were set up, together with a more relaxed objective of a maximum system latency. All these performance objectives offer major improvements over GSM and GPRS, as specified at that time, toward better serving the IoT verticals. One additional objective was that it should be possible to introduce the IoT features to the existing GSM networks through software upgrade. Building out a national network takes many years and requires substantial investment up front. With software upgrade, however, the well-established cellular network can be upgraded overnight to meet all the key performance requirements of the IoT market.

Among the solutions proposed to the *Cellular IoT study*, some were backward compatible with GSM/GPRS and were developed based on the evolution of the existing GSM/GPRS specifications. EC-GSM-IoT described in Chapters 3 and 4 is the solution eventually standardized in 3GPP Release 13.

Historically, the group carrying out the study, 3GPP TSG GERAN (Technical Specifications Group GSM/EDGE Radio Access Network), had focused on the evolution of GSM/GPRS technologies, developing features for meeting the need of GSM operators. Certain GSM operators, however, at that point considered refarming their GSM spectrum for deploying the *Long-Term Evolution* (LTE) technology as well as to LPWAN dedicated for IoT services. This consideration triggered the study on non–GSM backward compatible technologies, referred to as *clean-slate solutions*. Although none of the clean-slate solutions in the study were specified, it provided a firm ground for the *Narrowband Internet of Things* (NB-IoT) technology that emerged after study completion and was standardized in 3GPP Release 13. As described later in this chapter, the entire NB-IoT system is supported in a bandwidth of 180 kHz, for each of the downlink and uplink. This allows for deployment in refarmed GSM spectrum as well as within an LTE carrier. NB-IoT is part of the 3GPP LTE specifications and employs many technical components already defined for LTE. This approach reduced the standardization process and leveraged the LTE ecosystem to ensure a fast time to market. It also possibly allows NB-IoT to be introduced through a software upgrade of the existing LTE networks. The normative work of developing the core specifications of NB-IoT took only a few months and was completed in June 2016 [3].

Since its first release in 2016, NB-IoT up to 2018 has gone through two additional releases, i.e. 3GPP Releases 14 and 15. These later releases continued to improve device energy efficiency. Furthermore, features for improving system performance and for supporting new use cases and additional deployment options were also introduced. Table 7.1 provides a summary. These enhancements further improve NB-IoT's position as a superior LPWAN technology. Like before, all the Releases 14 and 15 features can be enabled through a software upgrade of the existing LTE or NB-IoT networks. In Release 15, 3GPP evaluated NB-IoT against a set of agreed *Fifth Generation* (5G) performance requirements defined for the *massive machine-type communications* (mMTC) use case [4]. As shown in Section 8.9, NB-IoT meets these requirements with margins and in all relevant aspects qualifies as a 5G mMTC technology.

The GSM Association, which is an organization that represents the interests of mobile network operators worldwide, tracks the status of commercial launches of NB-IoT. Since the completion of its first release in June 2016, there had been 80 NB-IoT launches in 45 markets as of June 2019, according to GSM Association [5]. On the device side, the Global Mobile

TABLE 7.1 New NB-IoT features introduced in 3GPP Releases 14 and 15.

Motivation	Release 14 (2017)	Release 15 (2018)
Device energy efficiency improvement	Category NB2 device (Sections 7.2.3, 7.2.4.6, 7.2.5.2) Support of 2 HARQ processes (Section 7.3.2.2.3)	Early data transmission (Section 7.3.1.7.3) Wake-up signal (Sections 7.2.4.8 and 7.3.1.4) Quick RRC release procedure Relaxed cell reselection monitoring (Section 7.3.1.3) Improved scheduling request (Section 7.3.2.5) Periodic buffer status reporting (Section 7.3.2.5) RLC unacknowledged mode
System performance improvement	Multicast (Section 7.3.1.11) Non-anchor carrier paging (Section 7.3.1.10) Non-anchor carrier random access (Section 7.3.1.10) Mobility enhancement DL channel quality reporting during random access procedure (Section 7.3.1.8)	Improved access barring (Section 7.3.1.9) Improved system information acquisition (Section 7.3.1.2.1) Improved measurement accuracy (Section 7.3.1.3) Improved UE power headroom reporting (Section 7.3.2.3.2) UE differentiation
New use case	14 dBm device power class (Section 7.1.2.1) Positioning (Section 7.3.2.6)	
Support for more deployment options		TDD Support (Section 7.2.8) Support of small cell deployment Extend cell radius up to 120 km (Section 7.2.5.1) Flexible uses of stand-alone carrier (Section 7.3.2.7)

Suppliers Association published a research report in 2018 [6] stating that as of August 2018, there were 106 modules supporting NB-IoT, 43 of which also support LTE-M. According to Ref. [7], the global NB-IoT chipset market is projected to have *compound annual growth rate* higher than 40% in the next decade. The device ecosystem has been well established and is expected to have strong momentum and growth in the coming years.

7.1.2 Radio access design principles

NB-IoT is designed for ultra-low-cost mMTC, aiming to support a massive number of devices in a cell. Low device complexity is one of the main design objectives, enabling low module cost. Furthermore, it is designed to offer substantial coverage improvements over GPRS as well as for enabling long battery lifetime. Finally, NB-IoT has been designed to give maximal deployment flexibility. In this section, we will highlight the design principles adopted in NB-IoT to achieve these objectives.

7.1.2.1 Low device complexity and cost

Device modem complexity and cost are primarily related to the complexity of baseband processing, memory consumption, and *radio-frequency* (RF) requirements. Regarding baseband processing, the two most computationally demanding tasks are synchronization during initial cell selection and demodulation and decoding during data reception. NB-IoT is designed for allowing low-complexity receiver processing in accomplishing these two tasks. For initial cell selection, a device needs to search for only one synchronization sequence for establishing basic time and frequency synchronization to the network. The device can use a low sampling rate (e.g., 240 kHz) and take advantage of the synchronization sequence properties to minimize memory requirement and complexity.

During connected mode, low device complexity is facilitated by restricting the downlink *transport block (TB) size* (TBS) to be no larger than 680 bits for the lowest device category and relaxing the processing time requirements compared with LTE. For channel coding, instead of using the LTE turbo code [8], which requires iterative receiver processing, NB-IoT adopts a simple convolutional code, i.e., the LTE *tail-biting convolutional code* (TBCC) [8], in the downlink channels. In addition, NB-IoT does not use higher-order modulations or multilayer *multiple-input multiple-output* transmissions. Furthermore, a device needs to support only half-duplex operation and is not required to receive in the downlink while transmitting in the uplink, or vice versa.

Regarding RF, all the performance objectives of NB-IoT can be achieved with one transmit-and-receive antenna in the device. That is, neither downlink receiver diversity nor uplink transmit diversity is required in the device. NB-IoT is designed for allowing relaxed oscillator accuracy in the device. For example, a device can achieve initial acquisition when its oscillator inaccuracy is as large as 20 *parts* per *million* (ppm). During a data session, the transmission scheme is designed for the device to easily track its frequency drift. Because a device is not required to simultaneously transmit and receive, a duplexer is not needed in the RF front end of the device. The maximum transmit power level of an NB-IoT device is either 20 or 23 dBm in 3GPP Release 13. Release 14 introduced a lower device power class with a maximum transmit power of 14 dBm. A device power level of 14 or

20 dBm allows on-chip integration of the *power amplifier* (PA), which can contribute to the device cost reduction. The 14 dBm device power class allows batteries with lower peak current draw, and thus facilitates the adoption of batteries of smaller form factors.

Economy of scale is yet another contributor to cost reduction. Thanks to its deployment flexibility and low minimum system bandwidth requirement, NB-IoT is already being globally available in many networks. This will help to increase the economy of scale of NB-IoT.

7.1.2.2 Coverage enhancement

Coverage enhancement (CE) is mainly achieved by trading off data rate for coverage. Like EC-GSM-IoT and LTE-M, repetitions are used to ensure that devices in coverage challenging locations can still have reliable communications with the network, although at a reduced data rate. Furthermore, NB-IoT has been designed to use a close to constant envelope waveform in the uplink. This is an important factor for devices in extreme coverage-limited and power-limited situations because it minimizes the need to back off the output power from the maximum configurable level. Minimizing the power backoff helps preserve the best coverage possible for a given power capability.

7.1.2.3 Long device battery lifetime

Minimizing power backoff also gives rise to higher PA power efficiency, which helps extend device battery lifetime. Device battery lifetime, however, depends heavily on how the device behaves when it does not have an active data session. In most use cases, the device actually spends the vast majority of its lifetime in idle mode as most of the IoT applications only require infrequent transmission of short packets. Traditionally, an idle device needs to monitor paging and perform mobility measurements. Although the energy consumption during idle mode is much lower compared with during connected mode, significant energy saving can be achieved by simply increasing the periodicity between paging occasions (POs) or not requiring the device to monitor paging at all. As elaborated on in Section 2.2, 3GPP Releases 12 and 13 introduced both *extended discontinuous reception* (eDRX) and *power saving mode* (PSM) to support this type of operation and optimize device power consumption. In essence, a device can shut down its transceiver and only keep a basic oscillator running for the sake of keeping a rough time reference to know when it should come out of the PSM or eDRX. The reachability during PSM is set by the *tracking area update* (TAU) timer with the maximum settable value exceeding 1 year [9]. eDRX can be configured with a cycle just below 3 h [10].

During these power-saving states, both device and network maintain device context, saving the need for unnecessary signaling when the device comes back to connected mode. This optimizes the signaling and power consumption when making the transition from idle to connected mode.

In addition to PSM and eDRX, NB-IoT also adopts *connected mode DRX* as a major tool for achieving energy efficiency. In Release 13, the connected mode DRX cycles were extended from 2.56 s to 10.24 s for NB-IoT [11].

7.1.2.4 Support of massive number of devices

NB-IoT achieves high capacity in terms of number of devices that can be supported on one single NB-IoT carrier. This is made possible by introducing spectrally efficient transmission scheme in the uplink for devices in extreme coverage-limited situation.

Shannon's well-known channel capacity theorem [12] establishes a relationship between bandwidth, power, and capacity in an *additive white Gaussian noise* channel as

$$C = W\log_2\left(1+\frac{S}{N}\right) = W\log_2\left(1+\frac{S}{N_0W}\right), \tag{7.1}$$

where C is the channel capacity (bits/s), S is the received desired signal power, N is the noise power, which is determined by the product of noise bandwidth (W) and one-sided noise power spectral density (N_0). The noise bandwidth is identical to the signal bandwidth if Nyquist pulse shaping function is used. At extreme coverage-limited situation, $\frac{S}{N} \ll 1$. Using the approximation $\ln(1+x) \approx x$, for $x \ll 1$, it can be shown that the channel capacity in the very low signal-to-noise-power ratio (SNR) regime is

$$C = \frac{S}{N_0}\log_2(e). \tag{7.2}$$

In this regime, the bandwidth dependency vanishes, and therefore channel capacity, in terms of bits per second, is only determined by the ratio of S and N_0. Thus, in theory the coverage for a target data rate $R = C$ depends only on the received signal power level, and not on the signal bandwidth. This implies that because the data rate at extreme coverage-limited situation does not scale according to the device bandwidth allocation, for the sake of spectral efficiency, it is advantageous to allocate small bandwidth for devices in bad coverage. NB-IoT uplink waveforms include various bandwidth options. While a waveform of wide bandwidth (e.g., 180 kHz) is beneficial for devices in good coverage, waveforms of smaller bandwidths are more spectrally efficient from the system point of view for serving devices in bad coverage. This will be illustrated by the coverage results presented in Chapter 8.

7.1.2.5 *Deployment flexibility*

To support maximum flexibility of deployment and prepare for refarming scenarios, NB-IoT supports three modes of operation, stand-alone, in-band, and guard-band.

7.1.2.5.1 Stand-alone mode of operation

NB-IoT may be deployed as a *stand-alone* carrier using any available spectrum with bandwidth larger than 180 kHz. This is referred to as the stand-alone deployment. A use case of the stand-alone deployment is for a GSM operator to deploy NB-IoT in its GSM band by refarming part of its GSM spectrum. In this case, however, additional guard-band is needed between a GSM carrier and the NB-IoT carrier. Based on the coexistence requirements in Ref. [13], 200 kHz guard-band is recommended, which means that a GSM carrier should be left empty on one side of an NB-IoT carrier between two operators. In case of the same operator deploying both GSM and NB-IoT, a guard-band of 100 kHz is recommended based on the studies in Ref. [2], and hence an operator needs to refarm at least two consecutive GSM carriers for NB-IoT deployment. An example is illustrated in Fig. 7.1. Here, NB-IoT bandwidth is shown as 200 kHz. This is because NB-IoT needs to meet the GSM spectral mask when deployed using refarmed GSM spectrum and the GSM spectral mask is specified according to 200 kHz channelization.

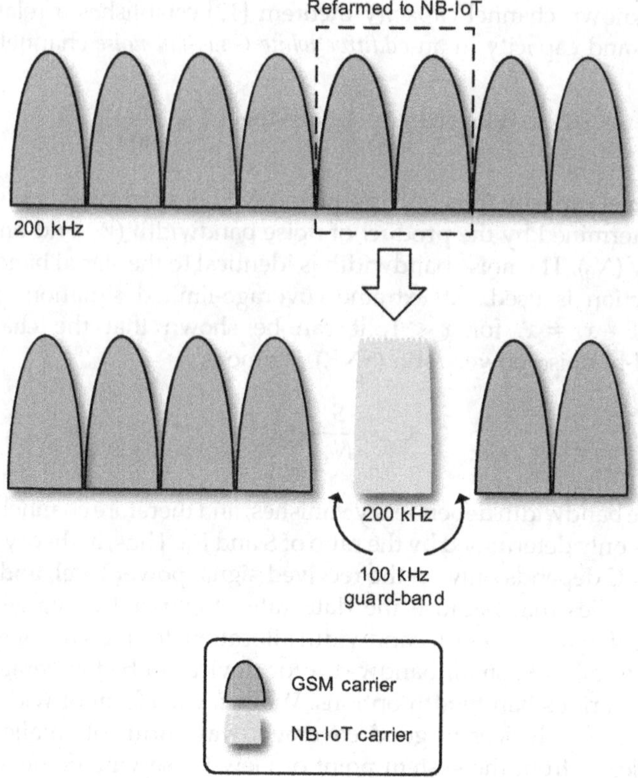

FIG. 7.1 NB-IoT stand-alone deployment using refarmed GSM spectrum.

7.1.2.5.2 In-band and guard-band modes of operation

NB-IoT is also designed to be possible for deployment in the existing LTE networks, either using one of the LTE *physical resource blocks* (PRBs) or using the LTE guard-band. These two deployment scenarios are referred to as in-band and guard-band deployments, respectively. An example is illustrated in Fig. 7.2. An LTE carrier with a number of PRBs is shown. NB-IoT can be deployed using one LTE PRB or using the unused bandwidth in the guard-band. The guard-band deployment makes use of the fact that the occupied bandwidth of the LTE signal is roughly 90% of channel bandwidth [14], leaving roughly 5% of the LTE channel bandwidth on each side available as guard-band. It is therefore possible to place an NB-IoT carrier in the guard band of LTE when the LTE carrier bandwidth is 5, 10, 15, or 20 MHz.

Yet another possible deployment scenario is to have NB-IoT in-band deployment on an LTE carrier that supports LTE-M features. The concept of *narrowband* used in LTE-M is explained in Section 5.2.2.2. Some of these LTE-M *narrowbands* are not used for transmitting LTE-M *system information (SI) Block Type 1* (SIB1) and thus can be used for deploying NB-IoT. More details about this deployment scenario are given in Section 7.2.1.1.

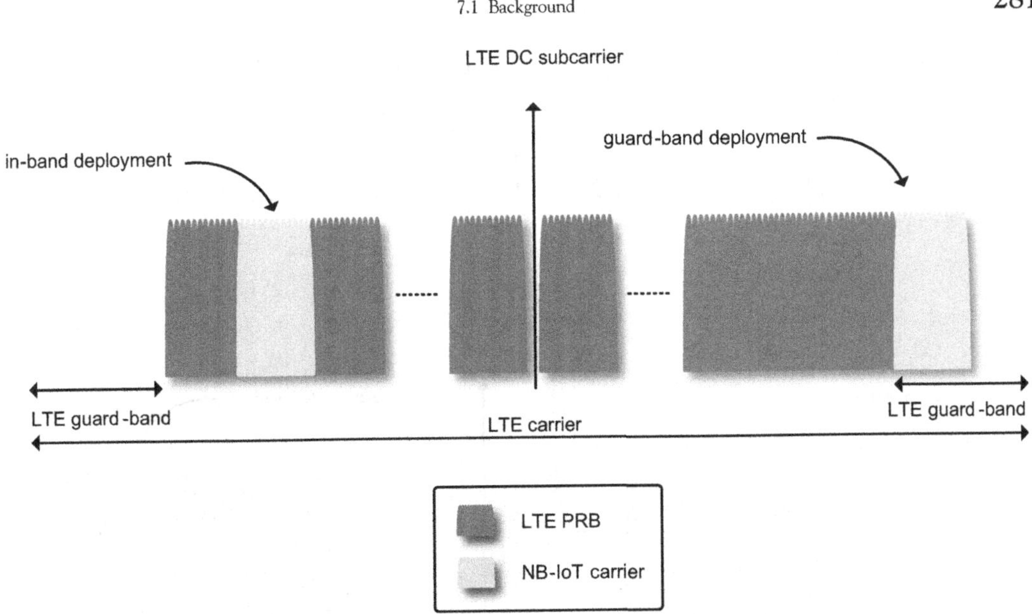

FIG. 7.2 NB-IoT deployment inside an LTE carrier, either using one of the LTE PRBs (in-band deployment) or using the LTE guard band (guard-band deployment).

7.1.2.5.3 Spectrum refarming

NB-IoT is intended to offer flexible spectrum migration possibilities to a GSM operator. An operator can take an initial step to refarm a small part of the GSM spectrum to NB-IoT as the example shown in the top part of Fig. 7.3. Thanks to the support of LTE in-band and guard-band deployments, such an initial migration step will not result in spectrum fragmentation to make the eventual migration of the entire GSM spectrum to LTE more difficult. As illustrated in Fig. 7.3, the NB-IoT carrier already deployed in the GSM network as a standalone deployment may become an LTE in-band or guard-band deployment when the entire GSM spectrum is migrated to LTE. This high flexibility was also envisioned to facilitate NB-IoT deployments when LTE is later refarmed to the 5G *New Radio* (NR) technology. Indeed, as it will be described in Section 7.4 NB-IoT can be deployed together with NR, and in such a deployment, NR and NB-IoT achieve superior coexistence performance.

7.1.2.6 Coexistence with LTE

When designing a new access technology, the degree of freedom is higher compared with when basing the design on an existing technology. Therefore NB-IoT could with limited restrictions be designed from ground-up with the intention to follow the radio access design principles outlined thus far.

The ambition to provide a radio access technology with high deployment flexibility and the capability to operate both in a refarmed GSM spectrum and inside an LTE carrier did, however, impose certain guiding principles onto the design of the technology.

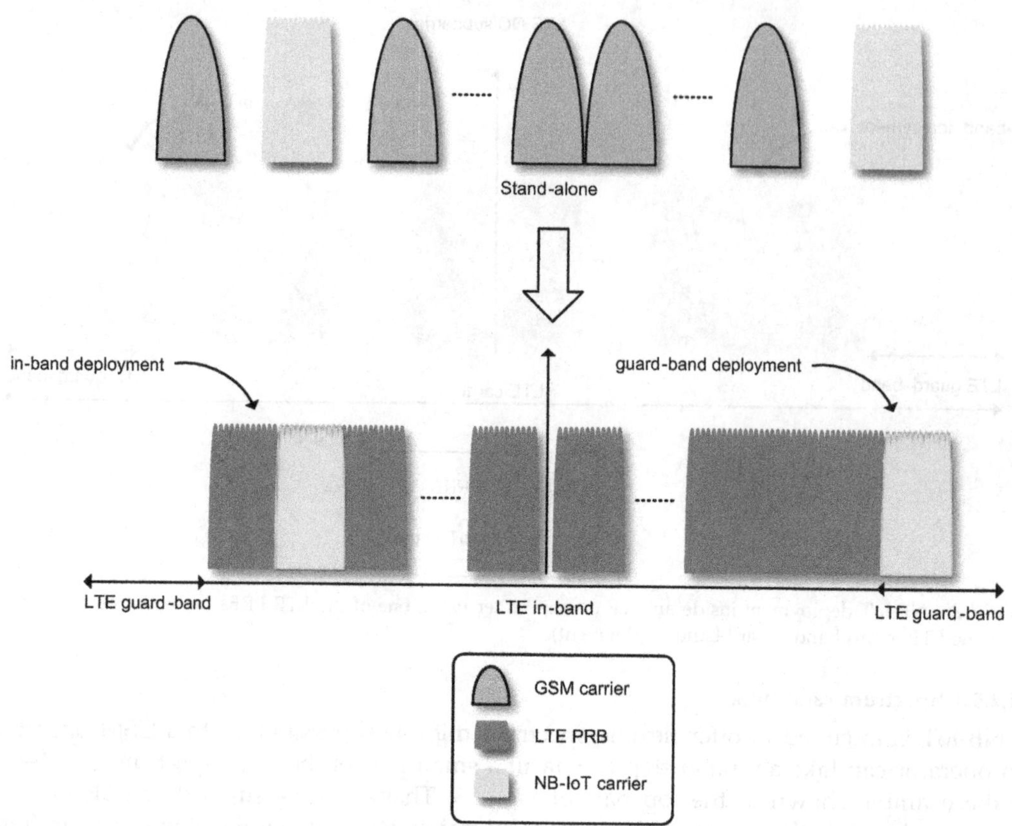

FIG. 7.3 Partial GSM spectrum to NB-IoT introduction as an initial spectrum refarming step, followed by eventual total migration to LTE.

While the stand-alone deployment in the GSM spectrum, discussed in Section 7.1.2.5, is facilitated by the introduction of a guard-band between the NB-IoT and GSM carriers, the expectations on close coexistence with LTE were higher. NB-IoT deployment inside an LTE carrier was hence required to be supported without any guard-band between NB-IoT and LTE PRBs. To minimize the impact on the existing LTE deployments and devices, this imposed requirements on the NB-IoT physical layer waveforms to preserve orthogonality with the LTE signal in adjacent PRBs. It also implies that NB-IoT should be able to share the same time-frequency resources as LTE the same way as different LTE physical channels share time-frequency resources. Last but not least, because legacy LTE devices will not be aware of the NB-IoT operation, NB-IoT transmissions should not collide with *essential* LTE transmissions.

Among the essential LTE transmissions are the physical channel signals transmitted in the downlink control region, including *Physical Control Format Indicator Channel* (PCFICH), *Physical Hybrid ARQ Indicator Channel*, and *Physical Downlink Control Channel* (PDCCH). PCFICH is used for indicating the number of *orthogonal frequency-division multiplexing* (OFDM) symbols in a subframe that can be used for PDCCH. In LTE, starting from the beginning of a

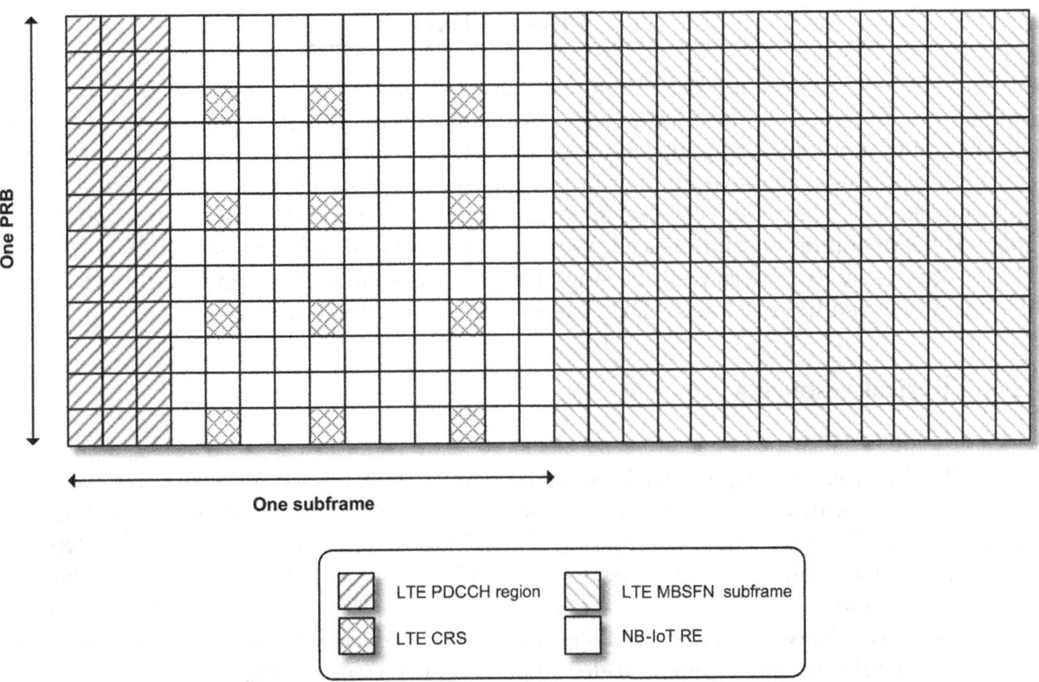

FIG. 7.4 An illustration of how NB-IoT physical channels of the in-band mode share REs with LTE in the downlink.

subframe, up to 3 OFDM symbols may be used for PDCCH transmissions. A PDCCH transmission may carry scheduling information, paging indicators, etc. As illustrated in Fig. 7.4, the *resource elements* (REs) of the first three OFDM symbols in an LTE subframe may therefore not be used by NB-IoT downlink channels.

Additional essential LTE physical channels and signals are the *Cell-specific Reference Signal* (CRS) and *Multicast-Broadcast Single-Frequency Network* (MBSFN) signal. The REs used by these channels and signals are also preserved and not mapped to any NB-IoT physical channels. MBSFN signal, for example, spans one subframe in the time dimension and all PRBs in the frequency dimension, and thus if one subframe is configured as an LTE MBSFN subframe, that subframe is not used by NB-IoT. Fig. 7.4 illustrates such an example.

Furthermore, NB-IoT also avoids using resources mapped to LTE *Primary Synchronization Signal* (PSS), *Secondary Synchronization Signal* (SSS), and *Physical Broadcast Channel* (PBCH). Since these transmissions use the middle six PRBs, in the case of 1.4, 10, and 20 MHz LTE carrier bandwidths, or middle seven PRBs, in the case of 3, 5, and 15 MHz LTE carrier bandwidths, an NB-IoT carrier cannot use any of these middle LTE PRBs in an in-band deployment.

Following these guiding principles will naturally mean that the physical layer to a large extent is inspired by LTE. At the same time, changes are required to meet the aforementioned design objectives.

Interestingly, the objective for achieving superior coexistence performance with LTE paves the way for NB-IoT to also achieve superior coexistence performance with 5G NR, after the LTE to NR migration. NB-IoT coexistence with NR will be described in Section 7.4.

7.2 Physical layer

In this section, we describe NB-IoT physical layer design with an emphasis on how these channels are designed to fulfill the objectives that NB-IoT targets, namely, deployment flexibility, ubiquitous coverage, ultra-low device cost, long battery lifetime, and capacity sufficient for supporting a massive number of devices in a cell. NB-IoT was initially designed to support *frequency division duplex* (FDD) operation. In 3GPP Release 15, support for *time division duplex* (TDD) operation was introduced. We focus on FDD design in this section. Many of the FDD designs apply to TDD as well, although there are also a few differences. The most critical differences between the FDD and TDD designs will be described in Section 7.2.8.

7.2.1 Physical resources

7.2.1.1 Channel raster

An NB-IoT carrier carrying essential physical signals that allow a device to perform cell selection is referred to as an *anchor* carrier. The carrier frequency of an NB-IoT anchor carrier is determined by the *E-UTRA absolute radio frequency channel number* [15]. The placement of an NB-IoT anchor carrier in frequency corresponding to a certain E-UTRA absolute radio frequency channel number is just as for LTE, based on a 100 kHz channel raster [15]. Contrary to LTE, an NB-IoT anchor carrier can, however, be located slightly off the 100 kHz channel raster. The need for this offset is justified by the required NB-IoT deployment flexibility as explained next.

The deployment scenario chosen by the operator, stand-alone, in-band, or guard-band mode of operation, should be transparent to a device when it is first turned on and searches for an NB-IoT carrier on its supported frequency bands. In case of stand-alone deployment, the NB-IoT anchor carrier can always be placed on a refarmed GSM 200 kHz channel, or as shown in Fig. 7.1, placed in the middle of refarmed 400 kHz GSM spectrum with 100 kHz guard band on both sides of the NB-IoT carrier. In either case, since GSM uses 200 kHz channel raster, it is easy to see that in the two scenarios mentioned above, the NB-IoT carrier will fall on the 100 kHz raster. However, for LTE in-band and guard-band deployments, NB-IoT needs to follow the same subcarrier grids used by LTE to maintain inter-subcarrier orthogonality. As a result, it is not possible to place the NB-IoT anchor carrier with a center frequency exactly on the 100 kHz raster grid. An example is given in Table 7.2 based on a 3 MHz LTE carrier, which has 15 PRBs. The PRB frequency offsets, i.e., the PRB center frequencies relative to the LTE DC subcarrier, are listed in the table. Because the LTE DC subcarrier is placed on the 100 kHz raster, the relative frequencies, listed in Table 7.2, give indications of the PRB centers relative to the 100 kHz raster. We will refer to this as raster offset in the discussion below.

TABLE 7.2 LTE PRB center frequencies relative to the DC subcarrier for 3 MHz LTE bandwidth.

PRB index	0	1	2	3	4	5	6	7
PRB frequency offset [kHz]	-1267.5	-1087.5	-907.5	-727.5	-547.5	-367.5	-187.5	0
PRB index	**8**	**9**	**10**	**11**	**12**	**13**	**14**	
PRB frequency offset [kHz]	187.5	367.5	547.5	727.5	907.5	1087.5	1267.5	

The PRB indexing starts from index 0 for the PRB occupying the lowest frequency within the LTE carrier.

As can be seen, no PRB, besides the center PRB, has a center frequency that exactly falls on the raster, i.e., has a PRB center frequency that equals a multiple of 100 kHz. As described in Section 7.1.2.6, the middle PRB on an LTE carrier cannot be used for NB-IoT deployment since it overlaps with LTE PSS, SSS, and PBCH. Therefore, although the center PRB has a center frequency that exactly falls on the raster, it cannot be used for NB-IoT deployment. From Table 7.2, PRB#2 and PRB#12 in this example have the smallest raster offset of 7.5 kHz. To facilitate an efficient initial cell selection for NB-IoT, in case of LTE in-band deployment, an anchor carrier is required to be configured on a PRB that has the smallest raster offset. It turns out that for LTE carrier bandwidth of 3, 5, and 15 MHz, the smallest magnitude of the raster offset is 7.5 kHz, whereas for LTE carrier bandwidth of 10 and 20 MHz, the smallest magnitude of the raster offset is 2.5 kHz. A full list of PRB indexes for in-band deployment is given in Table 7.3.

Because a 1.4 MHz LTE carrier has only six PRBs and very small guard-bands, in-band and guard-band deployments of NB-IoT are not supported on a 1.4 MHz LTE carrier. Furthermore, a 3 MHz LTE carrier also has a very small guard band, it does not support the guard-band deployments of NB-IoT.

Similar to the in-band deployment, an NB-IoT anchor carrier in the guard-band deployment needs to have a raster offset as small as possible. For LTE carrier bandwidth of 10 and 20 MHz, the smallest magnitude of the raster offset for the guard-band deployment is 2.5 kHz. An example is shown in Fig. 7.5, in which NB-IoT is deployed immediately adjacent to the edge LTE PRB, i.e., PRB#49 on a 10 MHz LTE carrier. The center of the NB-IoT guard-band PRB is 4597.5 kHz from the DC subcarrier, giving rise to a -2.5 kHz raster offset. For LTE carrier bandwidth of 3, 5, and 15 MHz, an anchor carrier cannot be placed immediately adjacent to the edge LTE PRB as the raster offset would become too large. However, shifting the NB-IoT anchor by additional three subcarriers away from the edge LTE PRB gives rise to a raster offset of ±7.5 kHz.

Multicarrier operation of NB-IoT is supported. Because NB-IoT requires only one anchor carrier for facilitating device initial cell selection, the additional carriers may be located with an offset of up to 47.5 kHz outside the 100 kHz raster grid (a total of 21 offset values are defined) [15]. These additional carriers are referred to as non-anchor carriers. A non-anchor carrier does not carry the physical channels that are required for device initial cell selection.

As mentioned in Section 7.1.2.5, NB-IoT can be deployed as an LTE in-band deployment together with LTE-M on the same LTE carrier. The PRB indexes that can be used for deploying

TABLE 7.3 Suitable PRB indexes for NB-IoT anchor carrier in the in-band deployment.

	Allowed PRB indexes for NB-IoT anchor carrier		
LTE bandwidth	**PRBs below DC subcarrier**	**PRBs above DC subcarrier**	**Magnitude of raster offset**
1.4 MHz	Not supported		Not applicable
3 MHz	2	12	7.5 kHz
5 MHz	2, 7	17, 22	7.5 kHz
10 MHz	4, 9, 14, 19	30, 35, 40, 45	2.5 kHz
15 MHz	2, 7, 12, 17, 22, 27, 32	42, 47, 52, 57, 62, 67, 72	7.5 kHz
20 MHz	4, 9, 14, 19, 24, 29, 34, 39, 44	55, 60, 65, 70, 75, 80, 85, 90, 95	2.5 kHz

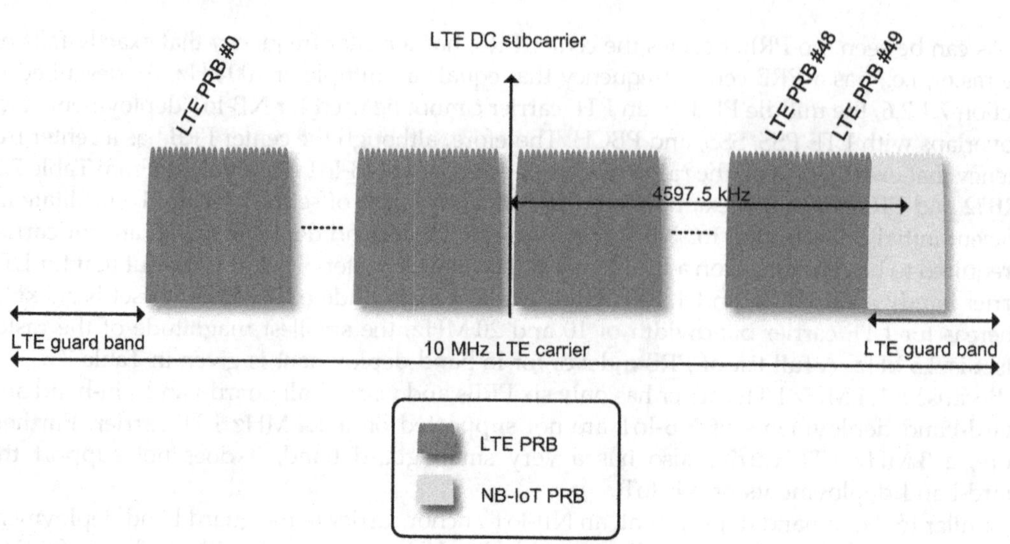

FIG. 7.5 Deployment of NB-IoT in the guard-band of a 10 MHz LTE carrier.

an NB-IoT anchor carrier for such scenarios without collisions with an LTE-M narrowband are shown in Table 7.4. The PRB indexes that can be used for deploying NB-IoT non-anchor carriers are also shown in Table 7.4. As indicated in Table 7.4, deploying both NB-IoT anchor carrier in-band and LTE-M on a 3-MHz LTE carrier is not possible without collision with an LTE-M narrowband. This is because LTE-M defines two narrowbands on a 3-MHz LTE carrier, as illustrated in Fig. 7.6, leaving one PRB on the center and both edge PRBs unused. However, none of these three PRBs can be used as an NB-IoT anchor. The edge PRBs cannot be used as an anchor carrier because they do not meet the raster offset requirements, but they can nevertheless be used as a non-anchor carrier. The center PRB cannot be used as an anchor carrier, nor as a non-anchor carrier, as it is used by LTE PSS, SSS, and PBCH.

7.2.1.2 *Frame structure*

The overall time frame structure on the NB-IoT access stratum (AS) is illustrated in Fig. 7.7. On the highest level, one hyperframe cycle has 1024 hyperframes each consisting of 1024 frames. One frame consists of 10 subframes, each dividable into two slots of 0.5 ms. Hyperframe and frame are each labeled with a *hyper system frame number* (H-SFN) and a *system frame number* (SFN), respectively. Each subframe can therefore be uniquely identified by the

TABLE 7.4 Suitable PRB indexes for NB-IoT in-band deployment on an LTE carrier that supports LTE-M.

LTE bandwidth [MHz]	Allowed PRB indexes for NB-IoT carrier
1.4	Not supported
3	Anchor carrier: none; non-anchor: 0 or 14
5	Anchor carrier: 7, 17; non-anchor carrier: 6, 7, 8, 16, 17, 18
10	Anchor carrier: 19, 30; non-anchor carrier: 0, 19, 20, 21, 28, 29, 30, 49
15	Anchor carrier: 32, 42; non-anchor carrier: 0, 31, 32, 33, 41, 42, 43, 74
20	Anchor carrier: 44, 55; non-anchor carrier: 0, 1, 44, 45, 53, 54, 55, 98, 99

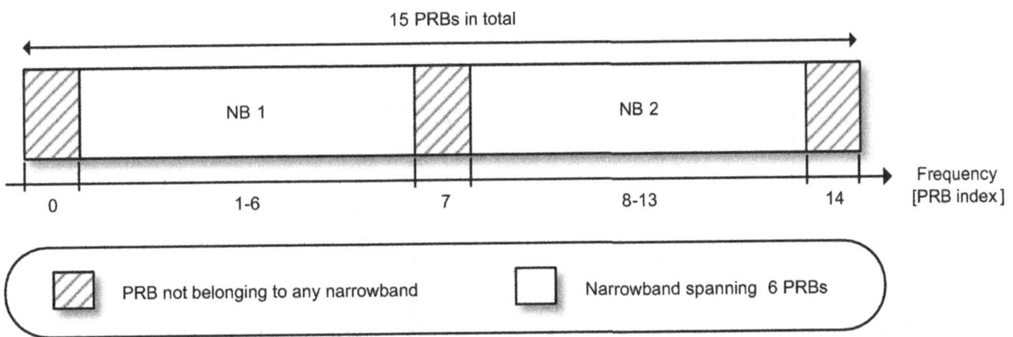

FIG. 7.6 Definition of LTE-M narrowband (NB) on a 3 MHz LTE carrier.

combination of an H-SFN, an SFN, and a *subframe number* (SN). The ranges of H-SFN, SFN, and SN are 0–1023, 0–1023, and 0–9, respectively.

In the downlink and uplink, the NB-IoT design supports a subcarrier spacing of 15 kHz, for which each frame contains 20 slots as illustrated in Fig. 7.7. In the uplink, the technology supports an additional subcarrier spacing of 3.75 kHz. For this alternative subcarrier spacing, each frame is directly divided into five slots, each of 2 ms as shown in Fig. 7.8.

7.2.1.3 *Resource grid*

In the downlink, the concept of PRBs is used for specifying the mapping of physical channels and signals onto REs. The RE is the smallest physical channel unit, each uniquely identifiable by its subcarrier index k and symbol index l within the PRB. A PRB spans 12 subcarriers over 7 OFDM symbols and in total $12 \times 7 = 84$ REs. One PRB pair is the smallest schedulable unit in most downlink cases, as of Release 15, and fits into two consecutive slots as shown in Fig. 7.9.

For the uplink, the *resource unit* (RU) is used to specify the mapping of the uplink physical channels onto REs. The definition of the RU depends on the configured subcarrier spacing and the number of subcarriers allocated to the uplink transmission. In the basic case where 12 subcarriers using a spacing of 15 kHz are allocated, the RU corresponds to the PRB pair in Fig. 7.9. In case of *sub-PRB* scheduling assignments of 6, 3, or 1 subcarrier, then the RU is expanded in time to compensate for the diminishing frequency allocation. For the single subcarrier allocation, also known as *single-tone* allocation, the NB-IoT RU concept supports an additional subcarrier spacing of 3.75 kHz. Section 7.2.5.2 presents the available formats of the RU for the different uplink transmit configurations in detail.

7.2.2 Transmission schemes

7.2.2.1 *Duplex modes*

NB-IoT since its introduction supports FDD operation. The support of TDD operation was introduced in 3GPP Release 15. A brief description on FDD and TDD operations is given in Section 5.2.2.1. Table 7.5 lists all the bands that NB-IoT support as of Release 15 [15]. An FDD NB-IoT device is only required to support *half-duplex FDD type B* operation, which allows single local oscillator for carrier frequency generation for both downlink and uplink. As explained in Section 5.2.2.1, half-duplex FDD operation means that the device is not required to simultaneously receive in the downlink and transmit in the uplink. Furthermore, HD-FDD type B allows ample time for the device to switch between downlink and uplink.

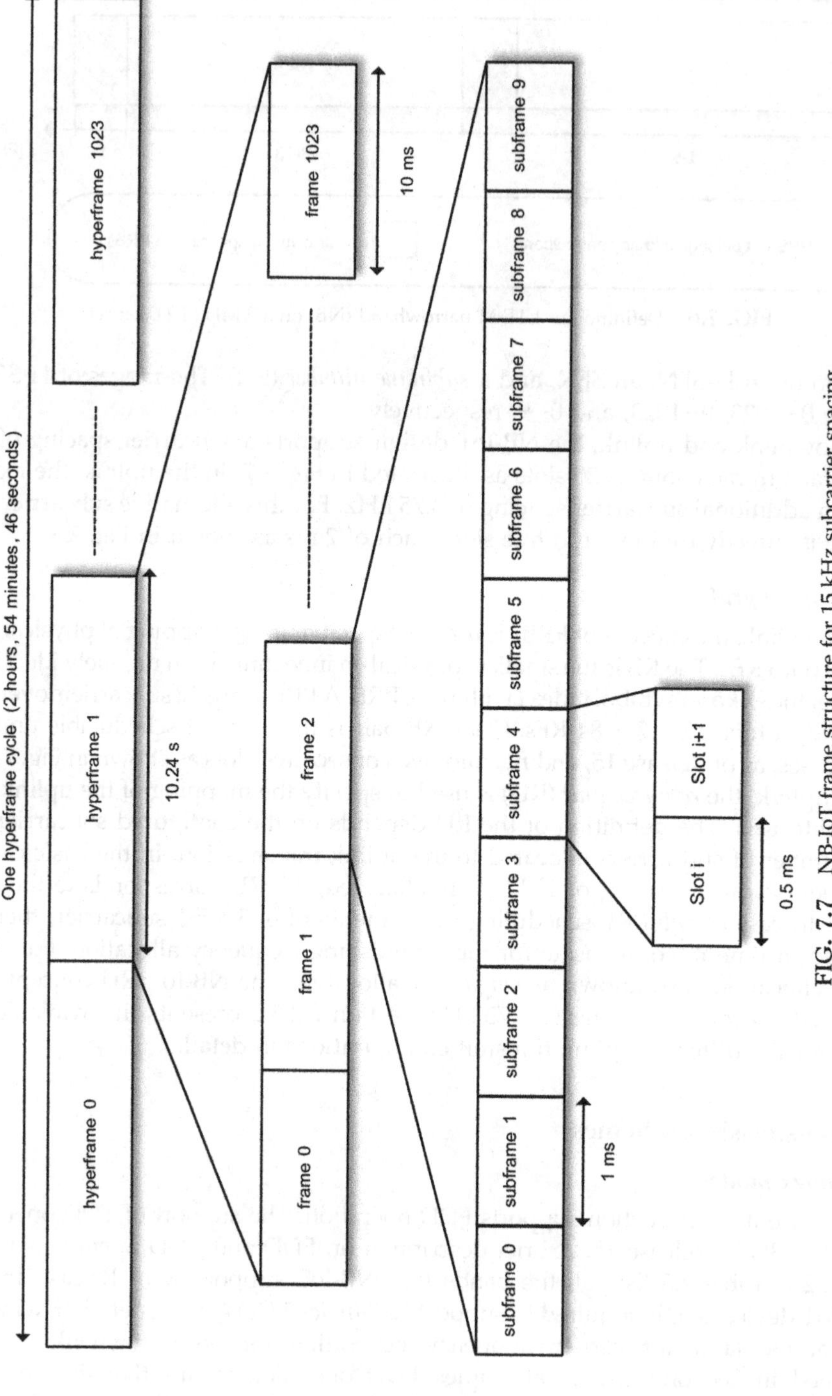

FIG. 7.7 NB-IoT frame structure for 15 kHz subcarrier spacing.

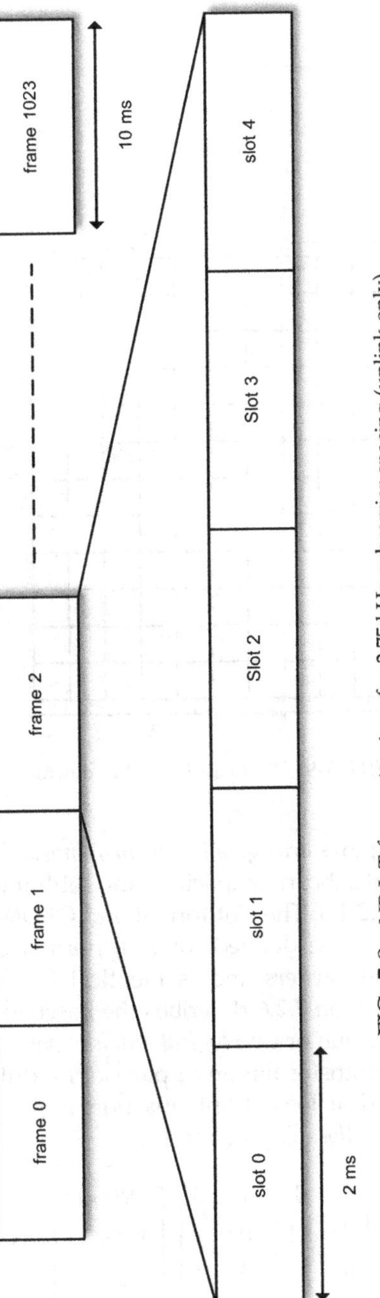

FIG. 7.8 NB-IoT frame structure for 3.75 kHz subcarrier spacing (uplink only).

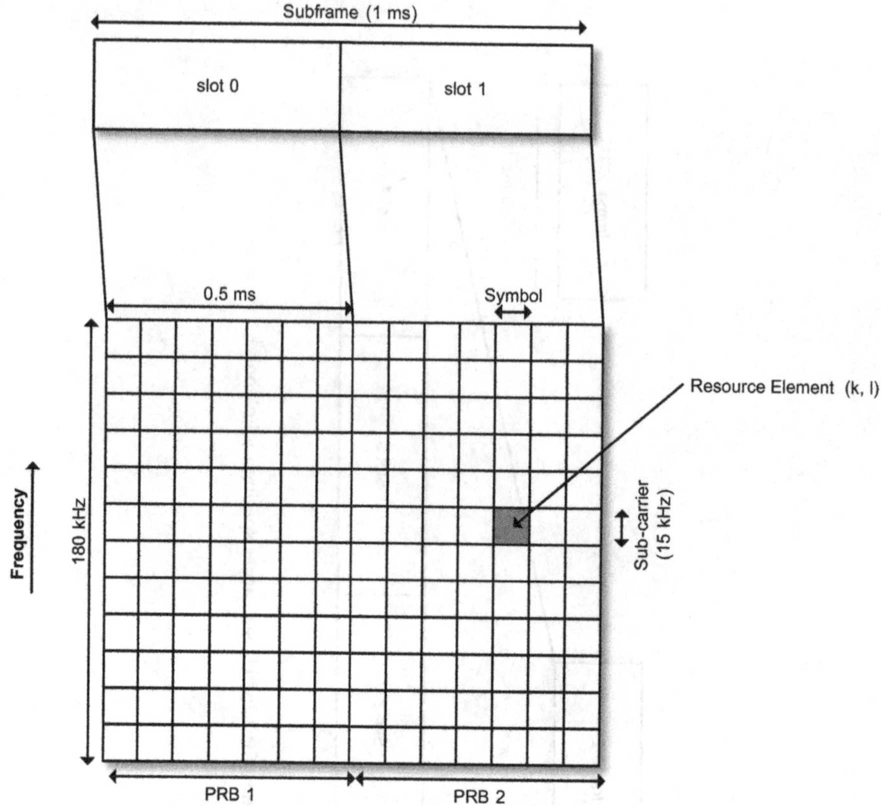

FIG. 7.9 PRB pair in NB-IoT downlink.

7.2.2.2 *Downlink operation*

In the downlink, NB-IoT employs *orthogonal frequency-division multiple-access,* using the same numerologies as LTE in terms of subcarrier spacing, slot, subframe, and frame durations as presented in Sections 7.2.1.2 and 7.2.1.3. The slot format and OFDM symbol and *cyclic prefix* (CP) durations are also identical to those defined for LTE normal CP, i.e., 4.7 μs. The downlink waveform is defined for 12 subcarriers and is identical for stand-alone, guard-band, and in-band modes of operation. Section 7.2.6 describes the baseband signal generation in detail.

In the downlink, support for one or two *logical antenna ports* has been specified. In case of two logical antenna ports, the transformation of pairs of modulated symbols s_{2i}, s_{2i+1} to sets of precoded symbols presented at logical antenna port p, $p=0$ or 1, is based on transmit diversity using *Space-Frequency Block Coding* [16,17]:

$$\begin{bmatrix} y_{2i}^{0} \\ y_{2i}^{1} \\ y_{2i+1}^{0} \\ y_{2i+1}^{1} \end{bmatrix} = \frac{1}{\sqrt{2}} \begin{bmatrix} 1 & 0 & j & 0 \\ 0 & -1 & 0 & j \\ 0 & 1 & 0 & j \\ 1 & 0 & -j & 0 \end{bmatrix} \begin{bmatrix} \Re(s_{2i}) \\ \Re(s_{2i+1}) \\ \Im(s_{2i}) \\ \Im(s_{2i+1}) \end{bmatrix} = \frac{1}{\sqrt{2}} \begin{bmatrix} s_{2i} \\ -s_{2i+1}^{*} \\ s_{2i+1} \\ s_{2i}^{*} \end{bmatrix} \tag{7.3}$$

TABLE 7.5 Frequency bands that are defined for NB-IoT.

Band	Duplex mode	Uplink [MHz]	Downlink [MHz]
1	FDD	1920–1980	2110–2170
2	FDD	1850–1910	1930–1990
3	FDD	1710–1785	1805–1880
4	FDD	1710–1755	2110–2155
5	FDD	824–849	869–894
8	FDD	880–915	925–960
11	FDD	1427.9–1447.9	1475.9–1495.9
12	FDD	699–716	729–746
13	FDD	777–787	746–756
14	FDD	788–798	758–768
17	FDD	704–716	734–746
18	FDD	815–830	860–875
19	FDD	830–845	875–890
20	FDD	832–862	791–821
21	FDD	1447.9–1462.9	1495.9–1510.9
25	FDD	1850–1915	1930–1995
26	FDD	814–849	859–894
28	FDD	703–748	758–803
31	FDD	452.5–457.5	462.5–467.5
41	TDD	2496–2690	2496–2690
66	FDD	1710–1780	2110–2200
70	FDD	1695–1710	1995–2020
71	FDD	636–698	617–652
72	FDD	451–456	461–466
73	FDD	450–455	460–465
74	FDD	1427–1470	1475–1518
85	FDD	698–716	728–746

Here, $\Re(s_{2i})$ and $\Im(s_{2i})$ represent the real part and imaginary part of complex symbol x. In essence, logical antenna port 0 transmits symbol pair (s_{2i}, s_{2i+1}) while logical antenna port 1 transmits symbol pair $\left(-s_{2i+1}^*, s_{2i}^*\right)$. Each symbol pair is mapped to two consecutively available REs within an OFDM symbol.

The mapping from the logical antenna ports to the base station physical antenna ports is up to implementation, and there is no restriction in the number of supported physical antennas. This is important for NB-IoT in-band operation when LTE may use four or more antenna ports. Many NB-IoT devices are expected to be stationary and equipped with a single antenna, i.e., offering low spatial diversity. The system is, in addition, supporting limited frequency diversity because of the narrow system bandwidth. The option of transmit diversity is therefore beneficial.

To support operation in an extended coverage range, mapping of a single *TB* over up to 10 subframes is supported in combination with repetition-based link adaptation. A consequence of this is that a single transmission time interval may range up to 20,480 subframes (20.48 s).

Finally, to allow the device receiver to perform coherent combining or cross-subframe channel estimation, it is expected that the downlink waveform is transmitted with a continuous and stable phase trajectory. Coherent combining and cross-subframe channel estimation improve receive performance and allow the device to detect a received signal far below the thermal noise floor.

7.2.2.3 Uplink operation

NB-IoT uplink uses *single-carrier frequency-division multiple-access* (SC-FDMA), also referred to as *discrete Fourier transform (DFT) spread OFDM* (DFTS-OFDM), with 15 kHz subcarrier spacing for multitone transmissions. In this case, the same numerologies as NB-IoT downlink are used. Multitone transmissions may use 12, 6, or 3 subcarriers. In addition, single-tone transmissions are supported, and in that case, the time–frequency resource grids can be based on 15 or 3.75 kHz subcarrier spacing.

The single-tone waveform has been designed with a close to constant envelope modulation to allow transmit operation without any power backoff to optimize coverage as well as PA efficiency. The 3.75 kHz subcarrier spacing allows for increased system capacity in the extended coverage domain where the data rate is power limited and not bandwidth limited.

Single-tone transmissions with 15 kHz subcarrier spacing use the same numerologies as multitone transmissions and thus achieve the best coexistence performance with multitone transmissions as well as with LTE. As for the downlink, the uplink supports mapping of a TB over multiple consecutive RUs in combination with an extensive set of repetitions to achieve operation in extended coverage.

SC-FDMA with single-tone is mathematically identical to OFDM, as the *DFT* precoding step can be omitted.

Just as in case of the downlink, also the uplink waveform is expected to be transmitted with a continuous and stable phase trajectory to allow the base station receiver to perform coherent combining of the received waveform or cross-subframe channel estimation at low signal-to-noise ratios.

7.2.3 Device categories and capabilities

NB-IoT defines two device categories, *Category NB1* (Cat-NB1) and *Category NB2* (Cat-NB2). They are differentiated by the parameters listed in Table 7.6. For each device category, a device can indicate the support of RF or baseband capabilities. A highlight of NB-IoT device capabilities is provided in Table 7.7. All the features associated with the device capability listed in Table 7.7 will be described throughout this chapter.

TABLE 7.6 Cat-NB1 and Cat-NB2 physical layer parameters.

Device category	NB1	NB2
Maximum UL transport block size [bits]	1000	2536
Maximum DL transport block size [bits]	680	2536
Total number of soft channel bits for decoding	2112	6400
Total layer 2 buffer sizes [bits]	4000	8000
Half-duplex FDD operation type	Type B	Type B

TABLE 7.7 A highlight of NB-IoT device capability.

Release	Device capability	Parameter value	Section
13/14	Device power class [dBm]	14, 20, or 23	7.1.2.1
13	Support of multi-tone NPUSCH transmissions	YES or NO	7.2.5.2
13	Multicarrier support	YES or NO	7.3.2.7
14	Support of two simultaneous HARQ processes	YES or NO	7.3.2.2.3
14	Support of non-anchor random access	YES or NO	7.3.1.10
14	Support of multicarrier paging	YES or NO	7.3.1.10
14	Support of interference randomization in connected mode	YES or NO	7.2.4.5 & 7.2.4.6
14	Support of SC-PTM	Not indicated	7.3.1.11
14	Support of downlink channel quality reporting	Not indicated	7.3.1.8
15	Support for wake-up signal	YES or NO	7.2.4.8 & 7.3.1.4
15	Support of NPRACH format 2	YES or NO	7.2.5.1
15	Support of additional SIB1-NB transmissions	YES or NO	7.3.1.2.1
15	Support of scheduling request through piggybacking NPUSCH F2	YES or NO	7.3.2.5
15	Support of scheduling request through NPRACH	YES or NO	7.3.2.5
15	Support of periodic NPUSCH for buffer status report	YES or NO	7.3.2.5
15	Supports of EDT	YES or NO	7.3.1.7.3
15	Support of multi-carrier operation with mixed operation mode	YES or NO	7.3.2.7
15	Support of enhanced random access power control	Not indicated	7.3.2.3.1
15	Support of enhanced power headroom reporting	Not indicated	7.3.2.3.2
15	Support for relaxed cell reselection monitoring	Not indicated	7.3.1.3
14	Support of NSSS-Based RRM measurements	Not indicated	7.3.1.3
15	Support of NPBCH-Based RRM measurements	Not indicated	7.3.1.3
15	TDD support	YES or NO	7.2.8

7.2.4 Downlink physical channels and signals

NB-IoT supports the set of downlink physical channels and signals depicted in Fig. 7.10. At a high level, the downlink physical channels and signals are time-multiplexed, except for *Narrowband Reference Signal* (NRS), which are present in every subframe carrying *Narrowband Physical Broadcast Channel* (NPBCH), *Narrowband Physical Downlink Control Channel* (NPDCCH), or *Narrowband Physical Downlink Shared Channel* (NPDSCH). Fig. 7.11 shows time-multiplexing of different mandatory downlink physical channels in a 20-ms period. The same pattern is repeated in subsequent periods. As shown, NPBCH and *Narrowband Primary Synchronization Signal* (NPSS) are transmitted in subframes 0 and 5 in every frame, respectively, and *Narrowband Secondary Synchronization Signal* (NSSS) is transmitted in subframe 9 in every other frame. The remaining subframes may be used to transmit NPDCCH, NPDSCH, *Narrowband Positioning Reference Signal* (NPRS), or *Narrowband Wake Up Signal* (NWUS).

7.2.4.1 NB-IoT subframes

Some of the subframes not carrying NPBCH, NPSS, or NSSS may be declared as *invalid subframes* according to the *Narrowband System Information Block Type 1* (SIB1-NB) described in Section 7.3.1.2. Those are not considered to belong to the set of NB-IoT subframes and are skipped (i.e. postponed) when the NB-IoT NPDSCH, NPDCCH and NWUS transmissions are mapped onto subframes. A device monitoring or receiving the downlink will skip these invalid subframes.

The notion of invalid subframe is especially useful when NB-IoT is deployed inside an LTE carrier having MBSFN subframes configured [14]. An MBSFN subframe uses all PRBs within the LTE carrier, making the REs in the subframe not available for NB-IoT. The notion of

FIG. 7.10 DL physical channels and signals used in NB-IoT.

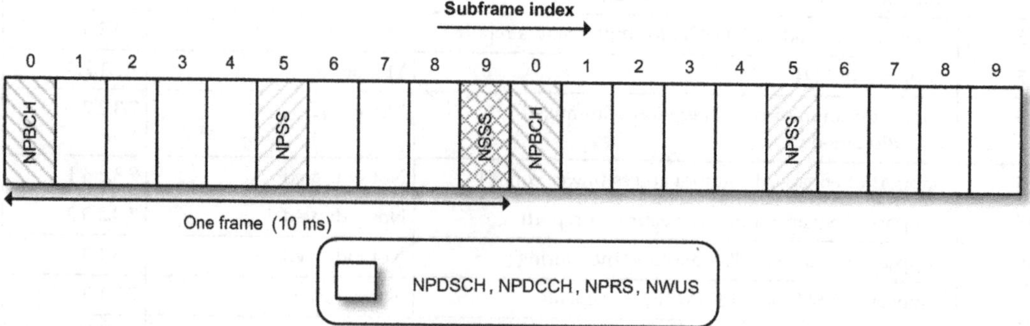

FIG. 7.11 Time-multiplexing of downlink physical channels on an FDD NB-IoT anchor carrier.

invalid subframe is also useful when NB-IoT is deployed in the guard-band of an LTE carrier configured with MBSFN subframes. An MBSFN subframe has a different subframe structure than a regular subframe, making it hard to ensure coexistence performance between NB-IoT and LTE MBSFN subframes on adjacent PRBs. This is due to that the CP used in an MBSFN subframe is longer than the normal CP, resulting in a different OFDM symbol duration compared with that of NB-IoT. In this case, it is convenient to simply declare the subframes used as LTE MBSFN subframes as invalid subframes for NB-IoT and avoid transmitting NB-IoT downlink signals in these subframes. The notion of invalid subframe is also useful for configuring the subframes for transmitting NPRS used for a device to perform *reference signal time difference* (RSTD) measurements as described in Section 7.3.2.6. An example of NB-IoT valid subframe bitmap is given in Fig. 7.12, in which subframes #7 are declared as invalid subframes and not available for NPDCCH, NPDSCH, NRS, or NWUS.

7.2.4.2 *Synchronization signals*

The NPSS and NSSS allow a device to synchronize to an NB-IoT cell. They are transmitted in certain subframes based on an 80-ms repetition interval as illustrated in Fig. 7.13. By synchronizing to NPSS and NSSS, the device will be able to detect the cell identity number and identify the framing information within an 80-ms NPSS and NSSS repetition interval.

NPSS and NSSS are designed to allow a device to use a unified synchronization algorithm during initial acquisition without knowing the NB-IoT operation mode. This is achieved by avoiding collision with REs used by LTE as much as possible. For example, numbering subframes within a frame from 0 to 9, LTE may use any of subframes 1, 2, 3, 6, 7, or 8 as an MBSFN subframe. However, during initial cell acquisition the device does not know the operation mode and whether any of these subframes is used as an MBSFN subframe. Collision avoidance with any possible LTE MBSFN subframe is achieved by using subframe 5 for NPSS and subframe 9 for NSSS as illustrated in Figs. 7.12 and 7.13. Furthermore, LTE may use up to the first three OFDM symbols in every subframe for PDCCH. To avoid potential collision with LTE PDCCH, the first three OFDM symbols are not used in the subframes that carry either NPSS or NSSS. This leaves only 11 OFDM symbols per subframe available for NPSS and NSSS (see Fig. 7.13). These symbols are generated from frequency-domain

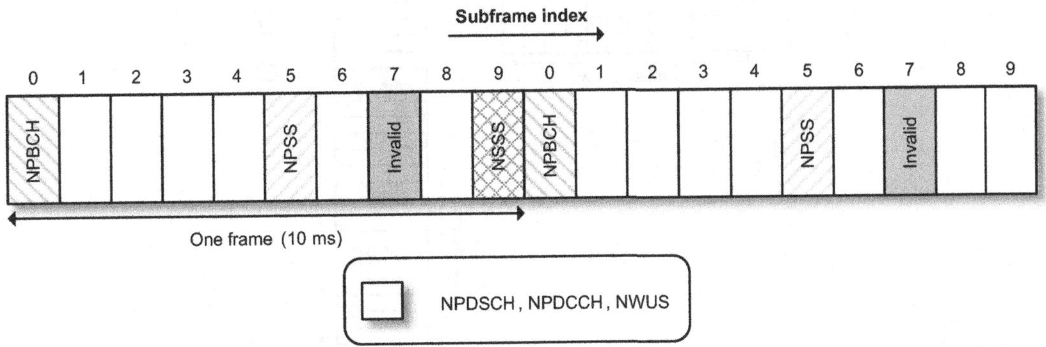

FIG. 7.12 Time-multiplexing of downlink physical channels, with invalid subframe declared through SIB1-NB (an FDD example).

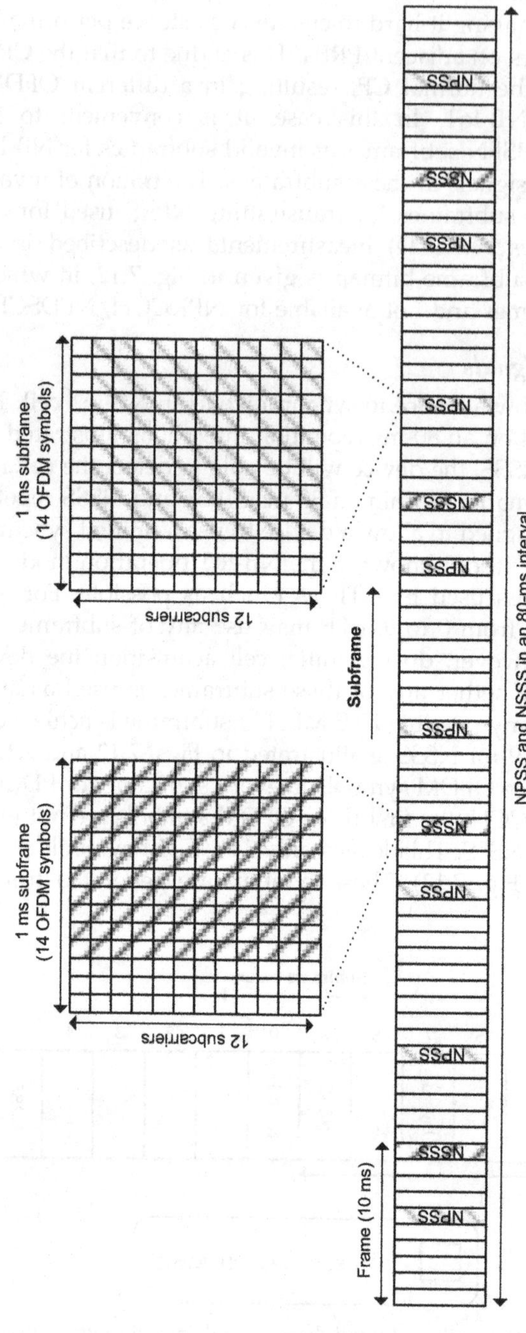

FIG. 7.13 NPSS and NSSS transmissions within an 80-ms repetition interval (an FDD example).

Zadoff-Chu (ZC) sequences, as described in next two subsections, and finally modulated into OFDM waveforms according to Section 7.2.6.2.

7.2.4.2.1 NPSS

Subframe	5
Subframe periodicity	10 ms
Sequence pattern periodicity	10 ms
Basic TTI	1 ms
Subcarrier spacing	15 kHz
Bandwidth	180 kHz
Carrier	Anchor

The NPSS is used by the device to achieve synchronization, in both time and frequency, to an NB-IoT cell. After the device wakes up from a prolonged period of deep sleep, the time base will no longer have a reliable reference, and the frequency base can be off by as much as 20 ppm (e.g., 18 kHz in a 900 MHz band) because of the limited accuracy of the low-power oscillator keeping track of time during periods of deep sleep. The NPSS hence needs to be designed so that it is detectable even with a very large frequency offset.

Because of the consideration of device complexity required for NPSS detection, all the cells in an NB-IoT network uses the same NPSS sequence. As a result, a device only needs to search for one NPSS sequence. In comparison, an LTE network uses three Primary Synchronization Signal sequences. NPSS is a hierarchical sequence generated based on a base sequence $\mathbf{p}$ and a binary cover code $\mathbf{c}$. The base sequence $\mathbf{p}$ is a length-11 frequency-domain ZC sequence of root index 5, whose nth frequency-domain element is given by

$$p(n) = e^{\frac{-j5\pi n(n+1)}{11}}, n = 0, 1, \ldots, 10 \tag{7.4}$$

The binary cover code is $\mathbf{c} = (1,1,1,1,-1,-1,1,1,1,-1,1)$.

Each of the 11 OFDM symbols in an NPSS subframe carries a copy of the base sequence, either $\mathbf{p}$ or $-\mathbf{p}$, based on the binary cover code. The same NPSS sequence is repeated in all subframes that are designated to transmit NPSS. The hierarchical sequence design reduces the device complexity in searching for an NPSS subframe.

Resource mapping within an NPSS subframe is shown in Fig. 7.14. For the in-band mode, a number of NPSS REs overlap with LTE CRS. The NPSS frequency-domain symbols on those REs will be punctured by the LTE CRS. However, the device performing NPSS detection does not need to be aware of such puncturing. For example, the device can simply correlate the received signal with an unpunctured NPSS. Although there is a mismatch between the NPSS transmitted from the base station and the device locally generated NPSS in the in-band mode, the impact on NPSS detection performance is small because there is only a small percentage of NPSS symbols punctured.

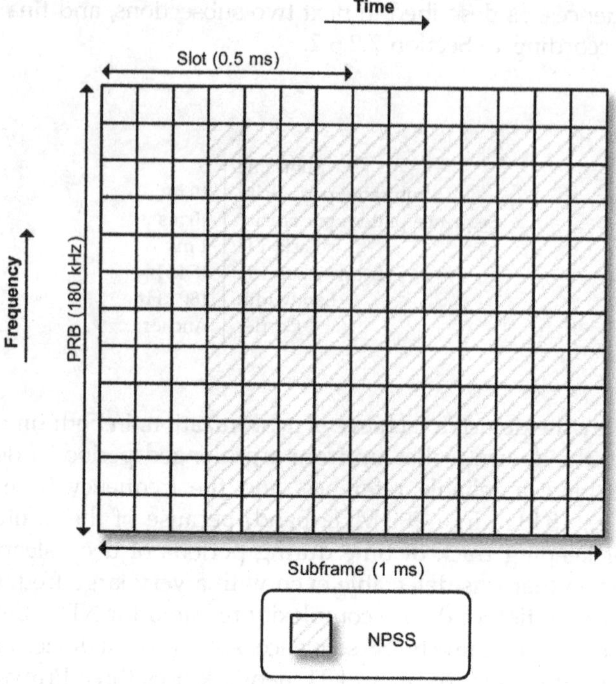

FIG. 7.14 Resource mapping in an NPSS subframe. (stand-alone and guard-band modes).

7.2.4.2.2 NSSS

Subframe	9
Subframe periodicity	20 ms
Sequence pattern periodicity	80 ms
Subcarrier spacing	15 kHz
Bandwidth	180 kHz
Carrier	Anchor

After the device has performed coarse synchronization in time and frequency when acquiring the NPSS, it turns to NSSS to detect the cell identity and acquire more information about the frame structure.

NB-IoT supports 504 unique *physical cell identities* (PCIDs) indicated by NSSS. NSSS has an 80-ms repetition interval, within which four NSSS sequences are transmitted as shown in Fig. 7.13. The four NSSS sequences transmitted in an 80-ms repetition interval are all different; however, the same set of four sequences are repeated in every 80-ms repetition interval. As explained earlier, only the last 11 OFDM symbols in an NSSS subframe are used to carry NSSS. However, compared with NPSS, NSSS is mapped to all 12 subcarriers of the PRB resulting in 132 REs in an NSSS subframe for NSSS.

The frequency-domain symbols for these 132 NSSS REs are determined according to the sequence $s(n)$ described below.

$$s(n) = b_q(n)e^{-j2\pi\theta_l n}\tilde{z}_u(n), n = 0, 1, \ldots, 131. \tag{7.5}$$

In essence, the NSSS for a cell with PCID k is determined by an extended ZC sequence $\tilde{z}_u(n)$, a binary scrambling sequence $b_q(n)$, and a phase shift θ_l. The extended ZC sequence $\tilde{z}_u(n)$ is obtained by first generating a length 131 ZC sequence of root u:

$$z_u(n) = e^{\frac{-j\pi un(n+1)}{131}}, n = 0, 1, \ldots, 130. \tag{7.6}$$

This sequence is extended to length 132 by repetition of the first element:

$$\tilde{z}_u(n) = z_u(n \; mod \; 131), n = 0, 1, \ldots, 131. \tag{7.7}$$

The root u is determined by the cell identity k as:

$$u = (k \; mod \; 126) + 3. \tag{7.8}$$

The binary scrambling sequence $b_q(n)$ is obtained based on a length-128 Walsh–Hadamard sequence [16] with the first four elements repeated at the end to become a length-132 sequence. The sequence index q is determined based on the cell identity k as:

$$q = \left\lfloor \frac{k}{126} \right\rfloor, \tag{7.9}$$

where $\lfloor x \rfloor$ is the floor function, which gives the largest integer less than or equal to x.

In a cell, all the NSSS transmissions share the same binary scrambling sequence and extended ZC sequence as these are determined by the cell identity k. Within an 80-ms NSSS repetition interval, the four occurrences, $l \in \{0,1,2,3\}$, of NSSS are differentiated by the phase shift θ_l defined as:

$$\theta_l = \frac{33l}{132}. \tag{7.10}$$

NSSS is designed to allow a device to unambiguously identify the cell identity k through matching the binary scrambling sequence and extended ZC sequence. It also supports frame synchronization within the 80-ms repetition interval through matching the phase shift term. Note that because the duration of a radio frame is 10 ms, by identifying the 80 ms framing information, the device essentially knows the three *least significant bits* (LSBs) of the SFN.

Resource mapping within an NSSS subframe is shown in Fig. 7.15. As in the case of NPSS, for the in-band mode, NSSS frequency-domain symbols mapped to the REs used by LTE CRS are punctured by CRS. Also like the case of NPSS, the device can simply correlate the received signal with an unpunctured NSSS. As there is only a small percentage of NSSS symbols punctured, the impact on NSSS detection performance due to puncturing is small.

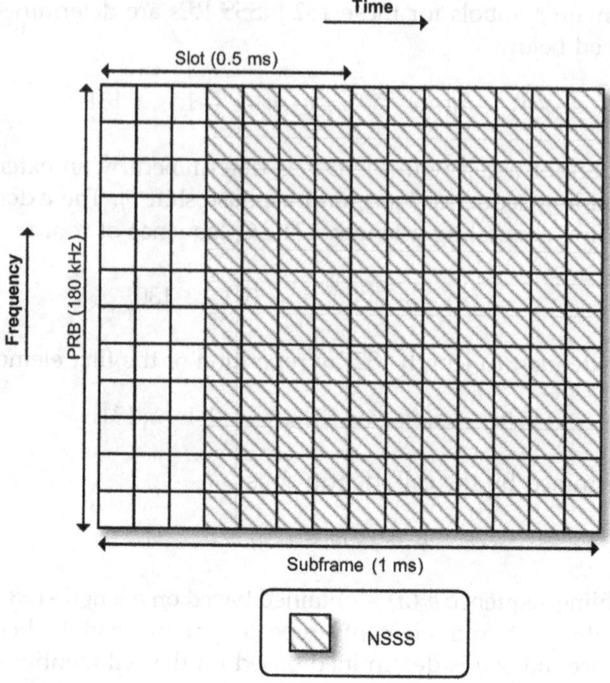

FIG. 7.15 Resource mapping in a NSSS subframe. (stand-alone and guard-band modes).

7.2.4.3 NRS

Subframe	Any
Basic *transmission time interval* (TTI)	1 ms
Sequence pattern periodicity	10 ms
Subcarrier spacing	15 kHz
Bandwidth	180 kHz
Carrier	Any

The NRS is used to allow the device to perform coherent demodulation of the downlink channels and perform downlink signal strength and quality measurements both in idle and connected mode procedures. It is mapped to certain subcarriers in the last two OFDM symbols in every slot within a subframe that carries NPBCH, NPDCCH, or NPDSCH. NRS may be transmitted also in subframes that do not have any NPDCCH or NPDSCH scheduled. For both channel estimation and downlink measurement, it is important that there is no ambiguity regarding which subframes that a device can assume the presence of NRS. In Ref. [16], NRS presence in different operation scenarios is elaborated. We highlight the important general rules below.

- In all operation modes, NRS is present in subframes 0 and 4 as well as in subframe 9 not containing NSSS.

- In stand-alone and guard-band modes, NRS is also present in subframes 1 and 3.
- In all operation modes, NRS is present in all valid NB-IoT downlink subframes (see Section 7.2.4.1)

Based on these rules, a device before acquiring any information about a cell can only assume the presence of NRS in subframes 0 and 4 as well as in subframe 9 not containing NSSS. Similarly, a device knowing the operation mode as stand-alone or guard-band, but without knowing the valid subframe configuration, can only assume NRS in subframes 0, 1, 3 and 4 as well as in subframe 9 not containing NSSS.

These general rules apply to most of the cases, but not all. Specifically, exceptions are applied to minimize unnecessary NRS transmissions on a non-anchor carrier. Starting from 3GPP Release 14, a non-anchor carrier can be configured for paging and random access (see Section 7.4.3). Since the load of paging and random access related downlink signaling might be very low in certain time periods, it is desirable to avoid NRS transmissions during a period when there is no paging or random access related downlink signals that need to be transmitted. Therefore, for a device monitoring paging and random access response (see Section 7.3.1.10) on a non-anchor carrier, it cannot assume NRS being transmitted in all NB-IoT downlink subframes before it has made the transition to the connected mode. In this case, only subframes that indeed carry information, either via NPDCCH or NPDSCH, would include NRS. In addition, to facilitate cross-subframe channel estimation and facilitate the device preparing its receiver for receiving NPDCCH or NPDSCH, 10 valid subframes before the start of NPDCCH transmission, or 4 valid subframes before the start of NPDSCH transmission, would also include NRS. Furthermore, 4 valid subframes following the end of NPDCCH or NPDSCH transmission would also include NRS.

The exact subcarriers that NRS is mapped onto depend on the cell identity and logical antenna port number. NB-IoT supports one or two logical antenna ports using space-frequency transmit diversity (see Section 7.2.2.1). An NRS resource mapping example is shown in Fig. 7.16, where NRS resource elements for the first antenna port and the second antenna port are illustrated. The NRS resource mapping pattern may shift up or down in the frequency domain depending on the cell identity. This is to allow adjacent cells to use orthogonal NRS resources to avoid mutual interference between NRS.

The NRS symbol sequence is generated based on cell identity and port number. In essence, a pseudorandom *quadrature phase shift keying* (QPSK) sequence is used for randomizing interference between cells. The NRS sequence repeats itself in every 10 ms radio frame.

7.2.4.4 NPBCH

Subframe	0
Periodicity	10
Basic TTI	640 ms
Subcarrier spacing	15 kHz
Bandwidth	180 kHz
Carrier	Anchor

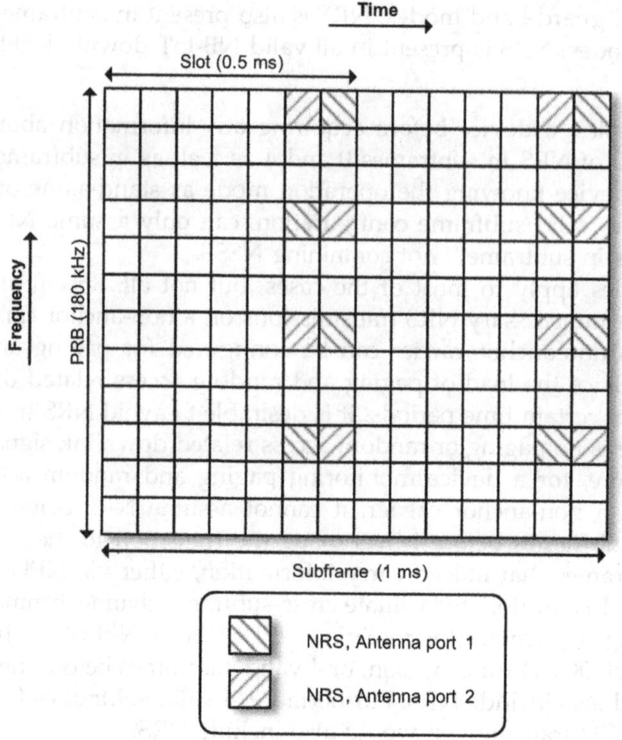

FIG. 7.16 A resource mapping example for NRS in a subframe carrying NPBCH, NPDCCH, or NPDSCH, or in a valid NB-IoT DL subframe.

The NPBCH is used to deliver the NB-IoT *Master Information Block* (MIB), which provides essential information for the device to operate in the NB-IoT network. NPBCH uses a 640 ms TTI, but within the TTI only subframe 0 is used in every radio frame. A resource mapping example is shown in Fig. 7.17. As mentioned earlier, the subcarriers used for NRS depend on NB-IoT cell identity. Thus, the NRS REs may shift up or down in the frequency domain. Some REs are reserved because they may be used by LTE, in case of in-band deployment. The LTE CRS may use REs outside of the potential LTE control region, i.e., the first three OFDM symbols. These REs may also shift up and down depending on the LTE cell identity. NB-IoT requires that the in-band deployment uses a cell identity that results in an NRS subcarrier set, e.g., subcarriers 2, 5, 8, and 11 in Fig. 7.17, which is identical to the subcarrier set used by CRS of the hosting LTE cell. Thus, when the device knows the NB-IoT cell identity, it also knows which REs are used for LTE CRS in case of in-band deployment.

As shown in Fig. 7.17, there are 100 REs available for NPBCH in a subframe. NPBCH uses QPSK modulation, and thus 200 encoded bits can be carried in an NPBCH subframe.

The TBS of NPBCH is 34 bits. A 16-bit *cyclic redundancy check* (CRC) is attached to the TB. Together, these 50 bits are encoded using the LTE TBCC [8] and rate-matched to generate 1600 encoded bits. The encoded bits are segmented into eight *code subblocks* (CSBs), each of

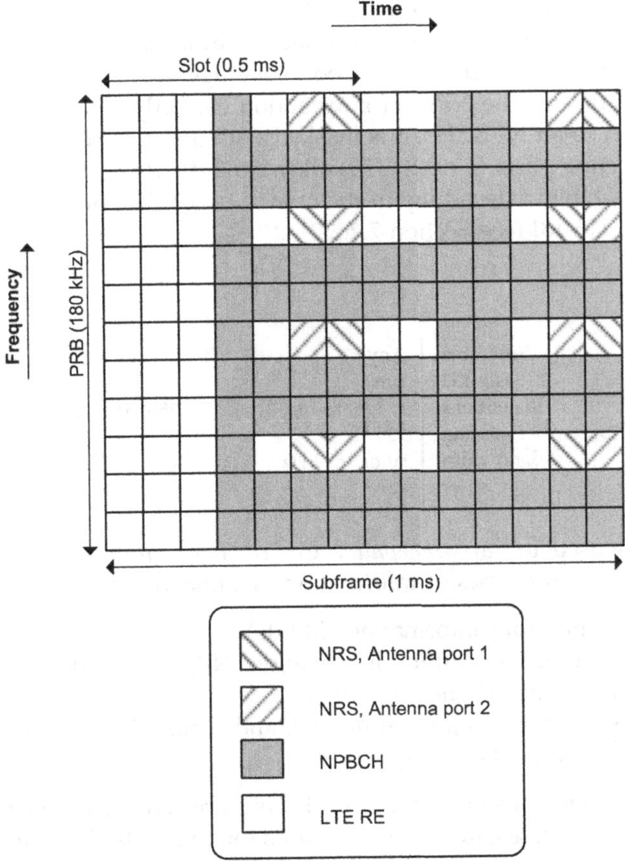

FIG. 7.17 A resource mapping example for NPBCH.

which is 200-bit long and mapped to 100 QPSK symbols. The OFDM modulation of the QPSK symbols is presented in Section 7.2.6.2 in detail.

On each CSB, a symbol-level scrambling is applied to provide robust protection against intercell interference, especially when the intercell interference happens to be dominated by the NPBCH signal from another cell. The scrambling pattern is dependent on the cell identity and SFN. This is achieved by re-initializing the scrambling pattern at the start of each radio frame with a seed that is determined by both the cell identity and SFN modulo 8. As a result, each CSB is scrambled to eight unique sets of 100 QPSK symbols mapped to subframe 0 in eight consecutive radio frames. The scrambling is implemented as symbol-level rotations that can easily be undone at the receiver side because the device after synchronizing to the NSSS knows the cell identity and the frame structure within an 80-ms interval. Thus, it knows how to descramble the NPBCH symbols and obtain the subframe symbol sequences over the multiple repetitions of a CSB. This facilitates, e.g., coherent combining of the repeated CSBs. Correlating repeated CSBs, after descrambling, is also a powerful tool for performing frequency offset estimation.

The transmission of the eight NPBCH subblocks in an NPBCH TTI is illustrated in Fig. 7.18. Each NPBCH subframe is self-decodable, but all the NPBCH subframes can also be jointly decoded. For certain devices in good coverage, a single transmission of one CSB may be sufficient to decode the NPBCH information correctly.

The CRC attached to an NPBCH TB is masked with a sequence that is dependent on the number of NRS antenna ports (1 or 2). This allows the device to detect the number of NRS antenna ports through blind decoding. In the case of two NRS ports, a *Space-Frequency Block Code* (SFBC) is used [16-18] (see Section 7.2.2.2).

7.2.4.5 NPDCCH

Subframe	Any
Basic TTI	1 ms
Repetitions	1, 2, 4, 8, 16, 32, 64, 128, 256, 512, 1024, 2048
Subcarrier spacing	15 kHz
Bandwidth	90 or 180 kHz
Carrier	Any

The NPDCCH is used to carry *Downlink Control Information* (DCI). A device needs to monitor NPDCCH for three types of information mentioned below.

- DCI format N0: Uplink grant information (23 bits).
- DCI format N1: Downlink scheduling information, NPDCCH order, and notification of changes in multicast control channel (23 bits)
- DCI format N2: Indicator of paging, SI update, and scheduling or notification of changes in multicast control channel (15 bits)

Based on the information bits in the DCI, 16 CRC bits are generated. In addition to providing error detection capability, these CRC bits are also used for differentiating between different information types for each DCI format and for identifying the device that the DCI is intended for. This is done by scrambling the CRC bits based on the different *radio network temporary identifiers* (RNTIs) associated with the different information types. More details are provided in Section 7.3.

3GPP Release 14 introduces an option for a Cat-NB2 device to support two hybrid automatic repeat request (HARQ) processes (see Section 7.3.2.2.3). To support the operation of two simultaneously active HARQ processes, one bit is added to DCI formats N0 and N1 to indicate the HARQ process number. In this case, DCI formats N0 and N1 become 24 bits long.

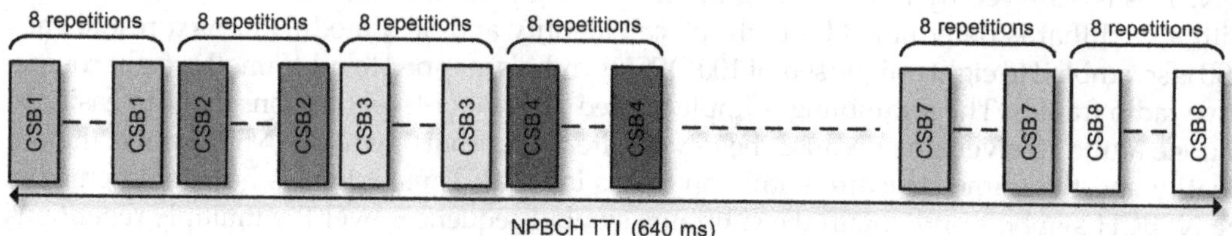

FIG. 7.18 Transmission of NPBCH CSBs.

An NPDCCH subframe is divided into two *narrowband control channel elements*. NCCE0 takes the lowest six subcarriers and NCCE1 take the highest six subcarriers. The number of REs available for one NCCE depends on NB-IoT operation modes and the number of logical antenna ports. For the in-band deployment, it further depends on the configuration of the LTE cell. Two examples are shown in Fig. 7.19.

After cell selection and *SI* acquisition, the device will understand the exact mapping of NPDCCH REs in an NPDCCH subframe (see Section 7.3.1.2.2). For example, for the in-band deployment, NPDCCH is not mapped to the first few OFDM symbols in a subframe. This is to avoid the LTE downlink control region. The index of the starting OFDM symbol within an NPDCCH subframe depends on the size of LTE downlink control region and is signaled to the device. Table 7.8 lists all possible number of REs per NCCE. It ranges from 50 to 80.

A DCI can be mapped to one NCCE, referred to as *aggregation level* (AL) 1, or both narrowband control channel elements in the same subframe, referred to as AL 2. As mentioned earlier, 16 CRC bits are generated. These CRC bits are scrambled with a sequence determined by an RNTI. The scrambled CRC bits are attached to the DCI. After the CRC attachment, TBCC encoding and rate-matching are used to generate a code word with a length matched to the number of encoded bits available. QPSK modulation is used for NPDCCH, and thus the code word length ranges from 100 to 160 for AL 1, or from 200 to 320 for AL 2. The baseband signal generation described in Section 7.2.6.2 takes the QPSK symbols as input and generates the baseband waveform.

NPDCCH AL 2 is used for increasing the coverage of NPDCCH. Using more REs for transmitting a DCI message gives rise to a higher energy level per information bit. Further coverage enhancement can be provided by subframe-level repetitions. Fig. 7.20 shows an NPDCCH transmission using AL 2 and 8 repetitions. The NPDCCH bit stream is scrambled before being mapped to symbols. The same scrambling sequence is used in sets of four consecutive subframes, which means that the same NPDCCH symbols are transmitted in each of the four consecutive NPDCCH subframes. This allows the devices to optimize reception performance through the use of coherent combining for received power estimation and for frequency offset estimation. The scrambling sequence is reinitialized after transmitting every four repetitions as indicated by the figure to provide randomization of the transmitted waveform. 3GPP Release 14 introduces an option to apply additional symbol-level scrambling for achieving further randomization and improved robustness to inter-cell interference. A QPSK-modulated NPDCCH symbol is rotated by *1*, *-1*, *j*, or *-j*, based on a pseudo-random scrambling mask. Unlike the bit-level scrambling mask, the optional symbol-level scrambling mask is reinitialized in every NPDCCH subframe. In this case, the device can remove the symbol-level scrambling mask to restore the identical set of NPDCCH symbols in each of the four consecutive subframes. From inter-cell interference mitigation point of view, reinitializing the scrambling mask in every NPDCCH subframe ensures a processing gain over the interfering signal from a neighbor cell.

NB-IoT allows the same NPDCCH to be repeated up to 2048 times. Thus, in the most extreme case, a DCI is transmitted in 2048 subframes. To avoid a DCI blocking other downlink NPDCCH/NPDSCH resources for an extended period of time, a transmission gap can be configured through *Radio Resource Configuration* (RRC) signaling. More details about downlink transmission gap will be discussed in Section 7.2.6.1.

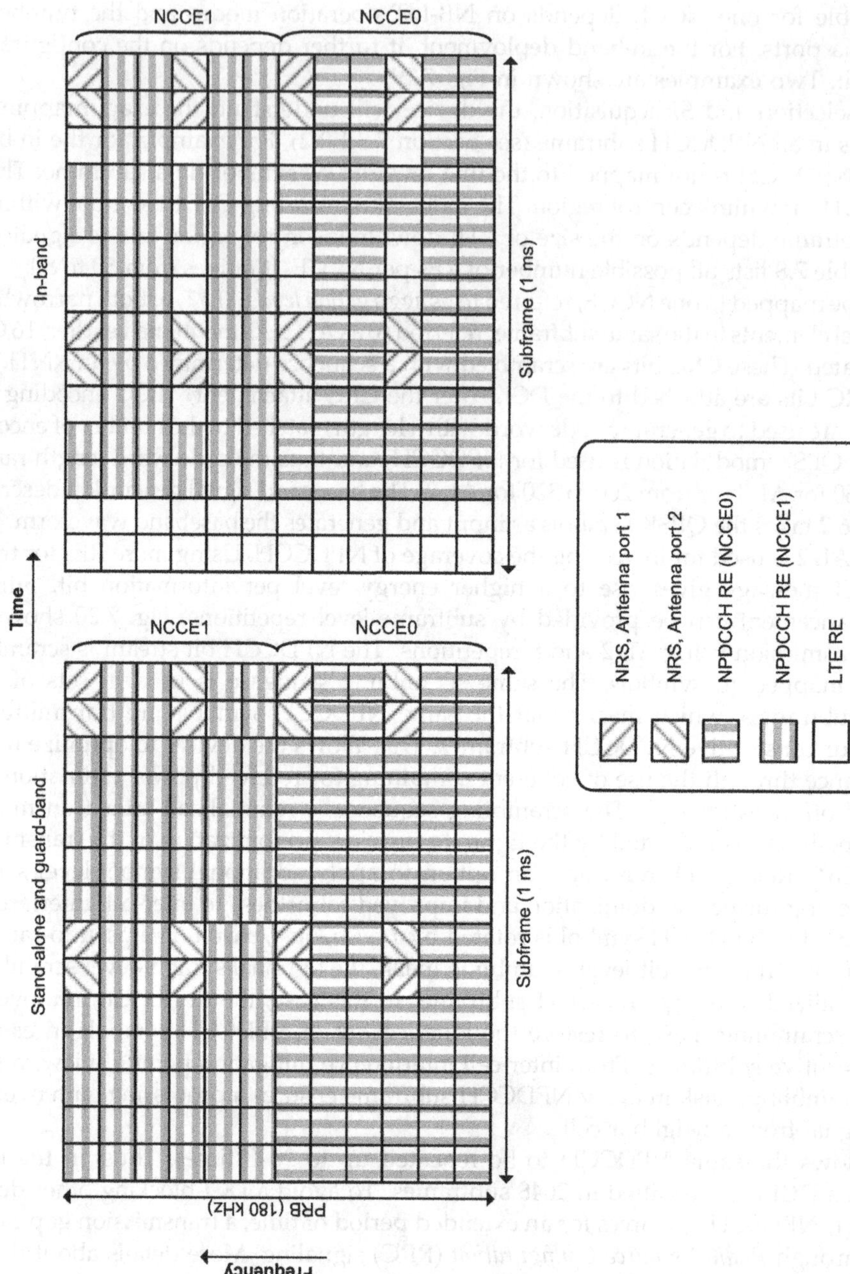

FIG. 7.19 Two NPDCCH resource mapping examples. On the left: Stand-alone and guard-band deployment with two NRS ports. On the right: In-band deployment with two NRS ports, four CRS ports, and two OFDM symbols for LTE downlink control region.

TABLE 7.8 Number of REs available per NCCE.

Operation mode	Number of LTE CRS antenna Ports	Number of OFDM symbols for LTE control Region	Number of NRS Antenna ports	Number of REs per NCCE
standalone, guard-band	N/A	N/A	1	80
standalone, guard-band	N/A	N/A	2	76
in-band	2	1	1	68
in-band	2	1	2	64
in-band	2	2	1	62
in-band	2	2	2	58
in-band	2	3	1	56
in-band	2	3	2	52
in-band	4	1	1	66
in-band	4	1	2	62
in-band	4	2	1	60
in-band	4	2	2	56
in-band	4	3	1	54
in-band	4	3	2	50

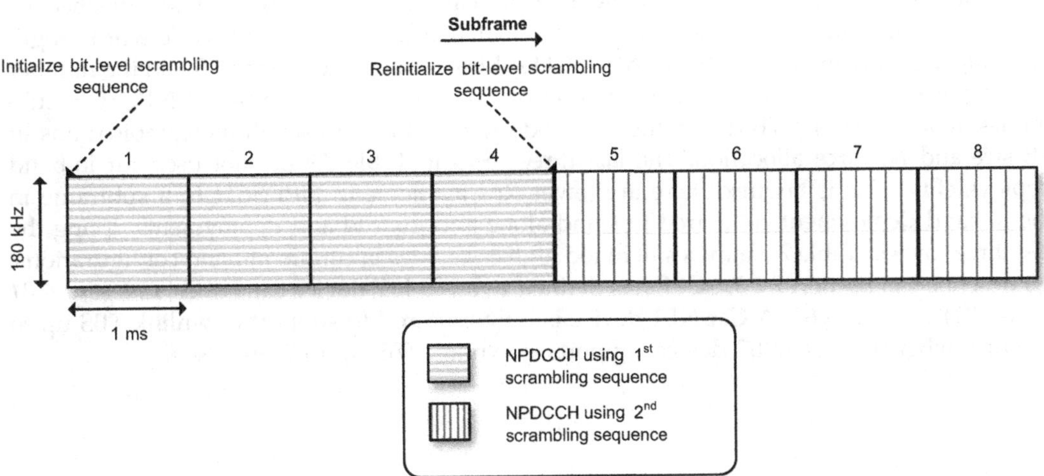

FIG. 7.20 An example of NPDCCH transmission configured with eight repetitions.

7.2.4.6 NPDSCH

Subframe	Any
Basic TTI	1, 2, 3, 4, 5, 6, 8, 10 ms
Repetitions	1, 2, 4, 8, 16, 32, 64, 128, 192, 256, 384, 512, 768, 1024, 1536, 2048
Subcarrier spacing	15 kHz
Bandwidth	180 kHz
Carrier	Any

The NPDSCH is used to transmit unicast data. A data packet from high layers is segmented into one or more TBs, and NPDSCH transmits one TB at a time. NPDSCH is also used to transmit broadcast information such as *SI* messages.

NPDSCH has a similar subframe-level resource mapping as NPDCCH. There are, however, two differences:

- NPDCCH may multiplex resources in a subframe to transmit two DCI messages. One NPDSCH subframe, however, can at most carry one TB. That is, the basic RU for NPDSCH is one PRB over one subframe.
- The starting OFDM symbol in an NPDSCH subframe may be different from that in an NPDCCH subframe in the in-band mode if the subframe is used to transmit SIB1-NB (see Section 7.3.1.2 for the description of SIB1-NB). Like in the case of NPDCCH, the starting OFDM symbol in an NPDSCH subframe in the in-band mode is determined based on the LTE control region size. Such information is carried in SIB1-NB. However, a device needs to be able to acquire SIB1-NB without knowing the LTE control region size. Therefore, if an NPDSCH subframe is used for transmitting SIB1-NB, the starting OFDM symbol position is always the fourth symbol in the subframe.

NPDSCH uses QPSK and supports a TB size up to 680 bits in case of the lowest device category. A TB is mapped to a number of NPDSCH subframes. The LTE TBCC is used as the only *forward-error correcting* code for NPDSCH. The base station processing of NPDSCH TB is as follows. First, a 24-bit CRC is calculated and attached to the TB. The CRC-attached TB is encoded using the TBCC encoder and rate-matched according to the code-word length determined jointly by the number of NPDSCH subframes allocated to the TB and the number of REs per subframe. Thus, the combination of TB size and the number of NPDSCH subframes allocated to the TB determines the code rate. Table 7.9 lists all the combinations of TB size and resource allocation. The last three rows in Table 7.9 are not used for in-band deployments. This is because there are fewer REs available to NPDSCH in a subframe in the in-band deployment compared to stand-alone and guard-band deployments. Using the last three rows in Table 7.9 may result in too high code rates in certain in-band configurations. As described in Section 7.2.3, 3GPP has defined two NB-IoT device categories, *Category NB1* (Cat-NB1) and *Cat-NB2*. A Cat-NB1 device is only required to support downlink TBS up to 680 bits, whereas a Cat-NB2 device supports downlink TBS up to 2536 bits.

TABLE 7.9 Combinations of NPDSCH TBS and number of NPDSCH subframes.

NPDSCH TBS index	Number of subframes [N_{SF}]							
	1	**2**	**3**	**4**	**5**	**6**	**8**	**10**
0	16	32	56	88	120	152	208	256
1	24	56	88	144	176	208	256	344
2	32	72	144	176	208	256	328	424
3	40	104	176	208	256	328	440	568
4	56	120	208	256	328	408	552	680
5	72	144	224	328	424	504	680	872
6	88	176	256	392	504	600	808	1032
7	104	224	328	472	584	680	968	1224
8	120	256	392	536	680	808	1096	1352
9	136	296	456	616	776	936	1256	1544
10	144	328	504	680	872	1032	1384	1736
11	176	376	584	776	1000	1192	1608	2024
12	208	440	680	904	1128	1352	1800	2280
13	224	488	744	1032	1256	1544	2024	2536

To limit the requirement on the receiver, only a single redundancy version is specified for the NPDSCH encoding.

For each of the combinations listed in Table 7.9, the code rate for a stand-alone deployment with two NRS ports is shown in Table 7.10. For a TB size, better coverage is achieved by allocating more NPDSCH subframes, giving a higher energy level per information bit and, in most cases, a higher coding gain as well.

As seen in Table 7.10, up to 10 NPDSCH subframes (i.e., $N_{SF} = 10$) can be used to carry one NPDSCH TB. Before mapping the bits of the encoded TB to QPSK symbols, the bits are scrambled. The scrambling is reinitialized every min(N_{REP}, 4) repetition of the code word, where N_{REP} is the number of configured repetitions. At most 2048 repetitions can be transmitted. After mapping the NPDSCH code word on a subframe, the subframe is repeated min(N_{REP}, 4) times before the mapping of the code word continues. Fig. 7.21 shows the transmission of a two-subframe TB configured for eight repetitions. The first subframe is repeated four times before the mapping continues to the second subframe. After the second subframe has been repeated four times, the scrambling is reinitialized and the procedure is repeated once to complete eight repetitions of the code word in total.

The repeated subframes do, just as in case of the NPDCCH, allow coherent combining for received power estimation and for frequency offset estimation. It also allows a device to attempt decoding of the code word before the transmission has completed. The example shown in

TABLE 7.10 NPDSCH code rates for stand-alone deployment with 2 NRS ports.

NPDSCH TBS index	Number of NPDSCH subframes [N_{SF}]							
	1	2	3	4	5	6	8	10
0	0.13	0.09	0.09	0.09	0.09	0.10	0.10	0.09
1	0.16	0.13	0.12	0.14	0.13	0.13	0.12	0.12
2	0.18	0.16	0.18	0.16	0.15	0.15	0.14	0.15
3	0.21	0.21	0.22	0.19	0.18	0.19	0.19	0.19
4	0.26	0.24	0.25	0.23	0.23	0.24	0.24	0.23
5	0.32	0.28	0.27	0.29	0.29	0.29	0.29	0.29
6	0.37	0.33	0.31	0.34	0.35	0.34	0.34	0.35
7	0.42	0.41	0.39	0.41	0.40	0.39	0.41	0.41
8	0.47	0.46	0.46	0.46	0.46	0.46	0.46	0.45
9	0.53	0.53	0.53	0.53	0.53	0.53	0.53	0.52
10	0.55	0.58	0.58	0.58	0.59	0.58	0.58	0.58
11	0.66	0.66	0.67	0.66	0.67	0.67	0.67	0.67
12	0.76	0.76	0.77	0.76	0.76	0.75	0.75	0.76
13	0.82	0.84	0.84	0.87	0.84	0.86	0.84	0.84

Fig. 7.21 does, for example, support decoding of the full code word already after eight subframes. Like the case of NPDCCH, 3GPP Release 14 introduces an option to apply additional symbol-level scrambling for achieving further randomization and improved robustness to inter-cell interference. This optional symbol-level scrambling mask is initialized in every subframe.

Because of the high number of supported repetitions, an NPDSCH TB may be mapped over 20,480 NPDSCH subframes when a TB is mapped to 10 subframes per repetition and configured with 2048 repetitions. To avoid one long NPDSCH transmission blocking other NPDCCH or NPDSCH transmissions, NPDSCH transmission gaps can be configured. The concept is similar to the NPDCCH transmission gap and will be described in more detail in Section 7.2.7.1.

As already indicated, the NPDSCH is also used for transmission of the SIBs. These transmissions are, however, following slightly different principles compared with those just described. For the SIB1-NB the TBS is always configured for $N_{SF} = 8$ subframes, and the available set of TBSs are limited to the set of {208, 328, 440, 680} bits. For the other SIBs, the available set of TBSs are limited to the set of {56, 120, 208, 256, 328, 440, 552, 680} bits. The two lowest TBSs use $N_{SF} = 2$ subframes while the six larger options are mapped to $N_{SF} = 8$ subframes.

The scrambling of the NPDSCH carrying a SIB is reinitialized for every repetition. The full code word is mapped to the full set of configured subframes N_{SF} before the repetition starts. This allows all devices in good coverage to decode a SIB already after the first transmission.

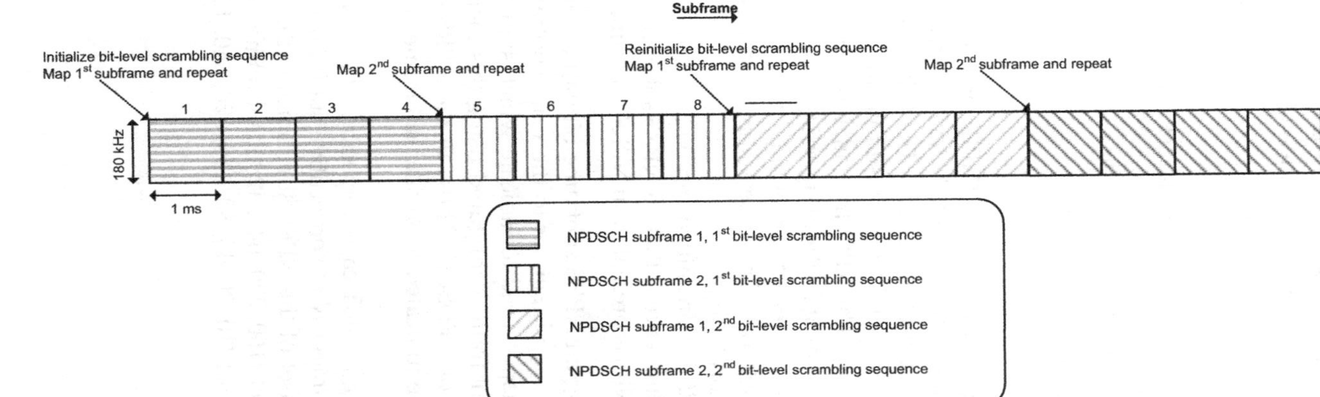

FIG. 7.21 Transmission of a TB mapped to two subframes and configured with eight repetitions.

The mapping and scheduling of the SIB repetitions on the downlink frame structure follows specific rules described in detail in Section 7.3.1.2.3.

The baseband generation of the NPDSCH signal uses the QPSK-modulated symbols as input and generates the transmitted waveform as described in Section 7.2.6.2.

7.2.4.7 NPRS

Subframe	Configured by LPP signaling
Basic TTI	1 ms
TTI	According to higher layer configuration
Subcarrier spacing	15 kHz
Bandwidth	180 kHz
Carrier	Any

The Narrowband Positioning Reference Signal (NPRS) was introduced in 3GPP Release 14 to improve the accuracy of positioning an NB-IoT device, beyond what cell identity based positioning methods can offer (see Section 7.3.2.6). The NPRS is a broadcast signal that enables *observed time difference of arrival* (OTDOA) based positioning estimation of a device. It is not like the other signals configured by RRC. Instead, it is configured by the *LTE Positioning Protocol* (LPP) [19]. The LTE Positioning Protocol signaling is device specific and is carried between a device and a positioning server known as the *Evolved Serving Mobile Location Center* (E-SMLC) located in the core network. The base station and E-SMLC negotiate the NPRS configuration via the LPPa protocol [20].

The subframes in which NPRS is transmitted can be indicated by using two configuration methods as follows. One main difference is that an NPRS subframe configured using *Method 1* does not include NRS, whereas NRS is present in an NPRS subframe configured using *Method 2*.

- Method 1: Configuration using an NPRS subframe bitmap of length 10 or 40. The concept of NPRS subframe bitmap is similar to the valid subframe bitmap described in Section 7.2.4.1. Here, the bitmap is used to indicate which subframes are configured as NPRS subframes. An NPRS subframe configured using this method is not available for transmitting NRS, NPDCCH, or NPDSCH. Thus, all the NPRS subframes configured using *Method 1* need to also be indicated as invalid subframes using the valid subframe bitmap.
- Method 2: Configuration using NB-IoT carrier-specific NPRS parameters, including a configuration period T, the number N of consecutive subframes where NPRS shall be transmitted, and a starting offset of the NPRS subframes within a configuration period. The period T belongs to the set {160, 320, 640, 1280} subframes, while the number of subframes N is selected from the set {10, 20, 40, 80, 160, 320, 640, 1280} subframes.

Note that in the case of Method 2, an NPRS subframe may be a valid NB-IoT subframe with NRS. Since Release 13 devices expect NRS in a valid NB-IoT subframe, it is therefore important to ensure that NPRS does not collide with NRS.

An NB-IoT cell may use the combination of both configuration methods described above. In this case, NPRS is transmitted in a subframe where both configuration methods indicate as an NPRS subframe, i.e. the set of NPRS subframes is the intersection of the two subframe sets indicated by the two configuration methods. For in-band operation it is mandatory to use at least Method 1 to facilitate coexistence with LTE. NPRS cannot be transmitted in subframes designated for NPBCH, NPSS, NSSS, or SIB1-NB.

Resource mapping in an NPRS subframe for the in-band mode operation configured by the Method 1 NPRS subframe bitmap is illustrated in Fig. 7.22. In the in-band mode, NPRS cannot be mapped to OFDM symbols that may be used by LTE PDCCH or CRS. Observe that the resource mapping patterns can be shifted up or down in the frequency dimension to create 6 different, orthogonal mapping patterns. These orthogonal mapping patterns can be used in neighboring cells in a synchronized network to avoid inter-cell interference. The mapping pattern used by a cell is determined by a configurable NPRS identity, with a default value equal to the cell identity.

In the stand-alone and guard-band modes, NPRS can be mapped to all the OFDM symbols if the Method 1 NPRS subframe bitmap is configured. If the cell is configured for stand-alone or guard-band operation and does not signal an NPRS subframe bitmap and only uses

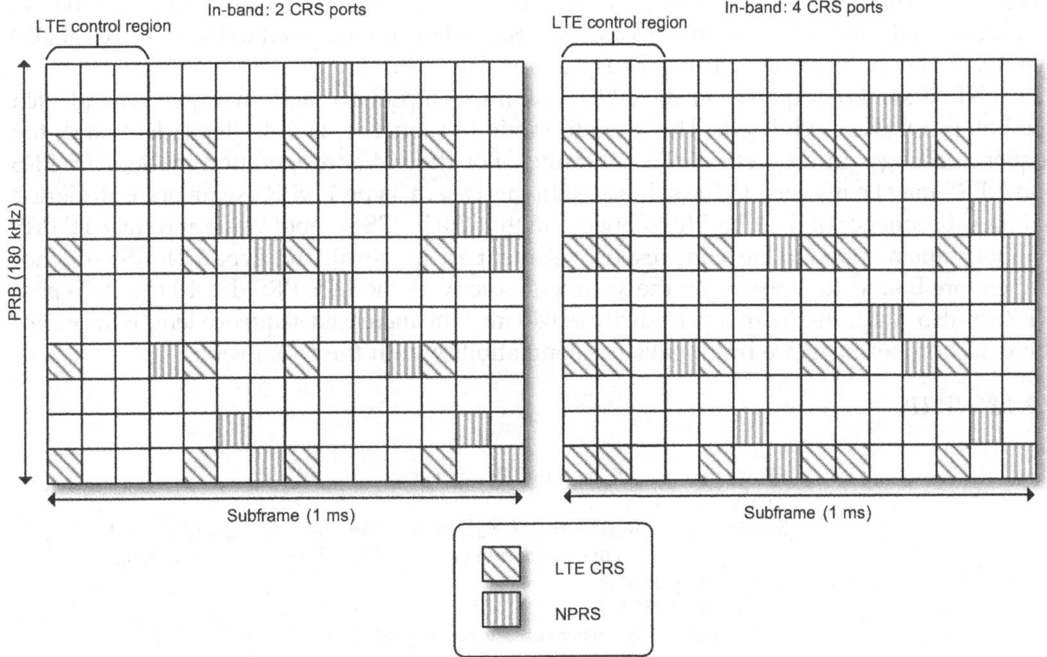

FIG. 7.22 An example of in-band resource mapping in an NPRS subframe configured by the NPRS subframe bitmap.

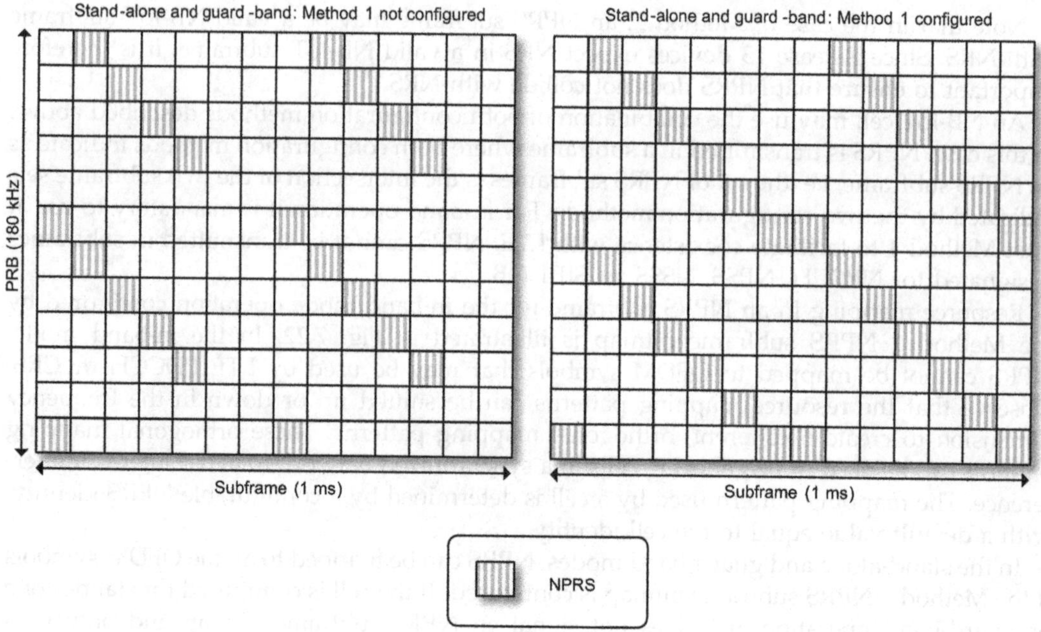

FIG. 7.23 An example of guard-band and stand-alone resource mapping in an NPRS subframe.

Method 2 to configure NPRS subframes, NRS in NPRS subframes are expected by all the devices. To avoid collision between NRS and NPRS, NPRS is not mapped to the last two OFDM symbols in every NPRS slot. This is illustrated in Fig. 7.23.

An NPRS symbol sequence in an NPRS subframe is a pseudo-random sequence, and each symbol is QPSK modulated. The pseudo-random sequence is cell dependent and the sequence changes in different NPRS subframes. For the in-band operation mode, LTE PRS and NPRS may be mapped to the same REs. In this case, a Type 1 NPRS sequence is designed to be a subsequence of the LTE PRS sequence in that the NPRS symbol value and the LTE PRS symbol value mapped to the same resource element are identical. The Type 1 NPRS sequence is therefore bound to repeat with the same periodicity as the LTE PRS, i.e. 10 ms. A Type 2 NPRS is also supported with a periodicity of 640 ms. The increased sequence length improves the detection performance but breaks the compatibility with the LTE PRS.

7.2.4.8 NWUS

Subframe	Configured by RRC signaling
Basic TTI	1 ms
Repetitions	1, $R_{max}/2$, $R_{max}/4$, $R_{max}/8$, $R_{max}/16$, $R_{max}/32$, $R_{max}/64$, $R_{max}/128$ R_{max}: maximum number of NPDCCH repetitions for paging
Subcarrier spacing	15 kHz
Bandwidth	180 kHz
Carrier	Any carrier configured for paging

The *NWUS* was introduced in 3GPP Release 15 for enhancing device energy efficiency. An NB-IoT device is most of the time in idle mode, during which it wakes up periodically to monitor a *PO* to check whether it is *paged*. More detailed description about NB-IoT idle mode procedures will be given in Section 7.3.1.4. For now, it suffices to note that a paging indicator is transmitted in NPDCCH by using DCI format N2, which has 15 information bits (see Section 7.2.4.5). As in most POs no paging indicator is sent, a device usually ends up waking up to look for a paging signal only to find that no paging indicator is sent. A much shorter NWUS can be transmitted before a UE would need to wake up to look for paging to indicate whether the device needs to stay awake during subsequent POs or can go back to sleep without monitoring those POs. More details on how NWUS is used are given in Section 7.3.1.4 when paging is discussed. In this section, we focus on the physical layer aspects.

NWUS resource mapping within a subframe is shown in Fig. 7.24. For in-band deployment, NWUS is mapped to the last 11 OFDM symbols in a subframe, which amount to 132 REs. However as illustrated in Fig. 7.24, LTE CRS and NRS *puncture* NWUS. The length-132 NWUS sequence in a subframe is generated based on an extended length-131 ZC sequence, exactly the same as Eqs. (7.5) and (7.6) used for generating the NSSS sequence. The extended ZC sequence is scrambled based on a pseudo-random scrambling mask. Each element in the scrambling mask takes values from {*1*, *-1*, *j*, *-j*}. The scrambling mask varies from subframe to subframe when the number of repetitions R_{NWUS} configured exceeds 1. It should be noticed

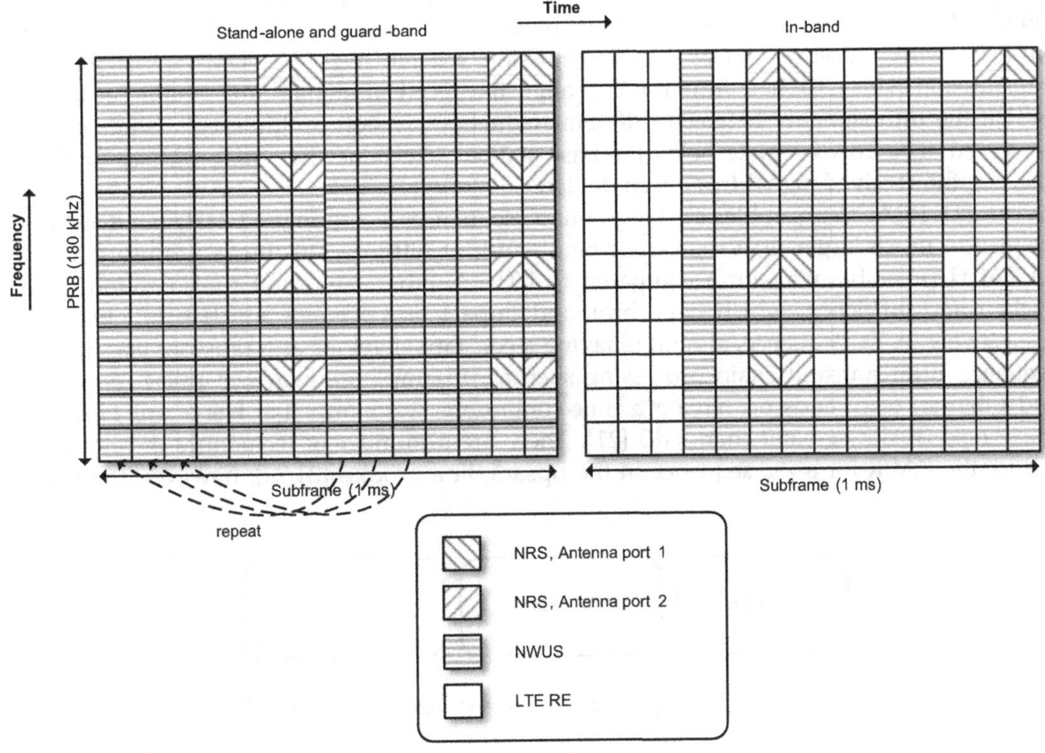

FIG. 7.24 NWUS resource mapping.

that R_{NWUS} is determined relative to the number of maximum repetitions R_{MAX} configured for the subsequent NPDCCH transmission indicating a paging signal. For stand-alone and guard-band deployments, the first 3 OFDM symbols in a subframe can also be used for NWUS. In these cases, OFDM symbols #7, #8, and #9 are repeated in OFDM symbols #0, #1, and #2, respectively.

7.2.5 Uplink physical channels and signals

NB-IoT supports the set of uplink physical channels and signals depicted in Fig. 7.25.

7.2.5.1 *NPRACH*

Subframe	Any
Basic TTI	5.6, 6.4, 19.2 ms (depending on NPRACH format)
Repetitions	1, 2, 4, 8, 16, 32, 64, 128
Subcarrier spacing	3.75 or 1.25 kHz (depending on NPRACH format)
Bandwidth	3.75 or 1.25 kHz (depending on NPRACH format)
Carrier	Anchor from 3GPP Release 13 Non-anchor from 3GPP Release 14

Like the *Physical Random Access Channel* (PRACH) in LTE, the NPRACH in NB-IoT is used by the device to initiate connection and allows the serving base station to estimate the *time of arrival* (ToA) of the received NPRACH signal. The ToA of the received NPRACH signal reflects the round-trip propagation delay between the base station and device. Because NB-IoT uplink employs OFDM-like transmission scheme (e.g., SC-FDMA or single-tone transmission with CP), it is important to align the received signals from multiple devices so that the orthogonality between different frequency-division multiplexed devices can be preserved. The ToA estimate helps the base station determine the *timing advance* (TA) for aligning the received signal from each device.

The LTE PRACH preambles are based on ZC sequences, spanning ~1 MHz bandwidth in frequency, much larger than the carrier bandwidth of NB-IoT. This motivates using a new NPRACH preamble waveform compared to LTE. Another important consideration is PA backoff and efficiency, which have profound impact on coverage and battery efficiency, respectively. A ZC sequence, although having a constant envelope as a time-domain discrete sequence, after transmit chain processing such as upsampling, *discrete-to-analog conversion*, and filtering, often ends up having a time-continuous waveform that has a *peak-to-average power ratio* (PAPR) greater than 3 dB [21]. There are a number of techniques developed to reduce the PAPR for a ZC sequence in the upsampling process [21,22]; however, the PAPR

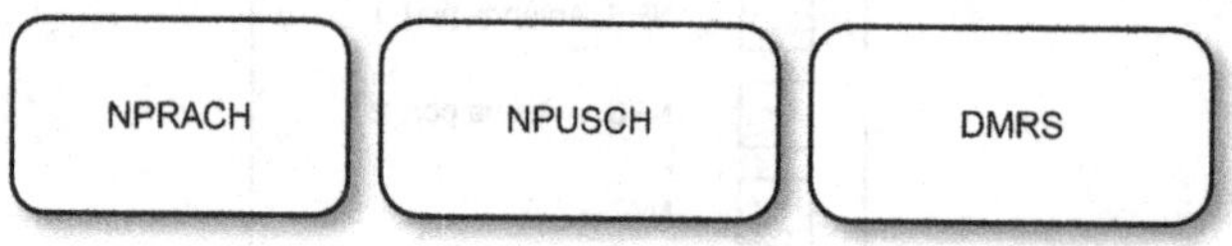

FIG. 7.25 uplink physical channels and signals used in NB-IoT.

with such techniques is still not close to 0 dB. This is undesirable as it is advantageous to keep power backoff as low as possible. Power backoff results in a compromise on uplink coverage as the PA cannot be used at its maximum configurable output power level. Power backoff also reduces PA efficiency, resulting in a compromise of battery lifetime. NPRACH preamble design targets a time-continuous waveform that has a PAPR close to 0 dB. As described in this section, NPRACH preambles are based on single-tone frequency-hopping waveforms, which indeed achieve a PAPR very close to 0 dB.

In NB-IoT, up to three NPRACH configurations can be used in a cell to support devices in different coverage classes. Different configurations are separated by using different time–frequency resources. Before we describe how NPRACH is configured, we first describe the waveform adopted for the NPRACH preambles.

NPRACH preambles use single-tone transmission with frequency hopping. A preamble repetition unit consists of multiple symbol groups and each symbol group consist of a CP and multiple symbols. There are three NPRACH formats in NB-IoT. NPRACH Format 0 and Format 1 were defined in 3GPP Release 13 while NPRACH Format 2 was introduced in Release 15. We start with the preamble structure for NPRACH Formats 0 and 1. The symbol groups for Formats 0 and 1 are illustrated in Fig. 7.26. As shown, one symbol group consists of a CP plus five single-tone symbols of tone frequency $n\Delta f_{\mathrm{NPRACH}}$, where n is an integer number, $n \in \{0, 1, \ldots, 11\}$, and $\Delta f_{\mathrm{NPRACH}}$ is 3.75 kHz, which is the NPRACH tone spacing. The value n is fixed within a symbol group. The resulting waveform of an NPRACH symbol group is thus a continuous phase sinusoidal of baseband frequency $n\Delta f_{\mathrm{NPRACH}}$. NPRACH Format 0 uses a CP duration 66.7 μs and NPRACH Format 1 uses a CP duration 266.67 μs for supporting cell radii up to at least 10 and 40 km, respectively.

When NB-IoT is deployed in rural or remote areas, support for a larger cell radius is desirable. In Ref. [23], it is shown that by exploiting the structure in Release 13 NPRACH Format 1, it is actually possible to support a 100 km cell radius using advanced algorithm at the base station.

3GPP Release 15, however, still introduced a new NPRACH format, as illustrated in Fig. 7.27, to support larger cell radii without resorting to advanced base station algorithms. This new NPRACH format is referred to as NPRACH Format 2, which uses a longer CP of 800 μs and supports a cell radius up to 120 km. A symbol group in this case consists of the CP and three single-tone symbols. Similar to Format 0 and Format 1, the waveform of each single-tone symbol is a sinusoidal waveform of frequency $n\Delta f_{\mathrm{NPRACH}}$, where n is an integer number. The tone spacing, $\Delta f_{\mathrm{NPRACH}}$, for Format 2, is 1.25 kHz, which results in a symbol duration of 800 μs.

The basic NPRACH repetition unit consists of four symbol groups for Formats 0 and 1, or 6 symbol groups for Formats 2, with special relationship between tone frequencies within a repetition unit as illustrated in Fig. 7.28. The example shown in Fig. 7.28 uses deterministic tone hopping pattern between the four symbol groups within a repetition unit; that is, the second symbol group uses a tone that is right above the tone used by the first symbol group, the third symbol group uses a tone that is six tones above the tone used by the second symbol group, and the fourth symbol group uses a tone that is right below the tone used by the third symbol group.

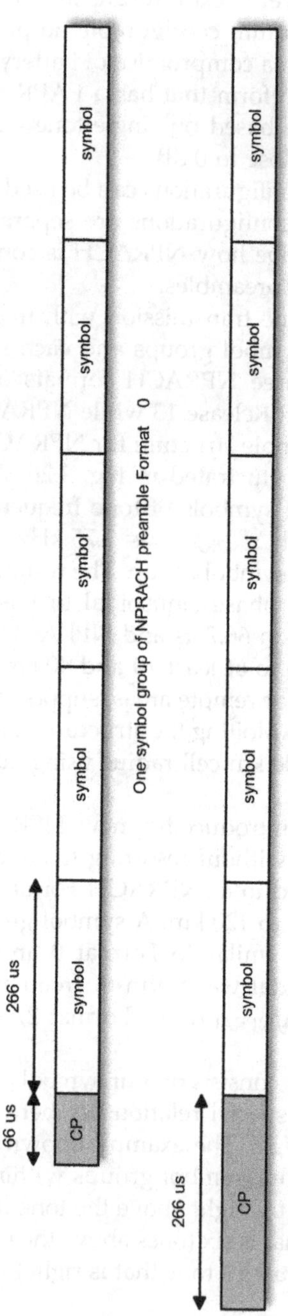

FIG. 7.26 NPRACH symbol groups for Format 0 and Format 1, respectively (FDD).

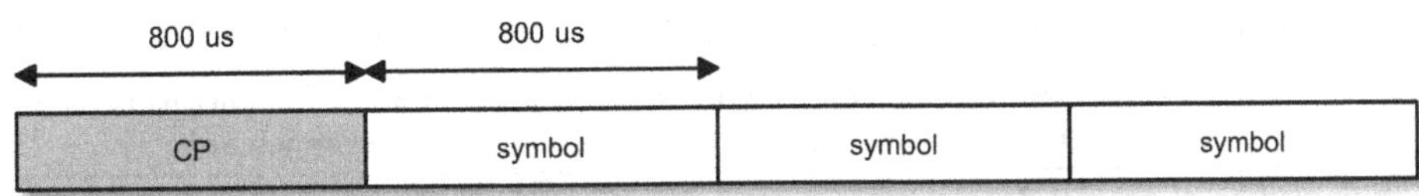

FIG. 7.27 NPRACH symbol group for Format 2 introduced in 3GPP Release 15 (FDD).

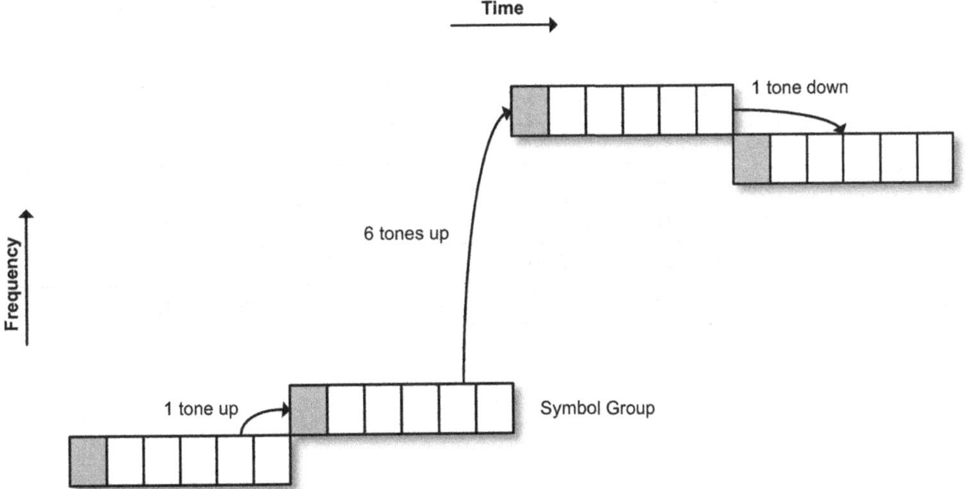

FIG. 7.28 One NPRACH preamble repetition unit (Format 1) and an example of the tone relationship between the four symbol groups.

The deterministic tone hopping pattern within a repetition unit is designed to facilitate the base station to estimate the TA in the presence of an unknown device residual frequency offset. As mentioned earlier, the same frequency tone is used within a symbol group. In essence, this time–frequency relationship between the four symbol groups allows the base station to solve for the two unknowns of TA and device residual frequency offset. For Formats 0 and 1, the hopping pattern uses a band of 12 contiguous tones, and the tones within the band can be indexed by 0, 1, …, 11. There are four possible deterministic frequency-hopping patterns within an NPRACH repetition unit. These are shown in Table 7.11. For Format 2, the hopping pattern uses a band of 36 tones, and the tones within the band can be indexed by 0, 1, …, 35. There are eight possible deterministic frequency-hopping patterns for Format 2 within an NPRACH repetition unit, as shown in Table 7.11.

These deterministic hopping patterns create a set of orthogonal NPRACH preambles within a repetition unit as illustrated in Tables 7.12 and 7.13. The baseband signal generation of the NPRACH preambles is explained in Section 7.2.6.1 in detail, which also illustrates the use of the parameters presented in Tables 7.12 and 7.13.

Using one NPRACH repetition unit alone, however, is not sufficient for meeting the aggressive NB-IoT coverage target. To ensure that NPRACH signals from devices in coverage

TABLE 7.11 Deterministic hopping patterns within a NPRACH repetition unit (FDD).

	Index of the tone used by the first symbol group	Deterministic hopping patterns within a repetition unit
Format 0 & Format 1	0, 2, 4	{+1, +6, −1}
	1, 3, 5	{−1, +6, +1}
	6, 8, 10	{+1, −6, −1}
	7, 9, 11	{−1, −6, +1}
Format 2	0, 6, 12	{+1, +3, +18, −3, −1}
	1, 3, 7, 9, 13, 15	{−1, +3, +18, −3, +1}
	2, 4, 8, 10, 14, 16	{+1, −3, +18, +3, −1}
	5, 11, 17	{−1, −3, +18, +3, +1}
	18, 24, 30	{+1, +3, −18, −3, −1}
	19, 21, 25, 27, 31, 33	{−1, +3, −18, −3, +1}
	20, 22, 26, 28, 32, 34	{+1, −3, −18, +3, −1}
	23, 29,35	{−1, −3, −18, +3, +1}

TABLE 7.12 An example of NPRACH Formats 0 and 1 preambles defined for a repetition unit (12 preambles) (FDD).

NPRACH preamble	Tone index $k(l)$ for symbol group l			
0	0	1	7	6
1	1	0	6	7
2	2	3	9	8
3	3	2	8	9
4	4	5	11	10
5	5	4	10	11
6	6	7	1	0
7	7	6	0	1
8	8	9	3	2
9	9	8	2	3
10	10	11	5	4
11	11	10	4	5

TABLE 7.13 An example of NPRACH Format 2 preambles defined for a repetition unit (36 preambles) (FDD).

NPRACH preamble	Tone index *k(l)* for symbol group *l*					
0	0	1	4	22	19	18
1	1	0	3	21	18	19
2	2	3	0	18	21	20
3	3	2	5	23	20	21
4	4	5	2	20	23	22
5	5	4	1	19	22	23
6	6	7	10	28	25	24
7	7	6	9	27	24	25
8	8	9	6	24	27	26
9	9	8	11	29	26	27
10	10	11	8	26	29	28
11	11	10	7	25	28	29
12	12	13	16	34	31	30
13	13	12	15	33	30	31
14	14	15	12	30	33	32
15	15	14	17	35	32	33
16	16	17	14	32	35	34
17	17	16	13	31	34	35
18	18	19	22	4	1	0
19	19	18	21	3	0	1
20	20	21	18	0	3	2
21	21	20	23	5	2	3
22	22	23	20	2	5	4
23	23	22	19	1	4	5
24	24	25	28	10	7	6
25	25	24	27	9	6	7
26	26	27	24	6	9	8
27	27	26	29	11	8	9
28	28	29	26	8	11	10
29	29	28	25	7	10	11
30	30	31	34	16	13	12
31	31	30	33	15	12	13

(Continued)

TABLE 7.13 An example of NPRACH Format 2 preambles defined for a repetition unit (36 preambles) (FDD).—cont'd

NPRACH preamble	Tone index $k(l)$ for symbol group l					
32	32	33	30	12	15	14
33	33	32	35	17	14	15
34	34	35	32	14	17	16
35	35	34	31	13	16	17

challenging locations can be detected reliably at the base station, NPRACH preambles may be configured with 1, 2, 4, 8, 16, 32, 64, or 128 repetition units. When the number of NPRACH repetitions increases, it is desirable to avoid persistent interference between NPRACH preambles in different cells. To achieve this objective, pseudorandom frequency hopping is introduced between different repetition units. This is done by applying a pseudorandom integer tone offset χ to all of the tone indexes in Tables 7.12 and 7.13. Because the hopping range needs to be within 12 or 36 tones, depending on the NPRACH format, the offset is applied in the modulo-12 or modulo-36 sense, keeping the tone indexes in the desired range. The pseudorandom tone offset χ is determined by both the cell identity and the repetition index. Thus, the NPRACH preambles in different cells will not end up having the same hopping pattern over the entirety of its transmission interval, thereby avoiding persistent interference.

An NB-IoT cell can configure up to three NPRACH configurations, each supporting a set of preambles and repetition units. Each configuration targets a specific coupling loss. The number of preambles corresponds to the number of supported access attempts. Information about NPRACH configurations is provided in the SI. Parameters in an NPRACH configuration include the number of repetitions, number of NPRACH preambles, the time periodicity by which they are reoccurring, a received signal-level threshold, etc. A few configurations are illustrated in Section 7.2.5.4, and the random access procedure including the meaning of these parameters is described in Section 7.3.1.6.

Finally, Table 7.14 summarizes the three NPRACH formats used in an FDD system.

7.2.5.2 NPUSCH

Subframe	Any
Basic TTI	1, 2, 4, 8, 32 ms
Repetitions	1, 2, 4, 8, 16, 32, 64, 128
Sub-carrier spacing	3.75, 15 kHz
Bandwidth	3.75, 15, 45, 90, 180 kHz
Carrier	Any

NPUSCH is used to carry uplink user data or control information from higher layers. Additionally, NPUSCH also carries *HARQ* acknowledgment for NPDSCH transmissions.

TABLE 7.14 NPRACH formats 0, 1, and 2 (FDD).

	Format 0	Format 1	Format 2
Supported cell radius	10 km	40 km	120 km
Cyclic prefix, T_{CP}	66.7 μs	266.7 μs	800.0 μs
Symbol group length, T	1400.0 μs	1600.0 μs	3200.0 μs
Number of symbol groups in a repetition unit	4	4	6
Subcarrier spacing	3.75 kHz	3.75 kHz	1.25 kHz
Tone hopping range	12 tones	12 tones	36 tones

3GPP Release 15 introduces a feature that further allows the device to signal a scheduling request using NPUSCH. The waveform adopted by NPUSCH is in principle the same as the LTE SC-FDMA waveform. However, in LTE, SC-FDMA supports device bandwidth with a granularity of one PRB, i.e., 12 subcarriers. A device may be scheduled for $12K$ subcarriers, where K is a positive integer. Thus, the minimum device-scheduled bandwidth allocation in LTE is one PRB or 12 subcarriers. However, because NB-IoT uses only one PRB, the maximum device-scheduled bandwidth can be only one PRB. The considerations below motivate NB-IoT to include lower device-scheduled bandwidth options:

- NB-IoT targets ultra-low-end IoT use cases, and it is envisioned that such use cases often have small data packets. Thus, in many cases, a device may not need to use the entire radio resources of one PRB (180 kHz).
- NB-IoT targets devices in coverage-limited scenarios. These devices operate in power limited regime, rather than in bandwidth limited regime, and thus do not benefit from having higher device bandwidth (see the discussion on this aspect in Sections 7.1.2.4 and 8.2.3).
- As mentioned in the NPRACH discussion above, having a low PAPR waveform is important for coverage and battery lifetime, mainly for devices at the edge of the network coverage. NPUSCH needs to include a waveform that has a PAPR close to 0 dB to best serve devices at the edge of coverage.

NPUSCH thus adds sub-PRB scheduled bandwidth options, including two special cases of single-tone transmissions, which have the advantages of having a close to 0 dB PAPR.

NPUSCH employs two transmission formats depending on the data it carries. Format 1 is used for uplink data transfer and uses the same turbo code as used in LTE [8] for error correction. The maximum TBS of NPUSCH Format 1 is 1000 bits in case of device category Cat-NB1 and 2536 for Cat-NB2. Format 2 is used for signaling HARQ feedback for an NPDSCH transmission as well as an uplink scheduling request. It uses a repetition code for error correction. Both Format 1 and Format 2 use SC-FDMA waveform involving DFT-precoding and CP insertion in the waveform generation process. As mentioned earlier, NPUSCH supports both multitone and single-tone device scheduling bandwidth options. Waveform generation in the single-tone case may omit the DFT-precoding. All the multitone NPUSCH transmissions are based on 15 kHz subcarrier spacing; however, the single-tone transmissions use either 15 or 3.75 kHz subcarrier spacing.

The slot formats applicable to NPUSCH Format 1 are illustrated in Fig. 7.29. The definition of CP and OFDM symbol durations for 15 kHz subcarrier spacing are identical to those in LTE, as described in Section 7.2.2.2. For single-tone transmissions with 3.75 kHz subcarrier spacing, NB-IoT introduces new numerologies as shown in Fig. 7.29. The slot duration in this case is 2 ms, consisting of seven SC-FDMA symbols and a guard period (GP) at the end. Each SC-FDMA symbol is 275 μs, including a CP of 8.33 μs. The GP is created to avoid collision with LTE *Sounding Reference Signal* (SRS), which is a signal transmitted by an LTE device to facilitate the base station to estimate the channel quality. An LTE device may be configured to transmit SRS on a PRB that is used as an NB-IoT carrier. SRS can only use the last OFDM symbol in an LTE subframe and can be configured with a periodicity as small as 2 ms and as large as 320 ms [11]. An SRS collision example is illustrated in Fig. 7.30. LTE devices 1 and 2 are configured with SRS transmission. The collision with the SRS transmitted by LTE device 2 is avoided by having the GP in the slot format of NPUSCH with 3.75 kHz subcarrier spacing. However, SRS collision may still occur on the fourth OFDM symbol, as shown in Fig. 7.30. Collisions between SRS and NPUSCH symbols are avoided by puncturing the NPUSCH symbols.

For 15 kHz subcarrier spacing, the middle OFDM symbol in each slot is used as *Demodulation Reference Signal* (DMRS), which allows the base station to estimate the uplink propagation conditions. The DMRS design is described in Section 7.2.5.3. For 3.75 kHz subcarrier spacing, the placement of the DMRS is shifted to the fifth OFDM symbol in the slot. This is also for the consideration of SRS avoidance. It can be seen in Fig. 7.30 that the fourth symbol in the slot may have collision with SRS.

NPUSCH Format 2 uses similar slot formats. The only difference compared to NPUSCH Format 1 is that three OFDM symbols are used as DMRS, leaving only four information-bearing symbols per slot, as illustrated in Fig. 7.31. Like in the case of NPUSCH Format 1, the placements of DMRS are designed to avoid collision with LTE SRS. For 15 kHz subcarrier spacing, using the middle 3 OFDM symbols avoids collision with LTE SRS, and for 3.75 kHz, using the first 3 OFDM symbols achieves the same.

The basic NPUSCH time scheduling unit is referred to as a *RU*. It is specified in terms of number of slots and is dependent on user bandwidth allocation and NPUSCH format. Table 7.15 summarizes the definition of RU for various NPUSCH configurations. A device may be scheduled with 1, 2, 3, 4, 5, 6, 8, or 10 RUs per repetition, resulting in a transmission interval per repetition as short as 1 ms, using one RU for 180 kHz scheduled bandwidth, and as high as 320 ms, using 10 RUs for 3.75 kHz scheduled bandwidth. The numbers of data symbols per RU for the various NPUSCH configurations are shown in Table 7.15.

The TBSs supported for NPUSCH Format 1 are shown in Table 7.16. The smallest TBS is 16 bits long and the largest is 2536 bits long. A Cat-NB1 device is only required to support uplink TBS up to 1000 bits. The TBSs given in the last three rows in Table 7.16 are used only for multitone transmissions because the code rates would have been higher than one in these cases. All the other rows are used for both single-tone and multitone transmissions. A 24-bit CRC is calculated and attached to a TB. Afterward, the LTE turbo code is used for encoding, based on the same LTE mother code and rate-matching scheme [8]. NPUSCH Format 1 supports incremental redundancy; however, only redundancy versions 0 and 2

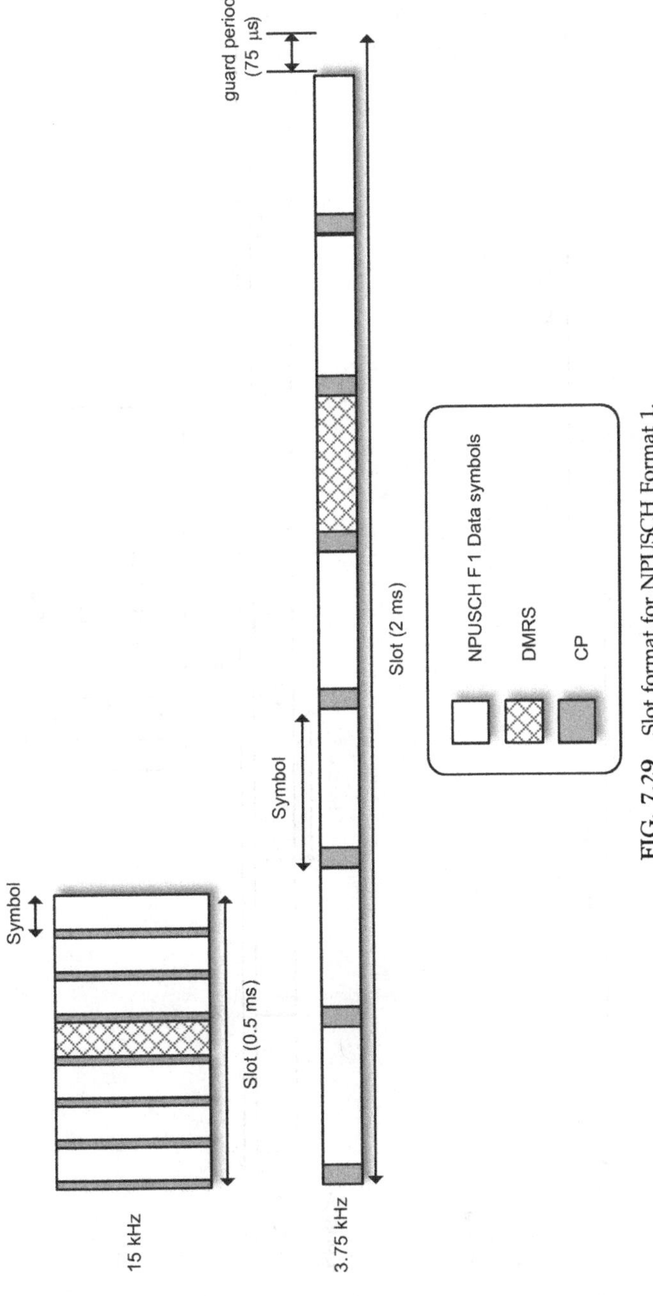

FIG. 7.29 Slot format for NPUSCH Format 1.

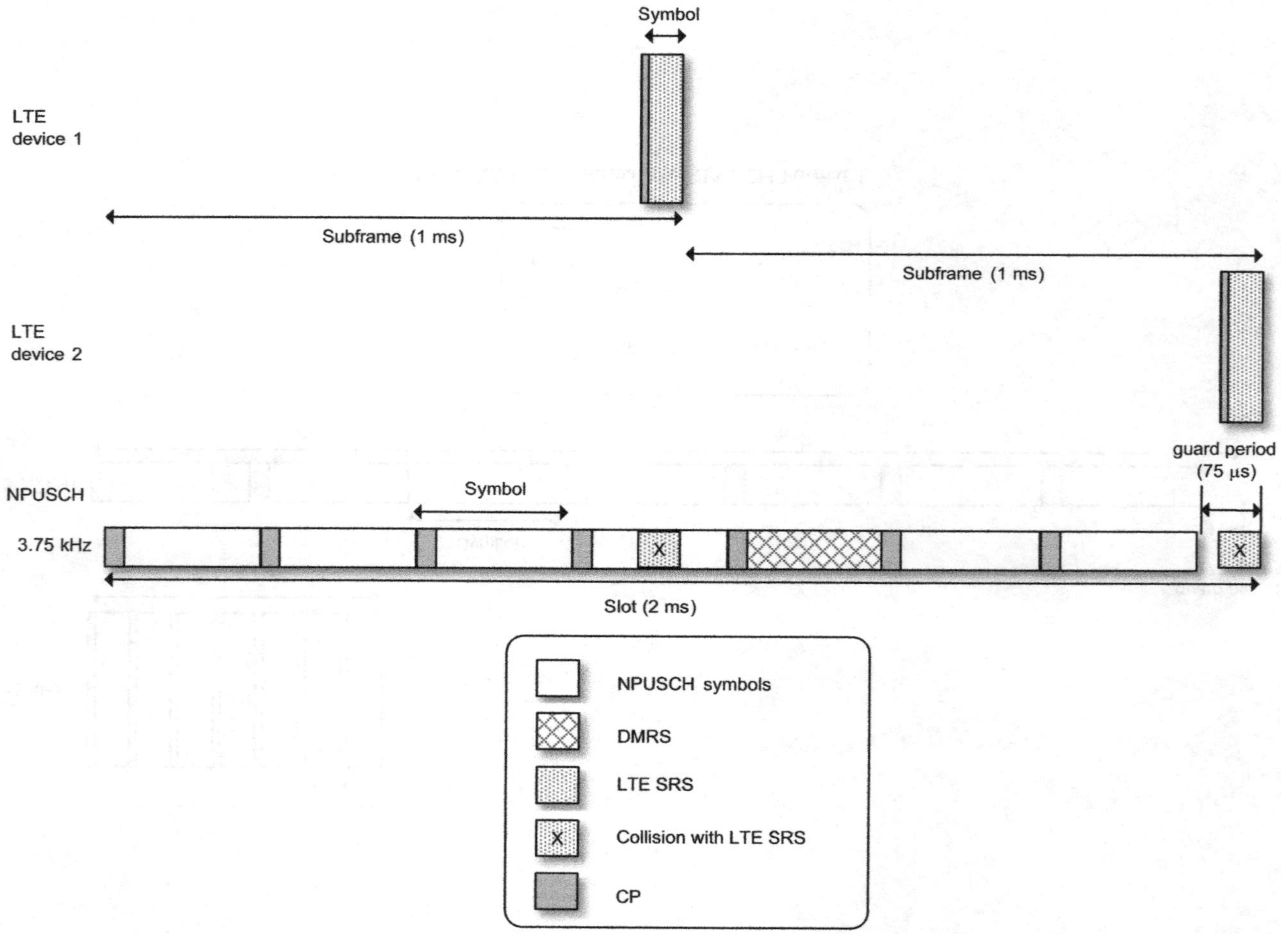

FIG. 7.30 An example of NPUSCH collision with LTE SRS.

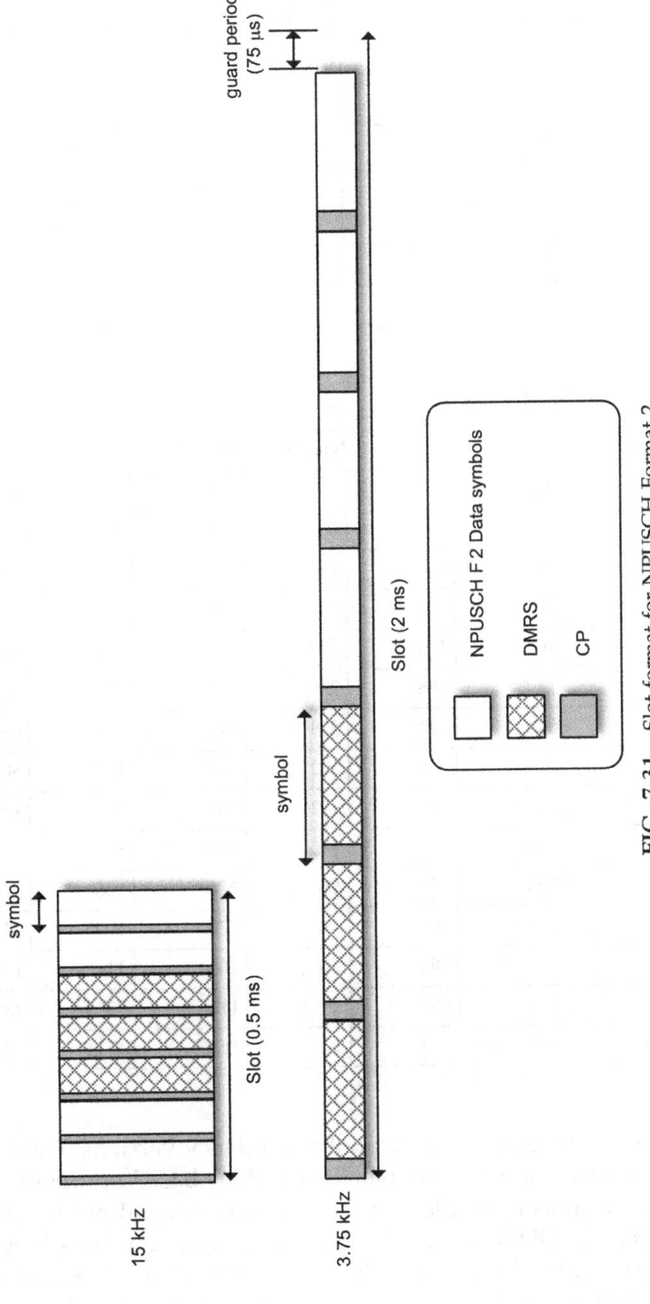

FIG. 7.31 Slot format for NPUSCH Format 2.

TABLE 7.15 Number of slots and number of data symbols per NPUSCH RU.

NPUSCH format	Device scheduled bandwidth [kHz]	Number of OFDM symbols per slot, N_{SYMB}	Length of slot [ms]	Number of slots per RU, N_{SLOTS}	Length of RU [ms]	Number of REs for data per RU
Format 1	180	7	0.5	2	1	144
	90	7	0.5	4	2	144
	45	7	0.5	8	4	144
	15	7	0.5	16	8	96
	3.75	7	2	16	32	96
Format 2	15	7	0.5	4	2	16
	3.75	7	2	4	8	16

TABLE 7.16 TBSs for NPUSCH format 1.

	Number of RUs [N_{RU}]							
TBS (I_{TBS})	**1**	**2**	**3**	**4**	**5**	**6**	**8**	**10**
0	16	32	56	88	120	152	208	256
1	24	56	88	144	176	208	256	344
2	32	72	144	176	208	256	328	424
3	40	104	176	208	256	328	440	568
4	56	120	208	256	328	408	552	680
5	72	144	224	328	424	504	680	872
6	88	176	256	392	504	600	808	1000
7	104	224	328	472	584	712	1000	1224
8	120	256	392	536	680	808	1096	1384
9	136	296	456	616	776	936	1256	1544
10	144	328	504	680	872	1000	1384	1736
11	176	376	584	776	1000	1192	1608	2024
12	208	440	680	1000	1128	1352	1800	2280
13	224	488	744	1032	1256	1544	2024	2536

as defined in LTE [8] are used. Combination of redundancy versions 0 and 2 gives a higher coding gain than versions 0 and 1. Rate-matching is done based on the modulation scheme and number of data symbols available in a transmission interval per repetition. For all the multitone transmissions, QPSK is used. For single-tone transmissions, both 15 and 3.75 kHz numerologies, either *binary phase shift keying* (BPSK) or QPSK, are used, depending on the TBS index. BPSK is used for $I_{TBS} = 0$, or 2, and all the other I_{TBS} values use QPSK.

The reason to extend BPSK to $I_{TBS} = 2$ is to allow BPSK to be used for higher uplink TBS, up to 424 bits for single-tone transmissions. Similarly, using QPSK for $I_{TBS} = 1$ allows QPSK to be extended for the use of smaller uplink TBSs for single-tone transmissions. For single-tone transmissions, the BPSK and QPSK modulations are later converted to $\pi/2$-BPSK and $\pi/4$-QPSK, respectively, in the baseband signal generation process (see Section 7.2.6.1). The $\pi/2$ and $\pi/4$ rotations for single-tone transmissions are intended for reducing the PAPR, aiming to improve the PA efficiency. For a TB size, better coverage is achieved by allocating more RUs, giving a higher energy level per information bit, and in most cases, a higher coding gain as well.

The processing of NPUSCH Format 2 differs from NPUSCH Format 1 in that there is no CRC attachment and it is only based on a simple repetition code, which is essentially repeating the HARQ feedback bit by 16 times as shown in Table 7.17. NPUSCH Format 2 uses only BPSK and like the case of Format 1 converted to $\pi/2$-BPSK in the baseband signal generation process to help reduce the PAPR. 3GPP Release 15 introduced an option of signaling a scheduling request using NPUSCH Format 2 (see Section 7.3.2.5). This is done by applying a cover code to the BPSK modulated symbol sequence corresponding to an NPUSCH Format 2 codeword. When the base station detects the presence of the cover code, it knows that the device in addition to signaling a HARQ feedback, has indicated a scheduling request.

Before modulating the bits of the encoded TB to BPSK or QPSK symbols, the bits are scrambled. If a single-tone transmission is used, then the scrambling is reinitialized for every repetition. For multitone transmission, the scrambling is reinitialized every min($N_{REP}/2$, 4) repetition of the code word, where N_{REP} is the number of configured repetitions. Also, the redundancy version is changed at the same time. At most 128 repetitions can be transmitted.

For multitone NPUSCH transmissions, after mapping the code word on a pair of slots, the pair of slots are repeated min($N_{REP}/2$, 4) times before the mapping of the code word continues. In case of single-tone transmissions, the mapping is done by first mapping the entire code word to consecutive slots before repeating.

Fig. 7.32 shows the transmission of a TB configured on 12 subcarriers, 2 RUs, and 8 repetitions. The first pair of timeslots 1, 2 are repeated four times before the mapping continues to the second pair of timeslots 3, 4. After 16 timeslots the full code word has been repeated four times and the scrambling is reinitialized and the redundancy version is updated. The procedure is then repeated once to complete eight repetitions of the TB in total.

The repeated slots do, just as in case of the NPDCCH and NPDSCH, allow for coherent combining for received power estimation and for frequency offset estimation. It also allows a base station to attempt decoding of the code word before the transmission has completed. The example shown in Fig. 7.32 supports decoding of the full code word already after 16 timeslots.

TABLE 7.17 Repetition code used for NPUSCH Format 2.

HARQ ACK bit	NPUSCH format 2 code word
0	0,0,0,0,0,0,0,0,0,0,0,0,0,0,0,0
1	1,1,1,1,1,1,1,1,1,1,1,1,1,1,1,1

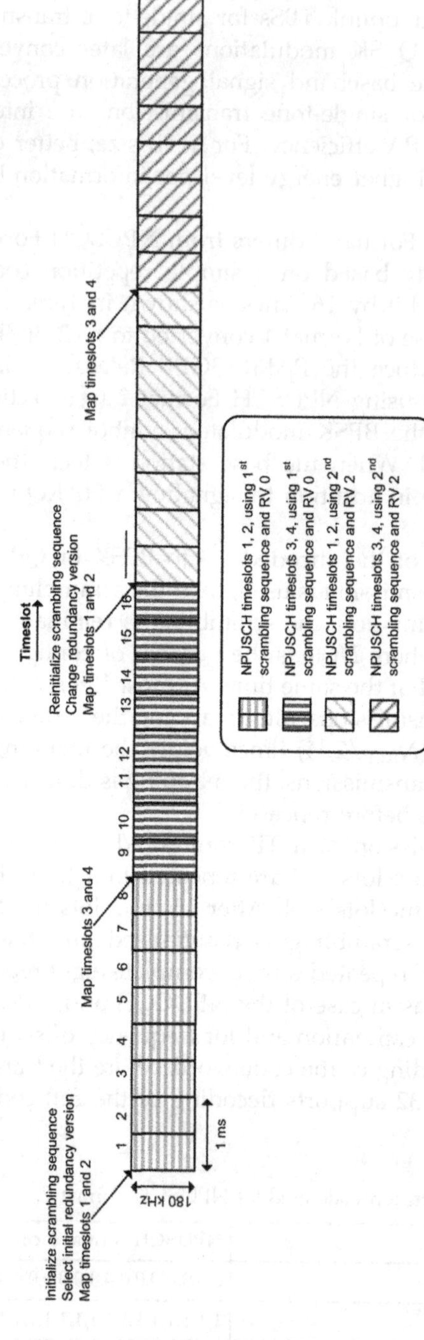

FIG. 7.32 NPUSCH F1 transmission configured with 12 subcarriers, two RUs and eight repetitions.

7.2.5.3 DMRS

Subframe	Any
TTI	Same as associated NPUSCH
Repetitions	Same as associated NPUSCH
Sub-carrier spacing	3.75, 15 kHz
Bandwidth	Same as associated NPUSCH
Carrier	Any

DMRS is always associated with NPUSCH, either Format 1 or 2, and is transmitted in every NPUSCH slot, as shown in Fig. 7.29 and 7.31. The bandwidth of DMRS is identical to the associated NPUSCH.

NB-IoT DMRS of 180 kHz bandwidth reuses the LTE DMRS sequences defined for one PRB [16]. New DMRS sequences are introduced to support NB-IoT DMRS with bandwidth smaller than 180 kHz. For all the multitone formats, the DMRS sequences are QPSK sequences, and for the single-tone formats, either 15 or 3.75 kHz bandwidth, BPSK sequences are used. We will use the NPUSCH Format 1 three-tone DMRS as an example to describe the key features of DMRS. Interested readers can refer to Ref. [16] for more details.

The base DMRS sequences for three-tone transmissions are associated with a length-3 frequency-domain sequence of the form $e^{j\phi(0)\pi/4}, e^{j\phi(1)\pi/4}, e^{j\phi(2)\pi/4}$ with $\phi(n)$ determined by one of the 12 base sequences shown in Table 7.18. Each element of the sequence is mapped to an RE in the frequency domain, i.e., to one of the three transmitted tones. Because $\phi(n) \in \{\pm 1, \pm 3\}$, each element in the basic sequence is essentially a QPSK symbol. A device is given a base sequence index through a higher layer parameter, or, if such a higher layer

TABLE 7.18 DMRS basic sequences for three-tone transmissions.

Base sequence index, u	$\phi(0)$	$\phi(1)$	$\phi(2)$
0	1	-3	-3
1	1	-3	-1
2	1	-3	3
3	1	-1	-1
4	1	-1	1
5	1	-1	3
6	1	1	-3
7	1	1	-1
8	1	1	3
9	1	3	-1
10	1	3	1
11	1	3	3

parameter is not given, the device determines the base sequence index by PCID, i.e., $u = \text{PCID mod } 12$.

A base sequence can be applied with cyclic shift α. The cyclic shift α therefore gives rise to a DMRS sequence of the form $\left(e^{\frac{j\phi(0)\pi}{4}}, e^{j\alpha}e^{\frac{j\phi(1)\pi}{4}}, e^{j2\alpha}e^{\frac{j\phi(2)\pi}{4}}\right)$. The value of α is given to the device through a higher layer parameter.

To randomize interference, DMRS sequence hopping can be optionally used when it is associated with NPUSCH Format 1. In that case, the base sequence index varies from slot to slot following a pseudorandom pattern. The pseudorandom pattern is cell specific and depends on PCID. If DMRS sequence hopping is not activated, the DMRS symbols are repeated across the full length of the NPUSCH Format 1 transmission.

7.2.5.4 NPRACH and NPUSCH multiplexing

The resources for NPRACH and NPUSCH are time and frequency multiplexed. An example is given in Fig. 7.33 where NPRACH resources reserved for coverage classes 0, 1, and 2 are identified. The starting tone index and number of allocated tones for each coverage class are signaled in SI. These parameters are used to configure the frequency portions of the NB-IoT carriers allocated for NPRACH. Coverage class 0 can be said to correspond to normal coverage, while coverage classes 1 and 2 are facilitating system access for users in extended and extreme coverage, respectively. Coverage class 0 may not need to provide NPRACH repetitions and is therefore of short time duration. Because most devices locate in normal coverage, it spans many tones, or preambles, and appears with a short periodicity to facilitate large capacity. Coverage classes 1 and 2 are configured for support of NPRACH repetitions to facilitate system access, for example, from deep indoors. They are expected to require fewer tones than coverage class 0 because of relatively fewer devices in the extended and extreme coverage domains.

Except for the resources carved out for NPRACH, most of the uplink resources are available for NPUSCH transmissions. The NPRACH resources are signaled in the SIBs, and thus the device knows exactly which resources are set aside for NPRACH. During an NPUSCH transmission, if an overlap with NPRACH resources occurs, NPUSCH transmission needs to be postponed until the first available uplink slot without overlap.

The NPRACH is further described in the context of the random access procedure in Section 7.3.1.6.

7.2.6 Baseband signal generation

7.2.6.1 Uplink

7.2.6.1.1 Multitone NPUSCH

The multitone NPUSCH baseband signal generation is based on the same principle that is used for LTE PUSCH. The data symbols are DFT precoded into a set of frequency-domain complex symbols $a(k, l)$ [16]. Elements of the DMRS sequence, however, are directly used as frequency-domain complex symbols $a(k, l)$. The symbol $a(k, l)$ defines the transmitted

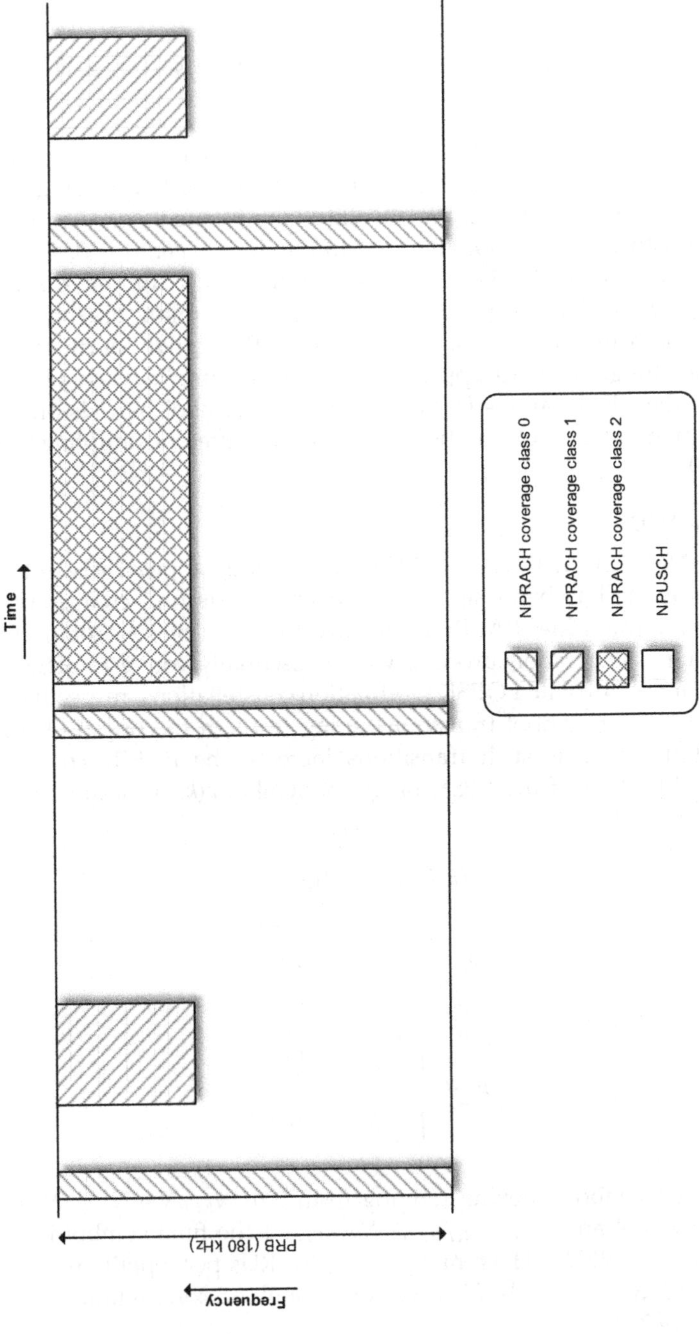

FIG. 7.33 Multiplexing of NPRACH and NPUSCH resources.

waveform as it modulates the k-th tone of the l-th OFDM symbol through the *inverse DFT* (IDFT):

$$s_l(t) = \sum_{k=0}^{11} a(k,l)e^{j2\pi(k-5.5)\Delta f(t-N_{cp}(l)T_s)}, \quad 0 \le t < (N + N_{cp}(l))T_s \tag{7.11}$$

where Δf is the subcarrier spacing of 15 kHz, $N_{cp}(l)$ is the number of samples for the CP of the l-th OFDM symbol, T_s is the *basic time unit*, and N is 2048. The frequency grid illustrated in Fig. 7.34 defines the tone-index k and the absolute tone frequency $(k-5.5)\Delta f$. Note that in Eq. (7.11) if a tone with index k is not allocated to the device, $a(k, l)$ is 0.

The basic time unit in Eq. (7.11) is specified as in LTE assuming a sampling rate of 30.72 MHz, i.e., $T_s = 1/30.72\ \mu s$. In practice though, NB-IoT uplink baseband signal generation can be based on a much lower sampling rate as the signal bandwidth is not higher than 180 kHz. One straightforward approach is to use a sampling rate of 1.92 MHz, as at this sampling rate the CP durations in different OFDM symbol periods amount to integer numbers. CP duration in terms of number of samples at different sampling rates is summarized in Table 7.19.

7.2.6.1.2 Single-tone NPUSCH

For single-tone NPUSCH with $\Delta f = 15$ kHz, one obvious approach is to take only one k term in Eq. (7.11) as the baseband signal. However, one desired characteristic for single-tone baseband waveform is low PAPR. To achieve this, $\pi/2$ or $\pi/4$ rotation is introduced in the baseband signal generation process, which essentially converts BPSK and QPSK to $\pi/2$-rotated BPSK and $\pi/4$-rotated QPSK modulation constellations, respectively. This allows NPUSCH to avoid having symbol transitions going through the origin of the *in-phase and quadrature-phase* (IQ) plane, as such transitions increase the PAPR. The rotation can be expressed as a modification of the BPSK or QPSK symbol $a(k, l)$ modulating the k-th tone of the l-th OFDM symbol:

$$\tilde{a}(k,l) = a(k,l)e^{j\varphi_{l'}}$$

$$\varphi_{l'} = \rho \times (l' \bmod 2) \tag{7.12}$$

$$\rho = \begin{cases} \frac{\pi}{2} \text{ for BPSK} \\ \frac{\pi}{4} \text{ for QPSK} \end{cases}$$

Here, we use l' for symbol indexing ranging from 0 to $N_{REP}N_{RU}N_{SLOT}N_{SYMB} - 1$, where 0 defines the first symbol and $N_{REP}N_{RU}N_{SLOT}N_{SYMB} - 1$ the final symbol in a transmission of N_{REP} repetitions of an NPUSCH TB mapped to N_{RU} RUs per repetition, each corresponding to N_{SLOT} slots of N_{SYMB} symbols. l is used for symbol indexing within a slot and because $N_{SYMB} = 7$, $l = l' \bmod 7$.

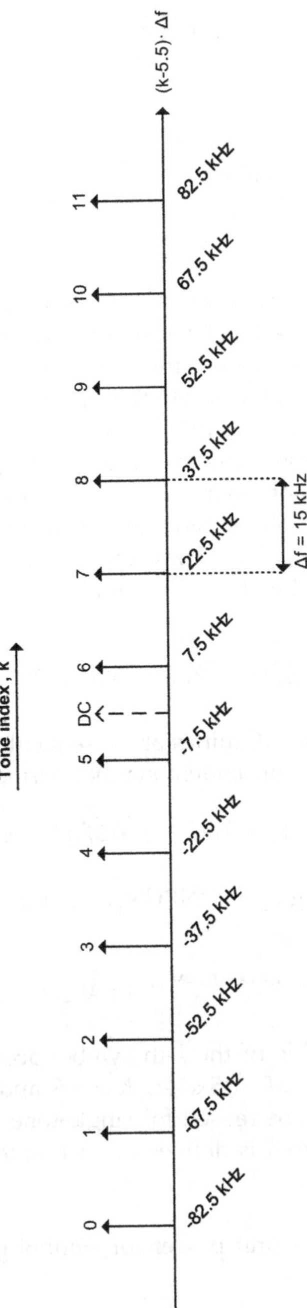

FIG. 7.34 Frequency grids used for uplink baseband signal generation (15 kHz subcarrier spacing).

TABLE 7.19 CP durations in terms of number of samples at different sampling rates for the 15 kHz numerology.

Parameter	CP of first OFDM symbol in a slot	CP of other OFDM Symbols in a slot
Time	5.21 μs	4.69 μs
Samples at 30.72 MHz	160 samples	144 samples
Samples at 1.92 MHz	10 samples	9 samples

The desired symbol transition statistical properties of $\pi/2$-BPSK and $\pi/4$-QPSK are preserved if the phase rotation in one OFDM symbol interval is exactly an integer multiple of π. However, because of the CP insertion, this is not the case. Thus, to preserve symbol transitions according to $\pi/2$-BPSK and $\pi/4$-QPSK modulations at an OFDM symbol boundary, an additional phase term is introduced in the baseband signal generation. From Eq. (7.11) for NPUSCH single-tone transmission using tone-index k, the phase term due to the sinusoid $e^{j2\pi(k-5.5)\Delta f(t-N_{cp}(l)T_s)}$ at the end of symbol 0 (i.e., $l=0$, $t = (N+N_{cp}(0))T_s$) is $\phi_e(0) = 2\pi(k-5.5)\Delta f N T_s$ and the phase term at the beginning of symbol 1 (i.e., $l=1$, $t=0$) is $\phi_b(1) = -2\pi(k-5.5)\Delta f N_{cp}(1)T_s$. Therefore, the additional phase term that needs to be introduced to compensate for the phase discontinuity at the boundary between symbols 0 and 1 is

$$\phi(1) = \phi_e(0) - \phi_b(1) = 2\pi(k-5.5)\Delta f(N+N_{cp}(1))T_s. \quad (7.13)$$

To compensate for the phase discontinuities at subsequent symbol boundaries, the phase term needs to be accumulated over subsequent symbol periods. Thus, $\phi(0) = 0$, and

$$\phi(l') = \phi(l'-1) + 2\pi(k-5.5)\Delta f(N+N_{cp}(l))T_s. \quad (7.14)$$

The baseband waveform for single-tone NPUSCH using tone-index k can be described mathematically as

$$s_{k,l'}(t) = \tilde{a}(k,l')e^{\phi(l')}e^{j2\pi(k-K)\Delta f(t-N_{cp}(l)T_s)}, \quad 0 \le t \le (N+N_{cp}(l))T_s \quad (7.15)$$

where $\tilde{a}(k,l')$ is the modulation value of the l'-th symbol according to either $\pi/2$-BPSK or $\pi/4$-QPSK modulation. In case of $\Delta f = 15$ kHz, $K=5.5$ and the tone-index k is defined according to Fig. 7.34. Eq. (7.15) can be reused for single-tone NPUSCH with $\Delta f = 3.75$ kHz, by using $K=23.5$ and the tone-index k is defined according to Fig. 7.35.

7.2.6.1.3 NPRACH

The NPRACH baseband signal of unit power for symbol group l with Format 0 or 1 is given by:

$$s_l(t) = e^{j2\pi(k'(l)-23.5)\Delta f(t-T_{cp})}, \quad 0 \le t \le T \quad (7.16)$$

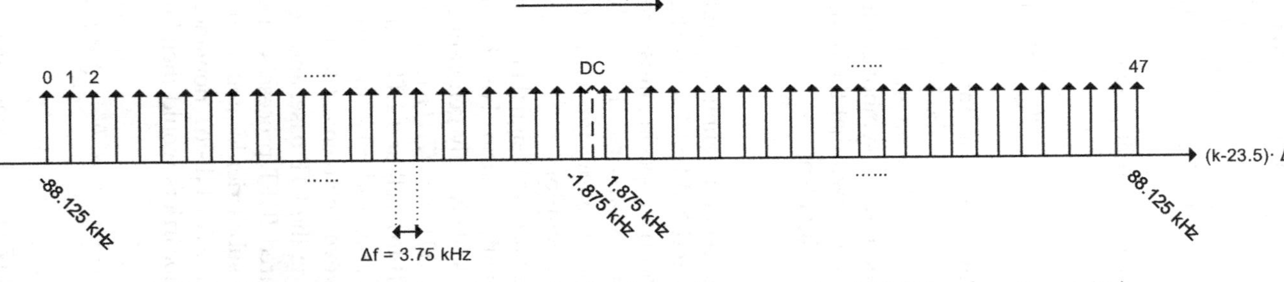

FIG. 7.35 Frequency grids used for uplink baseband signal generation (3.75 kHz subcarrier spacing).

where $k'(l)$ is the tone index according to the frequency grid depicted in Fig. 7.35 selected for transmission of the NPRACH symbol group l in an NPRACH repetition unit, Δf equals the NPRACH Formats 0 and 1 subcarrier spacing of 3.75 kHz, while the length T and CP duration T_{cp} are presented in Table 7.14. In Eq. (7.16), $k'(l)$ is determined by the hopping pattern $k(l)$ exemplified in Table 7.12, $k'(l) = k_{start} + k(l)$, where k_{start} is a parameter used for configuring the starting tone index of NPRACH. Thus while $k(l)$ ranges from 0 to 11, $k'(l)$ ranges from 0 to 47.

Similarly, NPRACH baseband signal of unit power for Format 2 can be generated using Eq. (7.17):

$$s_l(t) = e^{j2\pi(k'(l)-71.5)\Delta f(t-T_{cp})}, 0 \le t \le T \tag{7.17}$$

with Δf, T and T_{cp} based on the values given in Table 7.14.

When comparing Eqs. (7.16) and (7.17) with the NPUSCH definition in Eq. (7.15) it is seen that the additional phase term $e^{\phi(l)}$ is not implemented for NPRACH. For NPRACH preamble Formats 1 and 2, the phase rotation over a symbol group is equal to an integer multiple of 2π. For NPRACH Format 0, it can be shown that the transition between the symbol groups will not go through the origin of the IQ-plane.

7.2.6.2 *Downlink*

In the downlink case a unified baseband definition applies for the NPSS, NSSS, NPBCH, NPDSCH, NPDCCH, NRS, NPRS, and NWUS. The baseband signal generation is based on the frequency grid with 15 kHz subcarrier spacing illustrated in Fig. 7.34 with the DC carrier located in between the two center subcarriers. The baseband waveform is also the same across all three modes of operation to allow for a single device receiver implementation regardless of the network mode of operation.

The actual implementation of the baseband signal generation may, however, take different forms depending on the deployment scenario. For the stand-alone deployment scenario where the base station transmitter is likely only generating the NB-IoT signal, the baseband implementation may follow Eq. (7.11). Eq. (7.11) may also be used for guard-band and in-band deployments, when the NB-IoT cell and LTE cell are configured with different cell identities implying that there is no specified relation between the NRS and LTE CRS.

In the in-band scenario it is, however, convenient to generate the LTE and NB-IoT signals jointly. This can be achieved by using the LTE baseband definition to basically generate both LTE and NB-IoT signal with a shared IDFT process. Consider, e.g., a 20 MHz LTE carrier containing 100 PRBs plus one extra subcarrier located in the center of the carrier to eliminate the impact from the DC component. An NB-IoT non-anchor carrier located at PRB 0 and spanning the first 12 of the 1201 subcarriers would then be generated as:

$$s_l(t) = \sum_{k=0}^{11} a(k,l) e^{j2\pi(k-600)\Delta f(t-N_{cp}(l)T_s)}, \quad 0 \le t \le (N + N_{CP}(l))T_s \tag{7.18}$$

When comparing Eqs. (7.18) and (7.11), it becomes evident that the joint LTE and NB-IoT generation in Eq. (7.18) introduces a frequency offset $f_{NB-IoT} = (5.5 - 600)\Delta f$ for each tone k.

This is automatically compensated for when the NB-IoT carrier is upconverted to its RF carrier frequency. But the frequency offset also introduces a phase offset at start of each symbol l:

$$\theta_{k,l} = 2\pi f_{NB-IoT}\left(lN - \sum_{i=0}^{l} N_{CP}(i\ mod\ 7)\right)T_s. \tag{7.19}$$

To allow joint IDFT generating the LTE and NB-IoT signal, this phase shift must be compensated for as well. A detailed description of this compensation is found in Ref. [16] where the herein made example is generalized by defining f_{NB-IoT} as the center frequency of the NB-IoT PRB minus the LTE center frequency as illustrated in Fig. 7.36.

The phase compensation in Eq. (7.19) determines a fixed relation between NB-IoT NRS and LTE CRS symbols that is utilized by the system in case of in-band operation when the NB-IoT cell and LTE cell are configured with the same cell identity and the same number of logical antenna ports to allow devices to use the LTE CRS, in addition to the NRS, for estimating the NB-IoT downlink radio channel.

7.2.7 Transmission gap

7.2.7.1 *Downlink transmission gap*

Transmission gaps are introduced in the downlink and uplink for different reasons. In the downlink case, NPDCCH and NPDSCH transmissions serving a device in extreme coverage can take a long time. For example, if the maximum repetitions of 2048 are used for NPDCCH, the total transmission time is more than 2 s. For NPDSCH, it can be more than 20 s if one repetition requires 10 subframes. Thus, for NPDCCH and NPDSCH transmissions with large repetitions, it helps to define transmission gaps so that the network can serve other devices

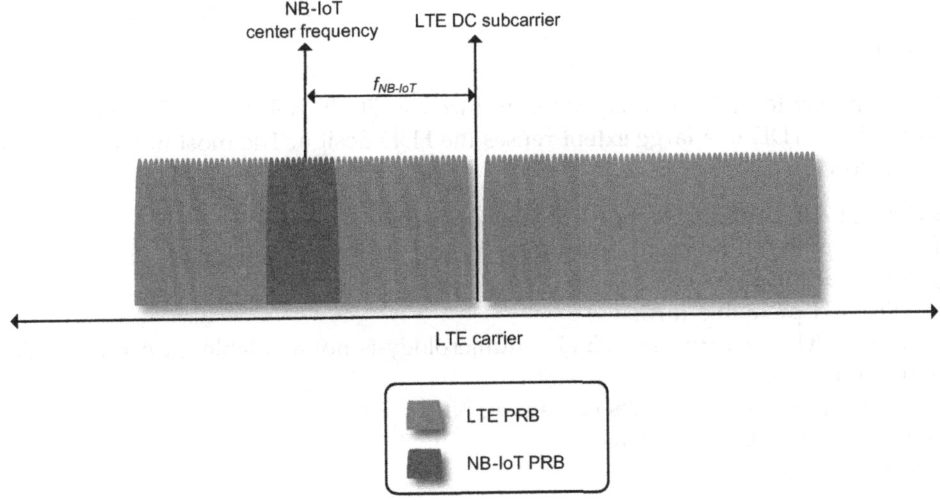

FIG. 7.36 Frequency offset between the centers of NB-IoT and LTE.

during a transmission gap. The higher layers can signal a threshold applied to the number of repetitions. If the number of repetitions is less than such a threshold, no transmission gaps are configured. An example is illustrated in Fig. 7.37. It is shown that transmissions gaps are configured for devices in extreme coverage as these devices require a large number of repetitions for NPDCCH and NPDSCH transmissions. In the example of Fig. 7.37, during a transmission gap, the NPDCCH or NPSDCH transmission to device 1 is postponed and the base station can serve device 2 and device 3 in normal coverage during the transmission gap of device 1.

The transmission gap parameters such as gap periodicity and gap duration are signaled by the higher layers. The transmission time between gaps can be calculated as the gap periodicity minus the gap duration. The transmission time, gap periodicity, and gap duration are all given in terms of NPDCCH or NPDSCH subframes. Therefore, subframes used by other downlink physical channels (e.g., NPSS, NSSS, NPBCH) or invalid subframes cannot be counted when mapping the NPDCCH or NPDSCH on the subframe structure.

7.2.7.2 *Uplink transmission gap*

In the uplink case, devices in extreme coverage are most likely scheduled with single-tone transmissions. Thus, they will not block the radio resources of the entire carrier because other subcarriers can be used to serve other devices. Low-cost oscillators that are not temperature-compensated can have frequency drifts resulting from self-heating. Thus, after certain continuous transmission time, self-heating will cause a frequency drift. Uplink transmission gaps therefore are introduced, not for the consideration of avoiding blocking other devices. Rather, they are introduced to allow a device operating in an FDD band to have an opportunity to recalibrate its frequency and time references by resynchronizing to the downlink reference signals, e.g., NPSS, NSSS and NRS. For NPUSCH transmissions, for every 256 ms continuous transmission, a 40 ms gap is introduced. For every 64 NPRACH preamble repetitions of Formats 0 and 1, or for every 16 repetitions of Format 2, a 40 ms gap is introduced. Illustration of NPUSCH and NPRACH Format 1 transmission gaps is shown in Fig. 7.38.

7.2.8 TDD

NB-IoT support for TDD bands was introduced in 3GPP Release 15. The physical layer design of NB-IoT TDD to a large extent reuses the FDD design. The most notable differences are listed below:

- LTE TDD UL-DL configurations 1 to 5 are supported
- Subframe mapping of different downlink physical channels and signals
- Usage of *special subframes*
- New NPRACH preamble formats
- Single-tone NPUSCH with the 3.75 kHz numerology is not available for certain TDD configurations
- Device assumption on subframes containing NRS
- System information transmissions
- Uplink transmission gaps

In this section, we highlight these main differences.

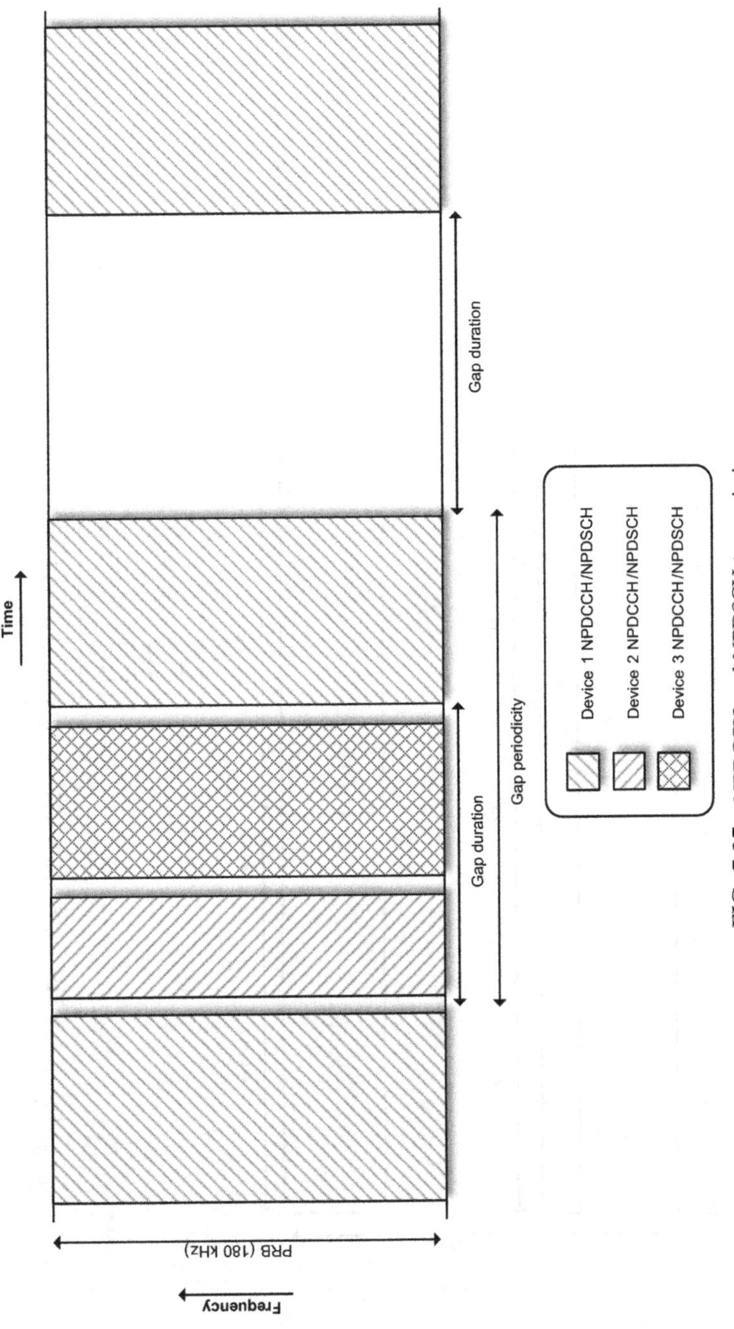

FIG. 7.37 NPDCCH and NPDSCH transmission gaps.

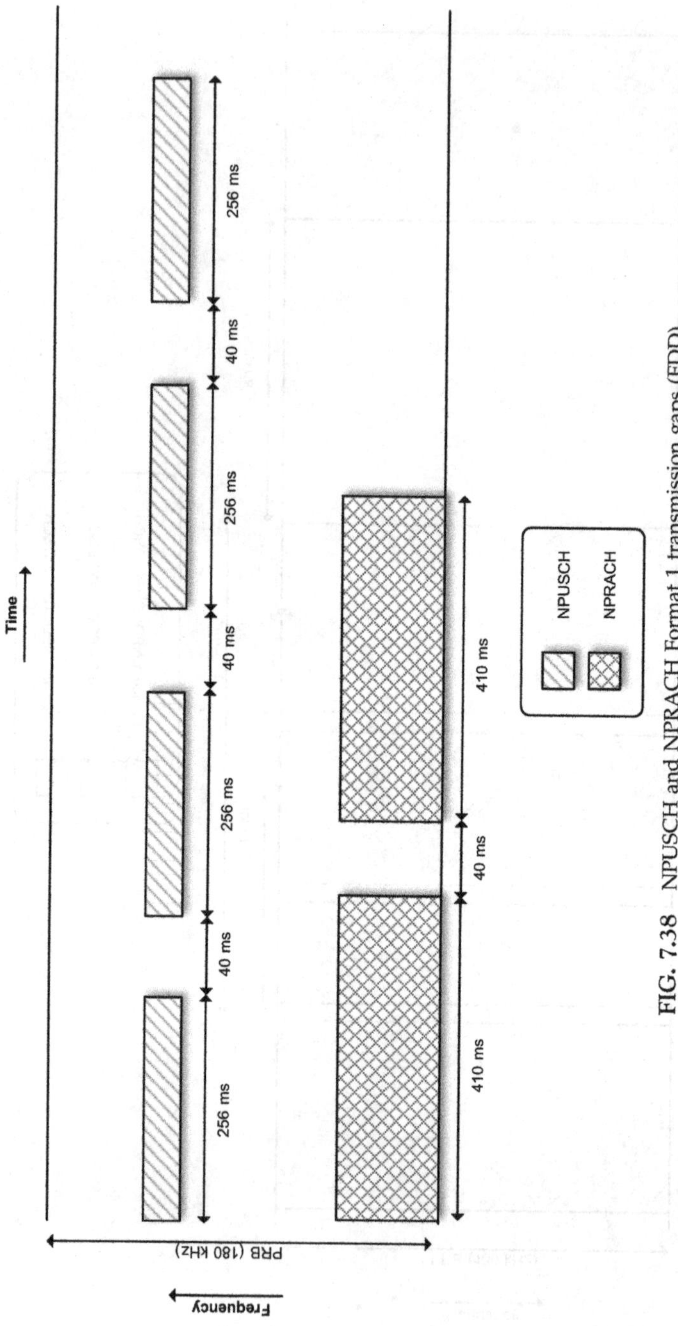

FIG. 7.38 NPUSCH and NPRACH Format 1 transmission gaps (FDD).

TABLE 7.20 LTE TDD configurations. (D: DL subframe; S: special subframe; U: UL subframe).

Configuration	Supported by NB-IoT	Subframe number									
		0	1	2	3	4	5	6	7	8	9
0	No	D	S	U	U	U	D	S	U	U	U
1	Yes	D	S	U	U	D	D	S	U	U	D
2	Yes	D	S	U	D	D	D	S	U	D	D
3	Yes	D	S	U	U	U	D	D	D	D	D
4	Yes	D	S	U	U	D	D	D	D	D	D
5	Yes	D	S	U	D	D	D	D	D	D	D
6	No	D	S	U	U	U	D	S	U	U	D

7.2.8.1 *Subframe mapping*

NB-IoT supports all LTE TDD UL-DL configurations, except for configurations 0 and 6 due to insufficient numbers of downlink subframes in a radio frame. As shown in Table 7.20, the supported TDD configurations have at least 4 downlink subframes in every radio frame. Also, as shown in Table 7.20, subframes #0, #5, and #9 are downlink subframes in all configurations supported in NB-IoT. Thus, these subframes are used to map the initial acquisition signals such as NPSS, NSSS, and NPBCH. Fig. 7.39 shows how these signals are mapped to different subframes. Compared to the mapping in FDD (see Fig. 7.11), the only difference is the swapping of NSSS and NPBCH subframe assignments. This results in different NPSS and NSSS time relationship. In the FDD case, an NSSS subframe is always 4 subframes after an NPSS subframe. In the TDD case, an NSSS subframe is always 5 subframes after an NPSS subframe. The difference in NPSS and NSSS subframe relationship allows the device to detect whether the NB-IoT carrier of interest is configured for FDD or TDD. Also shown in Table 7.20, subframe #1 is always a special subframe and subframe #2 is always an uplink subframe. Subframes #3, #4, #6, #7, #8 can be either a downlink, uplink, or special subframe depending on the TDD configuration. The information about TDD UL-DL configuration is provided in SIB1-NB.

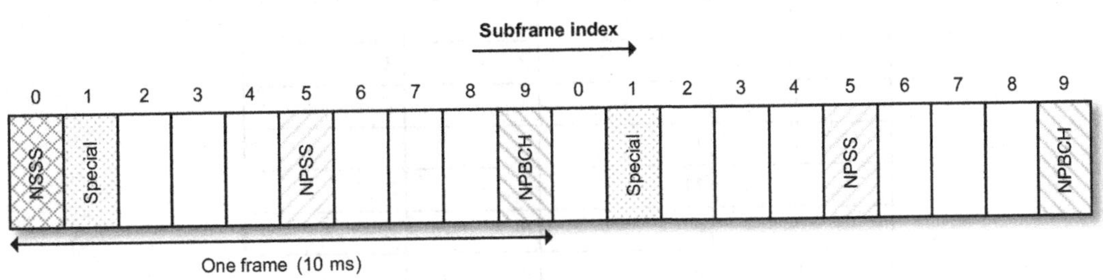

FIG. 7.39 Time-multiplexing of downlink physical channels on an NB-IoT TDD anchor carrier.

7.2.8.2 *Usage of special subframes*

As shown in Table 7.17, all the TDD configurations include *special subframes*, which are placed after a full downlink subframe and before a full uplink subframe. A special subframe includes three portions, *Downlink Pilot Time Slot* (DwPTS), *GP* and *Uplink Pilot Time Slot* as illustrated in Fig. 7.40. There are 11 configurations of special subframe defined, as shown in Table 7.21. The information about special subframe configuration is provided in SIB1-NB.

In NB-IoT, the Downlink Pilot Time Slot (DwPTS) portion of a special subframe can be used to map NPDCCH or NPDSCH symbols, but not NPRS. The GP and Uplink Pilot Time Slot portions are not used.

To support mapping NPDCCH and NPDSCH in a special subframe, NRS needs to be provided. The subcarriers that NRS is mapped onto in a special subframe are identical to those in a regular downlink subframe; and as explained in Section 7.2.4.3, these subcarriers are determined by the cell identity and logical antenna port number. The placement of NRS in a regular downlink subframe in the time dimension is fixed to the last two OFDM symbols in a slot. This is however not always true for special subframes. Fig. 7.41 shows the mapping of NRS in a special subframe with configuration #3. In this case, the NRS is mapped to the third and fourth OFDM symbol in each slot. The mapping of NRS onto OFDM symbols is adjusted for each special subframe configuration based on the number

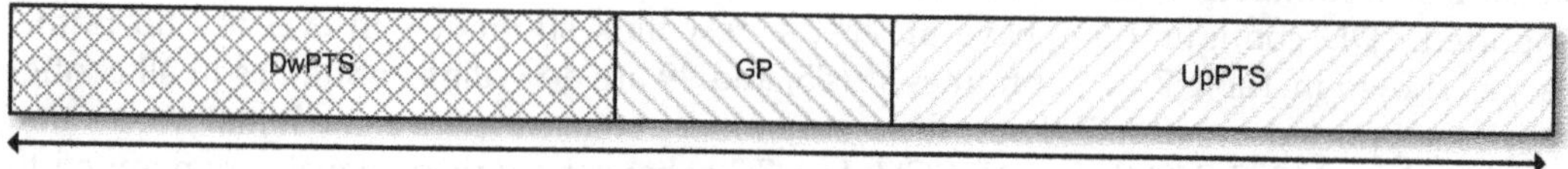

FIG. 7.40 Special subframe used in TDD.

TABLE 7.21 TDD special subframe configurations, durations of each portion in symbols.

Configuration	DwPTS [symbols]	GP [symbols]	UpPTS [symbols]
0	3	10	1
1	9	4	1
2	10	3	1
3	11	2	1
4	12	1	1
5	3	9	2
6	9	3	2
7	10	2	2
8	11	1	2
9	6	6	2
10	6	2	6

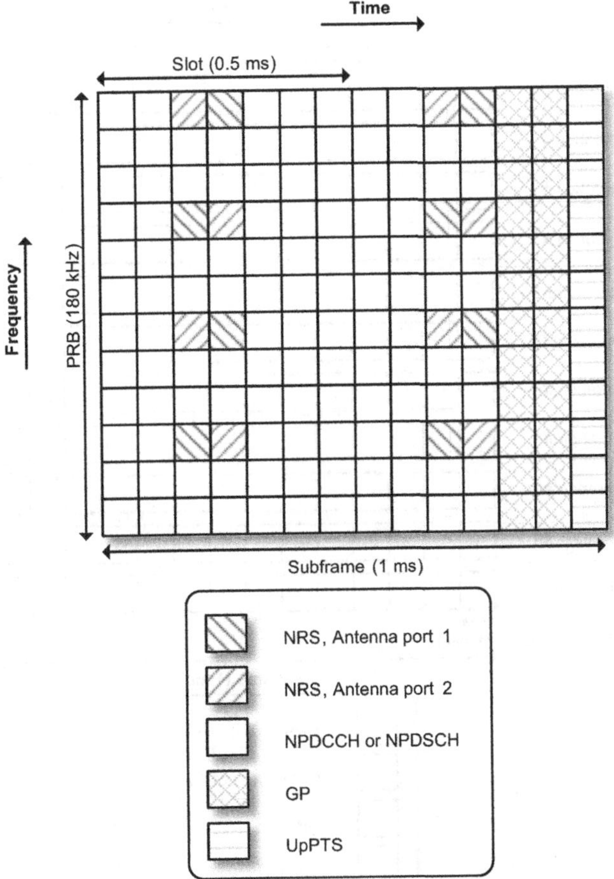

FIG. 7.41 Mapping of NRS in a TDD special subframe with special subframe configuration #3. (stand-alone deployment example).

of OFDM symbols in the DwPTS portion. This is illustrated in Fig. 7.42. As shown, no NRS is transmitted in a special subframe with configuration #0 or #5.

The mapping of NPDCCH and NPDSCH symbols in a special subframe depends on whether the repetition factor is greater than one or not. When NPDCCH or NPDSCH is configured with repetition factor equal to 1 (i.e. no repetition), all the available symbols in the DwPTS portion (i.e. symbols that are not mapped to NRS or reserved for LTE) are used; and NPDCCH or NPDSCH determines rate-matching parameters based on the number of available symbols. When the repetition factor is greater than 1, it is desirable to make it easy to combine the symbols repeated in regular and special subframes. Thus, when the repetition factor is greater than 1, a special subframe is used for repeating a regular downlink subframe and the mapping basically follows that in a regular downlink subframe, except that the symbols mapped to REs not available for NPDCCH or NPDSCH are punctured.

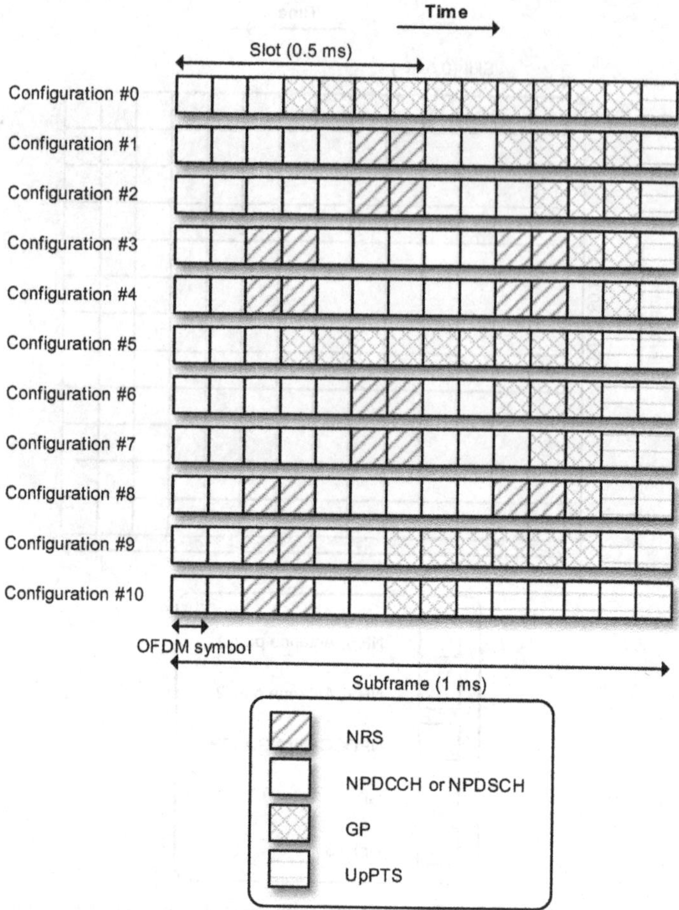

FIG. 7.42 Resource mapping of OFDM symbols in a TDD special subframe.

An example is illustrated in Fig. 7.43. As shown, some of the symbols in a regular subframe are not repeated in a special subframe as corresponding REs in the special subframe is either taken by NRS or not in the DwPTS portion.

7.2.8.3 *NPRACH for TDD*

In TDD, there are 5 NPRACH formats defined, as shown in Table 7.22. Which one of these formats is used in a cell depends on TDD configuration and the cell radius. The basic structure of these TDD NPRACH formats is the same as that of FDD formats described in Section 7.2.5.1, although the exact parameters are different. One major difference though is that while in the FDD case all the symbol groups within a repetition unit are time contiguous, the TDD cases are not. This is because, except in certain cases for NPRACH Format 0/0-a, it is not possible to fit an entire repetition unit to a contiguous uplink transmission interval due to the presence of downlink and special subframes.

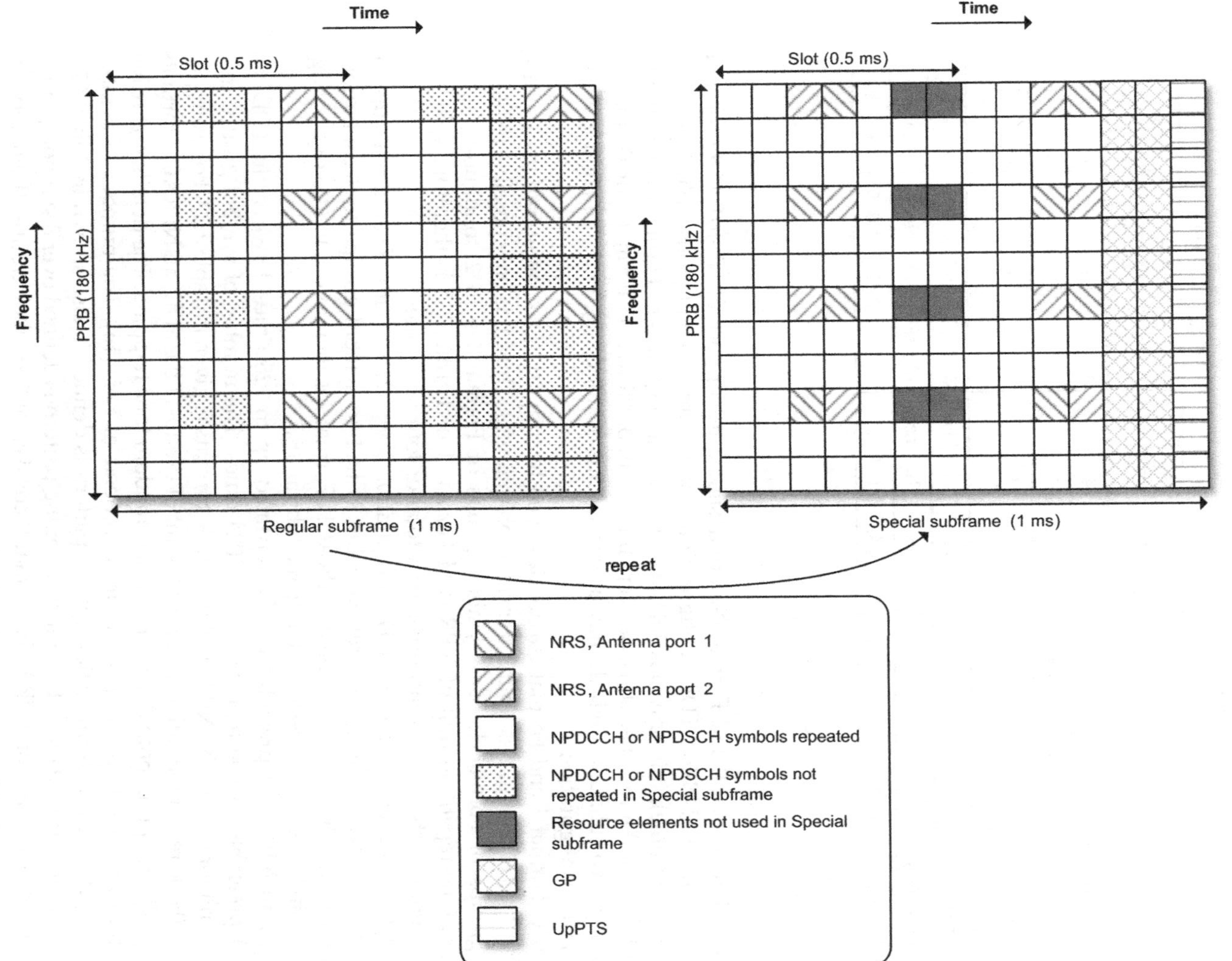

FIG. 7.43 An illustration of how NPDCCH or NPDSCH symbols in a regular subframe are repeated in a TDD special subframe with configuration #3 (stand-alone deployment example).

TABLE 7.22 NPRACH formats for TDD.

	Format 0	Format 1	Format 2	Format 0-a	Format 1-a
Supported cell radius	23.3 km	40 km	40 km	7.5 km	15 km
Cyclic prefix, T_{CP}	155.5 μs	266.7 μs	266.7 μs	50 μs	100 μs
Symbol group length, T	422.2 μs	800.0 μs	1333.3 μs	316.7 μs	633.3 μs
Number of symbol groups in a repetition unit	4	4	4	6	6
Number of time-contiguous symbol groups	2	2	2	3	3
Number of symbol groups over which deterministic hopping is defined	8	8	8	6	6
Subcarrier spacing	3.75 kHz	3.75 kHz	3.75 kHz	3.75 kHz	3.75 kHz
Tone hopping range	12 tones	12 tones	12 tones	12 tones	12 tones
TDD configurations supported	1, 2, 3, 4, 5	1, 4	3	1, 2, 3, 4, 5	1, 4

Observe that Format 0 and Format 0-a are designed to support all NB-IoT TDD configurations. Since for all NB-IoT TDD configurations the minimum contiguous uplink period is 1 ms (i.e., 1 subframe), it is only possible to fit two symbol groups of Format 0, or 3 symbol groups of Format 0-a to 1 ms. Thus, one repetition unit of NPRACH Format 0 or Format 0-a encompasses two subframes which may be contiguous or not contiguous, depending on the NB-IoT TDD configuration.

Similarly, Format 1 and Format 1-a are designed to support TDD configurations 1 and 4, which has 2 ms of contiguous uplink transmission intervals. Thus, it is only possible to fit two symbol groups of Format 1, or 3 symbol groups of Format 1-a to 2 contiguous uplink subframes. One repetition unit of NPRACH Format 1 or Format 1-a is therefore split into two contiguous parts, since it encompasses a total of four subframes.

Format 2 is designed to support TDD configuration 3, which has 3 ms of contiguous uplink transmission intervals. Thus, it is only possible to fit two symbol groups to 3 contiguous uplink subframes. One repetition unit of NPRACH Format 2 is therefore split into two contiguous parts since it encompasses a total of six subframes.

In FDD, an NPRACH preamble can be repeated up to 128 times; however in TDD, an NPRACH preamble can be repeated up to 1024 times. The number of symbol groups in a repetition unit for each of the NPRACH preamble formats is given in Table 7.22. For example, for Format 0 one repetition unit consists of 4 symbol groups. Like FDD, all NPRACH formats defined for TDD support both deterministic and random tone hopping. The details regarding the deterministic tone hopping pattern are described in Ref. [16]. The number of symbol groups over which the deterministic hopping pattern is defined is also shown in Table 7.22. For example, for Format 0 the deterministic hopping pattern is defined over 8 symbol groups. Since each repetition unit of NPRACH Format 0 consists of 4 symbol groups, the deterministic hopping pattern is therefore defined over two repetition units.

7.2.8.4 NPUSCH for TDD

Both Multi-tone and single tone transmissions are supported as in FDD, as well as NPUSCH Formats 1 and 2. All the FDD NPUSCH configurations with 15 kHz subcarrier spacing are supported in TDD. However, single-tone NPUSCH based on 3.75 kHz numerology has slot length of 2 ms, and thus is only supported by TDD configurations 1 and 4, which have exactly two contiguous uplink subframes.

7.2.8.5 Device assumption on subframes containing NRS

Before the device acquires the information about TDD configuration, it does not know which subframes are special subframes or uplink subframes, and therefore it can only assume subframes #0, #5, and #9 are downlink subframes according to Table 7.20. Since subframe #5 is used for NPSS, and subframe #0 in every other radio frame is used for NSSS, the device can therefore only assume that NRS is available in subframes #9 and in subframes #0 not containing NSSS. The procedure of SI acquisition is explained in Section 7.3.1. For now, it suffices to say that the device will at a certain point acquire the information about which subframes are used for transmitting the *Narrowband System Information Block Type 1* (SIB1-NB) described in Section 7.3.1.2. The device can therefore assume NRS is available in subframes carrying SIB1-NB. SIB1-NB contains the information about both the TDD configuration and the valid subframe bitmap. After knowing the TDD configuration and the valid subframe configuration, the device can assume NRS is present in all valid downlink subframes.

As described in Section 7.2.8.6, NB-IoT TDD supports an option of using a non-anchor carrier to transmit SIB1-NB. When a non-anchor carrier is configured for transmitting SIB1-NB, the device may assume NRS is available in subframes #0 and #5 on the non-anchor carrier.

7.2.8.6 System information transmissions

In an FDD system, all the system information is transmitted on an anchor carrier. Since some of the supported TDD configurations have very few downlink subframes per radio frame, it might be necessary to use a non-anchor carrier to transmit the system information. NB-IoT supports an option of transmitting system information using a non-anchor carrier. First, whether SIB1-NB is transmitted on an anchor or a nonanchor carrier is indicated in the MIB-NB. In the case of anchor carrier, SIB1-NB may be transmitted in subframes #0 not carrying NSSS, or in subframe #4, depending on the SIB1-NB scheduling information provided in the MIB-NB. In the case of non-anchor, subframes #0 and #5 are used for SIB1-NB transmissions and the location of the nonanchor carrier used for SIB1-NB transmissions is indicated in the MIB-NB.

The information about the carrier used for transmitting all the other types of system information blocks is provided in SIB1-NB.

7.2.8.7 Uplink transmission gaps

As mentioned in Section 7.2.7.2, in an FDD system, uplink transmission gaps are inserted during long NPUSCH and NPRACH transmissions to allow a device to synchronize to a downlink reference signal (e.g. NPSS, NSSS, or NRS) for re-calibrating its frequency and time references. In a TDD system, the UL-DL configurations listed in Table 7.20 all have a number of downlink subframes in every radio frame. The device can use some of these downlink subframes to perform frequency and time reference calibration whenever necessary. Therefore, uplink transmission gaps are not needed in a TDD system.

7.3 Idle and connected mode procedures

In this section, we describe NB-IoT idle and connected mode procedures and higher layer protocols, including all activities from initial cell selection to initiating a connection and transmitting and receiving data during the connection. This section shows how the physical layer described in Section 7.2 is utilized for supporting idle and connected mode procedures.

Like in the case of LTE-M, the idle mode procedures employed in NB-IoT include the initial cell selection, SI acquisition, cell reselection, paging procedures and PSM operation. The transition from idle to connected mode involves the procedures for random access, access control, connection establishment, and multicast. The connected mode operation includes procedures for scheduling, retransmission, power control and positioning. Idle mode procedures and connected mode procedures are described in Sections 7.3.1 and 7.3.2, respectively.

The NB-IoT radio protocol stack is inherited from LTE and is described in Section 2.2. Throughout this section, we will in most cases use an FDD network as an example to describe the radio access procedures.

7.3.1 Idle mode procedures

7.3.1.1 Cell selection

The main purpose of cell selection is to identify, synchronize to, and determine the suitability of an NB-IoT cell. Besides the *initial cell selection*, there are also the procedures of *noninitial cell selection* and *cell reselection*. For NB-IoT, cell reselection is used to support idle mode mobility and is described in Section 7.3.1.3. From a physical layer perspective, one of the main differences between initial and noninitial cell searches is the magnitude of carrier frequency offset (CFO) that the device has to deal with when synchronizing to a cell.

The initial cell selection is carried out by the device before it possesses any knowledge of the network and before any prior synchronization to the network has taken place. This corresponds to, e.g., the *public land mobile network* and cell search performed by the device on being switched on the first time. In this case, cell selection needs to be achieved in the presence of a large CFO because of the possible initial oscillator inaccuracy of the device. The initial oscillator inaccuracy for a low-cost device module may be as high as 20 ppm. Thus, for a 900 MHz band, the CFO may be as high as 18 kHz ($900 \cdot 10^6 \cdot 20 \cdot 10^{-6}$). Furthermore, as explained in Section 7.2.1, as the device searches for a cell on the 100 kHz raster grid, there is a raster offset of ±2.5 or ±7.5 kHz for the in-band and guard-band deployments. Thus, the magnitude of total initial frequency offset for in-band and guard-band deployments can be as high as 25.5 kHz. A brute-force search on the 100 kHz raster grid is however a painstakingly slow process. Release 15 introduces EARFCN [15] pre-provisioning, where the device is provided with information about a list of frequency candidates of NB-IoT anchor carriers. This information helps the device speed up the initial cell search process.

The noninitial cell selection, also known as *stored information cell selection*, is carried out by the device after previous synchronization to the network has taken place, and the device possesses stored knowledge of the network. After the device has synchronized to the network, it has, e.g., resolved the raster offset and has corrected its initial oscillator inaccuracy. In this case, the CFO may be smaller compared with that during initial cell selection. One example for the device to perform the noninitial cell selection procedure is when the device's connection to the current cell has failed and it needs to select a new cell or when the device wakes up from sleep.

The general steps in the NB-IoT cell selection procedure are as follows:

1. Search for the NPSS to identify the presence of an NB-IoT cell.
2. Synchronize in time and frequency to the NPSS to identify the carrier frequency and the subframe structure within a frame.
3. Identify the PCID and the three LSBs of the SFN by using the NSSS. The relative subframe between NSSS and NPSS is used to identify whether the cell is an FDD or TDD cell.
4. Acquire the MIB-NB to identify the complete SFN as well as the two LSBs of H-SFN, and resolve the frequency raster offset. MIB-NB further provides information concerning how the SIB1-NB is transmitted. In a TDD cell, the MIB-NB includes information whether the SIB1-NB is transmitted on the anchor or a non-anchor carrier. In the FDD case, the SIB1-NB is always transmitted on the anchor carrier.
5. Acquire the SIB1-NB to identify the complete H-SFN, the PLMN, tracking area, and cell identity and to prepare for verification of the cell suitability.

These procedures are described in the next few sections in detail.

7.3.1.1.1 Time and frequency synchronization

The first two steps in the initial cell selection procedure aim to time-synchronize to NPSS and obtain a CFO estimation. In principle, they can be combined into one step of joint time and frequency synchronization. However, joint time and frequency synchronization is more costly in terms of receiver complexity. For low-end IoT devices, it is easier to achieve NPSS time synchronization first, in the presence of CFO, and once the NPSS time synchronization is achieved, the device can use additional occurrences of NPSS for CFO estimation. As shown in Figs. 7.11 and 7.39, NPSS is transmitted in subframe #5 in every frame in both FDD and TDD systems, and by time synchronizing to NPSS the device detects subframe #5 and consequently all the subframe numbering within a frame. NPSS synchronization can be achieved by correlating the received signal with the known NPSS sequence or by exploiting the autocorrelation properties of NPSS. As described in Section 7.2.4.2.1, NPSS uses a hierarchical sequence structure with a base sequence repeated according to a cover code. The NPSS detection algorithm can be designed to exploit such structure to achieve time synchronization to NPSS. Interested readers can refer to Ref. [24] for more details. Once the device achieves time synchronization to NPSS, it can find NPSS in the next frame and use it for CFO estimation. In coverage-limited condition, these two steps may rely on accumulating detection metrics over many NPSS subframes.

As described in Section 7.2.1.1, for in-band and guard-band deployments, there is a frequency raster offset referred to as the frequency separation between the 100 kHz raster grid, which is the basis of device searching for an NB-IoT carrier, and the actual center frequency of an NB-IoT anchor carrier. This frequency raster offset is, however, unknown to the device before the initial cell selection. The CFO contributed by the oscillator inaccuracy

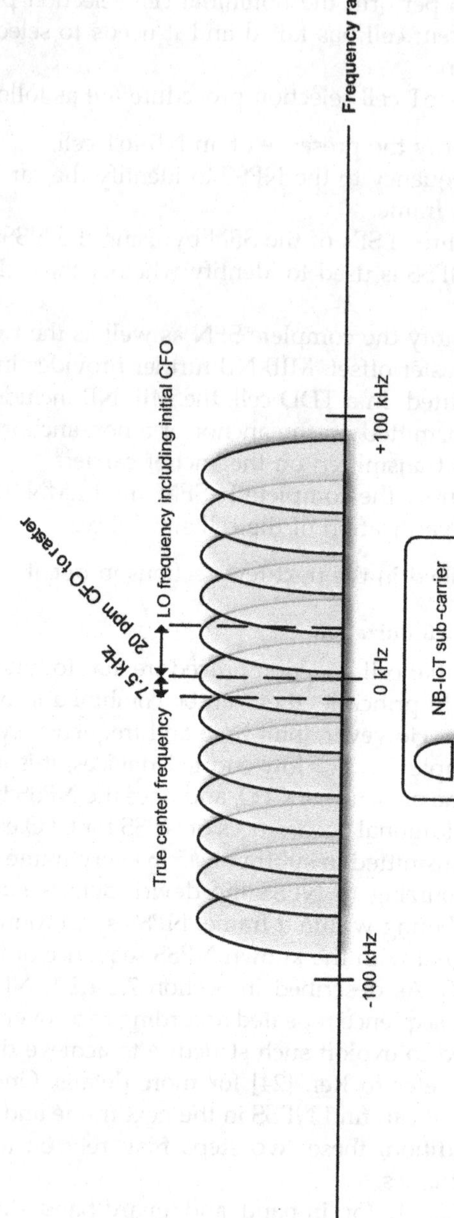

FIG. 7.44 Illustration of over-CFO estimation due to the raster offset in the in-band and guard-band deployments.

is, at this stage, relative to the raster grid. As illustrated in Fig. 7.44, this may result in a needed correction of the local oscillator exceeding the oscillator-induced CFO. The additional correction is equivalent to the raster offset and thus has a magnitude 2.5 or 7.5 kHz for in-band and guard-band deployments. This means that the initial cell selection algorithm needs to be robust in the presence of a raster offset up to 7.5 kHz.

7.3.1.1.2 Physical cell identification and initial frame synchronization

The NSSS transmissions are mapped to subframe #9 in an FDD cell, or subframe #0 in a TDD cell, in every even-numbered radio frame. After NPSS synchronization, the device knows where subframe #9 or #0 is, but does not know whether a frame is even- or odd-numbered. As described in Section 7.2.4.2.2, the NSSS waveform also depends on the SFN, i.e., SFN mod $8 = 0$, 2, 4, or 6. Furthermore, NSSS waveforms depend on the PCID. A straightforward NSSS detection algorithm is therefore to form $504 \times 8 = 4032$ hypotheses, where 504 equals the number of PCIDs used in a NB-IoT network. Each hypothesis corresponds to a hypothesized NSSS waveform during the NSSS detection period. Correlating the received signal with the NSSS waveforms based on each of these hypotheses would allow the device to detect the PCID and the three LSBs of SFN, essentially the 80 ms framing structure. In coverage-limited condition, NSSS detection may rely on accumulating detection metrics over multiple NSSS repetition intervals. For a device interested in searching for both FDD and TDD cells, it needs to repeat the NSSS detection process on subframes #9 and #0. When NSSS is successfully detected, the device knows whether the detected cell is an FDD or TDD cell.

7.3.1.1.3 MIB acquisition

After acquiring the PCID, the device knows the NRS placement within a resource block as the subcarriers that NRS REs are mapped to are determined by PCID. It can thus demodulate and decode NPBCH, which carries the NB-IoT MIB, often denoted as MIB-NB. One of the information elements carried in the MIB-NB is the four most significant bits (MSBs) of the SFN. Because the SFN is 10-bit long, the four MSBs of SFN change every 64 frames, i.e., 640 ms. As a result, the TTI of NPBCH is 640 ms. A MIB-NB is encoded to an NPBCH code block, consisting of eight CSBs. See Section 7.2.4.4. NPBCH is transmitted in subframe #0 in an FDD cell, or subframe #9 in a TDD cell, and each NPBCH subframe carries a CSB. A CSB is repeated in eight consecutive NPBCH subframes. Thus, by knowing the 80 ms frame block structure (after synchronizing to NSSS), the device knows which NPBCH subframes carry an identical CSB. It can combine these repetitions to improve detection performance in coverage extension scenarios. However, the device does not know which one of the eight CSBs is transmitted in a specific 80 ms interval. Therefore, the device needs to form eight hypotheses to decode a MIB-NB during the cell selection process. This is referred to as NPBCH blind decoding. In addition, to correctly decode the NPBCH CRC, the device needs to hypothesize whether one or two antenna ports are used for transmitting NPBCH. There is a CRC mask applied to the CRC bits of NPBCH, and the mask is selected according to 1 or 2 transmit antenna ports used for NPBCH, NPDCCH, NPDSCH transmissions. Therefore, in total 16 blind decoding trials are needed. A successful MIB-NB decoding is indicated by having a correct CRC.

When the device can successfully decode NPBCH, it acquires the 640 ms NPBCH TTI boundaries. Afterward, from MIB-NB the device also acquires the information listed below:

- Operation mode (stand-alone, in-band, guard-band).
- In case of in-band and guard-band, the frequency raster offset (±2.5, ±7.5 kHz).
- Four MSBs of the SFN.
- Two LSBs of the H-SFN.
- Information about SIB1-NB scheduling.
- SI value tag, which is essentially a version number of the SI. It is common for all SIBs except for *SI Block Type 14* (SIB14-NB) and *SI Block Type 16* (SIB16-NB).
- Access Barring (AB) information which indicates whether AB is enabled, and in that case, the device shall acquire a specific SI (i.e., SIB14-NB, see Section 7.3.1.2.3) before initiating RRC connection establishment or resume.

The operation mode information further indicates, in the in-band case, how the NB-IoT cell is configured compared with the LTE cell. Such an indication is referred to as *same PCI indicator*. If the *same PCI indicator* is set true, the NB-IoT and LTE cells share the same PCID, and NRS and CRS have the same number of antenna ports. One implication is that in this case, the device can also use the LTE CRS for channel estimation. As a result, when the *same PCI indicator* is set true, MIB-NB further provides information about the LTE CRS sequence. On the other hand, if the *same PCI indicator* is set false, the device still needs to know where the CRSs are in a PRB. For the in-band mode, there is a required relationship between the PCIDs of LTE and NB-IoT in that they need to point to the same subcarrier indexes for CRS and NRS. Thus, the device already knows the subcarrier indexes for CRS by knowing the PCID of the NB-IoT cell. However, it does not know whether the LTE CRS has the same number of antenna ports as NRS or it has four antenna ports. Knowing this is important as NPDCCH and NPDSCH are rate-matched according to the number of CRS REs in a PRB. The information concerning whether the LTE cell uses four antenna ports or not is carried in the MIB-NB.

7.3.1.1.4 Cell identity and H-SFN acquisition

After acquiring the MIB-NB, including the scheduling information about SIB1-NB, a device is able to locate and decode SIB1-NB. We will describe more about how the device acquires SIB1-NB in Section 7.3.1.2.1. From a cell selection perspective, it is important to know that the SIB1-NB carries the eight MSBs of the H-SFN, the PLMN, tracking area, and a 28-bit long cell identity, which is used to unambiguously identify a cell within a public land mobile network. Thus, after acquiring SIB1-NB, the device has achieved complete synchronization to the frame structure shown in Fig. 7.7. Based on the cell identifiers, the device is able to determine if it is allowed to attach to the cell. It will finally be able to evaluate the suitability of the cell against a pair of minimum required signal strength and signal quality threshold parameters broadcasted in SIB1-NB. A suitable cell is a cell that is sufficiently good to camp on, but it is not necessarily the best cell. The cell reselection procedure described in Section 7.3.1.3 supports the selection of the best available cell for optimizing link and system capacity.

Fig. 7.45 illustrates and summarizes how the device acquires complete framing information during the initial cell selection procedure.

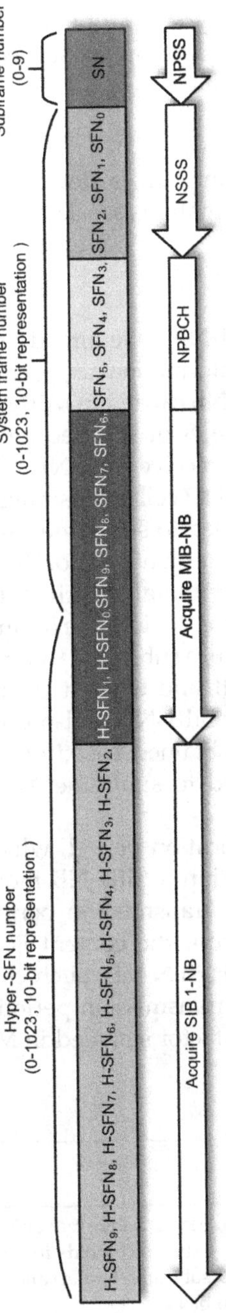

FIG. 7.45 Illustration of how the device acquires complete timing information during the initial cell selection.

After completing the initial cell selection, the device is expected to have a time-accuracy within a few microseconds and a residual frequency offset within 50 Hz. In essence, the device has achieved time and frequency synchronization with residual errors that will not result in significant performance degradation in subsequent transmission and reception during idle and connected mode operations.

7.3.1.2 SI acquisition

After selecting a suitable cell to camp on, a device needs to acquire the full set of SI messages. This procedure and the information associated with the eleven NB-IoT SI messages are presented in the next few sections.

7.3.1.2.1 System Information Block Type 1

The content and importance of SIB1-NB have already been indicated in Section 7.3.1.1.4. Table 7.23 presents in more detail the information acquired by the device on reading SIB1-NB.

The scheduling information of SIB1-NB carried in MIB-NB describes the TBS (208, 328, 440, or 680 bits) and number of repetitions (4, 8, or 16) used for SIB1-NB transmissions. With such information, the device knows how to receive SIB1-NB.

The transmission of SIB1-NB in an FDD cell is illustrated in Fig. 7.46. A SIB1-NB TB is carried in eight SIB1-NB subframes, mapped to subframe #4 in every other frame during a 16-frame interval. These 16 frames are repeated 4, 8, or 16 times. The repetitions are evenly spread over the SIB1-NB transmission interval, which is defined as 256 frames, i.e., 2.56 s.

To reduce SIB1-NB acquisition time for devices at extremely poor coverage locations, 3GPP Release 15 introduces an option to also use subframe #3 for SIB1-NB transmissions in an FDD cell. This option is indicated in MIB-NB, and when it is enabled every subframe #3 immediately preceding a subframe #4 carrying SIB1-NB is also used for SIB1-NB transmissions. The codeword mapped to subframe #3 is obtained based on a rate-matching process, which in essence extends the codeword mapped to subframe #4 by continuing the circular buffer reading process, see Ref. [8].

There is a notion of SIB1-NB modification period, which equals to 40.96 s. The SIB1-NB content is not supposed to change within a SIB1-NB modification period. Thus, the same SIB1-NB TB is repeated in all SIB1-NB transmission periods within a modification period. In the next SIB1-NB modification period, the content is allowed to change. However, in practice, excluding the changes of the H-SFN bits, such changes occur rarely.

The starting frame for SIB1-NB in a transmission period depends on the PCID as well as the aforementioned SIB1-NB repetition factor signaled in MIB-NB. The motivation of having

TABLE 7.23 System information block.

System information block	Content
SIB1-NB	Hyperframe information, network information such as PLMN, tracking area and cell identities, access barring status, thresholds for evaluating cell suitability, TDD configuration, valid subframe bitmap and scheduling information regarding other system information blocks.

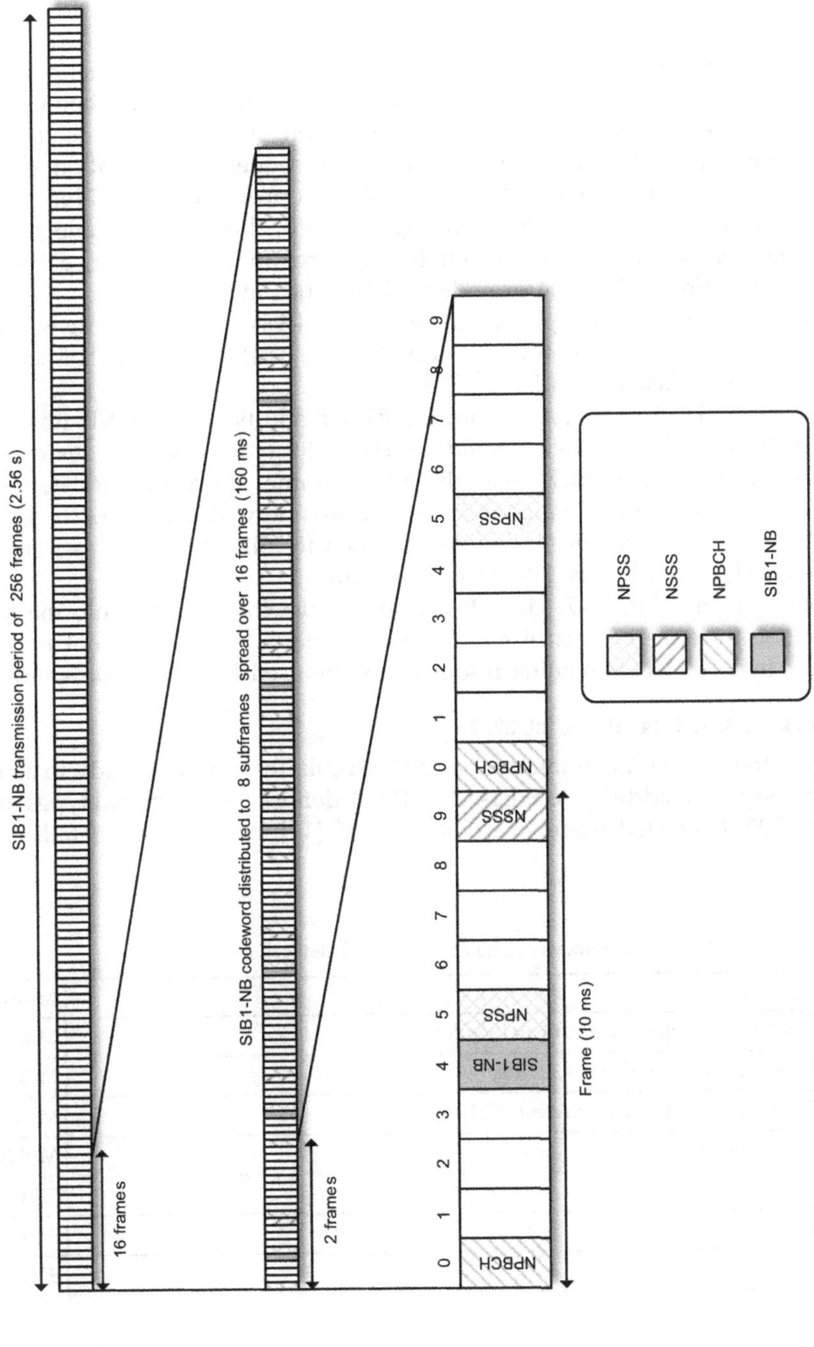

FIG. 7.46 Transmission of Narrowband System Information Block Type 1 (SIB1-NB) in an FDD cell.

different starting frames in different cells is to randomize interference and avoid persistent intercell interference between the transmissions of SIB1-NB in different cells.

7.3.1.2.2 Information specific to in-band mode of operation

For the in-band mode, an NB-IoT device needs to acquire certain LTE carrier parameters to be able to know which resources are already taken by LTE, and in cases when CRS may be used for assisting measurements and channel estimation, the device needs to know the relative power between CRS and NRS as well as the sequence information of CRS. As explained in Section 7.3.1.1.3, some of this information is provided by MIB-NB. Additional information is carried in SIB1-NB. By acquiring SIB1-NB, the device is able to have the complete information it needs to figure out the resource mapping within a PRB for the case of in-band operation. Table 7.24 provides a list of such information.

The valid subframe bitmap as explained in Section 7.2.4.1 is used to indicate which downlink subframes within a 10 or 40 subframes interval can be used for NB-IoT. This can be used, for example, to avoid collision with LTE MBSFN.

When a Release 15 cell configures subframe #3 for additional SIB1-NB transmissions, subframe #3 will be declared as an invalid subframe to avoid a legacy device mapping NPDCCH or its scheduled NPDSCH to a SIB1-NB subframe. However, a Release 15 device knows exactly which subframes #3 are used for SIB1-NB and which are not. Thus, a Release 15 device will still treat subframes #3 that are not used for SIB1-NB transmissions as valid, and map NPDCCH or its scheduled NPDSCH to them.

An example is given in Fig. 7.47. The NB-IoT devices use the provided information about LTE configuration in terms of control region, antenna ports, and CRS to not make use of resources taken by LTE. The remaining resources can be used for NB-IoT downlink.

7.3.1.2.3 SI blocks 2, 3, 4, 5, 14, 15, 16, 20, 22, 23

After reading the SI scheduling information in SIB1-NB, the device is ready to acquire the full set of SI messages. In addition to SIB1-NB, NB-IoT defines ten additional types of SIBs as listed in Table 7.25. Interested readers can refer to Ref. [11] for additional details regarding these SIBs.

TABLE 7.24 LTE configuration parameters signaled to NB-IoT devices.

Information	Message
Whether the NB-IoT cell is configured as in-band mode	MIB-NB
If the same PCID is used for NB-IoT and LTE cells	MIB-NB
Number of LTE antenna ports (in case different PCIDs)	MIB-NB
CRS sequence information	MIB-NB
Valid subframe bitmap	SIB1-NB
LTE control region size	SIB1-NB
NRS to CRS power offset	SIB1-NB

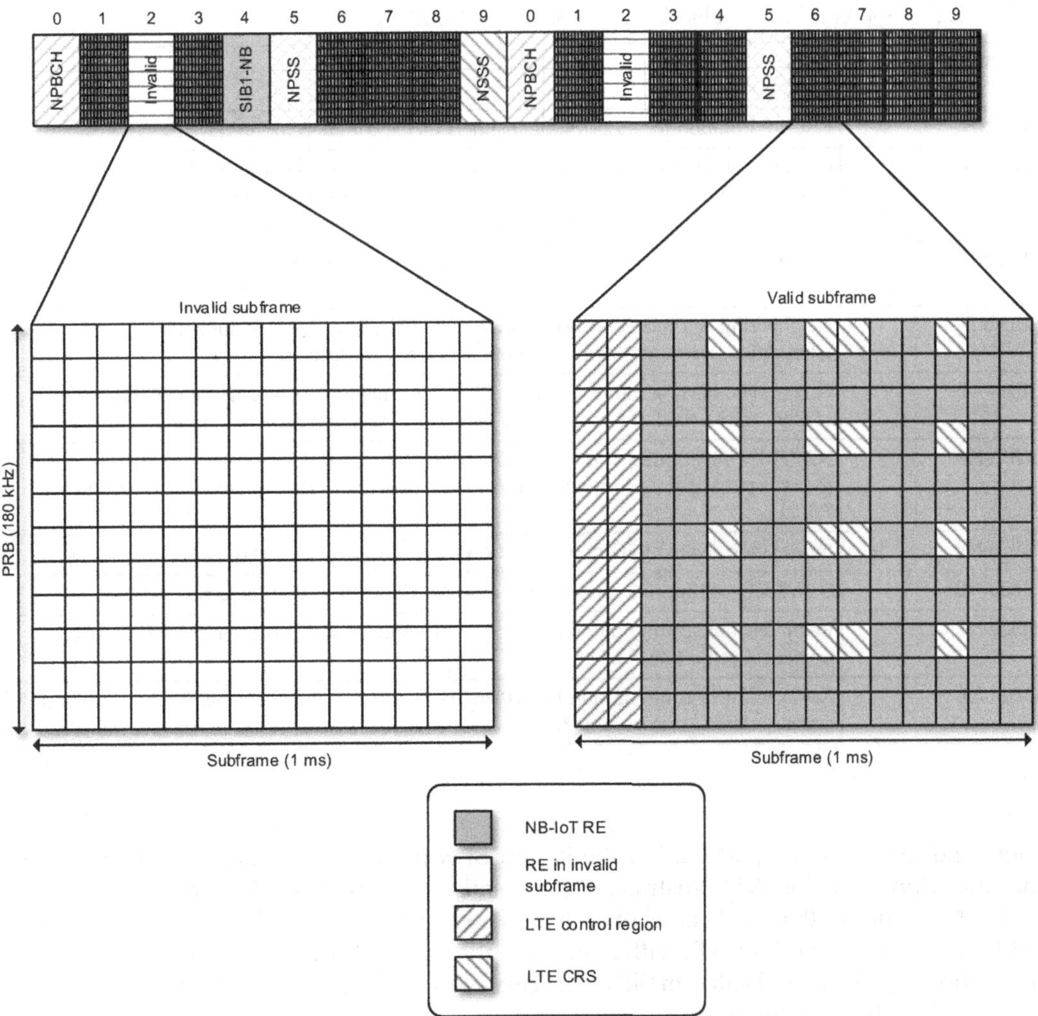

FIG. 7.47 An example of LTE configuration information provided to NB-IoT devices for determining the resource mapping at the physical layer.

SIBs 2, 3, 4, 5, 14, 15, 16, 20, 22, 23 are periodically broadcasted during specific time-domain windows known as the *SI windows*. SIB1-NB configures a common SI window length for all SIBs and schedules periodic and nonoverlapping occurrences of the SI windows.

To support a variable content across the SI messages as well as future extension of the messages, each SI message is configured with a TBS selected from the set of {56, 120, 208, 256, 328, 440, 552, 680} bits. While the two smallest sizes are mapped over two consecutive NB-IoT subframes, the six largest are mapped over eight consecutive NB-IoT subframes.

Furthermore, to support operation in extended coverage, a configurable repetition level of the SIBs is supported. Each SIB can be configured to be repeated every second, fourth, eighth, or 16th radio frame. The total number of repetitions depends on the configured SI window

TABLE 7.25 System information blocks 2, 3, 4, 5, 14, 15, 16, 20, 22, 23.

System information block	Content
SIB2-NB	RRC information for all physical channels that is common for all devices.
SIB3-NB	Cell reselection information that is common for intrafrequency and interfrequency cell reselection. It further provides additional information specific to intrafrequency cell reselection such as cell suitability related information.
SIB4-NB	Neighboring cell-related information, e.g., cell identities, relevant only for intrafrequency cell reselection.
SIB5-NB	Neighboring cell-related information, e.g., cell identities and cell suitability-related information, relevant only for interfrequency cell reselection.
SIB14-NB	Access class barring information per PLMN. Contains a specific flag for barring of a specific access class. It also indicates barring of exception reporting. (see Section 7.3.1.9)
SIB15-NB	SC-PTM related information, including a list of additional frequency bands applicable for the cells participating in the SC-PTM transmission. SC-PTM is described in Section 7.3.1.11.
SIB16-NB	Information related to GPS time and *Coordinated Universal Time* (UTC).
SIB20-NB	SC-MCCH configuration information. SC-MCCH is described in Section 7.3.1.11.
SIB22-NB	Radio resource configuration for paging and random access procedures on non-anchor carriers. (See Section 7.3.1.10)
SIB23-NB	Radio resource configuration for NPRACH resources using preamble Format 2 on non-anchor carriers. (See Section 7.2.5.1)

length and the repetition pattern. For the largest SI window of 160 frames, up to 80 repetitions are supported. Fig. 7.48 illustrates the transmission of an NB-IoT SI message.

SIB1-NB indicates the latest status of each of the scheduled SIBs. A change in the content of a SIB is indicated in the MIB-NB, with a few exceptions, as described in Section 7.3.1.2.4, and may optionally also be indicated in SIB1-NB. This allows a device to determine if a specific SI block needs to be reacquired.

7.3.1.2.4 SI update

The value tag (5 bits) provided in MIB-NB serves as content version number and is to be considered valid for 24 h by the devices. Different version numbers correspond to different SI contents. When the SI has changed, the network can explicitly notify the devices. A device in the idle mode with a DRX cycle shorter than the modification period has the option to monitor the paging messages for SI change notifications. Paging will be discussed in Section 7.3.1.4. There might be cases where the device has not been notified about the SI changes before it attempts to access the network, e.g., in case the device has been configured with a DRX cycle longer than the modification period, in case the device uses PSM (see Section 7.3.1.5) during idle mode, or in case the base station does not transmit any paging at SI change. Such a potential problem is addressed by requiring the device to always read the

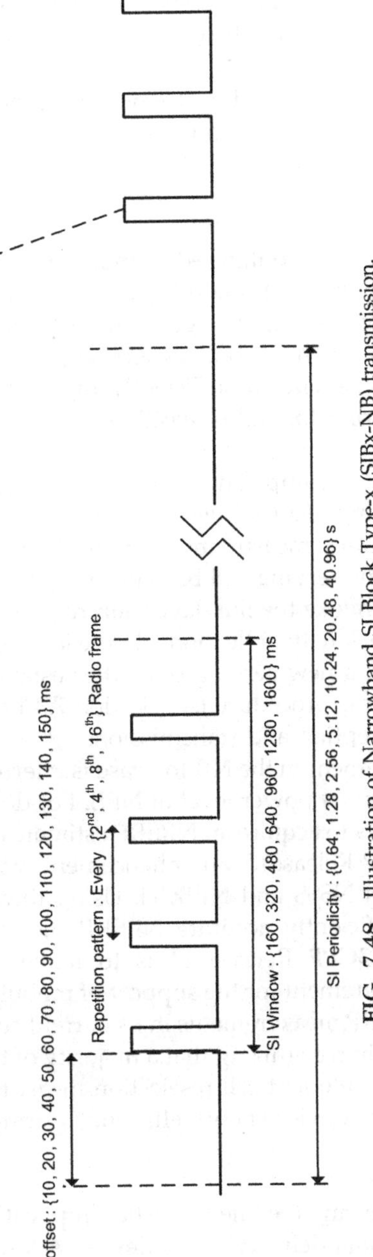

FIG. 7.48 Illustration of Narrowband SI Block Type-x (SIBx-NB) transmission.

MIB-NB on an access attempt. By reading the MIB-NB, the device will realize that there is an SI change through the SI value tag. This also allows the device to acquire information regarding the barring status as explained in Section 7.3.1.9. Note, however, that changes in SIB14-NB (AB parameters) and SIB16-NB (information related to GPS time and UTC) will not result in an SI value tag change. This is because the information carried in SIB14-NB and SIB16-NB changes more frequently, but has no implication or relationship to other system information. It is therefore advantageous to make such changes independent of the SI value tag.

7.3.1.3 Cell reselection

After selecting a cell, a device is configured to monitor up to 16 intrafrequency neighbor cells and up to 16 neighbor cells on an interfrequency. In simple words, in case the device detects that a neighbor cell has become stronger in terms of the *NRS received power* (NRSRP) than the currently serving cell, then the cell reselection procedure is triggered. Devices that are in good coverage, i.e., experience a sufficiently high NRSRP level in the serving cell, can be excluded from measuring for cell reselection. This helps improve the battery life of these devices.

Besides securing that a device camps on the best cell, the cell reselection procedure is the main mechanism for supporting *idle mode mobility*. During the connected mode, the device does not need to perform mobility measurements on the serving or on the neighboring cells. In case the signal quality of the serving cell becomes very poor, resulting in persistent link-level failures, the device will invoke the link-layer failure procedure, which in essence moves it from the connected mode back to the idle mode. In the idle mode, the device can use the cell reselection mechanism to find a new serving cell. After establishing a new serving cell, the device can start a random access procedure (see Section 7.3.1.6) to get back to the connected mode to complete its data reception and transmission.

The cell reselection measurement in the NB-IoT case is referred to as NRSRP measurement, as it intends to reflect the received power level of NRS. For devices in poor coverage, it may take many downlink subframes to acquire an NRSRP estimate of required accuracy. This consumes device energy. In 3GPP Release15, an enhancement was introduced to allow NRSRP measurements to be based on NSSS and NPBCH. Using these additional signals, the time required for obtaining a sufficiently accurate NRSRP measurement is reduced. Another enhancement introduced in 3GPP Release 14 is to make it possible for the device to completely skip NRSRP measurement for the support of mobility during 24 h after the device confirms the variation in NRSRP measurements in its current cell is less than a threshold. This enhancement was introduced in recognizing that a majority of the NB-IoT devices are stationary, and therefore performing frequent cell reselection measurement is not necessary.

These features help improve device energy efficiency during idle mode.

7.3.1.4 Paging, DRX and eDRX

The monitoring of paging during the idle mode has implications on device battery lifetime and the latency of downlink data delivery to the device. A key to determining the impact is how often a device monitors paging. NB-IoT does just as LTE use *search spaces* for defining paging transmission opportunities. The search space concept, including the *Type-1 common search space* (CSS) implementation used for paging indication, is covered in detail in Section 7.3.2.1.

For now, it is sufficient to note that a device monitors a set of subframes defined by the Type-1 CSS to detect an NPDCCH containing a DCI of format N2 that schedules a subsequent NPDSCH containing a paging message addressed to the device. The P-RNTI is the identifier used to address a device for the purpose of paging and is, as described in Section 7.2.4.5, used to scramble the NPDCCH CRC.

The starting subframe for a Type-1 CSS candidate is determined from the location of NB-IoT PO subframe, which is determined based on the configured DRX cycle [25]. If the starting subframe is not a valid NB-IoT downlink subframe, then the first valid NB-IoT downlink subframe after the PO is the starting subframe of the NPDCCH repetitions. The Type-1 CSS candidates are based on only NPDCCH AL 2 described in Section 7.2.4.5. A search space contains NPDCCH candidates defined for repetition levels R up to a configured maximum NPDCCH repetition level R_{max}. R_{max} is typically configured to secure that all devices in a cell can be reached by the paging mechanism, and the relation between the possible repetition levels R for a configured R_{max} is given by Table 7.26.

Fig. 7.49 illustrates possible paging configurations in NB-IoT. Either DRX or eDRX can be used. In case of DRX, the POs occur with a periodicity of at most 10.24 s. For eDRX, the longest eDRX period is 2 h, 54 min, and 46 s, which corresponds to one hyperframe cycle. After each eDRX cycle a *paging transmission window* starts during which downlink reachability is achieved through the configured DRX cycle.

NB-IoT when introduced in 3GPP Release 13 only supported paging on the anchor carrier. To enable NB-IoT to support mobile terminated reachability for a very high number of users, paging support was added to non-anchor carriers in 3GPP Release 14. Radio resource

TABLE 7.26 NPDCCH Type-1 common search space (CSS) candidates.

R_{max}	R
1	1
2	1, 2
4	1, 2, 4
8	1, 2, 4, 8
16	1, 2, 4, 8, 16
32	1, 2, 4, 8, 16, 32
64	1, 2, 4, 8, 16, 32, 64
128	1, 2, 4, 8, 16, 32, 64, 128
256	1, 4, 8, 16, 32, 64, 128, 256
512	1, 4, 16, 32, 64, 128, 256, 512
1024	1, 8, 32, 64, 128, 256, 512, 1024
2048	1, 8, 64, 128, 256, 512, 1024, 2048

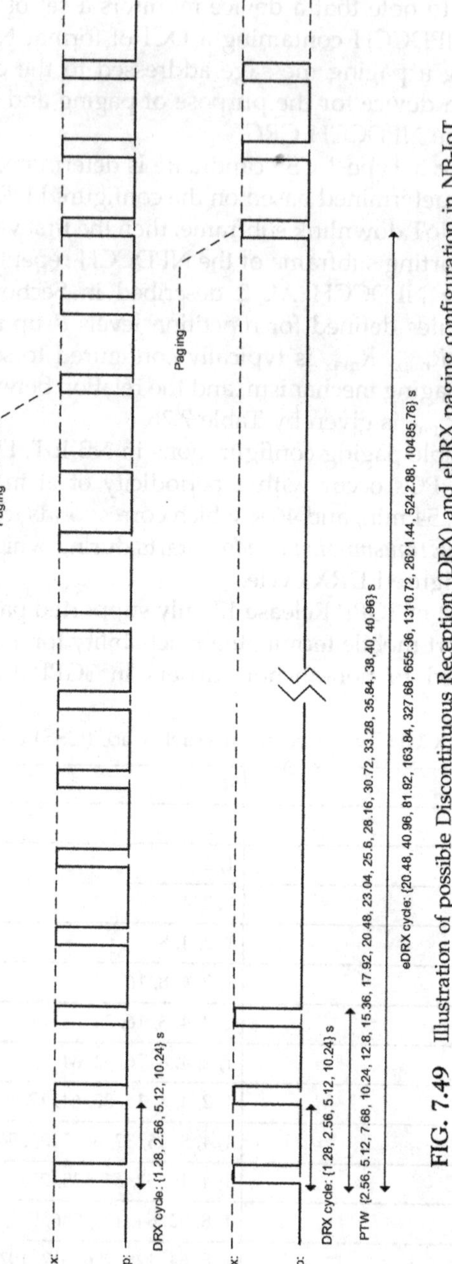

FIG. 7.49 Illustration of possible Discontinuous Reception (DRX) and eDRX paging configuration in NB-IoT.

configuration for paging on non-anchor carriers is provided in SIB22-NB. A device selects its paging carrier based on its device identity.

Device battery lifetime is one of the most important performance objectives that NB-IoT aims to achieve. To further improve device battery lifetime, 3GPP Rel-15 introduced NWUS (see Section 7.2.4.8). The NWUS facilitates device energy saving by indicating whether a paging indicator will be sent in an associated PO. As this in essence is a single bit information (i.e., paging indicator present or not present), the total transmission time required is much shorter than that required for NPDCCH with DCI format N2. The number of repetitions for NWUS, R_{NWUS}, is a fraction of the maximum number of repetitions configured for NPDCCH for paging, R_{max}:

$$R_{NWUS} = \max\{1, \alpha R_{max}\},$$

where $\alpha \in \left\{\frac{1}{2}, \frac{1}{4}, \frac{1}{8}, \frac{1}{16}, \frac{1}{32}, \frac{1}{64}, \frac{1}{128}\right\}$.

Thus, NWUS enables the device to go back to sleep sooner, thereby achieving energy saving. An NWUS can be associated with more than one PO for achieving even bigger energy saving. An example is illustrated in Fig. 7.50, where a device wakes up periodically during time durations P_0, P_1, P_2, and P_3 to monitor for paging, but only finds a paging indicator during duration P_2. In Release 15, an NWUS is used to indicate whether an associated PO will have a paging indicator present. As illustrated in Fig. 7.50, NWUS is transmitted in duration W_2 to indicate a paging indicator will be present in duration P_2. Thus, in this case, the device wakes up during durations W_0, W_1, W_2, and W_3 to look for NWUS, and as it detects NWUS during duration W_2, it goes on to receive the paging indicator during duration P_2. The total duration that the device wakes up in this case is much shorter than the combination of durations P_0, P_1, P_2, and P_3. The time gap between the NWUS and the associated paging occasion (where the NPDCCH starts) is configurable. When NWUS is used with either DRX or eDRX, a "short" gap length of 40, 80, 160 or 240 ms can be configured, but in the eDRX case it is also possible to configure a "long" gap length of 1 or 2 seconds.

The detection of NWUS is very straightforward, and thus it is feasible to have a dedicated, special baseband receiver wake up for NWUS detection and leave the rest of the hardware, including the ordinary receiver, in *deep sleep*. The aforementioned configuration of a longer gap between the NWUS and the associated paging occasion in the eDRX ensures that there is sufficient time for the device's ordinary receiver to start up and be ready to detect the NPDCCH once an NWUS was indeed detected. With such a *wake-up receiver* approach, additional energy efficiency improvement can be achieved. Furthermore, the design of the NWUS waveform and repetition allows a device to perform efficient re-synchronization after it wakes up. This can also contribute to energy efficiency improvement.

7.3.1.5 PSM

Some applications have very relaxed requirements on mobile terminated reachability (e.g. longer than a day) and then power consumption can be further reduced compared with using eDRX, which has a maximum eDRX cycle close to 3 h. For this type of devices, the most energy-efficient state of operation is the power saving mode (PSM) which is available for devices in RRC idle mode.

In PSM a device uses the smallest possible amount of energy. Essentially it only needs to leave its real time clock running for keeping track of time and scheduled idle mode events

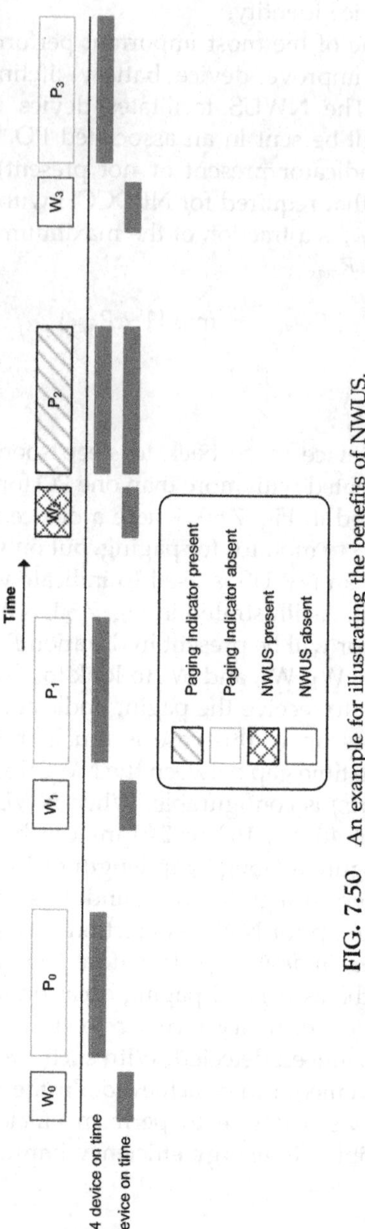

FIG. 7.50 An example for illustrating the benefits of NWUS.

such as the *TAU* timer. The device aborts energy consuming idle mode operations such as mobility measurements in PSM and will not transmit or monitor paging and need not keep an up-to-date synchronization to the network. The duration of PSM is configurable and may be more than a year in the most extreme configuration. The device exits PSM once it has uplink data to transmit or is mandated to send a TAU upon expiry of the TAU timer. The TAU is used as a tool for tracking a device. It informs the network of which cell the device camps on. After its uplink transmission, the device may enter DRX mode for a short period of time, configured by the active timer, for monitoring paging to enable mobile terminated reachability. After such a short duration in which the device monitors paging, the device enters the next PSM period. This procedure is illustrated in Fig. 2.6. PSM was introduced in 3GPP Release 12 as general improvement for devices with high requirements on long battery life and is described in more details in Section 2.3.3.

7.3.1.6 Random access in idle mode

NB-IoT random access procedure is generally the same as LTE. We will in this section mainly highlight aspects that are unique to NB-IoT.

The random access procedure is illustrated in Fig. 7.51. After synchronizing to the network and confirming that access is not barred, the device sends a random access preamble using NPRACH.

The device needs to determine an appropriate NPRACH configuration according to its coverage class estimation. Recall that SIB2-NB carries RRC information as explained in Section 7.3.1.2.3. One of the RRC information elements is the NPRACH configuration. The cell can configure up to two NRSRP thresholds that are used by the device to select the NPRACH configuration appropriate for its *coverage class*. An example is given in Fig. 7.52, in which two NRSRP thresholds are configured and therefore there are three NPRACH configurations for three CE levels, respectively. Essentially, the network uses these NRSRP thresholds to configure the coupling losses associated with each of the different CE levels. If the network does not configure any NRSRP threshold, the cell supports only a single NPRACH configuration used by all devices regardless of their actual path loss to the serving base station.

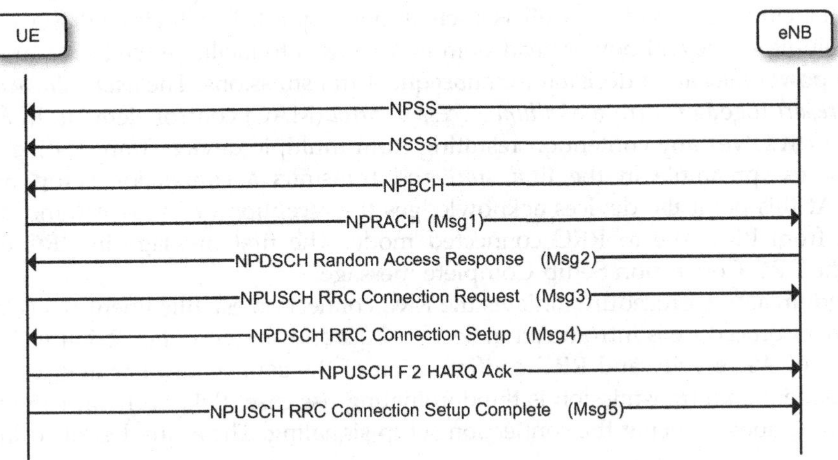

FIG. 7.51 NB-IoT random access procedure.

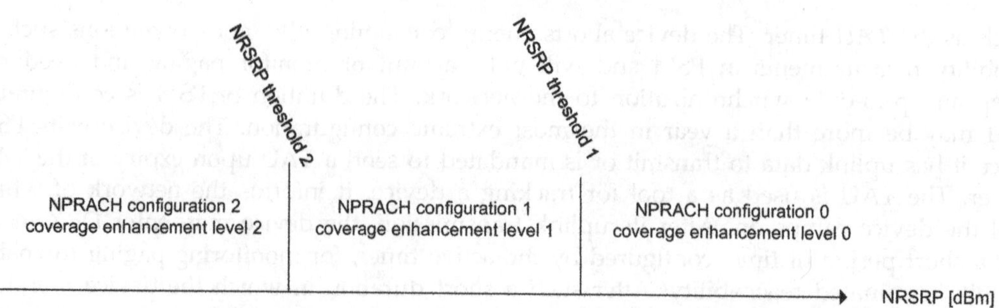

FIG. 7.52 NPRACH configurations and NRSRP thresholds.

The SIB2-NB NPRACH configuration information for each CE level includes the time–frequency resource allocation. The resource allocation in the frequency domain is a set of starting preambles. Each starting preamble is equivalent to the first NPRACH symbol group and associated with a specific 3.75 kHz or 1.25 kHz tone, depending on the NPRACH format configured. The set of starting preambles is determined by a subcarrier offset and a number of subcarriers. The time-domain allocation is defined by a periodicity, a starting time of the period, and the number of repetitions associated with the NPRACH resource. This is exemplified in Fig. 7.53 where an NPRACH configuration intended to support a high access load is illustrated.

This set of starting preambles may further be partitioned into two subsets. The first subset is used by devices that do not support multitone NPUSCH transmissions for *Message 3*, whereas the second subset is used by devices that are capable of multitone transmissions for *Message 3*. In essence, the device signals its NPUSCH multitone support by selecting the NPRACH starting preamble according to its capability.

If the base station detects an NPRACH preamble, it sends back a random access response (RAR), also known as *Message 2*. The RAR contains a *TA* parameter. The RAR further contains scheduling information pointing to the radio resources that the device can use to transmit a request to connect, also known as *Message 3*. Note that at this point the base station already knows the device multitone transmission capability, and thus resource allocation for *Message 3* will account for the device multitone transmission capability. In *Message 3*, the device will include its identity as well as a scheduling request. The device will always include its data volume status and power headroom in *Message 3* to facilitate the base station scheduling and power allocation decision for subsequent transmissions. The *data volume and power headroom report* together form a *Medium Access Control* (MAC) control element. In *Message 4*, the network resolves any contention resulting from multiple devices transmitting the same random access preamble in the first step and transmits a connection setup or resume message. At this point the devices acknowledges the reception of *Message 4* and makes the transition from RRC idle to RRC connected mode. The first message in RRC connected mode is the RRC Connection Setup Complete message.

The random access procedure initiates the RRC connection establishment. The classic LTE establishment procedure is introduced in Section 2.3.2 and shown in Fig. 2.5. It includes both negotiation of *AS* security and RRC configuration of the connection. For devices for which small infrequent data transmission is the dominating use case, Release 13 and 15 introduces three optimizations reducing the connection setup signaling. These are described in the next section.

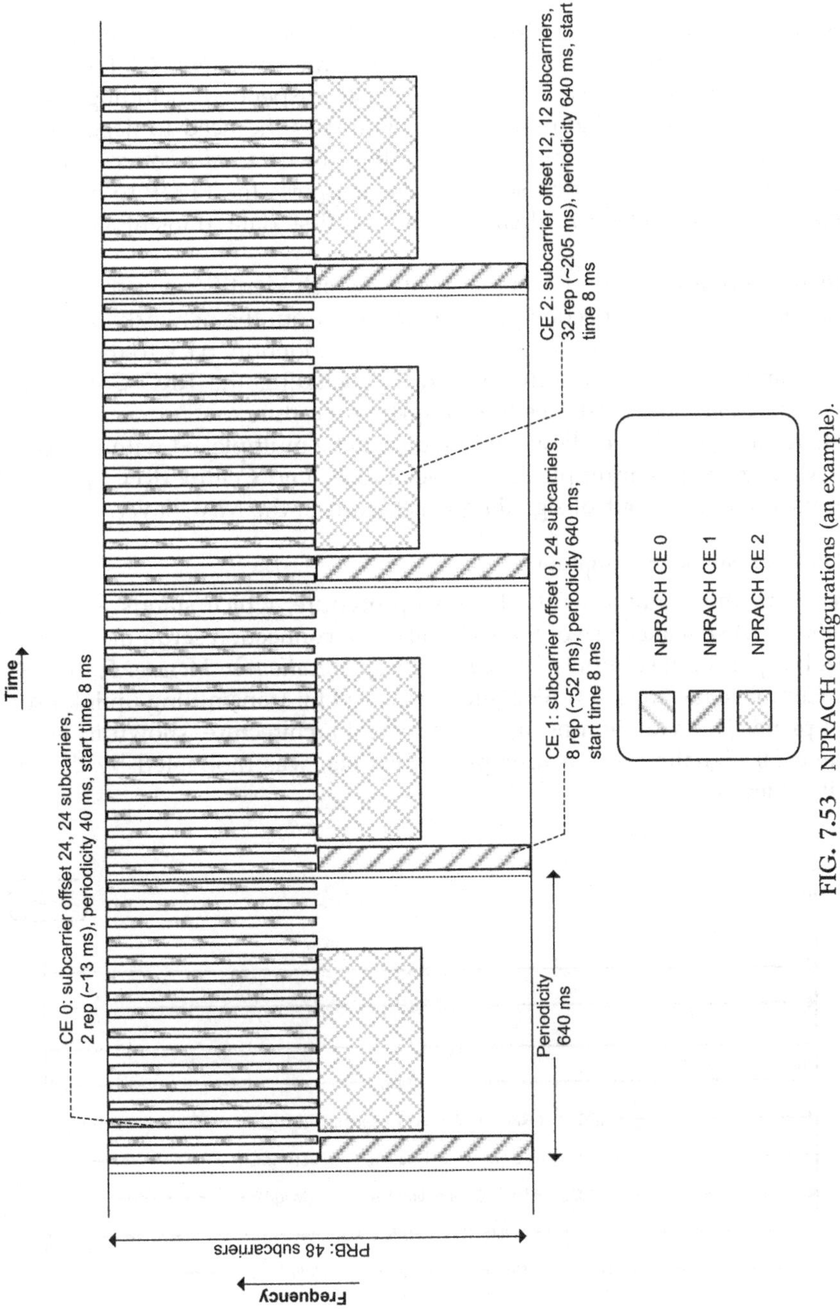

FIG. 7.53 NPRACH configurations (an example).

7.3.1.7 Connection establishment

7.3.1.7.1 RRC resume

The first method for signaling reductions is known as the RRC Suspend Resume procedure, or just the RRC Resume procedure. It is part of the *User Plane CIoT EPS optimizations.* It allows a device to resume a connection previously suspended including the PDCP state, the AS security and RRC configurations. This eliminates the need to negotiate AS security as well as configuring the radio interface, including the *data radio bearers* carrying the data over the air interface, at connection setup. It also supports the PDCP to make efficient use of its *Robust Header Compression* already from the first data transmission in a resumed connection.

This functionality is based on a resume identity which identifies a suspended connection. It is signaled in the RRC Connection Release message from the network to a device when a connection is suspended. The device signals the resume identity back to the network when it wants to resume a connection using *Message 3* including the RRC Connection Resume Request message. Fig. 7.54 illustrates the complete procedure.

The RRC resume procedure allows uplink data to be multiplexed with the RRC signaling already in Message 5. This multiplexing between *Radio Link Control* (RLC) packet data units containing user data and control signaling is achieved in the MAC layer.

7.3.1.7.2 Data over Non-access Stratum

The second method known as the DoNAS procedure, which stands for *Data over Non-Access Stratum (NAS),* makes use of the LTE Release 8 connection setup message flow shown in Fig. 7.51. It is part of the *Control Plane CIoT EPS optimization.* In *Message 5,* the RRC Connection Setup Complete message, a NAS container is used for transmitting uplink user data over the control plane. This violation of the LTE protocol architecture, shown in Fig. 2.3, with a control plane for signaling and a user plane for data, was motivated by the reduction in signaling it achieves.

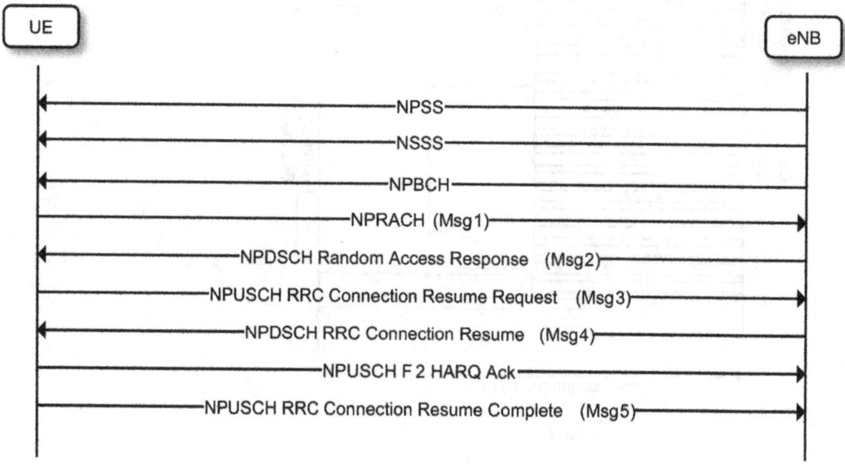

FIG. 7.54 The RRC resume procedure.

For the NAS interface, which terminates in the *Mobility Management Entity* (MME), security is negotiated when a device attaches to a network. This means that data sent in an NAS container is both integrity protected and ciphered. As a result, AS security needs not to be configured during connection establishment. Furthermore, as the NAS signaling is sent over a *signaling radio bearer* (SRB) with a default configuration, it also means that a data radio bearers for user plane data transmission needs not to be configured. To support uplink and downlink transmissions after the connection establishment procedure Release 13 specifies two RRC messages (*UL/DL information transfer*) that only carry a NAS container where user data can be inserted.

Since DoNAS does not provide a secure link over the AS, it does not support a reconfiguration of the RRC parameters configuring a connection. As a consequence, it is important to be able to correctly assess the radio quality and provide an accurate RRC configuration of the radio link already at connection setup. Since DoNAS transfers data over an SRB the quality of service frame work developed for LTE DRBs does also not apply. This simplification was motivated by the assumption that DoNAS is mainly intended for supporting small and short data transmissions.

For Cat-NB devices the support for DoNAS is mandatory. Support for data transmission over the user plane is an optional feature, but advisable to support for providing efficient operation for use cases beyond short infrequent data transmission.

7.3.1.7.3 Early Data Transmission

For a Release 13 or 14 device, uplink data and downlink data can at the earliest be delivered in *Message 5* and *Message 6*, respectively. This leaves room for further enhancements. 3GPP Release 15 introduces a feature called *Early Data Transmission* (EDT) that allows a device to transmit its uplink data in *Message 3*. In this case, the device can complete its data transmission in idle mode without a need to transition to connected mode.

EDT is however limited to small data payloads. Uplink EDT supports a TBS of 328, 408, 504, 584, 680, 808, 936, or 1000 bits. Although these TB sizes are not very large, they are nevertheless significantly larger than the TBS of 88 bits used in a conventional (non-EDT) *Message 3*. The EDT configuration in a cell, including the maximum allowed EDT TBS, is provided in SIB2-NB. A device can use the EDT procedure to transmit its uplink data only when the number of payload bits is less than the maximum TBS permitted. A device indicates its intent to use EDT by initiating a random access procedure with an NPRACH preamble randomly selected from a set of preambles configured for the EDT procedure. In this case, upon detecting the NPRACH preamble, the base station knows that the device attempts to transmit its uplink data through EDT, and therefore can include an EDT uplink grant for *Message 3* in *Message 2*. However, it is also possible for the base station to reject the device's EDT request. The EDT uplink grant includes information about the modulation and coding scheme (MCS) and number of repetitions (R_{max}) associated with the maximum TBS size (TBS_{max}).

As an option, the network may allow the device to select a TBS among a set of allowed TBS values. For example, if the configured maximum EDT TBS is 936 bits, the set of allowed TBS values can be {328, 504, 712, 936} [27]. From this set, the device can select a TBS for *Message 3* based on its data buffer status. This option reduces the amount of padding if the device data buffer has much fewer bits than the configured TBS_{max}. When the device selects a TBS less than TBS_{max}, it needs to reduce the number of repetitions compared to that provided in the uplink grant message. For example, if it selects a lower TBS, e.g. TBS_{max}/K, it reduces the number of repetitions to R_{max}/K. Depending on R_{max}, the reduced repetition factor R_{max}/K needs to be rounded up to either the nearest integer or the nearest integer which is a multiple of 4. To support the option of device-selectable TBS, the base station needs to blindly detect the TBS used in *Message 3* among a number of TBS hypotheses. Depending on the EDT configuration, the number of TBS hypotheses can be 2, 3, or 4.

EDT also support downlink data transmission in *Message 4*. This may be used to provide an application layer acknowledgment to the *Message 3* uplink transmission.

EDT is in Release 15 enabled for mobile originated access both for the User Plane CIoT EPS optimization procedure and Control Plane CIoT EPS optimization procedure. The user plane version builds on the RRC Resume procedure and makes use of the RRC Connection Resume Request in *Message 3*. *Message 4*, including potential downlink data, is defined by the RRC Connection Release message in case the connection is suspended or released immediately following the uplink data transmission. For the control plane solution, a pair of new RRC messages were defined for *Message* 3 and 4. Both of them include the needed NAS container carrying the data on an SRB over the NAS.

7.3.1.8 Channel quality reporting during random access procedure

As described in Section 7.3.1.6, a device initializing the random access procedure needs to determine an appropriate NPRACH configuration according to its coverage class estimation. The coverage estimation is based on NRSRP since the network set thresholds for determining the device coverage class are based on NRSRP. An example is given in Fig. 7.52. In the subsequent downlink messages during the remainder of the random access and connection establishment procedure, the coverage class of the NPRACH selected by the device is used by the network to determine the repetition factors of NPDCCH and NPDSCH.

There are a number of issues with this solution. First, there can be at most three NPRACH configurations. Thus, the network only knows the device's downlink received signal strength coarsely. Second, there might be a pronounced difference between the received signal strength and the downlink received *signal-to-interference-plus-noise ratio* (SINR), which in practice is more relevant for determining the number of repetitions required for NPDCCH and NPDSCH. Note that the downlink SINR can be also different from uplink SINR as the interference may be asymmetric between downlink and uplink. Not being able to more accurately determine the repetition levels needed for NPDCCH and NPDSCH may result in failures in downlink messages during the random access and connection establishment procedure. It also influences the ability of the network to configure the NPDCCH device specific search space periodicity (see Section 7.3.2.1), which will limit the achievable data rate in RRC Connected mode. The network may configure NPDCCH and NPDSCH with a more aggressive repetition factors, although this result in inefficiency in downlink resource utilization. Most NB-IoT devices are based on the Control Plane CIoT EPS optimization solution (see section 7.3.1.7.2) which does not support a reconfiguration of the radio link after the initial configuration. It is therefore of special importance in NB-IoT to start the connection with a correct link configuration.

To address these issues, 3GPP Release 14 introduced a feature that allows the device to include its downlink channel quality estimate as an RRC information element in *Message 3* or *Message 5* during the random access and connection establishment procedure. For *Message*

3 this quality reporting is provided by the device in terms of the number of repetitions needed for NPDCCH to secure reliable operation. The device can, e.g., obtain such an estimate when it receives the *RAR* in *Message 2*, as in that process the device knows how many repetitions it needs to decode the NPDCCH scheduling the RAR. Two reporting formats are supported. The first one allows the device to indicate one of the NPDCCH repetition factors, i.e. 1, 2, 4, 8, 16, 32, 64, 128, 256, 512, 1024, or 2048 (See Section 7.2.4.5). The device can also indicate a *null*, i.e. that no measurement was performed. This requires a 4-bit message. In addition, for some *Message 3* types with a limited space, a short format is used. With the short format, the device indicates its NPDCCH repetition estimate relative to the maximum NPDCCH repetition factor configured for the search space, R_{max}. The reported downlink channel quality can be $R_{max}/8$, R_{max}, or $4R_{max}$. The device can also indicate a *null*, i.e. no measurement reporting. Thus, the short channel quality reporting format requires a 2-bit message.

In *Message 5*, which is not constrained in size in the same way as Message 3, the downlink signal strength and downlink signal quality in terms of NRSRP and NRSRQ are reported by the device to the network. The intention behind this reporting is to allow the network to, in a long term, configure and plan the cells in a network in a suitable manner.

It should be noticed that the reporting described in this section is, together with the power headroom reporting, the HARQ ACK/NACK and the RLC Acknowledged Mode status reports, the only feedback sent from a NB-IoT device to the network.

7.3.1.9 Access control

Access Barring (AB) is an access control mechanism adopted in NB-IoT; it closely follows the Access Class Barring functionality described in Section 2.2 and allows PLMN-specific barring across 10 normal and 5 special access classes. It also supports special treatment of devices intending to transmit an *Exception report*. The Exception report concept was introduced in 3GPP Release 13 to allow a network to prioritize reports of high urgency transmitted by a device.

An AB flag is provided in the MIB-NB. If it is set false, then all devices are allowed to access the network. If the AB flag is set true, then the device must read the SIB14-NB before it attempts to access the network, which provides the just introduced access class–specific barring information. The device needs to check whether its access class is allowed to access the network. In case the device is barred, it should back off and then reattempt access at a later point in time.

A device in bad coverage locations requires a high repetition factor to be configured for its dedicated physical channels. During high network loads, it may be desirable to bar devices in bad coverage locations and use the available resources to serve many more devices in good coverage locations. 3GPP Release 15 introduces coverage-level-specific barring to prevent devices at certain coverage level, or worse, from accessing the network. This is enabled by providing an NRSRP threshold in SIB14-NB. If a device has measured NRSRP below this threshold, it is barred from accessing the network. It should back off and then reattempt access at a later point in time.

It should be noted that when the MIB-NB AB flag toggles, there is no impact on the SI value tag. See the discussion in Section 7.3.1.2.4.

7.3.1.10 System access on non-anchor carriers

The basic design of NB-IoT aimed at supporting at least 60,680 devices per km^2. From 3GPP Release 14 the target was increased to support 1,000,000 devices per km^2 as a consequence of a desire to make NB-IoT a 5G technology. Based on the traffic model described in Section 4.6, these loads can be assumed to correspond to 6.8 and 112.2 access arrivals per second, respectively. To keep the collision rate of competing preambles low, a sufficiently high amount of radio resources needs to be reserved for the NPRACH.

For *slotted* type of access schemes such as the NB-IoT NPRACH a collision probability between competing preambles can be estimated as [26]:

$$\text{Prob}(\text{collision}) = 1 - e^{-\gamma/L} \tag{7.22}$$

where L is the total number of random access opportunities per second and γ is the average random-access intensity. If a collision probability of 1% is targeted for a random-access intensity of 6.8 access arrivals per second, the total number of random access opportunities provided by the system needs to be set at 677 per second. NB-IoT can at most provide 1200 access opportunities per second by configuring a NPRACH resource for CE level 0 spanning 48 sub-carriers and reoccurring with a periodicity of 40 ms. It is clear that already this requires the NPRACH to consume a large portion of the available uplink resources. If a collision rate of 10% are acceptable, it is sufficient to configure 65 access opportunities per second. But in order to support 112.2 access arrivals per second, even a 10% collision rate will require the configuration of 1065 access opportunities per second. This in combination with the required support of devices in extended coverage motivated the introduction of random access on the non-anchor carriers. To also support mobile terminated reachability for a very high number of users, also paging support was added to the non-anchor carriers in 3GPP Release 14. The paging carrier is besides paging used for *Message* 2 and 4 transmissions, while the random-access carrier is used for both *Message* 1 and 3 transmissions. Radio resource configuration for paging and random access procedures on non-anchor carriers is provided in SIB22-NB (see Section 7.3.1.2.3).

One main difference between system access on a non-anchor carrier versus on an anchor carrier is the presence of NRS. More details on this aspect can be found in Section 7.2.4.3.

7.3.1.11 Multicast

In order to offer improved support of firmware or software updates, a group message delivery type of service known as *Single Cell Point to Multipoint* (SC-PTM) was introduced in NB-IoT Release 14. It builds on the *Multimedia Broadcast Multicast Service* architecture and provides the air interface for delivering single-cell broadcast and multicast transmissions. It shares design with LTE-M SC-PTM, described in Section 5.3.1.9.

SC-PTM defines two logical channels: *Single Cell Multicast Control Channel* (SC-MCCH) and *Single Cell Multicast Traffic Channel* (SC-MTCH). These logical channels are mapped on the NPDSCH. The NPDSCH is scheduled by the NPDCCH which is transmitted in two new common search spaces associated with the SC-MCCH and SC-MTCH. The *Type-1A Common Search Space* (CSS) contains NPDCCH candidates, with the CRC scrambled by the *Single Cell RNTI* (SC-RNTI) that schedules the NPDSCH carrying the SC-MCCH. As SC-MCCH is similar to paging, i.e. it is broadcasted to a group of devices in a cell, the design of

Type-1A CSS is based on Type-1 CSS, where the NPDCCH candidates can only start at the beginning of the search space.

Type-2A CSS contains NPDCCH candidates, with the CRC scrambled by the *Group RNTI* (G-RNTI), that schedules the NPDSCH carrying the SC-MTCH. To maintain scheduling flexibility for the SC-MTCH, the design principle of Type-2 CSS is followed by Type-2A CSS, where several starting points are defined for the NPDCCH candidates. The search space concept is described in details in Section 7.3.2.1.

The procedure of acquiring SC-PTM services is illustrated in Fig. 7.55. As shown in Fig. 7.55, if a device is configured by higher layer to receive an SC-PTM service, upon reading SIB1-NB, the device identifies the scheduling information of SIB20-NB. The SIB20-NB contains the information required to acquire the SC-MCCH configurations associated with transmission of SC-PTM in the cell. In SC-MCCH, a device can further find the configuration

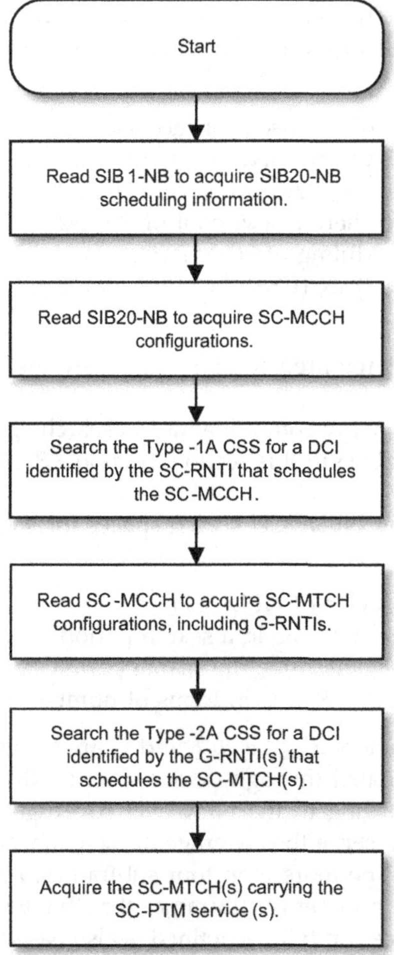

FIG. 7.55 Procedure for acquiring SC-PTM services.

information of the SC-MTCH that carries the multicast service a device is interested in. One SC-MCCH can be configured in a cell, and it may configure up to 64 simultaneous multicast and broadcast services transmitted over the SC-MTCH. Both anchor and non-anchor carriers can be used to carry SC-MCCH and SC-MTCH.

Retransmissions are not supported for the multicast packages, and therefore for each SC-PTM session, there is only a single transmission on the NPDSCH. Due to the high number of repetitions supported by the NPDCCH and NPDSCH, an MCL of 164 dB can still be expected to be supported by NB-IoT SC-PTM services.

7.3.2 Connected mode procedures

7.3.2.1 NPDCCH search spaces

A key concept related to connected mode scheduling as well as idle mode paging is the NPDCCH *search spaces*. A search space consists of one or more subframes in which a device may search for DCI addressed to the device. There are three major types of search spaces defined:

- *Type-1 Common Search Space (CSS),* used for monitoring paging. There is a variant of Type-1 CSS, namely Type-1A CSS, used for monitoring the scheduling of SC-MCCH.
- *Type-2 CSS,* used for monitoring RAR, *Message 3* HARQ retransmissions and *Message 4* radio resource assignments. There is a variant of Type-2 CSS, namely Type-2A CSS, used for monitoring the scheduling of SC-MTCH.
- *User equipment-specific search space* (USS), used for monitoring downlink or uplink scheduling information

The device however is not required to simultaneously monitor more than one type of search space.

Type-2 CSS and USS share many commonalities in search space configurations. Thus, we will focus on Type-2 CSS and USS in this section. Section 7.3.1.4 already presents the use of the Type-1 CSS.

Key parameters for defining NPDCCH search spaces for Type-2 CSS and USS are listed below:

- R_{max}: Maximum repetition factor of NPDCCH
- α_{offset}: Offset of the starting subframe in a search period
- G: Parameter that is used to determine the search period
- T: The search space period, $T = R_{max}G$ in terms of number of subframes

For Type-2 CSS, the parameters R_{max}, α_{offset}, and G are signaled in SIB2-NB, whereas for USS these parameters are signaled through device-specific RRC signaling. For Type-2 CSS, R_{max} should be adapted according to the NPRACH coverage class it is associated to. For USS, R_{max} can be optimized to serve the coverage of the connected device. There is a restriction that a search period must be more than four subframes, i.e., $T > 4$.

Within a search period, the number of subframes that the device needs to monitor is R_{max} and the number of search space candidates defined is also based on R_{max}. Note that the R_{max} subframes that the device needs to monitor within a search period have to exclude the subframes used for transmitting NPBCH, NPSS, NSSS, and SIB1-NB. Furthermore, these subframes need to be NB-IoT subframes according to the valid subframe bitmap described in Sections 7.3.1.2.1. Table 7.27 shows the USS candidates for different R_{max} values and NCCE ALs.

As described in Section 7.2.4.5, the CRC bits attached to DCI are also used for differentiating between different types of DCI and for identifying the device that the DCI is intended for. This is done by scrambling the CRC bits based on the different *Radio Network Temporary Identifiers* (RNTIs). Table 7.28 lists the combinations of NPDCCH search space, DCI format, and RNTI.

Describing the search space concept is done best with a concrete example. We will use a USS example to illustrate all the search space aspects we have discussed up to this point. Consider, for example, a device in coverage conditions requiring the NPDCCH to be transmitted with up to 2 repetitions. R_{max} will in this case be set to 2. It is further assumed that the scheduling periodicity is configured to be eight times longer than the maximum repetition interval, i.e., G is set to 8. Finally, an offset α_{offset} of 1/8 is selected.

With these parameter settings, the search period is $T = R_{max}G = 16$ subframes. Fig. 7.56 illustrates the search periods according to this reference case. For this reference case, the starting subframes are the ones satisfying (SFN $\times$ 10 + SN) mod $T = 0$, when the offset value is set to 0. According to the present example, the offset value is set to 1/8 of the search period, i.e., the starting subframe is shifted by two subframes.

According to Table 7.27, with $R_{max} = 2$, the search space may have NPDCCH repetition factor of $R = 1$ or 2. Furthermore, as indicated in Table 7.27, for the case of $R = 1$, AL 1 may be used and thus both NCCE0 and NCCE1 (see Fig. 7.19 in Section 7.2.4.5) are each individually a search space candidate. For AL 2, NCCE0 and NCCE1 are used jointly as a

TABLE 7.27 NPDCCH UE-specific search space candidates.

R_{max}	R	NCCE indices of monitored NPDCCH Candidates	
		AL = 1	AL = 2
1	1	{0},{1}	{0,1}
2	1	{0},{1}	{0,1}
	2	–	{0,1}
4	1	–	{0,1}
	2	–	{0,1}
	4	–	{0,1}
≥ 8	$R_{max}/8$	–	{0,1}
	$R_{max}/4$	–	{0,1}
	$R_{max}/2$	–	{0,1}
	R_{max}	–	{0,1}

TABLE 7.28 NPDCCH search spaces, RNTIs and DCI formats monitored by NB-IoT devices.

Mode	Search space	RNTI	Usage	DCI format
	–	SI-RNTI	Broadcast of system Information	–
Idle	Type-1 common	P-RNTI	Paging and SI update notification	N2
	Type-1A common	SC-RNTI	SC-PTM control channel scheduling	N2
	Type-2 common	RA-RNTI	Random access response	N1
		TC-RNTI, C-RNTI	Random access contention resolution with message 4	N1
		TC-RNTI, C-RNTI	Message 3 transmission and retransmission	N0
	Type-2A common	G-RNTI	SC-PTM traffic channel scheduling	N1
Connected	Device-specific	C-RNTI	Random access order	N1
		C-RNTI	Dynamic DL scheduling	N1
		C-RNTI	Dynamic UL scheduling	N0
		SPS-C-RNTI	Semi-persistent UL scheduling	N0

search space candidate. All the search space candidates are illustrated in Fig. 7.56, including the following set of seven candidates within a search period:

- Four candidates with $R = 1$ and AL = 1,
- Two candidates with $R = 1$ and AL = 2, and
- One candidate with $R = 2$.

It should be noted that the device needs to monitor a set of search space subframes that are not taken by NPBCH (subframe #0), NPSS (subframe #5), NSSS (subframe #9, in even-numbered SFN), and SIB1-NB.

The search space candidates shown in Table 7.27 to a large extent also applies to Type-2 CSS, with the only difference that Type-2 CSS candidates are only based on AL equal to 2. Furthermore, Type-2 CSS and USS share the same set of values for G, $G \in \{1.5, 2, 4, 8, 16, 32, 48, 64\}$ [11]. Considering the maximum repetition factor of NPDCCH is 2048, the values of the search period for Type-2 CSS and USS are $4 < T \leq 131{,}072$.

7.3.2.2 *Scheduling*

In this section, we describe how scheduling for uplink and downlink transmissions works. When the network needs to schedule a device, it sends a DCI addressed to the device using one of the search space candidates that the device monitors. The C-RNTI, masking the DCI CRC, is used to identify the device. The NPDCCH carries a DCI that includes resource allocation (in both time and frequency domains), *MCS*, and information needed for supporting the HARQ operation. To allow low-complexity device implementation, Release 13 NB-IoT adopts the following scheduling principles:

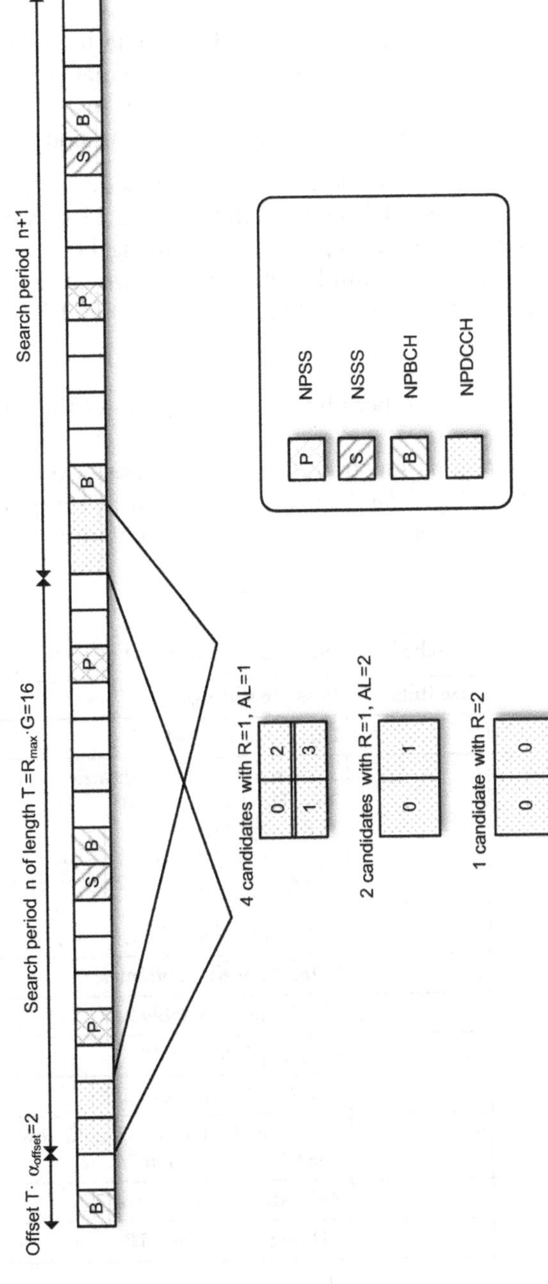

FIG. 7.56 An example of user equipment-specific search space (USS) configuration.

- A device only needs to support one HARQ process in the downlink.
- A device only needs to support one HARQ process in the uplink.
- The device need not support simultaneous transmissions of uplink and downlink HARQ processes.
- Cross-subframe scheduling (i.e., DCI and the scheduled data transmission do not occur in the same subframe) with relaxed processing time requirements.
- Half-duplex operation at the device (i.e., no simultaneous transmit and receive at the device) allows time for the device to switch between transmission and reception modes.

3GPP Release 14 introduces a device option of two simultaneously active HARQ processes in the downlink and uplink for Cat-NB2 devices, and Release 15 introduces TDD. We first describe the fundamental scheduling concept of NB-IoT using FDD networks and devices supporting only 1 HARQ process as examples. Aspects specific to devices supporting two HARQ processes and TDD are highlighted in Sections 7.3.2.2.3 and 7.3.2.2.4, respectively.

7.3.2.2.1 Uplink scheduling

DCI Format N0 is used for uplink scheduling. Table 7.29 shows the information carried in DCI Format N0.

An uplink scheduling example is illustrated in Fig. 7.57 based on the same USS search space configuration example in Fig. 7.56. We will highlight a few important aspects in this example and relate them to the scheduling information presented in Table 7.29.

TABLE 7.29 DCI Format N0 used for scheduling NPUSCH Format 1 (FDD, one HARQ process).

Information	Size [bits]	Possible settings
Flag for DCI format N0/N1	1	DCI N0 or DCI N1
Subcarrier indication	6	Allocation based on subcarrier index. 3.75 kHz spacing: {0}, {1}, …, or {47} 15 KHz spacing: 1-tone allocation: {0}, {1}, …, or {11} 3-tone allocation: {0, 1, 2}, {3, 4, 5}, {6, 7, 8}, {9, 10, 11} 6-tone allocation: {0, 1, …, 5} or {6, 7, …, 11} 12-tone allocation: {0, 1, …, 11}
NPUSCH scheduling delay	2	8, 16, 32, or 64 (subframes)
DCI subframe repetition number	2	The R values in Table 7.27
Number of RUs	3	1, 2, 3, 4, 5, 6, 8, or 10
Number of NPUSCH repetition	3	1, 2, 4, 8, 16, 32, 64, or 128
MCS	4	0, 1, …, or 13, for indexing the row of the NPUSCH TBS table (see Section 7.2.5.2)
Redundancy version	1	Redundancy version 0 or 2
New data indicator (NDI)	1	NDI toggles for new TB or does not toggle for same TB

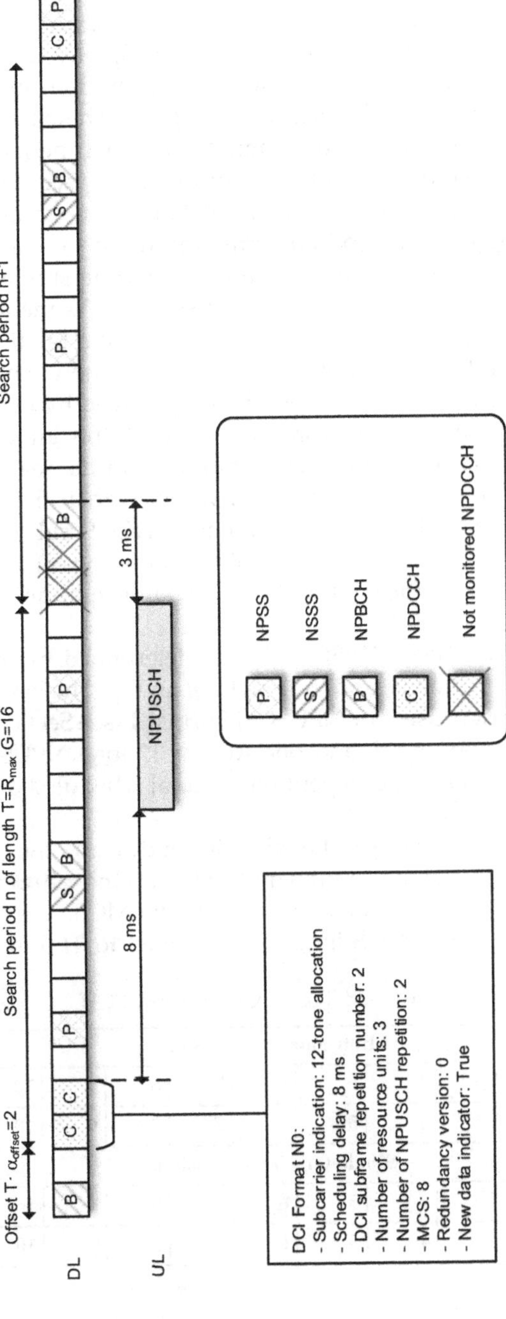

FIG. 7.57 An uplink scheduling example. (FDD, one HARQ process).

For uplink data transmissions, subframe scheduling with at least an 8 ms time gap between the last DCI subframe and the first scheduled NPUSCH subframe is required. This time gap allows the device to decode the DCI, switch from the reception mode to the transmission mode, and prepare the uplink transmission. This time gap is referred to as the scheduling delay and is indicated in DCI. After the device completes its NPUSCH transmission, there is at least a 3-ms gap to allow the device to switch from transmission mode to reception mode and be ready for monitoring the next NPDCCH search space candidate. This means that according to this example the network cannot use search period $n+1$ to send the next DCI to the device because the device will skip both search space candidates since they are both within the 3-ms gap from the end of its NPUSCH transmission. The network scheduler needs to follow this timing relationship in determining when to send the DCI to the device.

DCI Format N0 provides the information about the starting subframe as well as the total number of subframes of the scheduled NPUSCH resources. As mentioned earlier, the time gap between the last NPDCCH subframe carrying the DCI and the first scheduled NPUSCH slot is indicated in the DCI. However, how does the device know which subframe is the last subframe carrying the DCI? This potential problem is resolved by including the information of NPDCCH subframe repetition number in the DCI. With this information, if the device is able to decode the DCI using the first available subframe in the search space, it knows that the DCI will be repeated in one more subframe, which is then the last subframe carrying the DCI. Generally speaking, with the information of NPDCCH subframe repetition number, if the device is able to decode the DCI using any of the subframes of a search space candidate, it can unambiguously determine the starting and the ending subframes of this specific search space candidate.

The total number of scheduled NPUSCH slots is determined by the number of RUs per repetition, the number of repetitions, and the length of an RU. The length of an RU is inferred from the number of subcarriers used for NPUSCH Format 1 (see Section 7.2.5.2). According to the example, 12 subcarriers are used, and one RU for 12-tone NPUSCH Format 1 is 1 ms. Thus, with 2 repetitions and 3 RUs per repetition the total scheduled duration is 6 ms as illustrated in Fig. 7.57.

Modulation format is determined based on the MCS index, and the coding scheme is determined jointly based on the MCS index, number of RUs, and the redundancy version. According to the current example, the MCS index 8 is used. The MCS index is converted to a TBS index based on Table 7.30. Thus, MCS index 8 is mapped to TBS index 8, which, together

TABLE 7.30 Relationship between MCS and TBS indexes for NPUSCH.

Multi-tone transmissions														
I_{MCS}	0	1	2	3	4	5	6	7	8	9	10	11	12	13
I_{TBS}	0	1	2	3	4	5	6	7	8	9	10	11	12	13
Single-tone Transmissions														
I_{MCS}	0	1	2	3	4	5	6	7	8	9	10			
I_{TBS}	0	2	1	3	4	5	6	7	8	9	10			

with the information that each repetition uses three RUs, is used to determine TBS as 392 based on Table 7.16. The number of data symbols per RU is determined based on Table 7.15 and is 144 symbols in this case. With QPSK and three RUs per repetition, there are overall 864 coded bits available. With TBS of 392 bits, code-word length of 864 bits, and redundancy version 0, using the rate matching framework of LTE, as detailed in Ref. [8], the code word can be generated accordingly.

Because only one HARQ process is supported in Release 13, there is no need to signal the process number. From the device perspective, it only needs to know whether it needs to transmit the same or new TB. The HARQ acknowledgment is signaled implicitly using the *new data indicator* (NDI) in the DCI. If the NDI is toggled, the device treats it as an acknowledgment of the previous transmission.

7.3.2.2.2 Downlink scheduling

Scheduling of NPDSCH is signaled using DCI Format N1, which is shown in Table 7.31 including the parameter values of different information elements. Most of the general aspects

TABLE 7.31 DCI Format N1 for scheduling NPDSCH. (FDD, one HARQ process).

Information	Size [bits]	Possible settings
Flag for DCI format N0/N1	1	DCI N0 or DCI N1
NPDCCH order indication	1	Whether the DCI is used for NPDSCH scheduling or for NPDCCH order
Additional time offset for NPDSCH (in addition to a minimal gap of 4 subframes)	3	R_{max}<128: 0, 4, 8, 12, 16, 32, 64, or 128 (subframes) $R_{max} \geq$ 128: 0, 16, 32, 64, 128, 256, 512, or 1024 (subframes)
DCI subframe repetition number	2	The R values in Table 7.27
Number of NPDSCH subframes per repetition	3	1, 2, 3, 4, 5, 6, 8, or 10
Number of NPDSCH repetition	4	1, 2, 4, 8, 16, 32, 64, 128, 192, 256, 384, 512, 768, 1024, 1536, or 2048
MCS	4	0, 1, ..., or 13, for indexing the row of the NPDSCH TBS table (see Table 7.9)
NDI	1	NDI toggles for new TB or does not toggle for same TB
HARQ-ACK resource	4	15 kHz subcarrier spacing: • Time offset value: 13, 15, 17, or 18 • Subcarrier index: 0, 1, 2, or 3 3.75 kHz subcarrier spacing: • Time offset value: 13 or 17 • Subcarrier index: 38, 39, 40, 41, 42, 43, 44, or 45

of downlink scheduling are similar to those used for uplink scheduling, although the exact parameter values are different. For example, cross-subframe scheduling is also used for downlink scheduling, but the minimum time gap between the last DCI subframe and the first scheduled NPDSCH subframe is 4 ms. Recall in the uplink cross-subframe scheduling case, this gap is at least 8 ms. A smaller minimum gap in the downlink case reflects that there is no need for the device to switch from receiving to transmitting between finishing receiving the DCI and starting the NPDSCH reception.

An NPDSCH scheduling example is illustrated in Fig. 7.58, again based on the same NPDCCH USS configuration example from Fig. 7.56. First, the DCI indicates that there is no additional scheduling delay, and thus the scheduled NPDSCH starts after a minimum gap of 4 downlink subframes following the last subframe carrying the DCI. The DCI also indicates that one subframe is used per repetition of NPDSCH and there are two repetitions. Based on this information, the device knows that there are two subframes scheduled for its NPDSCH reception. These two subframes are the first two available *downlink subframes* from the scheduled starting point of NPDSCH. The available subframes are the ones not used by NPBCH, NPSS, NSSS, or SIB1-NB and not indicated as invalid subframes. As illustrated, the first subframe after the scheduled NPDSCH starting point is available, but the next two subframes need to be skipped as they are used for NSSS and NPBCH and therefore are not considered as *downlink subframes* as far as downlink scheduling is concerned. After the NPBCH subframe, the next subframe is available.

A major difference between downlink and uplink scheduling is that in the downlink case, the scheduler also needs to schedule NPUSCH Format 2 resources for the signaling of HARQ feedback. This information is provided in the DCI in the format of a subcarrier index and a time offset. The time offset is defined between the ending subframe of scheduled NPDSCH and the starting slot of NPUSCH Format 2. NB-IoT requires such a time offset to be at least 13 ms, giving a gap of at least 12 ms between the end of NPDSCH and the start of NPUSCH Format 2, as indicated in Fig. 7.58. This gap is to allow sufficient NPDSCH decoding time at the device, time for switching from reception to transmission, and time for preparing the NPUSCH Format 2 transmission. As described in Section 7.2.5.2, NPUSCH Format 2 uses single-tone transmissions, either with 15 or with 3.75 kHz numerology. The RU for NPUSCH Format 2 is 2 ms for the 15 kHz subcarrier numerology or 8 ms for the 3.75 kHz numerology. Whether the device uses 15 or 3.75 kHz numerology for signaling HARQ feedback is configured through RRC signaling. Furthermore, the NPUSCH Format 2 transmissions may be configured with multiple repetitions. This information, however, is not included in the DCI but signaled separately through higher-layer signaling [27]. The example in Fig. 7.58 shows that the device is configured to use four repetitions for NPUSCH Format 2 (assuming the 15 kHz subcarrier numerology).

After the device completes its NPUSCH Format 2 transmission, it is not required to monitor NPDCCH search space for 3 ms. This is to allow the device to switch from the transmission mode to getting ready for receiving the NPDCCH again. According to the example in Fig. 7.58, the first subframe of search period $n+2$ is within the 3-ms gap and thus cannot be used for signaling a DCI to the device. The subframe immediately after the NPSS subframe is another NPDCCH search space candidate and is not within the 3-ms gap. Therefore, a DCI may be sent using this subframe.

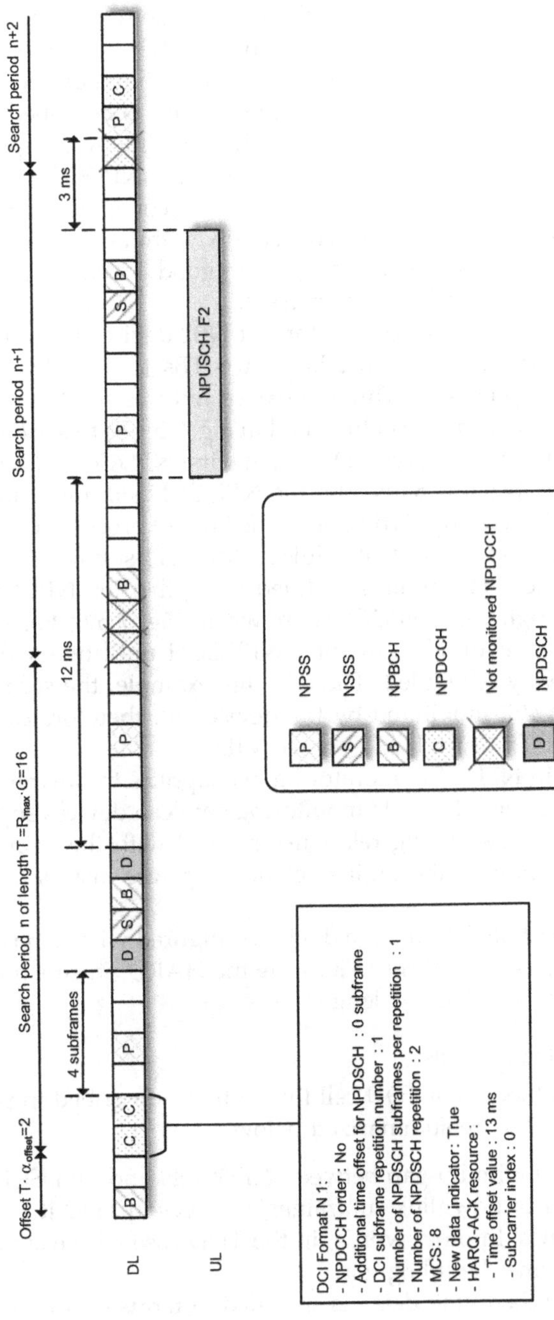

FIG. 7.58 A downlink scheduling example (FDD, one HARQ process).

7.3.2.2.3 Scheduling for Cat-NB2 devices supporting 2 HARQ processes

For a device in bad coverage, the physical layer throughput is limited by coverage in that its NPDCCH, NPDSCH, and NPUSCH all require large repetition factors and therefore it is the actual transmission time that limits the throughput. For a device in good coverage where repetition factor 1 is adequate, its throughput is then limited by the scheduling gaps that are designed for enabling low-cost device implementations as described in previous sections. As shown in Fig. 7.57 and 7.58, it takes a relatively long time to complete the delivery of a TB compared to the actual transmission time needed for NPDCCH, NPDSCH, or NPUSCH. Although NB-IoT is designed for allowing significant coverage extension, it should be noted that most of the devices in the network are actually within good coverage. It is thus desirable to improve the throughput for devices that are in good coverage. This may incentivize more capable devices to be introduced to the market.

3GPP Release 14 introduces an option for Cat-NB2 devices to support two simultaneously active HARQ processes. This option allows two TBs to be delivered in the two simultaneously active HARQ processes. This in essence reduces the time gap between delivering multiple TBs. An uplink example is illustrated in Fig. 7.59. In this example, the device receives uplink scheduling for HARQ process ID #0 in the first NPDCCH subframe in search period n. The scheduled NPUSCH starts 8 ms after the NPDCCH subframe according to the information in the DCI. For devices supporting only one HARQ process, it is not required to monitor NPDCCH until 3 ms after the last scheduled NPUSCH subframe. Thus, for single-HARQ-process devices, the next TB can be scheduled using the first NPDCCH subframe in search period $n+2$ (i.e. the rightmost NPDCCH shown in Fig. 7.59). For devices supporting two HARQ processes, it is required to monitor NPDCCH until two subframes before the start of the NPUSCH already scheduled. Thus, in this example, the second NPDCCH subframe in search period n is still monitored by the device and therefore can be used to scheduled NPUSCH for another TB using HARQ process ID #1.

Similar tighening in NPDCCH monitoring also applies to downlink scheduling. Fig. 7.60 illustrates the difference in NPDCCH monitoring between devices supporting 1 HARQ and 2 HARQ processes. The new timing relationships deal with the inter-process switching. For each HARQ process though, the timing relationships are just as those of Release 13 for 1 HARQ process.

As mentioned in Section 7.2.4.5, for a device configured with two HARQ processes, one bit is added to DCI formats N0 and N1 to indicate the HARQ process number. In this case, DCI formats N0 and N1 become 24 bits long.

7.3.2.2.4 TDD scheduling methods

The scheduling methods in a TDD cell follow those described in previous sections for an FDD cell. The differences are summarized below.

- For NPDSCH, the minimum gap between NPDCCH and NPDSCH in FDD is specified in terms of number of downlink subframes, whereas in TDD it is specified in terms of number of subframes. In other words, in the TDD case all the subframes are counted toward the minimum scheduling gap.
- For NPUSCH, the scheduling delay is specified in terms of number of subframes in the FDD case, whereas in TDD it is specified in terms of number of uplink subframes.

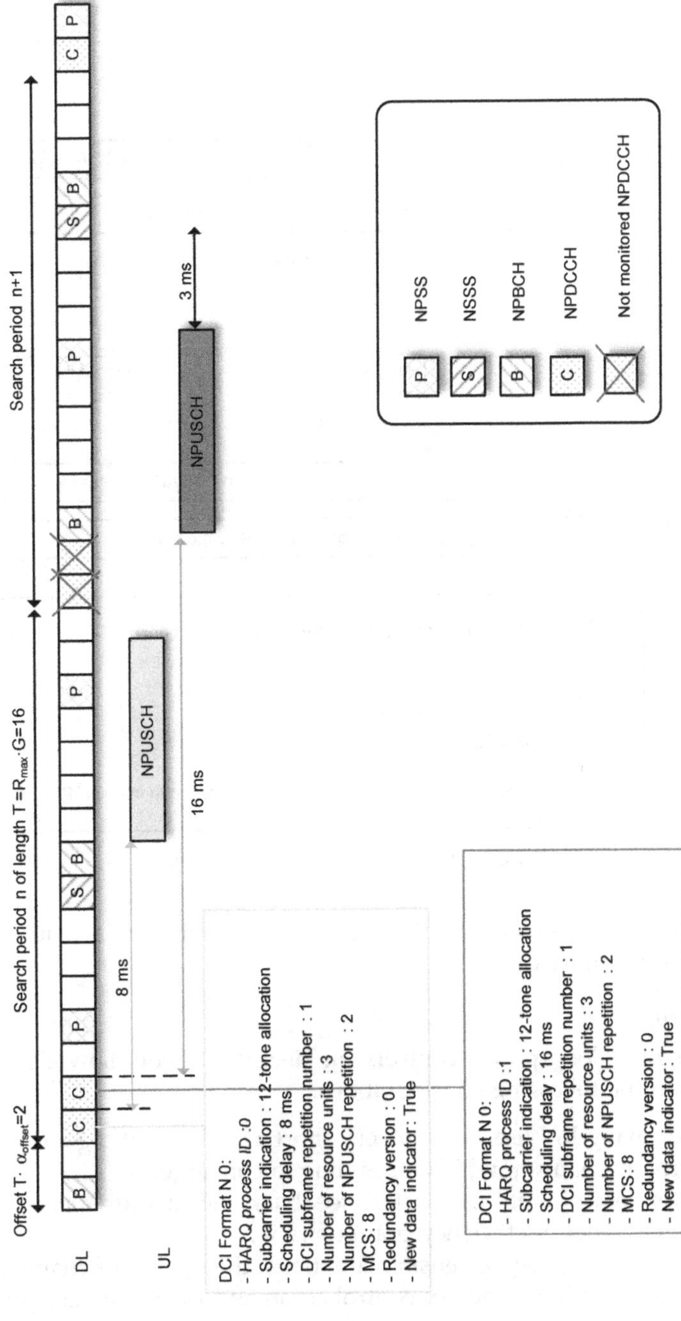

FIG. 7.59 An uplink scheduling example. (FDD, two HARQ processes).

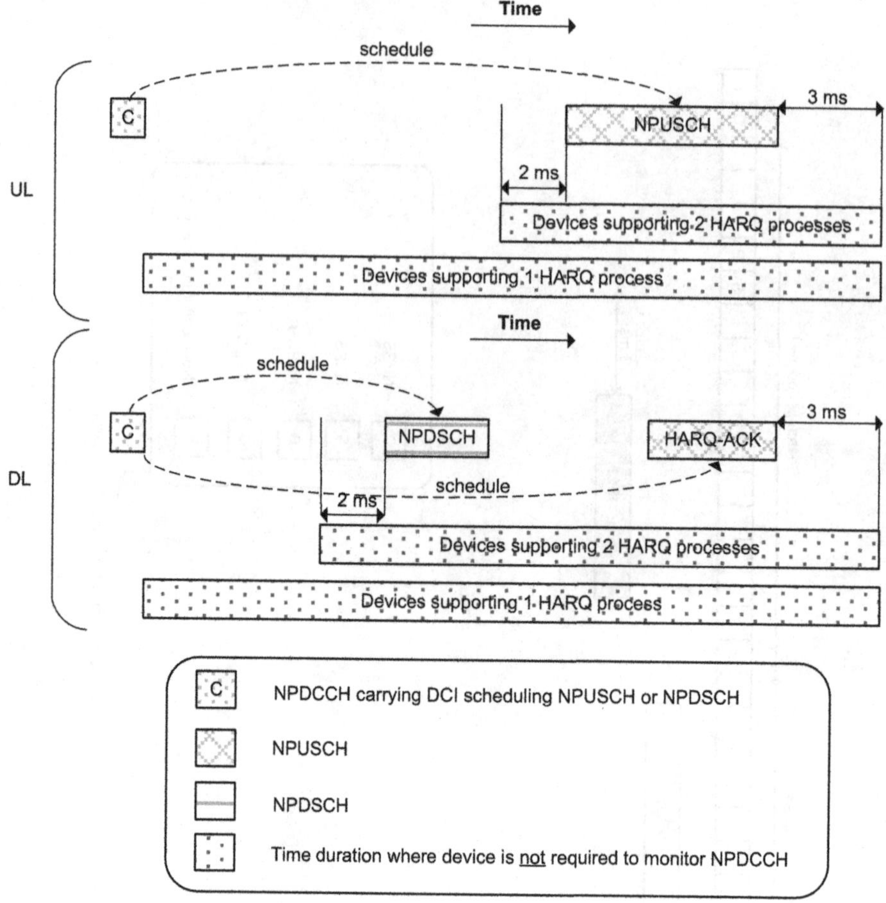

FIG. 7.60 NPDCCH monitoring restriction.

In other words, in the TDD case downlink and Special subframes are not counted toward the uplink scheduling delay.

7.3.2.3 *Power control*

NB-IoT supports open-loop power control. The decision to not allow closed-loop power control was based on the following considerations:

- A data session for many IoT use cases is very short and thus not a good match for a closed-loop control mechanism, which takes time to converge.
- Closed-loop power control requires constant feedback and measurements, which are not desirable from device energy efficiency point of view.
- For devices in extreme coverage extension situations, the quality of channel quality measurements and reliability of power control command might be very poor.

Instead, NB-IoT uses open-loop power control based on a set of very simple rules. For NPUSCH, both Format 1 and Format 2, if the number of repetitions is greater than 2, the transmit power is the maximum configured device power, P_{max}. The maximum configured device power is set by the serving cell. If the number of NPUSCH repetitions is 1 or 2, the transmit power is determined by

$$P_{NPUSCH} = \max\{P_{max}, 10\log_{10}(M) + P_{target} + \alpha L\} \text{ [dBm]}, \tag{7.20}$$

where P_{target} is the target received power level at the base station, L is the estimated path loss, α is a path loss adjustment factor, and M is a parameter related to the bandwidth of NPUSCH waveform. The bandwidth-related adjustment is used to relate the target received power level to target received SNR. The device uses the value of M according to its NPUSCH transmission configuration. The values of M for different NPUSCH configurations are shown in Table 7.32. The values of P_{max}, P_{target}, and α are provided by higher-layer configuration signaling.

The power control for NPRACH follows the same general principles. There may be multiple NPRACH configurations for supporting different coverage levels. The NPRACH preamble repetition levels may be different for these different NPRACH configurations. For NPRACH preambles not having the lowest repetition level, the maximum configured device power, P_{max}, is used in all transmissions. For NPRACH preambles having the lowest repetition level, the transmit power is determined based on the expression below.

$$P_{NPRACH} = \max\{P_{max}, P_{target} + L\}(\text{dBm}), \tag{7.21}$$

where P_{target} is the target NPRACH received power level, which is indicated by the higher layers [27]. If the device does not get a response and has not used the maximum configured device power, it can increase the target NPRACH received power level in its subsequent random access attempts until it reaches the maximum configured device power. This is referred to as power ramping.

7.3.2.3.1 Enhanced power control for transmitting random access preambles

As described earlier, for NPRACH preambles not having the lowest repetition level, the maximum configured device power, P_{max}, is used in all transmissions. This simple rule makes

TABLE 7.32 Bandwidth adjustment factors used in open-loop power control.

NPUSCH Configuration	NPUSCH bandwidth [kHz]	M
Single-tone, 3.75 kHz subcarrier spacing	3.75	1/4
Single-tone, 15 kHz subcarrier spacing	15	1
3-tone, 15 kHz subcarrier spacing	45	3
6-tone, 15 kHz subcarrier spacing	90	6
12-tone, 15 kHz subcarrier spacing	180	12

sense if a device uses a higher repetition level for sending an NPRACH preamble is for coverage extension.

However, NB-IoT allows CE-level ramping in the random access procedure. With this, a device can switch to an NPRACH preamble configured for a higher coverage level after it has reached the maximum number of attempts and maximum transmit power level using NPRACH preambles configured for the lowest coverage level. One cause for the CE-level ramping may very well be because the coverage situation was worse than initially estimated by the device. However, another possible cause for the CE-level ramping is NPRACH collision during high network loads. In such a scenario, allowing a device to transmit an NPRACH preamble configured for a higher coverage level at its configured maximum power level does not help resolve the collision problem. Worse, it may result in a significant increase in inter-cell interference. To address this problem, enhanced random access power control was introduced in 3GPP Release 14. For devices supporting enhanced random access power control, open loop power control according to Eq. (7.21) and power ramping are used for NPRACH preambles when the random access procedure begins in CE level 0 or 1. If the procedure begins in CE level 2, the configured maximum transmit power is still applied.

7.3.2.3.2 Power head room

The device can include a *power head room* (PHR) report in its uplink transmission in *Message* 3 described in Section 7.3.1.6. A PHR report indicates the difference between the device's configured maximum transmit power level and its estimated required transmit power level for NPUSCH Format 1 which is the $P_{target} + \alpha L$ term in Eq. (7.20) for a nominal NPUSCH bandwidth and coverage level. The information about PHR helps the base station scheduler adjust the scheduled bandwidth as well as the *MCS* for the device. When a device has a positive PHR, it benefits from a higher bandwidth allocation and a higher MCS. In 3GPP Release 13, a PHR is quantized to 4 levels and included in the MAC control element named DPR (Data volume status and Power headroom Report). The 4-level quantization however results in too coarse granularity and leaves room for further enhancement. 3GPP Release 15 introduced enhanced PHR to enable the device to report its power headroom with 16-level quantization, which gives finer granularity and more detailed information for the base station scheduler to optimize the scheduling decision.

7.3.2.4 *Random access in connected mode*

The base station can order a device to send an NPRACH preamble in connected mode by sending an NPDCCH order using DCI Format N1. This procedure can allow the base station and device to reacquire uplink time alignment. In this case, the exact resources for the device to transmit the NPRACH preamble are indicated in the DCI. Thus, the preamble is not subject to a collision. Table 7.33 lists the information included in an NPDCCH order.

7.3.2.5 *Scheduling request*

As NB-IoT was initially designed for use cases with devices transmitting or receiving infrequent, small data volume, there was no provisioning of scheduling request in connected mode when it was introduced. A scheduling request from the device can only be made by a random access procedure and including a data volume status report in *Message* 3 of the

TABLE 7.33 DCI Format N1 for NPDCCH order.

Information	Size [bits]	Possible settings
Flag for DCI format N0/N1	1	DCI N0 or DCI N1
NPDCCH order indication	1	Whether the DCI is used for NPDCCH order or for other purposes
Preamble format indicator	1	This information is included only for devices supporting the new NPRACH Format 2 introduced in 3GPP Release 15. The one bit information is to indicate whether the device should use Format 2 or not.
NPRACH configuration	2	Which NPRACH configuration to use. As mentioned in Section 7.2.5.4, each configuration is associated with a coverage class and has a specific repetition factor.
Subcarrier indication	6 for Formats 0 & 1 8 for Format 2	For NPRACH Formats 0 and 1, 3.75 kHz subcarrier spacing is used and thus subcarrier index might be 0, 1, …, 47. For Format 2 with 1.25 kHz subcarrier spacing, subcarrier index might be 0, 1, …, 143.
Carrier indication	4	Which NB-IoT carrier to use

random access procedure. To add to the versatility of NB-IoT and allow it to expand the use cases it can cater to, better supporting for long data sessions has been an interest. During a longer data session, the device may have bursty data arriving during connected mode. It is desirable for the device to be able to send its data volume status report without triggering a new random access procedure.

3GPP Release 15 introduced a few options supporting the transmission of a scheduling request. The first option allows a device in connected mode to send scheduling requests in the form of buffer status report through periodic NPUSCH F1 resources. Such periodic NPUSCH resources can be activated and deactivated through dynamic signaling on NPDCCH. The second option allows a device to modify its NPUSCH F2 transmission by the application of a cover code on top of the Release 13 NPUSCH F2 waveform. The presence of the cover code indicates a scheduling request to the base station. Yet another option is for the device to send a scheduling request using a NPRACH transmission that is specifically preconfigured for the device.

7.3.2.6 Positioning

The capability of obtaining the positioning of a device is attractive as it opens new business opportunities such as people and goods tracking. Today such services are associated with solutions based on GPS, but since low cost is highly important for IoT devices, it is desired to be able to support positioning using the NB-IoT cellular technology. This facilitates a low cost device based on a single chip for both wireless connectivity and positioning services. Also, although GPS based solutions work perfectly well outdoors, they may have limitations in determining the position of devices indoors, where GPS coverage is limited.

The first solution designed for positioning of a device was *enhanced cell identity* which basically is based on the estimated *TA* that determines the round trip time between the device and the base station. It can be translated to the distance between the base station and device and allows improved positioning compared with only positioning a device based on the identity of the serving cell. Fig. 3.42 illustrates the positioning accuracy of enhanced cell identity and the impact from the synchronization accuracy of the device and the base station.

The second solution specified for NB-IoT is *observed time difference of arrival* (OTDOA). It is based on a device measuring the *ToA* on a set of downlink NPRS transmitted from a set of time-synchronized base stations surrounding the device. The device reports the narrowband positioning *RSTD* to a positioning server. Each RSTD allows the positioning server to determine the position of the device to a hyperbola centered around the base stations transmitting the NPRS. If RSTDs between three or more base stations are reported, the positioning server will be able to determine multiple hyperbolas and fix the position of the device to the intersection of the hyperbolas. Fig. 7.61 illustrates this, and the accuracy of the RSTD measurement is reflected by the width of the hyperbolas. High timing accuracy and many positioning base stations allowing multiple hyperbolas to be determined lead to a better positioning estimate.

7.3.2.7 Multicarrier operation

To support a massive number of devices, NB-IoT also includes a multicarrier feature from its very first release. In addition to the *anchor carrier*, which carries synchronization and broadcast channels, one or more *non-anchor carriers* can be provided. The notion of anchor and non-anchor carriers is explained in Section 7.2.1.1.

As a non-anchor carrier does not carry NPBCH, NPSS, NSSS, and SI, a Release 13 device in idle mode camps on the anchor carrier, monitoring the paging messages on the anchor. When the device needs to switch from idle to connected mode, the random access procedure also takes place on the anchor carrier. The network can use RRC configuration to point the device

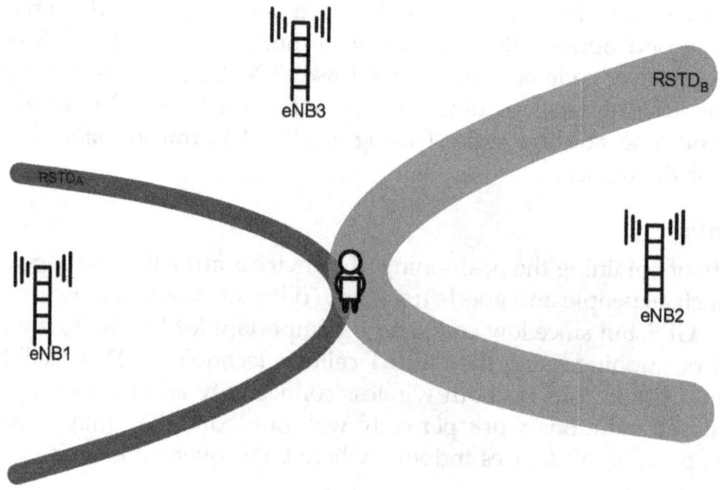

FIG. 7.61 OTDOA positioning.

to a non-anchor carrier. Essential information about the non-anchor carrier will be provided to the device using dedicated signaling. During the remaining duration of the connected mode, USS monitoring and NPDSCH and NPUSCH activities all take place on the assigned non-anchor carrier. After the device completes the data session, it comes back to camp on the anchor carrier during the idle mode.

3GPP Release 14 enhanced non-anchor carrier operation to allow a device to monitor paging, initiate random access, and monitor RAR on a non-anchor carrier. More details on system access on a non-anchor carrier is provided in Section 7.3.1.10. A device can also receive single-cell multicast on a non-anchor carrier. Up to 15 downlink and uplink non-anchor carriers can be configured.

The non-anchor carriers can be allocated adapting to the traffic load of NB-IoT. Many IoT use cases generate highly delay-tolerant traffic. Such traffic can be delivered during off-peak hours in the network. The multicarrier feature allows the radio resources normally reserved for serving broadband or voice to be allocated for NB-IoT when the load of broadband and voice traffic is low. For example, during the middle of the night many of the LTE PRBs may be configured as non-anchor NB-IoT carriers serving the IoT traffic. The network can, as an example, take advantage of the multicarrier feature to push firmware upgrades to a massive number of devices during the middle of the night. An example of replacing LTE PRBs with non-anchor NB-IoT carriers during off-peak hours of broadband and voice traffic is illustrated in Fig. 7.62.

In 3GPP Releases 13 and 14, a stand-alone anchor carrier cannot be configured together with an in-band or guard-band non-anchor carrier. Similarly, a stand-alone non-anchor carrier cannot be configured together with an in-band or guard-band anchor carrier. This restriction is removed in 3GPP Release 15.

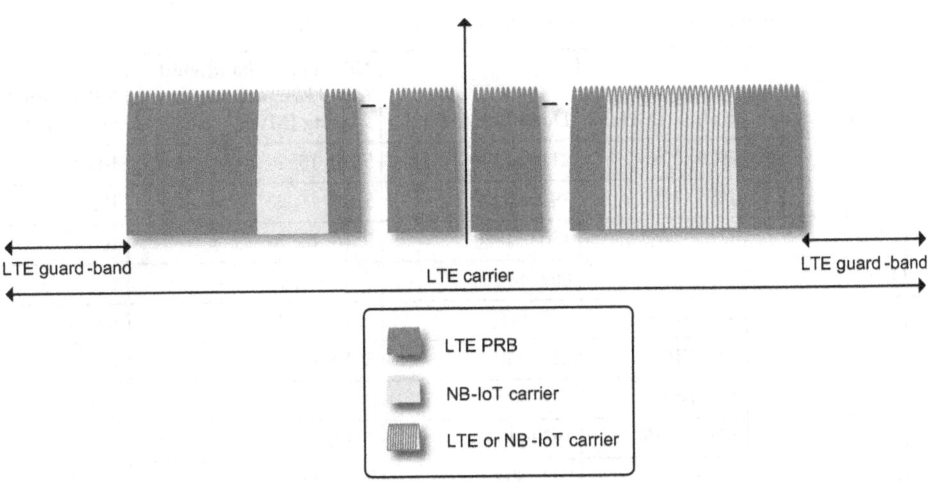

FIG. 7.62 An example of multicarriers, replacing LTE PRBs with non-anchor NB-IoT carriers during off-peak hours of broadband and voice traffic.

7.4 NR and NB-IoT coexistence

In 3GPP Release 15, a new radio access technology known as *NR* is introduced. NR offers performance advantages for both mobile broadband and *ultra-reliable low-latency communication* services, more deployment flexibility, and higher energy efficiency over LTE. NR also has superior scalability that makes it suitable for deployment in higher frequency bands such as millimeter wave bands where there is greater spectrum availability. Thanks to all these advantages, NR has attracted vast interest from existing and new operators. Network migration from LTE to NR has started in earnest in 2019 with US operators leading the way. Chapter 2.4 introduces NR, and the interested readers can refer to Ref. [28] for detailed description on NR.

Although NR is designed for enhancing the performance of mobile broadband and ultra-reliable low-latency communications services, it is not designed to be used for low-power wide-area IoT use cases. One main reason is that these use cases are already adequately addressed by existing 3GPP technologies such as LTE-M and NB-IoT. Thus, LTE-M and NB-IoT networks will continue to serve low-power wide-area IoT use cases even after the LTE-to-NR migration. In fact, as discussed in great details in Section 8.9, NB-IoT fully fulfills the *International Telecommunications Union* IMT-2020 and 3GPP *Fifth Generation* (5G) requirements on massive MTC, and therefore is a component of the 5G radio access technology. As such, it is important to ensure efficient and flexible coexistence between NB-IoT and other components of 5G radio access technology. In this section, we describe how these two important 5G component technologies, NR and NB-IoT, coexist.

NB-IoT is defined in LTE bands 1, 2, 3, 4, 5, 8, 11, 12, 13, 14, 17, 18, 19, 20, 21, 25, 26, 28, 31, 41, 66, 70, 71, 72, 73, 74 and 85 (see Section 7.2.2.1 for the frequency ranges of these bands). Many of these bands are also defined for NR [29]. Table 7.34 lists all the bands that are

TABLE 7.34 Frequency bands that are defined for both NR and NB-IoT.

Band	Duplex mode	Uplink [MHz]	Downlink [MHz]	NR channel bandwidth for 15 kHz subcarrier spacing [MHz]	NR channel raster [kHz]
1	FDD	1920–1980	2110–2170	5, 10, 15, 20	100
2	FDD	1850–1910	1930–1990	5, 10, 15, 20	100
3	FDD	1710–1785	1805–1880	5, 10, 15, 20, 25, 30	100
5	FDD	824–849	869–894	5, 10, 15, 20	100
8	FDD	880–915	925–960	5, 10, 15, 20	100
12	FDD	699–716	729–746	5, 10, 15	100
20	FDD	832–862	791–821	5, 10, 15, 20	100
25	FDD	1850–1915	1930–1995	5, 10, 15, 20	100
28	FDD	703–748	758–803	5, 10, 15, 20	100
41	TDD	2496–2690	2496–2690	10, 15, 20, 40, 50	15 or 30
66	FDD	1710–1780	2110–2200	5, 10, 15, 20, 40	100
70	FDD	1695–1710	1995–2020	5, 10, 15, 20, 25	100
71	FDD	636–698	617–652	5, 10, 15, 20	100
74	FDD	1427–1470	1475–1518	5, 10, 15, 20	100

defined for both NR and NB-IoT, as of 3GPP Release 15. These bands thus can be used to deploy both NR and NB-IoT. It is worth mentioning that in all these bands, NR can use 15, 30, or 60 kHz subcarrier spacing, and an NR device operating in these bands is required to support the 15 and 30 kHz subcarrier spacings.

There are many options for deploying NR and NB-IoT in the same band. We will describe these options in the subsections below. But before getting to that, it is worthwhile describing certain important aspects first. Obviously, such deployment needs to satisfy the following objectives:

- NB-IoT legacy devices can operate without knowing any NR-specific information.
- NR devices can operate regardless whether there is an NB-IoT carrier deployed in the same band.
- There is minimal mutual interference between NR and NB-IoT so the impact on NR and NB-IoT performance is negligible when they are deployed in the same band.

To satisfy the first objective, one fundamental aspect is that an NB-IoT device can identify an NB-IoT cell during the initial cell selection process (see Section 7.3.1.1.) To achieve this, the anchor NB-IoT carrier needs to be placed according to 100 kHz channel raster (see Section 7.2.1.1), i.e. the center frequency needs to be at $100N_{NR}$ kHz, where N_{NR} is an integer. As mentioned in Section 7.2.1.1, there is a raster offset in the case of in-band or guard-band deployment, and the raster offset is either 2.5 kHz, -2.5 kHz, 7.5 kHz, or -7.5 kHz. Furthermore, as NPSS, NSSS and NPBCH are the signals used by the device during cell search, they need to be preserved.

Similarly, to satisfy the second objective, the NR carrier needs to be placed according to NR channel raster shown in Table 7.34. In most cases, the NR channel raster is 100 kHz. There is an important difference in the location of channel raster between NR and NB-IoT. This is illustrated in Fig. 7.63. In the NB-IoT case, channel raster points to the center of the carrier, which is half-way (i.e. 7.5 kHz) between two subcarriers. NR channel raster however points to a subcarrier around the middle of the carrier. As shown, for an NR carrier with N resource blocks, there are $12N$ subcarriers; and indexing all the subcarriers from 0 to $12N-1$, the channel raster is mapped to subcarrier $6N$ [29].

Finally, the third objective suggests that if the NR carrier is configured with 15 kHz subcarrier spacing, it is desirable to align the NR and NB-IoT subcarriers on the same subcarrier grids, with the frequencies between an NR subcarrier and an NB-IoT subcarrier differing by an integer multiple of 15 kHz. With this, if the NR and NB-IoT networks are synchronized, NR subcarriers and NB-IoT subcarriers are mutually orthogonal. If the NR carrier is configured with subcarrier spacing other than 15 kHz, a guard band is needed between NR and NB-IoT to ensure minimal inter-subcarrier interference.

Before going further, it helps to clarify certain important terminology. As described in Section 7.1.2.5, there are three operation modes defined for NB-IoT, *stand-alone*, *in-band* and *guard-band* operation modes. These terms were introduced in Release 13 to describe the deployment of an NB-IoT carrier in its relationship with an LTE carrier. Going forward though, when the LTE spectrum is refarmed for NR deployment while the NB-IoT carrier continues to be in service, the meaning of the NB-IoT operation mode loses its relationship with the LTE carrier, as the LTE carrier may no longer exist. Is the NB-IoT operation mode still relevant when there is no LTE carrier? The answer is yes, and interestingly the

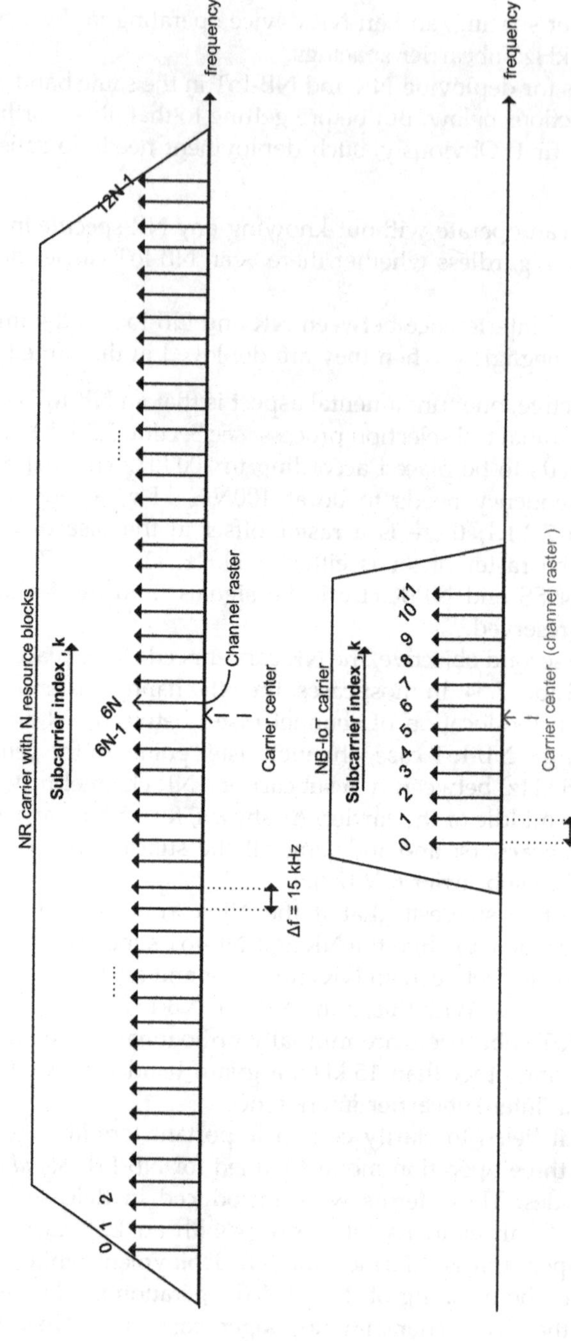

FIG. 7.63 Locations of NR and NB-IoT channel rasters. (assume the NR carrier is configured with 15 kHz subcarrier spacing).

deployment flexibility offered by these three operation modes is a useful tool for NR and NB-IoT coexistence. For NB-IoT devices, the NB-IoT operation mode means:

- Stand-alone operation mode: the NB-IoT carrier center is *exactly* at a 100 kHz channel raster. All REs in an NB-IoT subframe are available to an NB-IoT physical channel or signal.
- Guard-band operation mode: the NB-IoT carrier center is *not exactly* at a 100 kHz channel raster. The raster offset is -2.5 kHz, +2.5 kHz, -7.5 kHz, or +7.5 kHz. The exact raster offset is signaled in MIB-NB. All REs in an NB-IoT subframe are available to an NB-IoT physical channel or signal.
- In-band operation mode: the NB-IoT carrier center is *not exactly* at a 100 kHz channel raster. The raster offset is -2.5 kHz, +2.5 kHz, -7.5 kHz, or +7.5 kHz. The exact raster offset is signaled in MIB-NB. *Not all* REs in an NB-IoT subframe are available to an NB-IoT physical channel or signal. The device should expect some REs are taken by LTE CRS and PDCCH. The information about which REs taken by LTE is provided in MIB-NB and SIB1-NB. Although there is no LTE carrier, as far as the device is concerned, these resources are not available for NB-IoT.

The NB-IoT operation mode referred to in the following subsections can therefore be thought of as an indication to NB-IoT devices regarding channel raster location and resource element allocation in an NB-IoT subframe. It is *not about* the relationship between the NB-IoT carrier and the legacy LTE carrier which has been replaced by an NR carrier. It is also *not about* the relationship between the NB-IoT carrier and the new NR carrier that has replaced the LTE carrier.

7.4.1 NR and NB-IoT as adjacent carriers

The most straightforward option is to deploy NR and NB-IoT carriers as adjacent carriers as shown in Fig. 7.64. Depending on the subcarrier configuration of the NR carrier and whether there is tight synchronization between the two carriers, the guard band can be dimensioned accordingly to ensure minimal mutual interference. In the special case that NR and NB-IoT carriers are synchronized and the NR carrier is configured with 15 kHz

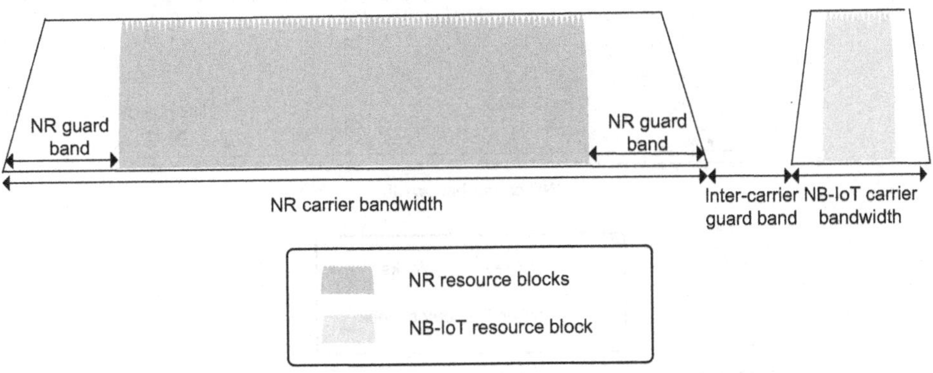

FIG. 7.64 NR and NB-IoT deployed as adjacent carriers.

subcarrier spacing, it is possible to achieve orthogonality between NR and NB-IoT without a guard band. As mentioned previously, the NR channel raster is mapped to a subcarrier. With a 100 kHz NR channel raster, the frequency of an NR subcarrier (in kHz) is:

$$f_{NR}(i) = 100N_{NR} + 15(i - 6N), \tag{7.23}$$

where N_{NR} is an integer which specifies the location of the NR channel raster, i is the NR subcarrier index. As shown in Fig. 7.63, the channel raster, which is at the frequency $100N_{NR}$, is mapped to subcarrier index $6N$. In comparison, NB-IoT channel raster is mapped to the midpoint between the middle two subcarriers, and thus the frequency (in kHz) of an NB-IoT subcarrier is:

$$f_{NB-IoT}(j) = 100N_{NB-IoT} + 15(j - 5.5), \tag{7.24}$$

where N_{NB-IoT} is an integer which specifies the location of the NB-IoT channel raster, j is the subcarrier index, $j = 0, 1, \ldots, 11$, and the channel raster is mapped to the midpoint between subcarriers 5 and 6, as illustrated in Fig. 7.63.

Eqs. (7.23) and (7.24) imply that perfect subcarrier alignment between NR and NB-IoT cannot be achieved as there is an additional offset of 7.5 kHz between the subcarrier grids of NR, Eq. (7.23), and those of NB-IoT, Eq. (7.24). This issue can be addressed by using the raster offset of NB-IoT. Let f_{offset} be the raster offset of NB-IoT, the frequency of NB-IoT subcarrier j then becomes

$$f_{NB-IoT}(j) = 100N_{NB-IoT} + 15(j - 5.5) + f_{offset}. \tag{7.25}$$

It is easy to show that with $f_{offset} \in \{\pm 2.5, \pm 7.5\}$ kHz, the subcarrier grids of NR and NB-IoT can be aligned, i.e., there exist N_{NR} and N_{NB-IoT} such that the frequencies of $f_{NR}(i)$ and $f_{NB-IoT}(j)$ differ by an integer multiple of 15 kHz, hence ensuring subcarrier orthogonality between NR and NB-IoT.

Thus, if the NB-IoT carrier in Fig. 7.64 is configured as guard-band or in-band operation mode with a raster offset of -2.5 kHz, 2.5 kHz, -7.5 kHz or 7.5 kHz, with proper choices of

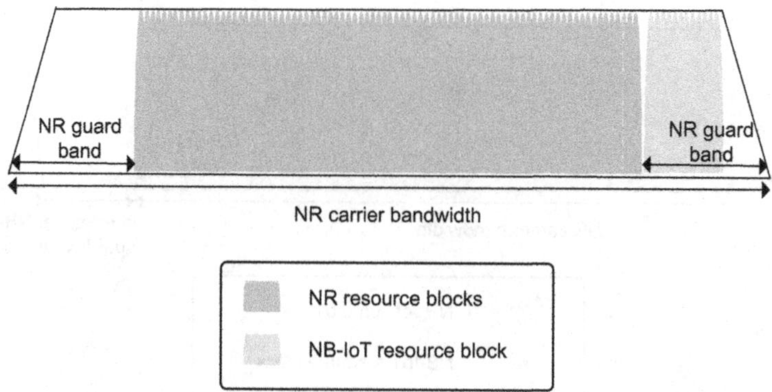

FIG. 7.65 NB-IoT deployed in the guard band of an NR carrier.

$N_{NB\text{-}IoT}$, its subcarriers will fall on the same subcarrier grid as the NR carrier. In this case, it is possible to achieve orthogonality between the NR and NB-IoT subcarriers.

One major advantage of this approach is that the inter-carrier guard band shown in Fig. 7.64 might not be needed at all. Obviously, among the two NB-IoT operation modes (guard-band and in-band), the guard-band operation mode is preferred since in such a scenario all the REs in an NB-IoT subframe can be made available to NB-IoT physical channels or signals.

7.4.2 NB-IoT in the NR guard band

As discussed in Section 7.4.1, there exists a solution to maintain orthogonality between NR and NB-IoT subcarriers. In fact, this makes it possible to deploy an NB-IoT carrier in the guard band of an NR carrier as illustrated in Fig. 7.65. The guard band of an NR carrier is smaller compared to that of an LTE carrier. Nevertheless, it may still be possible to accommodate an NB-IoT carrier. Table 7.35 shows the minimum guard bands for NR carriers with 15 kHz subcarrier spacing and frequencies below 6 GHz. Observe that in all cases the guard band is large enough to accommodate an NB-IoT carrier.

One important aspect for this deployment option is the NR unwanted emission requirements, which in most scenarios will limit the power level of the NB-IoT carrier.

Like the adjacent carrier scenario discussed in Section 7.4.1, also in this case it is preferred to configure the NB-IoT carrier with the guard-band operation mode as all the REs on the NB-IoT carrier can be made available to NB-IoT devices.

7.4.3 NB-IoT deployed using NR resource blocks

Yet another deployment option is to deploy NB-IoT within an NR carrier using an NR resource block (RB) as shown in Fig. 7.66. Similar to the NR guard band option described in Section 7.4.2, it is necessary to use the raster offset offered in NB-IoT in-band and guard-band operation modes to align the NB-IoT subcarriers with the NR subcarriers for achieving subcarrier orthogonality.

Compared to the NR guard band option described in Section 7.4.2, this option allows a higher NB-IoT transmit power level since the NB-IoT carrier is away from the carrier edge and therefore its power level has less impact on the fulfilment of the NR unwanted emission requirements specified in Section 6.6 of Reference [29]. This option however requires sharing of radio resources between NR and NB-IoT. The sharing can be configured semi-statically. NR has a feature called *reserved resources* intended for ensuring forward compatibility. An NR PDSCH time-frequency resource can be declared as *reserved* and not made available

TABLE 7.35 Minimum guard band for NR carriers with frequencies below 6 GHz. (15 kHz subcarrier spacing).

NR carrier bandwidth [MHz]	5	10	15	20	25	30	40	50
Guard band [kHz]	242.5	312.5	382.5	452.5	522.5	592.5	552.5	692.5

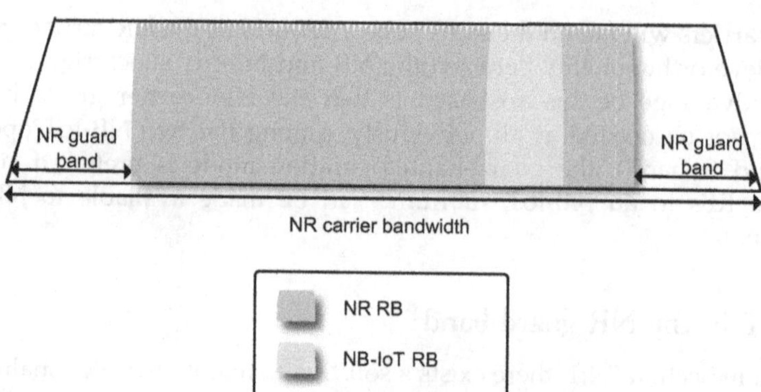

FIG. 7.66 NB-IoT deployed within an NR carrier by using an NR resource block.

to NR devices. NR reserved resource configuration can be done by using a frequency-domain bit map and a time-domain bit map as illustrated in Fig. 7.67. The frequency-domain bit map has granularity of resource block, with which each resource block can be individually indicated as reserved or not. The time-domain bit map has granularity of an OFDM symbol, with which each OFDM symbol can be individually indicated as reserved or not. When a resource element is indicated as reserved by both the frequency-domain and time-domain bit maps, it is not available to NR devices. Thus, the network can simply declare the resources used by NB-IoT carrier as reserved to enable the deployment of NB-IoT using one or more NR resource blocks. NR devices are required to rate match around the declared reserved REs.

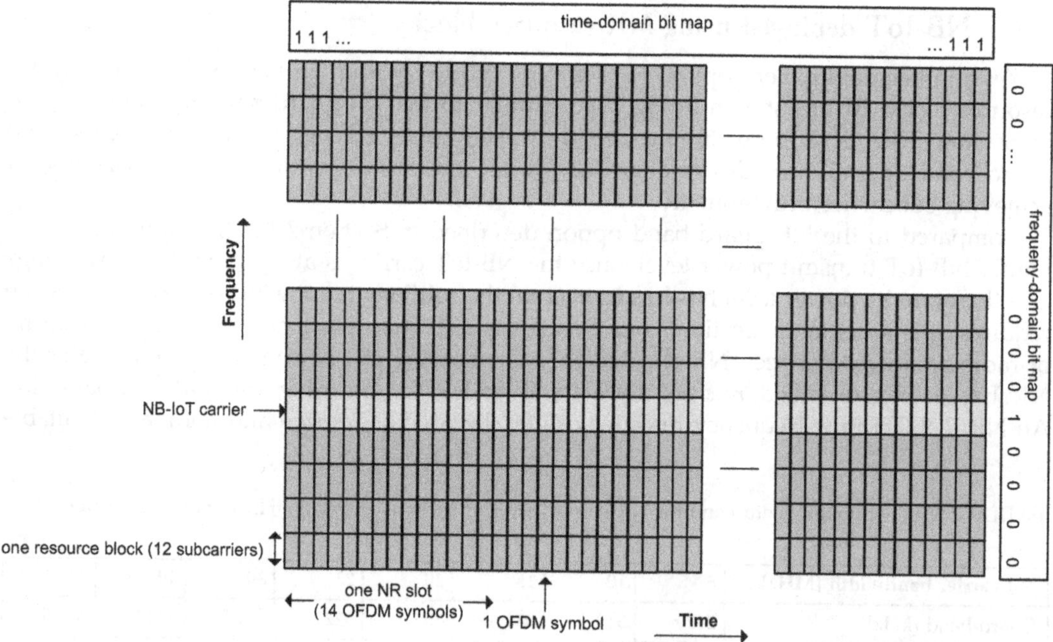

FIG. 7.67 Use of NR reserved resource configuration for supporting NB-IoT deployed within an NR carrier using an NR resource block.

Due to the raster offset requirement of NB-IoT and the different mappings of NR and NB-IoT channel rasters illustrated in Fig. 7.63, when deployed within an NR carrier, the NB-IoT anchor carrier may overlap with one or two NR resource blocks, depending on whether there is resource block alignment between the two. When the NB-IoT carrier overlaps with two NR resource blocks, the frequency-domain bit map needs to be configured for reserving two NR resource blocks. A detailed analysis on resource block alignment can be found in Ref. [30], and it is shown that it is possible to achieve resource block alignment between an NB-IoT anchor carrier and an NR resource block.

References

[1] LoRa Alliance. LoRaWAN R1.0 open standard released for the IoT, 2015 [Online]. Available from: https://www.businesswire.com/news/home/20150616006550/en/LoRaWAN-R1.0-Open-Standard-Released-IoT.

[2] Third Generation Partnership Project, Technical Report 45.820, v13.0.0. Cellular system support for ultra-low complexity and low throughput Internet of Things, 2016.

[3] 3GPP News. Standardization of NB-IOT completed, 2016 [Online]. Available from: http://www.3gpp.org/news-events/3gpp-news/1785-nb_iot_complete.

[4] Third Generation Partnership Project, Technical Report 37.910, v1.1.0. Study on self evaluation towards IMT-2020 submission, December 2018.

[5] GSMA. NB-IoT deployment guide to basic feature set requirements. Available from: https://www.gsma.com/iot/wp-content/uploads/2019/07/201906-GSMA-NB-IoT-Deployment-Guide-v3.pdf, June 2019.

[6] Global Mobile Suppliers Association. NB-IoT and LTE-M: global market status, August 2018.

[7] Persistence Market Research. NB-IoT chipset market to expand at a steady CAGR of 40.8% by 2028, February 2019. Available from: https://www.prnewswire.com/news-releases/nb-iot-chipset-market-to-expand-at-a-steady-cagr-of-40-8-by-2028-persistence-market-research-300800261.html.

[8] Third Generation Partnership Project, Technical Specification 36.212, v15.3.0. Evolved universal terrestrial radio access (E-UTRA) and evolved universal terrestrial radio access network (E-UTRAN); multiplexing and channel coding, 2018.

[9] Third Generation Partnership Project, Technical Specification 24.301, v15.5.0. Technical specification group core network and terminals; non-access-stratum (NAS) protocol for evolved packet system (EPS), 2018.

[10] Third Generation Partnership Project, Technical Specification 36.413, v15.4.0. Technical specification group radio access network; evolved universal terrestrial radio access network (E-UTRAN); S1 application protocol (S1AP), 2018.

[11] Third Generation Partnership Project, Technical Specification 36.331, v15.3.0. Evolved universal terrestrial radio access (E-UTRA) and evolved universal terrestrial radio access network (E-UTRAN); radio resource control (RRC); protocol specification, 2018.

[12] C. E. Shannon. Communication in the presence of noise. Proc. Inst. Radio Eng., 1949, Vol. 37, No. 1, 10−21.

[13] Third Generation Partnership Project, Technical Specification 37.104, v13.3.0. Group radio access network; E-UTRA, UTRA and GSM/EDGE; multi-standard radio (MSR) base station (BS) radio transmission and reception (release 13), 2016.

[14] E. Dahlman, S. Parkvall, J. Sköld. 4G: LTE/LTE-Advanced for mobile broadband. Oxford: Academic Press, 2011.

[15] Third Generation Partnership Project, Technical Specification 36.104, v15.5.0. Evolved universal terrestrial radio access (E-UTRA); base station (BS) radio transmission and reception, 2018.

[16] Third Generation Partnership Project, Technical Specification 36.211, v15.4.0. Evolved universal terrestrial radio access (E-UTRA) and evolved universal terrestrial radio access network (E-UTRAN); physical channels and modulation, 2018.

[17] S.M. Alamouti. A simple transmit diversity technique for wireless communications. IEEE J. Select. Areas Commun, October 1998, Vol. 16, No. 8, 1451−8.

[18] H. Bölcskei, A. J. Paulraj. Space−frequency coded broadband OFDM systems. Proc. IEEE Wireless Commun. Netw. Conf., Chicago, IL, USA, September 2000.

[19] Third Generation Partnership Project, Technical Specification 36.355, v15.2.0. Evolved universal terrestrial radio access (E-UTRA); LTE positioning protocol (LPP), 2018.
[20] Third Generation Partnership Project, Technical Specification 36.455, v15.2.0. Evolved universal terrestrial radio access (E-UTRA); LTE positioning protocol a (LPPa), 2019.
[21] Ericsson. R1-160094, NB-IoT – design considerations for Zadoff-Chu sequences based NB-PRACH, 3GPP TSG RAN1 Meeting NB-IoT#1, 2016.
[22] M. P. Latter, L. P. Linde. Constant envelope filtering of complex spreading sequences. Electron. Lett, August 1995, Vol. 31, No. 17, 1406–7.
[23] Ericsson. R1-1719368, NPRACH range enhancements for NB-IoT, 3GPP TSG RAN1 Meeting #91, 2017.
[24] Qualcomm. R1-161981, NB-PSS and NB-SSS design, 3GPP TSG RAN1 meeting NB-IoT#2, 2016.
[25] Third Generation Partnership Project, Technical Specification 36.304, v15.2.0. Evolved universal terrestrial radio access (E-UTRA) and evolved universal terrestrial radio access network (E-UTRAN); user equipment (UE) procedures in idle mode, 2018.
[26] Third Generation Partnership Project, Technical Report 37.868, v11.0.0. Study on RAN improvements for machine-type communications, 2011.
[27] Third Generation Partnership Project, Technical Specification 36.213, v15.3.0. Evolved universal terrestrial radio access (E-UTRA) and evolved universal terrestrial radio access network (E-UTRAN); physical layer procedures, 2018.
[28] E. Dahlman, S. Parkvall, J. Sköld. 5G NR: the next generation wireless access technology. Oxford: Academic Press, 2018.
[29] Third Generation Partnership Project, Technical Specification 38.104, v15.3.0. New radio (NR); base station (BS) radio transmission and reception, 2018.
[30] M. Mozaffari, Y.-P. E. Wang, O. Liberg, J. Bergman."Flexible and efficient deployment of NB-IoT and LTE-MTC in coexistence with 5G new radio," In Proc. IEEE Int. Conf. on Computer Commun., Paris, France, 29 April–2 May, 2019.

CHAPTER

8

NB-IoT performance

OUTLINE

Cellular Internet of Things, Second Edition
https://doi.org/10.1016/B978-0-08-102902-2.00008-X

© 2020 Elsevier Ltd. All rights reserved.

Abstract

This chapter presents the *Narrowband Internet of Things* (NB-IoT) performance in terms of coverage, data rate, latency, battery lifetime, connection density and positioning accuracy. All these performance aspects differ between the three operation modes of NB-IoT. Thus, in most cases, the performance of each NB-IoT operation mode is individually presented. It shows that all the three operation modes of NB-IoT meet the *Third Generation Partnership Project* (3GPP) performance objectives agreed for Cellular IoT. It further shows that NB-IoT Release 15 fulfills both *International Telecommunications Union* and 3GPP Fifth Generation (5G) performance requirements on massive machine type communications.

8.1 Performance objectives

When specified in 3GPP Release 13 (2016), Narrowband Internet of Things (NB-IoT) shared the same performance objectives as EC-GSM-IoT in terms of coverage, minimum data rate, service latency, device battery life, system capacity, and device complexity as presented in Section 4.1. NB-IoT Release 15 (2018) further achieves the *Fifth Generation* (5G) performance requirements for *massive machine type communications* (mMTC) set forth by *International Telecommunications Union* (ITU) and 3GPP. In addition, as mentioned in Section 7.1, NB-IoT aims to achieve deployment flexibility, including:

- A stand-alone mode for deployments using refarmed GSM spectrum as small as 400 kHz, facilitating one NB-IoT carrier plus a 100 kHz guard-band toward the surrounding GSM carriers.
- An in-band mode for deployments using a *Long-Term Evolution* (LTE) *physical resource block* (PRB).
- A guard-band mode for deployments using the guard-band of an LTE carrier. This mode of operation is also suitable for operation with an *New Radio* (NR) carrier as described in Section 7.4.

In this chapter, we will present the NB-IoT performance for each of these three operation modes.

In addition to discussing the performance achieved at the targeted *maximum coupling loss* (MCL) level of 164 dB, which represents 20 dB *coverage enhancement* (CE) compared to GSM/GPRS and matches the 3GPP 5G coverage requirement, this chapter discusses the following aspects:

- Best achievable performance for devices, which are in good coverage. The performance as such is then defined by the limitations set by the technology.
- Performance of devices in normal coverage, for example, within 144 dB coupling loss from the base station, which corresponds to the MCL of GSM/GPRS.
- Performance of devices with moderate requirements on extended coverage, i.e., 10 dB CE compared to GSM/GPRS.

Performance of NB-IoT, like any other system, depends heavily on the device and base station implementations and the evaluation assumptions. In the performance evaluations presented in Sections 8.2–8.7, unless mentioned otherwise, we adopt commonly used 3GPP Release 13 assumptions such as those described and referred to in Chapter 4. The assumptions used for ITU and 3GPP 5G performance evaluations are slightly different and will be described in Section 8.9. Any additional NB-IoT specific assumptions will be described as needed throughout this chapter.

Although NB-IoT performance results are abundant in 3GPP contributions and in the 3GPP Technical Report 45.820 *Cellular System Support for Ultra-low Complexity and Low Throughput Internet of Things* [1], most of these results were not evaluated exactly according to the Release 13 specifications. Therefore, the results presented in this chapter are to a large extent based on authors' own evaluations. For certain performance aspects where the performance results are available in the open literature, the results presented herein are well aligned with the archived results in the literature, e.g., References [2–4].

8.2 Coverage and data rate

Like EC-GSM-IoT and LTE-M, all three operation modes of NB-IoT aim to achieve coverage, in terms of MCL, up to 164 dB and at this coverage level a data rate of at least 160 bits per second (bps). In this chapter we start by describing the methodologies used to evaluate the coupling loss for each of the NB-IoT physical channels followed by actual performance evaluations and thereto corresponding results.

8.2.1 Evaluation assumptions

8.2.1.1 Requirements on physical channels and signals

For NB-IoT to meet the coverage requirement, all the physical channels must have *adequate performance* at 164 dB MCL. We will describe what is considered to be adequate performance for various physical channels and signals in the next few subsections.

8.2.1.1.1 Synchronization signals

The synchronization signals, *Narrowband Primary Synchronization Signal* (NPSS) and *Narrowband Secondary Synchronization Signal* (NSSS), need to be detected with a 90% detection rate. A successful detection of NPSS and NSSS includes identifying the physical cell identity, achieving time synchronization to the downlink (DL) frame structure with ~2.5 μs accuracy, and achieving frequency synchronization within ~50 Hz. The 2.5 μs time synchronization and 50 Hz frequency synchronization accuracy is considered *adequate* as these residual errors will only give rise to small performance degradation during the device's subsequent idle and connected modes operation. Note that after the cell selection, the device may employ a timing tracker and *automatic frequency correction* to further refine its time and frequency references.

Because NPSS and NSSS are transmitted periodically, in principle, the device can extend its synchronization time to achieve better coverage. This is where the latency requirement presented in Section 8.4 becomes relevant. The required synchronization time for achieving a 90% detection rate must allow an overall service latency requirement of 10 s, to be met. Efficient synchronization does also improve the device battery life evaluated in Section 8.5.

8.2.1.1.2 Control and broadcast channels

The Narrowband Physical Broadcast Channel (NPBCH) carries the *NB-IoT Master Information Block* (MIB-NB), which needs to be detected with 90% probability, i.e., support a 10% *block error rate* (BLER). Like in the case of the synchronization signals, a device can in principle compensate for a poor receiver implementation by repeated attempts to decode the NPBCH until the MIB-NB is received successfully. As in the case of the synchronization signals, this would, however, negatively impact the overall performance. Efficient NPBCH acquisition improves both latency and battery life.

The Narrowband Physical Downlink Control Channel (NPDCCH) carries *Downlink Control Information* (DCI), which is a control message containing information for *Narrowband Physical Uplink Shared Channel* (NPUSCH) and *Physical Downlink Shared Channel* (NPDSCH) scheduling. In these evaluations, the NPDCCH is evaluated at a BLER target of 1% to secure robust operation even at the most extreme CE levels.

The NPUSCH Format 2 (F2) carries *hybrid automatic repeat request acknowledgments* (HARQ-ACK) for NPDSCH transmissions. The requirement for such signaling is just as the NPDCCH targeting a 1% BLER.

For the *Narrowband Physical Radom Access Channel* (NPRACH), a 1% miss detection rate is targeted. A higher miss detection rate target may facilitate higher capacity on the NPRACH, which just as the EC-GSM-IoT *Extended Coverage Random Access Channel* and LTE-M *Physical Random Access Channel*, is a collision based channel. However, to secure a robust *timing advance*, estimated on the NPRACH, from the very start of the NB-IoT connection, a low miss detection rate target has been chosen. A timing advance estimation accuracy in the range of 3 μs is considered *adequate* as this will secure good NPUSCH reception performance at the base station.

8.2.1.1.3 Traffic channels

The NPDSCH and NPUSCH Format 1 (F1) carry downlink and uplink data for which the achievable data rate is a suitable criterion. For NB-IoT, a data rate of 160 bps is required at the MCL of 164 dB. In this chapter, the data rate is evaluated after the first HARQ transmission for which a BLER requirement of 10% is targeted. An erroneous transmission can be corrected through HARQ retransmissions. Thus, a 10% BLER for the initial transmission is more than adequate.

8.2.1.2 Radio related parameters

The radio related parameters used in the NB-IoT evaluations in Sections 8.2–8.7 are summarized in Table 8.1. These assumptions were agreed in 3GPP [1], and most of them are identical to those used for the EC-GSM-IoT evaluations as summarized in Section 4.2.1.

On the device side, the oscillator accuracy assumed is 20 *parts per million* (ppm), which reflects the expected accuracy of oscillators used in an ultra-low-cost device. For the 900 MHz band, such an oscillator accuracy gives rise to an initial frequency error up to 18 or -18 kHz.

TABLE 8.1 Simulation assumptions.

Parameter	Value
Frequency band	900 MHz
Propagation condition	Typical Urban (see Ref. [5])
Fading	Rayleigh, 1 Hz Doppler spread
Device initial oscillator inaccuracy	20 ppm (applied to initial cell selection)
Raster offset	Stand-alone: 0 Hz; in-band and guard-band: 7.5 kHz
Device frequency drift	22.5 Hz/s
Device NF	5 dB
Device antenna configuration	One transmit antenna and one receive antenna
Device power class	23 dBm
Base station NF	3 dB
Base station antenna configuration	Stand-alone: one transmit antenna and two receive antennas
	In-band and guard-band: two transmit antenna ports and two receive antennas
Base station power level	43 dBm (stand-alone), 35 dBm (in-band and guard-band) per 180 kHz
Number of NPDCCH/NPDSCH REs per subframe	Stand-alone 160; in-band: 104; guard-band: 152
Valid NB-IoT subframes	All subframes not carrying NPBCH, NPSS, and NSSS are assumed valid subframes

As mentioned in Section 7.2.1.1, for in-band and guard-band operations, there is also a frequency raster offset contributing to the initial frequency offset. The raster offset can be as large as ±7.5 kHz. Considering both device oscillator inaccuracy and a worst-case raster offset, the initial cell selection is thus evaluated with a frequency offset uniformly distributed within the interval of [7.5−18, 7.5 + 18] kHz, i.e., [−10.5, 25.5] kHz, or [−7.5−18, −7.5 + 18 kHz], i.e., [−25.5, 10.5] kHz. Such a frequency offset not only introduces a time varying phase rotation in the baseband signal, but also results in a timing drift. For example, a frequency offset of 25.5 kHz (including the raster offset) at 900 MHz translates back to a 28.3 ppm error, which means for every 1 million samples, the timing reference at the device may drift by 28.3 samples relative to a correct timing reference. For initial cell selection evaluation, the effect of timing drift is modeled according to the exact initial frequency offset. One of the objectives of the initial cell selection is to reduce such an initial frequency offset to a low value so that the reception, or transmission, of other physical channels in the subsequent communication with the base station will not suffer serious degradation. Considering that the oscillator used in a low-cost device may not hold on to a precise frequency, a frequency drift after initial frequency offset correction is also modeled. The assumed drift rate is 22.5 Hz/s. In addition to this, the *noise figure* (NF) assumed is 5 dB, and all the evaluations are done for devices equipped with one transmit antenna and one receive antenna using a maximum transmit power level of 23 dBm.

On the base station side, the NF assumed is 3 dB. Regarding antenna configuration and power level, the assumptions reflect different deployment scenarios for different operation modes. The stand-alone mode is intended for deployment in an existing GSM network. Most of the GSM base stations use one transmit antenna and 43 dBm transmit power and these assumptions are reused for NB-IoT stand-alone operation. The in-band and guard-band modes are intended for deployment in an LTE network. An LTE base station typically uses multiple transmit antennas or multiple antenna ports with each transmit branch associated with a power amplifier of maximum power level of 43 dBm. The notion of *antenna port* is a logical concept as each antenna port may be associated with multiple physical antennas. Fast fading channel coefficients of different antenna ports are generally assumed independent. Thus, the number of antenna ports is related to the degree of spatial diversity. 3GPP ended up using the assumption of two transmit antenna ports and thus a 46 dBm maximum base station total transmit power level when summed over the two transmit antenna ports. However, note that such a total transmit power level is over the entire LTE carrier bandwidth. For example, a total 46 dBm transmit power over 10 MHz LTE carrier bandwidth means that one PRB can only have 29 dBm if the power is evenly distributed over all the 50 PRBs within the LTE carrier. It was a common assumption during 3GPP standardization that PRB power boosting can be applied on an NB-IoT anchor carrier. For in-band and guard-band deployments within a 10 MHz LTE carrier, 6 dB power boosting has been assumed. This results in 35 dBm power for the anchor NB-IoT carrier. For LTE carrier bandwidth of 5, 15, and 20 MHz, power boosting of 3, 7.8, and 9 dB are assumed. This results in a 35 dBm transmit power level for an NB-IoT anchor carrier for all these LTE bandwidth variants. Note that power boosting is only assumed for the anchor carrier, and for a non-anchor carrier it is assumed that no power boosting is applied although this is a permitted configuration. Finally, for all modes of operations two base station receive antennas are assumed.

Furthermore, the amount of available resources for NB-IoT operation also depends on the operation mode. Obviously, for the in-band mode certain resource elements (REs) in the downlink are taken by the legacy LTE channels, giving rise to fewer resources available to NB-IoT. This has impact on performance. In a subframe, there are 12 subcarriers and 14 *orthogonal frequency-division multiplexing* (OFDM) symbols, giving rise to a total of 168 resource elements per subframe. The stand-alone operation assumes one *Narrowband Reference Signal* (NRS) antenna port, which takes 8 resource elements per subframe, see Section 7.2.4.3. Thus, the number of resource elements available to NPDCCH or NPDSCH per subframe is 160. For the guard-band operation, 2 NRS antenna ports are assumed, which need 16 resource elements for NRS per subframe, leaving 152 resource elements available to NPDCCH or NPDSCH per subframe. For the in-band operation, we study a case where the first 3 OFDM symbols in every subframe are taken by LTE *Physical Downlink Control Channel* (PDCCH) region and there are 2 *Cell-specific Reference Signal* (CRS) antenna ports. The total number of resource elements taken by LTE PDCCH and CRS per subframe is 48. In addition, the 2 NRS ports take 16 resource elements, leaving 104 resource elements available to NPDCCH or NPDSCH per subframe.

The assumptions described for in-band and guard-band modes of operation are also applicable for a NB-IoT carrier operating within a NR carrier. Most relevant for this deployment scenario is the NB-IoT guard-band mode of operation.

8.2.2 Downlink coverage performance

As discussed in Section 8.2.1, the base station transmit power level as well as antenna configuration and the number of resource elements for certain downlink channels depend on the operation mode. We will therefore present downlink coverage performance for all the three different operation modes. The downlink link budget for the 3 operation modes to achieve 164 dB MCL are presented in Table 8.2. As shown, the required *signal-to-interference-plus-noise power ratio* (SINR) is −4.6 dB for the stand-alone operation and −12.6 dB for the in-band and guard-band operations. We will therefore discuss the performance of all the downlink physical channels at these SINRs. For completeness, the performance at the edge of the normal coverage level, i.e., 144 dB coupling loss, and 10 dB CE level, i.e., 154 dB coupling loss, will also be shown. The required downlink SINRs for 154 and 144 dB coupling loss levels are 5.4 and 15.4 dB, respectively, for the stand-alone mode, and −2.6 and 7.4 dB, respectively, for both the in-band and guard-band modes.

8.2.2.1 Synchronization signals

Initial synchronization performance in terms of required synchronization time is shown in Tables 8.3–8.5 for stand-alone, in-band, and guard-band operations, respectively. Note that the stand-alone operation has the best performance thanks to a higher transmitted power level. The in-band and guard-band operations have the same transmit power level and

TABLE 8.2 NB-IoT downlink link budget for achieving 164 dB MCL in different operation modes.

#	Operation mode	Stand-alone	In-band	Guard-band
L1	Total base station Tx power [dBm]	43	46	46
L2	Base station Tx power per NB-IoT carrier [dBm]	43	35	35
L3	Thermal noise power spectral density [dBm/Hz]	−174	−174	−174
L4	Receiver NF [dB]	5	5	5
L5	Interference margin [dB]	0	0	0
L6	Channel bandwidth [kHz]	180	180	180
L7	Effective noise power [dBm] = (L3) + (L4) + (L5) + 10 log10(L6)	−116.4	−116.4	−116.4
L8	Required downlink SINR [dB]	−4.6	−12.6	−12.6
L9	Receiver sensitivity [dBm] = (L7) + (L8)	−121.0	−129.0	−129.0
L10	Receiver processing gain [dB]	0	0	0
L11	Coupling loss [dB] = (L2) − (L9) + (L10)	164.0	164.0	164.0

TABLE 8.3 NB-IoT initial synchronization performance for stand-alone operation (The 144 and 164 dB performances are based on Reference [2]).

Coupling loss	144 dB	154 dB	164 dB
Synchronization time [ms] for 50% devices	24	24	64
Synchronization time [ms] for 90% devices	84	104	264
Synchronization time [ms] for 99% devices	144	194	754
Average synchronization time [ms]	36	42	118

TABLE 8.4 NB-IoT initial synchronization performance for in-band operation (The 144 and 164 dB performances are based on Reference [2]).

Coupling loss	144 dB	154 dB	164 dB
Synchronization time [ms] for 50% devices	24	44	434
Synchronization time [ms] for 90% devices	84	124	1284
Synchronization time [ms] for 99% devices	154	294	2604
Average synchronization time [ms]	38	64	582

TABLE 8.5 NB-IoT initial synchronization performance for guard-band operation.

Coupling loss	144 dB	154 dB	164 dB
Synchronization time [ms] for 50% devices	24	44	354
Synchronization time [ms] for 90% devices	84	124	1014
Synchronization time [ms] for 99% devices	154	294	2264
Average synchronization time [ms]	38	60	470

thus have similar performance. The slightly longer synchronization time in the in-band case is because of that LTE CRS punctures NPSS/NSSS, giving rise to performance degradation due to puncturing. However, as indicated by the results in Tables 8.4 and 8.5, the impact of NPSS/NSSS puncturing is small.

8.2.2.2 *NPBCH*

NPBCH is carrying the MIB-NB. All the three operation modes share the same resource mapping and thus the performance of MIB-NB acquisition only depends on the transmit power level and coverage. The NPBCH has a 640 ms *transmission time interval* (TTI) in which every subframe 0 carries a code subblock as illustrated in Section 7.2.4.4. When the device is in good coverage, receiving one NPBCH subframe may be sufficient to acquire MIB-NB.

Recall that each code subblock is self-decodable and therefore the smallest MIB-NB acquisition time is only 10 ms, as the device can receive one NPBCH subframe in 10 ms. However, for devices in poor coverage, combining multiple NPBCH repetitions and multiple NPBCH code subblocks is necessary for the MIB-NB to be acquired. There are 64 NPBCH subframes in an NPBCH TTI. Thus, the device can jointly decode up to 64 NPBCH subframes. If the device still fails to acquire MIB-NB after jointly decoding 64 NPBCH subframes, the most straightforward approach is to reset the decoder memory and start a new decoding attempt in the next NPBCH TTI. With such an approach, the device simply tries again in the new TTI and sees if it can acquire MIB-NB. This is often referred to as the *keep-trying* algorithm. The results presented below are based on the *keep-trying* algorithm.

NB-IoT physical channel performance when channel estimation is required depends heavily on how the receiver acquires its channel estimate. In coverage limited scenarios, using NRS in one subframe may not be enough to ensure adequate channel estimation accuracy. Cross-subframe channel estimation, where NRS from multiple subframes are jointly used for channel estimation, is highly recommended. We will highlight the impact of channel estimation accuracy when the NPUSCH Format 1 performance is discussed (see Section 8.2.3.2). For now, it suffices to point out that for the NPBCH results presented below, cross-subframe channel estimation with NRS collected within a 20 ms window is used.

Tables 8.6 and 8.7 show the MIB-NB acquisition time required for achieving a success rate higher than 90%. The average acquisition time is also shown. Because the in-band and guard-band operations share the same transmit power level, they also share the same MIB-NB acquisition performance.

The keep-trying algorithm, although straightforward, is far from optimal. In MIB-NB, the information element that surely changes from one TTI to the next is the 6 bits used to indicate *system frame number* (SFN) and hyper SFN. This information changes in a predictable manner. If representing these bits as a decimal number n in one NPBCH TTI, the number in the next TTI is $n + 1 \bmod 64$. Exploiting this relationship, it is possible to jointly decode NPBCH subframes across TTI boundaries and significantly improve the performance of MIB-NB acquisition. Besides the SFN number, the 1-bit *access barring* (AB) indicator may change more often

TABLE 8.6 MIB-NB acquisition time for stand-alone operation.

Coupling loss	144 dB	154 dB	164 dB
Acquisition time [ms] required for achieving 90% success rate	10	20	170
Average acquisition time [ms]	10.3	16.2	83.9

TABLE 8.7 MIB-NB acquisition time for in-band and guard-band operation.

Coupling loss	144 dB	154 dB	164 dB
Acquisition time [ms] required for achieving 90% success rate	10	60	640
Average acquisition time [ms]	11.1	37.5	357.1

TABLE 8.8 NPDCCH performance on an anchor carrier for stand-alone operation.

Coupling loss	**144 dB**	**154 dB**	**164 dB**
Repetition required for achieving 1% BLER	1	8	128
Total TTI required for achieving 1% BLER [ms]	1	9	182
Average time required for device to receive DCI correctly [ms]	1.0	1.6	18.0

TABLE 8.9 NPDCCH performance on an anchor carrier for in-band operation.

Coupling loss	**144 dB**	**154 dB**	**164 dB**
Repetition required for achieving 1% BLER	2	32	256
Total TTI required for achieving 1% BLER [ms]	2	44	364
Average time required for device to receive DCI correctly [ms]	1.1	5.4	78.9

TABLE 8.10 NPDCCH performance on an anchor carrier for guard-band operation.

Coupling loss	**144 dB**	**154 dB**	**164 dB**
Repetition required for achieving 1% BLER	2	16	256
Total TTI required for achieving 1% BLER [ms]	2	22	364
Average time required for device to receive DCI correctly [ms]	1.0	3.3	51.8

than the other remaining information elements in MIB-NB. The MIB-NB decoder may need to hypothesize whether the AB indicator changes or not if it wants to jointly decode NPBCH subframes across TTI boundaries. Examples of advanced MIB-NB decoder can be found in Ref. [6]. It shows that it is possible to achieve a performance gain up to 2 dB, compared to the keep-trying algorithm, at a very low SINR region where the acquisition time might be longer than one NPBCH TTI.

8.2.2.3 NPDCCH

The performance of NPDCCH is shown in Tables 8.8–8.10. As described in Section 7.2.4.5, NPDCCH may use a repetition factor as low as one and as high as 2048. The repetition factors required for achieving a 1% BLER at DCI reception for the three coverage levels and the three operation modes are shown. Because each repetition of NPDCCH is mapped to one subframe, each repetition corresponds to 1 ms transmission time. However, not all subframes are available for NPDCCH. On the anchor carrier, NPDCCH transmissions need to skip subframes taken by NPBCH, NPSS, NSSS, and *NB-IoT System Information Block Type 1* (SIB1-NB). For example, for repetition factor 8, there are no 8 consecutive subframes available for NPDCCH. The shortest possible total TTI to fit in 8 NPDCCH subframes on an anchor carrier is 9 ms. This is illustrated in Fig. 8.1.

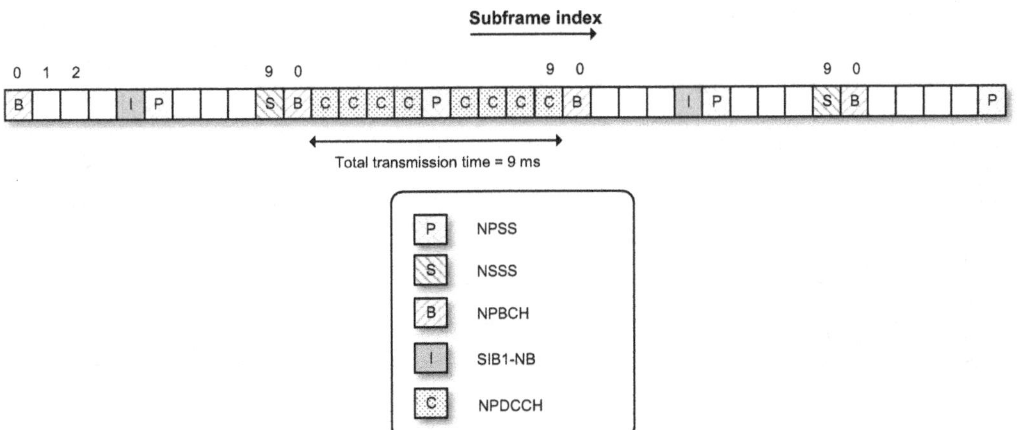

FIG. 8.1 Illustration of NPDCCH with repetition factor 8 will have at least 9 ms total transmission time on an anchor carrier.

8.2.2.4 NPDSCH

The performance of NPDSCH for different coverage levels and the three operation modes is shown in Tables 8.11–8.13. Here, we only show the performance for the maximum *transport block size* (TBS) of 680 bits as defined in 3GPP Release 13. In general, the entries on the same row of the TBS table (see Section 7.2.4.6) are expected to have approximately the same performance. Like NPDCCH, on the anchor carrier the subframes available for NPDSCH need to exclude those taken by NPBCH, NPSS, NSSS, and SIB1-NB. This results in the total transmission time longer than the total number of subframes needed for NPDSCH. We see that at 164 dB MCL, the data rates of NPDSCH measured over the total transmission interval of NPDSCH are 2.5, 0.47, and 0.62 kbps, for stand-alone, in-band, and guard-band operations, respectively. As described in Section 7.3.2.2, NPDSCH transmissions are scheduled by DCI transmitted in the NPDCCH and acknowledged by the HARQ-ACK bit carried in NPUSCH Format 2. There are also timing relationships between

TABLE 8.11 NPDSCH performance on the anchor carrier for stand-alone operation.

Coupling loss	**144 dB**	**154 dB**	**164 dB**
TBS [bits]	680	680	680
Number of subframes per repetition	4	6	6
Number of repetitions	1	4	32
Number of subframes used for NPDSCH transmission	4	24	192
Total TTI required [ms]	4	32	272
Data rate measured over NPDSCH TTI [kbps]	170	21.3	2.5
MAC-layer data rate [kbps]	19.1	8.7	1.0

TABLE 8.12 NPDSCH performance on the anchor carrier for in-band operation.

Coupling loss	144 dB	154 dB	164 dB
TBS [bits]	680	680	680
Number of subframes per repetition	10	8	8
Number of repetitions	1	16	128
Number of subframes used for NPDSCH transmission	10	128	1024
Total TTI required [ms]	12	182	1462
Data rate measured over NPDSCH TTI [kbps]	56.7	3.7	0.47
MAC-layer data rate [kbps]	15.3	2.4	0.31

TABLE 8.13 NPDSCH performance on the anchor carrier for guard-band operation.

Coupling loss	144 dB	154 dB	164 dB
TBS [bits]	680	680	680
Number of subframes per repetition	8	5	6
Number of repetitions	1	16	128
Number of subframes used for NPDSCH transmission	8	80	768
Total TTI required [ms]	9	112	1096
Data rate measured over NPDSCH TTI [kbps]	75.6	6.1	0.62
MAC-layer data rate [kbps]	15.3	3.8	0.37

these different channels in that specific timing gaps are defined between these physical channels. When all these factors are accounted for, the effective data rates are lower than if simply measured over the total NPDSCH transmission interval. In essence, such an effective data rate is the throughput perceived at the egress point of the *medium access control* (MAC) protocol layer. Therefore, for the rest of this chapter, we will simply refer to this data rate as the MAC-layer data rate. More details and examples will be given in Section 8.3. In Tables 8.11–8.13, the MAC-layer data rates are presented based on anchor carrier configuration when the overheads such as NPSS, NSSS, NPBCH, and SIB1-NB are accounted for.

8.2.3 Uplink coverage performance

Unlike the downlink, the UL link budget does not depend on the operation mode. Thus, the results presented in the next subsections, if the operation mode is not explicitly specified, apply to all three operation modes.

TABLE 8.14 NPRACH link budget.

#	Coupling loss	144 dB	154 dB	164 dB
L1	Total device Tx power [dBm]	23	23	23
L2	Thermal noise power spectral density [dBm/Hz]	−174	−174	−174
L3	Base station receiver NF [dB]	3	3	3
L4	Interference margin [dB]	0	0	0
L5	Channel bandwidth [kHz]	3.75	3.75	3.75
L6	Effective noise power [dBm] = (L2) + (L3) + (L4) + 10 log10(L5)	−135.3	−135.3	−135.3
L7	Required UL SINR [dB]	14.3	4.3	-5.7
L8	Receiver sensitivity [dBm] = (L6) + (L7)	−121.0	−131.0	−141.0
L9	Receiver processing gain [dB]	0.0	0.0	0.0
L10	Coupling loss [dB] = (L1) − (L8) + (L9)	144.0	154.0	164.0

8.2.3.1 NPRACH

The NPRACH link budget for the three coverage levels is shown in Table 8.14 based on the 23 dBm device power class. The link budget establishes the required SINR for NPRACH for each coverage level. Based on the required SINR, a suitable repetition factor is needed to achieve the required performance. A good description of NPRACH detection algorithm and its performance can be found in Ref. [7]. Table 8.15 shows the repetition level needed for each coverage level, and the achieved performance. The results in Table 8.15 are based on Reference [7].

8.2.3.2 NPUSCH format 1

We have mentioned that the performance of NB-IoT physical channels depend on the accuracy of the channel estimation. To achieve good channel estimation accuracy, cross-subframe channel estimation is highly recommended. An example is shown in Fig. 8.2. It can be seen that there is a substantial performance difference between single subframe (SF)

TABLE 8.15 NPRACH performance (based on NPRACH Format 1).

Coupling loss	144 dB	154 dB	164 dB
Number of repetitions	2	8	32
Total TTI required [ms]	13	52	205
Detection rate	99.71%	99.76%	99.16%
False alarm rate	0/100,000	0/100,000	13/100,000
TA estimation error [μs]	[−3, 3]	[−3, 3]	[−3, 3]

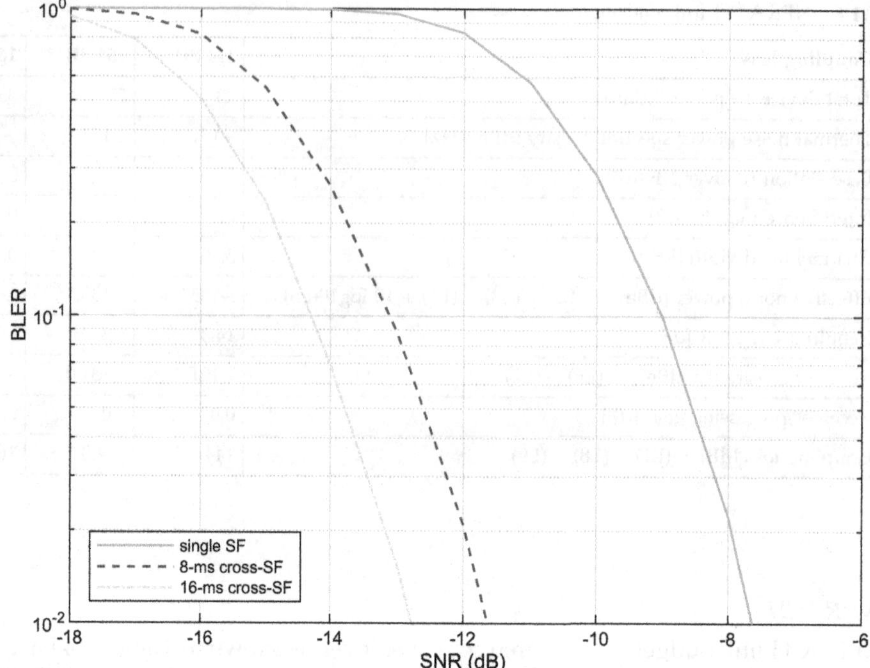

FIG. 8.2 NPUSCH Format 1 performance. Single-tone transmission with 15 kHz numerology, TBS 1000 bits, 80-ms transmission time per repetition, and 64 repetitions.

based channel estimation and cross-subframe channel estimation. All the NPUSCH results presented in the remainder of this chapter are based on cross-subframe channel estimation with an 8 ms estimation window, if not specified otherwise. For cases when the total NPUSCH TTI is lower than 8 ms, cross-subframe channel estimation is based on using all the NPUSCH subframes.

As described in Section 7.2.5.2, there are a few different transmission configurations for NPUSCH Format 1. Fig. 8.3 shows the performance of these different configurations. The data rates shown here are NPUSCH Format 1 data rates measured over the total NPUSCH transmission interval and do not account for the scheduling aspects. It can be seen that when coupling loss is high (e.g., greater than 150 dB), all these transmission configurations achieve approximately the same performance. In such an operation regime, it is thus advantageous to use a more spectrally efficient signal-tone transmission. However, for devices in good coverage, multi-tone transmission can be used to allow these devices to transmit at a higher data rate. As shown, when the coupling loss is lower than 141 dB, the 12-tone (i.e. using one full PRB) configuration achieves the highest data rate.

NPUSCH Format 1 performance at 144, 154, and 164 dB coupling loss is summarized in Table 8.16. Here we choose the most suitable transmission configuration for each of these coupling loss levels. Like the performance of NPDSCH shown in Section 8.2.2.4, two types of data rates are shown in Table 8.16. First, data rates measured over total NPUSCH Format

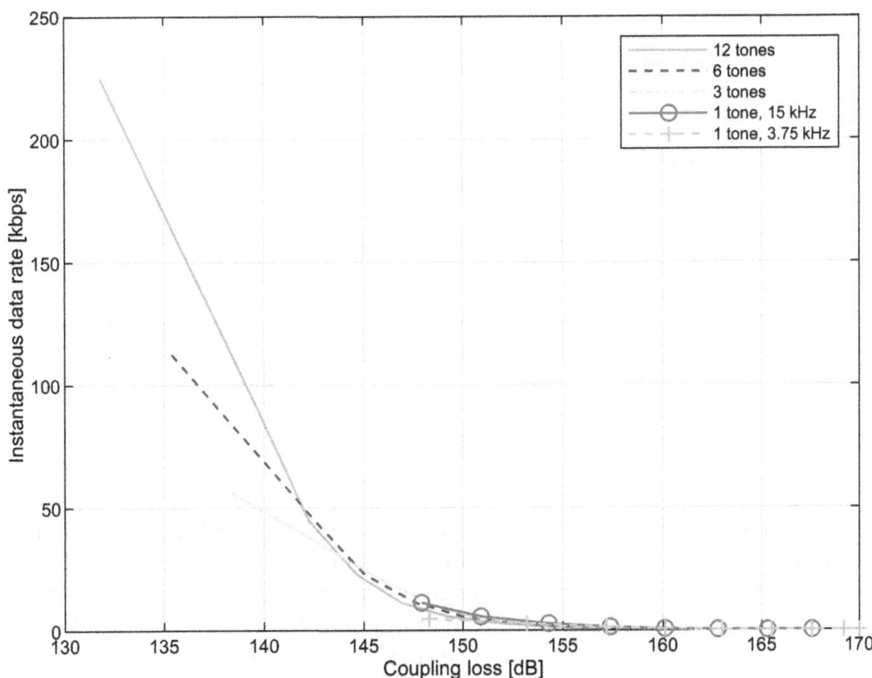

FIG. 8.3 NPUSCH Format 1 performance, physical-layer data rate versus coupling loss for various transmission configurations (TBS = 1000 bits, 8-ms cross-subframe channel estimation).

TABLE 8.16 NPUSCH Format 1 performance.

Coupling loss	**144 dB**	**154 dB**	**164 dB**
TBS [bits]	1000	1000	1000
Subcarrier spacing [kHz]	15	15	15
Number of tones	3	1	1
Number of resource units per repetition	8	10	10
Number of repetitions	1	4	32
Total TTI required [ms]	32	320	2560
Data rate measured over NPUSCH Format 1 TTI [kbps]	28.1	2.8	0.371
MAC-layer data rate, stand-alone [kbps]	18.8	2.6	0.343
MAC-layer data rate, in-band [kbps]	18.7	2.4	0.320
MAC-layer data rate, guard-band [kbps]	18.7	2.5	0.320

TABLE 8.17 NPUSCH Format 2 performance based on 1% BLER.

Number of repetitions	TTI [ms]	MCL [dB]
1	2	152.2
2	4	155.0
4	8	157.2
8	16	159.2
16	32	161.2
32	64	163.6
64	128	165.5
128	256	167.6

1 transmission interval are shown. In addition, we also show the MAC-layer data rates for which the time it takes to signal the scheduling grant, via NPDCCH, and the associated timing relationship are accounted for. Note that because NPDCCH performance depends on the operation mode, uplink MAC-layer data rates at different coverage levels also depend on the operation mode. To get the performance at 164 dB coupling loss as shown in Table 8.16, an improved channel estimation with 32-ms cross-subframe channel estimation is used. Using a higher degree of cross-subframe channel estimation requires a higher baseband complexity at the base station. Furthermore, performance improvement from more aggressive cross-subframe channel estimation will be more significant for devices of low mobility. According to the device coupling distribution in Ref. [1], there is only a small percentage of devices requiring an extreme CE level. Thus, the base station can afford to use more advanced channel estimator to help these devices improve performance. Furthermore, devices that need extreme CE are expected to be located indoors or underground. These devices are low mobility and indeed will benefit from more aggressive cross-subframe channel estimation.

8.2.3.3 NPUSCH format 2

NPUSCH Format 2 is used to transmit HARQ-ACK for NPDSCH and its performance is summarized in Table 8.17. It can be seen that the target of 164 dB coupling loss can be achieved with 64 repetitions.

8.3 Peak data rates

8.3.1 Release 13 Cat-NB1 devices

NB-IoT supports a range of data rates. For a device in good coverage, a higher data rate can be used, which leads to shortened reception and transmission times. This allows the device to return to a deep sleep state sooner, thereby improving the battery lifetime. In this section, we discuss the peak physical-layer data rates and MAC-layer data rates

TABLE 8.18 Peak physical-layer data rates (Release 13 Category NB1 devices).

	Stand-alone [kbps]	In-band [kbps]	Guard-band [kbps]
NPDSCH	226.7	170.0	226.7
NPUSCH multi-tone	250.0	250.0	250.0
NPUSCH single-tone (15 kHz)	21.8	21.8	21.8
NPUSCH single-tone (3.75 kHz)	5.5	5.5	5.5

achievable for Release 13 Cat-NB1 devices. The physical-layer data rate is measured over the NPDSCH or NPUSCH Format 1 TTI. The MAC-layer data rate in addition accounts for the time it takes to receive or transmit other control signaling, for example, DCI or HARQ-ACK, and the associated timing relationship. This corresponds to the data rate at which the MAC-layer *packet data units* are delivered.

The physical-layer data rates are purely determined based on the NPDSCH and NPUSCH configurations. For example, the maximum Release 13 NPDSCH TBS is 680 bits and can be mapped to 3 subframes, i.e., 3 ms, in the stand-alone and guard-band modes. The peak downlink physical-layer data rate is thus 226.7 kbps in these operation modes. For the in-band mode, the shortest time duration for NPDSCH carrying a 680-bit transport block is 4 ms, giving a peak physical-layer downlink data rate of 170 kbps. NB-IoT peak physical-layer data rates are summarized in Table 8.18. The peak physical-layer data rate indicates the optimal spectral efficiency that can be achieved by a technology, and is often used as a performance metric for cross-technology comparison.

The peak physical-layer data rates, do not account for the protocol aspects, and thus may not be a good indicator for the user experienced data rates. For NB-IoT, both NPDCCH search space configuration (see Section 7.3.2.1) and the timing relationship (see Section 7.3.2.2) affect the MAC-layer data rate. A non-anchor carrier scheduling example for a Cat-NB1 device is illustrated in Fig. 8.4. The device receives a series of 680-bit transport blocks. Each transport block is delivered in 3 NPDSCH subframes. This corresponds to the highest physical-layer data rate in the stand-alone and guard-band modes. Here, the NPDCCH search space is configured to have $R_{max} = 2$ and $G = 4$, giving a search period of $R_{max} \cdot G = 8$ ms. Recall that the device is not expected to monitor search space candidates after it receives a DCI and until 3 ms after it completes its HARQ feedback. For example, the search space candidate in the 24th subframe in Fig. 8.4 will not be monitored by the device. The earliest opportunity that the base station can send the second DCI to the device is the 25th subframe. The timing relationship illustrated is according to the minimum gaps between the NPDCCH DCI and NPDSCH and between the NPDSCH and NPUSCH Format 2 allowed by the specifications. The time between the NPDCCH DCIs can be thought of as the time it takes to transmit each transport block. As shown, it takes 56 ms to complete two full scheduling cycles delivering two transport blocks of 680 bits, giving rise to a MAC-layer data rate of 24.3 kbps. An anchor carrier example is more complicated as the presence of NPSS, NSSS, and NPBCH results in postponements of NPDCCH and NPDSCH transmissions. Such postponements may in some cases end up resulting in a slightly higher MAC-layer data rate. This

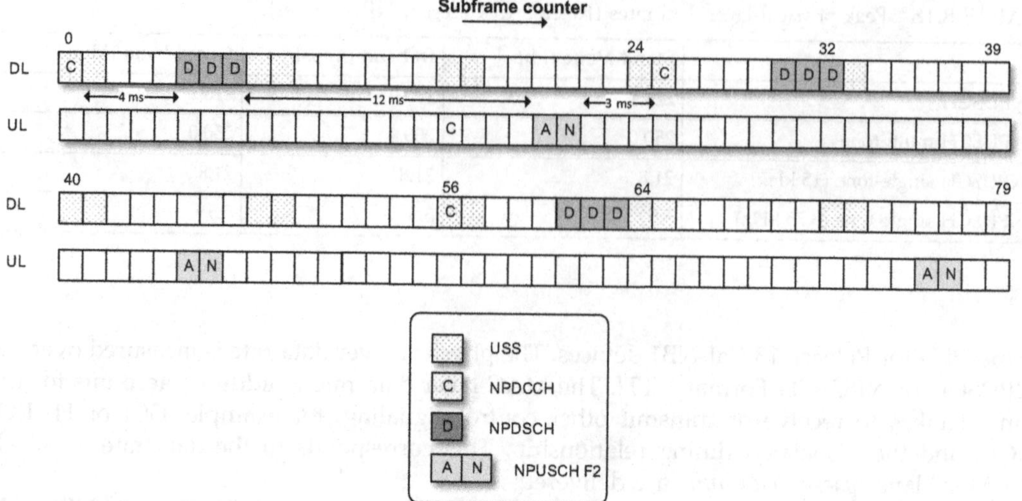

FIG. 8.4 Example of a Release 13 device receiving a series of transport blocks of 680 bits, each in 3 subframes on a non-anchor carrier (stand-alone and guard-band). NPDCCH search space configuration includes $R_{max} = 2, G = 4$.

TABLE 8.19 Peak MAC-layer data rate (Release 13 Category NB1 devices).

	Stand-alone [kbps]	In-band [kbps]	Guard-band [kbps]
NPDSCH	26.2	24.3	26.2
NPUSCH multi-tone	62.6	62.6	62.6
NPUSCH single-tone (15 kHz)	15.6	15.6	15.6
NPUSCH single-tone (3.75 kHz)	4.8	4.8	4.8

is due to that a postponement of NPDCCH extends the time for certain search space candidates, which in some cases makes it possible to send a DCI to the device without waiting for an extra NPDCCH search period.

Table 8.19 summarizes the peak MAC-layer data rate. For both the downlink and uplink, the MAC-layer data rate for a Cat-NB1 device is achieved by configuring the user-specific search space with $R_{\max} = 4$ and $G = 2$. Observe that the differences between the peak physical-layer data rates and MAC-layer data rate for the NPUSCH single-tone cases are relatively small. This is due to the actual NPUSCH transmissions take much longer time for the single-tone cases, and therefore the timing relationship and NPDCCH search space constraint have smaller impact on the MAC-layer data rates.

8.3.2 Cat-NB2 devices configured with 1 HARQ process

As described in Chapter 7, 3GPP Release 14 introduces a new device category – Cat-NB2, which supports a TBS up to 2536 bits. Additionally, Release 14 supports the configuration of

TABLE 8.20 Peak physical-layer data rates (Release 14 Category NB2 devices configured with 1 HARQ process).

	Stand-alone [kbps]	In-band [kbps]	Guard-band [kbps]
NPDSCH	258.0	174.4	258.0
NPUSCH multi-tone	258.0	258.0	258.0
NPUSCH single-tone (15 kHz)	21.8	21.8	21.8
NPUSCH single-tone (3.75 kHz)	5.5	5.5	5.5

TABLE 8.21 Peak MAC-layer data rates (Release 14 Category NB2 devices configured with 1 HARQ process).

	Stand-alone [kbps]	In-band [kbps]	Guard-band [kbps]
NPDSCH	79.3	54.3	79.3
NPUSCH multi-tone	105.7	105.7	105.7
NPUSCH single-tone (15 kHz)	18.1	18.1	18.1
NPUSCH single-tone (3.75 kHz)	5.2	5.2	5.2

a Cat-NB2 device with two simultaneously active HARQ processes, although this is an optional device feature. We discuss the peak data rates for Cat-NB2 devices configured with 1 HARQ process in this subsection. The peak data rates for devices configured with two simultaneous HARQ processes will be discussed in next subsection.

The peak physical-layer data rates and MAC-layer data rates of Cat-NB2 devices configured with 1 HARQ process are shown in Tables 8.20 and 8.21, respectively. Observe that increasing the maximum TBS does not really change the peak physical-layer data rates by much. This is because that a larger TBS requires a longer transmission time in most cases. For example, on the highest row of the TBS table (see, e.g. Table 7.9), doubling the TBS results in approximately doubling the transmission time required to deliver the transport block. However, doubling the TBS will not result in doubling the scheduling cycle, and therefore will give rise to a more significant increase in the MAC-layer data rate. This can be observed by comparing Tables 8.21 to Table 8.19.

8.3.3 Devices configured with two simultaneous HARQ processes

The peak MAC-layer data rate can be further increased if the device is configured with two HARQ processes, as shown in Tables 8.22. Fig. 8.5 illustrates how using two simultaneously active HARQ processes increases the MAC-layer data rate. Like the example in Fig. 8.4, a device is receiving a series of transports blocks of 680 bits. Each transport block is delivered using 3 subframes of NPDSCH, which corresponds to the peak physical-layer throughput for

TABLE 8.22 Peak MAC-layer data rates (Release 14 Category NB2 devices configured with two HARQ processes).

	Stand-alone [kbps]	In-band [kbps]	Guard-band [kbps]
NPDSCH	127.3	87.1	127.3
NPUSCH multi-tone	158.5	158.5	158.5
NPUSCH single-tone (15 kHz)	20.0	20.0	20.0
NPUSCH single-tone (3.75 kHz)	5.2	5.2	5.2

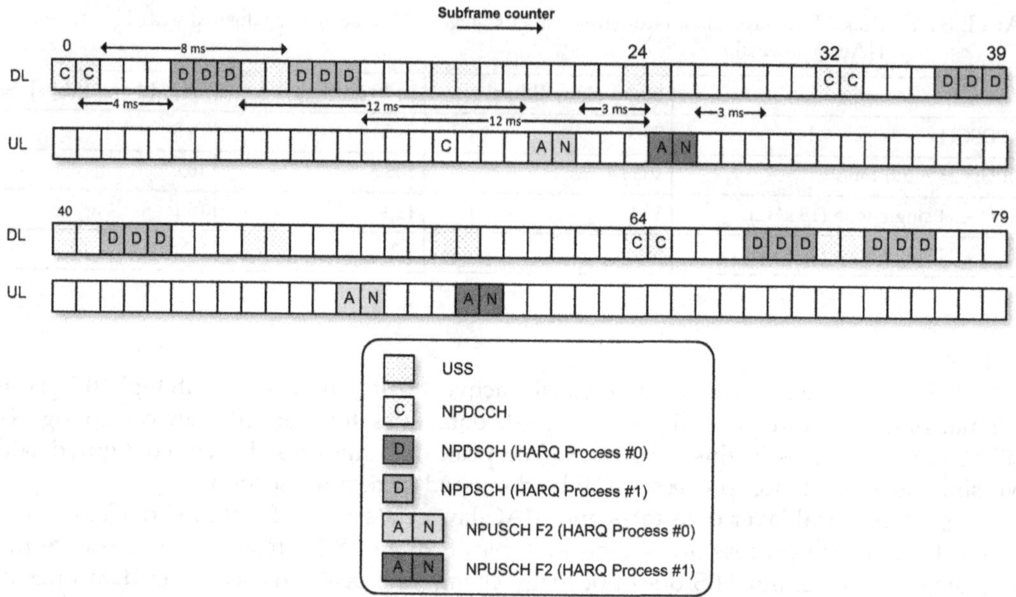

FIG. 8.5 Example of a Cat-NB1 device, configured with two simultaneously active HARQ processes, receiving a series of transport blocks of 680 bits, each in 3 subframes on a non-anchor carrier (stand-alone and guard-band operation modes). NPDCCH search space configuration includes $R_{max} = 2$, $G = 4$.

a Cat-NB1 device in the stand-alone and guard-band modes. As shown, there are two successive NPDCCH subframes used for scheduling the two HARQ processes, each specifying a timing relationship according to what is described in Section 7.3.2.2. Observe that compared to the example of Fig. 8.4, with two simultaneously active HARQ processes, the device can receive a total of four 680-bit transport blocks in every 64 subframes, giving rise to a MAC-layer data rate of 42.5 kbps

For single-tone NPUSCH transmissions however, using two active HARQ processes does not improve the peak MAC-layer data rates. As mentioned earlier, scheduling delay and timing relationship constraint do not have much impact on single-tone NPUSCH throughput; therefore there is hardly any benefit in using two active HARQ processes.

8.4 Latency

For NB-IoT an important use case is the transmission of small infrequent data. For this use case the latency to access the system and deliver a data packet is of higher relevance than the sustainable data rate supported in radio resource control (RRC) connected mode. In this section we focus on the NB-IoT latency performance achievable when using the Release 13 RRC resume procedure. At the time of designing NB-IoT in Release 13, the objective was to support a high priority report latency of at most 10 s.

8.4.1 Evaluation assumptions

The packet size definitions assumed for the NB-IoT latency evaluation is presented in Table 8.23. The protocol overhead from the application layer *constrained application protocol,* security layer Datagram transport layer security (DTLS) protocol, transport layer *user datagram protocol,* and *internet protocol* (IP) assumed are exactly the same as those assumed for EC-GSM-IoT in Section 4.4. It should be noted that the 40 bytes IP overhead can be reduced if Robust header compression is successfully applied at the packet data convergence protocol (PDCP) layer. The overheads from the radio protocols below the IP layer, i.e., *PDCP, radio link control* (RLC), and *MAC,* are the same as for LTE-M: There is 1 byte overhead at the PDCP layer; The overhead from RLC and MAC layers is dependent on the message type and the message content. As a rule of thumb, 4 bytes overhead from RLC/MAC headers can be assumed.

When accessing a cell, the *RRC connection resume request* message contains the information element *Establishment cause,* which may signal that a device requests radio resources for transmitting a so called *exception report* of high importance [8]. The process of transmitting such a report using the RRC resume procedure is depicted in Fig. 8.6. For simplicity, assignment

TABLE 8.23 NB-IoT packet definitions including application, security, transport, IP, and radio protocol overheads.

Type	UL report size [bytes]
Application data	20
CoAP	4
DTLS	13
UDP	8
IP	40
PDCP	1
RLC	2
MAC	2
Total	90

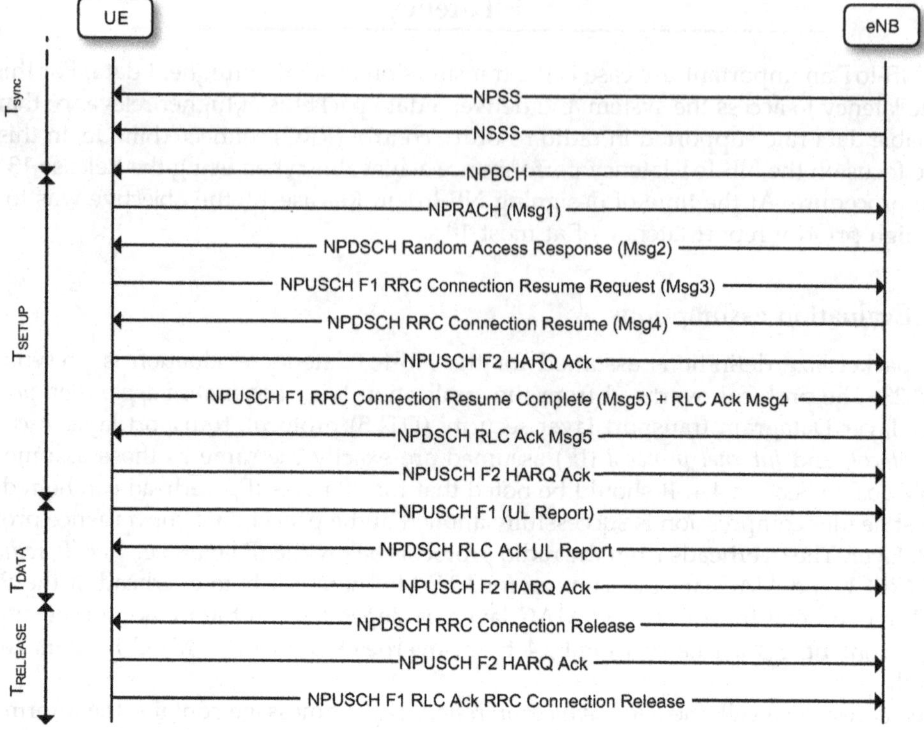

FIG. 8.6 NB-IoT data transfer based on the RRC resume procedure.

messages conveyed over the NPDCCH are omitted from the figure. The depicted procedure is based on the NB-IoT 3GPP Release 13 specifications where RLC acknowledge mode is mandatory. Here the uplink report is assumed to be transmitted after the connection has been resumed. In reality, it can be transmitted already in Message 5 together with the *RRC connection resume complete message*.

Four specific parts are identified in Fig. 8.6, namely:

- T_{SYNC}: The time to acquire synchronization.
- T_{SETUP}: The time to perform the random access procedure and setup the connection.
- T_{DATA}: The time to transmit the data.
- $T_{RELEASE}$: The time to release the connection.

Clear differences are seen in Fig. 8.6 compared to the EC-GSM-IoT message flow depicted in Section 4.4.1 where, for example, the data transmission starts already in the second uplink transmission. It can also be noted that the MIB-NB (carried in NPBCH) is read before accessing the system. This is needed to acquire frame synchronization, AB information, and system information status.

The transmission times for the NPDSCH and the NPUSCH Format 1 were based on a selected modulation and coding scheme that could carry the message at the targeted coupling

TABLE 8.24 NB-IoT exception report latency using the RRC resume procedure [9].

Coupling loss [dB]	Stand-alone	Guard-band [s]	In-band [s]
144	0.3	0.3	0.3
154	0.7	0.9	1.1
164	5.1	8.0	8.3

loss at a BLER of 10%. For the NPRACH, NPUSCH Format 2 and NPDCCH, a 1% BLER was assumed. For the NPBCH, NPSS, and NSSS the 90th percentile synchronization time was used in the evaluations. For the NPUSCH Format 1, a sub-PRB allocation was used in combination with cross subframe channel estimation to optimize the link budget at the 10% BLER target. The link level performance associated with these assumptions is presented in Section 8.2.

8.4.2 Latency performance

Table 8.24 summarizes the time to deliver the exception report described in Section 8.4.1. Results for stand-alone, guard-band, and in-band modes of operations are reported at coupling losses of 144, 154 and 164 dB. For in-band mode, the NB-IoT carrier is assumed to operate in a 10 MHz LTE carrier that limits the downlink power to 35 dBm per PRB when assuming a 6-dB power boosting. The number of available resource elements is also reduced due to LTE CRS and PDCCH transmissions. In case of guard-band mode, the downlink power is again restricted to 35 dBm due to the LTE *power spectral density* and the out of band emission requirements. In guard-band and stand-alone modes, the full set of resource elements are available for NB-IoT which improves downlink performance compared to in-band. For stand-alone mode, the full base station power can be allocated to the NB-IoT carrier which further improves the downlink performance. Using a higher power will improve downlink coverage and hence lower the latency needed to deliver the exception report. In good coverage conditions, the time to acquire synchronization, reading the MIB-NB and waiting for an access opportunity on the periodically occurring NPRACH time-frequency resource dominates the latency.

8.5 Battery life

NB-IoT is expected to support deployments of large number of devices. The ubiquity of the deployments requires NB-IoT devices to cope with all types of operating conditions, including non-rechargeable battery powered operation. Device battery life was therefore identified as a key aspect of the NB-IoT technology and a 10-year target was assumed in the Release 13 design. The next two sections describe battery life evaluations for NB-IoT including the expected performance.

TABLE 8.25 Packet sizes on top of the PDCP layer for evaluation of battery life [10].

Message type	UL report		DL application acknowledgment
Size	200 bytes	50 bytes	65 bytes
Arrival rate	Once every 2 h or once every day		

TABLE 8.26 NB-IoT power consumption [10].

Tx, 23 dBm	Rx	Light sleep	Deep sleep
500 mW	80 mW	3 mW	0.015 mW

8.5.1 Evaluation assumptions

To verify the battery life objective, a simple traffic model based on the packet sizes presented in Table 8.25 are considered. These packets sizes contain overheads from application, internet, and transport protocols presented in Table 8.23. To estimate the TBSs used over the air interface, additional overhead from PDCP, RLC, and MAC layers needs to be added to the values in Table 8.25.

It is assumed that a NB-IoT device can operate at the power consumption levels shown in Table 8.26. The transmitter power consumption assumes a 45% power amplifier efficiency. In addition to the 440 mW demanded by the power amplifier, a 60 mW power consumption is assumed for baseband and other circuitry. While transmitter and receiver power consumption depends on the characteristics of the supported access technology, the power levels when the device is in a sleep state, i.e. in RRC idle mode, or in RRC connected mode but not actively transmitting or receiving, are more related to the hardware architecture and design. Due to this, the assumptions on power consumption in light and deep sleep are aligned for NB-IoT, EC-GSM-IoT, and LTE-M.

Fig. 8.7 illustrates the uplink and downlink packet flow modeled in the battery life evaluations. It follows the RRC Resume procedure and includes the initial connection establishment, the uplink and downlink data transmissions and the connection release. In between the connections, the device is assumed to rest in power saving mode.

Not illustrated are NPDCCH transmissions and a 1 s period of light sleep in-between the end of the uplink report and the start of the downlink data transmission. A 20 s period of light sleep at the end of the connection, before the device goes back to power saving mode was assumed. This models the application of the Active timer during which the device uses a configured DRX cycle to facilitate downlink reach ability. Just as in the latency evaluations, a 10% BLER is targeted for the NPBCH, NPDSCH, and NPUSCH Format 1. A 1% BLER is assumed for the NPDCCH, NPUSCH Format 2, and NPRACH. For the synchronization time, the average NPSS and NSSS acquisition time is used and not the 90th percentile value used in the latency evaluations. The link level performance associated with these assumptions is presented in Section 8.2.

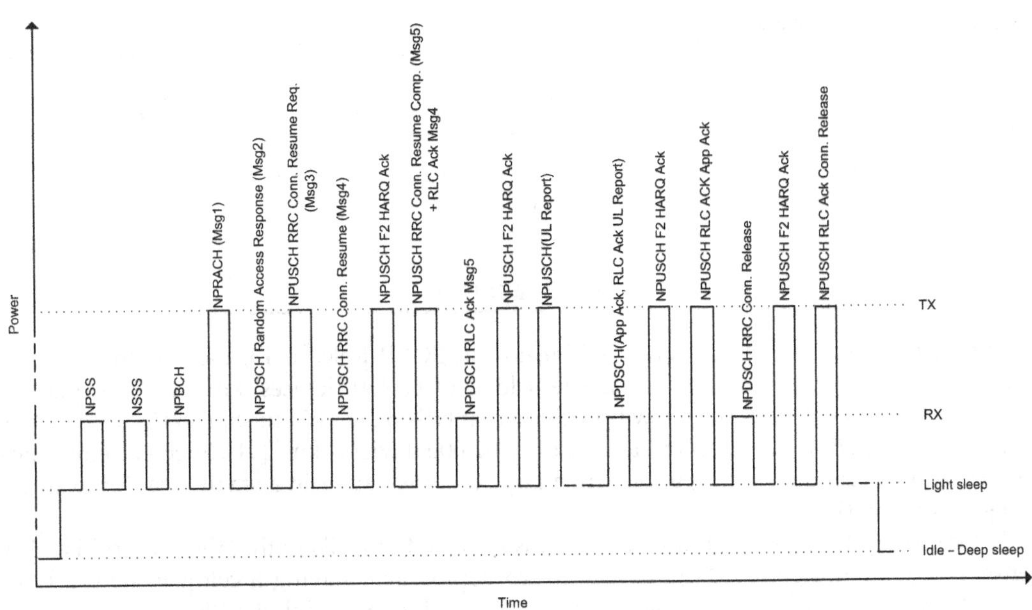

FIG. 8.7 NB-IoT packet flow used in the evaluation of battery life.

8.5.2 Battery life performance

The resulting battery lifetimes for the packet flow depicted in Fig. 8.7 are presented in Table 8.27. Slightly different performances are seen for the different modes of system operation. As the UL link level performance, which is decisive for the device power consumption, is not impacted by the mode of operation the differences are fairly small.

The general conclusion is that a 10-year battery life is feasible for a reporting interval of 24 h. It is also clear that the 2-h reporting interval is a too aggressive target when the device is at 164 dB MCL. Just as in the case of EC-GSM-IoT, see Section 4.5, these evaluations were performed assuming ideal battery characteristics.

TABLE 8.27 NB-IoT stand-alone (S), guard-band (G), and in-band (I) battery life [9].

Reporting interval [hours]	DL packet size [bytes]	UL packet size [bytes]	Battery life [years]								
			144 dB CL			154 dB CL			164 dB MCL		
			S	G	I	S	G	I	S	G	I
2	65	50	22.2	22.1	22.1	13	12.6	12.3	3.0	2.7	2.6
		200	20.0	20.0	20.0	7.9	7.8	7.7	1.4	1.3	1.3
24		50	36.2	36.1	36.1	33.0	32.8	32.6	19.3	18.4	18.0
		200	35.6	35.6	35.6	29.0	28.9	28.7	11.8	11.5	11.3

TABLE 8.28 3GPP assumption on required system capacity [11].

Assumed scenario	Devices/km^2	Devices/cell
3GPP Release 13 assuming 40 devices per home in central London	60,680	52,547
3GPP 5G requirement	1,000,000	865,970

8.6 Capacity

During the initial feasibility study of NB-IoT in 3GPP Release 13, the objective was to design a system with the ability to handle a load of 60,680 devices/km^2 [1], as explained in Section 4.6. Later in Release 14 it was decided that NB-IoT should meet also the 5G capacity requirement targeting a connection density of 1,000,000 devices/km^2 [11]. Such extreme loads are rare and can possibly be expected when large crowds gather in a small area, for example at big sport events.

At locations where this high network load is expected, it is likely that the cell grid is fairly dense, but, it cannot be ruled out that a macro deployment having a cell grid with a large inter-site distance may carry a significant part of the load. Table 8.28 summarizes the 3GPP Release 13 and 14 requirements under the assumption that base stations are located on a hexagonal cell grid with an intersite distance of 1732 m resulting in a cell size of 0.87 km^2.

8.6.1 Evaluation assumptions

To evaluate the NB-IoT system capacity, the traffic model described in Section 4.6.1.1 is used. It challenges mainly the uplink with up to 80% of the users autonomously accessing the system to send reports. But the downlink capacity is also tested, due to the network command messages. While EC-GSM-IoT in Section 4.6 is evaluated at the exact load of 60,680 users/km^2, NB-IoT is evaluated for a range of loads up to and beyond 100,000 devices per NB-IoT carrier.

The large- and small-scale channel models including the outdoor to indoor loss and fast fading characteristics are aligned with the assumptions described in Section 4.6.1. A macro deployment using as a hexagonal cell grid is assumed. In-band mode of operation within a 10 MHz LTE carrier is assumed. The base stations uses 46 dBm output power over 50 PRBs, implying 29 dBm per PRB. The PRB carrying the anchor carrier is assumed to be power boosted by 6 dB leading to a total output power of 35 dBm for the anchor carrier.

Table 8.29 summarizes the most relevant system simulation assumptions.

8.6.2 Capacity performance

The NB-IoT system capacity was simulated assuming a network load up to 110,000 users per carrier. This was done both for anchor carriers and non-anchor carriers. Fig. 8.8 shows how the DL NPDCCH and NPDSCH average resource utilization increases linearly on the

TABLE 8.29 System-level simulation assumptions [12].

Parameter	Configuration
Cell structure	Hexagonal grid with 3 sectors per site
Cell intersite distance	1732 m
Frequency band	900 MHz
LTE system bandwidth	10 MHz
Frequency reuse	1
Base station transmit power	46 dBm
Power boosting	6 dB on the anchor carrier
	0 dB on non-anchor carriers
Base station antenna gain	18 dBi
Operation mode	In-band
Device transmit power	23 dBm
Device antenna gain	−4 dBi
Device mobility	0 km/h
Pathloss model	$120.9 + 37.6 \times \log10(d)$, with d being the base station to device distance in km
Shadow fading standard deviation	8 dB
Shadow fading correlation distance	110 m
Anchor carrier overhead from mandatory downlink transmissions	NPSS, NSSS, NPBCH mapped to 25% of the downlink subframes.
Anchor carrier overhead from mandatory uplink transmissions	NPRACH mapped to 7% of the uplink resources.

anchor carrier as a function of the user arrival intensity. Depicted is only the resource utilization of the resources available for NPDCCH and NPDSCH transmission. Resources used for transmission of NPSS, NSSS, and NPBCH are not considered. Also, for the uplink, a linear increase in average NPUSCH Format 1 and Format 2 resource utilization is observed. For the uplink the resources allocated for NPRACH coupling loss 144, 154, and 164 dB are also included in the presented statistics.

The x-axis in Fig. 8.8 is defined in terms of device arrival rate in the system. Thanks to the traffic pattern described in Table 4.14, a linear relation between this rate and the absolute load is possible to derive. Two important reference points are:

- 6.8 arrivals per second correspond to the 3GPP Release 13 targeted load of 60,680 users/ km^2.
- 11.2 arrivals per second correspond to a load of 100,000 users/km^2.

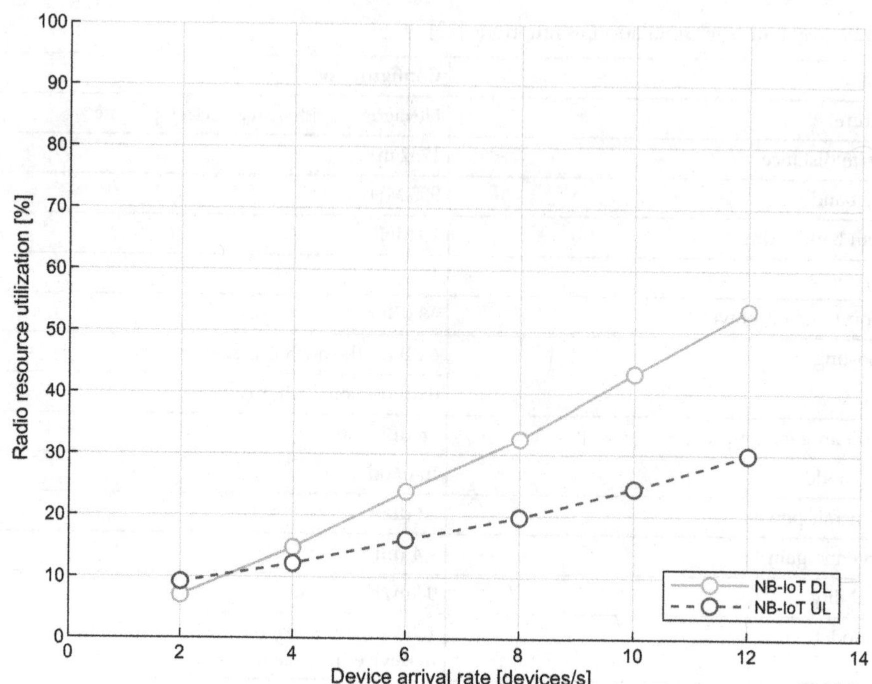

FIG. 8.8 NB-IoT anchor carrier average utilization of resources available for data transmission.

It may be surprising to see that despite the uplink heavy traffic model, it is downlink capacity that is limiting performance. This can be explained by the NPSS, NSSS, and NPBCH transmissions that takes up 25% of the available downlink subframes. Other contributing factors to the downlink load are the NPDCCH scheduling mechanism and the relatively signaling intensive connection setup and release procedures. It should also be noticed that mandatory narrowband system information transmissions, and load related to paging were not modeled. Taking also these transmissions into account would further increase the pressure on the anchor downlink capacity.

In the 3GPP work on 5G, it was required that 99% of all users need to be served at the peak capacity of the system [11]. Although it may seem like a stringent requirement, it is a reasonable objective for a system that was designed to reach the last mile. For the anchor carrier, a 1% *outage* was reached at 67,000 devices/km^2, corresponding to 7.5 user arrivals per second. Outage is here defined as the percentage of users not served by the system. Beyond this point the load increase seen in Fig. 8.8 comes at the expense of a rapidly increasing outage. It should also be clear that a more relaxed outage requirement would be directly translated into a higher system capacity.

For the non-anchor carriers, a load up to 110,000 devices/km^2 was achieved at the point where 1% outage was observed. This significant capacity increase compared to the anchor is explained by the fact that a non-anchor has no downlink overhead in terms of synchronization and broadcast channels. As a result, the resource utilization recorded on the non-anchor was fairly even between uplink and downlink.

TABLE 8.30 NB-IoT per carrier capacity.

Case	Connection density at 1% outage [devices/km^2]	Arrival rate at 1% outage [connections/s]
NB-IoT anchor	67,000	7.5
NB-IoT non-anchor	110,000	12.3

Table 8.30 summarizes the system capacity achievable on an anchor carrier and on a non-anchor carrier given the here made assumptions. Given these results, it is clear that NB-IoT has the potential to serve 1,000,000 devices/km^2 if 10 or more carriers are configured in a cell. It is important to note that the 3GPP Release 14 non-anchor support for random access as described in Section 7.3.1.10 was modeled in these simulations, which is necessary to cope with this high load.

8.6.3 Latency performance

When studying system capacity, it is relevant to consider also the achievable latency. Fig. 8.9 shows the latency recorded in the system simulations counted from the time a device accesses the system on the NPRACH to the point where the upper layer in the base station has correctly decoded the received transmission. Compared to the latency analysis performed in Section 8.4.2, it excludes the time to synchronize to the system and acquire the MIB-NB.

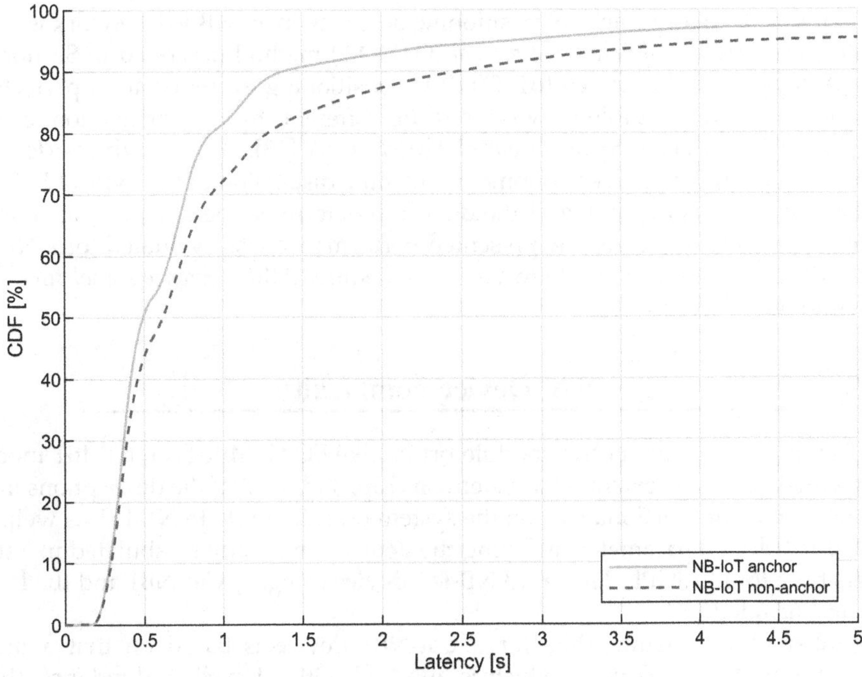

FIG. 8.9 Service latency for NB-IoT anchor and non-anchor carriers at the loads presented in Table 8.31.

TABLE 8.31 Service latency at extreme load extracted from Fig. 8.9

Service latency	NB-IoT anchor [s]	NB-IoT non-anchor [s]
At the 50th percentile	0.49	0.62
At the 90th percentile	1.4	2.5
At the 99th percentile	39	35

The results also include aspects such as scheduling delays, whereas the results in Section 8.4 assumes an absolute priority of the exception report, resulting in no delays due to scheduling queues.

Table 8.31 captures a few samples from the graphs in Fig. 8.9 including the latency at the 99th percentile not visible in the figure. At these extreme loads, the last 1% of all devices needs to accept a considerable latency. This is mainly due to the high load that creates queues for the available resources. In addition, it should be remembered that latency is one of the qualities that have been traded off for high coverage and high capacity in the design of NB-IoT. In case a low latency is required, it can be indicated to the network that the message is an exception report, as described in Section 8.4, enabling the network to prioritize its transmission.

8.7 Positioning

In this section, the performance of positioning accuracy in an NB-IoT network is presented based on the *observed time difference of arrival* (OTDOA) method described in Section 7.3.2.6.

Fig. 8.10 depicts simulated NB-IoT OTDOA positioning accuracy in a perfectly time-synchronized network assuming two types of different radio propagation conditions, namely, *Typical Urban* (TU) [5] or *Extended Pedestrian A* [13]. As seen, the performance is heavily dependent on the radio environment. The time dispersion of the Typical Urban channel significantly impacts the ability of the device to determine the line of sight channel tap, which impacts the performance. The presented performance was evaluated for a NB-IoT device using a high oversampling rate to be able to estimate the *reference signal time difference* (RSTD) with high accuracy [14].

8.8 Device complexity

NB-IoT aims to offer competitive module price. Like EC-GSM-IoT, an NB-IoT module can be implemented, to a large extent, as a system on chip, and most of the descriptions in Section 4.7 regarding different functionalities on the system on chip apply to NB-IoT as well. A summary of NB-IoT design parameters affecting the device complexity is tabulated in Table 8.32. Parameters both for the 3GPP Release 13 NB-IoT device category Cat-NB1 and the Release 14 Cat-NB2 are included.

The number of soft channel bits for a Cat-NB1 device is based on that a maximum NPDSCH TBS of 680 bits is used, which is attached with 24 cyclic redundancy check bits and then encoded by the LTE rate-1/3 *tail-biting convolutional code* (TBCC), giving rise to a

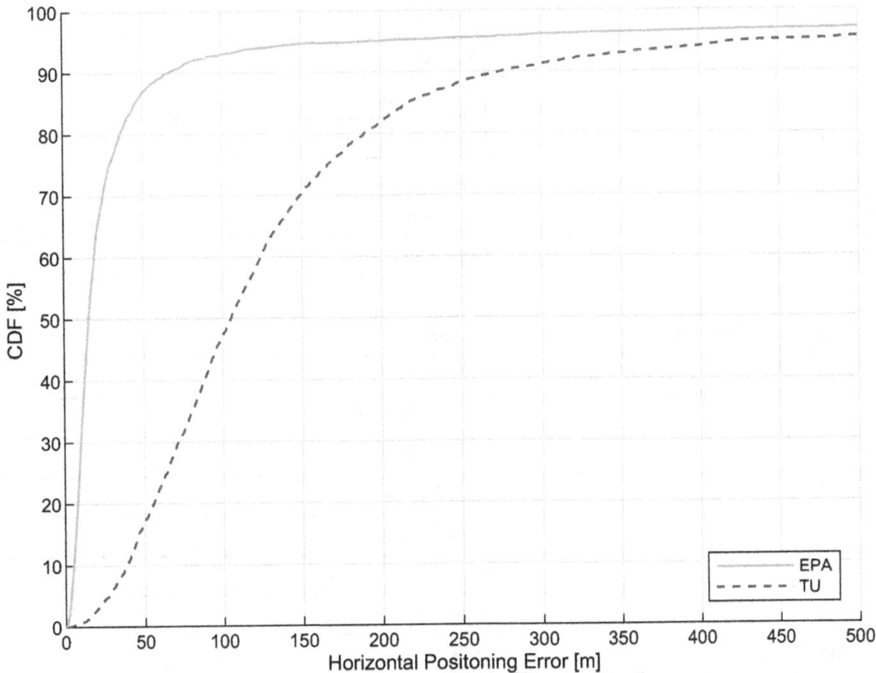

FIG. 8.10 Observed time difference of arrival positioning accuracy.

maximum 2112 coded bits. Rate-matching, either through puncturing or repetition, is used to match the actual number of encoded bits to the available set of bits according to the NPDSCH resource allocation. This results in a code rate lower or higher than the original code rate of 1/3. In case of a lower code rate than 1/3, there will be more encoded bits than 2112. However, rate-matching process can be handled at the device decoder to first combine the repeated bits. Thus, fundamentally, the decoder buffer for a Cat-NB1 device can be dimensioned based on the rate-1/3 TBCC to support 2112 soft bits as indicated in Table 8.32.

Regarding baseband complexity, the most noteworthy operations are the *fast fourier transform* (FFT) and decoding operations during connected mode, and the NPSS synchronization during cell selection and reselection procedures.

In NB-IoT, in principle only a 16-point FFT is needed. The complexity of N-point FFT is $6N\log_2 N$ real operations. There are 14 OFDM symbols per subframe, and therefore the complexity associated with FFT demodulation is approximately 5376 real-value operations per subframe.

The TBCC used is a 64-state code that can be decoded using 2 times trellis wrap-around [15, 16]. For a Cat-NB1 device, to decode the largest TBS of 680 bits, the device essentially needs to process $(680 + 24) \times 2 \times 64 = 90{,}112$ state metrics. The operation involved with processing each state is calculating the two path metrics merging onto the state and selecting a surviving path. This takes 3 real-value operations. Thus, the complexity of decoding a TBS of 680 bits is approximately 270k operations. Consider the most computationally demanding

TABLE 8.32 Overview of NB-IoT Cat-NB1 and Cat-NB2 device complexity.

Parameter	Value
Duplex modes	Half duplex
Half duplex operation	Type B
Number of device RX antennas	1
Power class	20, 23 dBm
Maximum bandwidth	180 kHz
Highest downlink modulation order	QPSK
Highest uplink modulation order	QPSK
Maximum number of supported DL spatial layers	1
Maximum DL TBS size	Cat-NB1: 680 bits
	Cat-NB2: 2536 bits
Number of HARQ processes	Cat-NB1: 1
	Cat-NB2: 1 or 2
Peak DL physical layer data rate	Cat-NB1: 226.7 kbps
	Cat-NB2: 258.0 kbps
DL channel coding type	TBCC
Physical layer memory requirement	Cat-NB1: 2112 soft channel bits
	Cat-NB2: 6,400 soft channel bits
Layer 2 memory requirement	Cat-NB1: 4,000 bytes
	Cat-NB2: 8,000 bytes

case for a Cat-NB1 device, the device receives TBS 680 bits in 3 NPDSCH subframes and has 12 ms before signaling HARQ-ACK. The complexity of FFT and TBCC decoding is approximately:

$$\frac{5376 \cdot 3 + 270000}{(3+12) \cdot 10^{-3}} \approx 19.1 \text{ millions of operations per second(MOPS)} \tag{8.1}$$

Regarding cell selection or reselection procedures, the complexity is dominated by NPSS detection, which requires the device to calculate a correlation value per sampling time interval. It was shown in Ref. [17] that the complexity of NPSS detection is less than 30 MOPS. Note also that, the device does not need to simultaneously detect NPSS and perform other baseband tasks.

Considering the most computationally demanding baseband functions as discussed above, NB-IoT device can be implemented with baseband complexity less than 30 MOPS.

8.9 NB-IoT fulfilling 5G performance requirements

Approximately two decades ago, the *Wideband Code-Division Multiple-Access* radio-access technology was developed according to the ITU IMT-2000 requirements. Wideband Code-Division Multiple-Access is often referred to as the *Third Generation* (3G) cellular technology, which extends cellular use cases from mainly voice communications and short text messages to video telephony and mobile internet. Fast forward another decade, the ITU brought forth a new set of requirements namely IMT-Advanced requirements, which set the development of *LTE* radio-access technology in motion. LTE is the *Fourth Generation* (4G) cellular technology, which significantly enhances the user experience in browsing the internet and streaming high resolution, high-fidelity multimedia content. Fast forward another decade, in November 2017, the ITU published a set of new requirements – IMT-2020 [18], which, in addition to further enhancing the mobile broadband services, aims to extend the cellular technology use cases to *machine type communications* (MTC) and *internet of things* (IoT). The ITU IMT-2020 requirements are generally referred to as the 5G requirements.

As described in Chapter 7, NB-IoT has since its first introduction added numerous features to enhance its performance in all relevant aspects. It has been well positioned to meet the 5G requirements on mMTC.

In this section, we show how NB-IoT fulfills 5G mMTC performance requirements. According to Ref. [18], a 5G mMTC radio access technology needs to meet the connection density requirement: one million devices per square kilometer (km^2). In addition to the connection density requirement, 3GPP also set the requirements on coverage, data rate, latency, and battery lifetime. In fact, these requirements are identical to the performance objectives used for the *cellular IoT* study in Release 13 [1], and described in the previous sections. The ITU and 3GPP 5G evaluations however do have slightly different assumptions compared to what were used in the *cellular IoT* study in Release 13. We will discuss the most notable differences in Section 8.9.1. The evaluation results are presented in Section 8.9.2.

8.9.1 Highlights of the differences in 5G mMTC evaluation assumptions

One main difference is the *NF* assumption on both the base station and device sides. In Ref. [1] and for all the results presented in the previous sections of this chapter, the NFs at the base station and device are 3 dB and 5 dB, respectively. In the 5G mMTC evaluation, NFs of 5 dB and 7 dB are assumed at the base station and device, respectively. The 2 dB higher NFs on both base station and device makes the requirements more challenging from radio performance point of view.

The *channel model* is another difference. The 5G evaluations are based on new channel models defined in Ref. [19]. For ITU defined IMT-2020 mMTC evaluation, the *urban-macro* (UMA) deployment scenarios, including both inter-site distances (ISD) of 500 m and 1732 m, are adopted [18, 20]. The defined deployment scenarios and large-scale channel models resulted in 4 different coupling gain distributions shown in Fig. 8.11. New small-scale fast fading models are also defined, with considerable shorter root-mean-square (rms) delay spread than the *Typical Urban* channel assumed in the previous sections.

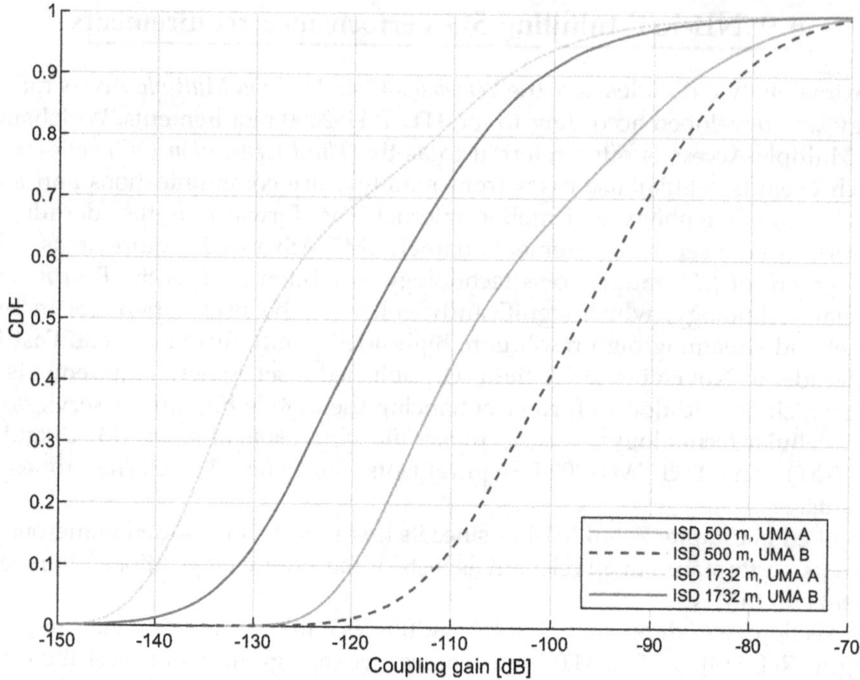

FIG. 8.11 The *cumulative distribution function* (CDF) of the coupling gain according to the ITU defined IMT-2020 mMTC evaluation scenarios.

All the evaluations are based on total transmit power of 35 dBm per NB-IoT carrier in the downlink and a 23 dBm device power level in the uplink.

8.9.2 5G mMTC performance evaluation

As described in Section 7.4, after LTE spectrum is fully refarmed to 5G *NR*, there are many options for NB-IoT and NR coexistence. In many coexistence scenarios, an attractive option is to configure the NB-IoT carrier with the guard-band operation mode. First, the raster offset options in the guard-band operation mode help achieve subcarrier grid alignment between NB-IoT and NR carriers. Second, the guard-band operation mode allows all the resource elements in an NB-IoT subframe to be fully utilized by NB-IoT. This makes sense when there is no longer an LTE carrier coexisting with NB-IoT, and therefore no resource elements need to be taken by LTE CRS or PDCCH. Thus, the results presented in this section are based on configuring the NB-IoT carrier as the guard-band operation mode.

8.9.2.1 Connection density

The ITU traffic model is that each user transmits a 32-byte layer-2 packet data unit every 2 h. The required connection density needs to be achieved with quality of service in such a way that at least 99% of the packets get a successful receipt within 10 s.

The evaluations are based on the user-plane RRC Resume procedure illustrated in Fig. 8.6, and the latency is defined as the time from the device initiates cell acquisition until the uplink packet is successfully received by the base station. The evaluation results based on dynamic

TABLE 8.33 Required number of NB-IoT non-anchor carriers for meeting the 5G mMTC connection density requirement of 1 million devices per km^2.

Inter-site distance [m]	Required number of NB-IoT carriers	
	UMA-A	UMA-B
500	1	1
1732	15	11

traffic modeling is shown in Table 8.33 [20]. For 500 m inter-site distance, one single NB-IoT non-anchor carrier is enough for meeting the required connection density. For more sparse base station deployment with 1732 m inter-site distance, 15 or 11 NB-IoT non-anchor carriers are required. In addition to the non-anchor carriers, an anchor carrier needs to be configured for NPSS, NSSS, NPBCH and system information transmissions.

8.9.2.2 Coverage

The 2 dB higher NFs at both base station and device, as mentioned in Section 8.9.1 pose a challenge in coverage. To counter this, we assume that the base station is equipped with 4 transmit and 4 receive antennas. In LTE network, already today, many base stations use 4 or more antennas. We also expect going forward, driven by mobile broadband capacity demand, advanced antenna features including the deployment of a higher number of antennas will be rolled out at more base stations. Thus, when NB-IoT shares the same base station radio equipment deployed for the mobile broadband services, as in most NB-IoT in-band and guard-band deployments today, it will also benefit from the deployment of more base station antennas. Although using more base station receive antennas benefits all the NB-IoT uplink channels, in the downlink, NPBCH, NPDCCH and NPDSCH are limited to using only 2 transmit antenna ports according to the NB-IoT specifications. Thus, for NPBCH, NPDCCH and NPDSCH, using more than two transmit antennas do not offer a higher degree of diversity. However, such a limitation does not apply to NPSS and NSSS, and thus the acquisition performance will be improved by using more than 2 transmit antennas. These limitations and potentials are accounted for in our coverage evaluations.

Based on the channel model specified in Ref. [19] with the assumption that the rms delay is 363 ns, and the Doppler spread is 2 Hz according to Ref. [21], NB-IoT coverage performance is presented in Table 8.34. All the physical channels meet the targeted MCL of 164 dB while achieving the desired synchronization or block error rate (BLER) performance.

8.9.2.3 Data rate

Based on the coverage results in Table 8.34, accounting for the timing relationship as discussed in details in Section 8.3, and assuming the transport block includes 5-byte overheads from the radio protocol, the MAC-layer data rates are 299 bps and 293 bps, in downlink and uplink, respectively. Excluding the 5-byte overheads, both downlink and uplink throughputs become 281 bps. Thus, the minimum data rate requirement of 160 bps is fulfilled.

8.9.2.4 Latency

For latency evaluation, both the RRC Resume protocol and the *early data transmission* (EDT) protocol are evaluated. The RRC Resume protocol latency evaluation follows the description

TABLE 8.34 NB-IoT coverage.

Performance/ Parameters	Downlink coverage				Uplink		
	NPSS/ NSSS	NPBCH	NPDCCH	NPDSCH	NPRACH	NPUSCH F1	NPUSCH F2
TBS [bits]	–	24	23	680	–	1000	1
Bandwidth [kHz]	180	180	180	180	3.75	15	15
Power [dBm]	35	35	35	35	23	23	23
NF [dB]	7	7	7	7	5	5	5
#TX/#RX	4/1	2/1	2/1	2/1	1/4	1/4	1/4
Transmission/ acquisition time [ms]	1280	1280	512	1280	205	2048	32
BLER	10%	10%	1%	10%	1%	10%	1%
SNR [dB]	-14.5	-14.5	-16.7	-14.7	-8.5	-13.8	-13.8
MCL [dB]	164	164	166.2	164.2	164.8	164	164

in Section 8.4. The EDT protocol, which is an enhancement introduced in 3GPP Release 15 is illustrated in Fig. 8.12. The EDT protocol allows an uplink data to be transmitted along with the RRC Connection Resume Request message in Msg3. Thus, in this case the latency is measured from the start of NPSS synchronization until when the base station receives the uplink data in Msg3. It is shown in Ref. [22] that the latency for delivering a MAC PDU of 105 bytes for the RRC Resume protocol is 9.0 s and for the EDT protocol is 5.8 s. Both are below the maximum allowed delay of 10 s.

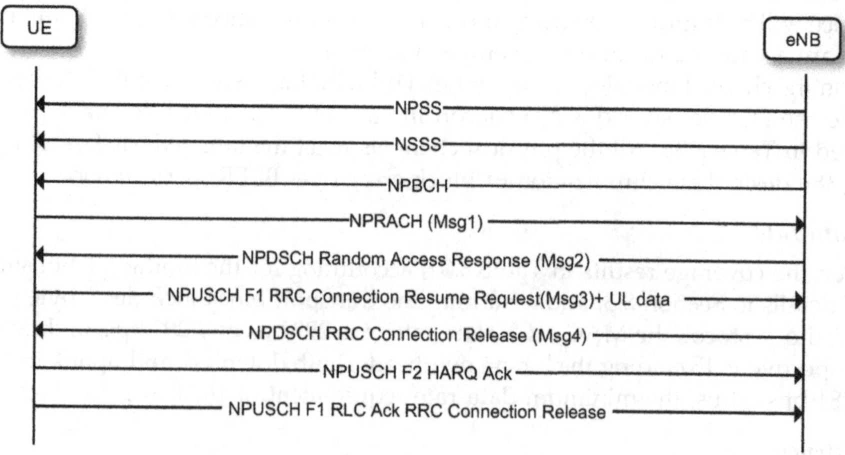

FIG. 8.12 The 3GPP Release 15 EDT protocol.

8.9.2.5 Battery life

The achievable battery life for a NB-IoT device was evaluated based on the user-plane RRC Resume procedure illustrated in Fig. 8.6, and the power consumption and traffic models described in Section 8.5. It is shown in Ref. [22] that for an assumed battery size corresponding to 5 Wh, a battery life of 11.8 years can be achieved. This fulfills the required 10 years battery life target.

References

[1] Third Generation Partnership Project, Technical Report 45.820, v13.0.0. Cellular system support for ultra-low complexity and low throughput internet of things, 2016.

[2] A. Adhikary, X. Lin, Y.-P. E. Wang. Performance evaluation of NB-IoT coverage, In: Proc. IEEE Veh. Technol. Conf., Montreal, Canada, September 2016.

[3] R. Ratasuk, N. Mangalvedhe, J. Kaikkonen, M. Robert. Data channel design and performance for LTE narrowband IoT. In: Proc. IEEE Veh. Technol. Conf., Montreal, Canada, September 2016.

[4] R. Ratasuk, B. Vejlgaard, N. Mangalvedhe, A. Ghosh. NB-IoT system for M2M communication, In: Proc. IEEE Wireless Commun. and Networking Conf., Doha, Qatar, April 2016.

[5] Third Generation Partnership Project, Technical Specifications 05.05, v8.20.0. Radio access network; radio transmission and reception, 2005.

[6] Ericsson. R1-1804159, System information acquisition time reduction for NB-IoT, 3GPP TSG RAN1 Meeting #92bis, April 2018.

[7] X. Lin, A. Adhikary, Y.-P. E. Wang. Random access preamble design and detection for 3GPP narrowband IoT systems. IEEE Wireless Commun. Letters, December 2016, Vol. 5, No. 6, pp. 640−643.

[8] Third Generation Partnership Project, Technical Specifications 36.331, v15.3.0. Evolved universal terrestrial radio access (E-UTRA) and evolved universal terrestrial radio access network (E-UTRAN); radio resource control (RRC); protocol specification, 2018.

[9] Ericsson. R1-1705189, Early data transmission for NB-IoT, 3GPP TSG RAN1 meeting #88, 2017.

[10] Ericsson. R1-1701044, On mMTC, NB-IoT and eMTC battery life evaluation, TSG RAN1 Meeting NR #1, 2017.

[11] Third Generation Partnership Project, Technical Report 38.913, v14.1.0. Study on scenarios and requirements for next generation access technologies, 2016.

[12] Ericsson. R1-1703865, On 5G mMTC requirement fulfilment, NB-IoT and eMTC connection density, 3GPP TSG RAN1 meeting #88, 2017.

[13] Third Generation Partnership Project, Technical Specifications 36.104, v15.5.0. Group radio access network; E-UTRA, UTRA and GSM/EDGE; multi-standard radio (MSR) base station (BS) radio transmission and reception, 2018.

[14] Ericsson. R1-1608698, OTDOA Performance for NB-IoT, 3GPP TSG RAN1 Meeting #87, 2016.

[15] Y.-P. Wang, R. Ramesh, A. Hassan, H. Koorapaty. On MAP decoding for tail-biting convolutional codes. In: Proc. IEEE Information Theory Symposium, 1997.

[16] Ericsson. R1-073033, Complexity and performance improvement for convolutional coding, 3GPP TSG RAN1 Meeting #49bis, 2007.

[17] Qualcomm. R1-161981, NB-PSS and NB-SSS design, 3GPP TSG RAN1 NB-IoT meeting#2, 2016.

[18] ITU-R, Report ITU-R M.2410-0. Minimum requirements related to technical performance for IMT-2020 radio interface(s), November 2017.

[19] Third Generation Partnership Project, Technical Report 38.901, v14.3.0. Study on channel model for frequencies from 0.5 to 100 GHz, December 2017.

[20] Third Generation Partnership Project, Technical Report 37.910, v1.1.0. Study on self evaluation towards IMT-2020 submission, December 2018.

[21] ITU-R, Report ITU-R M.2412-0. Guidelines for evaluation of radio interface technologies for IMT-2020, October 2017.

[22] Ericsson. R1-1903119, IMT-2020 self-evaluation: mMTC coverage, data rate, latency and battery life, 3GPP TSG RAN1 Meeting #96, March 2019.

CHAPTER

9

LTE URLLC

OUTLINE

Cellular Internet of Things, Second Edition
https://doi.org/10.1016/B978-0-08-102902-2.00009-1

© 2020 Elsevier Ltd. All rights reserved.

Abstract

This chapter describes the evolution of the LTE technology toward Ultra-Reliable-Low-Latency-Communications. It starts by providing a background as to why such a substantial step in evolving the LTE technology was taken. The background is followed by a detailed description of the changes both to the physical layer design and idle and connected mode procedures. This involves everything from how the physical channels are in detail designed to the resource allocation used. The chapter is written assuming the reader has some basic LTE knowledge, and describes mainly the changes introduced on top of 'legacy LTE'.

9.1 Background

In Release 15 of the 3GPP RAN specifications, the work on IoT expanded from being focused on the low-end part of the IoT spectra (ultra-low device cost, no stringent requirements on latency, etc) to also cater for more critical IoT applications. For more background on this development, see Chapters 1 and 2. Considering the global success of LTE, being the most advanced 3GPP technology (to that date), it was a natural choice to develop this technology toward more critical IoT applications. In parallel, work was also initiated in NR, see Chapter 2, targeting critical IoT applications, see Chapter 11.

To understand why a significant development of the LTE specification was needed to accommodate cMTC applications and services, let's first look at the abbreviation itself. URLLC stands for *Ultra-Reliable and Low Latency Communication*. From a RAN perspective, to make a radio link reliable, an improvement of the experienced radio signal quality (i.e. SINR) would quickly reduce the error rate of the transmission. However, if a too high SINR is required for the service to be considered ultra-reliable, the cellular network would contain many coverage holes where the reliability cannot be achieved. Furthermore, the use of time-based repetitions, as many of the techniques in this book utilizes to improve the reliability, will have a direct impact on the latency. Also, as described in Chapters 1 and 2, the extreme requirements on a low latency bound from the IMT-2020 requirements (upper bounded by 1 ms) would make the Release 14 LTE specification insufficient since the shortest possible transmission duration over-the-air already consumes the 1 ms latency budget (the IMT-2020 requirement also include e.g. alignment delay and processing time). It is however important to note that the aim of the LTE URLLC work was not only to cater for the extreme requirements of IMT-2020, but also to prepare LTE for other use cases with a less stringent latency bound and/or reliability requirement. The IMT-2020 requirements, however is the most stringent, and hence to a large extent what LTE is evolved toward fulfilling.

Knowing the limitations of the Release 14 LTE technology and the targeted URLLC performance, 3GPP Release 15 initiated work toward fulfilment of the IMT-2020 requirements for URLLC in two-stages.

- The first stage involved a lowering of the radio interface latency in the work item "Shortened TTI and processing time for LTE" [1]. This work involved both shortening the existing (up to Release 14) subframe-based transmission over the air and improving the device processing time in connected mode. Two new transmission durations were

defined based on *slot operation* and *subslot operation*, also referred to as *short TTI, sTTI*. The work was mainly motivated by improved user-throughput in TCP-like applications (where the start-up phase requires a fast turn-round time to ramp-up the throughput), but also to prepare LTE for the IMT-2020 URLLC requirements (see Chapter 1 and 2). The latter requirement is the focus of this book.
- The second stage was dedicated to improving the reliability, which was performed in the work item on "Ultra Reliable Low Latency Communication for LTE" [2].

It can be noted that improvements are limited to devices in connected mode and exclude any modification to PSS/Synchronization Signals (Primary and Secondary Synchronization Signals), Physical Broadcast Channel, Physical Control Format Indicator Channel and Physical Random Access Channel, the initial access procedures, SIB (System Information Block) signaling and Paging procedures. One exception is the work on reduced control plane latency, see Section 9.3.1.1, where processing times have been reduced to accommodate the IMT-2020 control plane latency requirements [3].

The following chapter will make no distinction from where a specific functionality originated, i.e. from which work item, but will describe the LTE URLLC functionality at the end of Release 15, including the work performed for both sTTI, URLLC work items and the work on improved control plane latency.

Throughout this chapter the terminology short transmission time interval (sTTI) will be used to refer to the functionality of shortening the air interface transmission duration to slot or subslot level.

The reader is assumed to have basic LTE knowledge and for a more complete coverage of LTE, the reader is referred to e.g. Ref. [4]. This chapter focuses on explaining the difference in the URLLC design compared to Release 14 LTE specifications. The latter is often referred to as 'legacy LTE'.

9.2 Physical layer

This section describes the physical layer of LTE URLLC. First, the reader is provided with the design principles in evolving the LTE technology toward URLLC, followed by a detailed review of the final design chosen.

In addition to the main thinking behind the LTE re-design, the reader is also provided with a detailed review of the changes to the physical resources, the downlink and uplink physical channels, and the impact on processing timelines and timing advance from shortening the transmissions time.

9.2.1 Radio access design principles

To vastly improve the air interface latency and to improve reliability has a significant impact to the overall system design, as the following sections will show. Hence, some design principles are needed to ensure proper system operation:

- **Backwards compatibility:** One principle is the co-existence between LTE URLLC and legacy LTE. That is, the introduction of URLLC operation to a network should be as seamless as possible from both an operator and an end-user point of view. The device penetration supporting URLLC will gradually increase in a network and hence to e.g. require separate resources for URLLC operation would have a huge impact to network operation. Furthermore, even if a device supports URLLC, it is not always the best mode of operation, considering for example control channel overhead when using the shorter TTIs over the air interface. That is, an sTTI capable device in an sTTI capable network would be expected to operate both in subframe operation and sTTI operation.

- **Fallback to subframe operation:** It is also important to allow a network to fallback to subframe operation from sTTI operation, e.g. to allow for a more robust operation, a less control signaling-heavy mode of operation and to allow RRC reconfiguration.
- **Connected mode improvements**: To improve the reliability in idle mode would have a great impact on the system design. In idle mode, broadcast/multicast channels are typically used which would directly impact all users in the system if for example the transmission duration is changed. Hence, in terms of latency, there were only improvements made to the initial access procedure (see Section 9.3.1.1) to fulfill the IMT-2020 requirement on control plane latency (see Chapter 2). All improvements related to reliability are exclusively targeting connected mode.

9.2.2 Physical resources

To fulfill the IMT-2020 latency requirement of 1 ms from the radio protocol layer (see Chapter 1), the latency over the air interface need to be significantly less than 1 ms. Considering that the subframe in LTE is already 1 ms which is the transmission duration used by the data channels (PDSCH/PUSCH), this had to be shortened. To limit the additional network complexity from URLLC, but allow enough flexibility, two additional transmission durations of Physical Downlink Shared Channel (PDSCH) and Physical Uplink Shared Channel (PUSCH) were defined: slot transmission and subslot transmission. A slot was already defined in LTE as the split of a subframe in two equal parts (each of 0.5 ms duration), but typical data channel transmission was still restricted to subframe duration, which changed with the introduction of sTTI. With the introduction of URLLC, a slot is further split into three subslots, each of two or three-symbol duration. The new radio frame in LTE is shown in Fig. 9.1.

As can be seen from the figure, two different subslot patterns have been defined. Their use depends on if the transmission takes place in the uplink or downlink, and in case of downlink, in which symbols the data transmission starts (in the 3GPP specifications. This symbol index is referred to as $l_{DataStart}$) and is typically dependent on how many symbols are used for PDCCH, see Table 9.1.

It could be noted that the 3GPP specification does not mention the use of different 'subslot patterns', but this is effectively how the design can be seen.

Fig. 9.2 shows the subslots available for data transmission depending on the direction and, for the downlink, the symbol index where the data starts. It can be seen that the subslot number in downlink starts either with '0' or '1' depending on the starting symbol index for data. One can note that from subslot number 2 onwards, the subslot borders are aligned.

This design might seem complicated, having different subslot lengths in the subframe, and also, using different subslot patterns depending on certain given conditions. The main reason

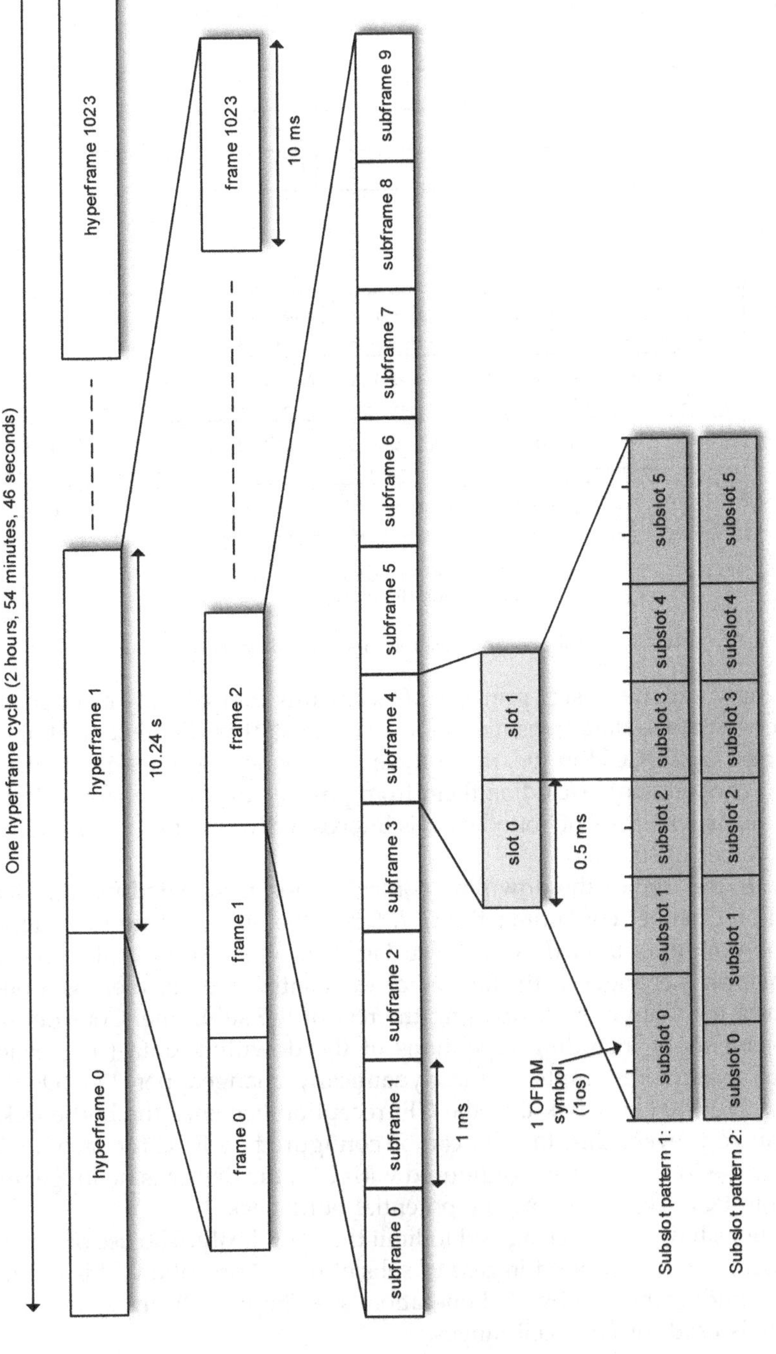

FIG. 9.1 Subslot patterns in the LTE frame structure.

TABLE 9.1 Subslot pattern used depending on data symbol start and transmission direction.

Direction	Starting index of data symbol in the subframe ($l_{DataStart}$)	Subslot pattern
Uplink	0	1
Downlink	1	1
	2	2
	3	1

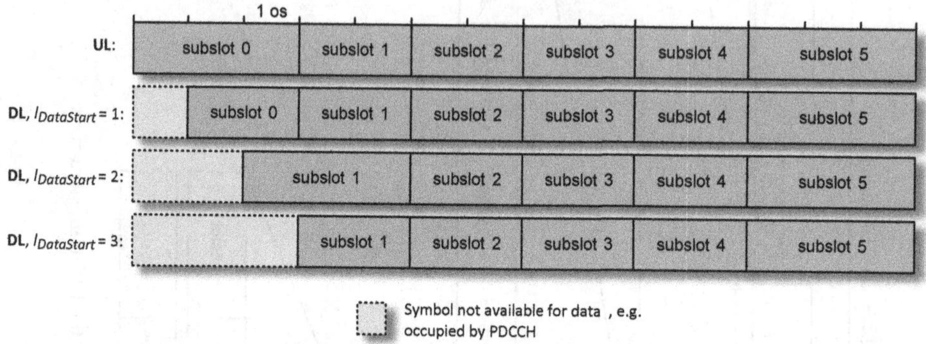

FIG. 9.2 Subslots available for data depending on direction.

for this is to comply with the design principle of backwards compatibility mentioned in Section 9.2.1. In the downlink, the data transmission in LTE can start in different symbol indices and could be occupied by PDCCH in the, up to three, first symbols (four symbol PDCCH is not supported in sTTI operation), excluding them from possible data transmission. Since the subslot boundary aligns with the slot boundary it is impossible to split the seven symbols of a slot into an even number of subslots.

In legacy LTE, the size of the downlink control region is signaled via Physical Control Format Indicator Channel containing the *CFI*. Since the subslot pattern is dependent on the start of the downlink data ($l_{DataStart}$), see Fig. 9.2, which follows the CFI, and, since the PDCCH mapping changes with the size of the control region, there is a need for the device to acquire the CFI before processing the rest of the subframe. Considering the reliability improvements of including repetitions of the downlink data (see Section 9.2.3.3) and the use of multiple attempts of the dynamically changed per TTI (DCI) decoding (see Section 9.2.3.3) there is a risk that the CFI reception becomes the bottleneck, in terms of performance. To prevent this, the CFI can be configured by RRC for each device (which then determines the $l_{DataStart}$). Once configured with CFI, the device is no longer expected to decode CFI from PCFICH, removing the potential bottleneck.

Some restrictions have also been agreed to limit the complexity. The use of extended cyclic prefix is for example not supported in case of subslot or slot operation. This is motivated by the reduced link budget implied by sTTI operation (see Chapter 10), and that extended cyclic prefix primarily is used for large cell ranges.

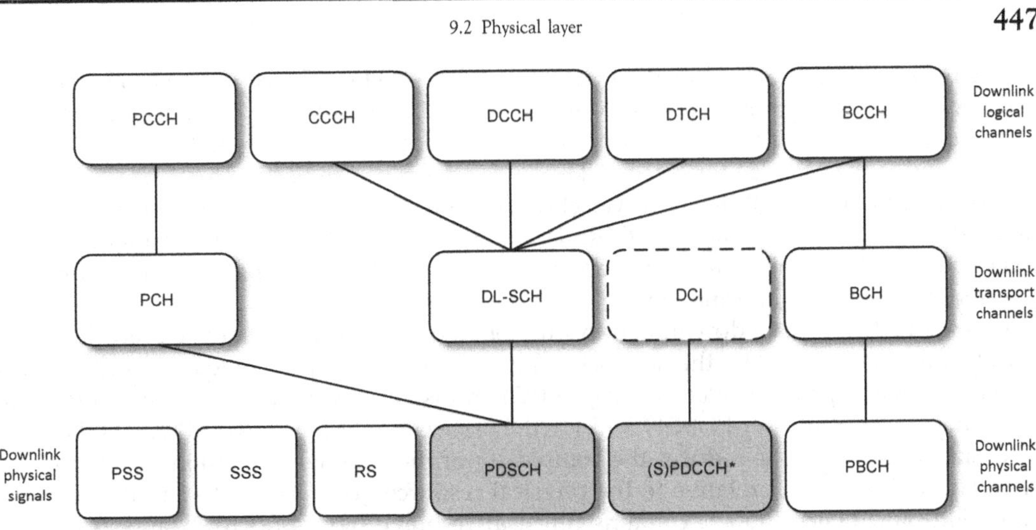

FIG. 9.3 Downlink physical channels in LTE URLLC (impacted channels compared to legacy LTE is filled).

9.2.3 Downlink physical channels and signals

The set of physical channels used in the downlink in case of sTTI operation are shown in Fig. 9.3. For a reference to the transport channels and logical channels that the physical channels are associated with, please refer to Section 5.2.4.

With the introduction of LTE URLLC, the only impacted channels on the downlink are the *Physical Downlink Shared Channel*, PDSCH and the *Short Physical Downlink Control Channel (SPDCCH)*, out of which SPDCCH is a completely new physical channel.

A PDSCH transmitted using slot or subslot duration are referred to as slot-PDSCH and subslot-PDSCH respectively. Similarly, for downlink control, slot-SPDCCH and subslot-SPDCCH is used when referring to a specific transmission duration.

9.2.3.1 Downlink reference signals

Subframe	DMRS - any CRS - not in MBSFN-subframes
Subcarrier spacing	15 kHz
CRS bandwidth	Full system bandwidth
DMRS bandwidth	Same as associated SPDCCH/PDSCH
CRS frequency location	According to Figure 5.9 in every PRB
DMRS frequency location	According to Fig. 9.4 (subslot) and 9.5 (slot) in affected PRBs

The reference signals used for the PDSCH and SPDCCH can be configured to be either a *Cell-specific Reference Signals* (CRS) or a device-specific *Demodulation Reference Signals* (DMRS).

The definition and mapping onto physical resources of the CRS is not changed in LTE URLLC and is the same as for legacy LTE, as also described in Section 5.2.4.3.

The mapping of the DMRS however changes with the use of slot and subslot transmissions. The reason to re-define the DMRS is mainly to avoid the delay and buffering at the receiver of OFDM symbols arriving earlier than the DMRS and to have, an as accurate as possible, channel state information.

In case of DMRS-based downlink transmission, the device can assume the precoder of the DMRS and payload used by the network to be the same over a predefined set of PRBs. This is referred to as a *precoder resource group* (PRG) and is always set to two resource blocks in frequency for subslot/slot-PDSCH and for the associated DMRS-based SPDCCH. Using a PRG size larger than one reduces the complexity of the channel estimation for the device. A given PRG is defined relative to the physical resource grid (not relative to the resource allocation) and hence PRG k, contains physical resource blocks (PRB) $2k$ and $2k+1$. The use of a PRG size of two opens up for a design of the DMRS pattern over two consecutive resource blocks in frequency. This helps to avoid channel estimation errors due to extensive extrapolation of the channel estimate for resource elements (REs) 'outside' the span of the DMRS resource in frequency, while at the same time keeping the DMRS density low (compared to having the DMRS pattern over a single resource block). For example, in Fig. 9.4, the REs that are outside the span of the DMRS is only one at high end of the frequency range for the baseline pattern and pattern $v2$.

The possible subslot-PDSCH DMRS patterns are shown in Fig. 9.4. Up to four antenna ports (where one port is used per layer in case of MIMO transmission) are defined. In case of 2-layer transmission, the two ports 7 and 8 are multiplexed using *orthogonal cover codes (OCC)*, over two OFDM symbols. As the name implies, the codes applied to the repeated symbols are orthogonal, see Ref. [4] for details.

In case of 4-layer transmission, the same OCC in time is used over a second pair of ports 9 and 10 placed in other positions in frequency.

Which one of the four DMRS patterns to use is selected based on pre-defined rules. The rules have been decided to avoid too many restrictions in network configurations and to ensure backwards compatibility.

To select a pattern, the rules below are followed:

The baseline pattern is selected:

- If the baseline pattern has no overlapping REs with neither CRS nor configured CSI-RS in the subslot.

Pattern v0, v1 and v2 is selected:

- In normal subframes, if the baseline pattern has overlapping REs with either CRS or configured CSI-RS. The applicable pattern to select is dependent on the parameter v_{shift} (the shift in frequency of the CRS pattern, see Ref. [5]) where $\mathrm{mod}(v_{shift},3) = x$ determines the shift to be used (v0, v1 or v2).

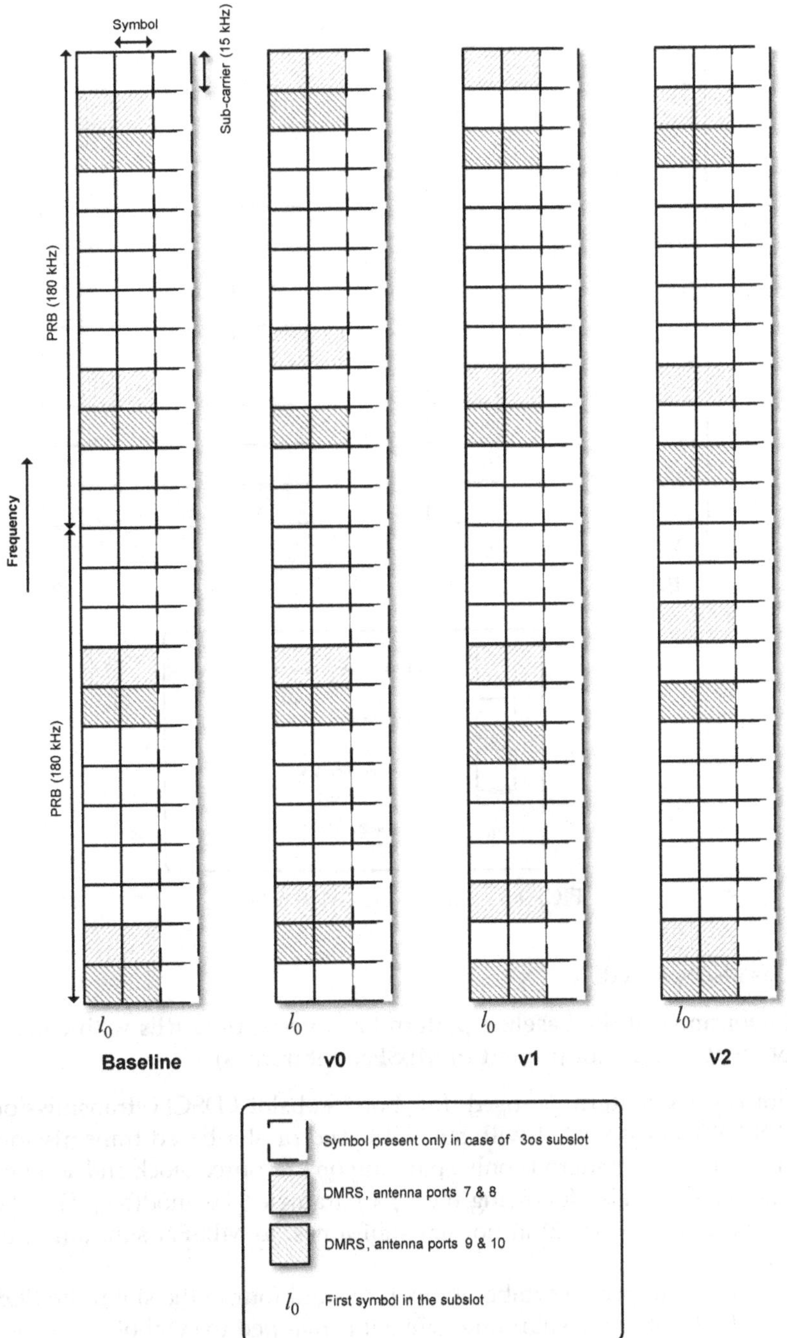

FIG. 9.4 Subslot-PDSCH DMRS pattern.

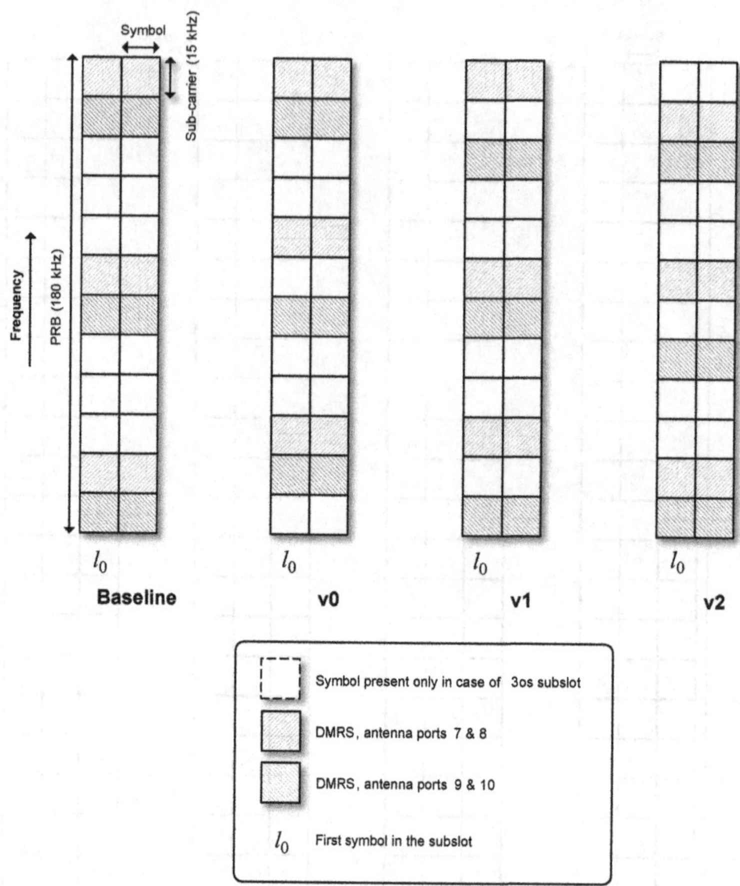

FIG. 9.5 Slot-PDSCH DMRS pattern.

Pattern v0 is also selected:

- In MBSFN subframes, if the baseline pattern has overlapping REs with configured CSI-RS (note that CRS is not present in MBSFN subframes).

The subslot-DMRS pattern is used for both subslot-PDSCH transmission and for DMRS-based SPDCCH associated with subslot-based or slot-based transmission.

The slot-PDSCH DMRS pattern is only spanning one resource block and shown in Fig. 9.5. In this case, the pattern is also following the v_{shift} parameter, i.e. mod(v_{shift},3) $= x$ determines the shift to be used (v0, v1 or v2) in normal subframes. In MBSFN subframes, the baseline pattern is used.

The DMRS pair is mapped to symbol index three and four in the slot if the PDSCH transmission is in the first slot of the subframe, while it is mapped to symbol index two and three of the slot if the PDSCH transmission is in the second slot.

9.2.3.2 *Slot/subslot-SPDCCH*

Subframe	Any
Duration	1 or 2 OFDM symbols (CRS)
	2 or 3 OFDM symbols (DMRS)
Repetitions	No
Subcarrier spacing	15 kHz
Bandwidth	Any - granularity of one PRB for CRS
	Any - granularity of two PRB for DMRS
Frequency location	Any, with the restriction of the DMRS to align with the PRG grid

9.2.3.2.1 General

Keeping the same downlink control channel for sTTI operation as for subframe operation would mean that the control channel is only present in the first symbols in a subframe. Hence, the periodicity of downlink assignments and uplink grants of data transmissions would only occur on a 1 ms basis. This would prevent reducing the alignment time before a device knows it is being scheduled, and, put an unnecessary bound on the achieved latency, see the upper figure in Fig. 9.6.

To ensure low latency a short periodicity of scheduling opportunities was introduced new scheduling information for each subslot or slot being scheduled, as illustrated by the lower figure in Fig. 9.6.

Adopting such a design, the downlink control channel had, to a large extent, be redesign for sTTI operation.

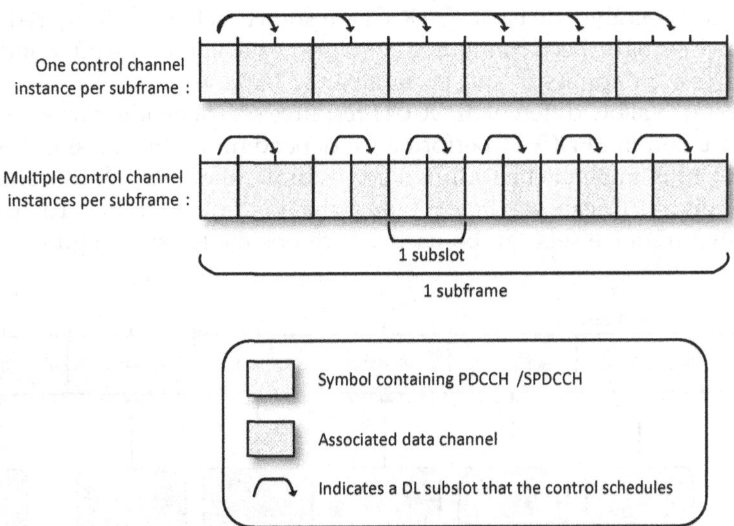

FIG. 9.6 Possible options for the downlink control channel occurrence (only the bottom approach was adopted).

In this clause, the new control channel, *SPDCCH*, is described.

It should however be noted that the control information associated with sTTI operation can be carried by both PDCCH and SPDCCH. Considering the mapping of PDCCH to the first symbols in the subframe, PDCCH is only used to schedule sTTI#0 (both in case of slot and subslot operation).

9.2.3.2.2 SPDCCH resource set

A SPDCCH is contained within a SPDCCH resource set, i.e. a set of possible resources that the SPDCCH can be mapped to.

The resource set is configured by RRC signaling with the following configuration options:

- The resource blocks in the frequency domain that the SPDCCH resource set occupy.
- If it is a DMRS-based or CRS-based reference signal:
- A DMRS-based resource set can be configured to apply to all subframes, only MBSFN subframes or only non-MBSFN subframes.
- A CRS-based resource set only applies to non-MBSFN subframes (there is no CRS transmission in MBSFN subframes). Furthermore, it can be configured to map to one or two symbols in time.
- If the SPDCCH resource set is distributed or localized.
- The number of SPDCCH candidates per aggregation level (AL) and the starting position of each AL.
- The rate-matching mode of the RB set.

The details of the configurations will be further described in this section.

A device can be configured with up to two SPDCCH resource sets in each subframe type (MBSFN and non-MBSFN) in the device specific search space. In case of subslot operation, the subslot numbers applicable to each SPDCCH resource set are configured. In Fig. 9.7, SPDCCH resource set 0 has been configured to apply to subslot number 0, 2 and 5, and hence SPDCCH resource set 1 applies to subslot number 1, 3 and 4.

The main reason to allow different SPDCCH resource sets depending on subslot number is the potential variation in SPDCCH performance depending on the subslot it is mapped to. Since the SPDCCH is rate-matched around for example the CRS, DMRS and CSI-RS, the code rate for a given AL can vary over time, which is not desirable. To compensate for this behavior, two resource sets can be used where one could be configured with a higher

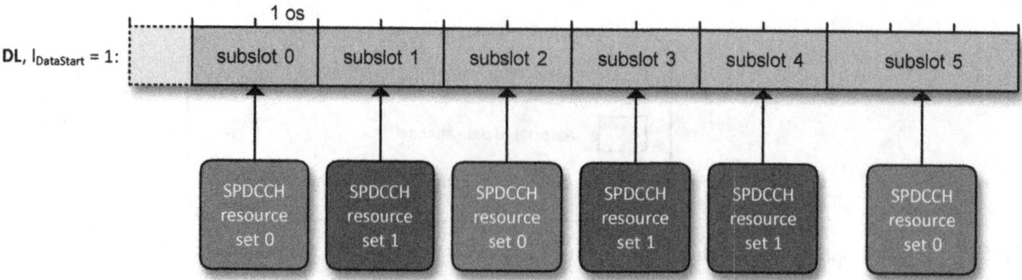

FIG. 9.7 SPDCCH resource sets.

AL and is configured to apply in subslots experiencing a larger overhead (a higher level of rate-matching).

9.2.3.2.3 Mapping to physical resources

Provided that we have configured the physical resources for the SPDCCH resource set, there are multiple ways on how to map a given SPDCCH to the resource set. As already mentioned, the mapping can be either distributed or localized. When to use which mapping can depend on multiple factors, such as service type and reference signal type. However, for cMTC services for fulfilling the URLLC requirements a distributed mapping using CRS-based channel estimation will provide a good trade-off between overhead and performance. Using a distributed mapping will increase diversity and hence improve reliability.

To assist the mapping, the concept of *short resource element groups* (SREG) and *short control channel elements* (SCCE) are introduced (similar to REGs and CCEs used in legacy LTE).

An SREG is defined over one OFDM symbol mapped to one resource block in frequency. That is, there are 12 REs that constitute a SREG. However, as already mentioned, not all 12 REs might be used for the SREG if e.g. CRS is mapped to the same physical resources. For a given SREG, the SPDCCH resources will be mapped to at most 12 REs.

An SCCE can be seen as a minimum building block of an SPDCCH. Each SCCE consists of a set of SREGs. The number of SREGs per SCCE is dependent on the SPDCCH resource set configuration, and are given in Table 9.2.

Each SPDCCH candidate belongs to an AL. For different ALs, different number of SCCEs are aggregated. The aggregation is carried out to achieve different level of coverage in the cell, considering that different UEs will experience different radio conditions.

The ALs defined for the SPDCCH candidates are 1, 2, 4 and 8. The AL corresponds to the number of SCCEs constituting the SPDCCH candidate, e.g. an SPDCCH of AL 4 consists of 4 SCCEs.

To be able to describe where an SPDCCH candidate is mapped onto the physical resources, the SREGs are first logically numbered. In the examples below, it is assumed that the SPDCCH resource set consists of 20 RBs (in reality, the size of the SPDCCH resource set is only limited by the system bandwidth).

The SREGs are numbered in different ways depending on the reference signal type of the SPDCCH:

- For CRS-based SPDCCH (see Fig. 9.8), the mapping follows a frequency first, time second mapping. This will allow a device to start decoding the SPDCCH candidates

TABLE 9.2 Number of SREGs per SCCE.

Reference signal type	Number of SREGs per SCCE
CRS	4
DMRS	4[a]
	6[b]

[a] *For a two-symbol long subslot, or for slot operation.*
[b] *For a three-symbol long subslot.*

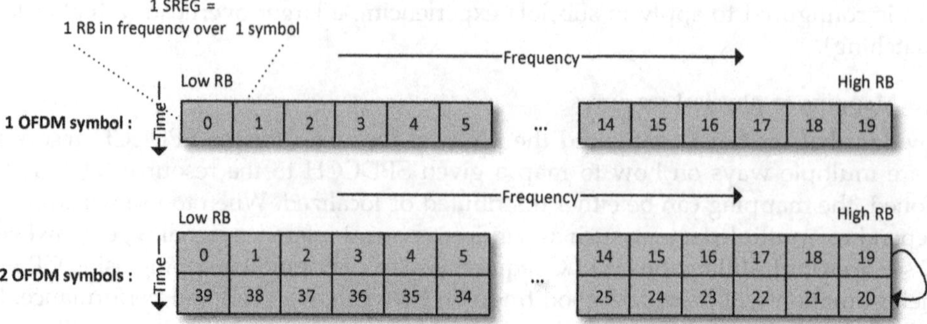

FIG. 9.8 SREG numbering for CRS-based SPDCCH.

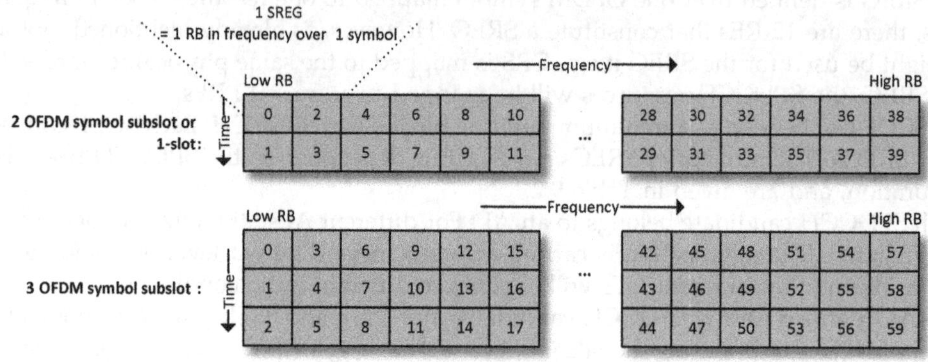

FIG. 9.9 SREG numbering for DMRS-based SPDCCH.

that map to a single symbol directly after receiving it and will hence help reducing the reception time. In case of a two symbol CRS-based SPDCCH, the numbering changes direction in the second symbol. This is to ensure that a SCCE mapped at the edge of an localized SPDCCH resource set and spanning two symbols, will be contained within the same resource blocks. For example, if an SCCE is mapped to SREGs 18–21, it will be mapped to the two highest resource blocks.

- For DMRS-based SPDCCH (see Fig. 9.9), the mapping follows a time first, frequency second mapping. Since the DMRS symbols used for channel estimation are mapped onto the full subslot, and, considering that the DMRS need to be received before performing channel estimation, there is no gain in using a frequency first, time second mapping. Using a time first, frequency second mapping will help in getting a condense mapping in frequency of the SPDCCH. This is useful since a DMRS-based SPDCCH is used when the eNB has knowledge about the channel and can perform frequency selective scheduling.

The next step in constructing the SPDCCH is the mapping of SREGs to SCCEs. There are four different configurations that have an impact on how the mapping is done:

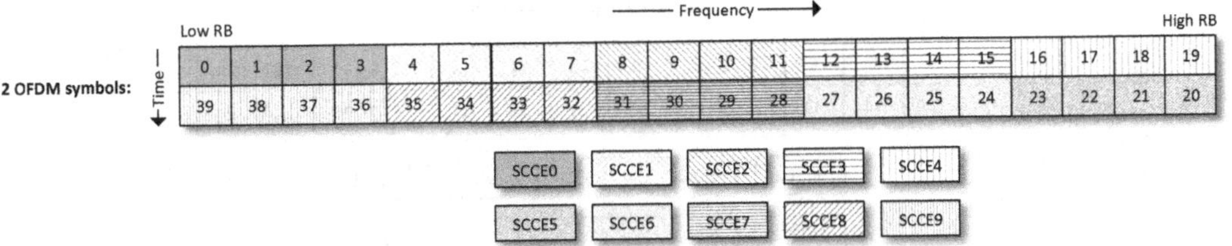

FIG. 9.10 SREG to SCCE mapping, Localized CRS-based SPDCCHillustrated for a 2 symbol SPDCCH resource set.

- CRS-based SPDCCH, localized mapping
- CRS-based SPDCCH, distributed mapping
- DMRS-based SPDCCH, localized mapping
- DMRS-based SPDCCH, distributed mapping

For CRS-based SPDCCH using localized mapping, the SREG index corresponding to SCCE index n is given by:

$$n \cdot N_{\text{SREG}}^{\text{SCCE}} + j \tag{9.1}$$

where

$$n = 0, \ldots, N_{\text{SCCE}} - 1$$

N_{SCCE} is the number of SCCE in the SPDCCH resource set,

$$j = 0, \ldots, N_{\text{SREG}}^{\text{SCCE}} - 1$$

$N_{\text{SREG}}^{\text{SCCE}}$ is the number of SREGs per SCCE.

The mapping is illustrated in Fig. 9.10 for a 2-symbol CRS-based SPDCCH.

For CRS-based SPDCCH using distributed mapping, the SREG index corresponding to SCCE index n is given by:

$$mod\left(n, \left\lfloor \frac{N_{RB}}{N_{SREG}^{SCCE}} \right\rfloor\right) + \left\lfloor \frac{n}{\left\lfloor \frac{N_{RB}}{N_{SREG}^{SCCE}} \right\rfloor} \right\rfloor \cdot N_{RB} + j \cdot \left\lfloor \frac{N_{RB}}{N_{SREG}^{SCCE}} \right\rfloor \tag{9.2}$$

where

$$j = 0, \cdots, N_{SREG}^{SCCE} - 1$$

N_{RB} is the number of resource blocks in frequency in the SPDCCH resource set.

This results in the SREG to SCCE mapping as shown in Fig. 9.11 for a two symbol CRS-based SPDCCH.

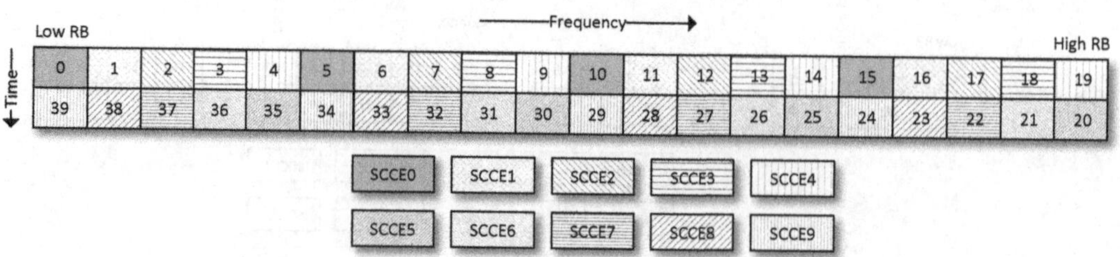

FIG. 9.11 SREG to SCCE mapping, distributed CRS-based SPDCCH, illustrated for a 2 symbol SPDCCH resource set.

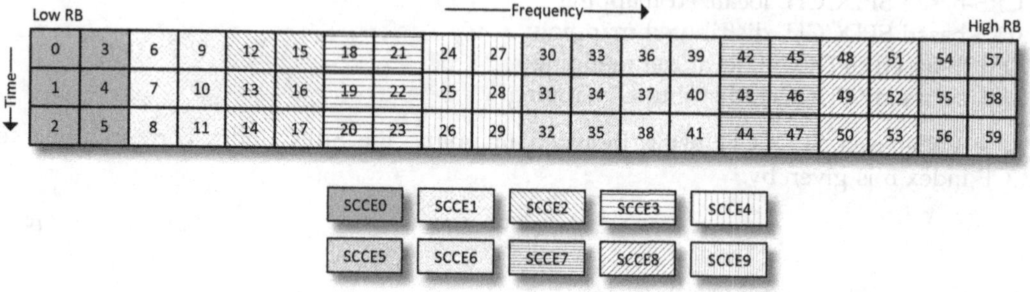

FIG. 9.12 SREG to SCCE mapping, DMRS-based SPDCCH, illustrated for a 3-symbol SPDCCH resource set.

For DMRS-based SPDCCH using either localized or distributed mapping, the SREG indexes corresponding to SCCE index n is given by the same equation as for CRS-based localized mapping (Eq. 9.1), but since the SREG numbering is performed in a time first manner, the SCCEs are formed more condense in frequency (compare Figs. 9.10 and 9.12), see Fig. 9.12.

It should be noted that, so far, the construction of SREGs, SCCEs and their mapping in the logical domain does not say anything about where a given SCCE index is mapped onto the physical resources. This is however easily derived for a given SPDCCH resource set. Consider in the examples above an SPDCCH resource set of 20 resource blocks. In this case, the mapping is simply done per resource block from low to high frequency. An example of mapping the DMRS based localized mapping to the physical resources in two different contiguous sub-bands is shown in Fig. 9.13.

Hence, it can be noted that although the mapping of SREGs to SCCEs is localized, the actual mapping to physical resources can still be distributed, depending on the SPDCCH resource set configuration.

One restriction that has been adopted related to the SPDCCH resource set configuration for DMRS-based SPDCCH is that the PRB need to be configured in no smaller entities than in contiguous pairs of resource blocks (i.e. the same PRG, see Section 9.2.3.1).

Let's now look at the final mapping stage of the SCCEs to a SPDCCH candidates. For an SPDCCH candidate of AL one, the mapping is already given (since a SPDCCH candidate of AL one consist of one SCCE, which we already have the mapping for), but for higher ALs, the CCE indices constituting a given SPDCCH need to be determined.

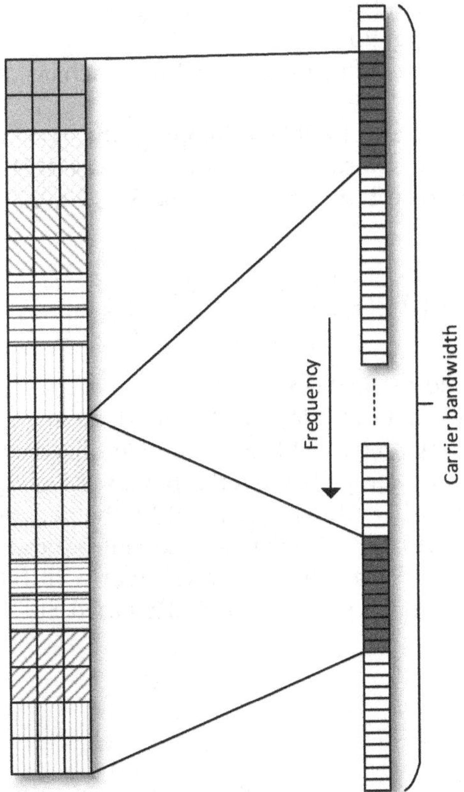

FIG. 9.13 Mapping of SCCEs to physical resources.

For CRS-based SPDCCH, the logical SCCEs indices corresponding to SPDCCH candidate m of the SPDCCH search space at AL L in a given slot/subslot are given by

$$mod\left(Y^{(L)} + L \cdot \left(mod\left(\frac{m \cdot N_{SCCE}}{L \cdot M^{(L)}}, \left\lfloor\frac{N_{SCCE}}{L}\right\rfloor\right) + i\right), N_{SCCE}\right) \tag{9.3}$$

where.

$Y^{(L)}$ is the RRC configured starting position for AL L

$$i = 0, \ldots, L-1$$

N_{SCCE} is the total number of SCCEs in the SPDCCH resource set

$$m = 0, \ldots, M^{(L)} - 1$$

$M^{(L)}$ is the number of SPDCCH candidates to monitor at AL L.

The resulting SPDCCH candidates for different aggregation levels (and different number of candidates) are shown in Fig. 9.14.

The localized mapping is not illustrated here but is achieved using the same SCCE to SPDCCH relation.

For DMRS-based SPDCCH and localized mapping, the same equation is used as in the mapping of SCCEs to SPDCCH candidates for CRS-based SPDCCH (Eq. 9.3). For the DMRS-based SPDCCH and distributed mapping, the SCCE to SPDCCH candidate mapping is performed according to:

$$mod\left(Y^{(L)} + \left\lfloor \frac{m \cdot N_{SCCE}}{L \cdot M^{(L)}} \right\rfloor, \frac{N_{SCCE}}{L}\right) + i \cdot \left\lceil \frac{N_{SCCE}}{L} \right\rceil \tag{9.4}$$

The mapping is illustrated in Fig. 9.15.

The mapping of the SPDCCH candidate to REs are done in a frequency-first-time-second manner. This implies that for CRS-based SPDCCH and localized mapping, the REs between a higher and a lower AL can be completely overlapping, see Fig. 9.16. Hence, there could be ambiguity for the device on what AL has been sent by the network.

To solve this, an interleaver on SREG level has been introduced. For each SPDCCH candidate, the SREGs are interleaved using a block interleaver with L (i.e. the aggregation level) rows and 4 (the number of SREGS per SCCE for CRS-based SPDCCH) columns.

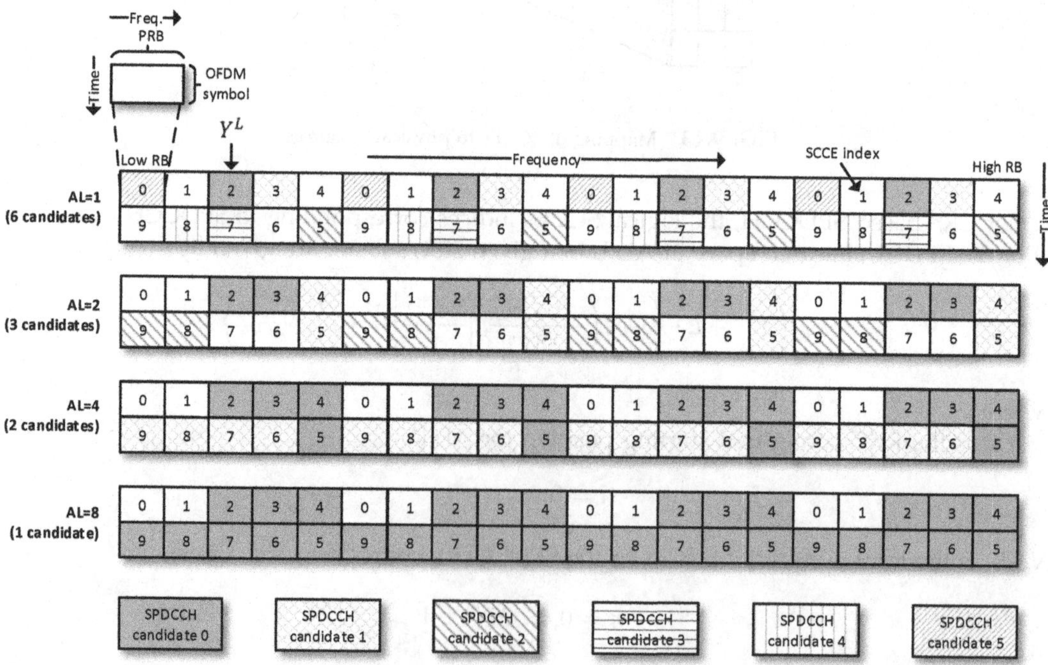

FIG. 9.14 SPDCCH candidates for CRS-based SPDCCH and distributed mapping.

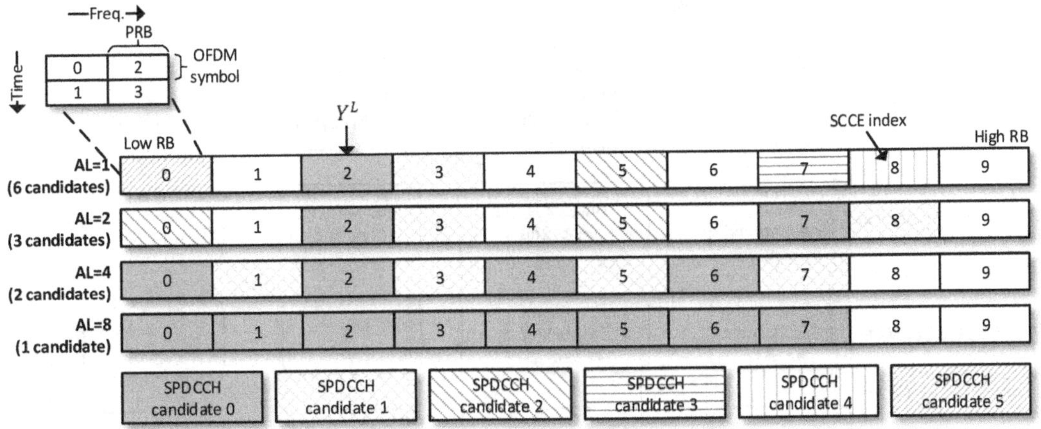

FIG. 9.15 SPDCCH candidates for DMRS-based SPDCCH and distributed mapping.

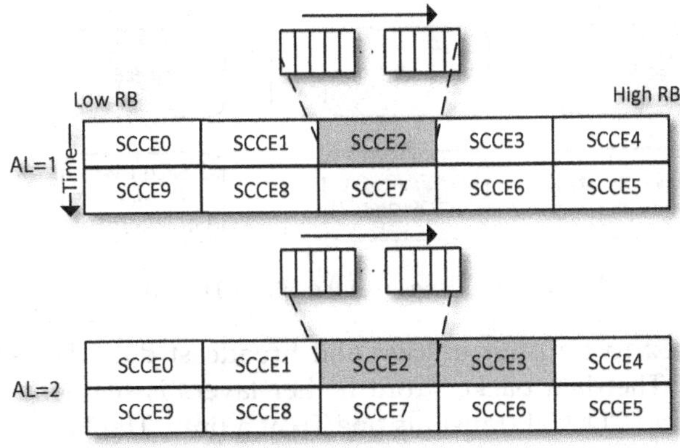

FIG. 9.16 Overlapping REs for different SPDCCH candidates for CRS-based SPDCCH and localized mapping without modified bit mapping.

9.2.3.2.4 Overview

To tie it all together, Fig. 9.17 shows an overview of the different steps in the creation of an SPDCCH candidate depending on the SPDCCH resource set configuration.

See Sections 9.2.3.2.2–9.2.3.2.3 for more details.

9.2.3.3 Slot/subslot-PDSCH

Subframe	Any
Basic TTI	Subslot, slot
Repetitions	Up to 6 transmissions
Subcarrier spacing	15 kHz
Bandwidth	See Section 9.3.2.9.1
Frequency location	See Section 9.3.2.9.1

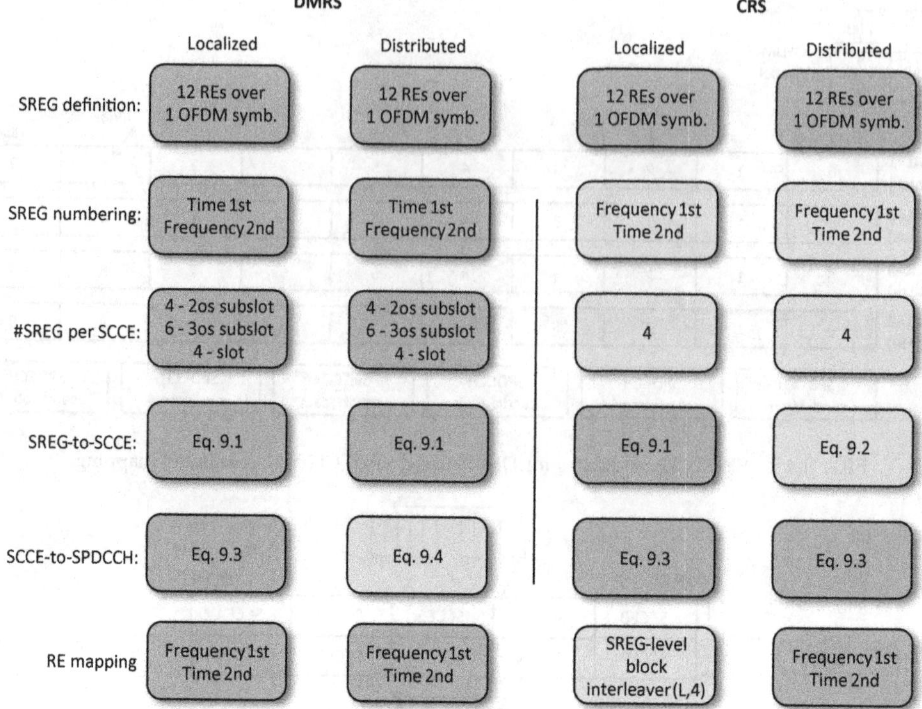

FIG. 9.17 Overview of the SPDCCH procedure.

The *PDSCH*, is used to transmit unicast and broadcast data. For sTTI the focus is on unicast transmission. The data packet from higher layers is segmented into one or more *transport blocks* (TB), and PDSCH transmits one TB at a time. The PDSCH for sTTI transmission is inherited from LTE and hence in this section, the differences introduced with sTTI will be the focus. For more details on the PDSCH design for LTE, an interested reader is referred to e.g. Ref. [4].

When shortening the transmission duration using subslot and slot operation, the available resources, for a given bandwidth, compared to subframe transmission is heavily reduced. To achieve roughly the same code rate for different *Modulation and Coding Schemes* (MCS), the information bits mapped to the subslot/slot need to be reduced, compared to subframe operation. To limit the impact to the specification, this was achieved by using the existing TBS tables in LTE, but scaling the TBS value by a fixed factor, α, depending on if the transmission duration is subslot or slot:

- Slot: $\alpha = 1/2$
- Subslot: $\alpha = 1/6$

After scaling, the resulting TBS value is rounded off to the closest valid TBS value in the existing table. It can be noted that this approach will result in a varying code rate over the different transmission opportunities, e.g. due to the overhead and subslot duration, but was chosen due to its simplicity in implementation.

TABLE 9.3 Transmission modes for slot/subslot-PDSCH.

Transmission mode	Description	Note
TM1	Single antenna transmissions	FDD, TDD
TM2	Transmit diversity	FDD, TDD
TM3	Open-loop codebook-based precoding/transmit diversity	FDD, TDD
TM4	Closed-loop codebook-based precoding	FDD, TDD
TM6	1-layer closed-loop codebook-based precoding	FDD, TDD
TM8	Non-codebook-based precoding, up to 2 layers	TDD
TM9	Non-codebook-based precoding, up to 4 layers	FDD; TDD
TM10	Non-codebook-based precoding, up to 4 layers	FDD, TDD

As for legacy LTE, a variety of transmission modes are supported using multiple antenna transmission and/or reception. Multiple-input multiple output, MIMO, transmission is supported with a maximum of four MIMO layers for both CRS- and DMRS-based transmission. Irrespective of the number of MIMO layers, a single codeword is used. Hence, only a single HARQ bit need to be reported from the device for each received downlink TTI, which improves Short Physical Uplink Control Channel (SPUCCH) performance.

The supported transmission modes are listed in Table 9.3. The downlink transmission modes for sTTI are configured separately from subframe operation.

For the 5G URLLC requirements (see Chapter 10), there is no high demand for high throughput but rather for a high reliability. Increasing diversity will generally increase reliability and hence transmit diversity is the expected mode of operation when reliability is of primary concern.

The mapping of the PDSCH to the physical resources follows the slot boundaries for slot operation and the subslot layout for subslot operation, see Fig. 9.2.

In case of slot operation in TDD, PDSCH transmission is supported in DwPTS for the first slot except in special subframe configuration 0 and 5, and in the second slot in case of special subframe configuration 3, 4 and 8, see Fig. 9.18. Each transmission in the two slots of DwPTS and UpPTS are scheduled independently.

For DMRS based PDSCH, when shortening the transmission duration, a natural consequence is that the DMRS overhead is increased since at least one DMRS symbol should be associated with each subslot/slot-based data transmission. However, when reducing the transmission duration to two or three symbols, as in the case of subslot operation, the overhead can become significant. To solve this, a sharing mechanism has been introduced to allow multiple subslots sharing the same DMRS. A DMRS can be shared between two consecutive subslots. Whether DMRS is included in a given subslot is indicated to the device dynamically via DCI.

9.2.3.3.1 Blind repetitions

To improve reliability of the PDSCH transmission, it is possible to configure *blind repetitions* (repetitions without HARQ feedback). This is a similar functionality as used in CIoT

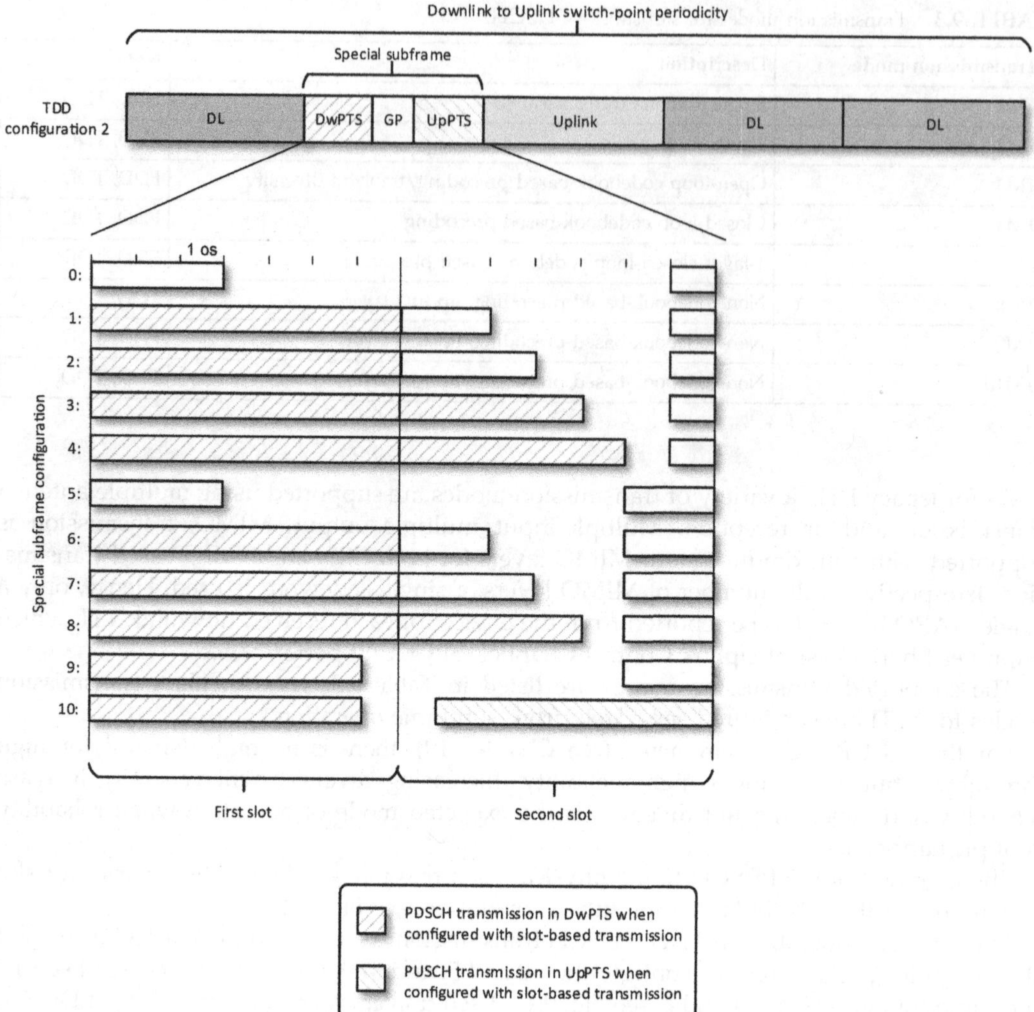

FIG. 9.18 Slot- PDSCH transmission in different special subframe configurations.

technologies to extend coverage (i.e. used in this case to lower the operating SNR at a given reliability). This allows for a fast reception at the device (not waiting for HARQ feedback). The drawback is the resource required. Compared to HARQ-based retransmissions where feedback is provided from the receiving node on which packets that need to be retransmitted, the number of blind repetitions will have to be selected to ensure proper reception in the worst-case scenario that the repetitions are dimensioned for. If the packet is received after a lower number of repetitions, the remaining repetitions is a waste of resources. This is however an acceptable price, considering the improved performance that is required for certain high-end segment cMTC services.

To realize the blind repetitions, a scheme has been adopted where the DCI indicates the number of repetitions used from that DCI onwards. The repetition value is decremented for each additional repetition. Hence, the device need only detect and decode one of these DCIs in order to understand the number of repetitions remaining. However, if the first DCI(s) are missed, the associated blind repetitions will also be missed. Hence, it is still of importance for the network to ensure good downlink control reception to increase the probability of the DCI being received in the *initial* transmission (ensuring reception of all blind repetitions). After decoding the DCI, the device assumed that the rate-matching and the resource allocation is the same for the remaining repetitions as in the TTI where the DCI was received.

The DCI can indicate number of transmissions 1,2,3, and, 4 or 6. Which one of {4,6} that is used is configured by RRC signaling.

An example is shown in Fig. 9.19 where four total transmissions are configured. The DCI decrements the number of blind repetitions with increasing TTI. On the left side of the figure, the behavior from the network is shown while to the right, the understanding of the device. It is assumed here that the device misses the DCI of the initial transmission but receives it in the first blind repetition. The DCI indicates that two repetitions remain. Based on this and the

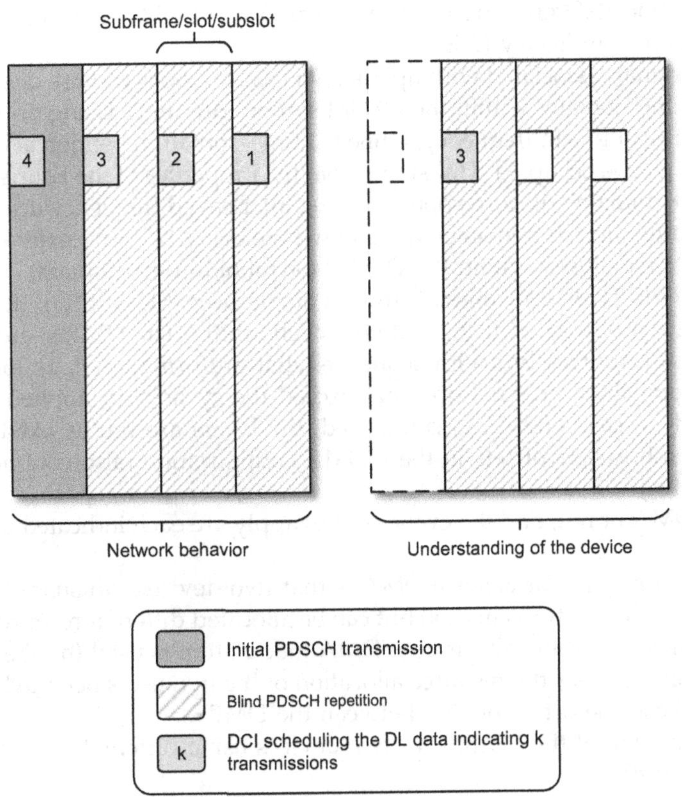

FIG. 9.19 Blind PDSCH repetitions.

knowledge that the resource allocation and the rate-matching for the remaining repetitions follow the decoded DCI, the remaining repetitions can be received and combined to achieve the processing gain required to improve the reliability.

Sharing of the DMRS between subslots to reduce overhead, cannot be used in case of blind repetitions.

In case of blind repetitions, the processing timeline (see Section 9.2.5) for the HARQ-ACK feedback to be reported by the device is based on the last repetition in the sequence.

9.2.4 Uplink physical channels and signals

9.2.4.1 Uplink reference signals

Subframe	Any
Subcarrier spacing	15 kHz
DMRS bandwidth	Same as associated PUSCH/PUCCH
DMRS frequency location	Same as associated PUSCH/PUCCH

The reference signals in the uplink span the bandwidth of the associated allocation, as for legacy LTE operation, whether it being uplink control or uplink data.

The Demodulation Reference Signals (DMRS) associated with the physical uplink control channel are re-used from legacy LTE.

The reference signals associated with uplink data transmissions can use one out of two repetition factors in time domain, within the OFDM symbol duration. Using no repetition factor, the DMRS occupies all REs in frequency, while if a time-domain repetition is used, every other REs in frequency will be occupied. This stems from the properties of the Fourier transform. The repetitions in time domain will correspond to using interleaved frequency domain multiple access (IFDMA) modulation in frequency domain (where every N^{th} subcarrier will be occupied, transforming to N repetitions within the ODFDM symbol in time domain).

When using multi-layer transmission from a single user (SU-MIMO), the DMRSs of the different layers are multiplexed using different cyclic shifts. The DMRSs are transmitted on the same physical resources but with sequences that are orthogonal, as long as the delay spread of the propagation channel does not exceed the cyclic shift applied in case IFDMA is not configured. In case IFDMA is configured, the layers are multiplexed using different cyclic shifts and subcarriers offsets in the IFDMA configuration (also commonly referred to as different 'combs').

The use of IFDMA or not, and the cyclic shift to apply are each indicated by a 1-bit field in the DCI.

One of the main benefits of using IFDMA is that two devices can share the same OFDM symbol for reference signal transmission but can be allocated different resources in frequency. A sharing of DMRS is possible also if no IFDMA modulation is used (by the use of different cyclic shifts) but in this case, the resource allocation of the two users need to be fully overlapping in order to keep the orthogonality between the DMRSs.

IFDMA transmission of the DMRSs for two devices using different resource allocations is illustrated in Fig. 9.20.

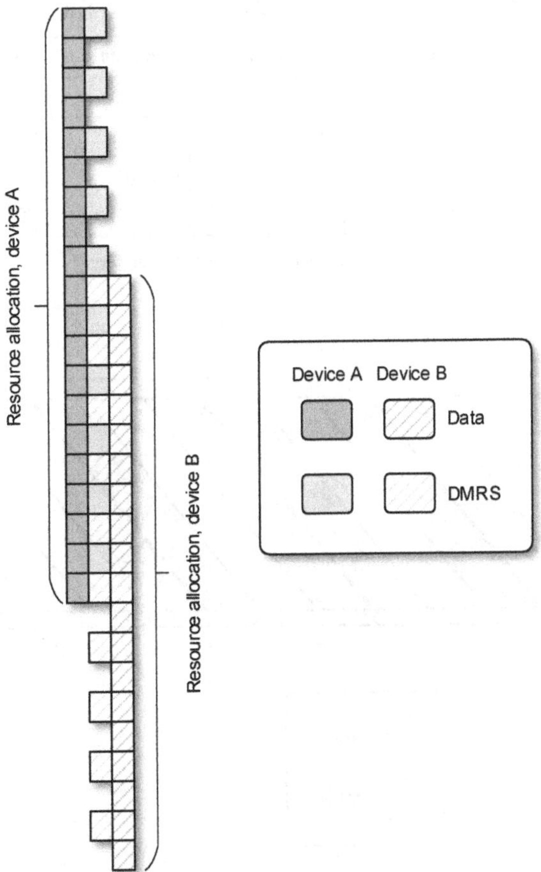

FIG. 9.20 Uplink IFDMA DMRS transmission.

As with subslot-PDSCH transmissions (see Section 9.2.3.3), to minimize DMRS overhead, DMRSs can be shared across subslots also in the uplink. As in the downlink, the sharing of DMRS is dynamically indicated to the device. However, instead of an indication of the DMRS present or not, two bits are used to provide flexibility of using different subslot allocations, providing different level of DMRS overhead and scheduling options. The, up to four, combinations per subslot in the subframe is shown in Table 9.4. The DMRS can be shared amongst at most the three consecutive subslots of a slot.

An illustration of the signaling is shown in Fig. 9.21.

In case of slot operation, the DMRS overhead does not vary and the DMRS is self-contained in each slot with the DMRS symbols positioned as for subframe operation in legacy LTE, see Fig. 9.22.

TABLE 9.4 PUSCH DMRS pattern for dynamic scheduling through uplink grant.

Bit value	Subslot#0	Subslot#1	Subslot#2	Subslot#3	Subslot#4	Subslot#5
00	R D D	R D	R D	R D	R D	R D D
01	D D R	D R	D D	D R	D R	
10		D D		D D |R	D D	
11		D D |R			D D |R	

'R denotes a DMRS symbol, 'D' denotes a data symbol, '\|' denotes the subslot boundary.

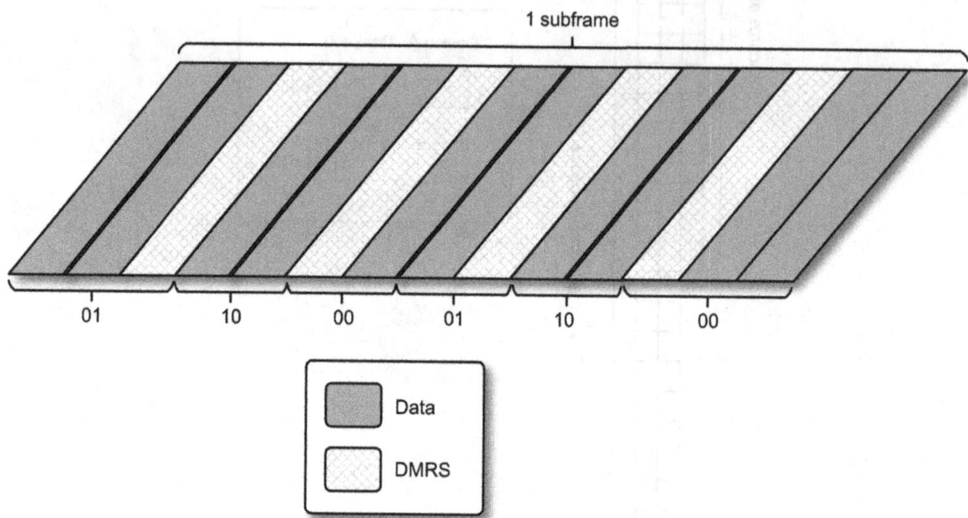

FIG. 9.21 Example of subslot PUSCH DMRS sharing including the DMRS pattern signaling.

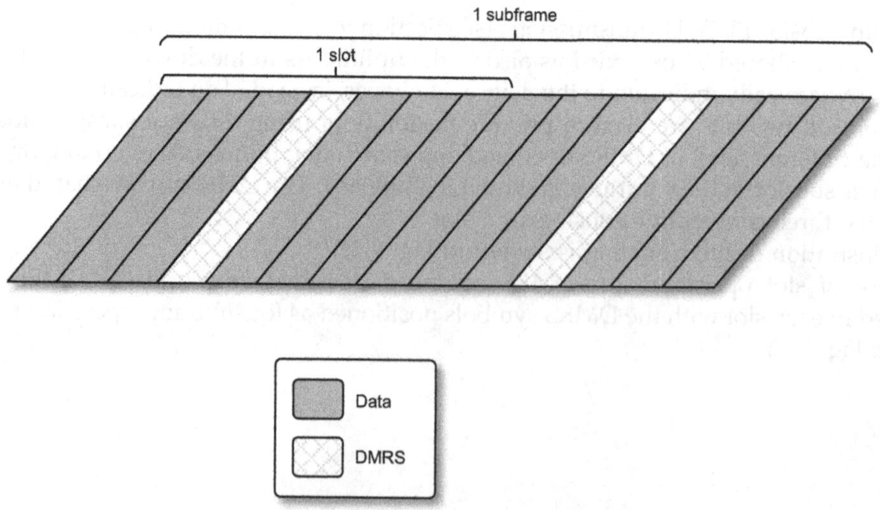

FIG. 9.22 DMRS positions in case of slot-operation.

9.2.4.2 *Slot/subslot-SPUCCH*

Subframe	Any
Basic TTI	Slot/subslot
Repetitions	No
Subcarrier spacing	15 kHz
Bandwidth	1 PRB (SPUCCH format 1/1a/1b/3) 1,2,3,4,5,6,8 PRB (SPUCCH format 4)
Frequency location	Any PRB
Frequency hopping	If applicable, between 2 PRB locations

9.2.4.2.1 General

The *SPUCCH*, is used to carry the following types of *Uplink Control Information* (UCI):

- Uplink scheduling request (SR)
- Downlink HARQ-ACK feedback

In contrast to other PUCCH designs in LTE, SPUCCH does not carry downlink *Channel State Information* (CSI), which is only carried by subslot-PUSCH and slot-PUSCH.

Different SPUCCH formats exist. The difference between the different formats is mainly the number of HARQ-bits it can carry and the multiplexing rate of the physical channel (i.e. how many UEs can be simultaneously transmitting on the same physical channel).

The SPUCCH format selection is based on the number of HARQ-ACK bits to be transmitted. The size of the HARQ-ACK bit field depends on multiple factors:

- The number of carriers configured, in case of carrier aggregation.
- Whether subslot operation is configured in the downlink and slot operation is configured in the uplink (see Sections 9.3.2.1 and 9.2.5), in which case the HARQ-ACK bits of the three downlink subslot are carried by an uplink slot-SPUCCH.
- Whether subframe HARQ-ACK bits are piggybacked on SPUCCH (see Section 9.3.2.5).
- Whether the number of HARQ-ACK bits is determined based on the carriers configured or dependent on the carriers where the downlink control is detected (usually referred to as fixed and dynamic codebook size respectively).

The device will determine the number of HARQ-ACK bits to be carried by a slot-SPUCCH/subslot-PUCCH considering the factors mentioned above and select the SPUCCH format accordingly.

An overview of the different SPUCCH formats is given in Table 9.5.

For slot operation, in case both SPUCCH format 3 and 4 are configured, SPUCCH format 4 is used from 12 bits onwards.

The channel coding used for SPUCCH is determined based on the payload size, irrespective of format 3 or 4 being used. A Reed-Muller block code is used for payload sizes of 3 bits up to 11 bits. For payload sizes between 12 and 22 bits, a dual Reed-Muller code is used (the same block code applied twice), and in case of payload size above 22 bits, a tail biting convolutional code with an 8-bit CRC is used.

TABLE 9.5 SPUCCH formats.

SPUCCH format	Applicability	HARQ Ack size [bits]	# Resource blocks	Frequency hopping
format 1/1a/1b	Subslot Slot	1 or 2	1	Yes Yes or No
format 3	Slot	3–11	1	No
format 4	Subslot Slot	≥3	1,2,3,4,5,6 or 8	No Yes

For all PUCCH formats, up to four time-frequency resources can be configured by RRC signaling. The resource to use is indicated to the device in a 2-bit field in the DCI. The same functionality exists in legacy LTE for PUCCH format 3.

9.2.4.2.2 SPUCCH format 1/1a/1b

9.2.4.2.2.1 Slot Slot-SPUCCH format 1/1a/1b is similar to PUCCH format 1/1a/1b in that it is based on coherent demodulation of modulated reference signal sequences. In the modulation of the sequences, either BPSK (SPUCCH format 1a) or QPSK (SPUCCH format 1b) is used.

However, in contrast to PUCCH format 1/1a/1b, the configuration of the time-frequency resources to be used is done through RRC signaling (as already mentioned in Section 9.2.4.2.1) and dynamically indicated in DCI which resources to use (see Section 9.3.2.7.1). This can be compared to PUCCH format 1a/1b, where the resources are implicitly given by the CCE index associated with the downlink data transmission.

Slot-SPUCCH format 1/1a/1b can be configured by RRC to either apply frequency hopping or not.

In case frequency hopping is not enabled, the format is identical to a single slot of PUCCH format 1/1a/1b (similar as the design choice of slot-SPUCCH format 3, see Section 9.2.4.2.3), including the use of *cyclic shifts* (CSs) and *OCC* to multiplex users.

In case of frequency hopping, the hop takes place after either the third symbol (for the first slot in a subframe) or after the fourth symbol (for the second slot in a subframe). The reason to have different hopping patterns is to align with the uplink subslot pattern, see Figs. 9.23 and 9.2.

How the cyclic shift and OCC-code applied varies over time is different depending on if frequency hopping is used or not. In case frequency hopping is not enabled, the cyclic shift and OCC is generated in the same way as for PUCCH format 1/1a/1b in legacy LTE. This allows users of slot-SPUCCH and PUCCH to be multiplexed as in the legacy case. In case frequency hopping is enabled, how the cyclic shift varies over time is instead taken from PUCCH format 2/2a/2b, as for the case of subslot-SPUCCH (see Section 9.2.4.2.2.2).

9.2.4.2.2.2 Subslot Subslot-SPUCCH format 1/1a/1b is the SPUCCH format that differs most from the subframe based PUCCH design in legacy LTE.

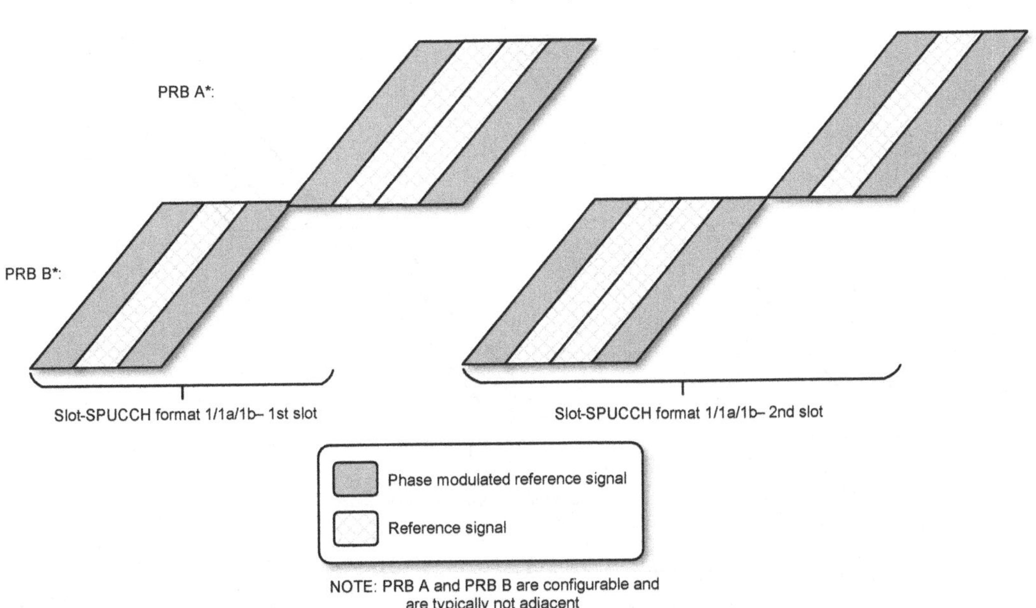

FIG. 9.23 Slot-SPUCCH format 1/1a/1b

The difference lies mainly in how the resources are configured (aligned with the configuration for slot-SPUCCH format 1, see Section 9.2.4.2.2.1) and how the SPUCCH format is detected at the receiver.

As for the detection, the format is based on *sequence selection* rather than DMRS based coherent demodulation. This means that there is no DMRS symbol used for phase reference/demodulation, but each symbol can instead be independently and non-coherently detected by the receiver.

The same sequence is used as for PUCCH format 1/1a/1b using cyclic shifts (CSs) to separate the sequences. Up to eight CSs can be allocated. The detected CS provides the HARQ feedback + (potential) SR information. The reason to use a sequence-based design was mainly to enable frequency hopping between the symbols in a subslot since using a design with coherent demodulation requires one reference symbol sequence and one modulated sequence transmitted transmitted on the same frequency resources to enable demodulation.

As seen from Fig. 9.24, the frequency hop in a 2 OFDM symbol (2 os) SPUCCH format 1/1a/1b design occurs naturally between the two symbols. For a 3 os SPUCCH format 1/1a/1b, the hopping occurs after the first symbol, which allows the 3 os subslot at the end of the subframe (see Fig. 9.2) to still provide frequency diversity, in the case of SRS transmission.

As with legacy PUCCH format 1, there is a cyclic shift variation over time to randomize interference. There is however no OCC applied since in each PRB being transmitted there is only one time-domain symbol, and hence OCC is not possible to apply (the orthogonality is lost in the frequency hop). Since the legacy randomization of cyclic shifts is more complex for PUCCH format 1/1a/1b, the randomization of cyclic shifts follows instead the one

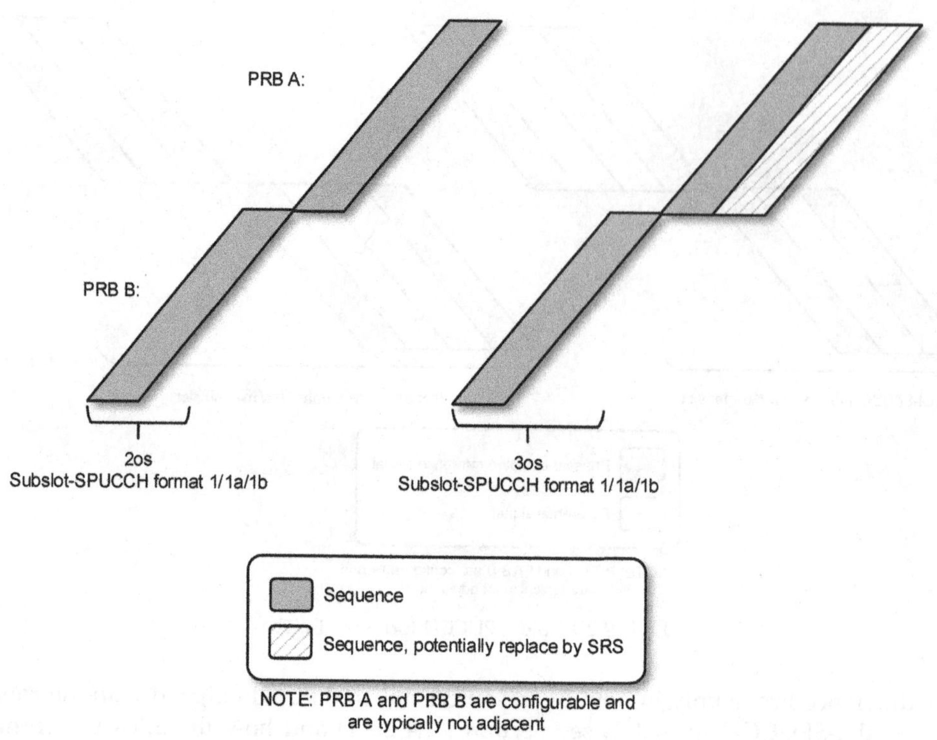

FIG. 9.24 Subslot-SPUCCH format 1/1a/1b.

used for PUCCH format 2/2a/2b. The consequence of this is that subslot-SPUCCH format 1/1a/1b cannot be easily multiplexed with legacy PUCCH format 1/1a/1b, but instead with PUCCH format 2/2a/2b. As for all (S)PUCCH formats, the format itself can be multiplexed to accommodate different users on the same physical resources by the use of different cyclic shifts.

9.2.4.2.3 SPUCCH format 3

SPUCCH format 3 is basically the same as the existing PUCCH format 3 with the transmission duration reduced to half (a slot instead of a subframe). PUCCH format 3 is using frequency hopping between the slots to increase diversity. However, since the transmission of SPUCCH format 3 is only over a single slot no frequency hopping is used. Similarly, PUCCH format 3 can carry up to 22 bits of payload, but since SPUCCH format 3 is only of half the duration, also the maximum payload size is halved to 11 bits. The format supports five simultaneous users transmitting on the same channel by the use of time domain OCC. As with PUCCH format 3, the format is mapped to a single PRB and uses SC-FDMA modulation, see Fig. 9.25.

Why is not frequency hopping used in this format? As with other formats, this would allow for improved performance. This is true. However, using frequency hopping would essentially convert slot-SPUCCH format 3 to slot-SPUCCH format 4 (see Section 9.2.4.2.4.2) since OCC can no longer be used, due to the frequency hop. Hence, slot-SPUCCH format

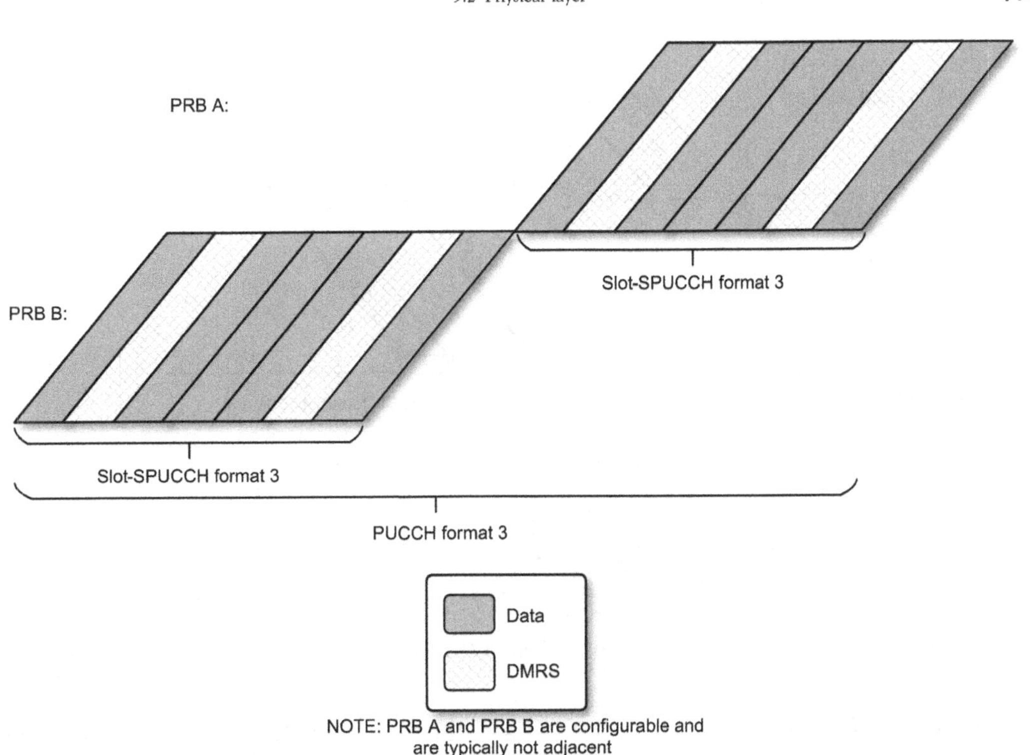

FIG. 9.25 Slot-SPUCCH format 3.

3 is an alternative to slot-SPUCCH format 4 (over the payload ranges where they overlap), where a higher user multiplexing rate is achieved instead of an improved link performance.

By using the same design as PUCCH format 3, multiplexing is also achieved with existing PUCCH format 3 users.

9.2.4.2.4 SPUCCH format 4

9.2.4.2.4.1 General As with PUCCH format 4 in legacy LTE, the slot/subslot-SPUCCH format 4 is similar to slot/subslot-PUSCH. It is based on coherent DMRS-based demodulation with all available bits used to provide an efficient channel coding rather than a high user multiplexing rate (only one user is intended for a given resource). The modulation used is QPSK and the DMRS sequence is generated in the same way as for PUCCH format 4.

The bandwidth allocated to SPUCCH format 4 can also vary depending on RRC configuration and dynamic signaling in the DCI, with allowed PRB allocations of 1,2,3,4,5,6 or 8.

9.2.4.2.4.2 Slot Slot-SPUCCH format 4 is illustrated in Fig. 9.26. As for slot-SPUCCH format 1/1a/1b, the frequency hopping pattern is dependent on the slot that the SPUCCH is transmitted in. One DMRS symbol is used in each sequence of OFDM symbols sharing the same frequency allocation, to enable coherent demodulation.

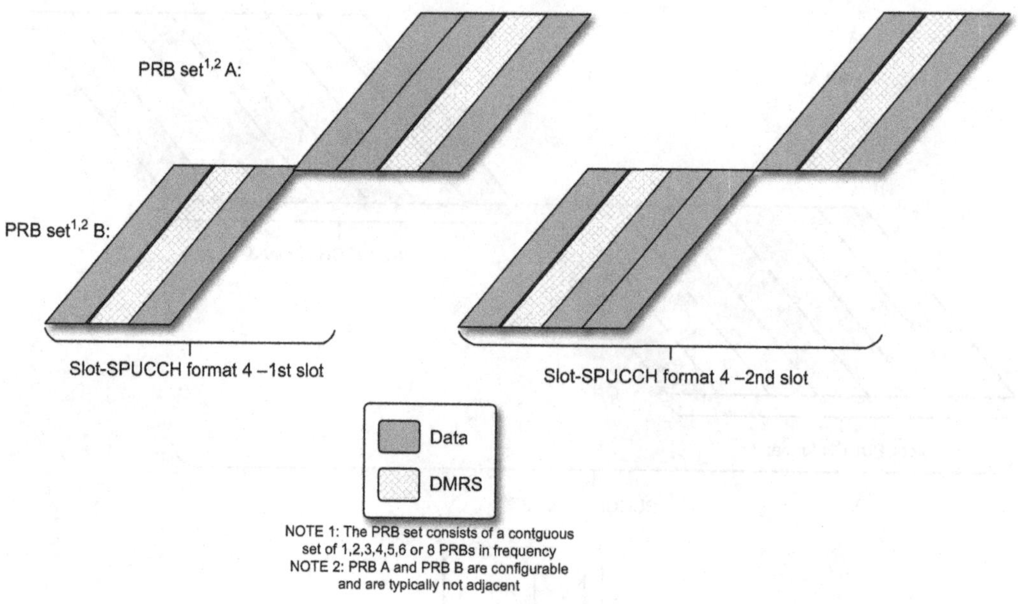

FIG. 9.26 Slot-SPUCCH format 4.

9.2.4.2.4.3 Subslot Subslot-SPUCCH format 4 is illustrated in Fig. 9.27. As can be seen, no frequency hopping is used due to the coherent demodulation. A front-loaded DMRS design (the DMRS is placed in the first symbol) is adopted which helps the receiver in performing a channel estimate early in the reception to allow for quick demodulation.

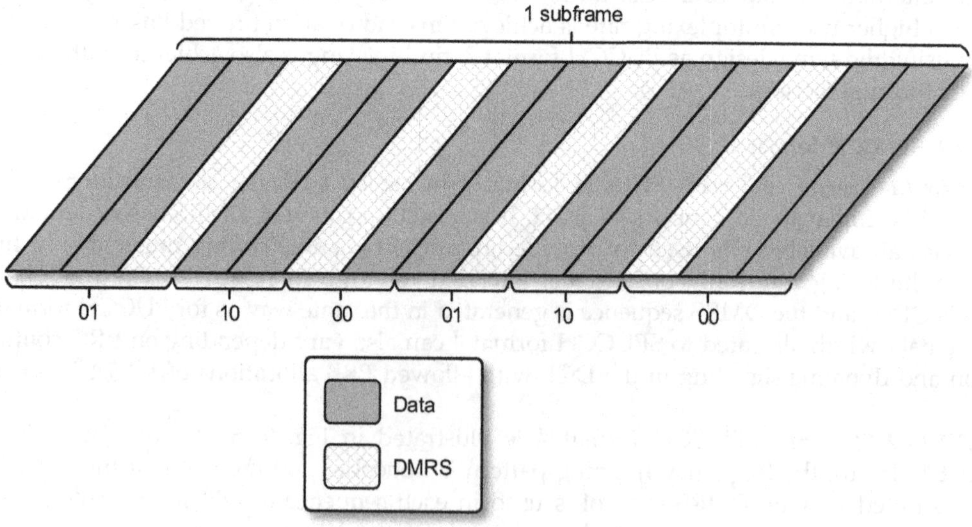

FIG. 9.27 Subslot-SPUCCH format 4.

9.2.4.3 Slot/subslot-PUSCH

Subframe	Any
Basic TTI	Slot/subslot
Repetitions	2, 3, 4 or 6 (SPS)
Subcarrier spacing	15 kHz
Bandwidth	See Section 9.3.2.9.2
Frequency location	See Section 9.3.2.9.2

The *PUSCH* is primarily used to transmit unicast data. The data packet from higher layers is segmented into one or more *TB*, and PUSCH transmits one TB at a time.

PUSCH is also used for transmission of UCI when aperiodic CSI transmission is triggered by setting the CSI request bit in DCI (see Section 9.3.2.7.1 and 9.2.3.10), or, to carry HARQ-ACK bits from PUCCH, in case of collision between PUSCH and PUCCH (and the device does not support, and is not configured with, simultaneous transmission of PUSCH and PUCCH).

As with the PDSCH, there is no new physical shared channel defined for the uplink, instead the transmission duration of the PUSCH is shortened when transmitting subslot-PUSCH or slot-PUSCH. The slot-PUSCH and subslot-PUSCH is inherited from LTE and hence in this section, the differences introduced with sTTI will be the focus. For more details on the PUSCH design for LTE an interested reader is referred to e.g. Ref. [4].

As with PDSCH, the TBS is scaled depending on subslot or slot transmission (see Section 9.2.3.3), using the same scaling factors of $\alpha = 1/2$ for slot, and using either $\alpha = 1/12$ or $\alpha = 1/6$ for subslot depending on if the subslot contains one or two data symbols respectively.

As for subframe-based PUSCH, slot-PUSCH and subslot-PUSCH can be configured with transmission mode 1 and 2, see Table 9.6. The downlink transmission modes for sTTI are configured separately from subframe operation. MIMO is also supported on the uplink with up to four MIMO layers configured. As for PDSCH (see Section 9.2.3.3), a single codeword is used, irrespective of the number of MIMO layers.

The mapping of the PUSCH to physical resources follows the slot boundaries for slot operation and the subslot layout for subslot operation, see Fig. 9.2. PUSCH can however not be mapped to symbols where the DMRS is mapped since the DMRS is transmitted over the full resource allocation, see Section 9.2.4.1.

In case of slot operation in TDD, PUSCH is only supported in UpPTS in special subframe configuration 10, see Fig. 9.18.

TABLE 9.6 slot/subslot-PUSCH transmission modes.

Transmission mode	Description
TM1	Single antenna
TM2	Transmit diversity

9.2.5 Timing advance and processing time

We have already seen how the time to transmit a packet over the air is improved by reducing the actual transmission time (to a subslot or slot operation). Another contributing factor to the overall latency is the processing time at the eNB and device. In case of eNB, the processing time is largely up to implementation, especially with the introduction of asynchronous operation with sTTI, see Section 9.3.2.6. On the device side however, a specified *processing timeline* needs to be strictly followed. This relates for example to the HARQ-ACK response to a downlink assignment and an uplink data transmission in response to an uplink grant. That is, for the network to properly schedule the uplink resources, and know when and where to expect what in the uplink, a processing timeline is required.

For cMTC services, the processing timeline of the device is an important factor since it determines the number of blind transmissions or number of HARQ retransmissions possible to perform within a given latency bound (see more details in Chapter 10), and is thus an important factor in reaching the 5G URLLC requirements.

The processing timeline is usually expressed in TTIs. For pre-release 15 devices, a processing timeline of n+4 is used. For example, if the downlink assignment is sent in subframe *n*, the associated HARQ-ACK response will be sent by the device in subframe *n*+4.

With shortened transmissions, the total amount of data to be processed, for a given bandwidth allocation, roughly scales with the transmission duration. However, other aspects, such as channel estimation will not. One of the more important processing related parameters that is independent of transmission duration is the timing advance, TA.

The timing advance is the advancement in transmission timing performed by the device on the uplink. The advancement is performed for the transmission to be received at the eNB aligned with the downlink frame timing. The amount of timing advance used by the device is set to correspond to twice the propagation time of the signal. This is since the downlink signal that the device synchronizes to, will be delayed by the same amount as the uplink signal received at the eNB. This is illustrated in Fig. 9.28.

The figure illustrates the case of a n+4 timing for two TTI durations (τ_T), with one being roughly 1/3 of the other one. As can be seen, since the propagation delay (τ_P) is not dependent on the transmission duration, the time remaining for processing (τ_{Proc}) by the device is greatly reduced, from roughly $7\tau_P$ to $1\tau_P$.

Hence, reducing the transmission time puts a high requirement on the device with regards to processing time, especially if the timing advance needs to be large. At the same time, allowing a too small timing advance will limit the propagation delay supported, and hence the cell size (including any fiber delay from the digital unit to the radio unit).

The above problem becomes most pronounced in the case of subslot transmission. In this case, the specification solves it by supporting multiple maximum timing advance values, each associated with a different processing timeline. The device can indicate its capability out of two possible processing timeline sets (Set 1 or Set 2). In case of Set 1, the timing advance values are associated with either n+4 or n+6 timing, while for Set 2, the timing advance values are associated with n+6 and n+8 timing, see Fig. 9.29.

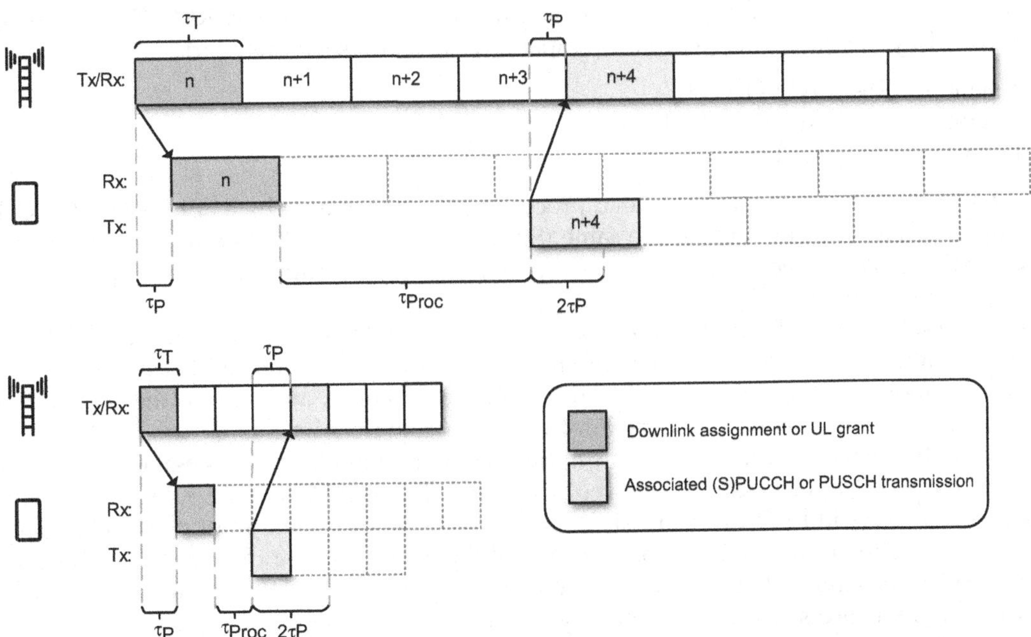

FIG. 9.28 Timing advance with different transmission duration but the same propagation delay (top figure: long transmission duration; bottom figure: short transmission duration).

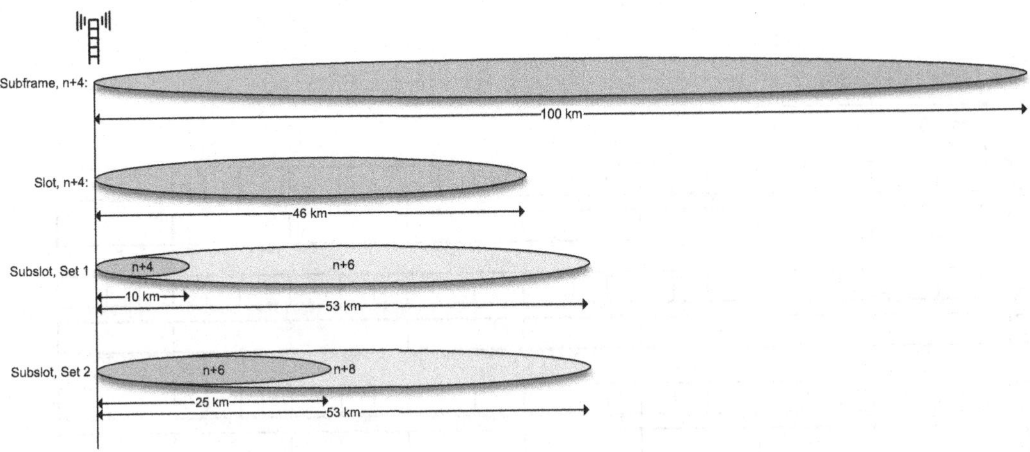

FIG. 9.29 Timing advance and cell size.

In case of slot operation, the processing timeline is not configurable and always equals n+4 slots.

The cell size supported for different configured transmission times and associated processing timelines is shown in Fig. 9.29.

In case of subslot operation, the processing timeline is configured by RRC. The TA set(s) supported by the device is indicated by capability signaling separately for:

- 1 symbol CRS-based SPDCCH
- 2 symbols CRS-based SPDCCH, and
- DMRS-based SPDCCH.

Only a single set, using an associated processing timeline is used by a device in a PUCCH group (a group of carriers using the same (S)PUCCH). Since Set 1 has a more strict timing than Set 2, it is assumed that if Set 1 is supported, the device implicitly also supports Set 2 for that configuration.

In case subslot is configured on the downlink and slot on the uplink (see Section 9.3.2.1) for FDD operation, the lowest maximum timing advance (for slot or subslot) is assumed.

Furthermore, in case subslot is configured on the downlink and slot on the uplink, the processing timeline follows the data direction (with some constraints on the fixed subslot and slot structures), e.g. if data is transmitted in the downlink, the processing timeline is based on the subslot processing timeline of the device. This is shown in Fig. 9.30 for downlink assignment to HARQ-ACK transmission, and in Fig. 9.31 for uplink grant to PUSCH transmission. Since the timing follows the direction of the data, there are three possible timelines for PDSCH, each associated with a different processing timeline, while only a single timeline for the uplink (since slot-transmission only supports n+4 timing).

The minimum processing timelines shown above for subslot and slot operation are for FDD. In case of TDD, the minimum delay of n+4 for slot operation cannot always be fulfilled depending on the restriction in the TDD configuration. This details of this are however not outlined in this book. Interested readers are referred to Ref. [7].

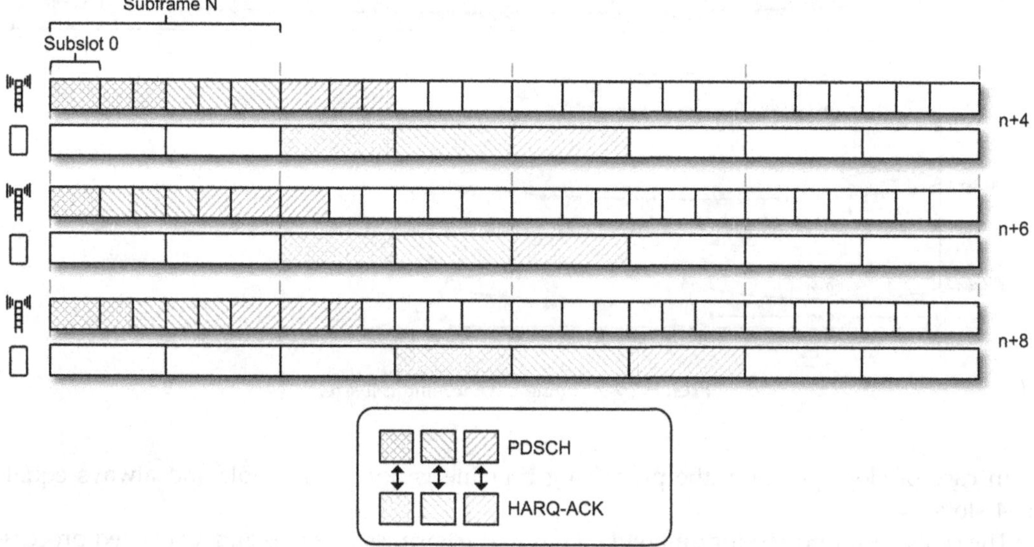

FIG. 9.30 Processing timeline for downlink subslot PDSCH to uplink slot HARQ ACK.

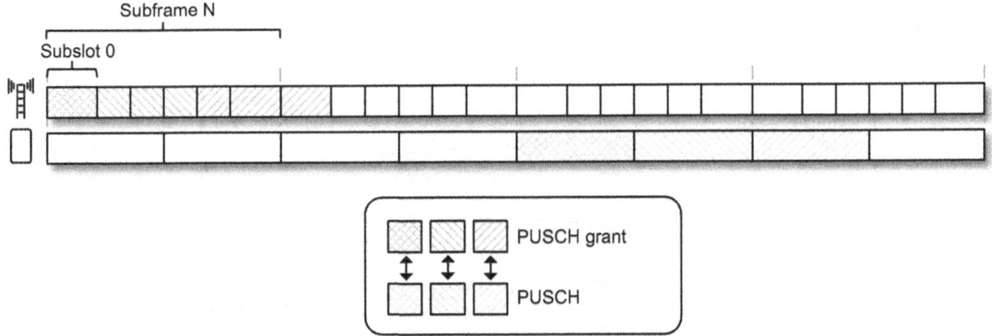

FIG. 9.31 Processing timeline for downlink subslot and uplink slot and PUSCH grant to PUSCH.

9.3 Idle and connected mode procedures

In this section, the idle and connected mode procedures are described. Since the work on LTE URLLC was mainly concerning user plane latency and reliability in connected mode, only a brief review on the reduced control plane latency to achieve the IMT-2020 requirements is provided. Related to connected mode procedures, the allowed configurations, multiplexing options between control and data, and the handling of collisions between physical channels are some of the aspects handled by this section.

9.3.1 Idle mode procedures

9.3.1.1 Control plane latency

As mentioned in Section 9.1, the main work in the LTE URLLC design is on the connected mode procedures. There is however one exception related to fulfilling the IMT-2020 requirements on the control plane latency, see Section 2.3, where a maximum of 20 ms shall be fulfilled.

Control plane latency is defined by ITU-R [3], as *"the transition time from a most "battery efficient" state* (e.g. *Idle state) to the start of continuous data transfer* (e.g. *Active state)"*. Hence, what needs to be analyzed is the initial access procedure where the device goes from a battery efficient idle mode state to an active state transferring data.

To achieve minimum specification impact, lowering the control plane latency was achieved by changes to device processing times and assumptions on processing times in the network.

A high-level signal diagram of the messages involved in case of the RRC Resume procedure is shown in Fig. 9.32.

As can be seen, the multi-step procedure involves multiple processing steps after receiving the messages transmitted, both at the device and the eNB.

Device processing delay after receiving the Random Access Response has been decreased by 1 ms, from 5 ms to 4 ms.

Also, the processing delay from the reception of the RRC connection resume, and reception of the uplink grant to the transmission of the associated connection resume complete has been

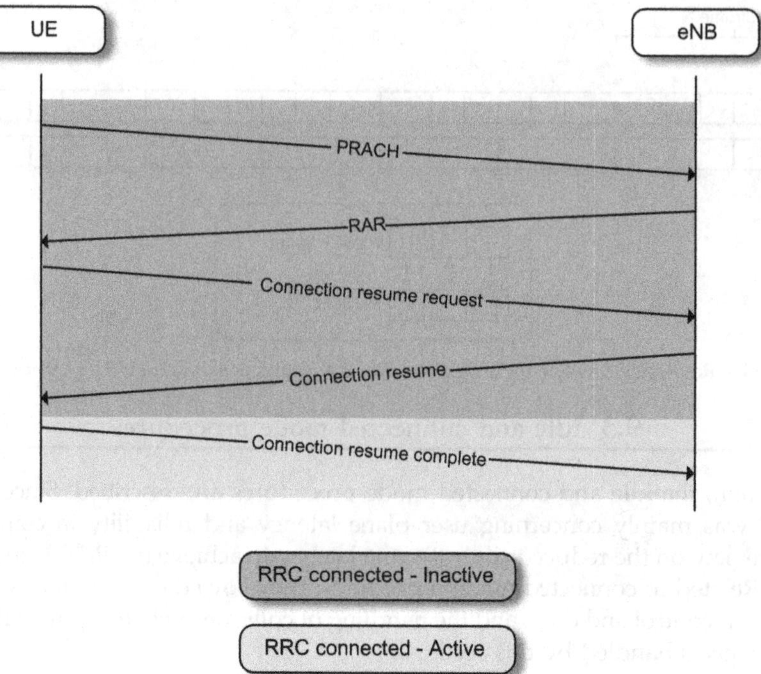

FIG. 9.32 Control plane latency.

reduced, assuming a typical 15 ms–7 ms. It is reasonable to assume that devices evolve with time and hence that some latency reduction can be expected compared to when the procedure was first introduced in Release 13. However, the significant reduction from 15 ms to 7 ms that has been specified comes at a cost, in order for the device to support it. The network is restricted in that the RRC message shall only include MAC and physical layer (re-)configurations and shall not include (re-)configurations of DRX, SPS, MIMO operation or to any secondary cell. Furthermore, the uplink grant of the connection resume complete must be sent on PDCCH DCI format 0 using the Common Search Space.

The support of a reduced control plane latency is an optional capability that the device can indicate. However, the network will not know that the device is supporting it when the device triggers the RACH preamble transmission. Hence, after detecting the Physical Random Access Channel and giving grant to the RRC connection resume request, the network will have to assume that the device either has 4 ms or 5 ms processing delay and hence allocate resource for both reaction times. In order for the network to control if a shorter processing is allowed to be used by the device or not, the activation of the reduced control plane latency is done via the broadcasted system information.

The resulting control plane latency is evaluated in 10.3.2 where the latency component of each step is outlined.

9.3.2 Connected mode procedures

9.3.2.1 *Configurations*

The more configurations that are allowed, the more complex the specification and implementation will be. This in turn leads to complex device and network implementations. One such specific consideration is for TDD operation where the existing TDD frame structure in LTE up to Release 14 was based on downlink subframes, uplink subframes and special subframes. To introduce subslots into the TDD structure without impacting the overall subframe structure would mean limited gain, and to change the overall frame structure has obvious impacts to the backwards compatibility and operator coexistence in the same geographical area. Thus, subslot operation is not allowed in TDD operation, and in case of slot operation, the same TDD configurations as in subframe operation applies (allowing some changes to the special subframes, see Fig. 9.18).

The allowed (per cell) configurations for slot and subslot operation are shown in Table 9.7.

The device will be configured to operate an sTTI configuration (i.e. subslot or slot transmission and reception) on a per PUCCH group basis. Hence, it is not possible to be configured with different sTTI lengths on different carriers within the same PUCCH group.

For sTTI (subslot or slot) operation, the network can, on a per-subframe basis, schedule the device interchangeably with sTTI and subframe transmissions. A typical implementation would however not be expected to switch frequently between the two, but e.g. use subframe operation for RRC reconfiguration, and when leaving a TCP slow-start (in which case a fast reaction time, using sTTI is primarily needed to ramp up the TCP throughput). The sTTI operation however, comes with a penalty in overhead both from the reference signals and the control channel, and is not advisable to be used, unless providing a performance benefit compared to subframe operation.

9.3.2.2 *Multiplexing of PDSCH and SPDCCH*

9.3.2.2.1 General

In legacy LTE, the downlink data (PDSCH) and the associated control (PDCCH) is not multiplexed in the same time domain OFDM symbols but are instead time-domain multiplexed. This changes with the introduction of sTTI. Not allowing any multiplexing of control and data in the same OFDM symbol in this case would severely restrict the data capacity since the control is sent in most of the symbols in the subslot. For some configurations, data transmission would even be prevented (e.g. for DMRS-based control signaling).

The multiplexing of control and data has been implemented on two levels, either by RRC configuration (semi-statically changed) or by signaling in the DCI.

TABLE 9.7 Allowed combinations of slot and subslot (sTTI) operation.

Downlink	Uplink	Frame structure type[1]
Slot	Slot	FDD, TDD
Subslot	Slot	FDD
Subslot	Subslot	FDD

FDD, TDD are in the 3GPP specifications also referred to as Frame Structure Type 1 and 2 respectively.

Irrespective of the configured mode, the device will always rate-match around its own DCI scheduling the downlink data.

9.3.2.2.2 RRC-based multiplexing

Four modes are defined for RRC-based multiplexing of data and control. The different modes are configured per SPDCCH resource set and relate to the device behavior on how the rate-matching for PDSCH is performed in association to the SPDCCH resource set:

- RRC mode 1: The device rate-matches around the DCI that schedules the slot or subslot transmission.
- RRC mode 2: The device rate-matches around the whole SPDCCH resource set
- RRC mode 3: The device rate-matches around the whole SPDCCH resource set, if the DCI that schedules the slot or subslot transmission was found in the resource set.
- RRC mode 4: The device rate-matches around the whole SPDCCH resource set, if the DCI that schedules the slot or subslot transmission was *not* found in the resource set.

It could be expected that different modes are used depending on the traffic situation in the cell. For example, in a light traffic load, when a single device is scheduled in a given TTI, there is no reason to have it rate-match around the whole SPDCCH resource set, but instead to rate-match around its own DCI (RRC mode 1). In a heavier traffic load, where multiple devices are expected to be scheduled, there is no functionality to communicate to a given device which DCIs resources are used by the other devices. In this case, rate-matching around the full set is recommended.

The different rate-matching behaviors are illustrated in Fig. 9.33, where it is assumed that both SPDCCH resource sets are configured with the same rate-matching RRC mode.

As can be seen, the different modes allow for a good level of flexibility. However, the rate-matching behavior cannot be changed frequently, due to the use of RRC signaling.

Hence, it is motivated to introduce a more flexible rate-matching behavior that can follow the momentary traffic behavior. This is solved by including information in the DCI in each sTTI about the rate-matching behavior of the device.

9.3.2.2.3 DCI-based multiplexing

DCI-based multiplexing of data and control can also be used instead of RRC-configured rate-matching.

DCI-based multiplexing is configured by RRC, and the device behavior is associated with three different possible modes.

- DCI mode 0: This mode is only used in case two SPDCCH resource sets are configured. In this case, one bit is allocated to each SPDCCH resource set, and, in case the bit is set, the device rate-matches around the whole SPDCCH resource set.
- DCI mode 1: Two bits are allocated to the first out of the two SPDCCH resource sets. No bits are allocated to the second SPDCCH resource set. If the most significant bit in the DCI is set, the rate-matching is applied around the first half of the SPDCCH resource set. If the least significant bit is set, the rate-matching is applied around the second half of the SPDCCH resource set. If both bits are set, the rate-matching is applied around the whole SPDCCH resource set. For the second set, no dynamic rate-matching is performed

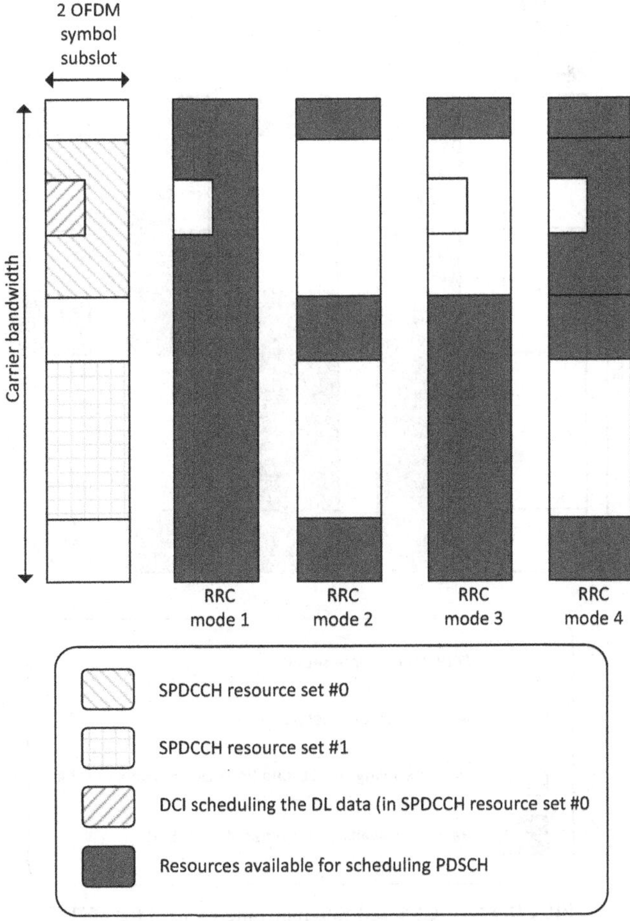

FIG. 9.33 RRC-based rate-matching modes for sTTI.

(no bits in the DCI associated with the SPDCCH resource set), and hence the RRC configured mode is applied.

- DCI mode 2: The mode is the same as DCI mode 1 with the difference that the DCI-based rate-matching is applied to the second SPDCCH resource set and the RRC-based rate-matching to the first SDCCH resource set.

As with the RRC-based rate-matching, different DCI modes can be used depending on the traffic situation and depending on if one or two SPDCCH resource sets are configured. For example, in case of high traffic load and large variations in the resource allocation over time between the two SPDCCH resource sets, DCI mode 0 could be advisable.

If only one SPDCCH resource set is configured, the use of DCI mode 1 clearly provides most flexibility (top part of Fig. 9.35).

The different modes are illustrated in Figs 9.34–9.36.

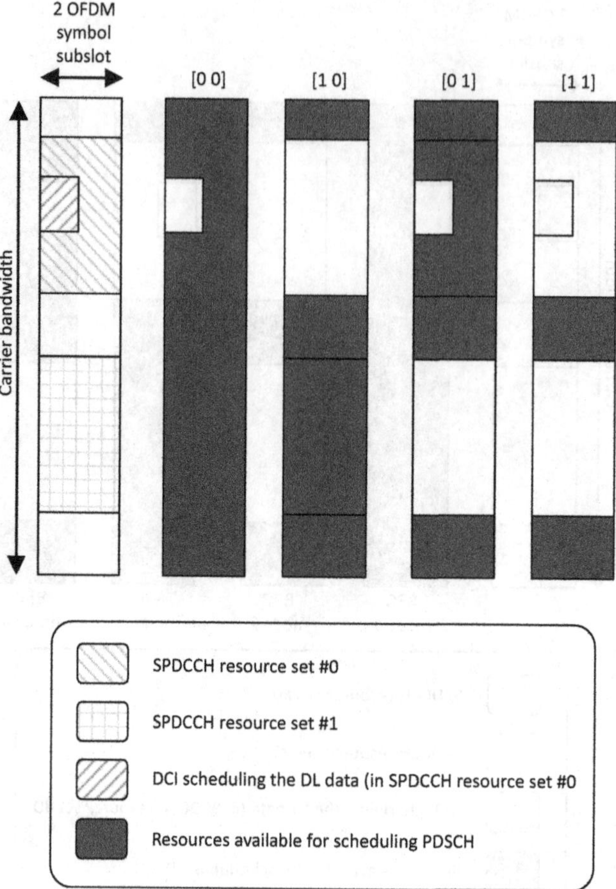

FIG. 9.34 DCI-based rate-matching mode 0 for sTTI.

9.3.2.3 *Scheduling request*

As for normal LTE operation, if the device wants to indicate the need for dynamic uplink scheduling from the network, it can send a *Scheduling Request* (SR). An SR-only transmission is carried by SPUCCH format 1. In case of HARQ and SR multiplexing in the same SPUCCH, the format can be adopted to carry both SR and HARQ bits. The multiplexing capacity of subslot-SPUCCH is reduced compared to PUCCH since no OCC can be used, and hence up to 12 cyclic shifts can be used for multiplexing.

To allow for as low latency as possible, the SR periodicities allowed to be configured to a given device is 1 sTTI (subslot or slot). The allowed range of SR periodicities for subslot operation and slot operation are listed in Tables 9.8 and 9.9 respectively.

9.3.2.4 *UCI on PUSCH*

UCI can be carried by (S)PUCCH or PUSCH, depending on the physical channel scheduled, and the collision between different physical channels, see Section 9.3.2.5.

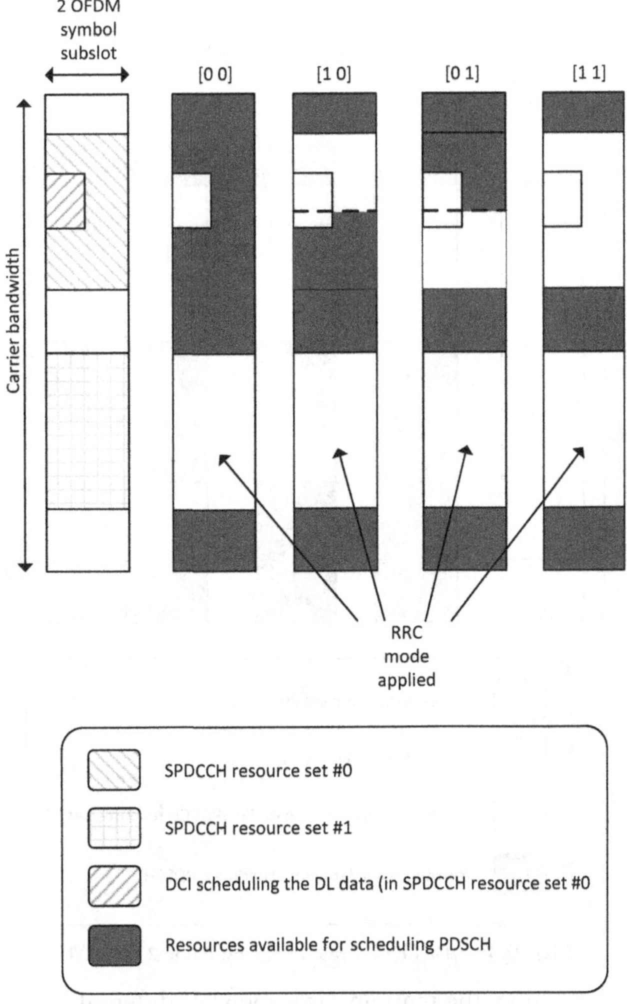

FIG. 9.35 DCI-based rate-matching mode 1 for sTTI.

In case of slot operation, the mapping of UCI onto PUSCH follows the same principles as for n+4 operation as shown in Fig. 9.37. The PMI/CQI bits related to the precoder and CSI are rate-matched with the data, and so is the rank indication (RI). The HARQ-ACK bits punctures the PUSCH data (instead of using rate-matching) to avoid the case that the downlink assignment associated with the HARQ-ACK is missed. If using rate-matching, the eNB and the device rate-matching would in this case be different (the device assuming HARQ-ACK bits not present). The HARQ-ACK bits are placed close to the DMRS to ensure a good channel estimate, and hence better performance.

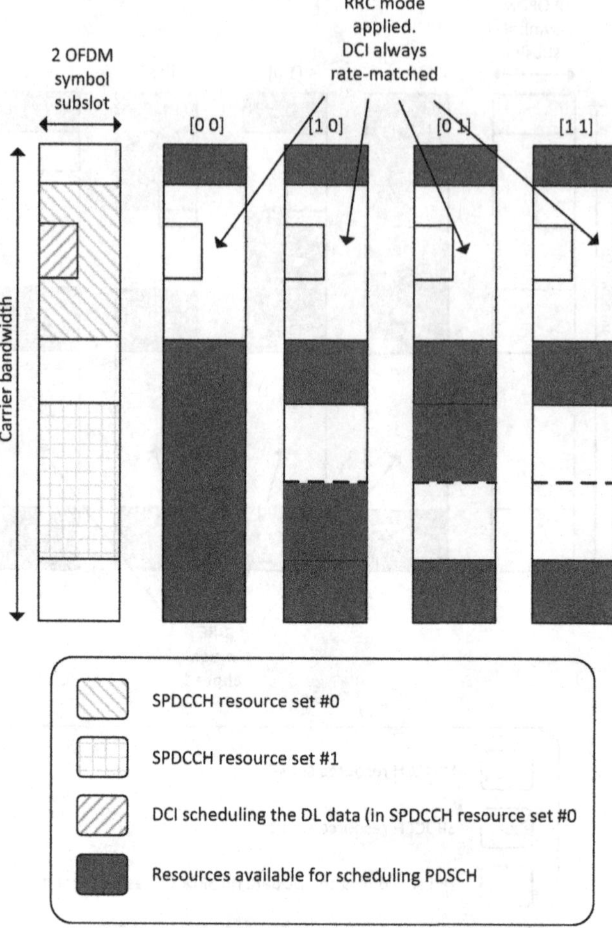

FIG. 9.36 DCI-based rate-matching mode 2 for sTTI.

In case of subslot operation, the mapping is somewhat different.

For a subslot containing two data symbols, the bits for:

- HARQ-ACK punctures PUSCH data from the end of the data symbol closest to the DMRS. In case of no DMRS (i.e. the subslot only contains two data symbols, see Section 9.2.4.1), the HARQ-ACK is mapped to the first symbol
- RI is rate-matched by PUSCH data and mapped from the end of the data symbol not mapped with HARQ-ACK bits.
- PMI/CQI are rate-matched by PUSCH data and mapped from the start of the data symbols in a time-first, frequency-second manner.

For a subslot containing one data symbols, the bits for:

- RI is rate-matched by PUSCH data and mapped from the end of the data symbol
- HARQ-ACK punctures PUSCH data and is mapped after the bits for RI
- PMI/CQI are rate-matched by PUSCH and mapped from the start of the data symbol

TABLE 9.8 Scheduling request periodicities – subslot.

Subslot SR periodicity
1 subslot
2 subslots
3 subslots
4 subslots
5 subslots
6 subslots (1 ms)
2 ms
5 ms
10 ms

TABLE 9.9 Scheduling request periodicities – slot.

Slot SR periodicity
1 slot
2 slots (1 ms)
2 ms
5 ms
10 ms

Fig. 9.38 illustrates the different mapping options for subslot-operation.

Furthermore, in case of subslot operation, the beta-offset (a scaling of the baseline code rate for the UCI) is allowed to take two values (configured by RRC) for HARQ-ACK and RI. This is to compensate for the potential degradation that can be experienced due to switching transients at the device. The switching transients stems from RF related imperfections at for example power ramp-up/ramp-down. In the switching transient period, the signal structure is not defined during (in the general case), up to 10 μs, and can thus assume to carry no information. A transient can occur (depending on scheduling) at the start of the first symbol and at the end of the last symbol. In case of a possible transient occurring (known to the network), a higher value of beta-offset can be indicated in the DCI to ensure proper HARQ-ACK reception.

The beta-offset values are independently configured for subframe-, slot- and subslot operation.

9.3.2.5 *Subframe and subslot/slot collisions*

As described in Section 9.3.2.1, the device can be scheduled with subframe and subslot/slot transmissions interchangeably in different subframes. Furthermore, since the processing

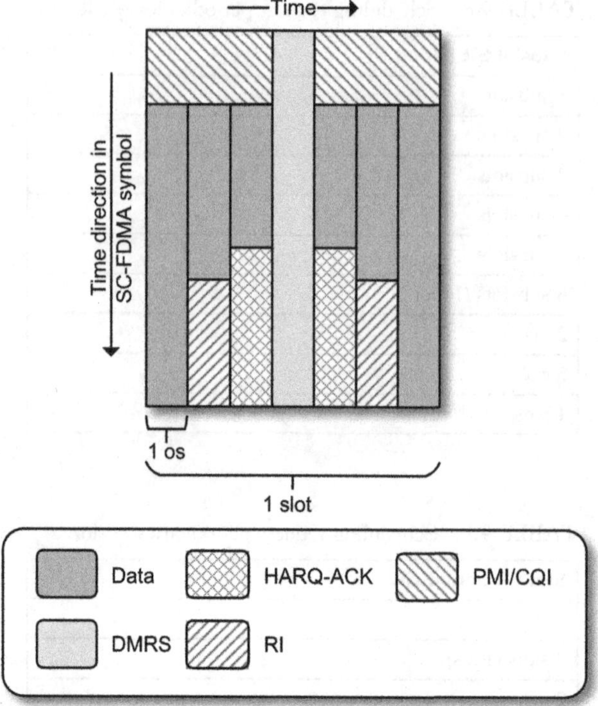

FIG. 9.37 UCI mapping for slot-PUSCH.

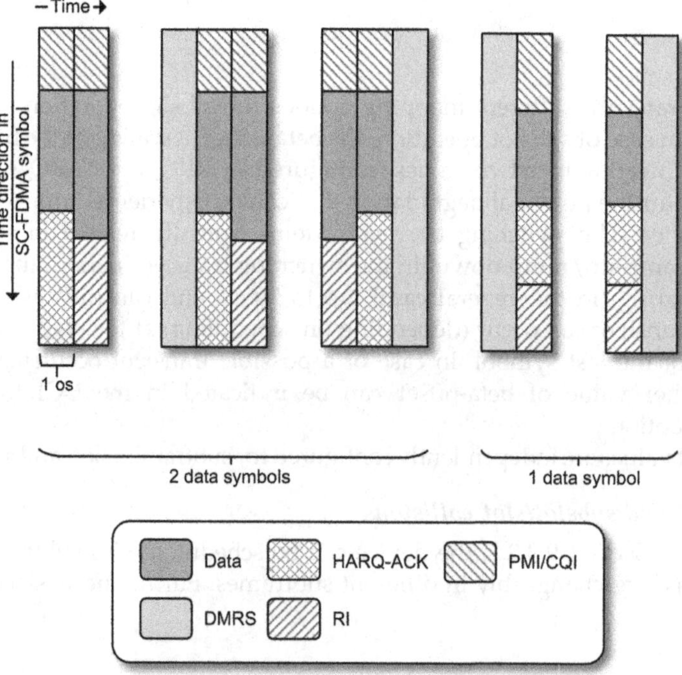

FIG. 9.38 UCI mapping for subslot-PUSCH.

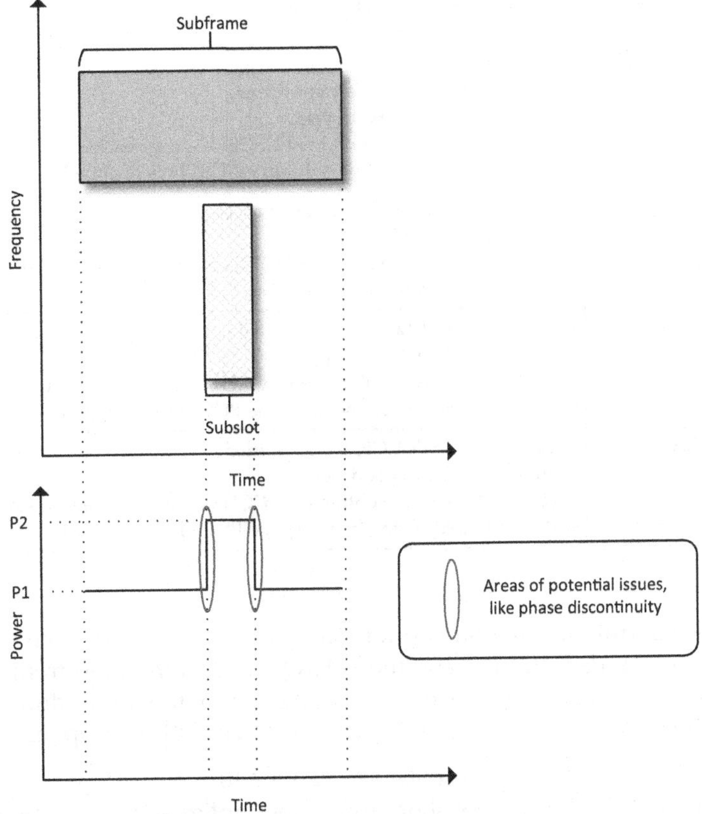

FIG. 9.39 Overlapping transmissions of different length in the same power amplifier.

timeline between subframe and subslot/slot differ (see Section 9.2.5), the network might schedule uplink transmission of different lengths on the same carrier to be simultaneously transmitted by the device.

However, transmitting two physical channels of different lengths on the same carrier using the same power amplifier will cause imperfections to the signal, including phase discontinuities, that prevents coherent demodulation at the eNB, see Fig. 9.39. Hence, a handling of collisions of physical channel with different duration on the same carrier is needed.

Some general principles are followed:

(i) The longer channel is dropped to transmit the channel of shorter duration. The main reason behind this is that the decision at the network to schedule the shorter channel is taken after the longer channel has been scheduled. Hence, the potential collision, and the associated device behavior, is known by the network when scheduling the channel with shorter duration. Furthermore, a shorter channel is more likely to be associated with a cMTC service with high reliability, and should hence be prioritized.

TABLE 9.10 Collisions between channels of different duration.

Subframe	Slot/subslot	Action
PUSCH	PUSCH	- Slot/subslot-PUSCH is transmitted. - Subframe-PUSCH is dropped - HARQ-ACK from subframe-PUSCH is piggy-backed on slot/subslot-PUSCH. - Potential CSI carried by subframe-PUSCH is dropped
PUSCH	SPUCCH	- Slot/subslot-SPUCCH is transmitted. - Subframe-PUSCH is dropped - HARQ-ACK from subframe-PUSCH is piggy-backed on slot/subslot-SPUCCH. - Potential CSI carried by subframe-PUSCH is dropped
PUCCH	PUSCH	- Slot/subslot-PUSCH is transmitted. - Subframe-PUCCH is dropped - HARQ-ACK from subframe-PUCCH is piggy-backed on slot/subslot-PUSCH. - Potential CSI carried by subframe-PUCCH is dropped
PUCCH	SPUCCH	- Slot/subslot-SPUCCH is transmitted. - Subframe-PUCCH is dropped - HARQ-ACK from subframe-PUCCH is piggy-backed on slot/subslot-SPUCCH. - Potential CSI carried by subframe-PUCCH is dropped

(ii) HARQ-ACK control information is prioritized to be transmitted, even for channels that are being dropped. In this case, the HARQ-ACK information from the channel with longer duration is piggy-backed on the channel with shorter duration.

(iii) UCI of less importance compared to HARQ-ACK (i.e. CSI) is dropped.

The different collision cases are described in Table 9.10.

Piggy-backing HARQ-ACK on the shorter channel implies a larger impact on the performance (less energy transmitted but the same number of bits carried). To minimize the negative impact on performance, bundling of the HARQ-ACK bits is supported. Bundling the bits implies that if one or more of the bits being bundled indicates a 'NACK', the bundled bit will indicate a 'NACK'. Bundling is performed over the spatial domain (i.e. the layers transmitted on the same carrier) and is always applied for subslot operation, while being configurable for slot operation.

All above collision cases are concerned with the collision on a given carrier. However, there are cases when different lengths can be transmitted on different carriers. For example, when falling back to subframe operation, when not configuring sTTI on all carriers, or when one PUCCH group is configured with subslot operation and the other PUCCH group with slot operation in the uplink. '

In this case, the device can either be capable, or not capable, of transmitting different transmission lengths on different carriers.

- In case the device is not capable, it will drop the channels with longer transmission duration.
- If the device is capable, the device can, as long as it is not power limited, transmit the different channels simultaneously. If it is power limited, channels are dropped based on priority. A general principle is followed that prioritizes channels carrying HARQ-ACK, and channels with shorter transmission duration. For more details of the channel priority specified, see Ref. [7].

9.3.2.6 HARQ

In legacy LTE, synchronous HARQ operation is supported on the uplink with a fixed timing between the initial transmission of PUSCH and its retransmission. In the downlink, asynchronous HARQ is used, i.e. the network does not have to follow a fixed timeline between initial transmission and retransmission.

For sTTI, asynchronous HARQ is adopted both in downlink and in uplink. This also implies that there is no longer a need to involve the PHICH, see Ref. [4].

The total number of HARQ processes have also been increased with sTTI operation, due to the shorter transmission duration, to be able to provide full throughput. A maximum of 16 processes regardless of slot or subslot operation is used.

To allow a smooth transition between sTTI and subframe operation, the HARQ processes are shared. That means that an initial transmission carried out in one transmission duration can be retransmitted using another transmission duration. The support is conditional of that the maximum TBS size for the respective transmission duration is respected.

9.3.2.7 Scheduling

In this section, we describe how scheduling for uplink and downlink transmissions works.

When the network needs to schedule a device, it sends a DCI addressed to the device in one of the PDCCH or SPDCCH candidates (see Section 9.2.3.2 for more details) in the search space that the device monitors. The *Cell RNTI*, masking the DCI CRC, is used to identify the device. The DCI includes resource allocation (in both time and frequency domains), modulation and coding scheme, and information needed for supporting the HARQ operation.

In case of slot#0 or subslot#0, the device monitors the search space in the PDCCH region, while for slot#1 and subslot#1–5, the SPDCCH search space is used.

As with PDCCH decoding, the more SPDCCH candidates the device need to search for, the more complex the decoding procedure becomes and the higher risk for a false SPDCCH decoding. Some limitations are therefore assumed.

For SPDCCH, the following applies:

(i) The search space over different ALs and SPDCCH resource sets is limited to:
- 16 SCCEs if the associated data is using subslot operation
- 32 SCCEs if the associated data is using slot operation

(ii) The SPDCCH candidates to be monitored by a device is limited per AL according to:
- ≤ 6 for AL 1 and 2
- ≤ 2 for AL 4 and 8

(iii) The blind decodes in a given TTI on a given carrier shall be:
- ≤ 6 for subslot operation
- ≤ 12 for slot operation

For PDCCH, the following applies:

(i) The overall search space is limited up to 28 CCEs for subslot operation.

(ii) The same restrictions as in (ii) and (iii) for SPDCCH also applies in case the DCI is mapped to PDCCH.

The DCI format for both downlink and uplink have been designed with the same number of payload bits. If the number of used bits for a given format in a given direction does not align with the format in the other direction, padding of bits is applied. Using an aligned DCI format will allow the use of a downlink/uplink flag in the DCI content and reduce the number of blind decodes required (same decoding for downlink and uplink DCI).

9.3.2.7.1 Dynamic downlink scheduling

The base station schedules downlink transmission on PDSCH dynamically using DCI Format 7-1A to 7-1F.

A list of the DCI fields for the baseline format DCI format 7-1A is provided in Table 9.11.

For downlink DCI formats other than DCI format 7-1A, the fields of DCI format 7-1A are included as baseline. Additional fields are included according to Table 9.12.

The other formats include additional information related to MIMO information, which is similar to legacy LTE operation. Interested readers are referred to Refs. [6,7].

TABLE 9.11 DCI format 7-1A.

Information	Size [bits]	Possible settings
Flag for DL/UL differentiation	1	0 - format 7-0A/B depending on the configured uplink transmission mode 1 - format 7-1X, where X is the DCI format transmitted
Resource allocation	variable	Different bit spaces depending on resource allocation type 0 or 2, see Section 9.3.2.9.1
Modulation and coding scheme	5	Modulation used together with an indicative code rate, see Ref. [7]
HARQ process number	4	See Section 9.3.2.6
New data indicator	1	The bit is toggled when indicating to the device that the soft-buffer is to be flushed and that new data is transmitted for the signaled HARQ process. See also [7]
Redundancy version	2	Indicates the set of bits, from the encoded set of bits, selected for transmission. See also [7]
TPC command	2	See Section 9.3.2.8
Downlink Assignment Index (DAI)	2 or 4	Assists the device in understanding the number of downlink assignments being transmitted by the eNB, even though only a subset is detected. Used to align the device's and eNB's understanding in the number of HARQ-ACK bits reported. See also [7]
Used/Unused SPDCCH resource indication	2	See Section 9.3.2.2
SPUCCH resource indication	2	See Section 9.2.4.2
Repetition number	2	See Section 9.2.3.3

TABLE 9.12 DCI format 7-1B to 7-1G.

Information	Size [bits]	DCI format 7-1						Possible settings
		B	C	D	E	F	G	
Precoding information	1 or 2	•						1 bit for transmission with 2 antenna ports 2 bits for transmission with 4 antenna ports
Precoding information	4 or 6		•					4 bits for 2 antenna ports 6 bits for 4 antenna ports
Precoding information	3 or 5			•				3 bits for 2 antenna ports 5 bits for 4 antenna ports
SRS request	0 or 1				•	•	•	For TDD operation, if the device has indicated the capability and is configured with SRS request
Scrambling identity	1				•			See Ref. [6]
Precoding information	2				•			One or two layers (with or without transmit diversity)
DMRS position indicator	1					•	•	See Section 9.2.3.1
Antenna port(s), scrambling identity and number of layers	1 or 3					•	•	See Ref. [6]
PDSCH RE Mapping and Quasi-Co-Location Indicator	2						•	See Ref. [7]

9.3.2.7.2 Dynamic uplink scheduling

The base station schedules uplink transmission on PUSCH dynamically using DCI Format 7-0A and 7-0B. DCI format 7-0B is used in case of MIMO transmission.

A list of the DCI fields is provided in Table 9.13.

9.3.2.7.3 Semi-persistent Scheduling

In addition to the dynamic scheduling in uplink and downlink, support has also been added for *Semi-Persistent Scheduling* (SPS) using slot/subslot transmission to reduce latency. A blind repetition-based scheme, similar to PDSCH (see Section 9.2.3.3), to support improved reliability has also been defined. SPS pre-allocates periodically (re)occurring resources using a configurable interval for the device to monitor and decode (on the downlink) and to transmit in (on the uplink). Using such pre-allocation of resources will for example eliminate the need for sending a SR to trigger scheduling in the uplink (reducing latency) and will also minimize the control overhead for the data transmission.

The use of SPS is also a key enabler for fulfilling the 5G requirements for URLLC as elaborated more upon in Chapter 10.

The SPS operation for sTTI is very similar to subframe-based operation in legacy LTE, see Ref. [4].

TABLE 9.13 DCI format 7-0A and 7-0B.

Information	DCI format 7-0A		DCI format 7-0B	
	Size [bits]	**Possible settings**	**Size [bits]**	**Possible settings**
Flag for UL/DL differentiation	1	0 - format 7-0A 1 - format 7-1A/B/C/D/E/F/G depending on the configured downlink transmission mode	1	0 - format 7-0B 1 - format 7-1A/B/C/D/E/F/G depending on the configured downlink transmission mode
Resource block assignment	variable	See Section 9.3.2.9	variable	See Section 9.3.2.9
Modulation and coding scheme	5	See Table 9.11 and [7]	5	See Table 9.11 and [7]
HARQ process number	4	See Section 9.3.2.6	4	See Section 9.3.2.6
New data indicator	1	See Table 9.11 and [7]	1	See Table 9.11 and [7]
Redundancy version	2	See Table 9.11 and [7]	2	See Table 9.11 and [7]
TPC command for scheduled PUSCH	2	See Section 9.3.2.8	2	See Section 9.3.2.8
DMRS pattern	2	See Section 9.2.4.1	2	See Section 9.2.4.1
Cyclic shift for DMRS and IFDMA configuration	1	See Section 9.2.4.1	1	See Section 9.2.4.1
UL index	2	In case of TDD, and for certain configurations, the UL index is used to allow scheduling of multiple PUSCHs from the same DCI. See also [7]	2	In case of TDD, and for certain configurations, the UL index is used to allow scheduling of multiple PUSCHs from the same DCI. See also [7]
Downlink Assignment Index (DAI)	2	See Table 9.11 and [7]	2	See Table 9.11 and [7]
CSI request	1, 2 or 3	See Section 9.3.2.10	1, 2 or 3	See Section 9.3.2.10
SRS request	0 or 1	For TDD operation, if the device has indicated the capability and is configured with SRS request	2	For TDD operation, if the device has indicated the capability and is configured with SRS request
Beta offset indicator	1	See Section 9.3.2.4	1	See Section 9.3.2.4
Cyclic shift field mapping table for DMRS	1	See Section 9.2.4.1	1	See Section 9.2.4.1
Precoding information and number of layers	–	–	3 or 6	See Section 9.2.4.3. Depending on the number of antenna ports at the device (2 or 4)

Information that is not carried in the DCI is configured by RRC. The DCI is then dynamically used to activate and release the SPS operation, where for example the selected MCS can be changed dynamically. For the device to understand that the DCI is related to SPS, it is scrambled by *SPS-C-RNTI*. SPS for sTTI is supported both on the downlink and the uplink.

The device monitors all subslots/slots in the subframe for possible SPS activation and release.

For downlink SPS, some behavior related to DCI operation need to be changed. As described in Section 9.2.4.2, the SPUCCH resources to be used for the HARQ-ACK response is indicated by a 2-bit field, indicating one out of four resources. For SPS operation, the activation of SPS will indicate one of these resources to be used until the SPS operation is reconfigured.

The power control loop for SPS is handled by the same DCI format as in legacy LTE, that is, DCI format 3/3A, using a processing timeline (see Section 9.2.5) of n+4 (subframes). The TPC-index configured to a device is separate for sTTI and subframe operation. As for PUSCH and PUCCH power control (see Section 9.3.2.8), the power control loops are independent for sTTI and subframe operation.

In contrast to dynamic downlink scheduling, there is no support for DMRS sharing across subslots in the downlink (see Section 9.2.3.3 and 9.3.2.7.1) when SPS is configured.

However, on the uplink where the overhead from DMRS can be up to 50% (in case of two-symbol subslot with one DMRS symbol), there are two configurations supported, one with DMRS contained in each subslot and one where the associated DMRS can be placed outside of the subslot boundary. The two possible configurations are shown in Table 9.14 (which can be compared to Table 9.4 used in case of dynamic uplink scheduling).

To allow for ultra-low latency, the SPS periodicity can be configured as low as 1 subslot or 1 slot for subslot and slot operation, respectively. The possible configurations are shown in Table 9.15.

Another difference to the legacy LTE operation is that the SPS configuration can be associated with a DMRS using a configurable cyclic shift and potentially IFDMA modulation (see Section 9.2.4.1). This allows users to be simultaneously multiplexed on the same physical resources, using Multi User MIMO, which compensates for the increase in resources used in sTTI, where the frequency allocation is typically larger, and for cMTC services, the SPS periodicity is typically shorter. However, for cMTC services, it should be noted that increasing the user multiplexing rate on the same physical resources will have an implication in the reliability achieved.

To improve reliability in uplink SPS operation, there is a possibility of configuring a device to transmit blind repetitions for subframe, slot and subslot operation. The number of repetitions possible to configure for uplink SPS repetitions are 2, 3, 4 or 6. To ensure a low waiting time from the time the packet is delivered to lower layers until it can be transmitted over the

TABLE 9.14 Uplink SPS DMRS configurations.

Bit value	Subslot#0	Subslot#1	Subslot#2	Subslot#3	Subslot#4	Subslot#5
0	R D D	R D	R D	R D	R D	R D D
1	R D D	D D |R	R D	D D |R	R D	R D D

'*R* denotes a DMRS symbol, '*D*' denotes a data symbol, '|' denotes the subslot boundary.

TABLE 9.15 Downlink and uplink SPS periodicities.

sTTI SPS periodicities		
sTTI [#]	**Subslot[a] operation [ms]**	**Slot operation [ms]**
1	0.2	0.5
2	0.4	1.0
3	0.5	1.5
4	0.7	2.0
6	1.0	3.0
8	1.4	4.0
12	2.0	6.0
16	2.7	8.0
20	3.4	10.0
40	6.7	20.0
60	10.0	30.0
80	13.0	40.0
120	20.0	60.0
240	40.0	120.0

[a]*Rough estimates since it should be noted that the periodicity will vary with CFI and subslot number, see Fig. 9.2.*

air, multiple SPS configurations can be used, each with a different starting position of the repeated PUSCH sequence. Up to six configurations can be used. The total number of transmission (the initial transmission and all repetitions) cannot exceed the periodicity configured since that would result in overlapping SPS transmissions.

An example is shown in Fig. 9.40, where four transmissions are used for each configuration (one initial transmission and three repetitions). Each configuration has a periodicity of four (s)TTIs and they each have different starting positions for the sequence of repetitions. In order to minimize the delay of the packet arriving, configuration#1 should be selected for the uplink transmission.

9.3.2.8 *Uplink power control*

The uplink power control for PUSCH and SPUCCH follow very similar principles as in legacy LTE operation.

9.3.2.8.1 PUSCH

The pathloss component (PL), target received power (P_0) and MCS related power offset (Δ_{TF}), all follow legacy LTE operation. The device is limited in the maximum power, using the same cell specific parameter as in legacy LTE operation, P_{CMAX}. Also, the compensation depending on the bandwidth allocation (in terms of resource blocks in frequency) is maintained (M_{PUSCH}). The power control equation for slot/subslot-PUSCH is shown in Eq. (9.5).

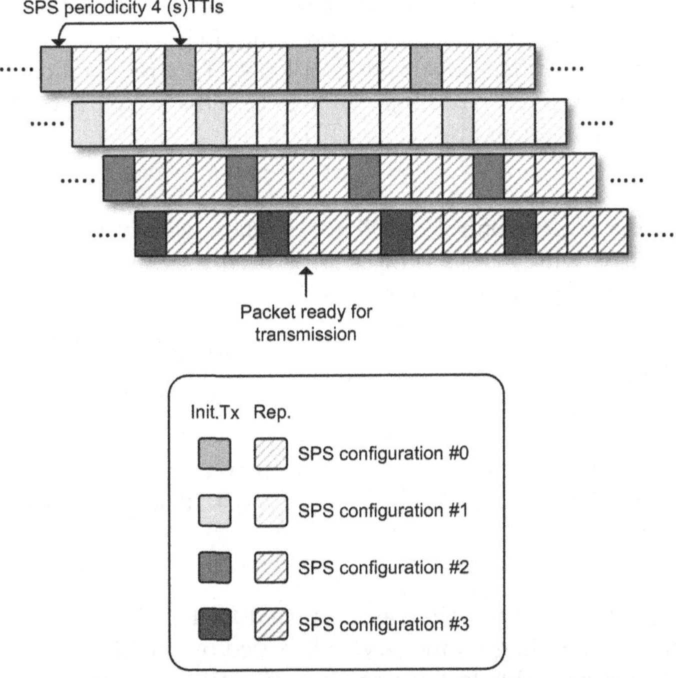

FIG. 9.40 Multiple uplink SPS configurations.

$$P_{PUSCH} = \min\begin{cases} P_{CMAX} \\ 10log_{10}(M_{PUSCH}) + P_0 + \alpha \cdot PL + \Delta_{TF} + f \end{cases} \tag{9.5}$$

The closed loop parameter (f) controlled by the TPC command in the DCI will only impact the power control loop related to subslot/slot transmission. That is, it will not have an impact to the power control loop used for subframe-based transmission, and hence they are independent.

The *power headroom report* (PHR), providing a rough indication to the network on the difference in power between the used transmit power and the maximum allowed power, is also similar to regular LTE operation.

A PHR can be transmitted by either subframe-based PUSCH or subslot/slot-PUSCH. The PHR is reported for all activated carriers. Both *Type 1* and *Type 2* PHR reporting is supported. As in legacy LTE, in case of.

- Type 1, the estimated power is based on transmitting PUSCH only
- Type 2, simultaneous transmission of PUSCH and PUCCH is assumed.

Furthermore, the power headroom for a given carrier can be based on either a scheduled transmission or non-scheduled transmission. In case the transmission is not scheduled, it is assumed that the resource allocation is one resource block (since the allocation is not known) and that there is no MCS-specific (since not known) contribution in the power estimation.

In case the PHR is transmitted on subframe-based PUSCH, the PHR is subframe-based for all carriers, regardless if the carrier is configured with sTTI or not.

In case the PHR is transmitted on subslot/slot-PUSCH, carriers not configured with sTTI follow the subframe-based PHR reporting, while in case the carrier is configured with sTTI, the power headroom is calculated based on sTTI.

9.3.2.8.2 SPUCCH

As for PUSCH, the power control of SPUCCH follow closely the corresponding power control for PUCCH. As with PUCCH, Eq. (9.6) is used for SPUCCH format 1/1a/1b and 3, while Eq. (9.7) is used for SPUCCH format 4.

$$P_{\text{PUCCH F1/3}} = \min\begin{cases} P_{\text{CMAX}} \\ P_0 + PL + \Delta_{TxD} + g + \Delta_{F\_\text{PUCCH}} + h \end{cases} \tag{9.6}$$

$$P_{\text{PUCCH F4}} = \min\begin{cases} P_{\text{CMAX}} \\ P_0 + PL + \Delta_{TxD} + g + \Delta_{F\_\text{PUCCH}} + 10\ \log_{10}(M_{\text{PUSCH}}) + \Delta_{\text{TF}} \end{cases} \tag{9.7}$$

- The pathloss estimate (PL) is the same as calculated for PUCCH operation,
- The compensations depending on the payload carried by the SPUCCH (h and Δ_{TF}),
- The change in power level due to the frequency allocation for SPUCCH format 4 (M_{PUSCH}), and,
- The compensation due to Tx diversity (Δ_{TxD})

are all the same as for SPUCCH.

However, to capture the difference in performance between PUCCH and SPUCCH, one of the parameters need to be modified, and this is done through the SPUCCH format dependent offset (Δ_{F_PUCCH}). A specific offset can be configured (within a specified range) for each subslot-SPUCCH/slot-SPUCCH transmission.

As for PUSCH, the closed loop power control parameter (g) updated through TPC in the DCI is independently applied for the PUCCH and the SPUCCH power control loops.

9.3.2.9 *Resource allocation*

9.3.2.9.1 Downlink

In the downlink, resource allocation type 0 and 1 is supported for subslot and slot transmissions. One of the two resource allocations is configured by RRC.

- Resource allocation type 0: The resource allocation can be non-contiguous and is indicated to the device via a bitmap in the DCI. The minimum contiguous resource allocation, or the *Resource Block Group* (RBG), has been increased compared to legacy LTE operation. The reason for doing so is motivated by the expected larger frequency allocation in case of short transmissions, and it also keeps the DCI size low improving the reliability. The RBG size for resource allocation 0 is shown in Table 9.16.

TABLE 9.16 RBG size for downlink resource allocation 0.

1,4 MHz	3 Mhz	5 MHz	10 MHz	15 MHz	20 MHz
1 RB	2 RBs	6 RBs	6 RBs	12 RBs	12 RBs

TABLE 9.17 RBG size and starting point granularity for downlink resource allocation 2.

	1.4 MHz	3 MHz	5 MHz	10 MHz	15 MHz	20 MHz
sRBG size	2 RB	2 RB	4 RBs	6 RBs	4 RBs	4 RBs
Starting point granularity	1 RB	1 RB	2 RBs	6 RBs	4 RBs	4 RBs

TABLE 9.18 RBG size and starting position granularity for uplink resource allocation 0.

	1.4 MHz	3 MHz	5 MHz	10 MHz	15 MHz	20 MHz
RBG size	1	1	4	4	4	4
Starting point granularity	1	1	4	4	4	4

- Resource allocation type 2: For resource allocation type 2, the allocation is contiguous in frequency, using an RBG size and starting point granularity according to Table 9.17 (also increased compared to legacy LTE).

If the RBG is not a multiple of the system bandwidth, the size of the last RGB is increased to avoid unscheduled resources.

9.3.2.9.2 Uplink

In the uplink, only resource allocation type 0 is supported. Similar to resource allocation type 2 on the downlink, this is a contiguous resource allocation with a starting position and a length. The granularity of the allocation and the starting position is shown in Table 9.18. As for the downlink allocation, the RBG size has been increased compared to legacy LTE.

9.3.2.10 CSI reporting

In legacy LTE operation, the *Channel State Information* (CSI) reporting from the device is done by performing measuring on the downlink resources, for the network to be able to apply channel dependent scheduling.

The CSI is either reported periodically or aperiodically, carried by PUCCH or PUSCH.

For sTTI operation, only aperiodic CSI reporting is supported, carried by PUSCH.

The triggering of CSI reporting from the DCI (see Section 9.3.2.7.1) follows the same processing timeline as subslot/slot-PUSCH and the downlink reference resources where the device is measuring is based on slot/subslot (depending on what is configured).

For more details on CSI reporting, the reader is referred to Ref. [4].

9.3.2.11 PDCP duplication

Increasing the reliability can be performed at different levels of the network to make LTE more suitable for cMTC services and fulfilling the URLLC requirements. We have for example already seen the use blind repetitions of the physical channels where the information sent over the air can be maximally combined in the receiver. Another scheme to increase diversity is the duplication of packets at the PDCP layer. The procedure is used both in LTE and NR with similar functionality and hence the reader is referred to Section 11.3.3.7 for more details.

References

[1] Ericsson, SouthernLINC Wireless, SK Telecom, T-Mobile USA, Orange, ITRI, OPPO, TELUS, Telstra, Sony, ETRI, Verizon, KDDI, CHTTL, Interdigital, Fujitsu, Spreadtrum Communicationsm, Nokia, Alcatel-Lucent Shanghai Bell, KT Corp, Sierra Wireless, Telecom Italia, TeliaSonera, Deutsche Telekom, Sprint, Sharp, NEC, CATT, Huawei, HiSilicon, AT&T, Intel, Samsung, ZTE, Qualcomm, LG Electronics. RP-171468, Work Item on shortened TTI and processing time for LTE, 3GPP RAN Meeting, Vol. 76, 2017.

[2] Alcatel-Lucent Shanghai Bell, Deutsche Telekom, Ericsson, Huawei, III, InterDigital, KT, LG Electronics, MediaTek, Nokia, Orange, Qualcomm, Samsung, SK Telecom, Softbank, SouthernLINC Wireless, Telecom Italia, Telefonica, Telenor, Telstra, Verizon, Vodafone, ViaviSolutions, Xilinx. RP-181259, Work item on ultra reliable low latency communication for LTE, 3GPP RAN meeting, Vol. 80.

[3] ITU-R, Report ITU-R M.2410. Minimum requirements related to technical performance for IMT-2020 radio interfaces(s), 2017.

[4] E. Dahlman, S. Parkvall, J. Sköld. "4G, LTE advanced pro and the road to 5G". Elsevier, 2018.

[5] Third Generation Partnership Project, Technical specification 36.211, v15.0.0. Evolved universal terrestrial radio access (E-UTRA); Physical channels and modulation, 2018.

[6] Third Generation Partnership Project, Technical specification 36.212, v15.0.0. Evolved universal terrestrial radio access (E-UTRA); Multiplexing and channel coding, 2016.

[7] Third Generation Partnership Project, Technical specification 36.213, v15.0.0. Evolved universal terrestrial radio access (E-UTRA); Physical layer procedures, 2016.

CHAPTER

10

LTE URLLC performance

OUTLINE

Abstract

This chapter shows the performance of LTE URLLC, comparing the technology to the existing 5G requirements put up by ITU (IMT-2020). Each 5G requirement applicable to URLLC is presented, together with the performance of LTE. Fulfilling the requirements involve both analytical calculations based on the existing design as well as system level and link level simulations. The performance evaluation includes user plane latency. Control plane latency and reliability evaluations. It is concluded that LTE can fulfill the 5G URLLC requirements.

In this chapter, the performance of the LTE URLLC design is presented. First the performance requirements on latency and reliability are reviewed. Following the requirements are the tools used to fulfill the requirements where both analytical calculations as well as system level and link level simulations have been used. Finally, link level simulations are provided showing how LTE fulfills the URLLC requirements of IMT-2020 to fulfill the 5G requirements.

Cellular Internet of Things, Second Edition
https://doi.org/10.1016/B978-0-08-102902-2.00010-8

© 2020 Elsevier Ltd. All rights reserved.

10.1 Performance objectives

As mentioned in Chapter 2, the URLLC requirements is a subset of the requirements set up by ITU-R used for a technology to declare itself 5G compliant. This section explains the requirements and how they apply to the LTE technology.

Out of the full set of 5G requirements, the ones of interest for URLLC are user-plane latency, control plane latency and reliability.

To give a better understanding to the reader on the actual requirements, they are described in separate sections below. For the detailed requirement specification, the reader is referred to see Ref. [2].

10.1.1 User plane latency

The requirement on user-plane latency in Ref. [2] is defined as from when a source node sends a packet to where the receiving node receives it. The latency is defined between layer 2 and layer 3 in both nodes. For LTE this would be on top of the PDCP protocol. It is assumed that the device is in an active state, assuming no queuing delays.

A 1 ms latency bound is targeted for URLLC. Both directions (uplink and downlink) have the same requirement.

10.1.2 Control plane latency

The requirement on control plane latency is 20 ms defined from a battery efficient state to when the device is being able to continuously transmit data.

In essence, the latency of interest is the point from where a device initiates a random access procedure connecting itself to the network. For more details on the different device states and the message transfer between the device and the network during the initial access, see Section 9.3.1.1.

10.1.3 Reliability

The requirement on reliability is defined as the probability of successful transmission of a packet of a certain size, within a certain latency bound, at a given channel condition. In other words, we want the system to guarantee that at a certain SINR we can deliver a certain packet reliably within a maximum time.

In case of IMT-2020, the requirement on the latency bound, packet size and reliability are 1 ms, 32 bytes and 99.999% respectively. The minimum SINR where this is achieved is technology specific and is evaluated according to the methodology described in Section 10.2.

10.2 Simulation framework

Simulations are required for the evaluation of the reliability, defined at the coverage edge of a macro scenario, designed for a wide area deployment. This section describes how the requirements on system level simulations and associated link level simulations from ITU has been interpreted by 3GPP in its evaluation toward IMT-2020 for LTE URLLC.

The simulation methodology is provided by the IMT-2020 documents [3] and involve the steps of:

(i) Run system level simulations in a macro scenario at full load.
(ii) Collect the SINR statistics from the simulation and register the fifth percentile SINR.
(iii) Use the fifth percentile SINR point in link simulations to show that the reliability is fulfilled.

One could think that allowing 5% of the devices in the network to be out of coverage, while requiring the devices in coverage to experience a reliability of 99.999% is not a consistent reasoning. One should however remember that assuming full traffic load of the network (see Table 10.1), i.e. that all resources in all cells are always occupied, is not realistic. Using instead a more realistic network load, the SINR distribution would shift toward higher SINR (assuming the network is interference limited, which is typically the case), which reduces the ratio of devices falling outside of the target SINR depicted in Fig. 10.1.

The relevant system simulation assumptions to produce the curve in Fig. 10.1 are shown in Table 10.1.

TABLE 10.1 Urban macro URLLC – Simulation assumptions.

Parameter	Setting
Carrier frequency for evaluation	700 MHz
Inter-site distance	500 m
Antennas, eNB	Two directional sector antennas with +45°, −45° polarization
Antenna elements, device	Two omni antennas with 0°, 90° polarization
Device deployment	80% outdoor, 20% indoor Randomly and uniformly distributed over the area
Inter-site interference modeling	Explicitly modeled
eNB noise figure	5 dB
Device noise figure	9 dB
Device antenna element gain	0 dBi
Traffic model	Full buffer
Simulation bandwidth	20 MHz
Device density	10 UEs per Transmission Reception Point
Device antenna height	1.5 m
eNB antenna height	25 m
Sites	3-sector
eNB antenna down-tilt	8°

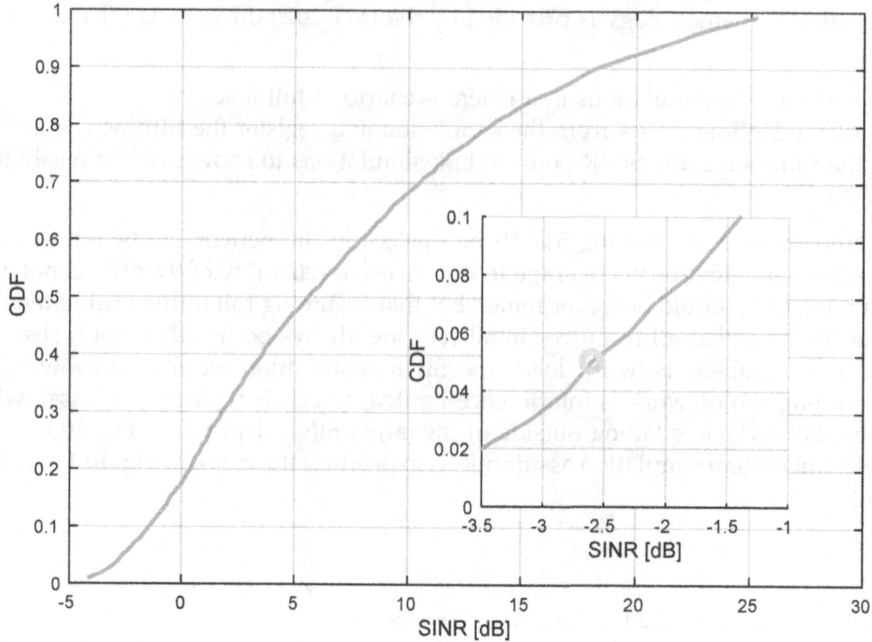

FIG. 10.1 PDSCH SINR distribution.

TABLE 10.2 Recorded SINR.

Physical channel	5th percentile SINR
Subslot-PDSCH(1)	-2.6 dB
Subslot-PUSCH(1)	-1.7 dB
Subslot-SPUCCH	-1.7 dB

NOTE 1: Same SINR value valid irrespective of transmission duration, i.e. also valid for e.g. slot operation.

The recorded SINR at the fifth percentile for PDSCH, subslot-PUSCH and subslot-SPUCCH are summarized in Table 10.2.

The main reason that the downlink SINR is worse than the uplink SINR is the use of power control on the uplink, which is not used on the downlink.

10.3 Evaluation

This section presents the performance evaluation for each URLLC requirement described in Sections 10.1.1–10.1.3.

10.3.1 User plane latency

For data transfer there are three different cases of interest:

- Downlink data transmission
- Uplink data transmission based on a configured grant, and,
- Uplink data transmission after SR

In case downlink data, the network has full control to schedule the packet after it has been prepared for transmission and put in the transmit buffer (assuming no queuing delay). Hence, only the processing delay and the alignment of the data transmission time with the air interface need to be considered.

For the uplink, the situation is somewhat different since the network will not know when the device has data in its buffer. In case of SR-based uplink transmission (see Section 9.3.2.3), there will be additional delay until the uplink grant is received by the device, compared to having a configured grant available when the data has been prepared. This is since the SR needs to be received by the eNB, followed by an uplink grant to the device, before transmission of data can start.

For services with ultra-low latency requirements, it can be assumed that Semi-Persistent Scheduling (SPS), see Section 9.3.2.7.3, is needed to remove the extra delay of sending SR and receiving uplink grant. SPS can also be assumed to be used in case of periodic traffic when the traffic pattern is predictable. Furthermore, subslot operation is required to get down to the IMT-2020 URLLC target of 1 ms latency bound.

To get the full picture of the user-plane latency, the tables below show performance for legacy LTE (subframe based operation with n+4 processing timeline, see Section 9.2.5), short processing time (subframe based operation with n+3 processing timeline, see Section 9.2.5), slot-based and subslot-based transmission.

Two cases are looked at:

- HARQ-based retransmissions: In this case, it is assumed that the data is retransmitted using HARQ. This will result in a spectrally efficient transmission since a packet is only retransmitted in case of error. However, the latency is negatively impacted by additional round-trip delays for each retransmission.
- Blind repetitions: When blind repetitions are used (here assumed consecutive in time), the only additional delay comes from the transmission time of each repetition. Compared to HARQ-based retransmissions however, the spectral efficiency is negatively impacted since extra resources are used without knowing if less resources would have been enough for delivering the packet.

The different delay components relevant for the latency calculation are described in Table 10.3.

It can be noted that additional delay due to the TDD frame structure has not been considered. This is since the structure of TDD, where downlink and uplink transmissions are multiplexed in time, will inherently increase the delay. Adding that TDD is only defined for slot

TABLE 10.3 Delay components.

Type	Comment
L1/L2 processing delay, $T_{L1/L2}$	For the L1/L2 processing performed in the eNB, a processing delay equal to the configured transmission time interval is assumed when both transmitting and receiving a packet. The same is assumed in the device. A three symbol subslot is assumed. At the receiver side, it is assumed that data can be delivered to higher layers after L1/L2 processing but before ACK feedback is transmitted.
Alignment delay, T_{Align}	The alignment delay is the wait time required after being ready to transmit until transmission can start on the air interface.
UE/eNB processing delay, T_{Proc}	The eNB delay between receiving a SR and transmitting an uplink grant, and between receiving downlink HARQ NACK and the next PDSCH retransmission is assumed to be the same as the timing of the device, i.e. the same as the time between PDSCH reception and downlink HARQ feedback transmission and uplink grant reception and PUSCH transmission. The shortest processing timeline is assumed to be configured for subslot operation, i.e. n+4, see Section 9.2.5. When applied, the processing delay includes the L1/L2 processing times. It could be noted that the values of eNB processing are not specified but vendor specific. The processing delay device on the other hand needs to be strictly followed, see Section 9.2.5.

operation, and not subslot operation, will make the resulting delay relatively far from the IMT-2020 requirements, which is the main focus of this performance chapter.

That is, the table and the following latency calculations assume FDD operation.

Fig. 10.2 shows the signal diagram for a downlink data transmission including the associated delays (ignoring propagation delay). It is considered that the packet can be delivered either in the first or the second attempt (depending on packet errors over the air).

The downlink data transmission delay can also be written as an equation, assuming k retransmissions.

$$T_{Tot} = T_{L1/L2,Device} + T_{L1/L2,eNB} + T_{Align} + 2kT_{Proc} + (1+2k)T_{Tx} \tag{10.1}$$

For one retransmission, assuming regular timing of n+4 ($T_{Proc} = 2\ TTI$), and that $T_{L1/L2} = T_{Align} = T_{Tx} = 1$ TTI, the delay becomes:

$$T_{Tot} = 4 + 8k \tag{10.2}$$

It can be noted that the processing timeline for the HARQ-ACK response, see Section 9.2.5, of the device is set by the specifications, while being implementation dependent for the eNB (in case of asynchronous HARQ). It is however assumed in Eqs. (10.1) and (10.2) that the same processing is assumed for both device and eNB. Furthermore, the L1/L2 processing time is different for different implementations in both network and device, and hence not defined by the specification.

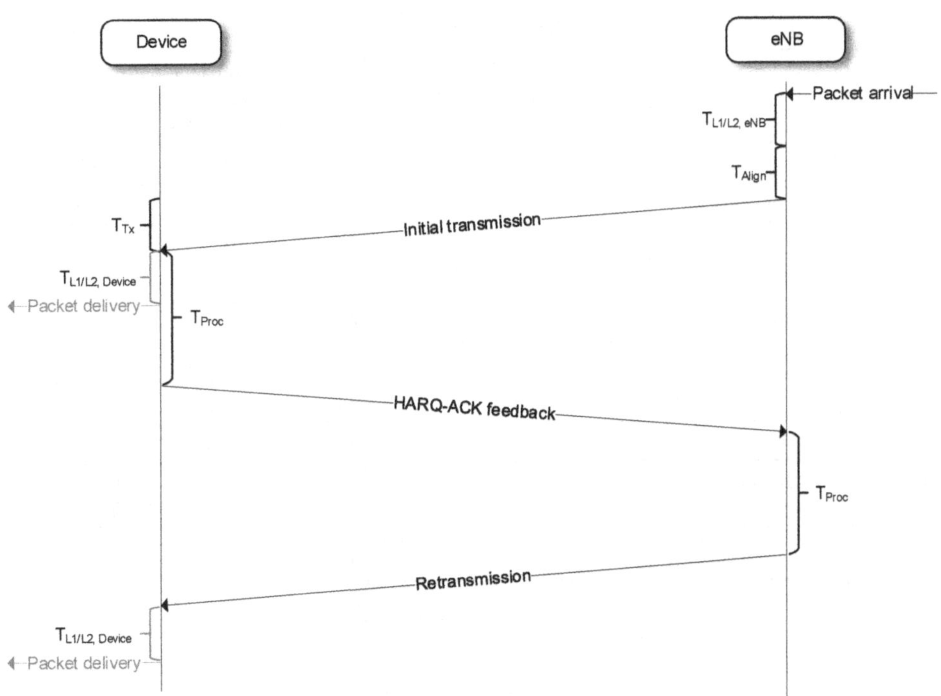

FIG. 10.2 Downlink data transmission – signal diagram.

The time to align with the frame structure will be given by the transmission granularity, i.e. how often a packet can be transmitted over the air. In case of LTE, this is based on a fixed structure using either subframe, slot or subslot granularity.

For uplink data using SR, the calculation of the delay is similar to the calculation of downlink data. The difference is that the packet is arriving at the device, and the packet is delivered to higher layers after reception at the eNB. Also, in this case, one round-trip time is consumed from SR transmission to the transmission of the PUSCH based on the uplink grant received. Hence, as expected, the latency for X uplink data retransmissions based on SR, is the same as the latency for X+1 uplink data retransmissions based on SPS. This is seen in Table 10.4 which presents the results from the user plane latency evaluations.

In all cases, it is assumed that the worst-case latency is used. This assumption has no impact on subframe based and slot-based operation due to the symmetry in the radio frame. For subslot operation however, the delay will vary over the subframe due to the irregular subslot structure, see Fig. 10.2. For subslot operation, the delays in Tables 10.4 and 10.5 are considering the largest delay of the five possible subslot starting numbers for the packet arrival. The three different possible PDCCH symbol configurations, one, two, or three, are considered. Using a single PDCCH symbol will decrease the overall latency since the

TABLE 10.4 User-plane latency – HARQ-based retransmissions.

Latency (ms)	HARQ	Rel-14	Rel-15 short processing time (n+3)	Rel-15 slot	Rel-15 subslot (1 PDCCH symbol)	Rel-15 subslot (2 or 3 PDCCH symbol)
DL data	1st transmission	4.0	4.0	2.0	**0.86**	**1.0**
	1st retransmission	12	10	6.0	2.1	2.4
	2nd retransmission	20	16	10	3.4	4.0
	3rd retransmission	28	22	14	4.9	5.4
UL data (SR)	1st transmission	12	10	6.0	2.1	
	1st retransmission	20	16	10	3.4	
	2nd retransmission	28	22	14	4.9	
	3rd retransmission	36	28	18	6.1	
UL data (SPS)	1st transmission	4.0	4.0	2.0	**0.86**	
	1st retransmission	12	10	6.0	2.1	
	2nd retransmission	20	16	10	3.4	
	3rd retransmission	28	22	14	4.9	

TABLE 10.5 User-plane latency – blind repetitions.

Latency (ms)	Blind repetitions	Rel-14	Rel-15 Short processing time (n+3)	Rel-15 slot	Rel-15 subslot (1 PDCCH symbol)	Rel-15 subslot (2 or 3 PDCCH symbol)
DL data	1st transmission	4.0	4.0	2.0	**0.86**	**1.0**
	1 repetition	5.0	5.0	2.5	**1.0**	1.1
	2 repetitions	6.0	6.0	3.0	1.1	1.3
	3 repetitions	7.0	7.0	3.5	1.3	1.4
UL data (SR)	1st transmission	12	10	6.0	2.1	
	1 repetition	14	12	7.0	2.4	
	2 repetitions	16	14	8.0	2.9	
	3 repetitions	18	16	9.0	3.1	
UL data (SPS)	1st transmission	4.0	4.0	2.0	**0.86**	
	1 repetition	5.0	5.0	2.5	**1.0**	
	2 repetitions	6.0	6.0	3.0	1.1	
	3 repetitions	7.0	7.0	3.5	1.3	

alignment time will be reduced. However, using less symbols for the DL control will have a direct impact on the DL control capacity, and hence this trade-off need to be considered when deciding the size of the DL control region.

Another influence on the latency due to the irregular subslot structure is the assumption on 1 TTI latency of L1/L2 processing. It should be noted that this is a simplified assumption and the actual timing need not be related to the over-the-air transmission duration. Still, in this evaluation, it is assumed that the L1/L2 processing assumes the value of 3 symbols, same as the longest subslot duration.

The user plane latency using HARQ-based retransmissions and using blind repetitions is shown in Tables 10.4 and 10.5 respectively using different number of retransmissions and repetitions.

If a 1 ms latency bound is to be respected (marked in bold in Tables 10.4 and 10.5), see Section 10.1.1, one can see that for:

- Downlink data and uplink data using grant-based transmission, either a single transmission is required, or one blind repetition. Using HARQ-based retransmissions will however violate the latency bound. It can be noted that using a blind repetition is only possible in the DL in case 1 PDCCH symbol is configured.
- Uplink SR-based transmission, no configuration fulfills the latency bound

10.3.2 Control plane latency

The background to the control plane latency and the improvements made to the 3GPP specifications are described in Sections 9.3.1.1 and 10.1.2.

To summarize, instead of changing existing initial access procedures in how information is transmitted over the air, which could imply implementation cost and backwards compatibility issues in existing networks, a simple approach was taken by shortening the processing time at both the device and the network in the initial access procedure.

The resulting control plane latency is shown in Table 10.6.

As can be seen, the target of 20 ms control plane latency, see Section 10.1.2, is reached.

10.3.3 Reliability

The latency calculations carried out in Section 10.3.1 is an analytical exercise assuming different number of retransmissions and no queuing effects at the scheduler (as assumed in the requirements, see Section 10.1.1). Still, it determines the configurations under which a certain latency bound can be met.

To determine the associated reliability, simulations are needed that model the radio performance under the conditions given by Section 10.3.1.

In Section 10.3.1 it was concluded that for downlink and uplink transmissions, to fulfill the 1 ms target, there is no time for HARQ retransmissions. Instead, blind repetitions need to be used to lower the latency while maintaining a high level of reliability. Furthermore, uplink transmissions need to be carried out by pre-configured uplink grant using SPS (not based on scheduling requests, which increase the overall latency by one round-trip time). Both the use of blind repetitions (instead of HARQ retransmissions) and pre-configured uplink grants

TABLE 10.6 Control plane latency.

Step	Description	Latency [ms]	
		Pre-release 15	Release 15
1	Transmission of RACH Preamble (uplink)	1	
2	Preamble detection and processing in eNB	2	
3	Transmission of RA response (downlink)	1	
4	Device processing delay	5	4
5	Transmission of RRC Connection Resume Request (uplink)	1	
6	Processing delay in eNB	4	3
7	Transmission of RRC Connection Resume (downlink)	1	
8	Processing delay in device of RRC Connection Resume and grant reception	15	7
9	Transmission of RRC Connection Resume Complete and user plane data (uplink)	–	
	Total delay	30	20

(instead of scheduling requests) will increase the usage of radio resources. This is a cost we need to be willing to pay to reach the ultra-reliable radio link performance at a low latency.

The reliability will be here estimated assuming a certain physical layer payload size to be transferred (32 bytes) assuming a certain latency bound (1 ms) achieving a target reliability (99.999%). These conditions are all given by the IMT-2020 requirements. Furthermore, these conditions are to be fulfilled at the cell-edge SINR assumed in the network, defined by Table 10.2.

10.3.3.1 Reliability of physical channels

To understand the reliability performance, we need to look at how the physical channels are designed. This is where most of the description in Chapter 9 comes in. Although it should be noted that the IMT-2020 use case of a latency bound of 1 ms at a reliability of 99.999% can be seen as an extreme case, this is what much of the LTE URLLC technology was designed and targeted for, and this is also the main use-case we evaluate in this chapter.

The possible physical channels involved in these evaluations are (S)PDCCH, SPUCCH, PDSCH and PUSCH, but as will be seen, not all channels need to be part of the evaluation.

Our task is now to determine an SINR where the overall block error rate, taking all channels involved in the data transfer into account, is below the target of 10^{-5}. To fulfill the overall reliability requirement, this determined SINR should be equal to, or below, the values listed in Table 10.2.

10.3.3.1.1 Downlink

For the downlink, the channels involved in a data transmission are (S)PDCCH and PDSCH. If HARQ retransmissions were to be considered, also SPUCCH would have to be included (i.e. carrying the HARQ-ACK feedback triggering the retransmission). There is however no time within the latency bound for this (see Section 10.1.1) and hence the uplink control channel can be excluded. The reason for writing "(S)PDCCH" is that the PDSCH can be scheduled by either PDCCH or SPDCCH depending on which subslot the data is scheduled on (See Section 9.2.2 and 9.2.3.2.1). However, in this simulation campaign, it is assumed that the downlink data is earliest mapped to symbol index three, implying that SPDCCH is always used. The reason for this choice is to model a worst-case situation where PDCCH is not used for scheduling and less blind transmissions can be performed within a given latency bound (compared to the case where the data can be mapped to symbol index 1).

The channels involved are shown in Fig. 10.3 for a two-antenna port CRS configuration, one symbol CRS-based SPDCCH, and a mapping of the data REs earliest in symbol index

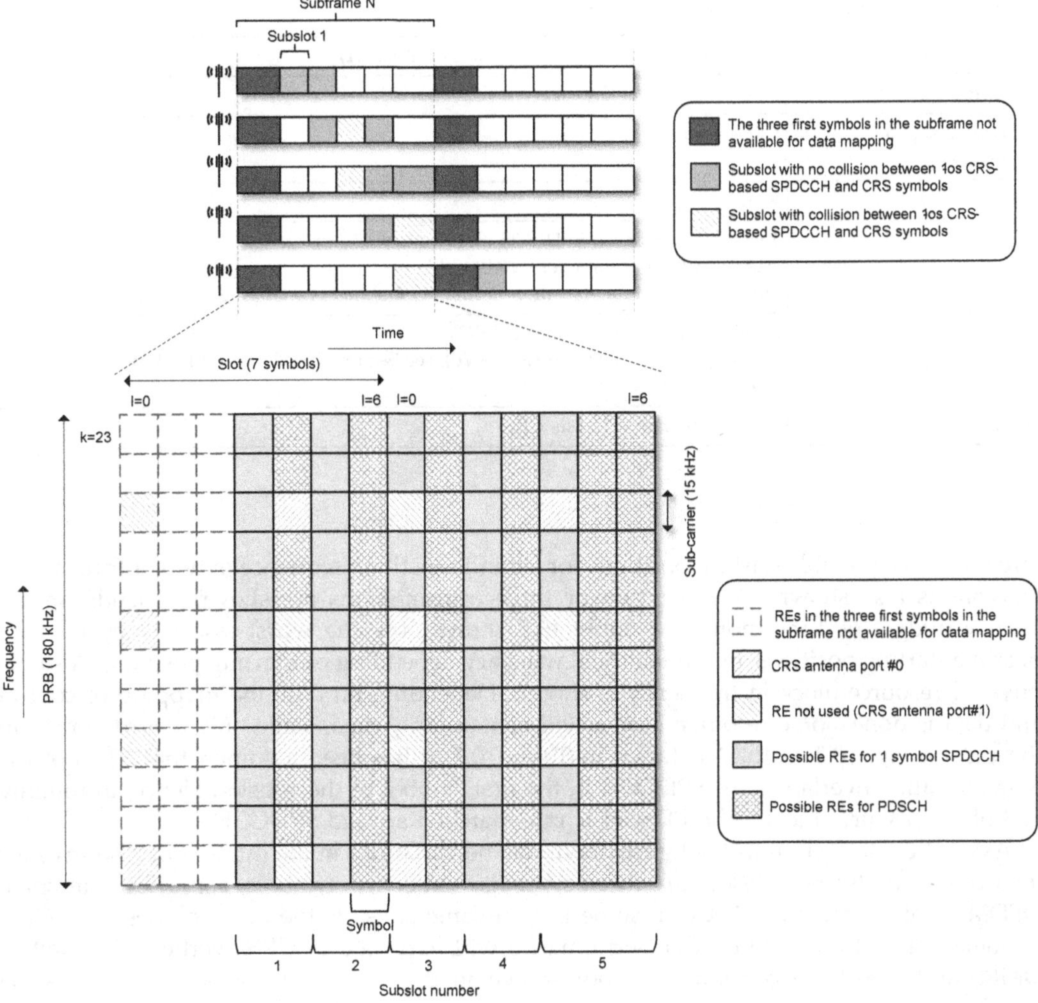

FIG. 10.3 Downlink data transmission from the five possible starting positions in the subframe using two blind repetitions.

TABLE 10.7 Link level simulation assumptions – downlink.

Parameter	Setting
Carrier frequency	700 MHz
Bandwidth	20 MHz (100 Resource Blocks)
Channel	TDL-C, 363 ns, see Ref. [4]
Device speed	30 km/h
TTI length	Subslot (2 or 3 OFDM symbols depending on where transmission is made)
Start of DL data	Symbol index 3 (See Section 9.2.2)
Payload	32 bytes
MCS	MCS-0 (occupying 55 resource blocks)
Resource allocation	Resource allocation type 0 (See Section 9.3.2.9.1)
Transmission mode	2TX, 2RX, 1-layer TX diversity
Reference signal transmission	CRS-based
Channel estimation	Realistic
SPDCCH	1 symbol CRS-based SPDCCH (See Section 9.2.3.2) AL8 (8 SCCE in total over bandwidth) (about 32% of data allocation in first symbol has overlapping SPDCCH)
Transmissions	2 (Blind repetitions without HARQ) The same redundancy version (RV0), see Section 9.3.2.7.1, used for both transmissions.
DCI size	40 bits payload + 16 bits CRC

three. The five possible starting positions for a blind repetition sequence of two or three transmissions is also shown. Whether two or three transmissions are shown depends on the latency budget of 1 ms (note that Table 10.5 shows only the worst-case latency over all possible starting positions, but in reality, it will vary depending on starting position). A single physical resource block in frequency is shown. Depending on how the mapping of control and data is done, some resource blocks will contain only data, some only control, or, as in the case of Fig. 10.3, both. As shown in Table 10.7, it has been assumed that 32 % of the data allocation overlaps with SPDCCH in the first symbol of the subslot. Hence, in roughly 1/3 of the resource blocks, the PDSCH is rate-matched around SPDCCH.

As can be also seen from the figure, there will be variations in the number of resources for data over time (some subslots are three symbols, some two symbols, some PRBs in some OFDM symbols contain SPDCCH, some not, in some subslots the data collides with CRS, in some not). Also, the SPDCCH performance will vary due to CRS-overhead (as well as DMRS and CSI-RS, if configured – not shown in this figure). These are all aspects the

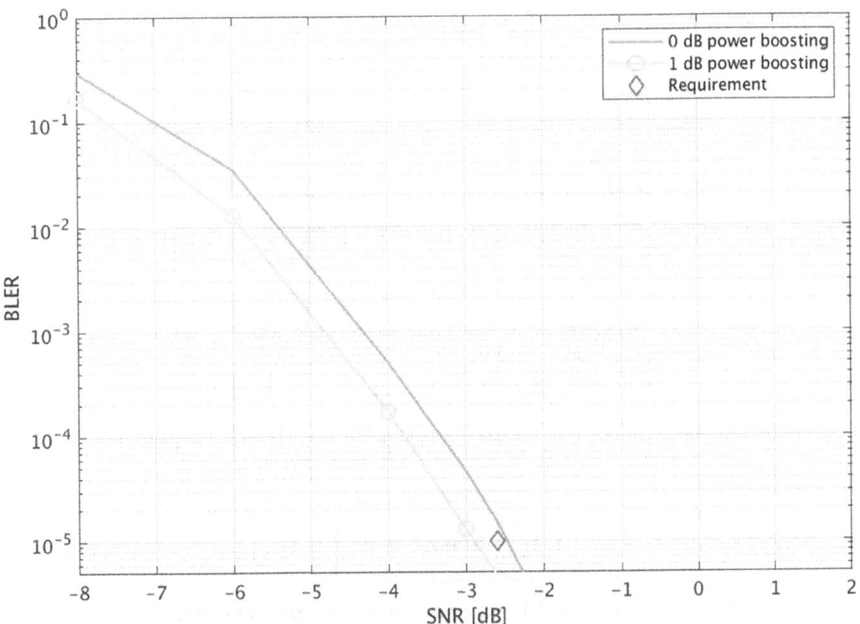

FIG. 10.4 PDSCH performance using two blind transmissions.

network has to consider when selecting the aggregation level for the device on the SPDCCH (see Section 9.2.3.2) as well as the MCSs selected for data transmission.

The simulations have assumed the worst-case possible configuration where one of the two SPDCCH transmissions are hit by CRS (fourth or fifth configuration in Fig. 10.3). This can be considered the limiting case in terms of performance for the SPDCCH. Regarding PDSCH performance, one of the subslots will have a three-symbol duration resulting in more REs available for data transmission (this does however not improve the SPDCCH performance, which is here configured to one symbol duration).

The simulation assumptions used are shown in Table 10.7.

The simulation results from the above simulation assumptions are shown Fig. 10.4.

As can be seen, the target performance of a reliability of 10^{-5} at SINR of -2.6 dB (see Table 10.2) is reached assuming a power boosting of 1 dB for the SPDCCH resources. The power boosting is necessary due to the performance imbalance between the data and control channel (the decoding of the data channel is conditioned by the decoding of the control channel). The implication of boosting the power of the control is that the remaining power left for the other resource elements in that OFDM symbol will be more restricted.

10.3.3.1.2 Uplink

For the uplink, the channel involved is only PUSCH. This is since we are only interested in the case of using a pre-configured grant (by SPS, see conclusion in Section 10.3.1), and it is

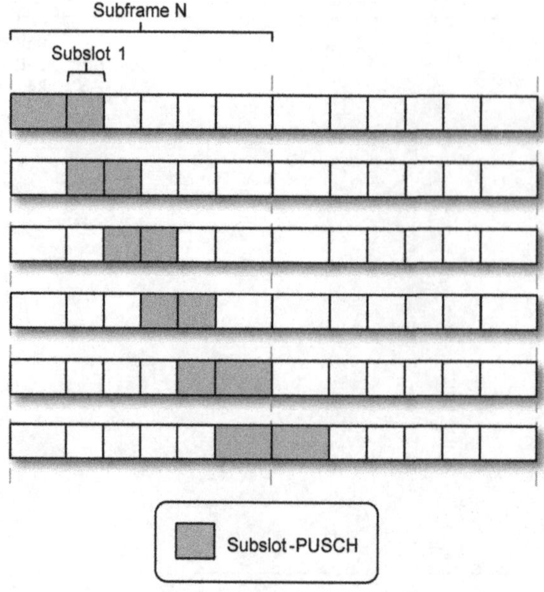

FIG. 10.5 Uplink data transmission using two blind repetitions.

TABLE 10.8 Link level simulation assumptions – uplink.

Parameter	Setting
MCS	MCS-1 (occupying 85 resource blocks)
Transmission mode	1TX, 2RX
Transmissions	1, 2 or 3 (Automatic repetitions without HARQ) The same redundancy version (RV0), see Section 9.3.2.7.1, used for all transmissions.
TTI length	Subslot (2 OFDM symbols)

assumed that this is pre-configured to the device. Since there is no time for HARQ-based retransmissions, the downlink control channel, SPDCCH, does not come into play.

From Table 10.5, it can be seen that the 1 ms bound can be fulfilled using two transmissions of PUSCH. The possible starting positions and the following consecutive transmissions are illustrated in Fig. 10.5. Compared to the downlink (see Fig. 10.3), the number of repetitions does not vary depending on where the repetition window starts. Using three transmissions will violate the latency bound.

The worst-case performance would be expected if the two transmissions are performed using three consecutive 2-symbol subslots (second, third and fourth configuration in

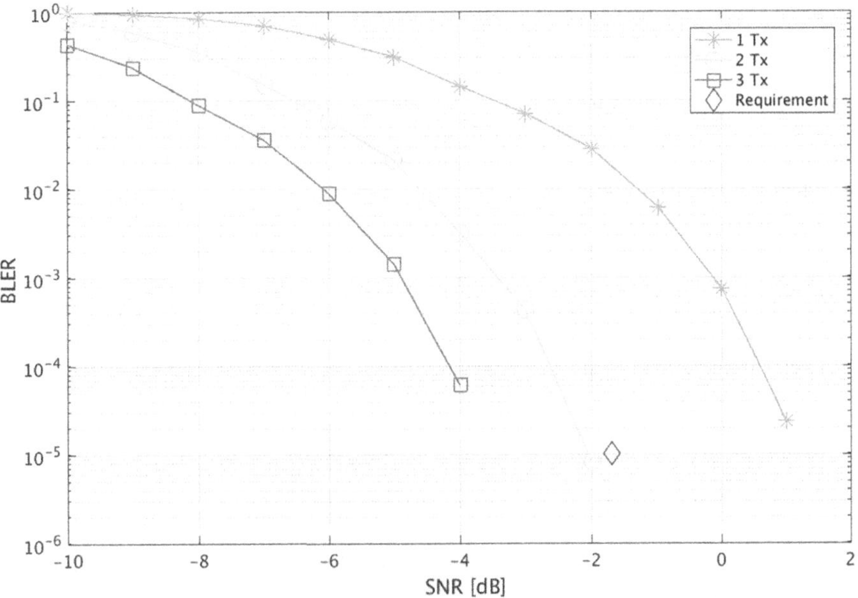

FIG. 10.6 PUSCH performance using one, two or three (HARQ-less) transmissions.

Fig. 10.5). This is since less resources are available for transmitting the data, and hence, this is what has been simulated (see Table 10.8).

The link simulation assumptions different from the ones for the downlink (see Table 10.7) are shown in Table 10.8.

The associated performance is shown in Fig. 10.6, see Ref. [1] for more details.

As can be seen, the PUSCH SINR target of -1.7 dB at a reliability of 10^{-5} is achieved with the use of two transmissions. However, it can be noted that the allocation consumes 85 out of 100 resource block in the carrier. Furthermore, it is assumed that this is a pre-configured resource to the device and that the scheduling interval is each subslot. This causes severe restrictions to the network operation (essentially one device taking up almost all of the resource in the uplink).

Using a higher MCS would lower the frequency allocation required. However, although the same overall energy would be transmitted with a smaller allocation (using a higher power spectral density, PSD), some additional degradation would be expected from the increased code rate of the MCS by reducing the resource allocation, less frequency diversity, and possibly from using a higher order modulation.

The results will depend on if the network is interference or noise limited, the fraction of cMTC users etc, but it could be expected that with a slightly relaxed SNR requirement, the impact to the use of the network resources can be alleviated. However, with the requirement of -1.7 dB, the performance margin is very limited for the case of two transmissions, as can be seen in Fig. 10.6.

References

[1] Ericsson. R1-1807301, URLLC techniques for uplink SPS. 3GPP TSG RAN1 Meeting #92bis, 2018.
[2] ITU-R, Report ITU-R M.2410, Minimum requirements related to technical performance for IMT-2020 radio interfaces(s), 2017.
[3] ITU-R, Report ITU-R M.2412, Guidelines for evaluation of radio interface technologies for IMT-2020, 2017.
[4] Third generation partnership project, Technical specification 38.901, v15.0.0. Study on channel model for frequencies from 0.5 to 100 GHz, 2018.

CHAPTER

11

NR URLLC

OUTLINE

Cellular Internet of Things, Second Edition
https://doi.org/10.1016/B978-0-08-102902-2.00011-X

© 2020 Elsevier Ltd. All rights reserved.

Abstract

In this chapter a general overview of NR is given with emphasis on the evolution from LTE. With this as base, the focus is set on the parts of the design of NR introduced to enable Ultra-Reliable Low-Latency Communications, going from the physical layer design up to procedures for idle and connected modes.

11.1 Background

In this section a brief overview of 5G and *New Radio* (NR) is given, its relation to LTE is discussed, and *Ultra-Reliable Low-Latency Communication* (URLLC) is introduced with related design principles.

11.1.1 5G system

The *3GPP 5G System* (5 GS) is a new generation cellular system that aims to broaden the usage of cellular systems. In addition to voice and Enhanced Mobile Broadband (eMBB), 5 GS is intended to support the use case areas of *massive Machine Type Communication* (mMTC) and *critical Machine Type Communication* (cMTC).

The 5 GS is based on the *5G Core* (5 GC) network and the *NR* radio access technology together aimed at meeting the 5G IMT-2020 requirements from ITU [1]. A first *non-standalone version* (NSA) of NR was standardized in 3GPP during 2017. It connects NR to the 4G Core Network (CN) *Evolved Packet Core* (EPC) through *Dual Connectivity* (DC) with LTE, effectively adding an NR cell to the existing LTE setup. The full *standalone* (SA) version, which connects NR to 5 GC, was finalized during 2018 as a part of 3GPP Release-15.

NR builds on the success of LTE and reuses well established concepts from its predecessor in the design of both the lower and higher layers. To that base, new features and modes of operation are added as motivated by the supported use cases. The main advances relate to the flexibility that is built into NR, increasing the scalability and adaptability compared to earlier systems GSM, WCDMA, and LTE. As an example a wider range of frequency bands spanning all the way up to 52.6 GHz is supported.

This chapter gives attention to the NR support for cMTC services, characterized by stringent requirements on service latency and reliability. In the following we will first give a brief introduction of NR and then continue by looking specifically at its *Ultra-Reliable Low-Latency Communication* (URLLC) features, designed for providing the needed support for critical services.

The process of standardizing the 5 GS as well as the basics of the NR physical layer (PHY) and higher layers are described in Chapter 2. An evaluation of the NR URLLC performance is presented in Chapter 11. For reference, the LTE version of URLLC is presented and evaluated in Chapter 9 and 10, and a comparison of the two technologies is given in Chapter 16.

11.1.2 URLLC

From the conception of 5G, it is designed to address three use case areas:

- *eMBB,*
- *Massive Machine Type Communication* (mMTC), as well as,
- *Critical Machine Type Communication* (cMTC).

These areas can be said to represent the performance dimensions of spectral efficiency, connection efficiency, and service quality, respectively. The structure of NR is flexible enough to provide support for all these use cases within one integrated system, by allowing shifting of the operation point toward one of the three dimensions. A mix of diverse services can thereby be provided over a single 5G radio interface, and even shared on the same carrier.

Ultra-reliable low-latency communication (URLLC) is the technology created to deliver on the cMTC use cases, which need communication links with very high requirements on *quality of service* (QoS) and availability. Examples of such use cases include protection services for power substations and factory automation with industry robots, as will be studied in Chapter 12, and also further applications such as high-precision remote control and tactile communication, as for example with the case of remote surgery.

To be able to provide URLLC in NR, the system must be able to deliver data packets with short latencies and with high reliability, and the service must be consistently guaranteed. This is achieved by a combination of techniques specified in the standard, and is secured by careful radio resource management in the deployed network. The toolbox of URLLC features must be flexible and powerful enough to make it possible to reach the requirements of the cMTC services. At the same time, since NR also should provide eMBB and mMTC services and connectivity for new verticals, the specific tools applied need to coexist well with the operation of other tools and features. This trade-off explains the design choices taken for URLLC in NR. Compared to LTE URLLC however, in NR URLLC there are no concerns with backwards compatibility, which makes the design less constrained.

11.1.3 NR as the successor of LTE

NR is the natural successor of LTE, and the two RATs share many features. At the basis, both operate on an *orthogonal frequency-division multiplexing* (OFDM) radio resource grid, and specify the same kinds of physical data and control channels. But some important differences in design should be highlighted:

- High bands: NR is designed to operate at higher frequencies (up to 52.6 GHz) and with higher carrier bandwidths (up to 400 MHz) compared to LTE, where a carrier is limited to 20 MHz and where the highest supported bands are at 5 GHz. The high frequency bands also facilitate small device form factors, thanks to the smaller antennas, which is important for many industrial applications. A scalable OFDM numerology in NR allows for efficient use of higher frequency bands.

- Lean design: Minimized always-on transmissions enable NR networks to have much improved energy performance through extended micro-sleep periods.
- Flexible design: The configuration and use of the NR time-frequency resources provides a high degree of flexibility, giving room for future enhancements and features.
- Low latency: NR offers flexible scheduling and shorter processing times, which are important tools for optimizing the service latency.
- Beam-centric: NR supports new and highly advanced antenna techniques, facilitating an antenna beam-centric design for enabling support of the new frequency range. Fast switching between beams belonging to different nodes is also supported.

At a glance then, NR offers the basic capabilities of LTE, while also being more forward-compatible by flexible design and enabling higher bands and more advanced transmission methods. The main step taken through NR is to be able to deliver on a wide range of use cases with a single integrated system. Using similar radio configurations we would expect roughly the same performance in LTE and NR in terms of spectral efficiency, but beyond the base NR comes with extra gears for higher data rates, lower latency and more options of operation option. Looking at network power consumption, NR is expected to be much more efficient due to the lean design with fewer obligatory signals always transmitted by the base stations.

11.1.4 Introduction of NR URLLC in existing networks

LTE is by now widely deployed on a large range of important spectrum bands ranging up to 5.9 GHz. NR on the other hand supports two *frequency ranges* (FR). FR1 corresponds to the existing LTE frequency range, while FR2 covers the range from 24 to 52.6 GHz. Traditionally, the *mobile network operators* (MNO) would stepwise re-farm parts of their existing spectrum to enable a roll-out of a new generation. However, since NR comes with a new set of bands not in current MNO use, an attractive alternative approach is to deploy NR in new rather than in current LTE bands. Therefore, the first NR deployments are expected to come as add-on higher frequency carriers to existing LTE deployments. The use of the FR2 bands, which support large system bandwidths and low latency, also allows NR to support demanding cMTC services with high traffic, such as factory automation, from day one. In addition, low latency can also be achieved on lower frequency bands thanks to short processing time and flexible scheduling with short transmissions.

Since 3GPP Release-15, the *Evolved Packet Core* (EPC) can, in addition to supporting LTE (called *Option 1*), also support NR as connected RAN. In this so-called *non-standalone* (NSA) NR mode (called *Option 3*), a carrier from an NR base station, called a *gNB*, provides service in a secondary cell in addition to the service provided in a primary cell by an LTE master eNB. This functionality is based on *Dual Connectivity* (DC), which in this case called *E-UTRA-NR Dual Connectivity* (EN-DC). Combined with the availability of new bands in FR2, EN-DC provides an attractive option for a seamless introduction of NR on top of existing LTE deployments.

With the deployment of *5G core networks* (5 GC), NR can instead be connected in the main *standalone* (SA) mode (called *Option 2*). LTE bands can then be re-farmed to become NR bands, or alternatively NR can be deployed in the data region of existing LTE carriers, by using the method of LTE-NR *dynamic spectrum sharing* (DSS). With DSS, NR dynamically uses spectrum from an LTE carrier, for instance to guarantee coverage in a lower band for an NR device. DSS can also be used to ensure sufficient resources for legacy LTE devices. For the sharing to work there is no requirement on upgrading nodes or devices to 5 GC-compliant LTE, making it more universally applicable as a means to migrate to NR.

Besides these main three options other NR-LTE combinations based on 5 GC are also possible. Using DC, SA NR and 5 GC-compliant LTE nodes can be connected, either with NR as master node and LTE as secondary node (called *Option 4*) in *NR-E-UTRA Dual Connectivity* (NE-DC), or with LTE as master node and NR as secondary node (called *Option 7*) in *NG-RAN-E-UTRA Dual Connectivity* (NGEN-DC).

Fig. 11.1 illustrates the main architectural options 1–3 for NR and LTE. With DC, the main node serving the primary cell supports both control plane signaling and user plane data transmissions. The secondary cell provides added capacity by means of data transmission over the user plane. In the pure SA form, NR is run as a separate system either on its own carriers or dynamically scheduled on LTE carriers using DSS.

11.1.5 Radio access design principles

With URLLC we want to transmit data packets with very high *quality of service* (QoS) requirements, but we still want to use the same system as is otherwise used for more relaxed requirements. This is possible through careful scheduling, using the knobs and levers connected to a set of mechanism specified in the NR standard. These adjustments make it possible to optimize NR for high QoS for the use cases where this is required. In this chapter we will mainly focus on these optimization features and the adjustments relevant for URLLC.

How do we achieve high reliability in varying radio conditions? Broadly speaking, the way to do it is through diversity: the information is duplicated in multiple copies so that at least one attempted copy can be received, or it is spread out in frequency or code domain to ensure successful decoding. But diversity can mean many things, and typically a combination of the following types of tools would be used for achieving reliability in NR:

- Code diversity. A lower code rate means that more coded bits are used to carry the information, and it is more likely that the received transmission conveys enough redundancy to decode the message. The code diversity can be achieved by use of a lower code rate, or by repeating the message, which effectively reduces the effective code rate.
- Frequency diversity. Spreading the message over a wider band, either contiguously or non-contiguously, increases the possibility that some part of the message is sent in good channel conditions. Applying frequency hopping is another way of achieving the same thing.
- Time diversity. Retransmissions of a message has both the effect of lowering the code rate after combining with previous attempts and in addition using the channel at different times, thereby increasing the chance of being in a better condition, in the case of a sufficiently time-varied channel. Attempts at different occasions can also enable interference diversity since different transmitters may be active at the time of transmission.
- Spatial diversity. The use of multiple transmit and receive antennas or transmission points improves the reception by sampling several spatial channels with low correlation.

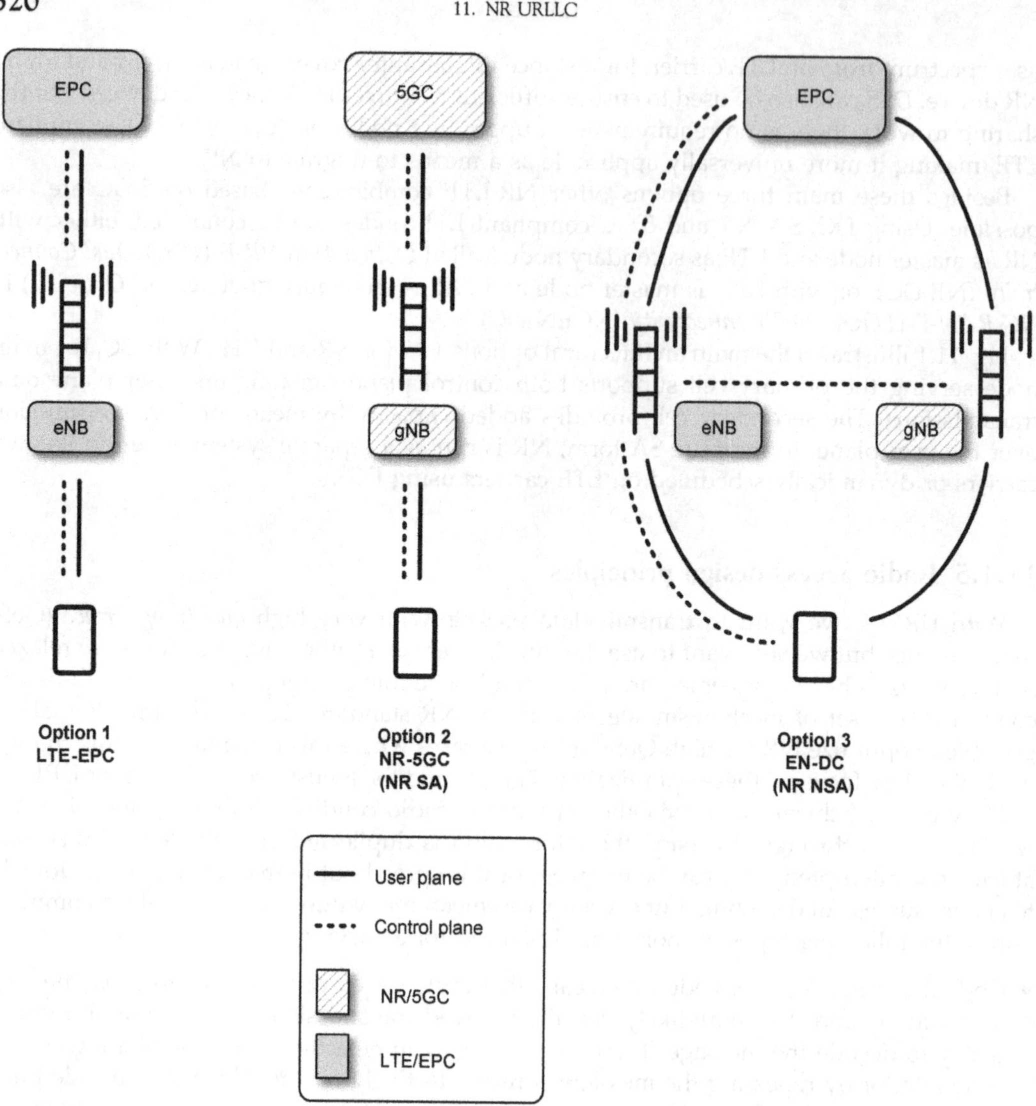

FIG. 11.1 LTE and NR connectivity options 1–3.

How do we then deliver a packet quickly? To answer this, we need to look at what actually constitutes the latency at RAN level when we transmit a packet. This includes delay for:

- Encoding a packet at the packet transmitter side
- Waiting for a transmission opportunity (alignment)
- Over the air transmission duration
- Decoding the packet at the packet receiver side
- Waiting to transmit feedback, and transmission of feedback
- Decoding the feedback at the packet transmitter side, preparing a retransmission, in case of packet failure

- Waiting for a retransmission opportunity
- Retransmission (transmission duration and decoding).

For uplink transmission with dynamic grant, additional steps for sending scheduling request (SR) and downlink control transmission are required before the uplink data transmission can start.

The delay contributions above are then the knobs to turn in order to reduce the delays. Correspondingly, the tools considered for achieving low latency in NR are:

- Enhanced packet encoding/decoding
- Reduced alignment delay by more transmission opportunities
- Reduced transmission time
- Reduced time between downlink data reception and transmission of feedback
- Reduced time between uplink grant and data transmission.

The additional challenge that comes with URLLC is that we want to achieve both qualities at the same time: a packet must be reliably delivered within a certain time. This requirement has the form of being able to deliver a payload of P bytes within a latency of L milliseconds with a probability or reliability of R. In those terms we can define a cMTC service requirement. Delivering the cMTC service in a real system would then be additionally characterized by a coverage, having the form of fraction or coverage C of devices in a certain population to which the service can be supplied. Ensuring the coverage of a service can be interpreted as guaranteeing a minimum SINR level across a certain deployment scenario, for instance in a factory hall or in a city. The available tools for ensuring coverage relate to the transmitting side's ability to provide a strong signal through beamforming and power control, and the receiving side's ability to filter out the interference.

11.2 Physical Layer

With these general design ambitions in mind we can identify a set of key technical components that have been introduced in NR. The intention is not to give an exhaustive review of the NR design and hence we restrict the description to an overview of only the essential features of NR, in addition to the more detailed review of URLLC specific functionality. For a fuller description of NR we recommend [2].

The material in this section is based on the 3GPP Release-15 specifications for the NR physical layer [3,4,5,6].

11.2.1 Frequency bands

As mentioned above, two sets of FR are defined for NR carriers; FR1 for the range 0.45–6 GHz, and FR2 for the range 24.25–52.6 GHz, as shown in Fig. 11.2. Especially the

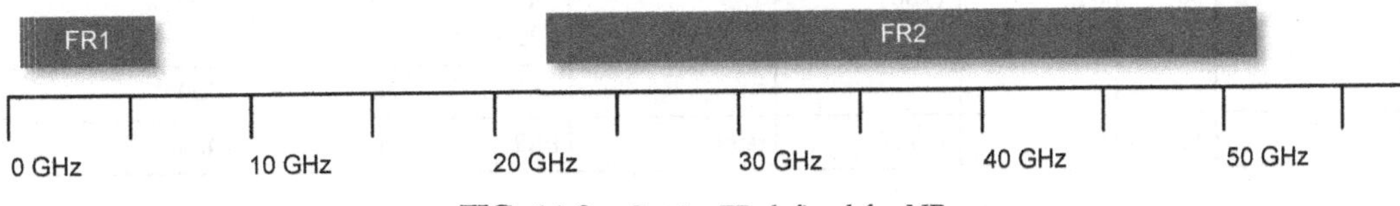

FIG. 11.2 Carrier FR defined for NR.

higher range opens many new spectrum opportunities supporting wider system bandwidths for cellular communication, owing to the fact that these bands are less occupied by incumbent radio services. This does not only enable eMBB higher data rates and increased capacity, it is also an important facilitator for URLLC. These higher carrier bandwidths are supported by the use of wider OFDM *subcarrier spacings* (SCS), which reduces symbol duration proportionally and hence also reduces latency.

Naturally, radio propagation is more challenging at the higher frequencies because of higher attenuation and reduced diffraction. One of the main achievements in NR is to provide the tools to cope with these more challenging conditions, and thereby enable also cMTC services.

11.2.2 Physical layer numerology

11.2.2.1 Flexible numerology

NR uses cyclic prefix-based OFDM modulation in both downlink and uplink. It is also possible to use the *digital Fourier transform* (DFT)-spread OFDM modulation in uplink, also known as *single-carrier frequency-division multiple access,* (SC-FDMA) could be introduced which is the only available modulation in LTE uplink. The availability of the two options for uplink in NR enables both the scheduling flexibility of OFDM and the possibility of improved coverage of DFT-OFDM, coming from its reduced *peak-to-average power ratio* (PAPR).

The OFDM subcarriers, which are grouped in sets of 12 into *resource blocks* (RBs), are separated by a *subcarrier spacing* (SCS) in frequency domain. NR can be configured with different SCS, defining different *numerologies*, while LTE only runs on 15 kHz SCS for data transmission. With a subcarrier spacing of W Hz, the duration of an *OFDM symbol* (OS), excluding the *cyclic prefix* (CP), is nominally $1/W$ s [7]. With the basic SCS of $W = 15$ kHz, 14 OS including CP takes 1 ms to transmit. One subcarrier during one OS is called a *resource element* (RE), as in LTE, and can carry one modulated symbol. In NR, the SCS can be set to N^*W for $N = 2^{\mu}$, with numerology $\mu = \{0, 1, 2, 3\}$, meaning that *14*N* OSs are transmitted per millisecond. The shorter symbol durations mean that a packet can be transmitted faster, at the cost of using more bandwidth. In NR, with higher frequencies considered, wider frequency bands also become available, so this tradeoff becomes reasonable and interesting for low-latency applications.

The set of available subcarrier configurations are listed in Table 11.1, together with the applicable FR and the supported maximum system bandwidths.

TABLE 11.1 NR numerologies for data transmissions.

Numerology μ	SCS [kHz]	Slot duration [ms]	Symbol duration [ms]	Normal cyclic prefix [μs]	Frequency range	Maximum nr. of RBs	Maximum bandwidth [MHz]
0	15	1	1/14	4.7	FR1	270	50
1	30	0.5	1/28	2.3	FR1	273	100
2	60	0.25	1/56	1.2	FR1 (Optional)	135	100
					FR2	264	200
3	120	0.125	1/112	0.59	FR2	264	400

At the highest subcarrier spacing intended for wide areas, 60 kHz, the symbol duration is short enough for the delay spread of the radio channel to become an issue in large-distance scenarios (i.e. the delay spread exceeds the CP which implies that the orthogonality between symbols is degraded and in worst case lost). To reduce the harmful effect of inter-symbol interference the CP duration can be set to a longer value (*extended CP*) at the cost of using fewer data symbols per second, from *14*N* symbols/*ms* to *12*N* symbols/*ms*.

11.2.2.2 Frame structure

In time domain the basic unit is the *subframe*, being 1 ms long, irrespective of used numerology. Depending on the chosen numerology, which determines the symbol duration, each subframe consist of a certain number of *slots*, each in turn being a set of 14 (normal-CP) or 12 (extended-CP) OS, as illustrated in Fig. 11.3. The basic transmission duration is a slot, but a transmission can occupy also part of a slot, a so-called *non-slot transmission* or commonly *mini-slot*, as illustrated in Fig. 11.4, where the frequency domain and different SCS are illustrated. In downlink, a mini-slot can start in any symbol and be 2, 4, or 7 OS long, whereas in uplink it can start in any symbol and be of any length up to the slot length, i.e. 1 to 14 OS long.

In NR the slot structure in terms of uplink and downlink symbols is flexible and there is therefore no principled difference between FDD and TDD slots, as described further below.

11.2.3 Transmissions schemes

11.2.3.1 Beam-based transmissions

All physical channels in NR, data as well as control channels, can be beamformed for enhanced coverage and higher rates. This motivates the label *beam-based* that is commonly

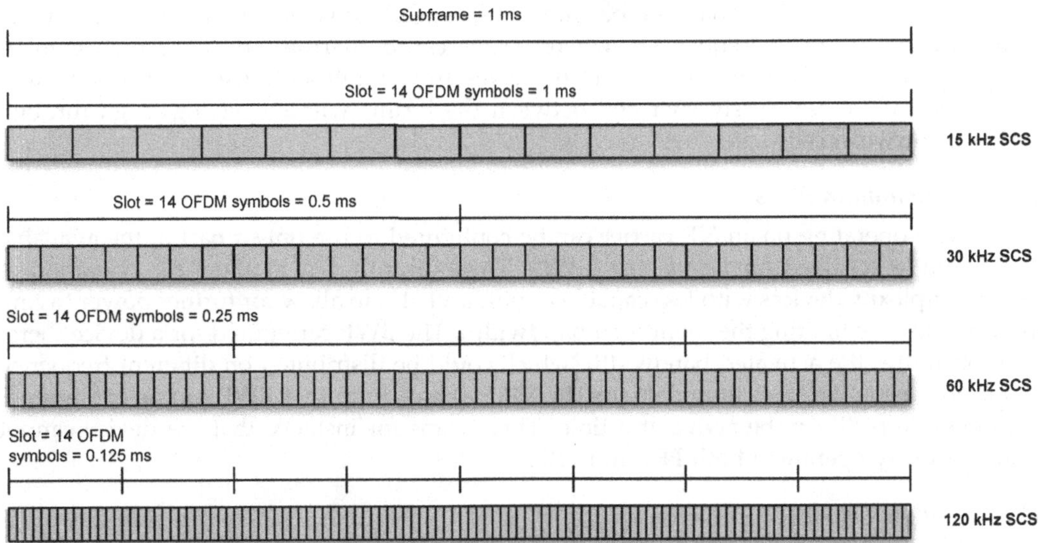

FIG. 11.3 Slot structure in NR with normal-CP.

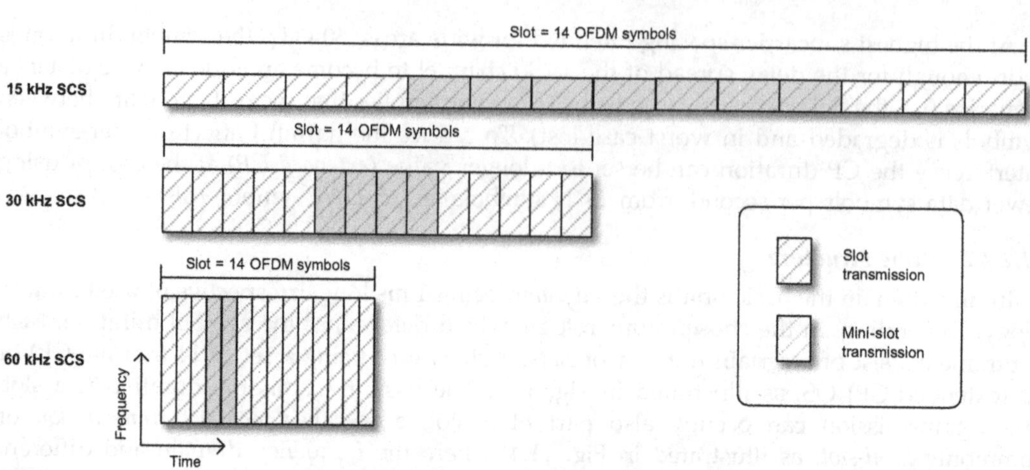

FIG. 11.4 Slots and example of mini-slots with different SCS.

attributed to NR. The beamforming can either be analogue (performed after *digital-to-analogue* (D/A) conversion) usually steering the beam for an entire carrier at a time, or digital (performed before D/A conversion) capable of simultaneously forming multiple beams at different frequencies. In the case of digital beamforming, a precoder matrix is used to form the beam from multiple antenna elements to and from the device. The precoder can be selected from a predefined set of matrices (*codebook-based*) by reporting the index of the best matrix (for uplink and downlink), or selected by the device based on downlink measurements (*non-codebook-based*), assuming channel reciprocity on a TDD carrier, for uplink beamforming. With analogue beamforming the task is instead to select a transmit and a receive beam out of a fixed set of time-multiplexed beams, with the selection based on strongest received reference signal. Analogue beamforming is well suited for the many antenna elements needed at higher frequencies, where beams can be narrow and have high gain. A cell can transmit and receive a large set of beams from multiple transmission points, and the device can be seamlessly switched between the beams without having to go through handover between cells.

11.2.3.2 Bandwidth parts

A device operating on an NR carrier can be configured to use only a part of the available bandwidth, a so-called *bandwidth part* (BWP). This is introduced to allow the operation of lower-complexity devices with less capable radios, and also to allow for further power saving in the devices by limiting the monitored bandwidth. The BWP configured for a device therefore constitutes the activated bandwidth, which could be distributed on different frequency bands and using different numerologies. In NR Release-15 up to 4 BWP can be configured, but only one BWP can be active at a time. This means for instance that the device cannot simultaneously operate in both FR1 and FR2.

11.2.3.3 Duplex modes

As mentioned above, there is one single frame format in NR supporting both TDD and FDD configurations. The device looks for downlink control messages in configured search

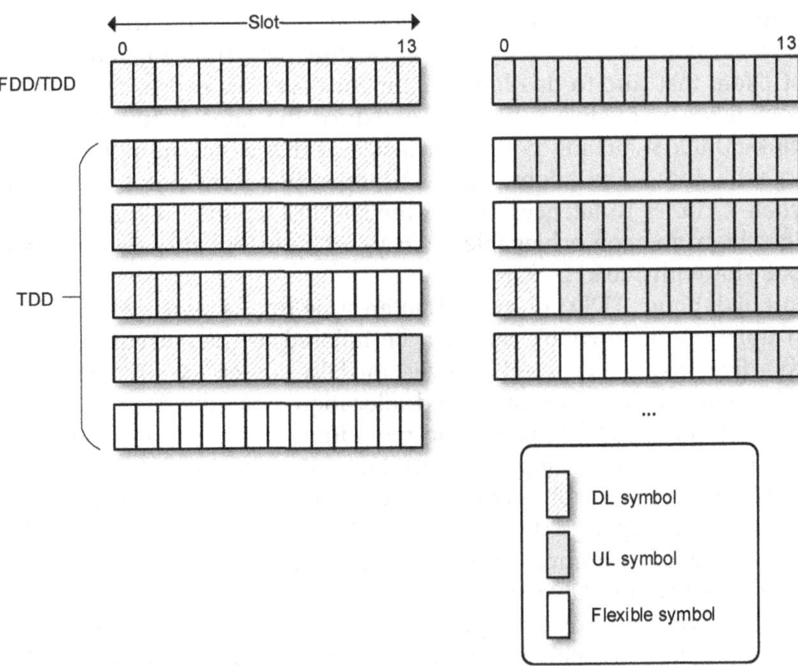

FIG. 11.5 Examples of NR slot formats for FDD and TDD use.

spaces (contained in CORESETs, as described below), and uses the slot for downlink reception or uplink transmission as dynamically indicated by the gNB. The slot can be partly used for downlink and partly for uplink, with a gap period around the switching point to avoid cross-link interference. A set of *slot formats* are defined with a sequence of downlink symbols, a set of flexible symbols, and a set of uplink symbols, see examples in Fig. 11.5. The gap period between downlink and uplink is thus taken from the flexible symbols. With TDD operation in a downlink slot the device expects that downlink and flexible symbols can be indicated for data reception, and in an uplink slot the flexible and uplink symbols can be used for transmission. The slot format can either be configured over RRC, dynamically indicated to each device in the DCI, or indicated to a set of devices in a special downlink control information (DCI) called *slot format indicator* (SFI). The sequence of slot formats indicated to the device the constitutes the TDD pattern of downlink and uplink slots. The fact that all symbols can be indicated as flexible allows for fully *dynamic TDD*, where the device is allocated downlink and uplink symbols on a per slot basis.

The mixed downlink-uplink slot formats also support so-called *self-contained* slots, where the Hybrid Automatic Repeat Request (HARQ) feedback is sent from the device toward the end of the downlink slot. This setup can enable very short roundtrip times with a high number of retransmissions possible in short time – a key feature for URLLC services. However, given a necessary gap period and allowing for processing on the device side, the number of symbols that can be used for data may be reduced, as further discussed below.

Comparing TDD with FDD, it is obvious that the alignment delay is necessarily longer when the direction is fixed, since we need to wait for the next downlink or uplink period and not only the next slot or mini-slot as in FDD. This is true even if the gNB can specify the configuration on a slot basis, since the uplink or downlink transmission opportunities are once per slot. Note that due to this long alignment delay we expect no significant latency reduction from using mini-slots in TDD. To reduce the latency further we would need more downlink-uplink switching points per slot, but this is not supported in NR Release-15. But even with these limitations, NR TDD operation constitutes a big improvement from LTE when it comes to latency. In LTE the shortest uplink-downlink switch period is 5 ms, placing a high floor on achievable latency, while in NR Release-15 it can be as low as 1/8 ms (one slot with 120 kHz SCS).

The support of dynamic TDD indication is interesting from a system efficiency perspective. Resources in uplink and downlink can then be tailored to the current need, which is not possible with a fixed slot allocation. However, for URLLC services dynamic TDD does not bring significant latency gains above a fixed allocation since the format is anyway set on slot basis with only one switching point, and more importantly the reliability may be risked since cells may independently change configurations resulting in so called *cross-link* (uplink-to-downlink and downlink-to-uplink) interference, which can have disastrous effects on link quality. It is therefore expected that dynamic TDD will in practice be most useful in controlled environments with limited number of cells and low interference.

When running URLLC in TDD bands, we can instead select a fixed configuration with as high frequency of switching points as possible, for instance by using a sequence of mixed downlink-uplink slot formats with a high SCS for short latency, as is illustrated in Fig. 11.6.

11.2.3.4 Short transmissions

The slot is the basic scheduling interval in NR, meaning that as a baseline the downlink control channel is sent once per downlink slot. This basic type of scheduling in blocks of slots

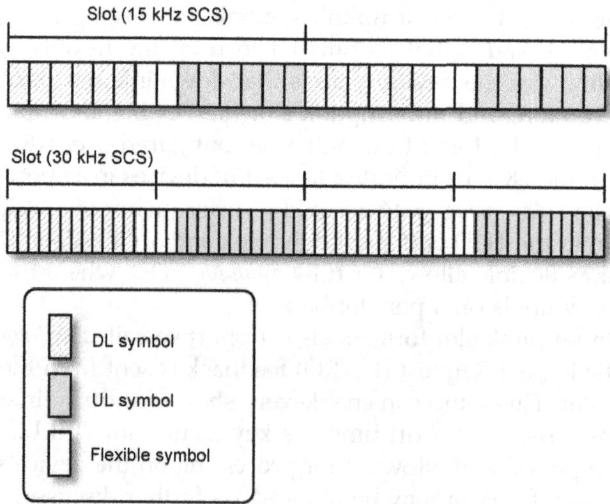

FIG. 11.6 Examples of possible low latency TDD configurations. Top: a lower SCS with mixed downlink-uplink slots. Bottom: a higher SCS with alternating downlink and uplink slots.

is called *Type A mapping*, which in downlink can start in symbol 0–3 and have a duration of 3–14 symbols, as shown in Fig. 11.7, and in uplink start in symbol 0 and have a duration of 4-14 OS. As mentioned above it is also possible to use shorter transmissions, which is what is referred to as non-slots or mini-slots, that can be scheduled with a shorter interval. These shorter transmissions can be scheduled in both downlink and uplink through *Type B mapping*, starting in any symbol and having a duration in the range 1–14 OS in uplink, but in downlink starting in symbol 0–12 and limited to {2, 4, 7} symbols in the downlink, as was exemplified in Fig. 11.4. Allocations of both types are not allowed to span over a slot border, as is seen in Fig. 11.7. This means that transmissions can be very flexibly scheduled, but not always with OS granularity. Of course, the flexibility comes at a cost; in downlink the devices need to monitor more downlink control occasions in order to see if they are scheduled.

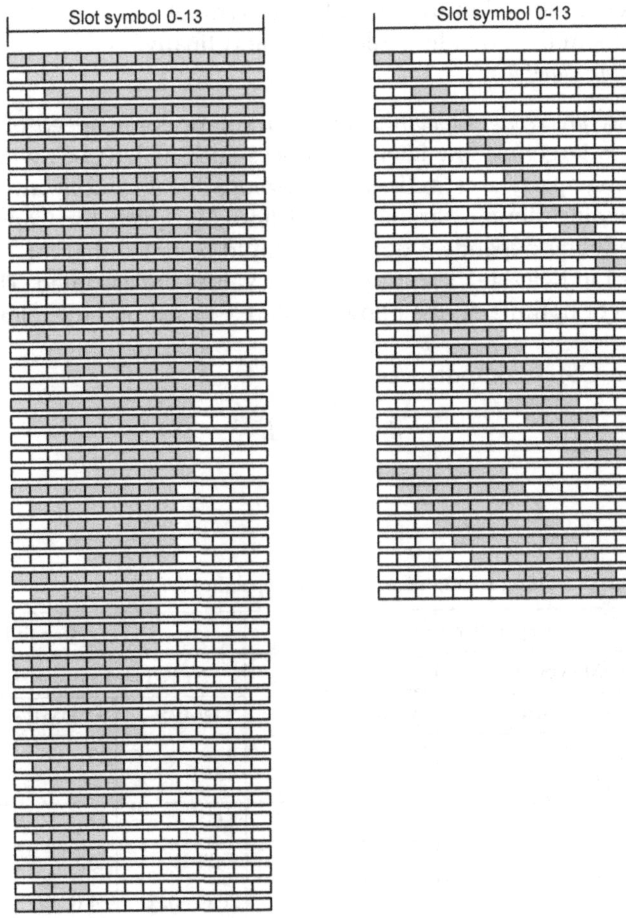

FIG. 11.7 Downlink allocations in time domain.

11.2.3.5 *Short processing time*

When it comes to timing, NR specifies the minimum scheduled timing that the device should manage for two cases:

- Processing time between the end of receiving a downlink data transmission and transmitting corresponding HARQ feedback, d_1
- Processing time between the end of receiving an uplink grant and transmitting the corresponding uplink data transmission, d_2

The length of these delays depends on a set of scheduling factors:

- How many demodulation reference signals (DMRS) symbols are used for Physical downlink shared channel (PDSCH) data transmission,
- If physical downlink control channel (PDCCH) and PDSCH symbols overlap in the downlink allocation,
- If DMRS and data are mixed in the first PUSCH symbol,
- The number of OSs in the downlink allocation, and finally,
- The *device processing capability*.

The factor of *device processing capability* reflects the processing speed of the device, out of which Capability 2 is targeting low-latency services such as URLLC, whereas Capability 1 is the basic processing speed. The defined expressions for d_1 and d_2 are set as the sum of the minimum processing timing parameters N_1 and N_2, given for Capability 1 and 2 in Table 11.2, and scheduling dependent delays. It should be noted that for 120 kHz SCS Capability 2 is not defined in Release-15, so Capability 1 values are assumed. The range of values reflects the intention of optimizing the allowed latency based on how computationally difficult the tasks are.

For d_1 the sum (in OS) is computed as:

$$d_1 = N_1 + d_{1,1}$$

TABLE 11.2 Minimum processing timing parameters for capability 1 and 2.

	Capability 1			Capability 2		
	N_1 [OFDM symbols]		N_2 [OFDM symbols]	N_1 [OFDM symbols]		N_2 [OFDM symbols]
SCS	1 DMRS	>1 DMRS		1 DMRS	>1 DMRS	
15 kHz	8	13 (14 if in OS #12)	10	3	13	5
30 kHz	10	13	12	4.5	13	5.5
60 kHz	17	20	23	9 (FR1)	20	11 (FR1)
120 kHz	20	24	36	20 (Cap. 1)	24 (Cap. 1)	36 (Cap. 1)

where

$$d_{1,1} = \begin{cases} 7 - i \text{ [Mapping type } A,\ i < 7] \\ 3 \text{ [Mapping type} B\text{, Capability 1, allocation 4 OFDM symbols]} \\ 3 + d \text{ [Mapping type } B\text{, Capability 1, allocation 2 OFDM symbols]} \\ d \text{ [Mapping type } B\text{, Capability 2, allocation 2 or 4 OFDM symbols]} \\ 0 \text{ [Otherwise].} \end{cases}$$

where i is the index of the last symbol in the PDSCH allocation, and d is the number of overlapping PDCCH and PDSCH symbols. As seen in the expressions the placement of downlink and uplink control impacts the processing delay. The intention here is to give the device some extra headroom for handling parallel processing tasks.

For d_2 we have the sum (in OS):

$$d_2 = N_2 + d_{2,1}$$

where

$$d_{2,1} = \begin{cases} 0 \text{ [Only DMRS in first PUSCH symbol]} \\ 1 \text{ [Otherwise].} \end{cases}$$

Here the device is allowed an extra symbol to perform the channel estimation from the DMRS before starting the decoding.

From the expressions above we can calculate the processing delays for some interesting URLLC configurations, with the assumption of Capability 2 processing in the device, see Table 11.3. Here it is assumed that 1 DMRS is configured in the PDSCH, that PDCCH and PDSCH overlap in 1 symbol for 4 and 2 symbol transmissions in downlink, and that DMRS is mixed with data on the first PUSCH symbol. It should be noticed here that processing delays do not decrease with shorter allocations, perhaps contrary to expected.

With this specified minimum timing in mind, the network explicitly indicates the uplink timing to the device in the DCI (see Section 11.3.3.1.2) for either the downlink HARQ feedback or the uplink data.

TABLE 11.3 Example of processing delays with capability 2 for 15 kHz and 120 kHz SCS.

	d_1 [OFDM symbols]		d_2 [OFDM symbols]	
Allocation	**15 kHz SCS**	**120 kHz SCS (Cap. 1 values)**	**15 kHz SCS**	**120 kHz SCS (Cap. 1 values)**
Slot [14 symbols]	3	20	6	37
7 symbols	3	20	6	37
4 symbols	4	21	6	37
2 symbols	4	21	6	37

11.2.3.6 Downlink multi-antenna techniques

There are many possibilities to improve the signal quality in the downlink. Perhaps the most obvious way is to equip the device with multiple cross-polarized antennas that receive the same transmission. Coherent processing of the signal gives receiver-side directivity, leading to suppression of noise and interference, and thereby improved quality. It is reasonable to expect that more high-end NR URLLC devices targeting cMTC services have at least 4 receive antennas, i.e. 2 pairs of cross-polarized elements.

Also without equipping the device with more antennas it is possible to utilize powerful beamforming on the gNB side, using the larger arrays. With increasing frequency, the antenna elements become smaller and it is feasible to have reasonably sized arrays with many elements. Since the size of the antenna elements are proportional to the wavelength (typically in the range of $\lambda/2$), a doubling of carrier frequency leads to half the array size, which would allow for twice as many elements if the same array size is kept. At the same time, the maximum directional gain of the antenna can be expected to scale with the number of elements. One can therefore typically enable much higher directional gains at the higher carrier frequencies used in NR.

Since the device-specific *DMRS* in NR are inserted in the downlink transmission, the beamforming can be made transparent to the device and doesn't need to be indicated. The DMRS are beamformed in the same way as data, and when the device demodulates based on the reference signals the beamforming is automatically handled.

For the antenna array the gNB can select the beam in two basic ways; either based on a precoder matrix (known as *digital beamforming*) or based on transmission weights (known as *analogue beamforming*). In the analogue setup the beam is a main lobe formed after *digital-analog* (D/A) conversion, and is typically chosen by sweeping a set of beams in a grid and letting the device report the received signal strength. This is rather robust and low-overhead option, but has the drawback of being restricted to one beam at a time.

The analogue beamforming with fixed beams is well suited for higher carrier frequencies when the focus is to achieve a high beamforming gain. This is because of the higher number of elements required and the thereby increasing processing complexity. A cell can have multiple associated fixed beams associated to it, each with its own set of synchronization and system information transmitted to the device. The device associates to the best beam during random access (*initial beam establishment*) and can then be moved between beams within the same cell on the physical layer (*beam adjustment*) based on measurement reports from signals in the beams. This change does not involve the handover procedure as in a cell change.

Using digital beamforming with a precoder matrix, the device is configured to receive pilot symbols, and try out a set of preconfigured precoders from a codebook. The index of the codebook giving the highest quality is indicated in a *precoding matrix indicatorprecoding* (PMI) reported by the device. Using precoders makes it possible for the gNB to use multiple beams to different devices at the same time, and it also enables MIMO transmissions with multiple data layers. With URLLC, however, we do not aim to achieve high spectral efficiency but rather robustness. Still, precoder-based beamforming can be a good choice for URLLC as it can be faster and more precise than grid-of-beams beamforming. The drawback is the extensive need for pilots and measurement reports, as well as the risk of using wrong precoder when conditions change.

11.2.3.7 *Uplink multi-antenna techniques*

For uplink transmissions a device with multiple antenna elements can use beamformed MIMO methods to improve the channel quality. Both non-codebook based (only in the case of downlink-uplink reciprocity over TDD, giving the full precoder matrix) and codebook-based (using an index corresponding to a precoder matrix) precoding with up to 4 antenna ports is supported. For codebook-based precoding the device needs to have the capability of performing fully coherent transmissions and thus be able to control the relative phase of the antennas. The gNB estimates the channel from *sounding reference signal* (SRS) (see Section 11.2.5.1.2) transmissions from the device, and then indicates a precoder in the uplink grant. The indication of precoder and the additional SRS transmissions means an increased downlink and uplink overhead. On the other hand, the use of beamforming should lead to improved SINR and thereby lead to higher reliability, but only as long as the right precoder is selected. A high rate of SRS transmissions may be needed to ensure the use of the right beam.

In addition to beamforming, the network can use *coordinated multi-point* reception to improve uplink quality. This implies that the signal is received in multiple locations (e.g. cells belonging to the same three-sector site) and processed jointly. To enable this, no additional signaling is required.

11.2.4 Downlink physical channels and signals

In downlink the most important aspects for URLLC is to ensure high reliability of the control channel PDCCH and the data channel PDSCH. Since the gNB can immediately react to an incoming data packet addressed to the device, scheduling in downlink is faster than in uplink, and the main concern is to provide scheduling opportunities with short intervals.

11.2.4.1 *Synchronization and broadcast signals*

Slot	Periodic 5 ms–160 ms
Subcarrier spacing	{15, 30, 120, 240} kHz
Bandwidth	240 subcarriers
Frequency location	Predefined raster locations

To enable low energy consumption through long sleep cycles, NR is defined with one unified broadcast message, the *synchronization signal block* (SSB). This transmission contains both system information and synchronization signals, both used during initial access. By using a long period and differently beamformed SSB, both good energy performance and good coverage can be achieved.

SSB consists of three channels: the *primary synchronization channel* (PSS), the *secondary synchronization channel* (SSS), and the *physical broadcast channel* (PBCH). PSS and SSS are 127 subcarriers wide and PBCH is 240 subcarriers wide, surrounding SSS in frequency and time domain. By placing the three channels in one limited time-frequency region, as shown in Fig. 11.8, and using a longer period (up to 160 ms) it is possible for the network to reduce

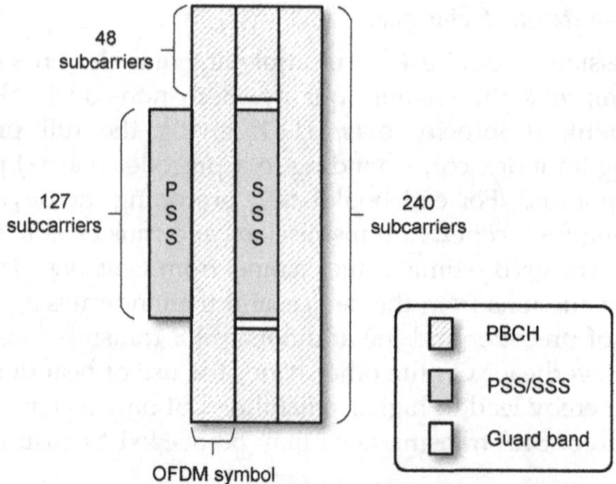

FIG. 11.8 SSB consisting of PSS, SSS, and PBCH.

its power consumption through micro-sleep. Also, the device power consumption can be reduced by using a limited set of frequency domain locations of the SSB.

From PSS and SSS the device derives the basic timing of the cell as well as the *physical cell identity*. Over PBCH the device receives the *master information block* (MIB) containing information for how to read the remaining system information, transmitted in *system information block one* (SIB1) on the downlink data channel.

11.2.4.2 Reference signals

11.2.4.2.1 DMRS

Slot	Any
Subcarrier spacing	{15, 30, 60, 120, 240} kHz
Time location	{1, 2, 3, 4} symbols per slot
Frequency location	On {4, 6} subcarriers per resource block

For both data and control in the downlink, device-specific *DMRS* are used. The DMRS is, just as in LTE, a predefined sequence known by the device from which it can estimate the channel state. Demodulation of up to 12 different antenna ports can be supported depending on configuration as given in Table 11.4, by applying different *orthogonal cover codes* in time and frequency over the DMRS resources. The DMRS are inserted within the data or control regions and are only transmitted when the device is receiving information.

TABLE 11.4 Number of supported antenna ports for different DMRS configurations.

#Ports	Single	Double
Type 1	4	8
Type 2	6	12

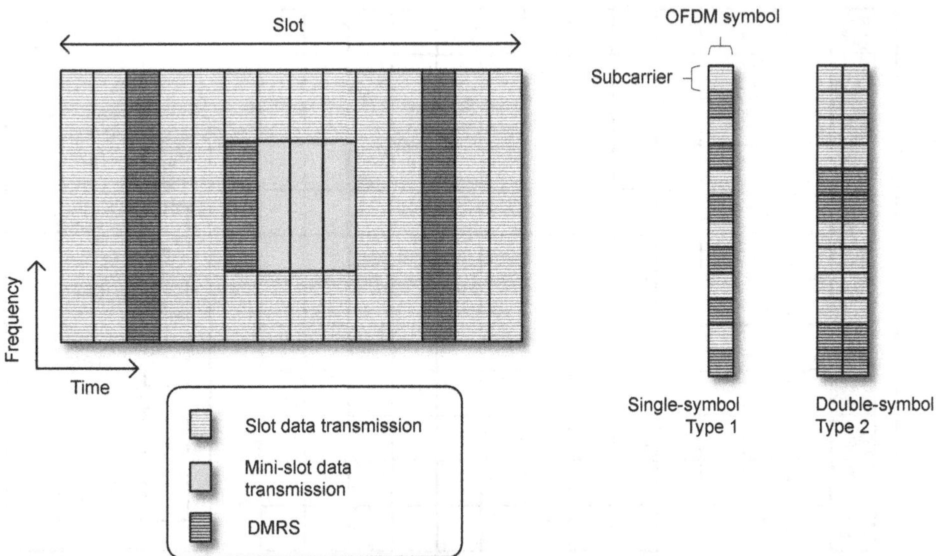

FIG. 11.9 Left: DMRS placement for PDSCH slots (Mapping A, Type 1, 2 single DMRS) and mini-slots (Mapping B, Type 1, 1 single DMRS). Right: examples of Type 1 and Type 2 DMRS.

For PDCCH (Section 11.2.4.3) the DMRS is inserted on every fourth subcarrier in the region where PDCCH is sent (see Fig. 11.8). PBCH (Section 11.2.4.1) also incorporates its separate DMRS for demodulation.

Two different time-domain mappings and two different frequency-domain types of DMRS are defined for PDSCH (Section 11.2.4.4). The used mapping and type are indicated in the downlink control DCI message.

Mapping A is intended for slot length transmission. In this case the first DMRS is inserted after the downlink control in the third or fourth symbol of the slot, independent of where the data is located.

Mapping B is intended for mini-slot transmission. Here the first DMRS is inserted in the first symbol of the mini-slot allocation, a configuration known as *front-loaded* DMRS. In case of overlap with a CORESET, DMRS is moved to the first symbol after CORESET.

On top of this basic mapping, extra DMRS can also be inserted for support of e.g. higher speeds, and double DMRS symbols can be used to support more antenna ports. An illustration of the mapping is shown in Fig. 11.9. Up to 4 single or 2 double DMRS symbols can be configured with both Mapping A and B.

On top of the mapping, there are two DMRS types: Type 1 DMRS is placed on every other subcarrier in the symbol, while Type 2 DMRS is placed on 4 out of 12 subcarriers in a RB over a symbol, also shown in Fig. 11.9. Both types can be sent as single or double DMRS symbols.

11.2.4.2.2 PT-RS

Slot	Any
Subcarrier spacing	{15, 30, 60, 120} kHz
Time location	Every {1–14} symbol per slot
Frequency location	6 subcarriers per resource block, every {2, 4} resource block

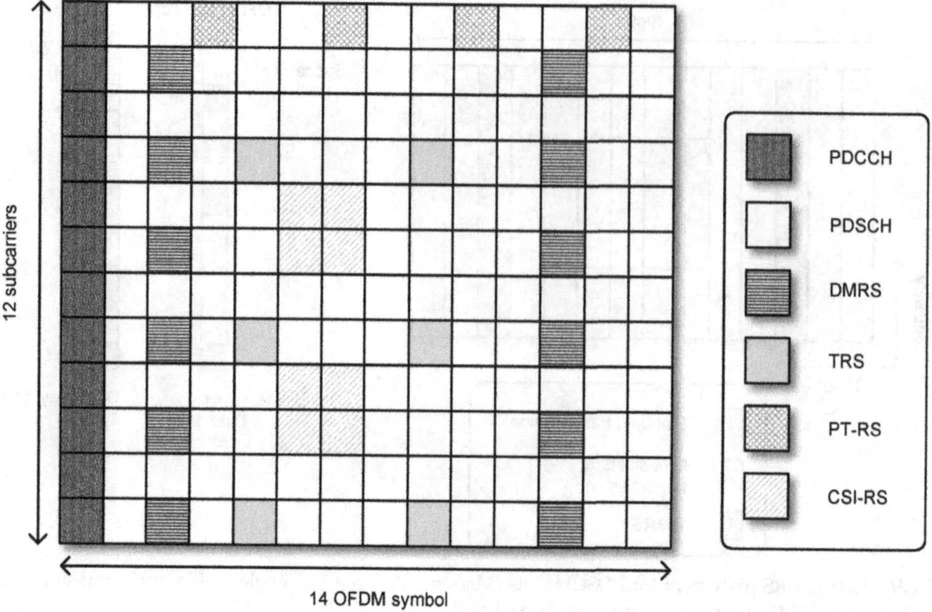

FIG. 11.10 Example configuration of physical channels in 1 RB during one downlink slot, with CSI-RS for 8 antenna ports.

As a complement to the DMRS an additional *phase-tracking reference signal* (PT-RS) can be configured in uplink and downlink. This consists of a DMRS transmitted on a sparser frequency grid of one subcarrier per second or fourth resource block, but on a shorter periodicity of every *n*th OS for accurate measurement of the phase. The main reason for configuring PT-RS is to assist the decoding in the higher frequencies of FR2 by better tracking of oscillator phase noise. An example configuration of PT-RS is shown in Fig. 11.10.

11.2.4.2.3 CSI-RS

Slot	Every 4th to 640th slot
Subcarrier spacing	{15, 30, 60, 120} kHz
Time location	Any symbol
Frequency location	Any subcarrier, every {1, 2} resource block

For sounding of the downlink channel quality, *channel state information (CSI) reference signal* (CSI-RS) is used. This consists of pilot sounding symbols transmitted in the downlink carrier on device-specific configured resources. The CSI-RS pattern is not overlapping with DMRS, SSB or PDCCH, and can be configured for sounding of the quality for up to 32 different antenna ports. The periodicity of *periodic* (configured with RRC, see Section 11.3.1) or *semi-persistent* (RRC and Medium Access Control (MAC) CE configured) *CSI resource sets* can be set in a range from 4 to 640 slots. In addition to the periodic sounding, CSI-RS can be aperiodically sent and indicated with DCI (see Section 11.3.3.1.2). The device measures on CSI-RS and reports in associated configured CSI reporting resources in the uplink, as described in Section 11.3.3.6. A device can be configured with both *non-zero-power* CSI-RS resources, with pseudo-random sequences transmitted by the gNB for sounding of channel quality, and *zero-power* CSI-RS resources without transmission that are used for interference measurements. An example of configured CSI-RS resources for 8 antenna ports is shown in Fig. 11.10.

11.2.4.2.4 TRS

Slot	Every {10, 20, 40, 80} ms
Subcarrier spacing	{15, 30, 60, 120} kHz
Time location	2 symbols in 2 consecutive slots
Frequency location	3 subcarriers per resource block, every {1, 2} resource block

An additional *tracking reference symbol* (TRS) can be configured in downlink in order to follow variations in the oscillator. TRS is essentially a configuration of two consecutive slots with 4 CSI-RS resources, separated by 4 subcarriers in frequency and 4 symbols in time. This pattern is repeated every 10, 20, 40, or 80 ms. The TRS is used by the device to improve the precision in time and frequency tracking. An example of a TRS configuration is shown in Fig. 11.10.

11.2.4.3 PDCCH

Slot	Any
Subcarrier spacing	{15, 30, 60, 120} kHz
Time location	1-3 symbol duration, any symbol
Bandwidth	2-96 resource blocks

As in LTE, the downlink control channel in NR is called *PDCCH,* and carries instructions to the device on how to read and transmit data. The PDCCH in NR can be flexibly configured in time and frequency, using a periodicity of once per slot or more, and can span 1–3 OSs. A device is configured with one or several search spaces, defined within *control resource sets* (CORESETs), where it blindly detects the presence of *DCI,* described in Section 11.3.3.1.2, which are coded on control channel elements (CCEs) described below. So PDCCH consists of DCI transmitted in CCEs located in CORESETs. The channel coding used for PDCCH is based on Polar codes. The DCI is encoded with a *radio network temporary identifier* (RNTI), which is the address of the device in the cell, by scrambling of an appended *cyclic redundancy check* (CRC).

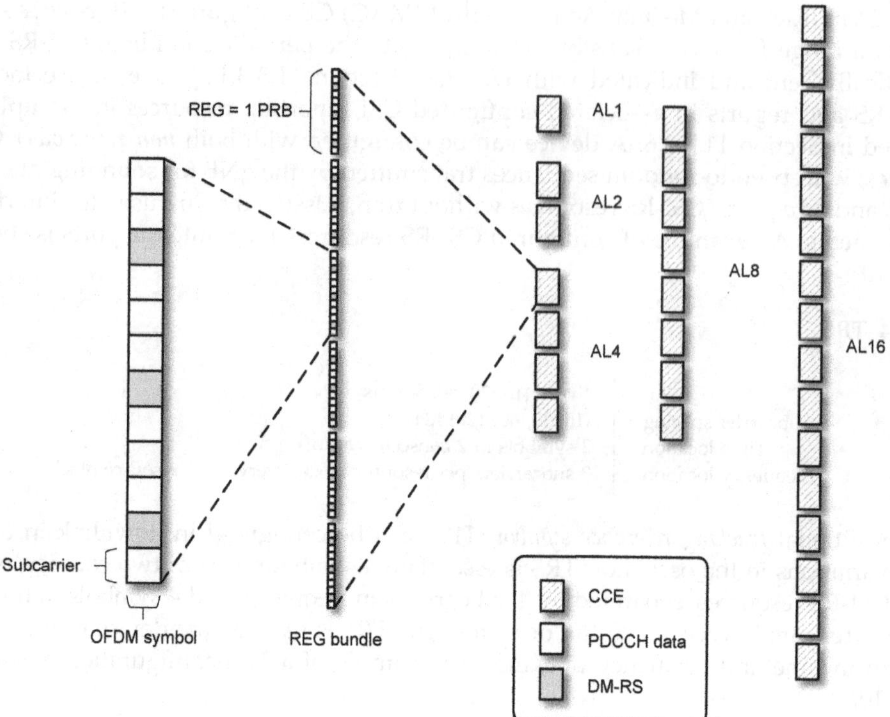

FIG. 11.11 PDCCH with DMRS placement, aggregation levels and REGs.

11.2.4.3.1 CCE

The resources used to transmit the coded DCI symbols are defined as *CCEs*, consisting of bundles of 6 *RE groups* (REGs), each being one RB in one OS. With DMRS inserted on every fourth subcarrier and QPSK modulation, a CCE can carry 108 bits of coded information. In NR one PDCCH message (a DCI) can be sent using 1, 2, 4, 8, or 16 CCEs (in LTE only up to 8), forming different *aggregation levels* (ALs), see Fig. 11.11, making it possible to use a very low code rate. Higher aggregation levels make PDCCH very robust, which of course is necessary for URLLC, since without decoding PDCCH the device will not be able to attempt decoding of downlink data or know the dynamic uplink resource.

PDCCH can be sent either in a *non-interleaved* manner, meaning on a contiguous allocation, or *interleaved* manner, where the REGs are distributed for increased diversity. In the non-interleaved case, the REGs are bundled in CCE sizes of 6 as described above and mapped without gaps in frequency domain, while in the interleaved case the CCEs are spread out in frequency using REG bundles of 2 or 6 REGs for 1–2 symbol PDCCH, or 3 or 6 REGs for 3 symbol PDCCH.

11.2.4.3.2 CORESET

A device can be configured with multiple *CORESETs*, which constitute configured resources on the time-frequency grid where the device is to perform blind decoding of

CCEs to find DCIs addressed to it. The location of the basic set, CORESET-0, is signaled in the master information block on PBCH, as described in Section 11.2.4.1, and additional CORESETs are later configured over Radio Resource Control (RRC). These resources are then the possible scheduling occasions for a device, and to achieve low alignment latency in NR URLLC the CORESETs should be configured with short interval. A CORESET can be 1–3 OS long (3 only when DMRS Type A is used in the fourth symbol of the slot, see Section 11.2.4.2.1), and be placed anywhere in the slot with multiple device-specific locations. CORESETs can overlap with each other but not with SSB.

Short intervals between CORESETs enable low latency, but having many CORESETs configured comes with some obvious costs. First, the device has more locations to monitor which increases power and processing needs. Second, more CORESETs likely means more blind decoding attempts performed by a device, with increased risk of falsely detecting downlink control from noise and thereby missing a real transmission. Third, we either need to reserve some resources for potential downlink control (which is expensive from a resource efficiency perspective), or we may be forced to interrupt a data transmission to send downlink control (which disturbs the data transmission). The latter alternative is supported by the feature of downlink pre-emption discussed in Section 11.2.4.4.3. As for sTTI in LTE, the unused CORESETs configured for a device can be reused for PDSCH transmission by indicating this in the DCI.

For 15/30/60/120 kHz SCS, the maximum number of blind decodes a device can be configured with per slot is 44/36/22/20, and the maximum number of CCEs is 56/56/48/32, respectively, with no restriction related to aggregation level.

A CORESET can be said to represent a possible start of a slot or mini-slot. By allowing CORESETs to be configured in any symbol of the NR slot we enable mini-slots to start anywhere, and the location is a matter of scheduling. Given that the available downlink transmission lengths are 2, 4, 7, or 14 symbols in NR Release-15, the gNB can then configure CORESETs with a matching periodicity to enable many scheduling opportunities, as illustrated in Fig. 11.12 for 4 symbol transmissions.

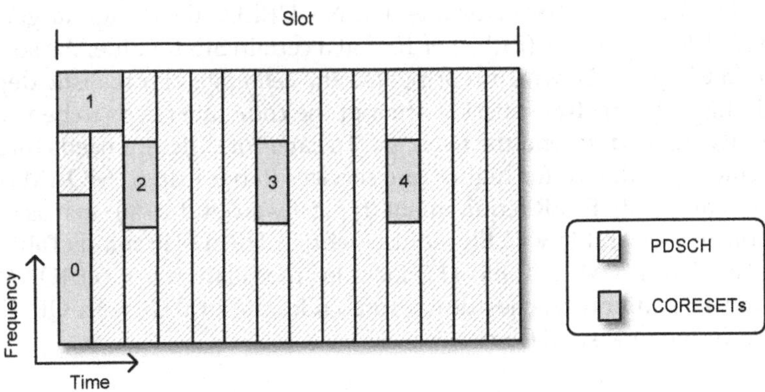

FIG. 11.12 Illustration of CORESET placements for slot and 4-symbol mini-slot transmissions.

11.2.4.4 PDSCH

Slot	Any
Subcarrier spacing	{15, 30, 60, 120} kHz
Duration	1-14 symbols
Bandwidth	Any (see Table 11.1)

PDSCH data is coded with defined *low density parity codes* (LDPC) according to code rates given in three *modulation and coding scheme* (MCS) tables. When a device receives data, it first reads a DCI message addressed to it on PDCCH. If the DCI is correctly decoded the device knows which MCS table to use, which table entry to use, and on which resources the data is found, and it can then start decoding. The data location is indicated as a start symbol and a duration of the message. This very flexible setup allows for locating the message practically anywhere, except that a transmission cannot spill over into the next slot.

Smaller downlink transport blocks (<3824 bits) are protected with a CRC of 16 bits and larger blocks with a CRC of 24 bits, appended before scrambling with the RNTI of the device.

By replacing the Turbo code used for data in LTE with LDPC improved performance for higher payloads is achieved. The codes used in NR are constructed with the help of two optimally constructed matrices, called *base-graphs,* one for smaller packets (up to 3840 bits) and one for larger packets (up to 8448 bits). Larger transport block (TB)s are segmented into *code blocks*(CBs), each with a 24-bit CRC added, before being encoded with LDPC and thereafter concatenated to PDSCH. The size of a set of CBs, a *CB Group* (CBG), can be defined over RRC, and retransmission can be handled separately per CBG for improved efficiency.

Similar to LTE, data is rate-matched using a circular buffer and sent with *incremental redundancy* for increased efficiency. A *redundancy version* (RV) in the sequence {0,2,3,1} is indicated for each data transmission, and the device can then place them in a receive buffer and efficiently combine the different sets of coded bits from multiple copies of a received transport block (from retransmission).

11.2.4.4.1 MCS table for low code rate

In LTE the lowest code rate in MCS-0 is around 0.094, meaning that the code word is about 10 times longer than the information bits. The extra redundancy of using a longer code gives robustness and leads to a reduced error rate. For NR URLLC the design target in Release-15 was to reach a *block level error rate* (BLER) of 10^{-5} at a certain SINR value, the so-called *Q-value* (see discussion in Chapter 12), which represented the cell-edge of a realistic deployment scenario. To reach this with one transmission attempt the code rate needs to be very low, meaning that many REs are used to send the message. To cater for different needs three MCS tables have been specified in NR, two for higher rate services (table 1 up to 64QAM and table 2 up to 256QAM) with moderate BLER requirement (typically around 10%) such as eMBB in mind, and one with lower rates and low BLER requirement for URLLC in mind (Table 11.3), shown in Table 11.5. This URLLC MCS Table 11.3 includes modulation up to 64QAM using a few higher rates, and in addition reaches down to a code rate of 0.03 with QPSK modulation, which is a third of the lowest LTE rate.

TABLE 11.5 Low rate MCS table (corresponding to table 5.1.3.1–3 in Ref. [6]).

MCS index	Modulation	Target code rate R x [1024]	Spectral efficiency
0	QPSK	30	0.0586
1	QPSK	40	0.0781
2	QPSK	50	0.0977
3	QPSK	64	0.1250
4	QPSK	78	0.1523
5	QPSK	99	0.1934
6	QPSK	120	0.2344
7	QPSK	157	0.3066
8	QPSK	193	0.3770
9	QPSK	251	0.4902
10	QPSK	308	0.6016
11	QPSK	379	0.7402
12	QPSK	449	0.8770
13	QPSK	526	1.0273
14	QPSK	602	1.1758
15	16QAM	340	1.3281
16	16QAM	378	1.4766
17	16QAM	434	1.6953
18	16QAM	490	1.9141
19	16QAM	553	2.1602
20	16QAM	616	2.4063
21	64QAM	438	2.5664
22	64QAM	466	2.7305
23	64QAM	517	3.0293
24	64QAM	567	3.3223
25	64QAM	616	3.6094
26	64QAM	666	3.9023
27	64QAM	719	4.2129
28	64QAM	772	4.5234
29–31	Reserved		

To tell the device which table to use, different RNTI encoding of the DCI is used. Devices are normally configured over RRC to expect DCI addressed to a default high-rate table RNTI, and for URLLC devices an additional configuration of a low rate table RNTI is made. The network can then dynamically switch between the two tables depending on the requested service, and the device will attempt decoding with both RNTIs out of which one is correct.

11.2.4.4.2 Downlink repetitions

An additional possibility beyond using the low-rate MCS table is to let the TB be repeated, using so-called *slot aggregation*. In terms of performance this is roughly equivalent to applying an extra repetition code on top of the basic LDPC, and thereby has the effect of lowering the effective code rate and adding robustness beyond what the MCS delivers. To make the device read a repeated message, the gNB signals a *K-factor* in the DCI. If $K > 1$ the allocation is repeated and sent in total K times. The *RV* of the nth transmission in the K-repetition sequence will be set mod(n,4) indices from the RV indicated in the DCI, following the basic RV sequence {0,2,3,1}. After each repetition in the sequence the device can attempt decoding, and therefore the packet latency increases only in the case of failed decoding attempts. In NR Release-15, repetitions can however only be done on slot level, meaning that the allocation will be repeated in K subsequent slots, see illustration in Fig. 11.13. To achieve the lowest latency with repetition we would need to repeat mini-slots *within* a slot, which would give low alignment delay, but this is not supported until Release-16 is specified.

Compared to HARQ retransmissions these repetitions are not triggered by feedback and are therefore sometimes called *blind, automatic*, or *HARQ-less* repetitions. The fact that they are always transmitted regardless of the decoding success makes repetition much less spectrally efficient compared to HARQ retransmissions. Moreover, the fact that the DCI is only transmitted once is a weak spot: if the device fails to decode the downlink control message the multiple copies of data don't matter. However, by using high enough AL for downlink control this problem can be avoided.

11.2.4.4.3 Downlink pre-emption

Unless we have enough critical traffic it is wasteful resource-wise to dedicate a certain frequency band for URLLC transmissions. This is true also for downlink data, where we would need to avoid scheduling longer transmissions in a part of the carrier to give room for shorter transmissions with short notice. And since the downlink control can be sent just before the downlink data, there is no latency gain from using configured grants (see Section 11.3.3.4), in downlink called *semi-persistent scheduling*. A new mechanism, *downlink preemptionindication*, was therefore introduced in NR to remove the issues connected to direct insertion of short data transmissions within ongoing transmissions. This solves both problems: direct access to downlink resources and using the whole carrier for longer transmissions.

If the whole downlink carrier is already used for slot-based transmissions and the gNB receives a high priority packet in the buffer, it can simply pre-empt the high priority data in the middle of the slot by using the next upcoming configured CORESET. Naturally, this means taking resources from the first transmission which will then likely fail. Moreover, it would not only fail, but when a retransmission is later sent to fix the first packet, the receive buffer would be contaminated with other data and the decoding would likely continue to fail. To fix this, a control message, a *pre-emption indication* (PI) DCI, is sent to the device that received the interrupted message, see illustration in Fig. 11.14. The PI informs the device

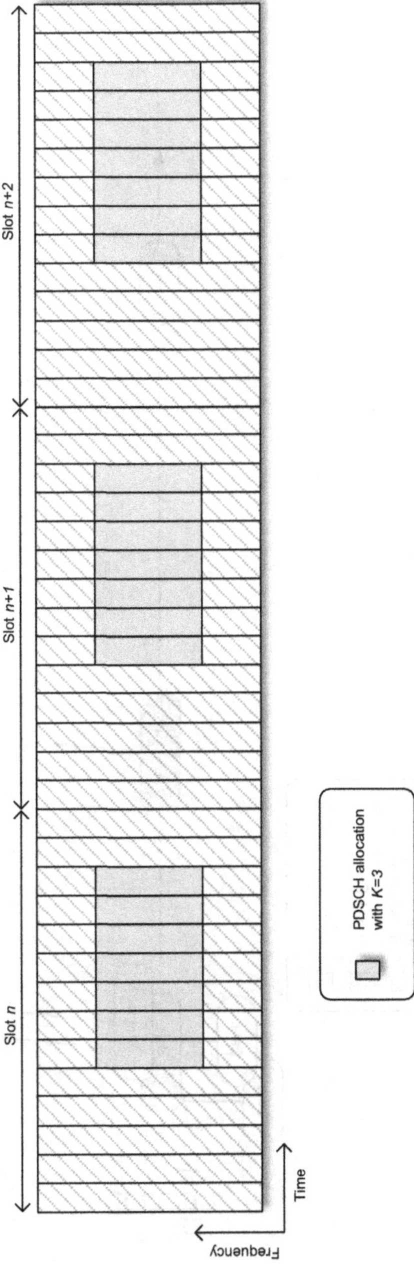

FIG. 11.13 Illustration of downlink repetition ($K = 3$) of a 7-symbol data allocation.

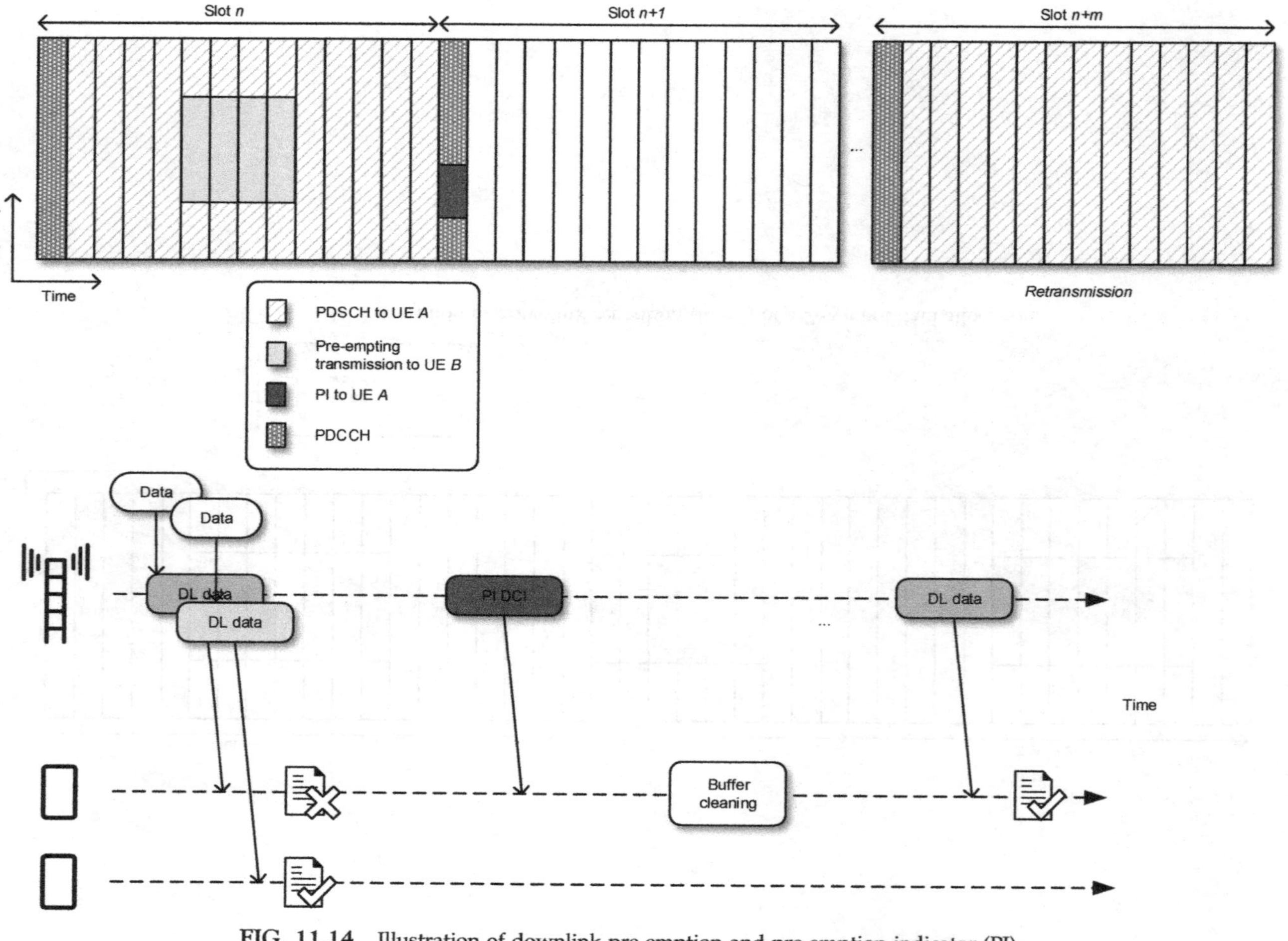

FIG. 11.14 Illustration of downlink pre-emption and pre-emption indicator (PI).

that the last attempt was pre-empted in an indicated resource region, and the device can then clean the receive buffer from the corrupted bits, and wait for the next retransmission attempt, which may be complete or only cover the impacted CBGs.

Downlink pre-emption thus offers quick downlink resources with affordable overhead. As long as the URLLC traffic is low compared to other traffic, the pre-emptied packets will not cause a big disturbance on the other traffic.

11.2.5 Uplink physical channels and signals

In the uplink, we face two main challenges with URLLC compared to downlink. First, the scheduling is controlled by the gNB while the transmit buffer status is known by the device, meaning that resources either needs to be requested or pre-allocated with cost on either latency or resource efficiency. Second, devices have a limited range of transmission power, fewer antenna elements and smaller antenna spacing, making the uplink more sensitive to poor radio conditions.

11.2.5.1 Reference signals

11.2.5.1.1 DMRS

Slot	Any
Subcarrier spacing	{15, 30, 60, 120} kHz
Time location	{1, 2, 3, 4} symbols per slot
Frequency location	{4, 6} subcarriers per resource block

Since the uplink is also based on OFDM, DMRS is used in the same way as for downlink (see Section 11.2.4.2.1), configurable with the same options of Type 1 and 2 placement in frequency domain and Mapping type A and B in time domain. An example configuration of DMRS resources is shown in Fig. 11.15, and an uplink RB pattern example is given in Fig. 11.16. But in uplink it is also possible to run with DFT-precoded OFDM for improved coverage. In this case the DMRS spans an entire symbol across the uplink frequency allocation, as in the LTE case.

11.2.5.1.2 SRS

Slot	Any
Subcarrier spacing	{15, 30, 60, 120} kHz
Time location	{1, 2, 4} symbols within last 6 symbols of slot
Frequency location	Every {2, 4} subcarrier per resource block

The sounding of channel quality in uplink is based on the *SRS*, which are extended Zadoff-Chu sequences that can be sent in the last 6 OS of an uplink slot. Just like CSI-RS, the SRS can be periodically (RRC-configured) or aperiodically (DCI-indicated) triggered. An SRS resource can be 1, 2, or 4 symbols long and be sent on every second (2-comb) or every fourth (4-comb)

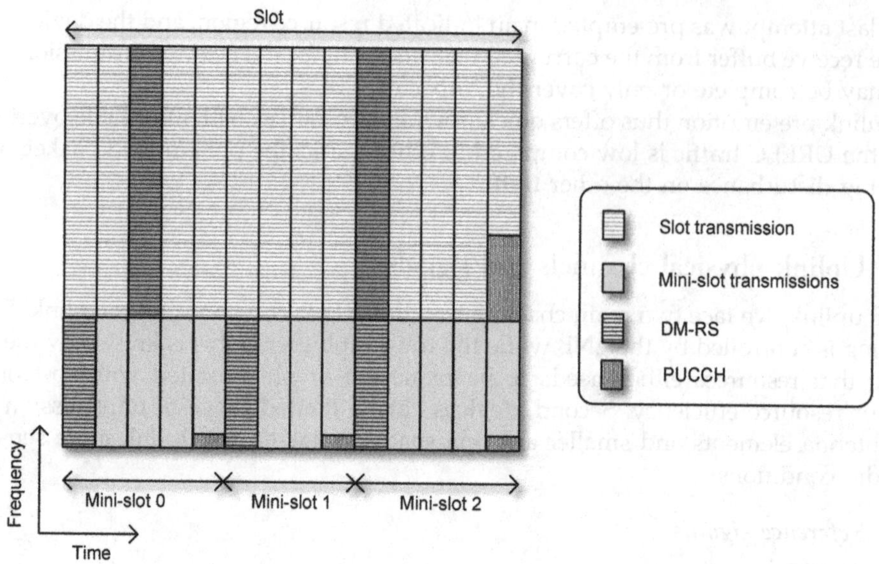

FIG. 11.15 Example uplink with slot transmission (DMRS Mapping A, Type 1, 2 DMRS, PUCCH Format 2) and mini-slot transmissions (DMRS Mapping A, Type 1, 1 DMRS, PUCCH Format 0).

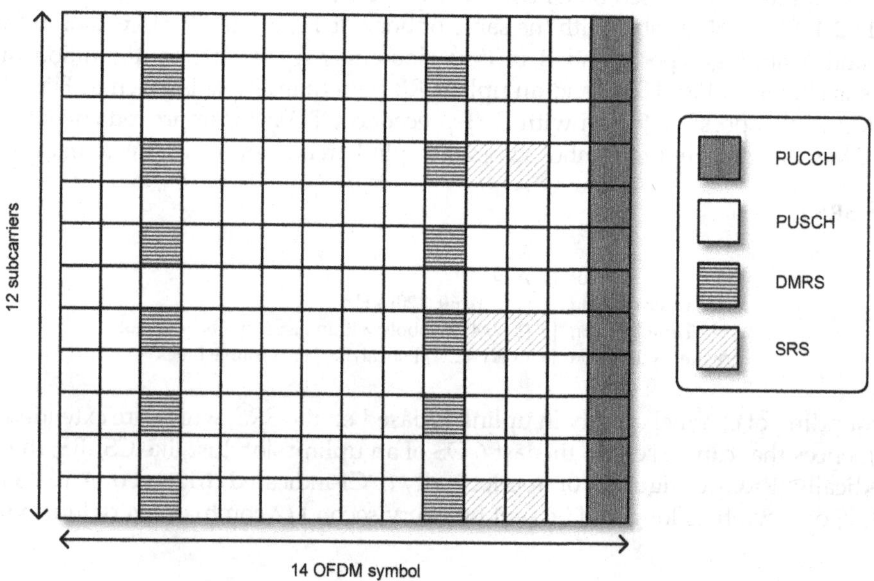

FIG. 11.16 Example pattern of uplink configuration of physical channels in 1 RB during one slot, with 2-symbol 4-comb SRS, short PUCCH, and 2 DMRS.

subcarrier. The combs allow for multiplexing of SRS from up to 4 devices on the same OS. By applying phase shifts of the SRS sequences up to 4 antenna ports can be sounded on one SRS resource. These ports can be configured to correspond to the different uplink beams used for PUSCH. An example configuration with 2 SRS symbols is shown in Fig. 11.16.

11.2.5.2 PRACH

Slot	Configurable slot set with 10–160 ms period
Subcarrier spacing	{1.25, 5, 15, 30, 60, 120} kHz
Duration	{1, 2, 4, 6, 12} symbols plus CP
Bandwidth	{139, 839} subcarriers, in blocks of {1, 2, 3, 6, 7}

When the device has acquired the system information it can register to a gNB by initiating random access. The first step in random access is to select and transmit a preamble over the *physical random access channel* (PRACH), followed by a *random access response* from the network containing a first *RNTI* and an uplink grant. PRACH is sent on RACH resources in the uplink, which are derived from the system information, and consists of a Zadoff-Chu sequence-based preamble. There are two main types of preambles defined: four long preamble formats (sequence length 839) with a SCS of 1.25 or 5 kHz and only in FR1, and nine short preamble formats (sequence length 139) for 15 or 30 kHz SCS in FR1 and 60 or 120 kHz SCS in FR2 (same as for data). By the use of different preamble formats different CP and repetition can be applied for improved coverage, so that the total duration can be set between 2-12 OS (short formats) and 1-4.3 ms (long formats). Depending on the combination of PUSCH and PRACH SCS the formats will occupy between 2 and 24 RBs on PUSCH, and for short formats always 12 RB. In frequency domain a RACH resource can consist of up to 7 blocks of consecutive copies of the sequence mapped to integer RBs.

11.2.5.3 PUCCH

Slot	Any
Subcarrier spacing	{15, 30, 60, 120} kHz
Duration	{1, 2, 4–14} symbols
Bandwidth	From 1 resource block

Since the uplink data transmission duration can be flexibly set in the wide range 1–14 symbols, we must be capable of adjusting the length of the *physical uplink control channel* (PUCCH) as well. To handle this, NR allows several formats and durations of PUCCH, which is used to send *uplink control information* (UCI) when it's not transmitted together with data on PUSCH. Five different PUCCH formats (0–4) are defined, see Table 11.6. Up to 4 PUCCH resource sets, mapping to different payload ranges, can be configured for a device, with each set consisting of at least 4 PUCCH configurations, each specifying a format and a resource. The configuration to be used within the payload-determined set is then indicated in the DCI, as described in Section 11.3.3.1.2. Naturally, for URLLC the shorter lengths are

TABLE 11.6 NR PUCCH formats.

PUCCH format		Length (OFDM symbols)	Payload (bits)	Bandwidth (RBs)	Intra-slot hopping possible	Multiplexing capacity (devices)	Coding
Short PUCCH	0	1–2	1–2	1	Yes (for 2 symbols)	3–6	Sequence selection
	2	1–2	>2	1–16	Yes (for 2 symbols)	1	Reed-Muller (3–11 b), Polar (>11 b)
Long PUCCH	1	4–14	1–2	1	Yes	1–7	Block
	3	4–14	>2	1–16	Yes	1	Reed-Muller (3–11 b), Polar (>11 b)
	4	4–14	>2	1	Yes	1–4	Reed-Muller (3–11 b), Polar (>11 b)

more interesting choices since they can be sent more often and can be decoded faster, thereby reducing latency. However, for power limited devices, a longer transmission duration is key to achieve a good enough SINR.

Besides delivering channel quality index (CQI) values to be used for scheduling, PUCCH delivers two critical pieces of information: the SR and the HARQ ACK/NACK (A/N). Both of these messages must be delivered fast and reliably, and in uplink this is especially challenging. This is because devices have a strict transmit power limitation of 0.2 W which could leave them power limited at locations where the downlink would still work fine. A power limited device doesn't benefit from a wider allocation with lower code rate, since the power is then split over more resources. What helps instead to improve the quality is to accumulate more energy per transmitted from extending the transmission duration, i.e. with repetition. But this directly contradicts the low latency target, and we are facing a tough trade-off: how to be fast and get high enough quality.

11.2.5.3.1 Long PUCCH

The three PUCCH formats 1, 3, and 4 are often together referred to as *long PUCCH*. They can be flexibly configured in the range 4–14 symbols. PUCCH format 1 can carry 1–2 bits information over base-12 sequence BPSK or QPSK symbols sent on every other OS, see Fig. 11.17, and can be multiplexed by using orthogonal sequences. This format has a rather high reliability since the code rate is low. The reliability can also be improved further with intra-slot frequency hopping. The number of UCI data symbols for different length and intra-slot hopping is given in Table 11.7. The number of orthogonal sequences that can be used for multiplexing is equal to the number of UCI data symbols. Resource efficiency and reliability are of course attractive aspects, but since it requires multiple symbols there is a latency drawback both in the form of alignment latency (waiting for the next PUCCH opportunity) and reception latency (waiting to the end to decode the message). Therefore, the long PUCCH format is not well suited for URLLC, except when we can anyway ensure a low latency e.g. through using a high SCS.

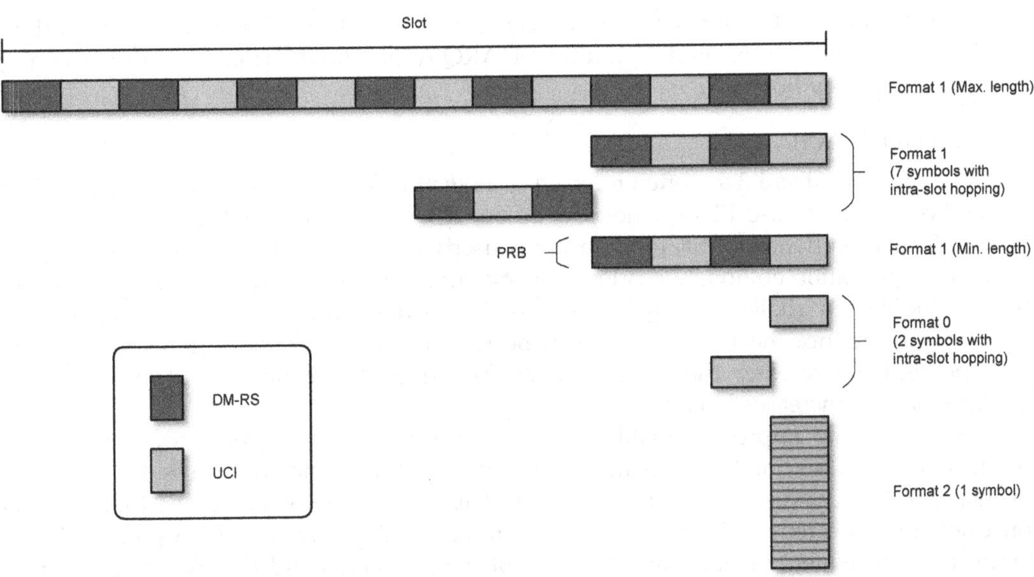

FIG. 11.17 PUCCH formats 0, 1, and 2.

TABLE 11.7 Number of UCI data symbols in PUCCH Format 1.

PUCCH length	#UCI data symbols		
	No intra-slot hopping	Intra-slot hopping	
		First part	Second part
4	2	1	1
5	2	1	1
6	3	1	2
7	3	1	2
8	4	2	2
9	4	2	2
10	5	2	3
11	5	2	3
12	6	3	3
13	6	3	3
14	7	3	4

The other long PUCCH formats 3 and 4 can carry more than 2 bits information and are used for CSI reports and for sending multiple HARQ A/N. These formats can also be configured with frequency hopping.

11.2.5.3.2 Short PUCCH

PUCCH formats 0 and 2 are often referred to as *short PUCCH* formats. Format 0 is 1–2 OS long and consists of a base-12 sequence sent on one RB that is phase rotated to convey up to 2 bits of information (when QPSK modulation is used), just enough for one A/N and one SR bit. This configuration obviously has low latency but since it uses sequence selection and no channel coding it's challenging to achieve high reliability for low-coverage devices. To improve performance the transmission can be repeated in a row. This increases the total signal power in the receiver, thereby improving the sequence detection, but of course extends the duration and increases latency.

A technique that improves quality beyond plain repetition is frequency hopping. By consecutively transmitting the same message on different frequency resources, the frequency diversity of the channel comes to the help. This way, a two-symbol PUCCH can be constructed from a repeated Format 0, see example in Fig. 11.17. A 1 RB resource is used during two subsequent symbols at both ends of the available band to send PUCCH. If the coherence bandwidth of the channel is small enough the attempts are uncorrelated, and we therefore expect a much improved success probability.

Still, with the short duration of a two-symbol PUCCH it can be hard to achieve high reliability. This is especially critical for SR, which is a device's only way to indicate that it has uplink data without grant. It doesn't matter how good the downlink control and uplink data qualities are, if the gNB doesn't know that it should send an uplink grant, nothing gets done (a solution to this problem is to use configured uplink grants as discussed in Section 11.3.3.4). For SR, a combination of the two methods can be used: two-symbol PUCCH with frequency hopping, and repetition of that sequence. Alternatively, if the latency allows, a longer PUCCH can be used to improve reliability.

Also HARQ A/N is a critical information element. Without it, the gNB does not know that a transmission failed and that it should do a retransmission. Even if there is time to perform multiple retransmissions and the combined reception of those would suffice, the chain is broken if a NACK is not delivered. However, compared to SR the A/N doesn't need quite the same level of reliability. This is because a failed data transmission in downlink would be quite rare to begin with, so a retransmission is only required in these rare cases. Some error is therefore acceptable on the HARQ A/N without jeopardizing the total reliability. However, one should keep in mind that the errors on downlink and uplink are likely to be somewhat correlated due to long-term fading dips, so it would often be in just these cases when the NACK is needed that the PUCCH transmission would fail.

In NR Release-15 there is a limitation of 2 PUCCH and 1 HARQ A/N transmission occasions that can be used by a device in a slot. If a short PUCCH is used, e.g. two-symbol long, it would be possible in theory to transmit 7 times per slot. But with the limitation, only two of these occasions can be used and only one for A/N, which effectively places a restriction on the rate of low latency packets a device can receive in downlink.

PUCCH format 2 is also 1–2 symbols long, but can carry a payload of >2 bits, which is required for CSI reports and multiple HARQ A/N feedback. In the range 3–11 bits a

Reed-Muller code is used, and for more than 11 bits, Polar code is used after CRC insertion. A DMRS sequence is mixed in on every third subcarrier. The number of RBs used depends on the total payload of the UCI. An example of this format is shown in Fig. 11.17.

11.2.5.4 PUSCH

Slot	Any
Subcarrier spacing	{15, 30, 60, 120} kHz
Duration	1-14 symbols
Bandwidth	Any (see Table 11.1)

In the same way as PDSCH, the uplink data channel in NR, *physical uplink shared channel* (PUSCH), is sent with OFDM waveforms, and is coded with LDPC with possible CB segmentation using the same CRC attachment. As an option DFT-spread OFDM is available. This is different from LTE, where only DFT-spread OFDM is used in uplink. The reason for using DFT-OFDM and not plain OFDM is to reduce the so-called *cubic metric,* which is a measurement of the amount of power back-off that needs to be done in the power amplifiers to handle the signal variation over time. High back-off means that the transmit power is reduced and the device becomes more power limited, and as an effect uplink coverage is reduced. When DFT-OFDM is used for the uplink for low back-off, transmissions need to be contiguous, and cannot be spread out over the band. This means less frequency diversity is enabled, and in addition that PUSCH and PUCCH cannot be transmitted at the same time.

As in PDSCH, a CRC is added to the uplink data TB, and additional CRCs in the case of segmented CBs, which is then each scrambled with a RNTI, and DMRS is sent within the data transmission on PUSCH, either in a front-loaded (Mapping type B) or later position (Mapping type A).

Similarly to PDSCH, the duration of PUSCH can be dynamically indicated. But for PUSCH there is no restriction on start and duration other than that an allocation cannot span across a slot border. Thus, PUSCH can start in symbol 0–13 and have a duration of 1–14 symbols.

The same principles of DCI informing the device of where it should transmit works also for uplink, and also for PUSCH the gNB can indicate that the allocation should be done with K slot repetitions (*slot aggregation*) for increased reliability, in the same way as was described for PDSCH in Section 11.2.4.4.2. However, no mechanism corresponding to the low latency feature of downlink pre-emption (Section 11.2.4.4.3) is supported for uplink in Release-15.

Uplink control information (UCI) can be sent on PUSCH if the device has UCI at the same time as a PUSCH data transmission. This is to avoid simultaneous PUSCH and PUCCH transmissions which can be challenging for power limited devices. The mapping of UCI into PUSCH follows a predetermined pattern, as in LTE, and the UCI is coded with rates set by the so-called *beta-factors* that the gNB signal in advance over RRC. The value of the beta factor is indicated with an index, see Table 11.8. For HARQ-ACK the device is configured with three indexes $I_{offset\ 1}^{HARQ-ACK}$ for 1–2 bits, $I_{offset\ 2}^{HARQ-ACK}$ for 3–11 bits, and $I_{offset\ 3}^{HARQ-ACK}$ for >11 bits payload, respectively. For CSI part 1 (CQI, RI, CRI) and CSI part 2 (PMI, and in some cases additional CQI) the device is configured with $I_{offset\ 1}^{CSI-1}$ and $I_{offset\ 1}^{CSI-2}$ for 1–11 bits payload, and $I_{offset\ 2}^{CSI-1}$ and $I_{offset\ 2}^{CSI-2}$ for >11 bits payload, respectively. The uplink scheduling DCI (Format 0–1) can be configured to contain a beta-offset indicator field, and in this

TABLE 11.8 Beta factors and indices.

Index	Beta factor for HARQ-ACK	Beta factor for CSI-1 and CSI-2
0	1.000	1.125
1	2.000	1.250
2	2.500	1.375
3	3.125	1.625
4	4.000	1.750
5	5.000	2.000
6	6.250	2.250
7	8.000	2.500
8	10.000	2.875
9	12.625	3.125
10	15.875	3.500
11	20.000	4.000
12	31.000	5.000
13	50.000	6.250
14	80.000	8.000
15	126.000	10.000
16	Reserved	12.625
17	Reserved	15.875
18	Reserved	20.000
19–31	Reserved	Reserved

case 4 indices are configured for each of the 7 parameters mentioned above. In the DCI the network can then dynamically indicate which of the 4 indices should be used.

11.3 Idle and connected mode procedures

In this section the idle and connected mode procedures in NR are described, with focus on the features and aspects relevant for URLLC, which are those related to scheduling and duplication.

11.3.1 NR protocol stack

Similar to LTE, NR defines the protocol layers PHY, *MAC*, *Radio Link Control* (RLC), and *Packet Data Convergence Protocol* (PDCP). On top of this a new layer *Service Data Adaptation Protocol* (SDAP) is added, with the role of *Quality of Service* (QoS) flow handling for *Internet Protocol*. The NR user plane and control plane protocol stack on the device side is shown in Fig. 11.18.

In the user plane (UP), *data radio bearers* can be of three types, *Master Cell Group* (MCG) bearer (where user plane follows the path of the control plane through the master node), *Secondary Cell Group* (SCG) bearer (where user plane data is mapped to the secondary node), or split bearer where data is split on PDCP layer and mapped two RLC entities in both master and secondary nodes.

In the control plane (CP), NR defines the protocol *Radio RRC*, closely related to that of LTE. This has the role of conveying control configurations between gNB and the device and to interface to the *Non-Access Stratum* in the CN.

Which protocols and procedures are relevant for URLLC? Obviously all are necessary, but two protocols ensure that packets are delivered: *RLC retransmission* on RLC level keeping track of data delivery, and HARQ on MAC level, triggering PHY retransmission attempts. Out of these, RLC can be excluded from a URLLC discussion since the time scale that RLC retransmission typically operates at is much longer than what is relevant for cMTC services. For ensuring delivery at short time scales we therefore rely on the HARQ protocol, or alternatively no protocol at all (just blind retransmission attempts).

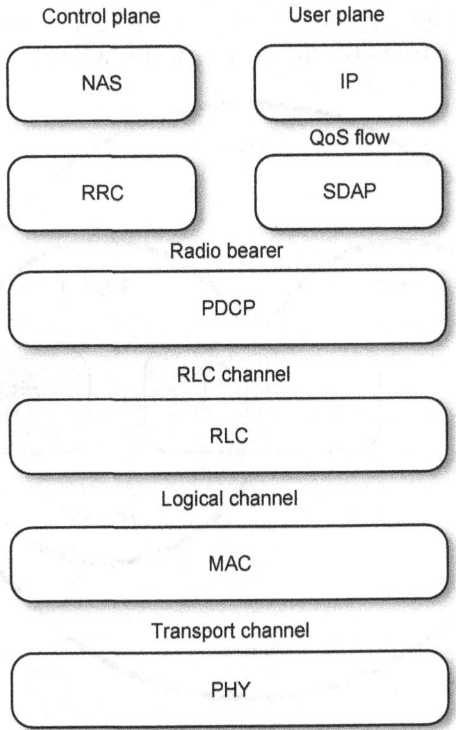

FIG. 11.18 Protocol stack in the device and interfaces between layers.

We also rely on uplink scheduling protocols for URLLC. As noted above, scheduling is straight-forward for downlink since the gNB is in control of resources, but for uplink data the device needs to either request resources (with SR) or be pre-allocated resources (configured uplink grants). Both these ways of getting an uplink resource have benefits and costs, as discussed below (Section 11.3.3.4).

In addition to diversity on the PHY and to multiple transmission attempts, we can also utilize the diversity from different links. This higher-layer diversity is enabled by branching a packet at the PDCP layer and then following the two legs through, as described further below (Section 11.3.3.7).

11.3.1.1 RRC state machine

One main step of NR is taken in the RRC state machine, which defines the three states IDLE, INACTIVE, and CONNECTED, shown with corresponding transitions in Fig. 11.19. The RRC states also correspond to different CN states. Here, the INACTIVE state is introduced to reduce the signaling and latency for devices with intermittent traffic, by keeping the device context in the gNB and staying in CN CONNECTED state. At the same time, the INACTIVE state allows the device to save battery by sleeping. This scheme is similar to the RRC suspend/resume mechanism that was introduced in LTE Release-13/14 to reduce signaling when going from IDLE to CONNECTED, but the NR solution avoids IDLE altogether and can thereby achieve even lower latency.

To begin with, we should clarify the focus we have here when it comes to procedures. Whether a device is in CONNECTED or IDLE mode will have significant impact on the service latency. A device first needs to wake up and synchronize to the network before data can be received or transmitted, and this process will inevitable take time since it involves the exchange of several messages. This is true also for devices in INACTIVE mode, but the transition is

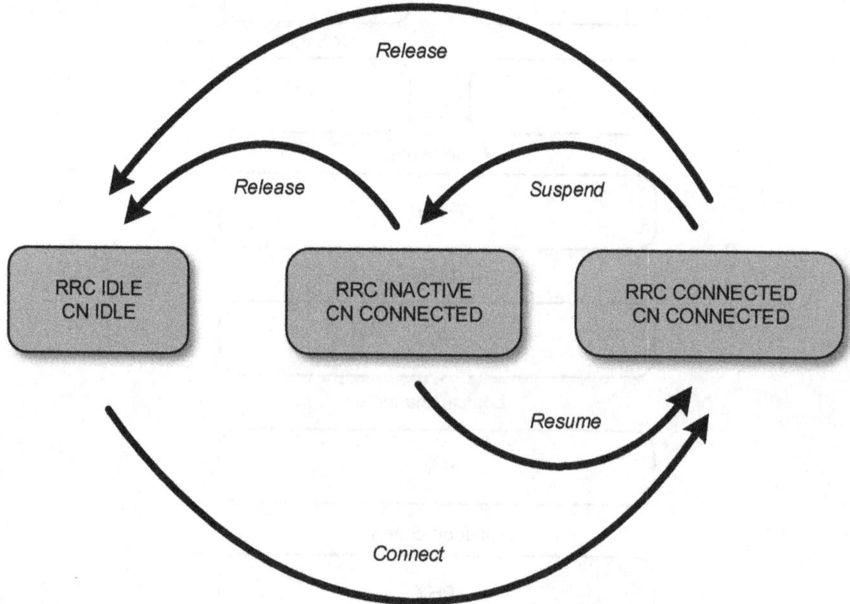

FIG. 11.19 RRC states and transitions.

simpler and faster. Even with the enhancements and speed-ups to the CP signaling introduced in NR, we can simply assume that when we speak of URLLC services we assume a latency that is low enough to exclude devices in IDLE mode. In future releases of NR, it is quite possible that the state transition can become faster or that first data can be transmitted earlier, but in Release-15, the CP signaling latency will be significantly higher compared to the UP (user plane) data latency, as shown in Chapter 12. Therefore, in the following we focus on CONNECTED mode procedures, and only discuss CP procedures in the context of handover.

11.3.2 Idle mode procedures

11.3.2.1 Control plane signaling

In cases when a device changes between different states or cells, or during the first attach to the network, it needs to set up the Control Plane (CP) through signaling before it can start to send or receive data over the User Plane (UP). As this CP signaling will take time, it is not obvious that a device that does not have the UP set up already can be counted on for a cMTC service. It therefore makes sense to say, as we did above, that only devices in the CONNECTED state are applicable for URLLC, and that devices need to be kept in this state and can't be allowed to go to a sleep state. However, this strict view can be relaxed a bit by considering handover and less latency critical services that still require high reliability.

For a device moving between cells, a handover is required to attach to the new cell. The handover means that the CP needs to switch to the new cell with RRC signaling. As already said, the signaling to make the transition happen will induce some latency. But the CP signaling is significantly faster in NR compared to LTE. To a large part this is because the processing is faster leading to shorter delay between messages and thereby faster round-trip. As with data, it's possible to reduce the time further with shorter transmissions and higher SCS, at the cost of using more bandwidth.

11.3.3 Connected mode procedures

11.3.3.1 Dynamic scheduling

The basic form of scheduling a device for downlink and uplink data transmissions is by dynamic scheduling using downlink control indication. While this method is resource efficient (a device is given resources only when needed) it has limitations in terms of reliability (SR must be received for uplink, and downlink control must be decoded) and latency in the case of uplink data (transmitting SR and receiving uplink grant).

11.3.3.1.1 Scheduling timeline

In NR the transmission timing can be dynamically indicated to the device by specifying three separate delay parameters, as illustrated in Fig. 11.20:

- K_0 for the delay between start of PDCCH and the start of PDSCH, in the range 0 OS to 32 slots.
- K_1 for the delay between end of PDSCH and start of HARQ feedback on PUCCH/PUSCH, in the range 0 OS to 15 slots.

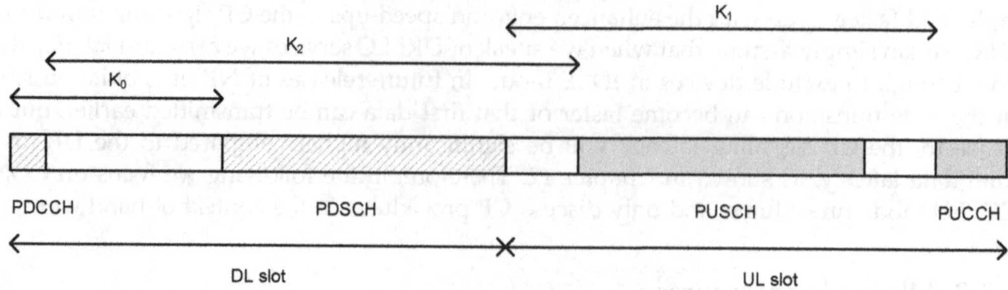

FIG. 11.20 Example of scheduling timeline parameters.

- K_2 for the delay between end of uplink grant in PDCCH and start of PUSCH, in the range 0 symbols to 32 slots.

The range of the parameters are RRC configured to indices and the indication is done in the DCI. This setup allows for high flexibility in scheduling.

11.3.3.1.2 DCI

Eight different DCI formats of four different sizes, that in turn depends on the system configurations, are defined, as shown in Table 11.9.

The fallback formats use only basic features and have the same smaller size compared to the full scheduling formats for downlink and uplink, see content in Table 11.10. Of these formats the last four (SFI, PI, two types of UPCI) also have the same size. The DCI is protected with a *CRC*, which is like in LTE a sequence added before scrambling with the RNTI, the configured device address to which the DCI is directed. The CRC for PDCCH is longer in NR, 24 bits compared to 16 bits in LTE. Out of the 24 bits, 3 bits are used to assist the decoding of the used Polar code. This helps reducing the false detection rate that arises from random false CRC checks, which have the approximate incidence of 2^{-N} for N bits CRC. For URLLC services false detection can be extra problematic, since it can lead to unavailability and buffer contamination and thereby reduced reliability.

TABLE 11.9 DCI formats.

Fallback DCI	Format 0-0 (uplink)
	Format 1-0 (downlink)
Non-fallback DCI	Format 0–1 (uplink)
	Format 1-1 (downlink)
Slot format indicator (SFI)	Format 2-0
Pre-emption indication (PI)	Format 2-1
Uplink power control indicator (UPCI)	Format 2-2 (for PUSCH/PUCCH)
	Format 2–3 (for SRS)

TABLE 11.10 DCI fields and size in bits for data transmissions addressed to C-RNTI.

DCI field	Format 0-0 (uplink fallback)	Format 0–1 (uplink scheduling)	Format 1-0 (downlink fallback)	Format 1-1 (downlink scheduling)
Uplink/downlink identifier	1	1	1	1
Frequency domain resource assignment	Formula	Formula	Formula	Formula
Time domain resource assignment	4	0–4	4	0–4
Frequency hopping flag	1	0–1	1	1
Modulation and coding scheme	5	5	5	5
New data indicator	1	1	1	1
Redundancy version	2	2	2	2
HARQ process	4	4	4	4
TPC command	2	2	2	2
Uplink/SUL indicator	1	1	–	–
Carrier indicator	–	0–3	–	0–3
Bandwidth part indicator	–	0–2	–	0–2
Downlink assignment index 1	–	1–2	–	0–4
Downlink assignment index 2	–	0–2	–	–
SRS resource indicator	–	Formula	–	–
Precoding information and number of layers	–	0–6	–	–
Antenna ports	–	2–5	–	4–6
SRS request	–	2	–	2–3
CSI request	–	0–6	–	–
CBG transmission information	–	0–8	–	0–8
PTRS-DMRS association	–	0–2	–	–
Beta offset indicator	–	0–2	–	–
DMRS sequence initialization	–	0–1	–	0–1
Uplink-SCH indicator	–	1	–	–
VRB-to-RB mapping	–	–	1	0–1

(Continued)

TABLE 11.10 DCI fields and size in bits for data transmissions addressed to C-RNTI.—cont'd

DCI field	Format 0-0 (uplink fallback)	Format 0–1 (uplink scheduling)	Format 1-0 (downlink fallback)	Format 1-1 (downlink scheduling)
PUCCH resource indicator	–	–	3	3
HARQ feedback timing	–	–	3	0–3
RB bundling size indicator	–	–	–	0–1
Rate matching indicator	–	–	–	0–2
ZP CSI-RS trigger	–	–	–	0–2
Transmission configuration indication	–	–	–	0–3
CBG flushing indicator	–	–	–	0–1

11.3.3.2 HARQ

HARQ is the protocol handling the default method for triggering retransmissions in the PHY layer. Compared to automatically repeating a message, as with downlink and uplink repetitions described above, the fact that retransmissions are conditionally triggered in HARQ ensures a much-improved resource efficiency. A retransmission is triggered when an ACK is not received in HARQ, and since we target high reliability, most of the packets will indeed result in an ACK, and a retransmission will therefore only rarely be needed. HARQ can thus be seen as a way to use a higher code rate first and then gradually lowering it to the needed level. Incremental redundancy using a circular buffer is used (as in LTE) for this purpose of combining retransmissions into one longer code. This is indeed very resource efficient, but has a drawback of longer latency and the dependency on receiving the feedback.

In NR the HARQ feedback is transmitted in UCI in the uplink for downlink data, and implicitly through a DCI in the downlink for uplink data. The latter is only true for NACK: an unsuccessful uplink transmission results in an uplink grant DCI for a retransmission, while a received uplink transmission results in no feedback on downlink. In NR, there is no equivalent to the LTE PHICH channel giving A/N feedback for uplink data transmissions. The main reason for this is that *synchronous* HARQ operation with hard-coded timing is not supported in NR, and both uplink and downlink operate using *asynchronous* HARQ with indicated timing.

If the TB is segmented into CBs the HARQ feedback and retransmission of the CBGs is handled separately by using multiple indication bits in the UCI and DCI, which can improve resource efficiency since only the erroneous parts of a transmission are resent.

NR HARQ is fully dynamic, meaning that there are no fixed connections between transmission and feedback. For downlink data the UCI where to send HARQ A/N is identified from the DCI, and for uplink data a DCI for retransmission is identified from the HARQ process index and a new data indicator, similar to LTE. There are 16 HARQ processes to use for a device.

With HARQ, we can trigger retransmission to ensure high reliability at low resource cost, but only as long as the chain of feedback is not broken. Therefore, we cannot set an arbitrarily high error rate target (using high code rate) for transmissions and rely on many retransmissions, since at some point the feedback may not be delivered (it can also take a long time). However, the reliability of HARQ NACK doesn't need to be on the same level as that of data since it will be rare that the data transmission fails, requiring a retransmission.

For both reasons of latency and reliability there is therefore a limit on how many HARQ retransmissions we can use for URLLC data. The more we can reliably use, the better for efficiency, which for uplink data appears a solid choice, but for downlink data we then rely on safe delivery of UCI which is not obviously achievable when the UE is power limited. This is discussed further in Chapter 12.

11.3.3.3 SR

When a device has URLLC data in its transmit buffer, it triggers a SR if it doesn't have a valid grant. This is often referred to as *dynamic* uplink scheduling. If the device only operates using one service, this is rather straight-forward: it will use its configured SR resource on PUCCH and the gNB will then know what resource the device needs. But in the case when a device is operating with multiple services (e.g. eMBB for video and cMTC for position) it needs to separate between them.

An SR is configured for a logical channel which is in turn connected to a radio bearer with a certain QoS requirement defined by the network. By using different SR resources, with up to 8 possible to configure in NR, the device can indicate which type of uplink data resource it needs. For cMTC services, we can then assume that the SR is configured on a short PUCCH with short periodicity to allow for low latency. As discussed in Section 11.2.5.3.2 this fast option has a drawback in the form of coverage limitations. Besides the PHY repetition (with frequency hopping in between) it is also possible to do a higher layer repetition of SR. This is enabled by configuring a delay parameter to 0, which has the effect of repeating the SR until the device receives the requested uplink grant. This is then a way to achieve reliability at the cost of latency (we can assume that there will be no additional resource cost from a device using its allocated PUCCH resources). For dynamic uplink scheduling, this possibility is crucial as already noted: until the gNB has received the SR no uplink data can be transmitted.

11.3.3.4 Uplink configured grant

As discussed above, the basic way to perform an uplink transmission is through the so-called *dynamic scheduling* sequence of SR - uplink grant - uplink data. For URLLC, this scheme has two main weaknesses: latency due to the extra round-trip of control signalling before data transmission can start, and the reliability of SR, as mentioned above. An alternative way of scheduling uplink transmissions is with *configured grant* (CG) scheduling. Here, the device is informed over RRC that a certain grant is configured, meaning that it is persistent and reoccurs with some defined periodicity. There are two different ways to set up the CG, see illustration in Fig. 11.21.

CG Type 1 only consists of the RRC configuration, meaning that all of the information typically sent in DCI is specified there instead and no additional control information is needed.

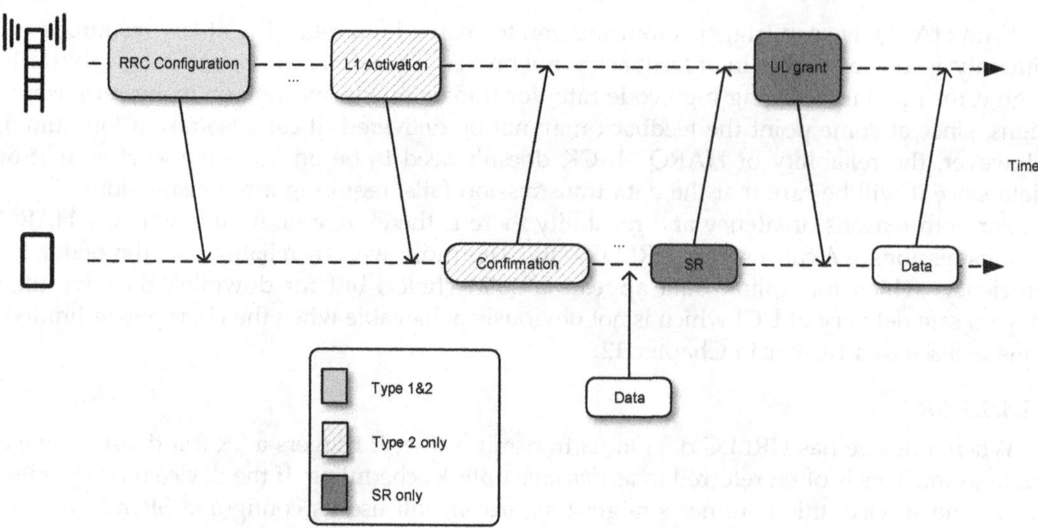

FIG. 11.21 Configured uplink grant Type 1 and 2, and SR-based access for uplink transmission.

CG Type 2 is similar to *semi-persistent scheduling* in LTE and consist of a configuration pointing out a certain RNTI address and a repetition pattern. A DCI scrambled with that RNTI is then interpreted by the device as an uplink grant that is reoccurring with the configured periodicity. Compared to Type 1, this type is more flexible since the gNB can at any time update the parameters of the uplink transmission in PDCCH, such as MCS and allocation.

With a CG in place, the device can in principle be given an uplink data transmission opportunity in every symbol, that it can readily use if it has data in its buffer. If it has no uplink data, it doesn't transmit anything on the CG. In practice, the shortest transmission duration would likely be two OS to give room for UCI and DMRS, which means that while avoiding overlapping allocations the device can start its transmission at 7 positions in the uplink slot without any further delay. Compared to dynamic scheduling, this offers a dramatic reduction of latency. The cost, however, is that uplink resources become "locked up" for a device, in a way that dynamic scheduling avoids. Unless the device has periodic data of the same period as the CG (this is indeed a possibility) there will be a waste of unused resources. At least this is the case if we want to uniquely assign uplink resources to ensure good SINR. If devices are allowed to share resources the resource management is much more effective, but we then run into the risk of colliding uplink transmissions. For URLLC data this is not an option, since the reliability would be jeopardized.

After a first transmission has taken place on the uplink CG the data will either be successfully received or not. If the transmission failed (but the presence of DMRS was detected) the gNB will issue a DCI for retransmission. Thus, from the first retransmission onwards dynamic grant scheduling will be used (of course the gNB is free to indicate the CG for retransmission also).

If the device has been indicated overlapping dynamic and configured uplink grants it will drop the CG and use the dynamically indicated one.

11.3.3.4.1 HARQ operation

The NR supported HARQ operation is asynchronous, meaning that the *process ID* (PID) must be indicated in the DCI for the device to know which buffer to use. After activation, the configured uplink grant resources will be numbered with a PID starting from 0 up to a maximum value configured by RRC, after which the numbering restarts at 0. This way the gNB and the device will both know which transmission a given PID refers to. When a retransmission is triggered the device is thereby dynamically scheduled with DCI even if the configured resource can be indicated for retransmission.

11.3.3.4.2 Repetition

It is possible to configure repetition of a TB over the CG, also in the case of mini-slot resources, just as for dynamic PUSCH allocations as mentioned in Section 11.2.5.4. However, as in the case of downlink repetition (illustrated in Fig. 11.13), the repetition in uplink will only take place on slot level, meaning that the same resource will be used for sending a TB in K subsequent slots. This will lead to increased reliability on shorter time compared to HARQ-based retransmission. For lowest latency it would require repetition on mini-slot resources with a slot, but this is not supported in NR Release-15. Over the K repetitions on the uplink CG a *RV* pattern is applied according to selecting the $(\text{mod}(n\text{-}1,4)+1)^{\text{th}}$ value of an RV sequence at repetition n. The configurable RV sequences are {0,2,3,1}, {0,3,0,3}, and {0,0,0,0}, and the K-sequence can start at any occasion when the RV from the configured RV sequence is 0. This means that when a varying RV sequence is configured, there will be an additional alignment delay to wait for the start of the repetition sequence.

11.3.3.5 Uplink power control

For determining the power used for PUSCH transmission a formula is used:

$$p_{\text{PUSCH}} = \min\{p_{\text{cmax}},\ p_0 + \alpha \cdot PL + 10\log_{10}(2^{\mu} \cdot M_{rb}) + \Delta_{TF} + \delta\}$$

where p_{cmax} is the maximum allowed power per carrier, p_0 and α are configurable parameters, PL is the pathloss estimate, μ is the numerology ($SCS = 2^{\mu} \cdot 15kHz$), M_{rb} is the uplink allocation, Δ_{TF} is calculated from the MCS, and finally δ is the closed-loop parameter. The *open-loop* components ($p_0 + \alpha \cdot PL$) are configured and handled by the device, while the *closed-loop* component δ is indicated in an uplink power control DCI. The PUCCH power control is the same as the PUSCH one, but for PUCCH α is fixed to 1.

11.3.3.6 CSI measurement and reporting

For the gNB to know which MCS to use for a device in downlink it is critical that it knows the status of the channel conditions. The device measures the channel quality on the pilot sounding signals CSI-RS (described in Section 11.2.4.2.3) and then reports a *CQI* in the *CSI* reporting, as part of the UCI together with HARQ A/N and SR. The reporting is based on measurement on configured CSI-RS resources and can in a similar way be triggered periodically (on PUCCH), semi-statically (on configured PUCCH or PUSCH resources) or aperiodically (on PUSCH) from a DCI indication.

The value of the CQI is connected to the MCS in such a way that the expected BLER using a certain MCS, when the corresponding CQI is reported, is at a target level. In LTE this

TABLE 11.11 CQI table for 0.001% error rate (corresponding to Table 5.2.2.1–4 in Ref. [6]).

CQI index	Modulation	Code rate x 1024	Efficiency
0	out of range		
1	QPSK	30	0.0586
2	QPSK	50	0.0977
3	QPSK	78	0.1523
4	QPSK	120	0.2344
5	QPSK	193	0.3770
6	QPSK	308	0.6016
7	QPSK	449	0.8770
8	QPSK	602	1.1758
9	16 QAM	378	1.4766
10	16 QAM	490	1.9141
11	16 QAM	616	2.4063
12	64 QAM	466	2.7305
13	64 QAM	567	3.3223
14	64 QAM	666	3.9023
15	64 QAM	772	4.5234

expected BLER level is set at 10%, meaning that this is the expected failure rate when a device is scheduled with a MCS corresponding to the reported CQI. In NR, addressing the needs of URLLC and the fact that there is a separate MCS table intended for high reliability, the device can be configured with two different BLER levels: 10% error mapping to CQI table 1 (up to 64 QAM) or CQI Table 11.2 (up to 256 QAM), and 10^{-5} error mapping to CQI Table 11.3, shown in Table 11.11. The configuration is semi-static, meaning that it is made over RRC, and not dynamically indicated. It can therefore be so that the CQI reported and the MCS table used do not match, but it is then up to the gNB to translate between the two sets.

In the case of only one antenna port used and one configured CSI-RS resource, the CSI report consists only of CQI. With more antenna ports and CSI-RS resources, the report can also include the codebook index with *precoding matrix indicator* (PMI), *rank indicator, layer indicator*, and also a *CSI-RS indicator* indicating the preferred beam. The reporting can be done for *type I single-panel* for simple multi-port antennas, *type I multi-panel* codebook for composite antennas each consisting of multiple ports, or for ***type II codebook*** with higher granularity intended for multi-user MIMO.

11.3.3.7 PDCP duplication

As in LTE, data can be duplicated in the PDCP layer for increased reliability from the redundancy over two transmission paths. The procedure is setup over RRC for a split radio bearer and defines an additional RLC entity connected to PDCP by adding an additional logical channel. In the PDCP layer, the incoming data packet in the form of a PDCP PDU is duplicated and sent to the two RLC entities; the *master cell group* (MCG) and the *secondary cell group* (SCG), as illustrated in Fig. 11.22. If the MCG and the SCG in turn belong to the same MAC entity the duplication will be over two different carriers, meaning *carrier aggregation* (CA), and if they belong to different MAC entities the duplication is done over two different cells, meaning *dual connectivity* (DC). In the case of CA, logical channel mapping restrictions are used to prevent the packets from being sent on the same carrier (which would not give as good redundancy).

On the receiving side the packets are separated all the way up to the PDCP layer, where a duplicate check is done so that only one PDCP PDU is delivered to the data buffer.

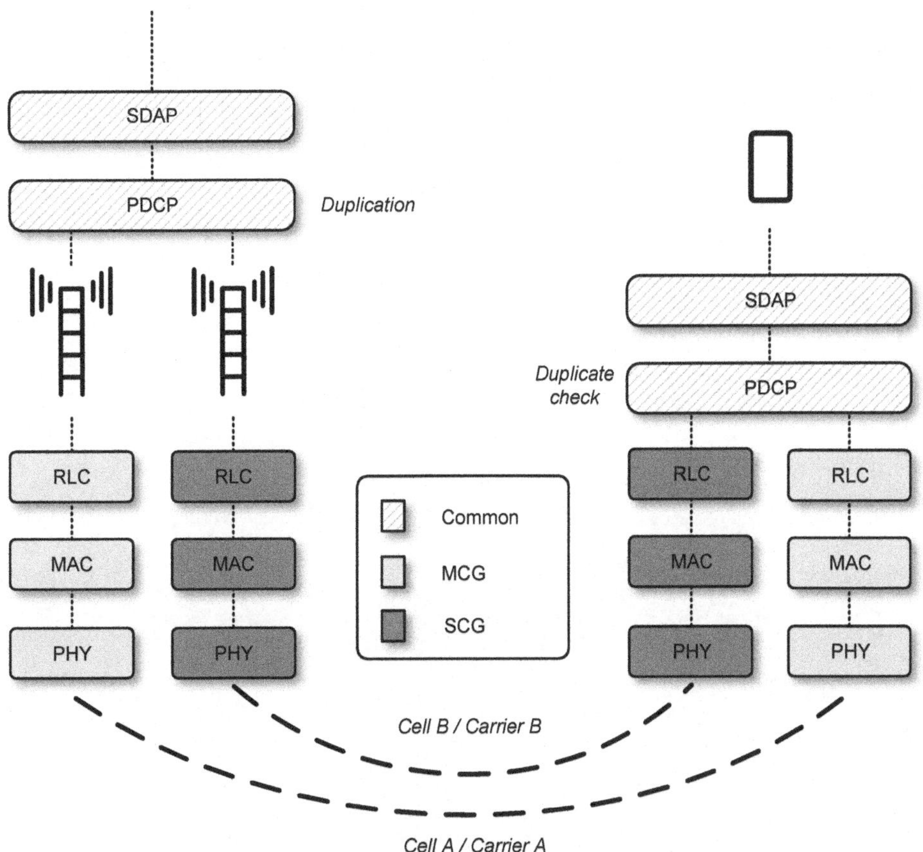

FIG. 11.22 PDCP duplication of data with Dual Connectivity.

References

[1] ITU-R. Report ITU-R M.2412-0, Guidelines for evaluation of radio interface technologies for IMT-2020, 2017.
[2] E. Dahlman, S. Parkvall, J. Sköld. 5G NR: the next generation wireless access technology. Academic Press, 2018.
[3] Third Generation Partnership Project, Technical specification 38.211, v15.2.0. Physical channels and modulation, 2018.
[4] Third Generation Partnership Project, Technical specification 38.212, v15.2.0. Multiplexing and channel coding, 2018.
[5] Third Generation Partnership Project, Technical Specification 38.213, v15.2.0. Physical layer procedures for control, 2018.
[6] Third Generation Partnership Project, Technical Specification 38.214, v15.2.0. Physical layer procedures for data, 2018.
[7] D. Tse, P. Viswanath. Fundamentals of wireless communication. Cambridge University Press, 2005.

CHAPTER

12

NR URLLC performance

OUTLINE

Abstract

This chapter shows the performance of NR URLLC, comparing the technology to the existing 5G requirements put up by ITU. The performance evaluation includes user plane and control plane latency, reliability evaluations, and spectral efficiency.

In this Chapter we will study the performance of NR URLLC, which was presented in detail in Chapter 11, and assess if it lives up to the strict requirements that it was designed for. First, a general evaluation of latency and reliability is presented, based on requirements from IMT-2020 specifications. This is followed by two use case studies of cMTC services

Cellular Internet of Things, Second Edition
https://doi.org/10.1016/B978-0-08-102902-2.00012-1

© 2020 Elsevier Ltd. All rights reserved.

where the URLLC technology is applied to enable new wireless solutions. For explanations of technical terms the reader should refer to the description in Chapter 11.

12.1 Performance objectives

The requirements on URLLC in 5G are set by ITU-R in the IMT-2020 specifications [1], which means that a radio technology that wants to have the 5G certification must fulfill these conditions. Out of the full set of 5G requirements, the ones of interest for URLLC are user plane (UP) latency, control plane (CP) latency and reliability. In the first few sections below, the assessment of these values is done for an NR system. The chosen setup follows the evaluations that have been done in 3GPP [2] as a preparation for submitting NR as a 5G RAT to ITU.

12.1.1 UP latency

The requirement on UP latency is defined from when a source node sends a packet to when the receiving node receives it. The latency is defined from and to the interface between layer 2 and layer 3 in both nodes, corresponding to QoS flows to and from the SDAP layer in NR. It is assumed that the device is in an active state, and therefore assuming no queuing delays.

A 1 ms latency bound is targeted for URLLC, and 4 ms for eMBB. Both directions (uplink and downlink) have the same requirement.

12.1.2 CP latency

The requirement on CP latency is 20 ms, defined from a battery efficient state to when the device is being able to continuously transmit data. Further reduction to 10 ms is encouraged but is not a requirement from ITU.

For NR the interpretation of a battery efficient state is chosen to be RRC INACTIVE, and the CP latency is therefore taken as the transition time from INACTIVE to CONNECTED state.

12.1.3 Reliability

The requirement on reliability is defined as the probability of successful transmission of a packet of a certain size, within a certain latency bound, at a given channel condition. Although it may sound complicated, it is simply that we want the system to guarantee that at a certain SINR we can deliver a certain packet reliably within a latency bound.

For the case of IMT-2020, the requirement on the latency bound, packet size and reliability are 1 ms, 32 bytes and 99.999%, respectively. The minimum SINR where this is achieved is technology specific and depends on the agreed evaluation scenario. However, it is specified that the reliability target as given above should be achievable at the cell-edge of the evaluation scenarios, defined as the fifth percentile of the device SINR distribution.

12.2 Evaluation

This section presents the NR URLLC performance evaluation for each URLLC requirement described in Section 12.1.1–12.1.3, and also an additional evaluation of spectral efficiency for NR URLLC.

12.2.1 Latency

The latency for a cMTC service can be directly evaluated based on the specification of the NR standard. But it is important to remember that in a real system we would need to look beyond the radio interface and account for possible additional delays from scheduling, transport, and core network functionality. We also have to make assumptions on what processing delay is needed at the gNB, as this is not defined in the standard. A simple and useful assumption is that the processing delays are equal in the device and gNB. Again, in a real system this may not be entirely correct, especially in a highly loaded network due to the higher processing load required in the gNB coming from scheduling and handling many devices.

Since we are looking at the latency achievable in critical systems we can confine the study to the maximum latency in a given scenario, meaning that we e.g. make a worst-case assumption when it comes to the waiting time until the next transmission opportunity (alignment delay).

12.2.1.1 Processing delay

Processing delay in the RAN domain is caused at the transmitting node by preparation, e.g. from protocol headers, ciphering, encoding and modulation, of transmission and in the receiving node from e.g. equalizing, decoding and deciphering. For downlink data transmissions the processing delay in the device upon reception of the downlink data on PDSCH includes the reception and decoding procedure. For uplink data transmission on PUSCH with dynamic scheduling, there is a processing delay in the device due to reception and decoding of the downlink control on PDCCH, carrying the uplink grant. In the gNB there is also processing delay as in the device, with the addition that the processing delay in the gNB also needs to comprise delay caused by e.g. scheduling and link adaptation.

The response timing in the gNB between receiving a *scheduling request* (SR) and transmitting a PDCCH containing an uplink grant, and between downlink HARQ reception and downlink PDSCH retransmission, is assumed in integer units of *transmission time interval* (TTI), meaning the used slot or mini-slot duration (the time between PDCCH occasions). For higher *subcarrier spacings* (SCS) and fewer symbols in the mini-slot, the TTI duration is shorter, and more TTIs are needed for processing. The assumed delays for the gNB side are given in *OFDM symbols* (OS) in Table 12.1. This processing consists of three main components:

- Reception processing for uplink (PUSCH data processing, PUCCH control processing for SR/HARQ-ACK)
- Scheduling and protocols for downlink
- Physical layer processing for PDSCH and PDCCH

TABLE 12.1 gNB processing time assumptions in nr. of OFDM symbols (OS).

Timing	15/30 kHz SCS				60/120 kHz SCS			
TTI [OS]	14	7	4	2	14	7	4	2
gNB processing time t_b [OS]	14	7	4	4	14	14	12	10

For simplicity we refer to gNB processing time (t_b) as the total processing time and further that the processing time is equal always. For example, the same processing time is assumed for scheduling first transmission and re-transmission. Same processing time is also assumed for downlink transmission and uplink reception.

The minimum response timing in the device between PDSCH reception and PUCCH downlink HARQ transmission, and between reception of a PDCCH containing an uplink grant and PUSCH transmission was discussed in Section 11.2.3.5. In downlink, the device processing time is according to the d_1 value (Table 12.2) while for uplink the device processing time is according to d_2 value (Table 12.3) for device capability 2. For 120 kHz SCS there's no Capability 2 values agreed so we use the Capability 1 values. For d_1 it is assumed here that PDCCH and PDSCH overlap on 1 OFDM symbol, and that 1 DMRS symbol is used in PDSCH. For d_2 it is assumed that the first PUSCH symbol is only used for DMRS.

12.2.1.2 UP latency

For data transfer we can single out three cases:

- Downlink data transmission,
- Uplink data transmission based on a *configured grant* (CG), and,
- Uplink data transmission based on SR (SR-based).

Of these, it is natural to expect that downlink data transmission is the fastest since the PDSCH data can (at least in theory) be sent by the gNB in the next available transmission opportunity in direct connection to the downlink control on PDCCH. In case of uplink data transmission on PUSCH, we should expect a higher latency, either due to the additional signaling and delay from SR and uplink grant, or because the gNB have configured scheduling opportunities with a higher interval compared to downlink opportunities. However, in the latter case it is possible to match the scheduling opportunities with the data periodicity

TABLE 12.2 PDSCH processing time for device Capability 2.

	d_1 [OFDM symbols]			
Allocation	**15 kHz SCS**	**30 kHz SCS**	**60 kHz SCS**	**120 kHz SCS (Cap. 1 values)**
Slot (14 symbols)	3	4.5	9	20
7 symbols	3	4.5	9	20
4 symbols	4	5.5	10	20
2 symbols	4	5.5	10	20

TABLE 12.3 PUSCH processing time for device Capability 2.

Allocation	d_2 [OFDM symbols]			
	15 kHz SCS	**30 kHz SCS**	**60 kHz SCS**	**120 kHz SCS (Cap. 1 values)**
Slot (14 symbols)	5	5.5	11	36
7 symbols	5	5.5	11	36
4 symbols	5	5.5	11	36
2 symbols	5	5.5	11	36

(for periodic traffic) or give the device an opportunity to transmit in every TTI (for random traffic). Here, we will study this more optimally configured case for configured uplink data. Similarly, for SR-based uplink data transmission we assume that the device has an SR opportunity on PUCCH in every uplink TTI.

The components of UP latency for uplink and downlink data are shown in a signaling flow chart in Fig. 12.1, including the processing delay discussed in the previous section, the alignment delay from the TTI structure, and the delay from the transmission duration itself.

In the following we will analyze the worst-case UP latency after a first transmission and up to three HARQ retransmissions. We will follow the ITU definition for UP latency as outlined

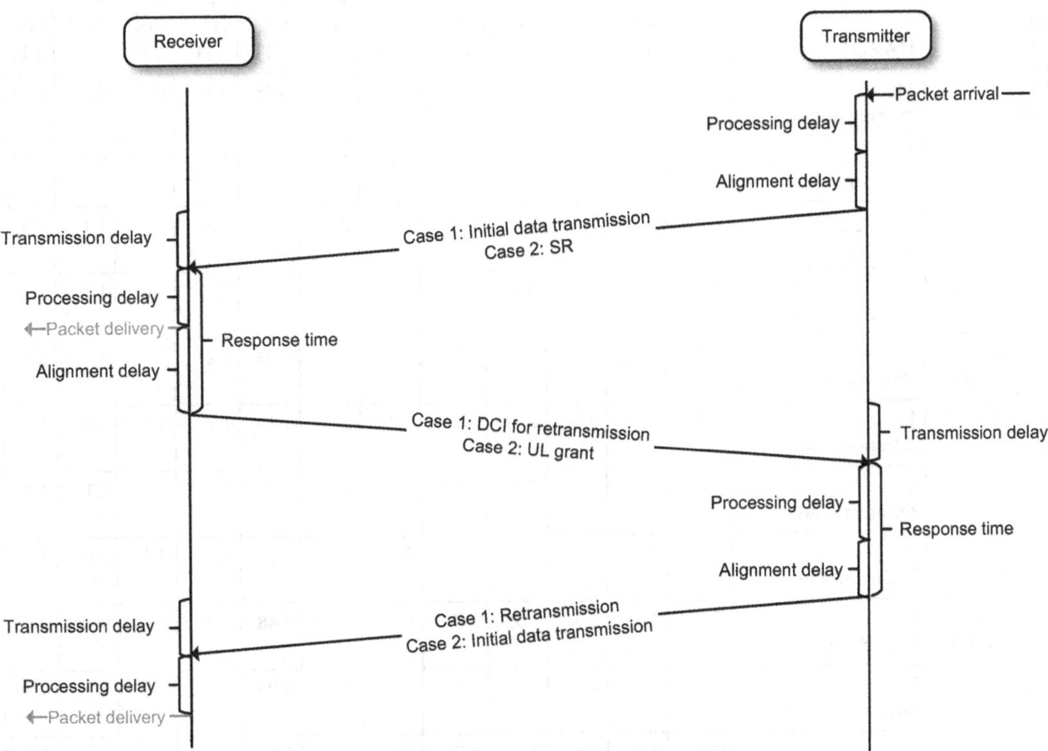

FIG. 12.1 Signaling flow and latency components showing downlink data or uplink CG-based data (Case 1) with one retransmission, and uplink SR-based initial data transmission (Case 2).

in Section 12.1.1. Since 60 kHz is an optional SCS for FR1 in NR Release-15 we here focus on evaluating the UP latencies for 15, 30, and 120 kHz SCS, which will anyway show the full span of achievable latency.

For downlink HARQ and SR we assume a 2-symbol PUCCH (Format 0) placed at the end of the uplink TTI. PDCCH opportunities and PUCCH opportunities are assumed to be present in every scheduled TTI.

The alignment delay is the time required after being ready to transmit until a transmission can start. We assume the worst-case latency, meaning the alignment delay is assumed to be the longest possible given the transmission opportunities.

On the device side we assume the full processing delays d_1 and d_2, also for decoding the downlink data without transmitting HARQ feedback, and when preparing for a first CG-based uplink data transmission.

12.2.1.2.1 Data latency in FDD

The resulting UP latency for a SCS of 15, 30 and 120 kHz in FDD is shown in Table 12.4. As can be seen, the 1 ms requirement can be reached for SCS 15 kHz and higher depending on

TABLE 12.4 FDD UP one-way latency for data transmission with HARQ-based retransmission. Bold numbers indicate meeting the 1 ms URLLC requirement.

		15 kHz SCS				**30 kHz SCS**				**120 kHz SCS**			
Latency [ms]	**HARQ**	**14-OS TTI**	**7-OS TTI**	**4-OS TTI**	**2-OS TTI**	**14-OS TTI**	**7-OS TTI**	**4-OS TTI**	**2-OS TTI**	**14-OS TTI**	**7-OS TTI**	**4-OS TTI**	**2-OS TTI**
Downlink data	1st transmission	3.2	1.7	1.3	**0.86**	1.7	**0.91**	**0.70**	**0.48**	**0.55**	**0.43**	**0.38**	**0.31**
	1 retransmission	6.2	3.2	2.6	1.7	3.1	1.6	1.3	**0.96**	1.1	**0.87**	**0.76**	**0.63**
	2 retransmissions	9.2	4.7	3.6	2.6	4.7	2.4	2	1.5	1.6	1.3	1.1	**0.96**
	3 retransmissions	12	6.2	4.6	3.4	6.1	3.1	2.7	2	2.1	1.7	1.5	1.3
Uplink data (SR)	1st transmission	5.5	3	2.5	1.8	2.8	1.5	1.3	**0.93**	1.2	1.1	**1**	**0.89**
	1 retransmission	9.4	4.9	3.9	2.6	4.7	2.4	2	1.4	1.9	1.7	1.6	1.3
	2 retransmissions	12	6.4	4.9	3.5	6.2	3.2	2.6	1.9	2.6	2.3	2.1	1.8
	3 retransmissions	15	7.9	5.9	4.4	7.7	3.9	3.3	2.3	3.2	2.8	2.6	2.2
Uplink data (CG)	1st transmission	3.4	1.9	1.4	**0.93**	1.7	**0.95**	**0.70**	**0.48**	**0.70**	**0.57**	**0.52**	**0.45**
	1 retransmission	6.4	3.4	2.6	1.8	3.2	1.7	1.4	**0.93**	1.3	1.1	1.1	**0.89**
	2 retransmissions	9.4	4.9	3.9	2.6	4.7	2.4	2	1.4	1.9	1.7	1.6	1.3
	3 retransmissions	12	6.4	4.9	3.5	6.2	3.2	2.6	1.9	2.6	2.3	2.1	1.8

mini-slot configuration. With uplink CG the latency is considerably reduced compared to SR-based scheduling.

12.2.1.2.2 Data latency in TDD

With TDD, there are additional alignment delays caused by the sequence of downlink and uplink slots. Depending on when the data arrives in the transmit buffer the latency may be the same or longer than the FDD latency. It should be noted that the pattern of downlink and uplink is fixed in slot periods of 14 symbols, and having mini-slot scheduling periods will not impact the direction: one downlink slot may consist of 2 or more mini-slots but no uplink mini-slot, as is illustrated in Fig. 12.2. This means that the alignment delay from the downlink/uplink pattern is only moderately reduced with mini-slots.

For a DL-UL-DL-UL slot pattern the resulting latency is as indicated in Table 12.5. As can be seen in the table, the 4 ms target for eMBB data can be reached with a SCS of 15 kHz for 7-symbol mini slot, while 30 kHz SCS is possible also with slot length transmission. The 1 ms target for URLLC data can be reached with 120 kHz SCS and mini-slots for downlink data and uplink data with CG. With a more downlink-heavy slot pattern as DL-DL-DL-UL the latency increases for uplink data due to the extra alignment delay, as is seen in Table 12.6.

12.2.1.3 CP latency

As was outlined in Section 12.1.2 the study of CP latency is taken to be the study of the latency for the transition from INACTIVE to CONNECTED state. This delay represents passing through the random access sequence and will be present during handover between cells,

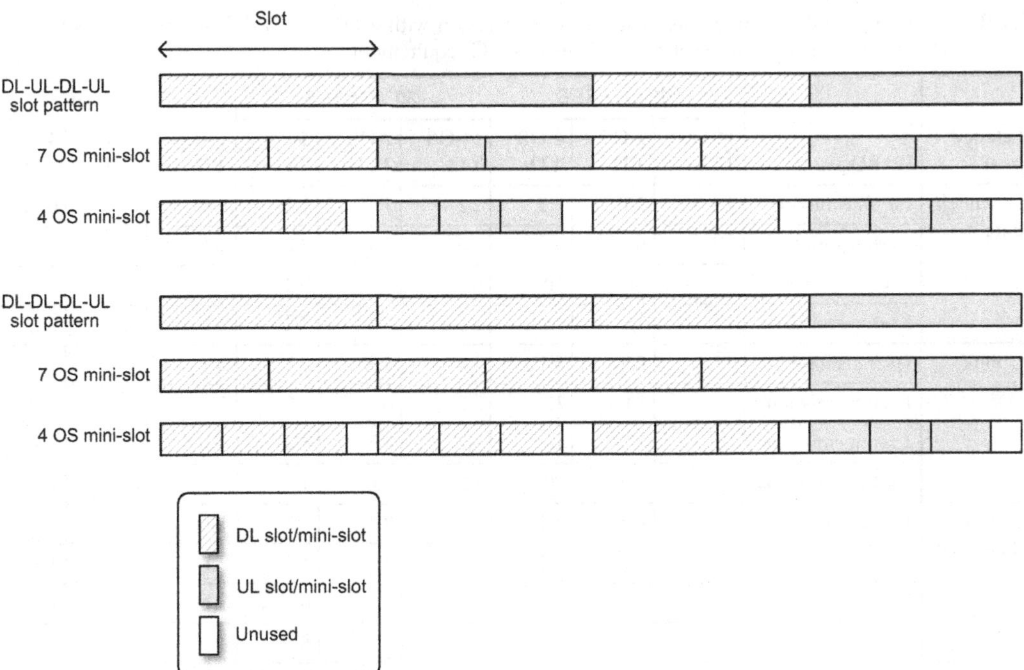

FIG. 12.2 Studied TDD slot patterns with mini-slots.

TABLE 12.5 TDD UP one-way latency for data transmission with alternating DL-UL-DL-UL slot pattern. Bold numbers indicate meeting the 1 ms URLLC requirement.

		15 kHz SCS			**30 kHz SCS**			**120 kHz SCS**		
Latency [ms]	**HARQ**	**14-OS TTI**	**7-OS TTI**	**4-OS TTI**	**14-OS TTI**	**7-OS TTI**	**4-OS TTI**	**14-OS TTI**	**7-OS TTI**	**4-OS TTI**
Downlink data	1st transmission	4.2	2.7	2.3	2.2	1.4	1.2	**0.68**	**0.55**	**0.51**
	1 retransmission	8.2	4.7	4.3	4.1	2.4	2.2	1.4	1.1	1
	2 retransmissions	12	6.7	6.3	6.2	3.4	3.2	2.2	1.6	1.5
	3 retransmissions	16	8.7	8.3	8.1	4.4	4.2	2.9	2.1	2
Uplink data (SR)	1st transmission	7.5	4.5	4.1	3.8	2.3	2.1	1.5	1.2	1.2
	1 retransmission	12	6.9	6.4	6.2	3.4	3.2	2.3	1.9	1.7
	2 retransmissions	16	8.9	8.4	8.2	4.5	4.2	3.1	2.5	2.2
	3 retransmissions	20	11	10	10	5.4	5.2	3.8	3.2	2.7
Uplink data (CG)	1st transmission	4.4	2.9	2.4	2.2	1.4	1.2	**0.82**	**0.70**	**0.64**
	1 retransmission	8.4	4.9	4.4	4.2	2.5	2.2	1.6	1.3	1.2
	2 retransmissions	12	6.9	6.4	6.2	3.4	3.2	2.3	1.9	1.7
	3 retransmissions	16	8.9	8.4	8.2	4.5	4.2	3.1	2.5	2.2

TABLE 12.6 TDD UP one-way latency for data transmission with a DL-DL-DL-UL slot pattern. Bold numbers indicate meeting the 1 ms URLLC requirement.

		15 kHz SCS			**30 kHz SCS**			**120 kHz SCS**		
Latency [ms]	**HARQ**	**14-OS TTI**	**7-OS TTI**	**4-OS TTI**	**14-OS TTI**	**7-OS TTI**	**4-OS TTI**	**14-OS TTI**	**7-OS TTI**	**4-OS TTI**
Downlink data	1st transmission	4.2	2.7	2.3	2.2	1.4	1.2	**0.68**	**0.55**	**0.51**
	1 retransmission	9.2	6.7	6.3	4.6	3.4	3.2	1.4	1.2	1.1
	2 retransmissions	13	11	10	6.7	5.4	5.2	1.9	1.7	1.6
	3 retransmissions	17	15	14	8.6	7.4	7.2	2.4	2.2	2.1
Uplink data (SR)	1st transmission	9.5	8.5	8.1	4.8	4.3	4.1	2	1.5	1.4
	1 retransmission	14	13	12	7.2	6.4	6.2	3.1	2.4	1.9
	2 retransmissions	18	17	16	9.2	8.5	8.2	4.1	3	2.4
	3 retransmissions	22	21	20	11	10	10	5.1	3.9	2.9
Uplink data (CG)	1st transmission	6.4	4.9	4.4	3.2	2.4	2.2	1.1	**0.95**	**0.89**
	1 retransmission	10	8.9	8.4	5.2	4.5	4.2	2.1	1.5	1.4
	2 retransmissions	14	13	12	7.2	6.4	6.2	3.1	2.4	1.9
	3 retransmissions	18	17	16	9.2	8.5	8.2	4.1	3	2.4

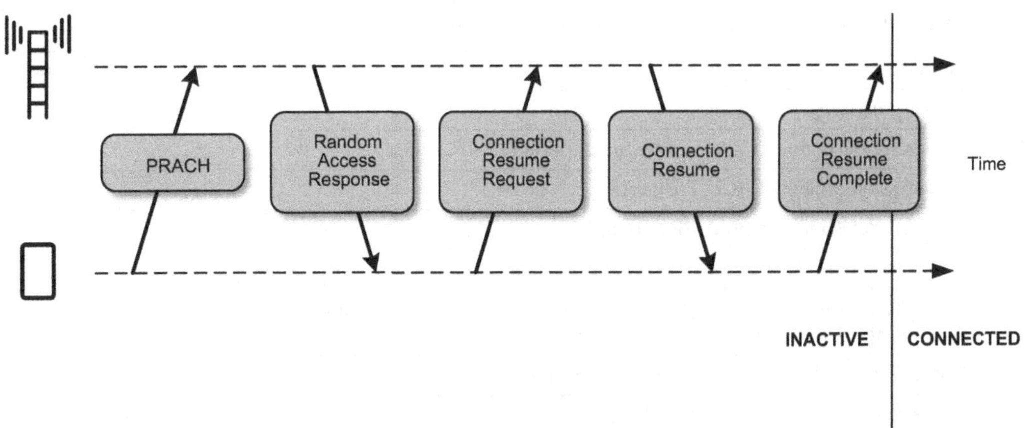

FIG. 12.3 Illustration of CP signaling during transition from INACTIVE to CONNECTED state.

and also during RRC reconfiguration for state transition and uplink synchronization, as was mentioned in Chapter 11. The sequence of signals exchanged during the state transition is illustrated in Fig. 12.3, and we assume that the latency covers the time from waiting for a PRACH opportunity in the device until RRC Connection Resume Request is processed in the gNB.

The latency associated with the CP signaling can be estimated from the same delays used for UP data, i.e. assuming the same processing (t_b and d_2) and slot alignment delays as in Section 12.2.1.2. But we must also take the significant processing associated with RRC updates, which is assumed to be a 3 ms additional delay on both device and gNB sides [3]. The different signaling steps are presented in Table 12.7, together with values for the processing components. For PRACH it is assumed here that a short preamble is used with a format fitting within the TTI (with or without repetition), so that one entire TTI is used for transmitting it.

We study the latency for slot (14 OS) and 7 OS mini-slot. With different numerologies and therefrom depending TTI duration the CP latency will differ strongly due to the many steps, as is seen in Table 12.8 for FDD, where the accuracy in the calculation is on OFDM symbol level instead. The latency levels required (20 ms) and encouraged (10 ms) by ITU are seen to be comfortably fulfilled for all (20 ms)/most (10 ms) of the possible configurations in NR.

Also with TDD these targets for CP latency can be reached, but the levels are higher than with FDD, as expected from the additional downlink-uplink alignment delay. The values for an alternating DL-UL-DL-UL slot pattern are given in Table 12.9, and for a downlink-heavier DL-DL-DL-UL slot pattern in Table 12.10, also here with a symbol level accuracy in the calculation.

12.2.2 Reliability

What we are interested in when it comes to reliability is that a certain data packet (of a certain size) is delivered to the receiver with high probability, considering an upper latency bound, as outlined in Section 12.1.3. The total reliability of the packet delivery can be seen as

TABLE 12.7 Control plane signaling steps and assumed latency.

Step	Description	Latency
0	Device processing	d_2
1	Worst-case delay due to RACH scheduling period (1 TTI period)	1 TTI
2	Transmission of Short RACH Preamble	1 TTI
3	Preamble detection and processing in gNB	t_b
4	Transmission of RA response	1 TTI
5	Device Processing Delay (decoding of scheduling grant, timing alignment and C-RNTI assignment + L1 encoding of RRC Connection Request)	d_2
6	Transmission of RRC Connection Resume Request	1 TTI
7	Processing delay in gNB (L2 and RRC) **3 ms extra processing assumed**	t_b + 3 ms
8	Transmission of RRC Connection Resume (and uplink grant)	1 TTI
9	Processing delay in the device (L2 and RRC) **3 ms extra processing assumed**	d_2 + 3 ms
10	Transmission of RRC Connection Resume Complete (including NAS Service Request)	1 TTI
11	Processing delay in gNB (Uu to S1-C)	t_b

TABLE 12.8 FDD CP latency.

CP latency (ms)	15 kHz SCS	30 kHz SCS	60 kHz SCS	120 kHz SCS
Slot (14-symbol TTI)	15.4	10.7	8.4	7.9
Mini-slot (7-symbol TTI)	10.9	8.4	7.9	7.7

TABLE 12.9 TDD CP latency, assuming DL-UL-DL-UL slot pattern.

CP latency (ms)	15 kHz SCS	30 kHz SCS	60 kHz SCS	120 kHz SCS
Slot (14-symbol TTI)	18.4	12.2	9.2	8.3
Mini-slot (7-symbol TTI)	12.9	9.4	8.4	7.9

TABLE 12.10 TDD CP latency, assuming DL-DL-DL-UL slot pattern.

CP latency (ms)	15 kHz SCS	30 kHz SCS	60 kHz SCS	120 kHz SCS
Slot (14-symbol TTI)	20.4	13.2	9.7	9.1
Mini-slot (7-symbol TTI)	18.4	12.2	9.4	8.4

increasing over time with an increasing number of transmission attempts (HARQ retransmissions). In other words, the task is then to be able to estimate the probability of success in the receiver as a function of attempts within the latency bound. While doing this we should take the potential failure of all essential physical channels into account. In this evaluation we will only consider FDD, since reaching the reliability target with one configuration is sufficient.

12.2.2.1 Reliability of physical channels

First, we need to study the success rate of the physical channels we described in Chapter 11: PDCCH, PDSCH, PUSCH, and PUCCH. This is done through link simulations where the SNR level is swept by changing the noise within the interesting range for mobile communication. By collecting statistics of many simulated transmissions, we can find the error rate as a function of SNR for the relevant channels. Since we know that the success rate will depend not only on SNR but also on code rate, the study is done for a set of modulation and coding schemes (MCSs) for data, and a set of *aggregation levels* (ALs) for the downlink control in PDCCH.

The assumptions for the link level simulations are given in Table 12.11. Three separate simulation data sets are used to represent data, downlink control, and uplink control. For

TABLE 12.11 Assumptions for the link-level simulations.

	Value	
Assumption	**Configuration A (mid band)**	**Configuration B (low band)**
Channel model	TDL-C [12.1] with 300 ns delay spread	
Carrier	4 GHz	700 MHz
Bandwidth	20 MHz	
Subcarrier spacing	30 kHz	
Antenna configuration	2TX 2RX (data), 1TX 2RX (control)	
TX diversity	Rank 1 (TX diversity precoding based on CSI reports with 5 slots periodicity).	
Speed	3 km/h	
Channel estimation	Realistic, 4 OS mini-slot − 1 OS front-loaded DMRS type 2	Realistic, 7 OS mini-slot − 2 OS front-loaded DMRS type 2
Frequency allocation	Frequency allocation type 1 (contiguous)	
Time allocation	4 OS allocations, type B	7 OS allocations, type B
PUCCH	1 A/N bit, PUCCH format 0 with 2-symbol duration and frequency hopping between band edges 1% probability of detecting a NACK from noise (D2N) Simulated at 4 GHz	
PDCCH	Polar codes, 40 bits payload excl. CRC. CCEs distributed over the carrier AL {4, 8, 16} Simulated at 700 MHz	
PDSCH	LDPC base-graph 2, 256 bits transport block, MCS table 3 (MCS-1 to MCS-6) (see Section 11.2.4.4.1)	

downlink (PDSCH) and uplink (PUSCH) data the same link evaluation is used, which is possible since the coding, channel estimation, and MCS tables are the same for these channels. For PDCCH a downlink control information (DCI) message of size 40 bits excluding CRC is assumed. For the uplink control on PUCCH we have assumed 1 bit *uplink control information* (UCI) carried by PUCCH format 0 with 2 OS duration and frequency hopping.

Resulting *block-level error rates* (BLER) as function of SNR for the control channels are shown in Fig. 12.4, for failed PDCCH and two types of PUCCH errors: N2A, where a NACK is interpreted as an ACK, and N2D where a NACK is not detected. It should be noted that delivering a HARQ ACK is not required for the data to be correctly received, so this step is not taken into account in the latency and reliability calculation. For downlink and uplink control the same evaluation data is used to represent both configurations, which is a reasonable assumption given low expected impact from doppler spread in a short PUCCH transmissions, and since the PDCCH is spread out over the same bandwidth in both cases.

With PDSCH and PUSCH, for first transmission attempt with data in the two configurations, the BLER as function of SNR for the highest MCS in the relevant range are shown in

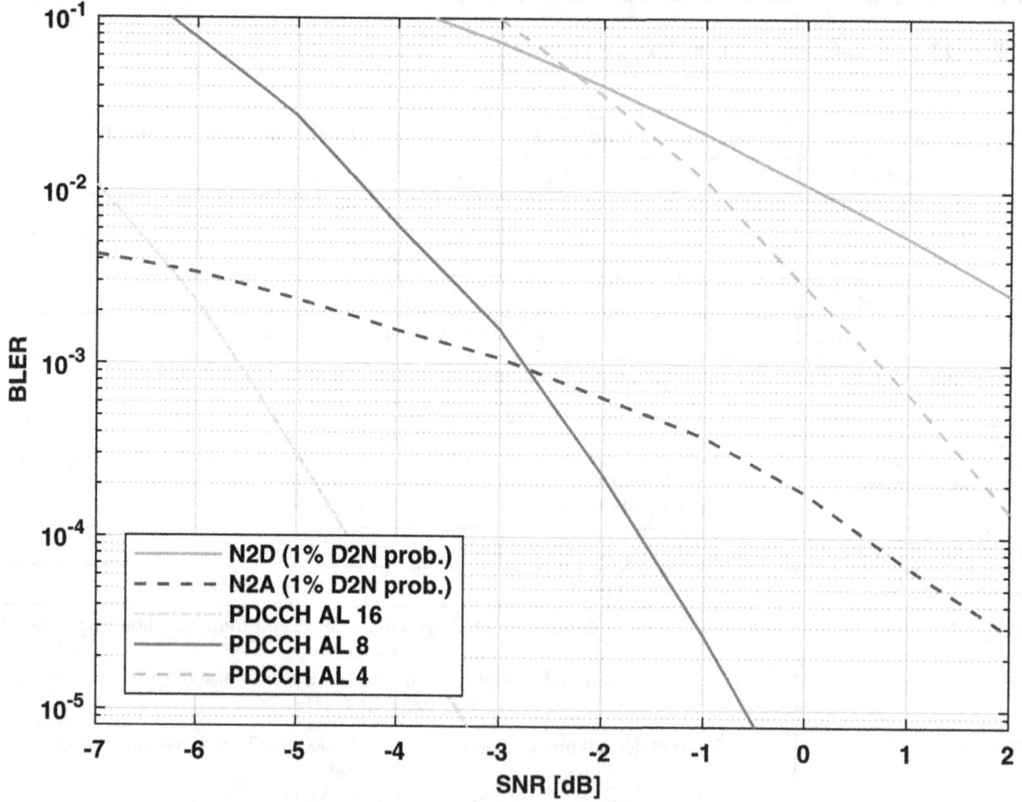

FIG. 12.4 PUCCH Format 0 and PDCCH BLER as function of SNR.

Fig. 12.5. The relevant SNR range is found from lower percentiles of the system simulations presented below.

12.2.2.2 SINR distributions

The assumptions for the system level simulations, aimed at finding the SINR distributions for a deployment scenario, are given in Table 12.12, with values that have been aligned with the 3GPP calibration campaign for the evaluation toward the 5G requirements.

The scenario was simulated for the Urban Macro URLLC configuration A (4 GHz) and configuration B (700 MHz) [1], and was evaluated separately for channel models UMa A and UMa B [1]. For Configuration A, the gNB antenna was set to be an array of 2 × 8 vertical x horizontal (VxH) panels each consisting of 4 × 1 V × H cross-polarized antenna elements, while in Configuration B it was set to be an array of 4 × 4 V × H panels, each consisting of 2 × 1 V × H cross-polarized antenna elements. Each array panel corresponds to one antenna port per polarization (P).

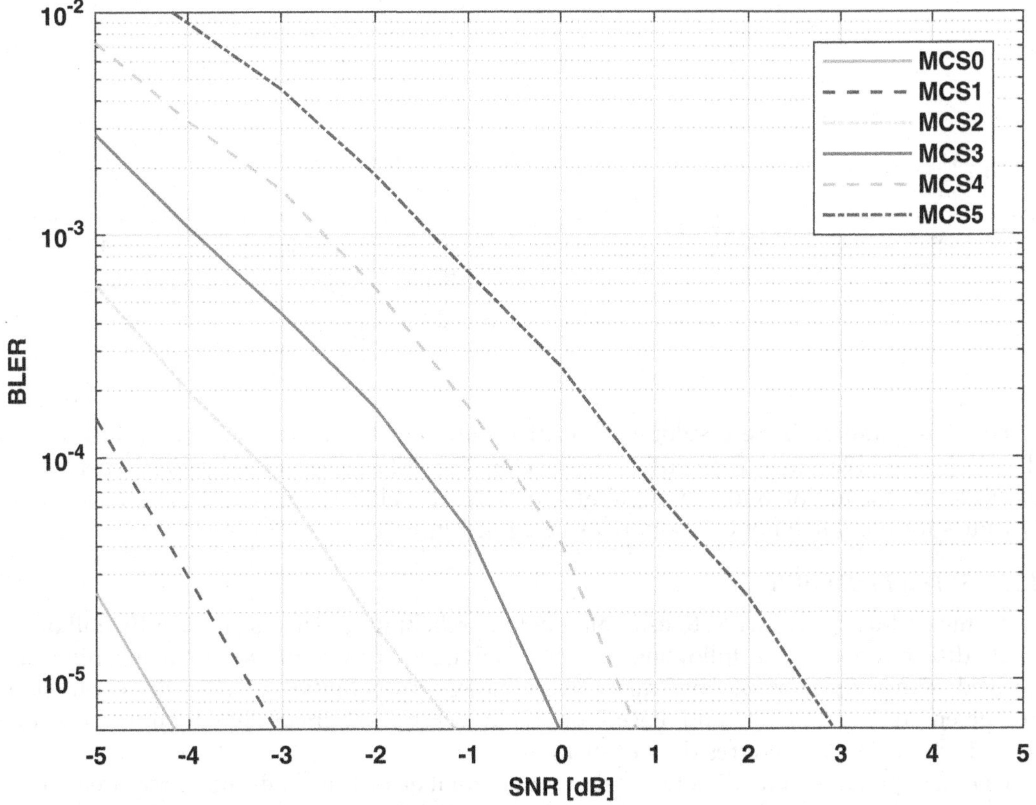

FIG. 12.5 Data BLER for QPSK with different MCS as function of SNR for Configuration A (top) and Configuration B (bottom).

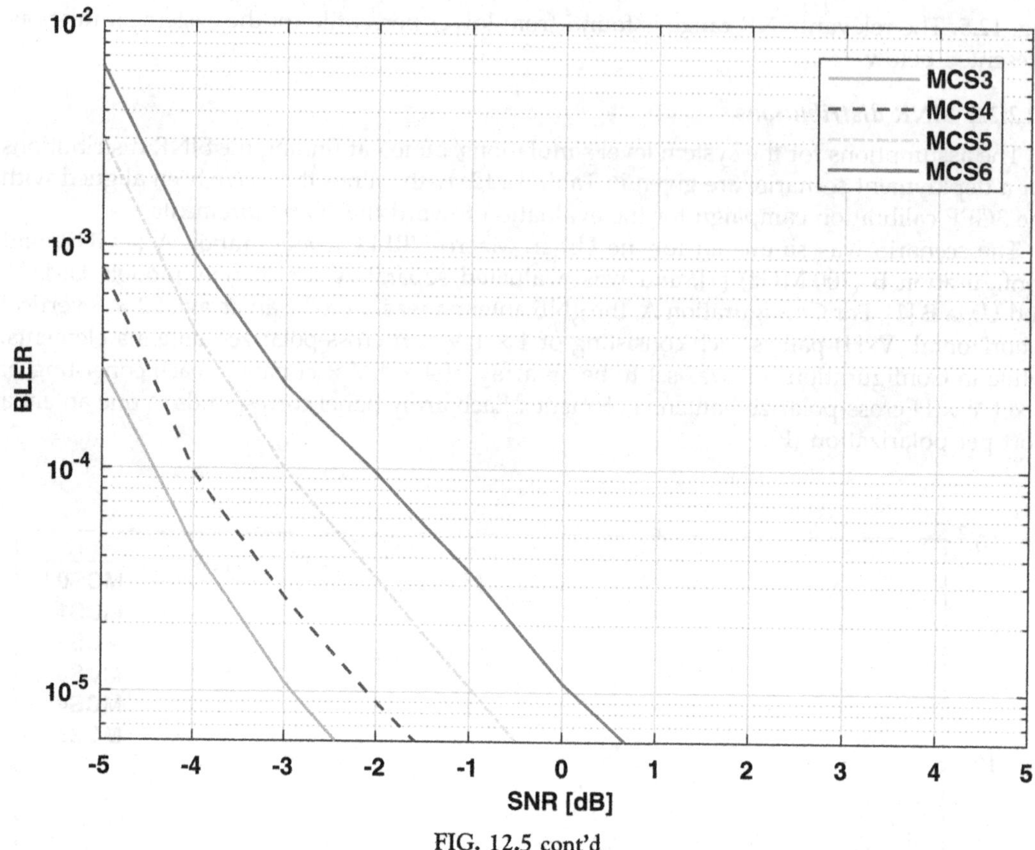

FIG. 12.5 cont'd

For configuration A the resulting SINR distribution at full load (100% cell utilization) is drawn in Fig. 12.6, and for configuration B in Fig. 12.7. The cell-edge (fifth percentile) SINR values for uplink and downlink are collected in Table 12.13, and these are the target Q-values for the reliability evaluation, as discussed in Section 12.1.3.

12.2.2.3 *Total reliability*

As the end result, we want to estimate the total reliability or success rate, at the tail of the SINR distribution. In the following section, analytical expressions for total reliability are derived. The definitions of physical channel success probabilities used in the expressions are given in Table 12.14, and based on these we will estimate the total success rate $p_t = 1 - \varepsilon$, where ε is the residual error rate.

One may perhaps immediately expect that the total error translates into a requirement on all physical channels, such that they should have a reliability exceeding p_t. But this is a simplification and an exaggeration, which is only relevant for the simplest case of having only one

TABLE 12.12 Assumptions for the system-level simulations.

Configuration parameters	Configuration A (mid band)	Configuration B (low band)
Carrier	4 GHz, FDD	700 MHz, FDD
Subcarrier spacing	30 kHz	
Base station Antenna Height	25 m	
Inter-site distance	500 m	
Sectors per site	3	
Bandwidth	20 MHz (50 RB)	
Device deployment	80% outdoor, 20% indoor	
Number of device antenna elements	4	
device noise figure	7	
device power	23 dBm	
Path loss model	UMa A and B	
gNB antenna VxH panel (VxHxP elements)	2 × 8 (4 × 1 × 2)	4 × 4 (2 × 1 × 2)
gNB transmit power	49 dBm	
gNB noise figure	5	
Electrical down tilt	9°	
Traffic model	Full buffer	
Uplink power control	Alpha = 1, P0 = -106dBm	
Uplink allocation	5 RB (10 devices sharing 50 RB)	

transmission attempt. In this case, for downlink transmission data, under the assumption of independent errors, we can expect the combined success rate of PDCCH and PDSCH to be:

$$p_t = p_1 p_2$$

This expression can then describe the success rate of a *one-shot attempt*, where we do not consider feedback or subsequent attempts.

How can we formulate the total success rate in the case of multiple attempts? First, we must have a model for how the attempts differ from each other. There are two extremes that are useful to consider here:

- Uncorrelated transmissions. In this case, the attempts are independent of each other and can be treated as separate probability processes. This could correspond to the case of two transmissions either separated in time more than the coherence time (e.g. when performing retransmissions while moving at high speed) - which is less relevant for URLLC - or separated in frequency more than the coherence bandwidth. It could also correspond to a duplication of the packet sent over two independent channels from e.g. different nodes. Two different attempts would therefore experience different channel conditions and therefore different SINR.

- Completely correlated transmissions. In this case the channel characteristics (i.e. the experienced SINR) are exactly the same in all transmission attempts. The gain from having more than one attempt of data transmission is then that the accumulated code rate can be reduced (using incremental redundancy), giving a higher success rate at the given SINR.

In an actual scenario, two transmission attempts would be partially correlated, meaning that the channel conditions will have changed somewhat between attempts, but the main characteristics of the channel will likely largely remain. By tracking the channel in time and frequency between attempts we can capture this correlation. The effect a channel correlation will have is that success rates become coupled to earlier success rates and therefore to the number of attempts.

In downlink, we need to take PDCCH and PUCCH control into account besides PDSCH data, and we can formulate the total reliability for N transmission attempts as (using the definitions in Table 12.14):

$$p_t = \sum_{n=1}^{N} \sum_{i=1}^{n} \left\{ \binom{n-1}{n-i} [(1-p_1)p_4]^{n-i} p_1 p_{2,i} \prod_{j=1}^{i-1} p_1 p_3 (1 - p_{2,j}) \right\}$$

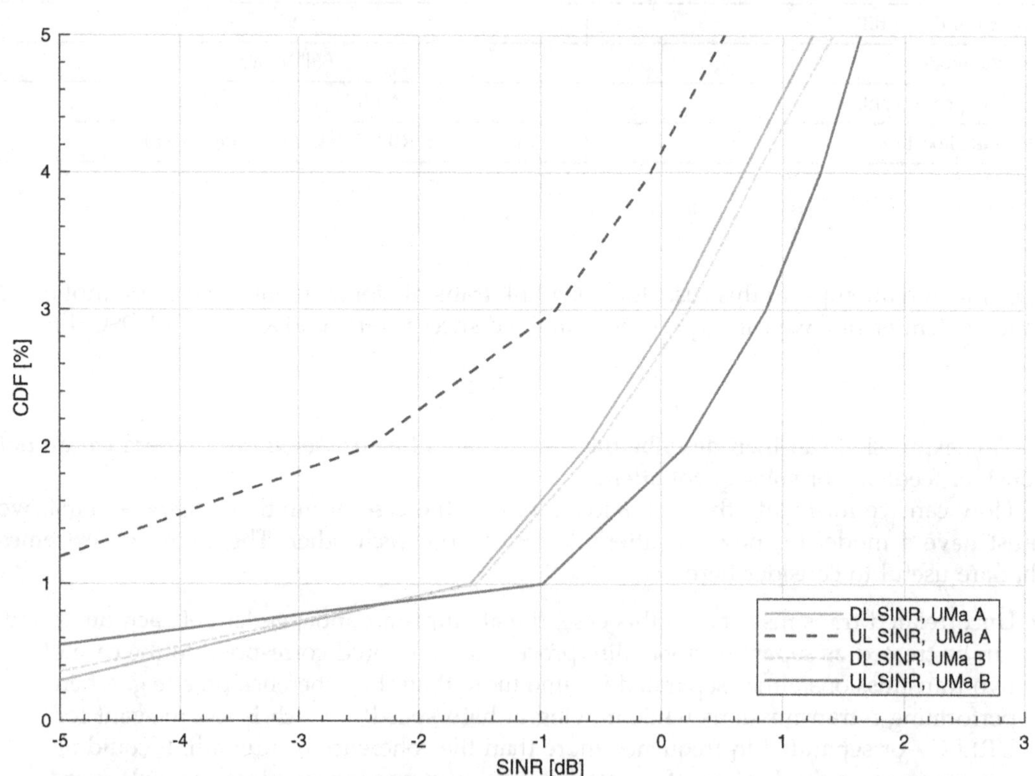

FIG. 12.6 SINR at full load for configuration A (4 GHz band). Full distribution (top) and fifth percentile (bottom).

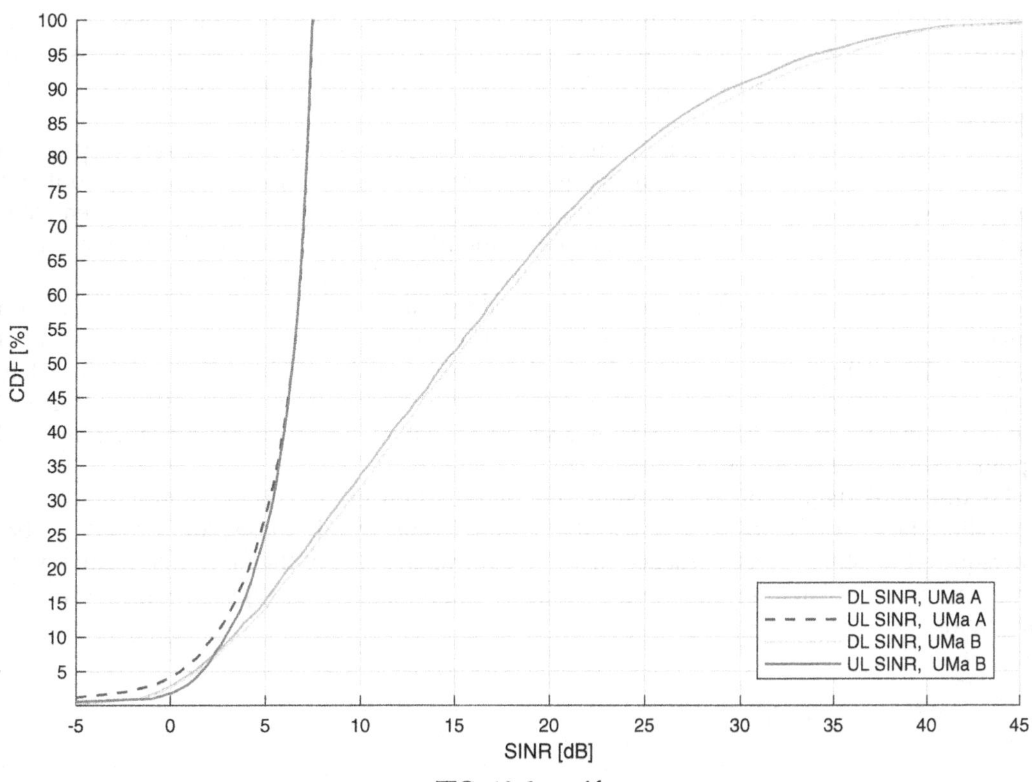

FIG. 12.6 cont'd

where for any positive integer k, $p_{2,k}$ is the probability of a data block being correctly received after exactly k transmissions have been soft combined. In this expression the downlink and uplink control transmissions are seen as uncorrelated with each other and with data. This is an approximation, but it can be motivated by e.g. shifting the frequency placement of PDCCH between attempts. The downlink data attempts are correlated with each other according to the tracking of the used channel model.

With SR-based uplink transmissions we also need to take both PDCCH and PUCCH performance in to account besides PUSCH performance. For m SR attempts followed by N uplink data transmission attempts the total reliability would be:

$$p_t = \left(1 - (1 - p_0)^m\right) \sum_{n=2}^{N} p_1 p_{2,n} \prod_{i=2}^{n-1} \left(1 - p_1 p_{2,i}\right)$$

With CG-based uplink scheduling instead we remove the SR step over PUCCH and the first downlink control for uplink grant, and the total reliability after N uplink data transmission attempts can be described as:

$$p_t = p_{2,1} + \left(1 - p_{2,1}\right) \sum_{n=2}^{N} p_1 p_{2,n} \prod_{i=2}^{n-1} \left(1 - p_1 p_{2,i}\right)$$

Here the PDCCH reliability only comes in starting from the first retransmission. In line with expectations, perfect energy detection performance on the PUSCH resource is assumed, meaning that gNB will always correctly identify the device from the first uplink transmission based on scheduled allocation.

With these expressions at hand we can construct the total reliability plots from the physical channel BLER plots, as function of the SINR experienced in the studied scenario (here only for the UMa B path loss model). As was mentioned in Section 12.1.3, the IMT-2020 requirement is that the reliability target should be achieved at the fifth percentile of the SINR distribution, given by Table 12.13. To achieve a simple plot, we make the assumption that the percentile of uplink and downlink SINR is equivalent, meaning that we assume that a user is on the fifth percentile of both the uplink and downlink SINR distributions at the same time. This is of course a strong simplification which we do not expect to generally hold, but not completely unrealistic and it allows us to study the main trends and more importantly the fulfilment of the requirement. For downlink control it is assumed that AL 8 is used. In both downlink and uplink frontloaded DMRS is assumed.

For downlink data the total reliability for Configuration A and B is given in Fig. 12.8, and for uplink data with CG the total reliability is given in Fig. 12.9. In these plots we can see that the required reliability can be reached below the fifth percentile of the SINR distribution with only one transmission attempt, using a range of MCSs for both configurations, and in both directions.

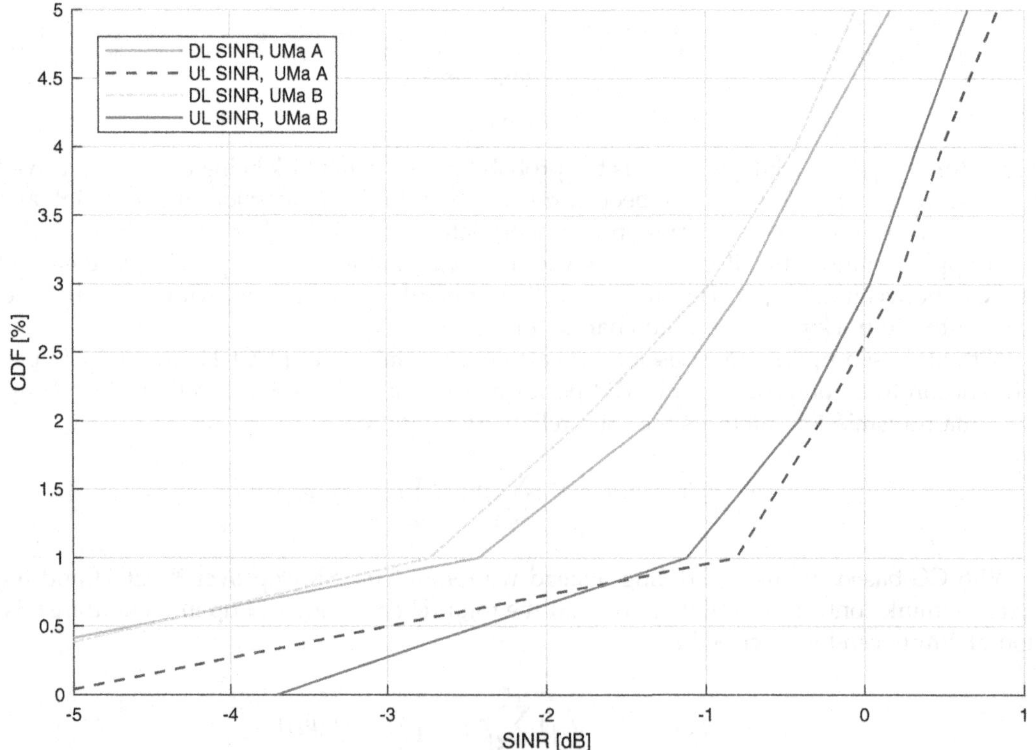

FIG. 12.7 SINR at full load for configuration B (700 MHz band). Full distribution (top) and fifth percentile (bottom).

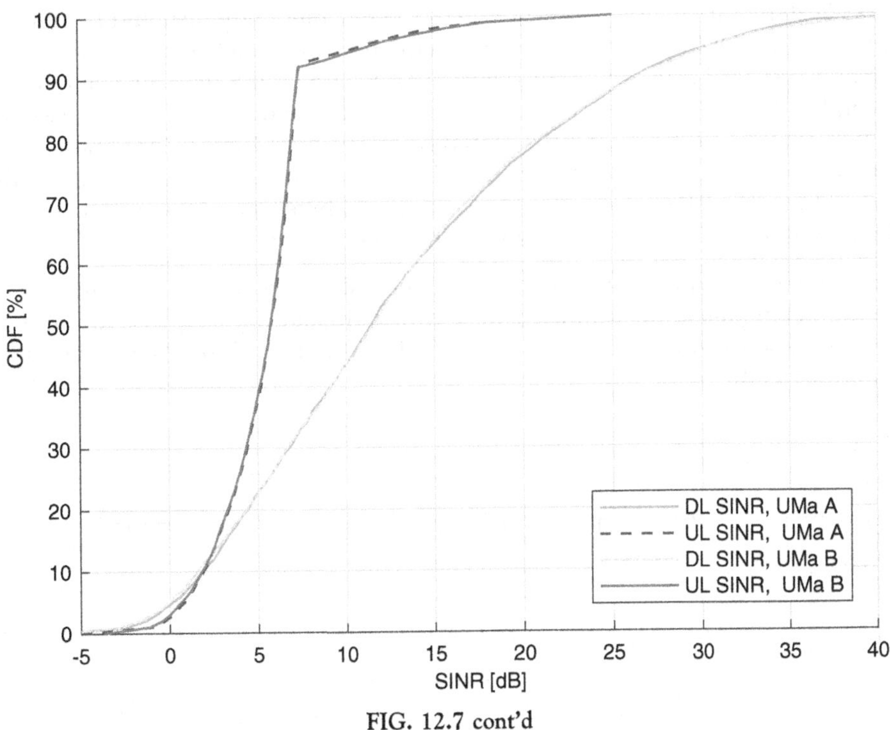

FIG. 12.7 cont'd

TABLE 12.13 Fifth percentile SINR values (Q-values) for configuration A and B and pathgain UMa A and B.

	Configuration A		Configuration B	
	UMa A	**UMa B**	**UMa A**	**UMa B**
Downlink SINR [dB]	1.2	1.4	0.16	-0.06
Uplink SINR [dB]	0.52	1.6	0.83	0.65

TABLE 12.14 Success probabilities for calculating total reliability.

Probability of successful reception	Description
p_0	PUCCH SR
p_1	PDCCH
p_2	PDSCH/PUSCH
p_3	PUCCH NACK detection
p_4	PUCCH DTX detection

The IMT-2020 requirement states further that the reliability target should be fulfilled for a packet of 32 bytes, within 1 ms. With QPSK modulation and a code rate according to MCS-1 to MCS-6 and overhead from downlink control (AL-8) and DMRS (one full symbol in uplink, every second subcarrier in one symbol in downlink), the total required transmission allocation in *resource blocks* (RB) per such transmission is given in Table 12.15. These allocations should then be compared with the carrier bandwidth of 20 MHz, giving 50 RB with 30 kHz SCS. Here, the TBS is assumed to be exactly 32 bytes, and CRC is not considered.

Studying now specific configurations we can check if the total requirement is fulfilled:

- In Configuration A with UMa B with a 4 OS mini-slot, one transmission attempt with MCS-6 fulfills the reliability requirement in downlink and uplink (Figs. 12.8 and 12.9). This would require 40 RB in downlink and 31 RB in uplink (Table 12.15) and take 0.7 ms (Table 12.4).

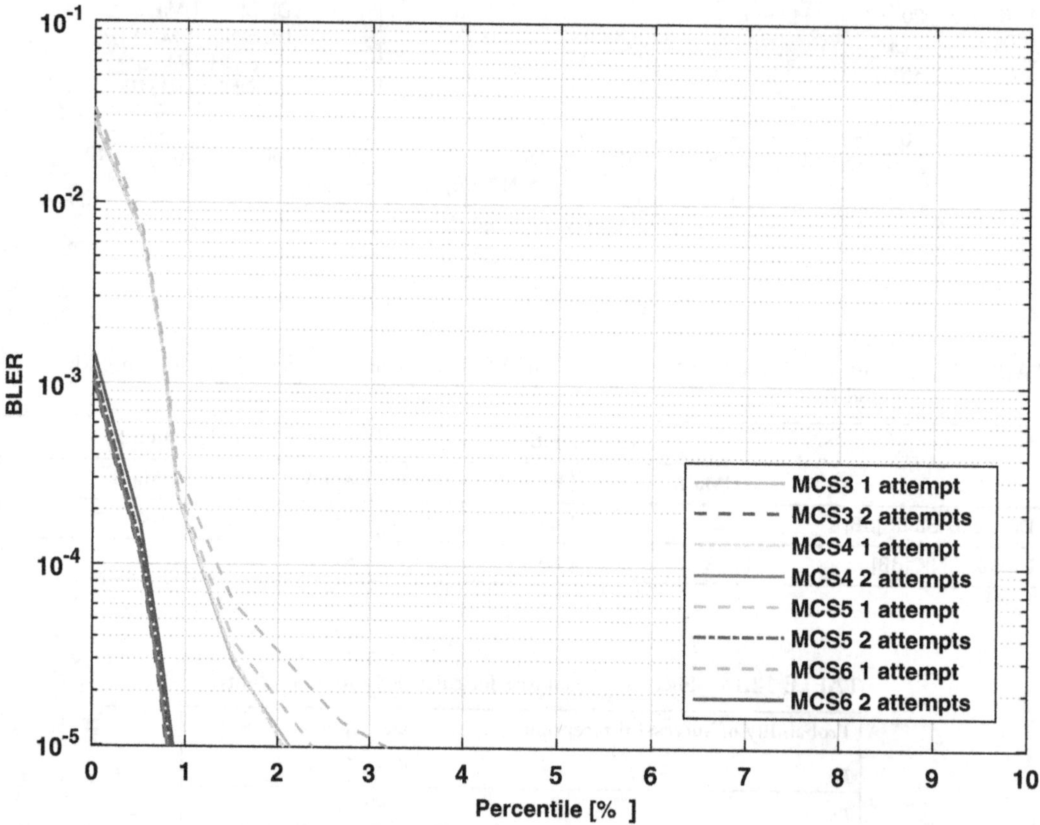

FIG. 12.8 Total reliability for downlink data as function of uplink and downlink SINR percentile, for Configuration A (top) and Configuration B (bottom) with UMa B.

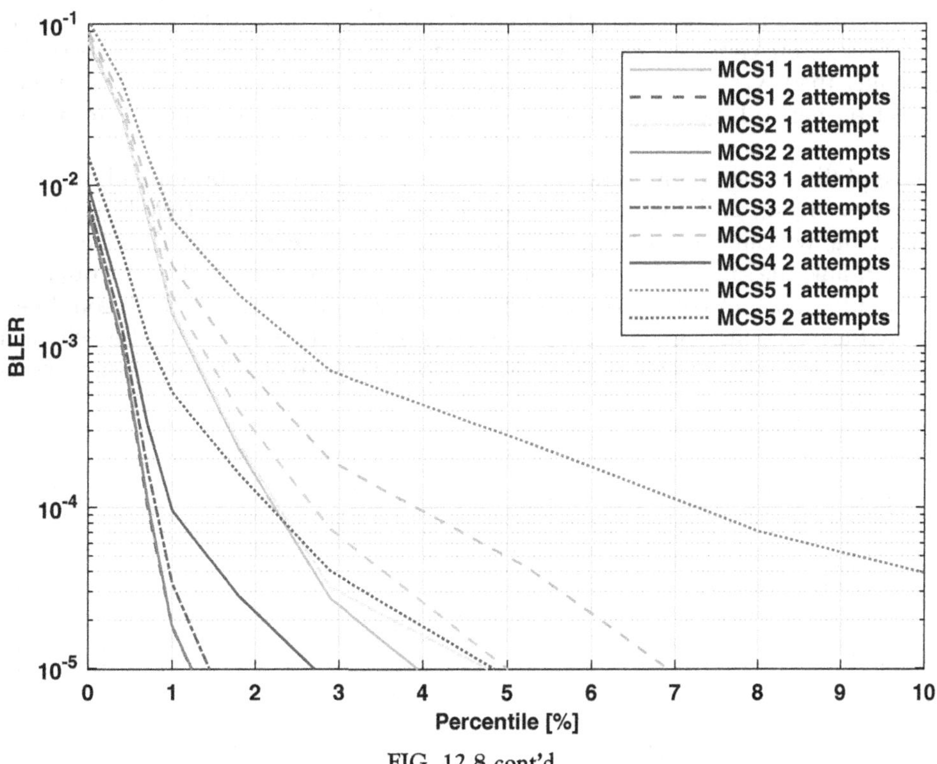

FIG. 12.8 cont'd

- In Configuration B with UMa B with a 7 OS mini-slot, one transmission attempt with MCS-3 in downlink and MCS-4 in uplink fulfills the reliability requirements (Figs. 12.8 and 12.9). This would require 34 RB in downlink and 24 RB in uplink (Table 12.15) and take 0.9 ms (Table 12.4).

Since the reliability, latency, and payload requirements can be met we can therefore conclude that the IMT-2020 requirements for URLLC in 5G are fulfilled with NR URLLC.

12.2.3 Spectral efficiency

Apart from the required performance on latency and reliability, another key value for any RAT is the spectral efficiency: the achievable bitrate per spectrum unit. The expectations on achievable URLLC rates should be set low from start, at least when compared to eMBB. This is because we know that reliability has additional cost arising from redundancy, which naturally means lower rates. While eMBB data packets can be sent with high code rate, high modulation, and using multiple MIMO layers, the URLLC data would be sent with robust transmission choices. On the other hand, the URLLC packets are expected to be much smaller than typical eMBB packets, and traffic is expected to be lower, and therefore the impact on network performance can therefore be expected to be moderate despite the lower rates.

Because of the smaller data packets, the relative overhead from control and reference signals becomes higher, and this also reduces the spectral efficiency of cMTC services. For downlink data, we can count on one DCI per data packet regardless of the size of the data packet. In uplink, the same is true for dynamic scheduling, but with configured uplink grants there is no additional downlink control overhead. However, in practice there could be a significant inefficiency with CGs if these are over-provisioned compared to the actual uplink data, meaning that we reserve transmission opportunities even if we are not sure the device will use them.

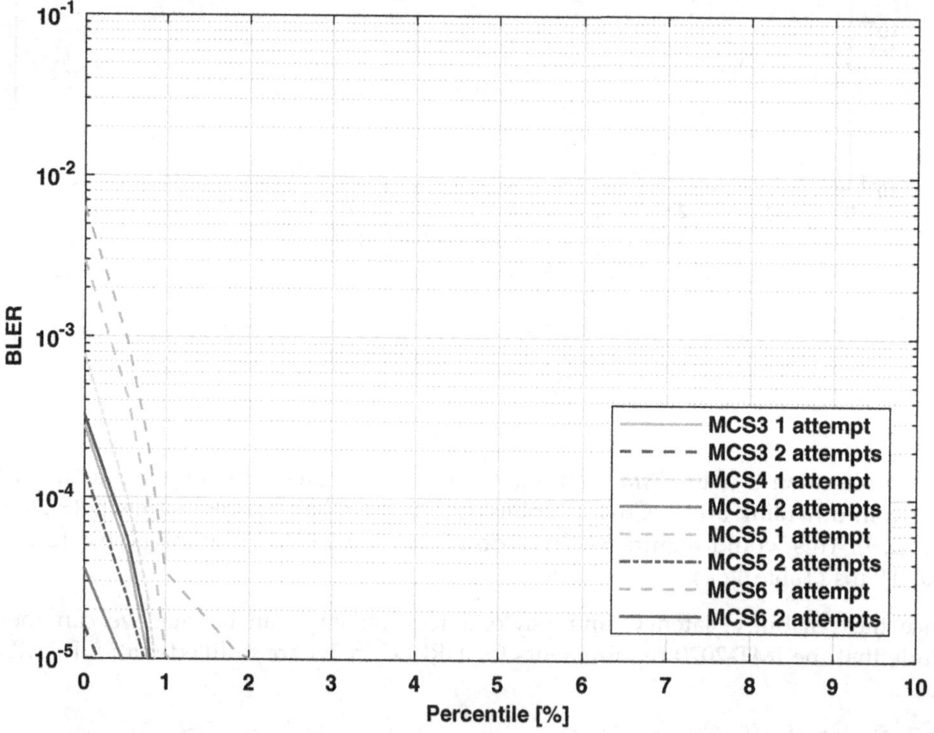

FIG. 12.9 Total reliability for uplink data (configured grants) as function of uplink and downlink SINR percentile, for Configuration A (top) and Configuration B (bottom) with UMa B.

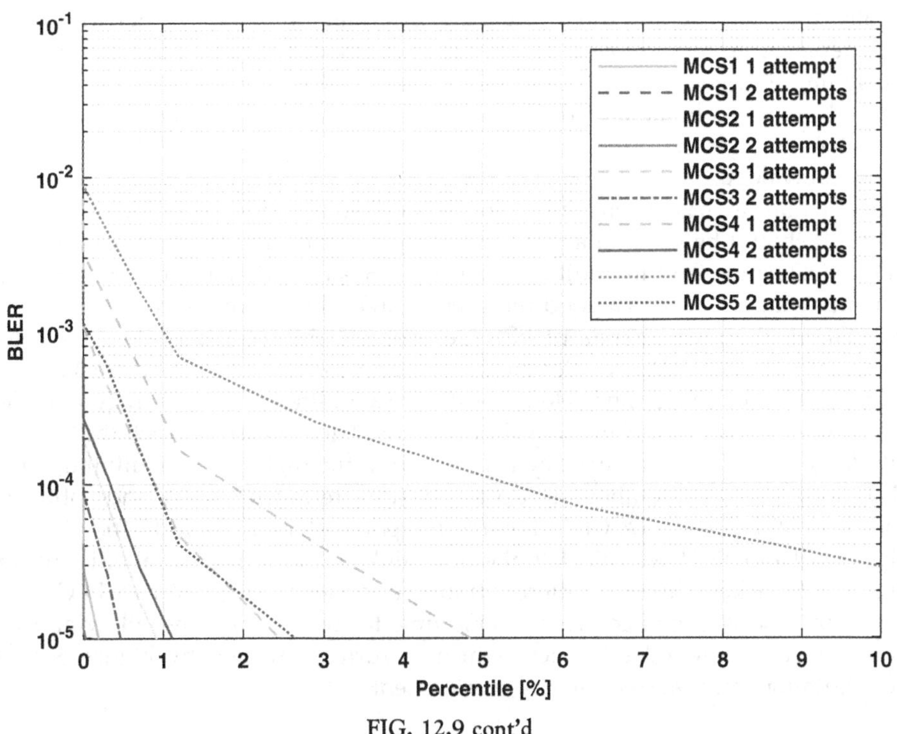

FIG. 12.9 cont'd

TABLE 12.15 Required #RBs for 32 B packet with different MCS.

	14-os TTI		7-os TTI		4-os TTI		2-os TTI	
Allocation size [RB]	**DL**	**UL**	**DL**	**UL**	**DL**	**UL**	**DL**	**UL**
MCS-1	24	22	50	46	92	92	215	274
MCS-2	20	17	41	37	77	73	178	219
MCS-3	17	14	34	29	63	57	146	171
MCS-4	14	11	29	24	54	47	126	141
MCS-5	12	9	25	19	46	37	106	111
MCS-6	11	8	22	16	40	31	93	92

Naturally, the observed spectral efficiency of a certain cMTC service depends on the scenario it is deployed in. Good coverage and high SINR means higher MCS can be used with higher resulting efficiency. We can estimate the spectral efficiency for a certain MCS at a certain SINR point. To be able to derive a simple plot we assume the HARQ feedback is sent on a channel with the same SINR as the data, that is the downlink and uplink SINR is set to be equivalent.

The evaluation is based on link simulation results with parameters according to Table 12.16. It should be noted that for the assumed carrier frequency of 3.5 GHz there are no defined FDD bands, and in addition 120 kHz SCS is only defined for FR2. Nevertheless, this frequency choice serves as a good representation of the performance of FDD in FR1 and 120 kHz TDD in the lower bands of FR2. To evaluate the total reliability, we follow the description in Section 12.2.2.3 for downlink data and for uplink data with CGs, using the same assumptions for scheduling. Here we study two different latency requirements, 1 ms and 2.5 ms, allowing for retransmissions in the higher latency case to reach the 99.999% total reliability target. The spectral efficiency is found from the highest MCS fulfilling the latency and reliability targets, while allowing up to 2 HARQ retransmissions, and subtracting for overhead from DMRS and PDCCH following the parameters in Table 12.16.

In Figs. 12.10 and 12.11 we calculate the downlink and uplink spectral efficiency, respectively, for a 32 B packet with four different transmission lengths: 2, 4, 7, and 14 OS. From the results we can observe the effect from having time for retransmissions, allowing for higher MCS which gives improved efficiency, which in turn is enabled by higher SCS, shorter mini-slot allocation, and relaxed latency requirement.

TABLE 12.16 Link simulation parameters and assumptions for spectral efficiency study.

Parameter	Value	
Carrier	3.5 GHz, 40 MHz bandwidth, {FDD, TDD}	
Subcarrier spacing	{30, 120} kHz	
Scheduling configurations	30 kHz SCS FDD {4, 7} OS TTI	120 kHz SCS TDD DL-UL-DL-UL slot pattern {7, 14} OS TTI
Channel model	TDL-C [1]	
Delay spread	300 ns	
Antenna configuration	2 TX 2 RX (data), 1 TX 2 RX (control)	
PDSCH/PUSCH	LDPC, 32 B, MCS {0,3,6,7,11,15,19} (MCS table 3), 1 frontloaded DMRS, 1-3 HARQ retransmissions	
PUCCH	2 OS with frequency hopping, 1 bit	
PDCCH	Polar code, 40 bits excl. CRC, AL-8	

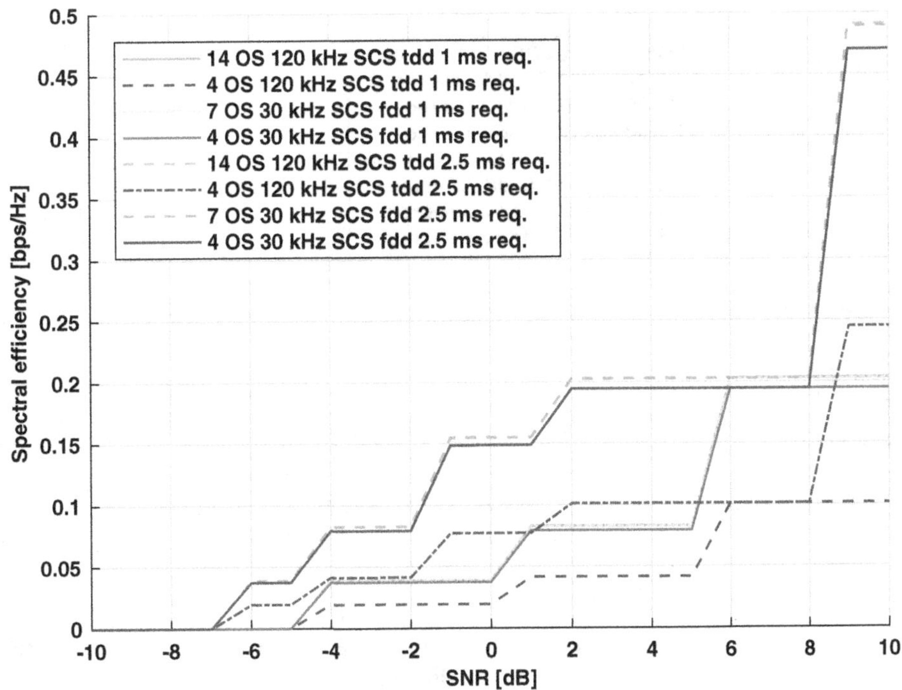

FIG. 12.10 Spectral efficiency for downlink data.

12.3 Service coverage

Defining a certain cMTC service from the requirement *"the delivery of a payload of P bytes within a latency of L milliseconds with a success probability or reliability of R"* we are now ready to study where such a service can be delivered. As a supplement to the above requirement we can then define service coverage as *"the fraction of devices in a certain population to which the service can be supplied"*. We saw in Section 12.2.2 that the fulfilment or not of the service requirement will depend on the SINR at the device location, and then the service coverage will be equivalent to studying how likely it is to have a SINR higher than a threshold value, the lowest at which the service can be delivered.

A device's experienced SINR depends on many factors, such as used transmit power, antenna configurations, channel condition and active interference. Hence, the SINR level experienced will be dependent on the load in the network, i.e. on the level of traffic. In an unloaded/low loaded system, the performance will be noise-limited (the thermal noise in the receiver is the dominant factor in determining the performance), and with growing traffic, it will become interference-limited.

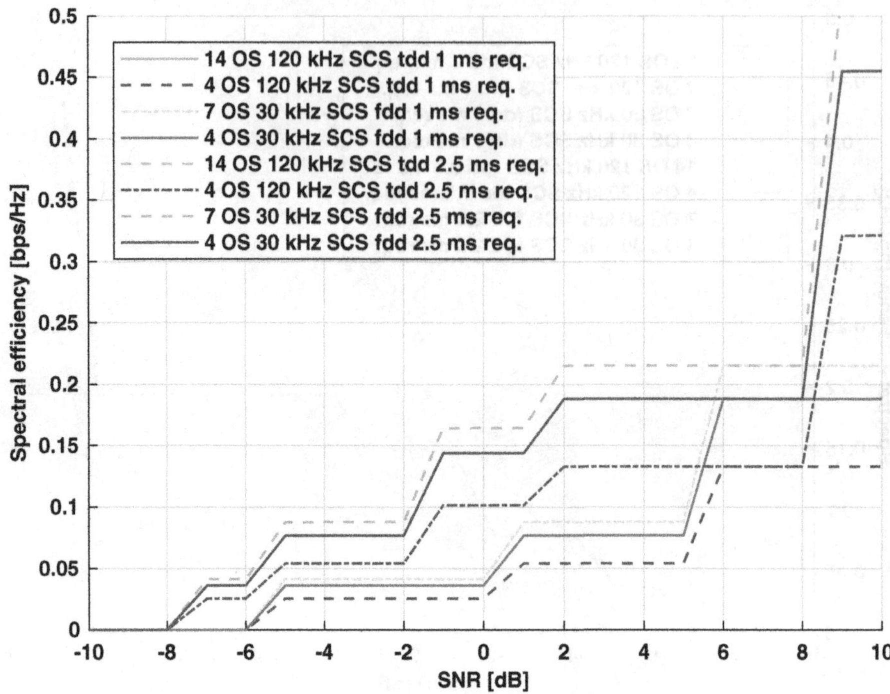

FIG. 12.11 Spectral efficiency for uplink data with CGs.

12.3.1 A wide-area service example: substation protection

For illustration of the expected cMTC service coverage, we can select a representative wide-area use case: power substation protection.

In this scenario power transformers and substations, connected on the electrical power grid, exchange sample values of current and voltage over 5G. The situation is schematically illustrated in Fig. 12.12.

In each power node, a protection unit compares the received values from the other end of the power link with its own values, and quickly breaks the power if the values diverge. In our example we define the cMTC service per NR radio link as sample values taken every 1 ms of packets of 100 bytes in uplink and downlink, which should be delivered in the gNB within 4 ms with a total reliability of 99,999%. With this dimensioning of the service, we can ensure a delivery at the other end's protection unit within 20 ms with 99,99% reliability, also taking inter-gNB transfer (assumed to take <12 ms) into account. For the deployment scenario, with parameters given in Table 12.17, we assume the *URLLC Urban Macro* case with UMa B channel [1], with all nodes placed outdoors, giving the total path gain distribution as in Fig. 12.13, including path gain, beamforming gain, and antenna gain.

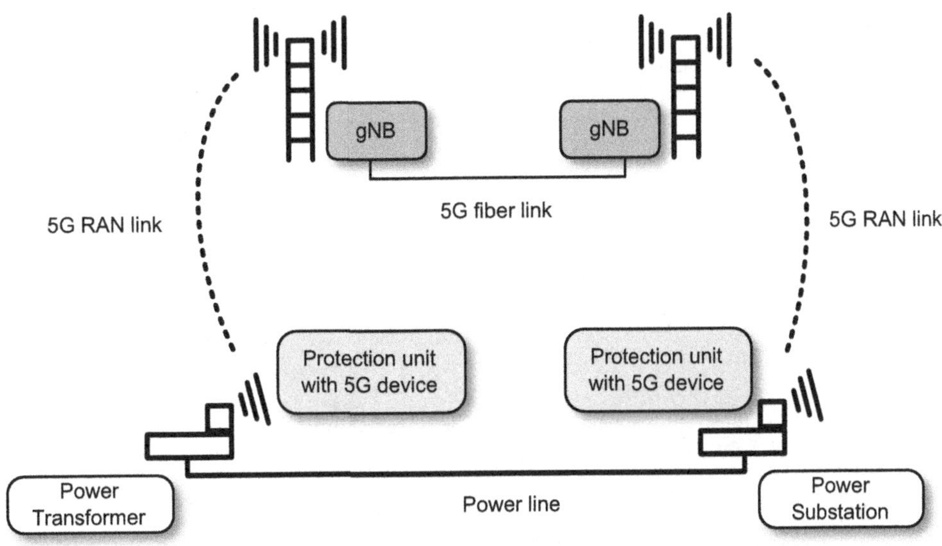

FIG. 12.12 Illustration of power substation scenario setup.

TABLE 12.17 System simulation parameters.

Parameter	Value
Carrier frequency and bandwidth	3.5 GHz, 40 MHz
5G gNB inter-site distance	500 m
Site configuration	3-sector (3 cells/site)
gNB height	25 m
gNB antenna power	40 W (per cell)
gNB noise figure	5 dB
gNB antenna down tilt	5°
Propagation model	ITU Uma B [1]
cMTC service traffic	100 B packet every 1 ms
cMTC service latency requirement	4 ms
cMTC service reliability requirement	99.999%
Device height	1.5 m
Device noise figure	7 dB
Uplink power control	10 dB SNR target, $\alpha = 0.8$
URLLC device deployment	100% outdoors Random, average 1/cell, 5000 positions
eMBB device deployment	80% indoor, 20% in-car, 5000 positions
URLLC device antenna	Omni, 1 W
eMBB device antenna	Isotropic, 0.2 W

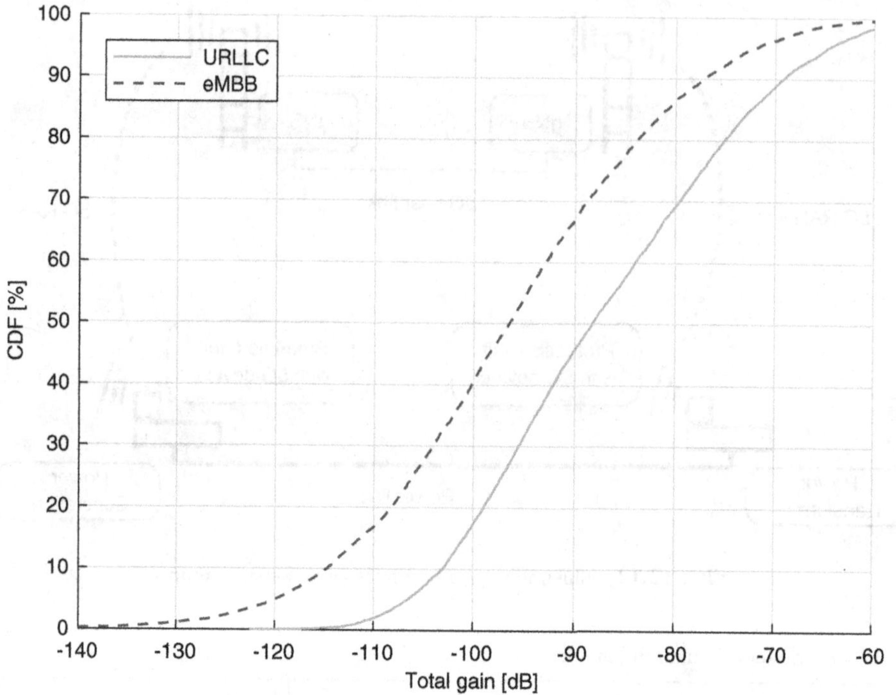

FIG. 12.13 Gain distribution for eMBB and URLLC devices.

In this case, "*URLLC*" is defined as the devices connected to protection units of the stationary power substations placed outdoors, while "*eMBB*" represents mobile devices that can be either inside buildings or outdoors.

The NR system configuration is set according to Table 12.18, assuming a realistic TDD pattern for eMBB services. In the used 3.5 GHz band, the SINR distribution in an unloaded respective fully loaded case is shown Fig. 12.14.

As can be seen, the impact on the interference is significant when there is a high load in the system, pushing the median of the SINR distribution by roughly 20 dB in the downlink and roughly 3–4 dB in the uplink. The lower impact on uplink is due to efficient power control that limits the interference.

The activity level in the system, the *utilization,* will mainly be set by the traffic of eMBB users in the system, and will therefore vary. At the same time we expect the service to work well regardless of traffic level and without having to coordinate transmissions between cells to minimize interference.

We are now ready to study the performance of the substation protection service on the selected NR system in the chosen scenario. Calculating the total reliability as function of SINR as in the previous sections, using link simulation with parameters as in Table 12.19,

TABLE 12.18 RAN system parameters.

Parameter	Value
gNB antenna (Vert. × Horiz. elements)	4-col. (8 × 4)
Number of antenna ports	32
5G RAN config.	NR TDD (downlink-downlink-downlink-uplink slot pattern)
Subcarrier spacing	30 kHz
TTI length	0.5 ms (14 OFDM symbols)
Carrier frequency	3.5 GHz
Carrier bandwidth	40 MHz

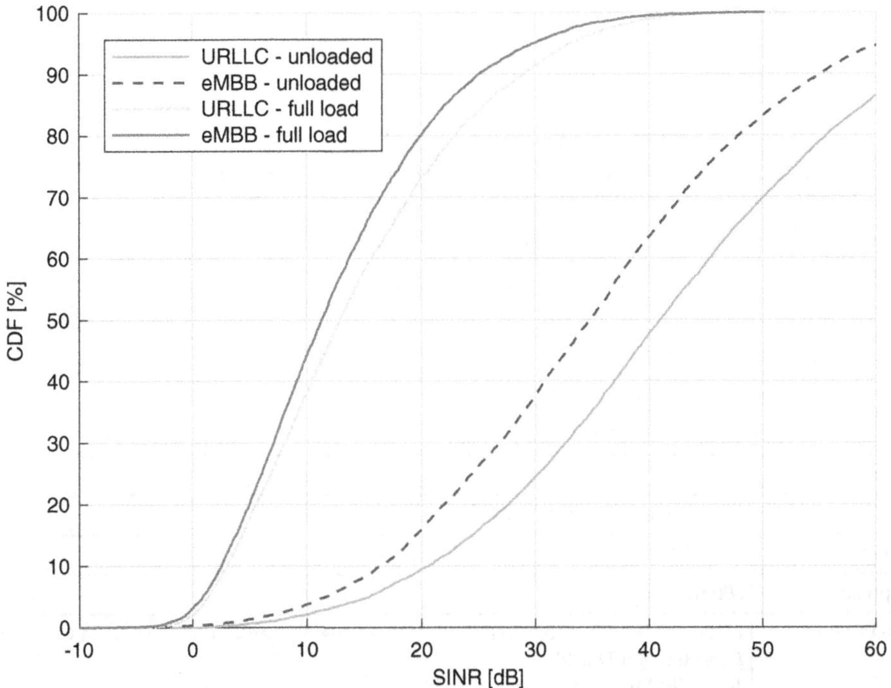

FIG. 12.14 Uplink and downlink SINR for eMBB and URLLC data for an unloaded (blue) (gray in print version) and highly loaded scenario (red) (light gray in print version).

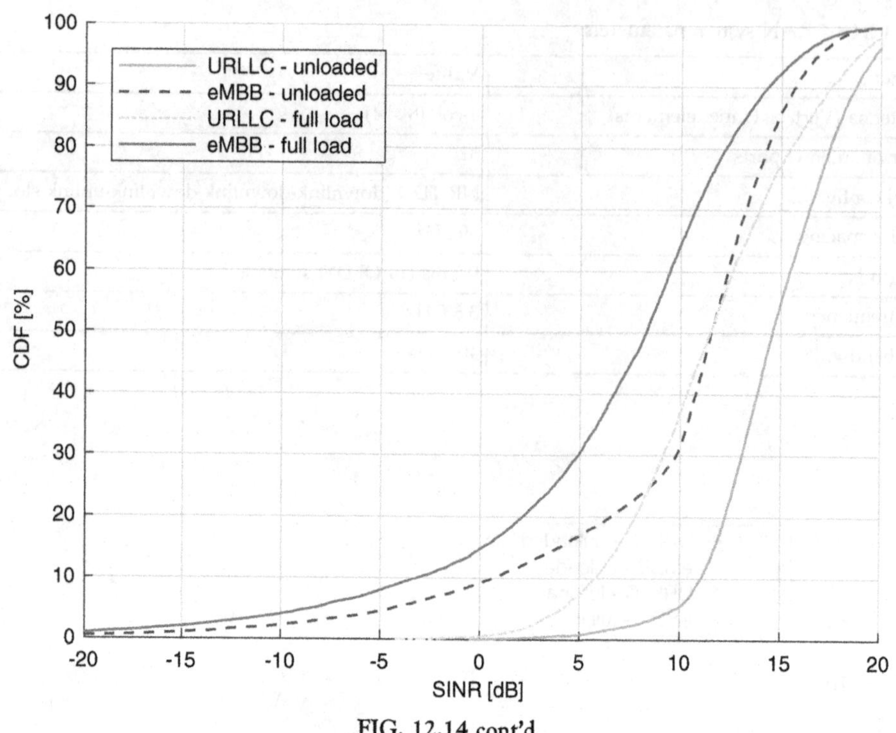

FIG. 12.14 cont'd

TABLE 12.19 Link simulation parameters.

Parameter	Value
Carrier	3.5 GHz, 40 MHz bandwidth
Channel model	TDL-C [1]
Delay spread	300 ns
PDSCH/PUSCH	LDPC, 32 B, MCS {0,3,6,7,11,15,19} (MCS table 3), transmission length 14 OFDM symbols, 1 frontloaded DMRS, 1-3 HARQ retransmissions
PUCCH	2 OS with frequency hopping, 1 bit
PDCCH	Polar code, 40 bits excl. CRC, AL8

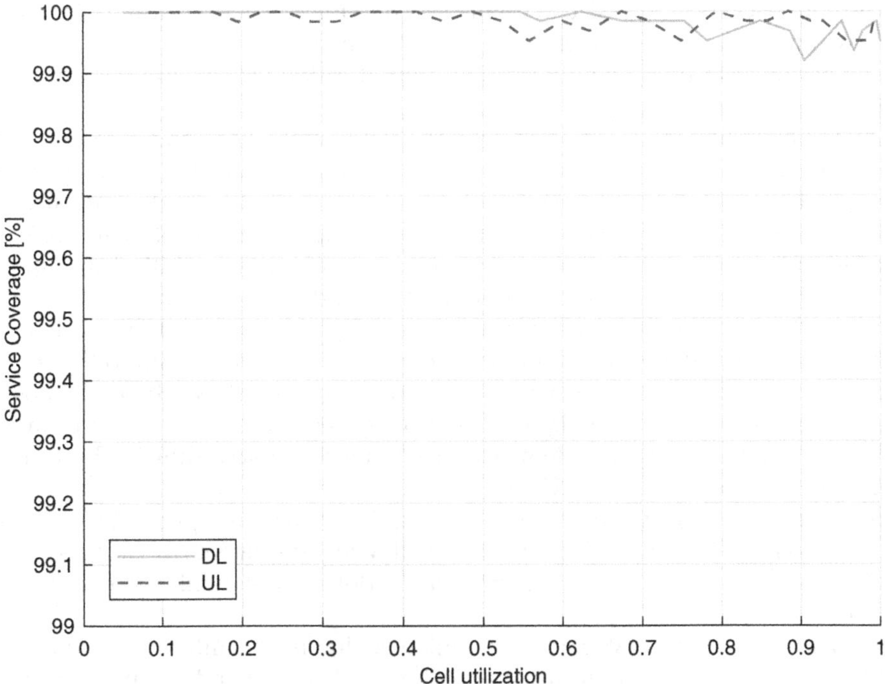

FIG. 12.15 cMTC protection service coverage for the studied substation scenario.

and comparing with the service definition we find the service coverage as of Fig. 12.15. The results indicate that the cMTC protection service can be consistently delivered by NR URLLC to more than 99.9% of studied device locations at high cell loads (up to 90% cell utilization), implying that only one in 1000 substation locations can't be reliably protected with the studied setup. Since the substations are stationary we would expect the coverage to be rather consistent over time, and for substations close to the coverage edge, extra measures such as directional antennas could be taken to improve coverage.

12.3.2 A local-area service example: factory automation potential

We also want to study whether NR URLLC has the potential to handle more challenging cMTC scenarios. To find out we can look at the use case of factory automation, and specifically wirelessly providing motion commands and sensor updates to and from industry robots. In this example we define the cMTC service as delivering sensor and actuator data every 1 ms with packets of 32 B size in uplink and downlink. These critical packets should be delivered within 1 ms with a reliability of 99,999%. The industry robots equipped with NR devices are assumed to be randomly moving around in the factory and performing critical tasks and full coverage is therefore expected.

The factory environment is modeled around 10 assembly lines and a central aisle. The assembly lines are surrounded by 3 m high and 0.2 m wide metallic fences, separated by 0.2 m in rows. Other types of equipment and objects are modeled as random metallic blockers distributed within the factory volume, as illustrated in Fig. 12.16. The blockers are rectangles with a uniform height distribution in the range 1–3 m, a uniform width distribution in the range 1–2 m, having a random rotation around the vertical axis, and placed randomly in the factory with a density of 0.1 blockers/m^2. Including blockers and fences in the model has an impact on the radio coverage in the factory, adding more realism to the scenario.

In the ceiling of the factory hall, different arrangements of NR base stations can be mounted. In this scenario, the configurations investigated are 1, 2, 4 or 6 base stations, see illustration in Fig. 12.17. Other system parameters used are according to Table 12.20 and link parameters according to Table 12.21.

The used carrier is 30 GHz, where a downlink-uplink-downlink-uplink TDD slot pattern is run with 120 kHz SCS and 7 OS mini-slots. Following the latency evaluation in Section 12.2.1.2.2 we find that 1 transmission attempt can be done in downlink and uplink, assuming CGs, within 1 ms.

Based on the resulting total gain distribution including beamforming gain, shown in Fig. 12.18, we can derive the SINR distribution for the 30 GHz band, in an unloaded respective fully loaded case, shown in Fig. 12.19.

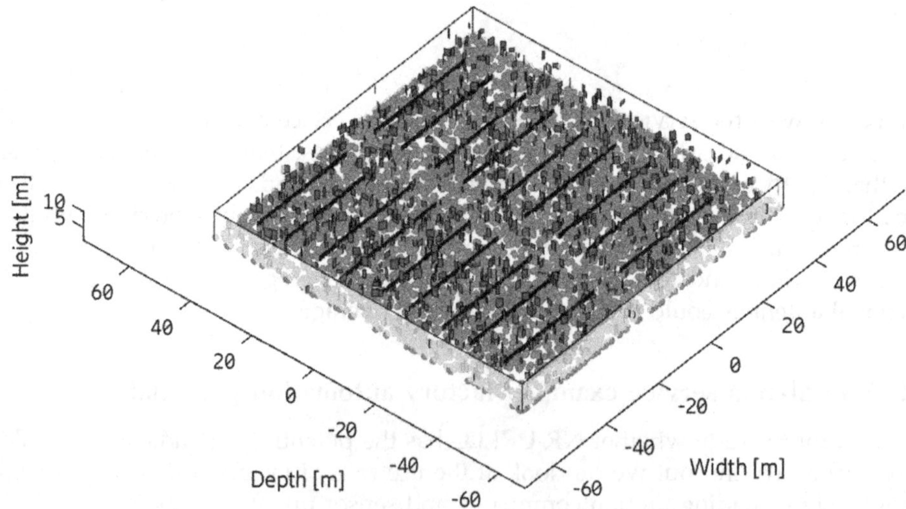

FIG. 12.16 Factory hall model with metallic fences (black) (dark gray in print vesrion), random blockers (red) (light gray in print version), and device positions (blue) (gray in print version).

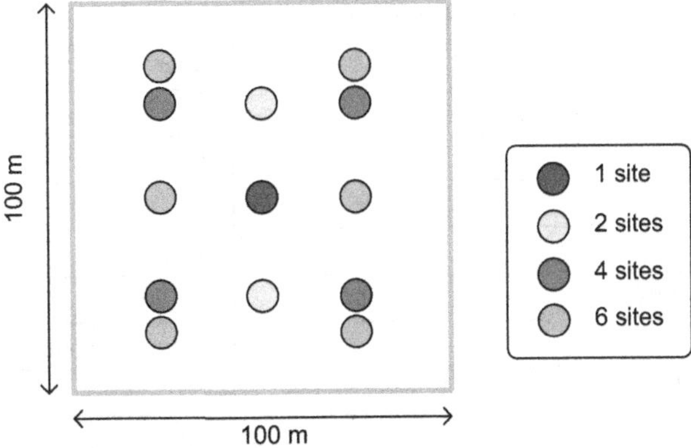

FIG. 12.17 Studied gNB site deployments in the factory ceiling.

TABLE 12.20 System simulation parameters for factory scenario.

Parameter	Value
Frequency [GHz]	30
RAT	NR
Bandwidth [MHz]	200
Duplex	TDD, downlink-uplink slot pattern
Scheduling	7 OS mini-slot in downlink and uplink Configured uplink grant with 1 TTI periodicity
Site configuration	3-sector (3 cells/site)
gNB Transmit Power [dBm]	33
gNB Antenna element gain [dBi]	8
gNB Antenna Array VxH x (VxHxP)	4 × 8,1 × 2, 2 × 2, 1 × 4, 2 × 4, 4 × 4, 2 × 8, and 4 × 8 panels of (2 × 1 × 2) antenna elements
gNB noise figure [dB]	7
Antenna tilting	Optimized to improve capacity
Device Transmit Power [dBm]	23
Device Antenna Gain [dBi]	9
Device Antenna configuration	Omni-directional
Device noise figure [dB]	10
Uplink power control	SNR based: Target SNR = 10 dB, $\alpha = 0.8$

TABLE 12.21 Link simulation parameters for factory scenario.

Parameter	Value
Frequency [GHz]	30
Subcarrier spacing [kHz]	120
OFDM symbols per TTI	7
TTI length [μs]	63
Device processing delay [OS]	$N_1 = 20$, $N_2 = 36$
Message Payload [B]	32 (downlink and uplink)
Latency requirement [ms]	1
Reliability requirement	99.999%
Channel model	TDL-D [1] with line-of-sight conditions
Delay spread [ns]	30
Device speed [km/h]	3
Traffic	Periodic
Modulation schemes	QPSK, 16 QAM, 64QAM
Code rates	{30, 64, 120, 251, 340, 438, 449, 490, 567}/1024

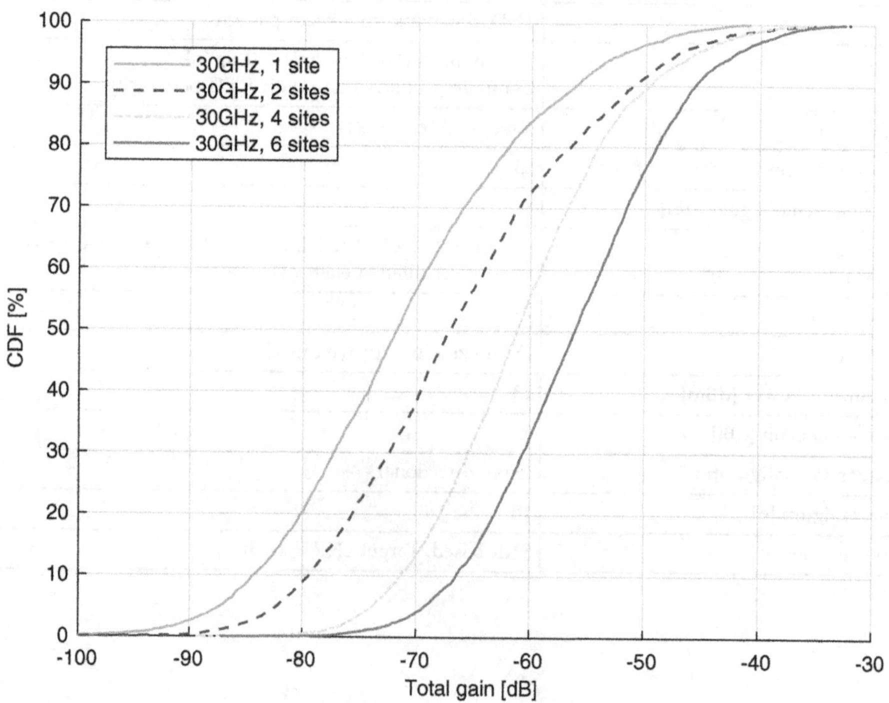

FIG. 12.18 Total gain including beamforming gain for factory scenario.

Starting from the SINR distributions we can then study the performance of the factory automation cMTC service running on the NR systems in the studied scenario. Calculating the total reliability as function of SINR using the methods outlined in the previous sections, and comparing with the service definition (payload, reliability, and latency requirements) we can find the service coverage, meaning the fraction of the factory positions where the cMTC service can be delivered. Requiring 100% service coverage for the cMTC service in the factory, we can then find the maximum traffic that the system supports in terms of a total number of data packets per second. This is shown in Fig. 12.20 for the two studied bands. Clearly, full coverage of the cMTC service can be delivered by NR URLLC in both directions in the factory, and a high rate of packets can be supported.

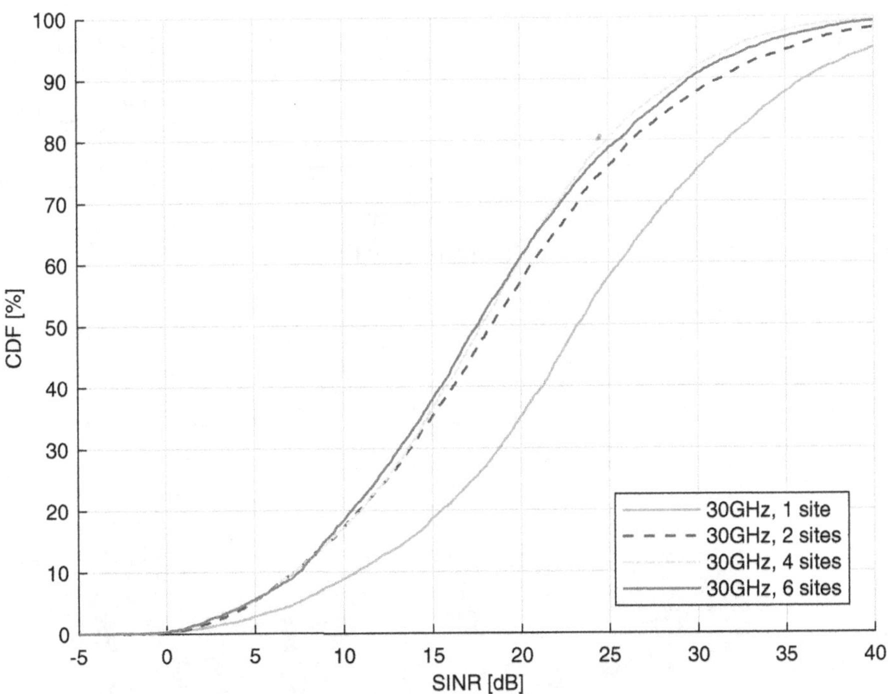

FIG. 12.19 Full load downlink and uplink SINR for factory scenario with different gNB site deployments.

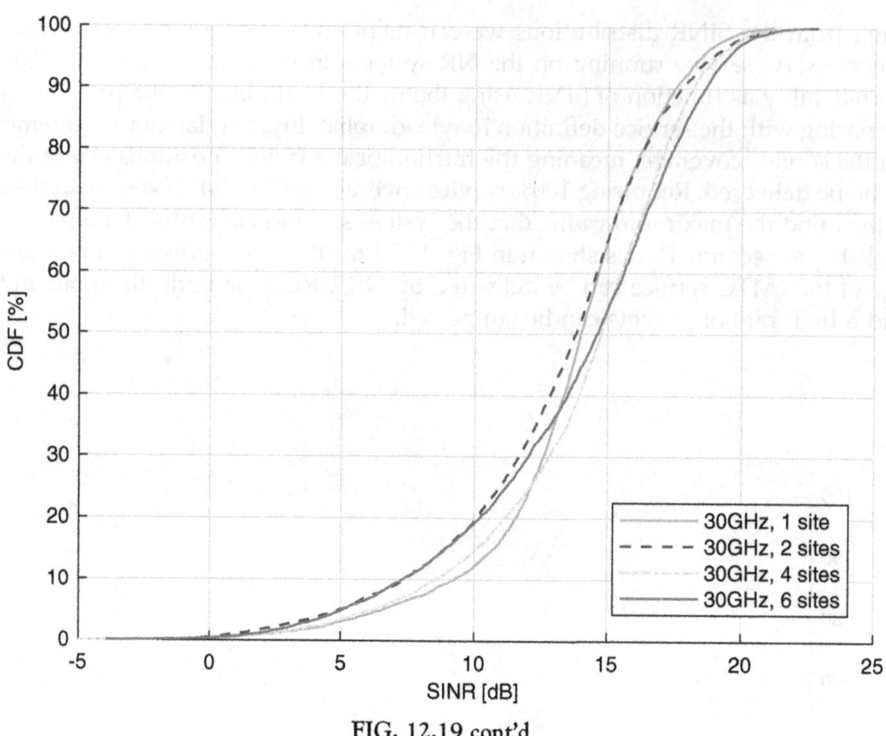

FIG. 12.19 cont'd

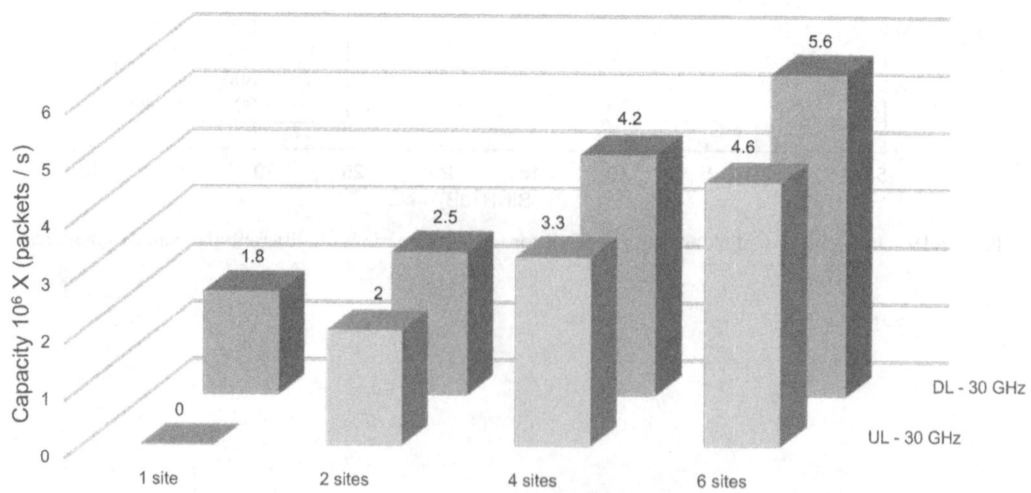

FIG. 12.20 Downlink and uplink system capacity at 100% service coverage for factory scenario with gNB densification.

References

[1] ITU-R M.[IMT-2020.EVAL].
[2] 3GPP TR37.910, "study on self evaluation towards IMT-2020 submission".
[3] 3GPP R2-1802686, "RRC device processing time for Standalone NR", Ericsson.

CHAPTER

13

Enhanced LTE connectivity for drones

OUTLINE

Abstract

This chapter presents new features introduced in 3GPP LTE Release 15 for enhancing wide-area connectivity for drones beyond visual line-of-sight. It starts with descriptions on different propagation channel characteristics for drones at a high altitude, and compare them to conventional terrestrial network channel models. These different propagation channel characteristics present certain unique challenges for providing cellular connectivity to drones from both device and network perspectives. This chapter continues on and describes how these challenges are addressed by the new features introduced in 3GPP LTE Release 15.

13.1 Introduction

Drones, also known as unmanned aerial vehicles (UAV), have in recent years gone much beyond being toys of hobbyists to becoming the centerpieces of many innovative use cases

Cellular Internet of Things, Second Edition
https://doi.org/10.1016/B978-0-08-102902-2.00013-3

© 2020 Elsevier Ltd. All rights reserved.

which have a potential to bring significant social-economic benefits. For example, drones are increasingly used for aiding search, rescue, and recovery missions during or in the aftermath of natural disasters like tornados [1], wildfires [2], hurricanes [3], etc. The commercial use cases of drones are also developing rapidly, including package delivery, inspection of critical infrastructure, surveillance, agriculture, etc. [4]. Wide-area connectivity is considered as one of the most critical technological components for fully realizing the potential of drone applications. An important aspect is command-and-control communications with drones beyond *visual line-of-sight* (VLOS). Reliable command-and-control communications are critical for drone operation safety and for enforcing aviation rules. Furthermore, many of the aforementioned drone use cases require connectivity, e.g., for delivering real-time imagery or video.

Recognizing an opportunity to expand the use cases of cellular networks for the *flying things*, 3GPP started a study on *Enhanced LTE Support for Aerial Vehicles* in 2017 [5]. The study focused on new aspects that are associated with connecting the flying things, in comparison with connecting the terrestrial devices. These aspects include.

- Propagation channel characteristics
- Downlink interference - aerial vehicles as victims
- Uplink interference - aerial vehicles as aggressors
- Handover performance
- Identification of aerial vehicles

All of these aspects are discussed in this chapter.

The study produced a comprehensive technical report [6] and concluded that LTE networks prior to Release 15 are already capable of serving flying drones; however, as the number of flying drones increases, certain challenges emerge. 3GPP then took action in Release 15 to introduce enhancement features for serving drones more efficiently and for better managing the impact on terrestrial devices should more flying drones are connected to LTE networks in the future. In Section 13.4, these enhancement features are described. Release 15 targets aerial vehicles of maximum height 300 m *above ground level* (AGL) and maximum speed 160 km/h [6]. Throughout this chapter, we will use drone, UAV, and aerial vehicle interchangeably. However, most descriptions in this chapter apply generally to cellular connectivity for all flying things which meet the aforementioned AGL and speed criteria.

13.2 Propagation channel characteristics

The propagation channel characteristics experienced by a drone is very different, compared to a terrestrial device. Therefore, understanding the difference in propagation channel characteristics provides a great insight in the potential drone communication problems that need to be addressed.

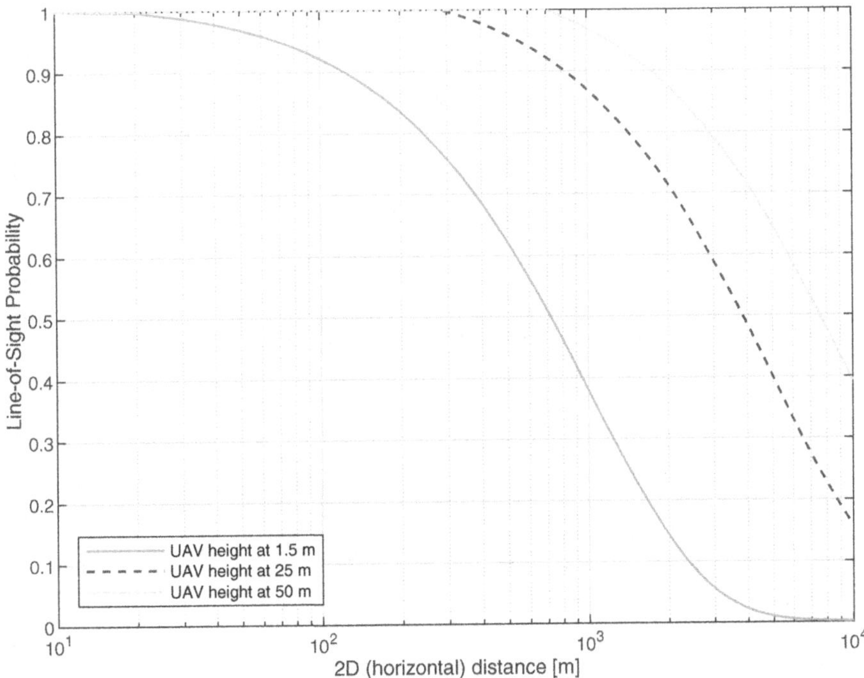

FIG. 13.1 Line-of-sight propagation channel probability for drones at different heights (rural macro-cell environment).

When a drone is at a height above buildings and trees, it has a much greater line of sight (LOS) probability to the base station antenna. Fig. 13.1 shows LOS probability in a rural macro-cell environment at various heights as a function of the horizontal distance to the base station, based on the channel model in Ref. [7]. It can be seen that the LOS probability increases as the drone flies higher. As the drone moves farther away from the base station, the elevation angle between the drone and base station antenna decreases. This increases the probability of having an object on the signal path, and therefore decreases the LOS probability. This phenomenon can also be interpreted from Fig. 13.1.

With a LOS propagation channel, there is no object on the signal path causing penetration, refraction, diffraction, and reflection loss. As a result, the overall path loss is smaller than in a non-line-of-sight (NLOS) channel when compared at the same distance. An example based on the channel model in Ref. [7] is shown in Fig. 13.2. Comparing the path loss between a drone at 50 m height with a LOS channel and a terrestrial device with NLOS channel in a rural macro-cell environment further illustrates this point. Observe that the path loss slope is smaller in the drone LOS case. The terrestrial device at 680 m from the base station experiences a 110 dB path loss. In comparison, the same path loss is experienced by the drone at 7380 m from the base station.

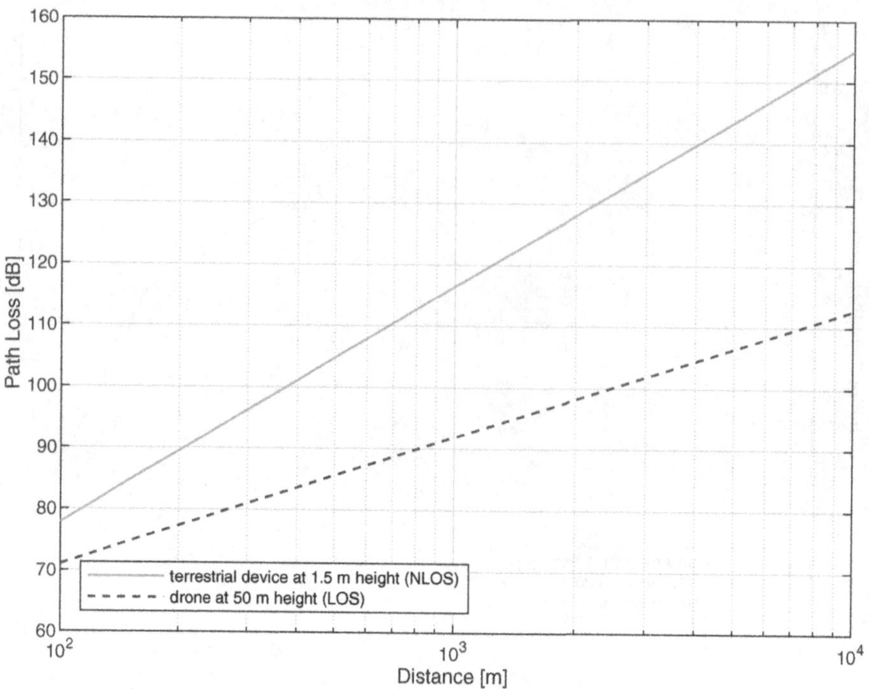

FIG. 13.2 Comparison of path loss in rural macro-cell environment with carrier frequency at 700 MHz between a terrestrial device at 1.5 m height with an NLOS propagation channel and a drone at 50 m height with an LOS propagation channel.

To complete the whole picture of drone propagation channel characteristics, it is also important to examine the base station antenna pattern. In a cellular network, a base station antenna is typically down-tilted to provide good terrestrial coverage. Antenna down-tilting also helps in alleviating inter-cell interference. Antenna down-tilting, however, is far from optimal in terms of aerial coverage. An example of base station antenna pattern is shown in Fig. 13.3 and 13.4 based on a column of 8 antenna elements. Fig. 13.3 shows antenna gain over azimuth angle ϕ. The antenna pattern in the example is a sector antenna, so the antenna pattern is designed to have sufficient antenna gains in the desired sector coverage area, which spans horizontally over certain azimuth angles centered at the boresight (0°). The sector antenna pattern is also designed to have an overlapped region between neighboring sectors in the sense that the antenna gain at the sector border is not too low. This is desirable for a terrestrial device to be handed over to an adjacent sector before the signal strength of its current serving sector drops too low. The antenna pattern example in Fig. 13.3 is for a 120-degree sector. The antenna gain at the sector border, i.e. 60° from the boresight, is approximately 7 dB below the peak.

Fig. 13.4 shows the antenna gain pattern over a range of elevation angles θ. In this example, the antenna is down-tilt by 3° and therefore the boresight points to the direction of 87-degree elevation angle. (The reference directions are 0 and 90° for the ground and

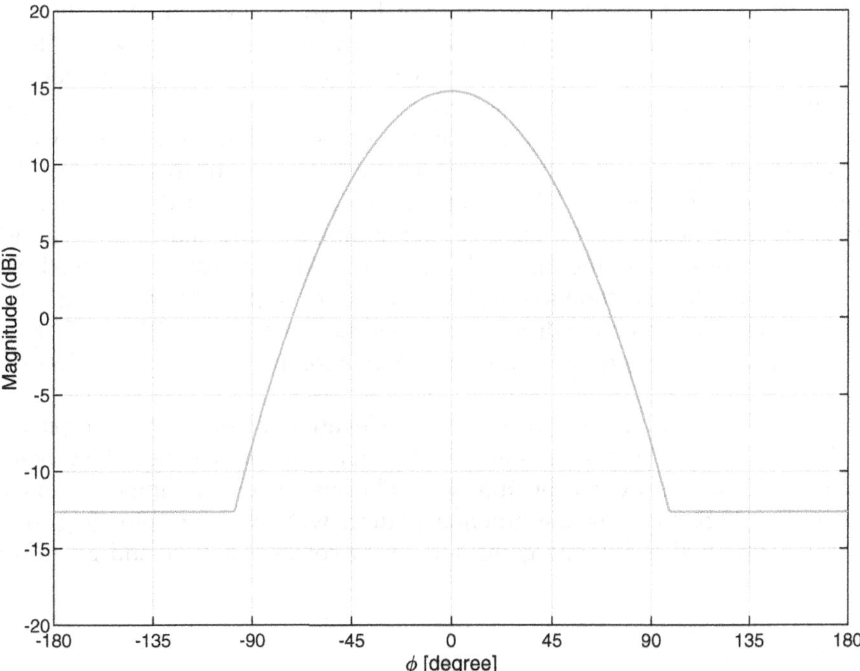

FIG. 13.3 Horizontal (azimuth) BS antenna pattern (sector antenna with a column of 8 antenna elements).

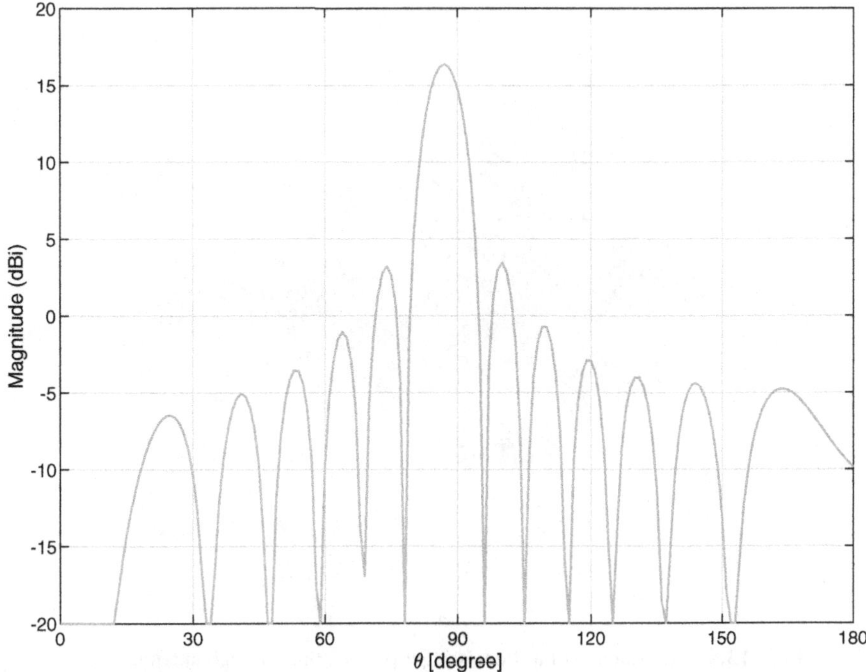

FIG. 13.4 Vertical BS antenna pattern (sector antenna with a column of 8 antenna elements).

the horizon, respectively.) Observe that less than 10-degree elevation angle above the boresight, there is a null, and in that direction, the antenna gain is very small-more than 35 dB lower compared to the peak antenna gain. Further increasing the elevation angle from the null, the antenna gain increases compared to the null, although still more than 15 dB lower than the highest antenna gain in the boresight direction. The elevation angles between the nulls, but not within the main-lobe in the boresight direction, can be said to be the antenna vertical sidelobes. These vertical sidelobes allow a drone flying above the base station antenna height to be covered. Although the antenna gain in a sidelobe is 10–25 dB lower compared to the main-lobe, the reduced path loss experienced by a drone with LOS can offset such reduction in antenna gain. For example, observe in Fig. 13.2 that the difference in path loss between the drone and terrestrial device is 24 dB at a 1 km distance from the base station. This difference can very much compensate the reduction in antenna gain.

The nulls of the antenna pattern at certain elevation angles however post a bigger challenge. When a drone flies at a constant height over a certain distance, the elevation angle changes, and the drone may enter the antenna null of its current serving cell. An example is illustrated in Fig. 13.5, showing the antenna pattern with warmer color (e.g. orange and yellow) representing higher antenna gains and cooler colors (e.g. blue and green) representing lower antenna gains.

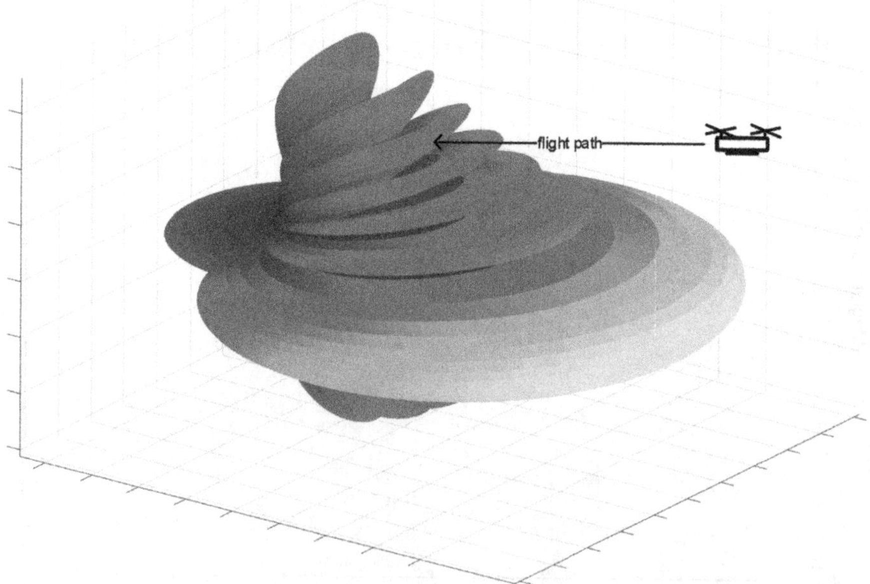

FIG. 13.5 An example of a drone's flight path cutting through antenna nulls.

13.3 Challenges

The propagation channel characteristics as described in Section 13.2 gives rise to challenges in connecting drones to cellular networks.

- Due to higher LOS probability and reduced path loss, a drone flying above a certain height may receive downlink signals with good signal strength from a greater number of cells. One of these signals would correspond to the serving cell; however, all the other signals would contribute to interference. Thus, a drone could be a victim of pronounced inter-cell interference, which posts challenges to maintaining good signal-to-noise-plus-interference ratio (SINR) in the downlink. An example is shown in Fig. 13.6. It shows the distributions of the SINR for drones and for terrestrial devices, respectively. The scenario is set up according to the rural macro-cell scenario described in Section A.1 of Reference [6]. Furthermore, a worst-case scenario where all the cells are fully loaded is assumed. It can be seen that the SINRs experienced by drones are much lower compared to those of terrestrial devices.
- Also due to higher LOS probability and reduced path loss, the uplink signal from a drone flying above a certain height may reach a greater number of cells, causing interference to many cells. Thus, in the uplink, the drone may be an aggressor. As some of the drone use cases require drone transmitting a rather large uplink payload, e.g. video

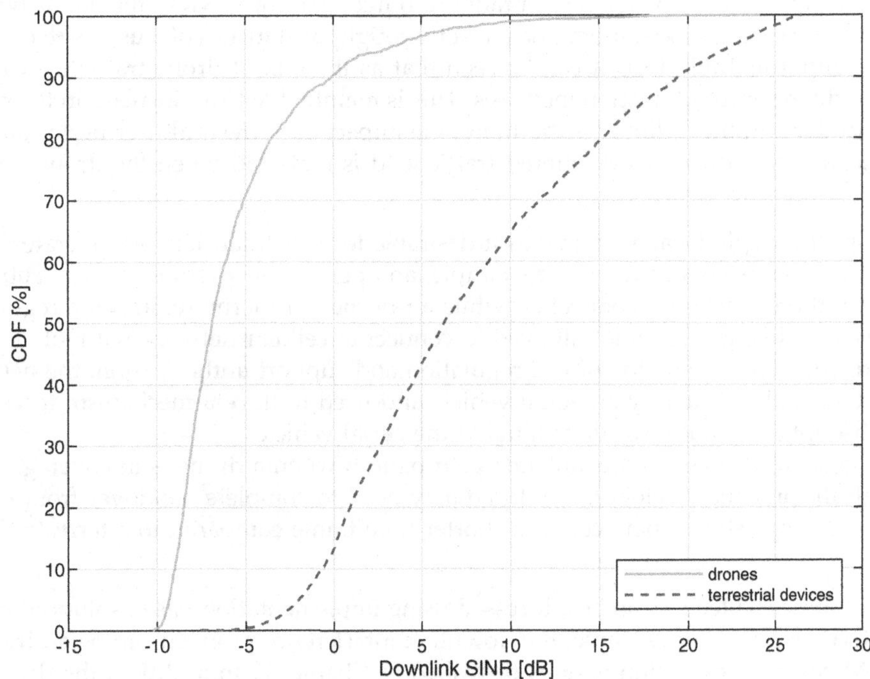

FIG. 13.6 Downlink SINR distribution in a rural macro-cell environment.

TABLE 13.1 Impact of drones on terrestrial device uplink performance in a rural macro-cell scenario defined in Ref. [6]. (Case 1: no drones in the network; Case 2: 1 drone in every 10 sectors; Case 3: 1 drone per sector; Case 4: 3 drones per sector).

Offered traffic per cell	**3.85 [Mbps]**				**7.45 [Mbps]**		
Drone density	Case 1	Case 2	Case 3	Case 4	Case 1	Case 2	Case 3
Resource Utilization [%]	20.00	20.00	22.14	28.30	50.00	51.06	70.27
5th-percentile terrestrial user throughput [Mbps]	7.59	7.59	6.64	4.50	2.91	2.71	1.11
medium terrestrial user throughput [Mbps]	20.20	20.00	18.22	13.74	10.80	10.32	5.73
Mean terrestrial throughput [Mbps]	18.17	18.06	16.69	13.32	11.79	11.37	7.28

traffic, the interference generated by a drone in the uplink could significantly reduce the uplink capacity in the network.

The results shown in Table 13.1, taken from Section D.2.1 in Ref. [6], illustrate such phenomenon. The density of drones in the network is increased from *Case 1* (no drones in the network) to *Case 4* (3 drones per sector). However, the offered traffic per cell is fixed at either 3.85 Mbps or 7.45 Mbps. Thus, as the drone density increases, the traffic mix is more from drones.

The offered traffic represents the average rate of the traffic arriving at a cell, averaging over time and over all the cells in the network. When a packet arrives, how fast the packet can be delivered depends on the availability of radio resources and the SINR of the link between the device and the serving base station. The packet throughput statistics of a user is referred to as user throughput in Table 13.1. It can be seen that as the mix of drone traffic increases, the network radio resource utilization increases. This is mainly due to an increase in the network interference level in the uplink. Furthermore, the impact on terrestrial user performance can be significant when the network offered traffic load is high and when the drone density is high.

- The impact on uplink capacity makes it desirable for a cellular network operator to regulate air-borne connectivity. For example, an operator may require special subscription to authorize network connection while air-borne. Furthermore, in some regulatory regions, aerial vehicles are not allowed to connect to cellular network without authorization. To be able to enforce regulation and support authorization, the network needs to be able to identify an aerial vehicle and need to have a mechanism to verify the subscription or authorization status of the aerial vehicle.
- Due to a sharp drop-off in the antenna gain pattern when a drone is in coverage through the antenna sidelobes, the drone may need to complete handover from its current serving cell to a new cell in a shorter time frame compared to a terrestrial device.

Some of these challenges can be addressed using implementation-based solutions or using existing LTE features. For example, the downlink interference problem can be addressed by the LTE-M coverage extension feaures described in Chapter 7. In addition, the drone radio receiver can implement receiver beamforming to focus its receive beam to the direction of the desired signal. With this, the interference from other cells, not in the same direction as the desired signal, will not cause pronounced performance degradation. The uplink interference problem can be addressed using power control techniques described in Ref. [8] and *coordinated multi-point* based approaches described in Ref. [9].

Aerial vehicle identification problem can also be addressed through implementation-based solutions. Examples can be found in Ref. [10], where machine learning approaches are applied to device measurement reports to classify whether the device is in flying mode or not. The radio measurements can be, e.g., *received signal strength indicator* or *reference signal received power*.

In the next section, we describe the LTE features specified in 3GPP Release 15 for addressing the challenges described in this section.

13.4 LTE enhancements introduced in 3GPP Rel-15

13.4.1 Interference and flying mode detection

To assist the network in identifying situations where a drone is generating significant uplink interference in *radio resource control* (RRC) connected mode, LTE in Release 15 added two reporting events H1 and H2.

- Event H1: the aerial vehicle height is above a threshold
- Event H2: the aerial vehicle height is below a threshold

The threshold is configured by the network. 3GPP Release 15 supports the height threshold as high as 8880 m above the *sea level*. The two new events can trigger a drone reporting its height and location information. The drone may additionally include its horizontal and vertical speed information in the report. Such reporting helps identify a drone above or below the network set height threshold. Obviously, this feature can be directly used for flying mode detection. In addition, as discussed in Sections 13.3, when the drone is flying above a certain height it may be a victim of inter-cell interference in the downlink and an aggressor for uplink interference. This feature can therefore also be used for interference detection.

In addition to the two newly introduced events H1 and H2, extensions were also introduced to existing events such as A3, A4, and A5 to help interference and flying mode detection. Events A3, A4, and A5 are mobility events defined for RRC connected mode. The definitions of these events are described below.

- Event A3: A neighbor cell becomes better than the primary serving cell
- Event A4: A neighbor cell becomes better than a threshold
- Event A5: The primary serving cell becomes worse than a first threshold and a neighbor cell becomes better than a second threshold

The Release 15 extension is that the network can further configure a threshold *N* in terms of the number of neighbor cells satisfying one of these events. For example, if the number of neighbor cells satisfying event A3 is greater than N, the drone reporting will be triggered. Such a triggering condition corresponds to drone experiencing that a greater number of neighbor cells having better signal strength, e.g., in terms of *reference signal received power*, or quality, e.g., in terms of *reference signal received quality*, than the primary serving cell. This is an indication that the drone is experiencing significant interference from many neighbor cells, which can very well imply that the uplink signal from the drone will generate interference at these neighbor cells. Similarly, when event A4 or A5 is met by many neighbor cells for a drone, the drone may have an interference problem. Again, when a drone is experiencing an interference problem, the drone is mostly likely air-borne. Thus, this feature can be used for flying mode detection.

Both features described in this section allow the network to detect scenarios where a drone is experiencing or generating significant other-cell interference. By detecting which drones having interference problems, the network can take appropriate measures to address the interference problem. One example of such actions is to set appropriate device-specific power control parameters based on the features described in Section 13.4.4.

13.4.2 Flight path information for mobility enhancement

To improve mobility performance in RRC connected mode, 3GPP Release 15 introduces a feature that supports the network to request a drone to provide its flight path information to the network via RRC signaling. The flight path information can be used by the network to determine a suitable new serving cell for the drone.

The flight path information includes a list of waypoints along the drone's planned flight path. Up to 20 waypoints can be included. The flight path information may further include the time stamp of the planned arrival times at each waypoint.

13.4.3 Subscription-based UAV identification

To support the authorization of LTE connectivity to a drone, 3GPP Release 15 introduces subscription-based UAV identification. The signaling of UAV subscription is illustrated in Fig. 13.7. The subscription information is stored at the *home subscriber server (HSS)*. Home subscriber server provides the subscription information to the *Mobility Management Entity*, which keeps the record of which base station is serving the drone and therefore can forward the subscription information to the base station that the drone is connected to. Based on the subscription information, the base station may deny the service to the drone if it determines that the drone is in the flying mode but has no UAV subscription. For X2-based handover, the existing base station can forward the subscription information via the X2 interface to the new serving base station.

13.4.4 Uplink power control enhancement

It is mentioned in Section 13.3 that a drone in flying mode may generate significant uplink interference in many cells, and therefore may have pronounced impact on network capacity. This problem is addressed in 3GPP Release 15 with the introduction of uplink power control enhancement.

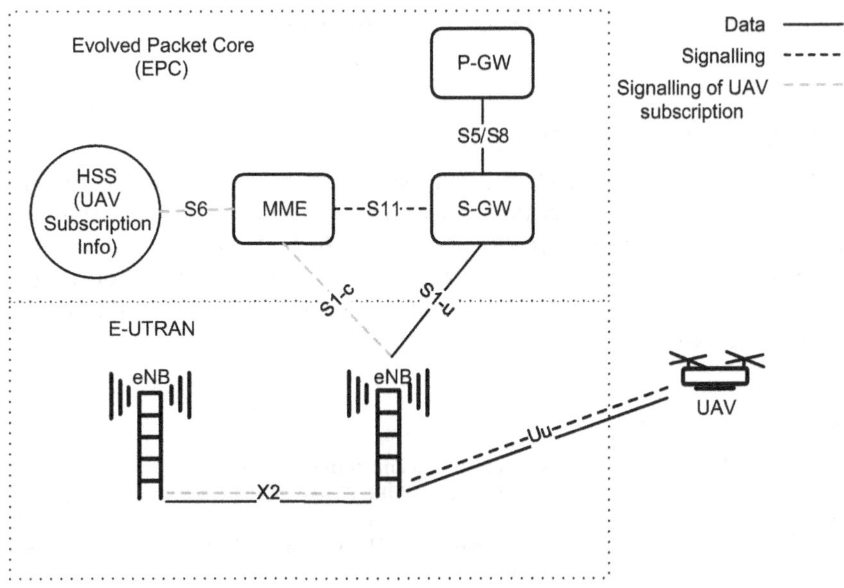

FIG. 13.7 Signaling of UAV subscription information.

LTE open-loop power control for *Physical Uplink Shared Channel* (PUSCH) is described in Section 9.3.2.8.1. The open-loop power control determines the transmit power level of PUSCH based on the parameters below.

- The bandwidth of PUSCH. The higher the PUSCH bandwidth is, the higher the transmit power.
- Adjustment for *modulation-and-coding scheme* (MCS) used for PUSCH. MCS of higher spectral efficiency requires higher transmit power.
- A parameter that is related to the target received power level at the base station. We will refer to this parameter as P_0.
- A pathloss compensation factor. If the path loss between the serving base station and the device is estimated to be L in dB, the pathloss compensation is αL dB, where α is referred to as the *fractional pathloss compensation factor*.

In LTE, up to 3GPP Release 14, the setting of the fractional pathloss compensation factor α is cell-specific, but not device-specific.

Motivated by the observation that a flying drone tends to have a different pathloss slope (see Fig. 13.2) and is likely to generate significant uplink interference compared to a terrestrial device, 3GPP Release 15 makes it possible to configure the fractional pathloss compensation factor α on a device-specific basis.

The P_0 parameter prior to Release 15 can already be configured on a device-specific basis. This is achieved by signaling a device-specific P_0 adjustment factor. However, its value range is from -8 to +7 dB. To increase the flexibility of PUSCH power control for a drone, Release 15 extends the value range of the device-specific P_0 adjustment factor to be from -16 to +15 dB.

The device-specific P_0 adjustment factor and α are included in a dedicated RRC signaling message.

13.4.5 UE capability indication

All the UAV features introduced in 3GPP Release 15, as described in this section, are optional for general LTE devices. A device can indicate whether it supports these features in the RRC information element *UE-EUTRA-Capability* [11]. However, for devices which have UAV subscription as described in Section 13.4.3, supporting the reporting of the two new height related events (H1 and H2) as well as the extension of mobility events A3, A4, A5 described in Section 13.4.1 is mandatory [12].

References

[1] Drones in disasters: Eyes in the air play key role in Alabama tornado zone – and beyond. USA Today, March 6, 2019. https://www.usatoday.com/story/news/nation/2019/03/06/drones-disasters-alabama-tornado-recovery-relief/3077948002/.

[2] California's fires face a new, high-tech foe: Drones. CNET, August 27, 2018. https://www.cnet.com/news/californias-fires-face-a-new-high-tech-foe-drones/.

[3] These drones and humans will work together in hurricane florence recovery efforts. Forbes, September 16, 2018. https://www.forbes.com/sites/jenniferhicks/2018/09/16/these-drones-and-humans-will-work-together-in-hurricane-florence-recovery-efforts/#30703b38b714.

[4] X. Lin, V. Yajnanarayana, S. D. Muruganathan, S. Gao, H. Asplund, H.-L. Maattanen, M. Bergstrom, S. Euler, Y.-P. E. Wang. "The sky is not the limit: LTE for unmanned aerial vehicles". IEEE Commun. Mag., July 2017, Vol. 56, No. 4, pp. 204–210.

[5] NTT DOCOMO INC., Ericsson. RP-170779, New SID on enhanced support for aerial vehicles, 3GPP TSG RAN meeting #75, 2017.

[6] Third generation partnership project, Technical Report 36.777, v15.0.0. "Study on enhanced LTE support for aerial vehicles", 2017.

[7] Third generation partnership project, Technical Report 38.901, v15.0.0. "Study on channel model for frequencies from 0.5 to 100 GHz", 2018.

[8] V. Yajnanarayana, Y.-P. E. Wang, S. Gao, S. Muruganathan, X. Lin. "Interference mitigation methods for unmanned aerial vehicles served by cellular networks,". 2018 IEEE 5G World Forum (5GWF), Silicon Valley, CA, 2018, pp. 118–122.

[9] H.-L. Määttänen, K. Hämäläinen, J. Venäläinen, K. Schober, M. Enescu, M. Valkama. System-level performance of LTE Advanced with joint transmission and dynamic point selection schemes. EURASIP J. Appl. Signal Process, November 2012, Vol. 2012, No. 1, 247.

[10] H. Rydén, S. B. Redhwan, X. Lin. "Rogue drone detection: a machine learning approach", 2019 IEEE Wireless Commun. and Networking Conf., Marrakech, Morocco, April 15–19, 2019.

[11] Third generation partnership project, Technical Specifications 36.331, v15.3.0. "Evolved universal terrestrial radio access (E-UTRA) and evolved universal terrestrial radio access network (E-UTRAN); radio resource control (RRC); protocol specifications", 2018.

[12] Third generation partnership project, Technical Specifications 36.306, v15.3.0. "Evolved universal terrestrial radio access (E-UTRA) and evolved universal terrestrial radio access network (E-UTRAN); user equipment (UE) radio access capabilities", 2019.

CHAPTER

14

IoT technologies in unlicensed spectrum

OUTLINE

Abstract

This chapter provides an overview of connectivity solutions for Internet of Things (IoT) applications and services. The focus lies on technologies that operates in unlicensed spectrum. The characteristics of using unlicensed spectrum are presented together with an introduction of short-range radio and low power wide area technologies that operate in unlicensed spectrum.

Cellular Internet of Things, Second Edition
https://doi.org/10.1016/B978-0-08-102902-2.00014-5

© 2020 Elsevier Ltd. All rights reserved.

14.1 Operation in unlicensed spectrum

14.1.1 Unlicensed spectrum regulations

Cellular Internet of Things (IoT) networks, such as EC-GSM-IoT, LTE-M, NB-IoT, LTE URLLC and NR URLLC operate in licensed spectrum. This means that mobile network operators have acquired long-term spectrum licenses from regulatory bodies in the country/region after, for example, an auction process. Such licenses provide an operator with exclusive spectrum usage right for a carrier frequency range. Such spectrum licenses may also be combined with an obligation to build out a network and provide network coverage and communication services in a certain area within a certain time frame. This obligation in combination with the cost of the license motivates mobile network operators to invest upfront into a network infrastructure. This exclusive spectrum usage right provides the prospect of good financial returns on the investment obtained via communication services within the lifetime of the license. There are also other spectrum bands, which do not abide to the rules of licensed spectrum. In unlicensed or license-exempt spectrum any device is entitled to transmit as long as it fulfills the regulation without requiring any player from holding a license. These regulatory requirements have the objective to harmonize and ensure efficient use of the spectrum.

Unlicensed spectrum bands differ for different regions in the world. In the following an overview of the usage of unlicensed spectrum is provided for two bands, one at around 900 MHz and one at 2.4 GHz. These are of particular relevance due to their ability to cater for IoT services and popular wireless communication standards such as Wi-Fi and Bluetooth have been specified for these bands. The sub-GHz range around 900 MHz provides attractive propagation characteristics in terms of facilitating good coverage. The 2.4 GHz range is interesting because it is considered to be a global band, which is important for systems targeting a global footprint. While the 2.4 GHz band is globally harmonized, the sub-GHz range has regional variations. However, most regions have some unlicensed spectrum allocation even if they differ in their specifics. A more detailed description is here provided for the US unlicensed spectrum at 902–928 MHz and the European unlicensed spectrum at 863–870 MHz. In Europe, some differences in the allocations of the 863–870 MHz band exist on a per country basis. There has been significant market traction for IoT connectivity solutions operating in the unlicensed frequency domain in these two regions. In other regions, the unlicensed spectrum allocation in the sub-GHz range varies for different countries. For example, the allocations in Korea and Japan are overlapping with the US spectrum region, and China has an allocation that is below the European spectrum allocation, see e.g., Ref. [1]. Radio technology standards that are addressing the unlicensed spectrum around 900 MHz, such as, IEEE 802.11ah, are typically designed in a way, in which they provide a common technology basis for different channelization options in this spectrum range; the detailed channelization is then adopted to the region where it is deployed, see e.g., Ref. [1] for the channelization of IEEE 802.11ah. IEEE 802.11ah is the basis upon which Wi-Fi HaLow is built.

For the United States, the usage of unlicensed spectrum for communication devices is regulated by the Federal Communications Commission and it is specified in Title 47 Code of Federal Regulations Part 15 [2]. For Europe, the spectrum rules are specified by the European Conference of Postal and Telecommunications Administrations (CEPT), which is a coordination body of the telecommunication and postal organizations within Europe. As of today, 48 countries are members of CEPT [3]. The CEPT recommendation for usage of short-range devices in unlicensed spectrum is described in Ref. [4]. This recommendation implements the European Commision decision on frequency bands used for short range devices [50]. It serves as the basis for European Telecom Standards Institute (ETSI) harmonized standards, which specify technical characteristics and measurement methods for devices that can be used by device implementers to validate their devices for conforming with the regulator rules. Such ETSI standards are as follows:

- ETSI standard EN 300 220 for short-range devices operating in the range 25 MHz–1 GHz [5,6],
- ETSI standard EN 300 440 for radio equipment to be used in the 1–40 GHz frequency range [7,8],
- ETSI standard EN 300 328 for data transmission equipment operating in the 2.4 GHz ISM band and using wide band modulation techniques. Direct-sequence spread spectrum (DSSS), frequency hopping spread spectrum, and Orthogonal Frequency-Division Multiplexing are considered to be wide band modulation techniques [9].

Some of the more relevant spectrum bands in the unlicensed spectrum range 863–870 MHz in Europe are given in Table 14.1 and for 902–928 MHz in the United States in Table 14.2. The tables present, e.g., the maximum allowed radiated power and requirements for interference mitigation. While the ETSI regulations in the band 863–870 MHz mandate the power in terms of Effective Radiated Power (ERP), i.e., the radiated power assuming a half-wave dipole antenna, the Federal Communications Commission sets requirements in terms of conducted power in combination with allowed antenna gain. Here we have converted these requirements to *Equivalently Isotropically Radiated Power* (EIRP), i.e., the radiated power assuming an isotropic antenna, according to the following equation:

$$\mathrm{EIRP} = \mathrm{ERP} + 2.15\ \mathrm{dB} \tag{14.1}$$

The 2.15 dB offset stems from the different reference antennas, i.e., dipole and isotropic, assumed for EIRP and ERP. It can be noticed that the permitted radiated power is higher for the unlicensed spectrum in the United States than in Europe. The power requirement is in some cases defined as a peak power requirement, and in some cases as an average power requirement. For a system using a modulation with a significant peak-to-average power ratio the peak power requirement will impact the average output power and the system coverage. This aspect is in detail discussed in Sections 15.2.3.3 and 15.3.3.2.

The European recommendations rely on duty cycle limitations, i.e., the percentage of time a transmitter may be active within a defined time span, while the US regulations defines maximum dwell times, i.e., the maximum continuous time by which a transmitting device may use a specific radio resource (e.g., a specific hopping channel). Regardless whether the limitations are on *duty cycle* or *dwell time*, these limits aim to avoid persistent interference. A short allowed dwell time can be understood to limit the coverage of a system, whereas a strict duty cycle requirement may limit coverage and system capacity.

TABLE 14.1 European unlicensed spectrum at 863–870 MHz, for more details see Refs. [4,50].

Spectrum band	EIRP	Mitigation requirement	Bandwidth	Other
863–870 MHz	16.1 dBm	<0.1% duty cycle or LBT		FHSS
863–870 MHz	16.1 dBm	<0.1% duty cycle or LBT + AFA		DSSS and other non-FHSS wideband techniques
863–870 MHz	16.1 dBm	<0.1% duty cycle or LBT + AFA	≤ 100 kHz, for 1 or more channels; modulation bandwidth ≤ 300 kHz	
868–868.6 MHz	16.1 dBm	<0.1% duty cycle or LBT + AFA		
868.7–869.2 MHz	16.1 dBm	<0.1% duty cycle or LBT + AFA		
869.4–869.65 MHz	29.1 dBm	<10% duty cycle or LBT + AFA		
865.6–865.8 MHz 866.2–866.4 MHz 866.8–867.0 MHz 867.4–867.6 MHz	29.1 dBm	10% duty cycle for network access points ≤ 2.5% duty cycle for other SRDs	≤ 200 kHz	Adaptive power control
869.7–870 MHz	9.1 dBm 16.1 dBm	For 4.8 dBm: No requirements; For 11.8 dBm: ≤ 1% duty cycle or LBT + AFA		

TABLE 14.2 US unlicensed spectrum at 902–928 MHz, for more details see Ref. [2].

Spectrum band	EIRP	Mitigation requirement	Bandwidth	Other
902–928 MHz	36 dBm	Dwell time per hopping channel: < 0.4 s/20 s	≤ 250 kHz	Frequency hopping with ≥50 hopping channels
902–928 MHz	30 dBm	Dwell time per hopping channel: < 0.4 s/10 s	200–500 kHz	Frequency hopping with ≥25 hopping channels
902–928 MHz	36 dBm		≥ 500 kHz	Digitally modulated

The duty cycle also limits the availability of downlink and uplink transmission opportunities which has a negative impact on service latency.

European recommendation allows to deviate from duty cycle limitations in case *listen-before-talk* (LBT) is used, i.e., when a transmitter first observes if a channel is not used by other devices before starting to transmit. Sometimes LBT needs to be combined with *adaptive frequency agility*, which is a method for avoiding transmission on already occupied channels (see e.g., Ref. [1]). LBT introduces an uncertainty in the systems transmissions that makes it difficult to provide high reliability and bounded latency.

The different rows in Tables 14.1 and 14.2 can be understood to cater for different types of applications and different types of equipment. For more details please see Refs. [2,4].

The spectrum usage recommendation for unlicensed spectrum at 2400–2483.5 MHz in Europe is given in Table 14.3 and the rules for the United States in Table 14.4. Also in this band the maximum allowed radiated power is higher in United States than in Europe. The European power limitation is given as EIRP. Higher radiated power is allowed for wide band transmission such as DSSS modulation, FHSS, and Orthogonal Frequency-Division Multiplexing under the condition of using a mitigation method such as LBT.

14.1.2 Coexistence in unlicensed spectrum

Different types of radio equipment and technologies can transmit on any frequency of the unlicensed spectrum. As depicted in Fig. 14.1, the simultaneous transmissions of different devices interfere with each other. As a result, this interference may lead to that one or both of the transmissions depicted in the figure fail. Spectrum coexistence mechanisms

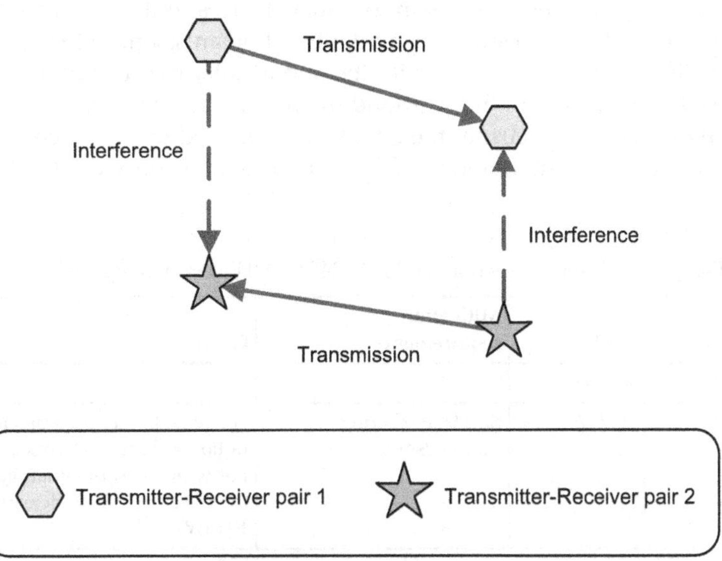

FIG. 14.1 Transmission of two different unlicensed devices to their corresponding receivers.

are mechanisms that limit the interference that a transmitter may cause on other nearby devices.

The simultaneous usage of unlicensed spectrum can occur between homogenous or heterogeneous types of devices, which means devices that use the same or different wireless communication technologies. A wireless communication technology designed for unlicensed spectrum, often, has some mechanism, which specifies how the spectrum is shared among different devices of that communication technology, so that each device has good transmission opportunities, while at the same time minimizing the interference to other devices. The coordination scheme typically follows the spectrum regulation introduced earlier but may also have additional technology specific features. While such a coordination scheme can be applied within one wireless communication technology, it can typically not provide the same level of interference mitigation in-between different wireless communication technologies.

The spectrum regulation for unlicensed spectrum provides requirements on devices, which shall provide technology neutral coexistence, i.e., independent of a particular wireless communication technology being used. Spectrum regulation has thereby mainly two types of communication devices in mind: *adaptive devices* and *nonadaptive devices*.

Nonadaptive devices are considered to transmit in unlicensed spectrum, while staying unaware of the other types of devices. To limit the amount of interference that can be generated, the devices are limited as follows:

- In the radiated power they may use, see Tables 14.1–14.4, sometimes further limited by an allowed power spectral density,
- By a duty cycle or dwell time, see Tables 14.1 and 14.2, which limit what fraction of time a device may transmit on a channel.

By these means the total amount of interference a device may cause on others is restricted.

In contrast, an adaptive device is aware of other devices in its vicinity, which also make use of the same channel. As a result, it can adapt its transmission, reducing the interference to other devices. This provides a device with the opportunity to transmit for a longer amount of time if there are few devices in the surrounding. In contrast, if many devices in the vicinity want to use the unlicensed spectrum, the adaptive device reduces the frequency of its transmissions and provides less interference. The common way to provide this type of adaptive

TABLE 14.3 European unlicensed spectrum at 2400–2483.5 MHz, for more details see Ref. [9].

Spectrum band	EIRP	Mitigation requirement	Other
2400–2483.5 MHz	10 dBm		
2400–2483.5 MHz	20 dBm	Spectrum sharing mechanism like LBT	For wide band data transmission and radio local area networks. For wide band modulations other than FHSS, the maximum EIRP-density is limited to 10 mW/MHz

TABLE 14.4 US unlicensed spectrum at 2400–2483.5 MHz, for more details see Ref. [2].

Spectrum band	EIRP	Mitigation requirement	Bandwidth	Other
2400 –2483.5 MHz	30 dBm	Dwell time per hopping channel: < (0.4 s * X), X = number of hopping channels	≤ 250 kHz	Frequency hopping with ≥15 and < 75 non-overlapping hopping channels
2400 –2483.5 MHz	36 dBm	Dwell time per hopping channel: < (0.4 s * X), X = number of hopping channels	200–500 kHz	Frequency hopping with ≥75 non-overlapping hopping channels
2400 –2483.5 MHz	36 dBm and max 8dBm/3 kHz PSD		≥ 500 kHz	Digitally modulated

device is LBT, where a device before a transmission listens on the radio channel and observes if other devices are communicating, and after a *clear channel assessment* it transmits for a limited time. In addition, technical standards consider a fair use of spectrum among devices, and thus consider that at a high utilization of the spectrum an adaptive device transmits less than in case of a low utilization.

The most common form of unlicensed spectrum usage is for short-range communication. One reason is that short-range communication provides a certain level of interference robustness. The amount of interference that a receiver is exposed to depends very much on the location of the interferer. Assuming that different transmitters in unlicensed spectrum use similar output powers, e.g., the maximum that is allowed by regulation, an interferer can be considered as a *strong interferer* if it is located closer to the receiver than the intended transmitter. Devices that jointly form a local network typically have some coordination or coexistence functionality provided by the wireless communication standard they are using, which avoids or reduces interference within the local network. However, other unlicensed radio technologies are typically not part of this interference coordination. In Fig. 14.2, it is depicted how different groups of devices use various unlicensed communication technologies. If these devices are separated in space, the interference is limited because it is typically significantly below the power levels of the communication within the group. However, if different unlicensed radio technologies operate at the same location, significant interference can occur.

Fig. 14.3 shows the challenge of long-range communication in unlicensed spectrum. With long-range communication it becomes more likely that an interferer is located closer to the receiver than the intended transmitter. The example of the figure shows a long-range system that is designed to cover a large path loss for transmission over several kilometers. There may exist several other local unlicensed networks using the same spectrum in vicinity of

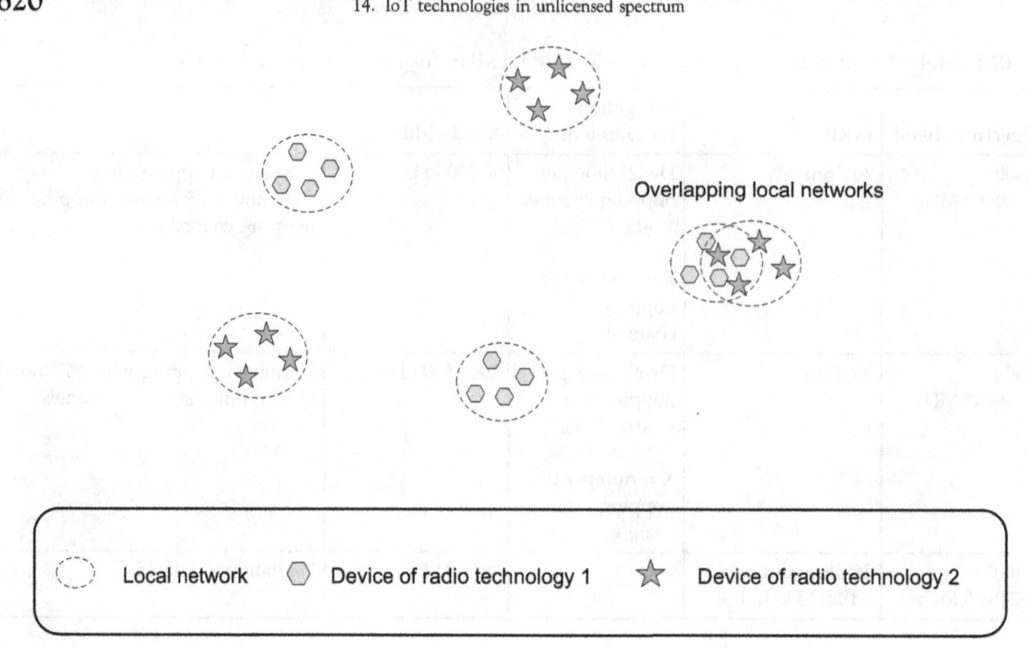

FIG. 14.2 Coexistence among different groups of unlicensed devices. Inter-system interference is most severe if different groups are overlapping in space.

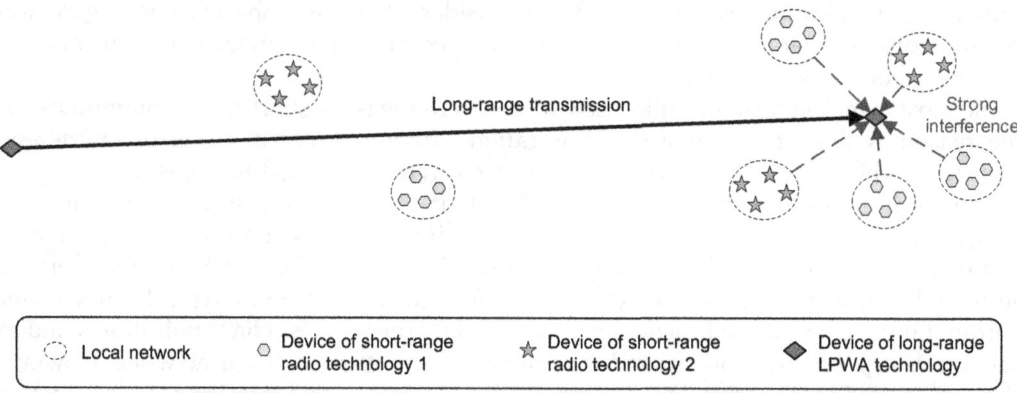

FIG. 14.3 Coexistence between long-range and short-range devices.

the long-range receiver. If the long-range receiver is, e.g., placed on the roof of a building, there may be some local unlicensed networks used in the same or neighboring buildings, e.g., for home automation. Because these devices are significantly closer to the long-range receiver, they may cause interference at the location of the long-range receiver, which is significantly higher, by e.g., several 10's of dB, than the strongly attenuated signal of the long-range transmitter, which is coming from far away. Furthermore, if we assume that the devices in the local network are adaptive devices, which e.g., use LBT to avoid interfering with other devices, this operation is likely to fail to adapt to long-range transmitters that are far away because the long-range signal is so strongly attenuated that it is below a sensitivity threshold used for clear channel assessment. As long as unlicensed spectrum is barely used, such interference situations may be unlikely. If it is anticipated that unlicensed IoT use cases (and other use cases) will drive the deployment of various local area networks using unlicensed spectrum the inference in unlicensed spectrum will increasingly play a role; long-range unlicensed radio technologies are more exposed to this interference.

One aspect that is worth to mention for unlicensed spectrum, is about the maximum transmit power that may be emitted by the device. The maximum transmit power provided by spectrum regulation is typically given with respect to a certain reference configuration of emission. The reason is that different antenna configurations have different power emission patterns, which leads to that the antenna has different gains in different directions. Often maximum power levels are defined as either effective isotropically radiated power or as EIRP, as shown above for Tables 14.1–14.4. For EIRP an ideal isotropic antenna is assumed, whereas for ERP a half-wave dipole antenna is assumed, which has a 2.15 dB antenna gained compared with an isotropic antenna in direction of the highest antenna gain as specified in Eq. (14.1). When the maximum transmit power is specified as EIRP or ERP, a transmitter with any antenna configuration should not emit power in any direction that would exceed the maximum power that could be emitted in this direction with an isotropic or dipole antenna, respectively. A practical result of this is that if a real antenna has an antenna gain of X dB with a maximum allowed radiated power of Y dBm EIRP, then the transmitter must limit its conducted power at the antenna port to Y-X dBm. One consequence is that downlink performance can be substantially limited compared with uplink, if a base station antenna with significant antenna gain is used, see e.g. Ref. [51]. In uplink direction, the base station can use the antenna gain at the receiver side to improve the link budget. In the reverse downlink direction, the base station has to compensate the antenna gain by a reduced transmit power resulting in that the base station antenna gain does not improve the actual link budget, such as maximum path loss. This is one significant difference for a communication system operating in unlicensed spectrum compared with licensed spectrum, where the antenna gain could also be used in downlink direction and where higher maximum radiated powers are permitted in downlink.

Another cause for asymmetry of uplink and downlink in unlicensed spectrum can be found regarding capacity for nonadaptive devices that are limited by a duty-cycle. As an example, we assume a large number of N devices connected to a single base station, each device transmitting at a rate R_u and being limited by a maximum duty cycle of D_u. The maximum achievable uplink capacity is then limited by the maximum data rate and the duty cycle limitation for each device. Assuming an ideal situation of all transmissions being successful by neglecting all possible collisions and interference situations, the upper bound of maximum achievable uplink capacity C_u becomes in this case

$$C_u = N \cdot D_u \cdot R_u, \tag{14.2}$$

In downlink, the capacity of the base station is limited by its own maximum duty cycle. This duty cycle has to be used for the transmission to all N devices, in contrast to uplink where a duty cycle is valid per device. The upper bound of maximum achievable downlink capacity C_d under ideal error free transmissions becomes then

$$C_d = D_d \cdot R_d, \tag{14.3}$$

for a downlink data rate R_d and a maximum duty cycle of D_d. Assuming the same uplink and downlink parameters, the downlink capacity is thus by a factor of N smaller compared with the uplink capacity, which leads to a significantly lower achievable effective downlink data rate per device. To some extent this can be compensated by configuring downlink transmissions to certain subbands that allow higher duty cycles (increasing D_d), see e.g., Table 14.1.

14.2 Radio technologies for unlicensed spectrum

Unlicensed spectrum enables immediate market access for any new radio technology with minimal regulatory requirements. As a result, a very large number of different Machine-to-Machine (M2M) connectivity solutions have been developed and brought to the market making use of this spectrum. This section presents a set of established solutions for both long and short range communications. In addition, Chapter 15 provides an indepth description of LTE-M-U and and NB-IoT-U which are designed by the Multefire alliance to provide long range communication in unlicensed spectrum.

14.2.1 Short-range radio solutions

In this section we provide an overview of the most promising unlicensed short-range radio communication technologies for the IoT, which are IEEE 802.15.4, Bluetooth Low Energy (BLE) and Wi-Fi HaLow. The choice to focus on these is made based on the following properties of those technologies: they address the communication requirements of IoT, they target an IP-based IoT solution, and they are based on open standards and are expected to reach a substantial economy of scale. We also address the capillary network architecture where any of these solutions may be used to provide the short-range connectivity within a larger network context.

14.2.1.1 IEEE 802.15.4

IEEE 802.15.4 was standardized in 2003 at a time of intense research on wireless sensor networking technologies [13,14], and it is applicable for a wide range of IoT use cases, ranging from office automation, connected homes to industrial use cases. IEEE 802.15.4 was one of the early standards taking advantage of the adaptation framework IPv6 over low power wireless personal area networks (6LoWPAN) [10–12] which was specified in IETF to enable IP communication over very constrained wireless communication technologies (see also Chapter 17). Several application-specific protocol stacks have been developed, which build on parts of the IEEE 802.15.4 standard (mostly the physical layer and to some extent the medium access control (MAC)) [15], including ZigBee, WirelessHART, ISA-100, and Thread.

IEEE 802.15.4 has been specified for the three frequency bands of 868 MHz (for Europe), 915 MHz (for United States), and 2.4 GHz (global) [16–18]. In the 2.4 GHz band, IEEE 802.15.4 has 16 channels available, each of 2 MHz bandwidth. It uses offset quadrature phase shift keying with DSSS and a spreading factor of 8. A gross data rate of 250 kbps is achievable, see Refs. [16,18]. In 868 MHz one channel of 600 kHz is available, which uses differential binary phase shift keying modulation and DSSS with a spreading factor of 15. The achievable data rate at 868 MHz is 20 kbps [16,18]. In the 915 MHz band 10 channels are available with a gross data rate of 40 kbps, see Ref. [18].

IEEE 802.15.4 uses *carrier-sense multiple access with collision avoidance* (CSMA-CA) for access to the radio channel; this can be complemented with optional Automatic Repeat Request retransmissions. Typical coverage ranges are in the order of 10–20 m [11]. Two different topologies are supported: star topology and mesh (or peer-to-peer) topology. Two types of devices are defined: full-function devices that provide all MAC functionality and can act as network coordinator of the local network, and reduced function devices that can only communicate with a full-function device and are intended for very simple types of devices. The network can operate in beaconed mode, which allows a set of devices to synchronize to a superframe structure that is defined by the beacon transmitted by a local coordinating device. The channel access in this case is slotted CSMA-CA. In nonbeacon mode, unslotted CSMA-CA is applied. In case of direct data transmission, a device transmits data directly to another device. In indirect data transmission, data is transferred to a device, e.g., from a network coordinator. When beacon transmission is active, the network coordinator can indicate the availability of data in the beacon; the device can then request the pending data from the network coordinator. In nonbeaconed transmission, the network coordinator buffers data and it is up to the device to contact the network coordinator for pending data.

In Ref. [18] the performance of IEEE 802.15.4 has been evaluated in an experiment with an ideal link and devices placed at 1 m distance. It has been found that for a configuration at 2.4 GHz with a theoretical gross data rate of 250 kbps, a net data rate of 153 kbps was measured for direct transmission (from the device), and a net data rate of 66 kbps for indirect transmission (toward the device). Furthermore, it has been shown that the effective data rate and delivery ratio decrease with an increasing number of devices.

A major step for broader relevance of IEEE 802.15.4 for the IoT has been to address end-to-end IP-based communication. To this end the IETF working group 6LoWPAN has been chartered in 2005 and it has developed IETF standards for header compression and data fragmentation. The maximum physical layer payload size of 802.15.4 is limited to 127 bytes, which is further reduced by various protocol headers and optional security overhead and can leave as little as 81 bytes available for application data within an IEEE 802.15.4 frame. IETF has developed standards that provide header compression and IP packet fragmentation that enable the transmission of IPv6 over 802.15.4 networks [17,19–22]. In addition, the RPL routing protocol has been developed to enable IP mesh routing over IEEE 802.15.4 [17,22,23]. In 2014 the Thread group was formed with the objective to harmonize the usage of IEEE 802.15.4 together with 6LoWPAN for home automation.

IEEE 802.15.4 has been extended in IEEE 802.15.4g to address smart utility networks with an objective to improve coverage and support higher data rates [24–26]. To this end multiple new physical layers have been defined which can be used from a common MAC layer. Physical layer implementations are multirate frequency shift keying, multirate orthogonal frequency-division multiplexing, and multirate offset quadrature phase shift keying. Multirate frequency shift keying has a benefit of good transmit power efficiency because of constant signal envelop, MR-OFDM enables higher data rates for frequency selective fading channels, and multirate offset quadrature phase shift keying has the benefit of the original IEEE 802.15.4 modulation with cost-effective and easy design. A physical layer agnostic management protocol is based on a common signaling mode to allow a network configuration with interference coordination among multiple IEEE 802.15.4 transmitters [26].

14.2.1.2 BLE

Bluetooth has been developed as a technology for wireless short-range connectivity [27,28] and has established itself as a leading technology for personal area networking. With the release of the Bluetooth core specification 4.0 [29] in 2010 a novel transmission mode called Bluetooth Low Energy (BLE) was introduced, which considerably reduces power consumption compared with Bluetooth classic. BLE has been a significant first step to expand the Bluetooth ecosystem toward IoT.

BLE uses the 2.4 GHz ISM band. The spectrum is divided into 40 channels, with 2 MHz channel spacing, of which 37 are data channels and 3 are used as *advertising channels*. Frequency hopping is applied to mitigate the impact of interference. The modulation is based on *Gaussian Frequency Shift Keying* and a data rate of up to 1 Mbps can be achieved over-the-air. A master-slave architecture has been adopted to assign asymmetric roles to devices; peripheral devices perform only a minimum amount of functions to enable ultra-low power consumption, while central devices perform coordination functions. BLE has short connection setup and data transfer times so that applications can transfer authenticated data within a few milliseconds. BLE allows connection-oriented or connectionless communication. It supports fragmentation and reassembly of large data packets into small radio frames, which are then transmitted over the radio interface. This enables BLE to support data services with large packets (e.g., IP packets).

An analysis of BLE for building automation use cases has been performed in Refs. [16,30–32]. With a single-hop deployment, the range for BLE in an indoor deployment setup is in the order of 10 m, and around five BLE gateways are needed to provide coverage in a 1000 m^2 office floor [32].

In 2014 the Bluetooth Special Interest Group (BT SIG), the standardization forum for Bluetooth, published the Internet Protocol Support Profile [33], which enables IP connectivity for BLE devices. Further, IETF has developed a standard for end-to-end IPv6 connectivity over BLE [34], including header compression. This enables that end-to-end IP-based IoT services can be provided via BLE systems [35].

A further evolution of BLE has occurred with the launch of Bluetooth 5, the Bluetooth core specification 5.0 [36,37]. It comprises quadrupling of the communication range at low data rates (i.e., 125 kbps) and the doubling of the peak data rates (to 2 Mbps). Another important extension of BLE has been the introduction of mesh networking in 2017, which can significantly increase the range of BLE [38].

14.2.1.3 Wi-Fi

Wi-Fi based on IEEE 802.11 is one of the most used unlicensed radio access technologies and its focus is on providing high data rate services to a range of mainly consumer electronics devices. Very long battery life has not been in focus. Also the scalability of IEEE 802.11 has mainly addressed being able to provide a high total throughput to a number of connected devices in an area; from the early Wi-Fi version IEEE 802.11b to the version IEEE 802.11ac the theoretically achievable physical layer peak data rates have increased from 11 Mbps to 6.9 Gbps [39]. The typical operation of IEEE 802.11 is in the 2.4 and 5 GHz unlicensed spectrum bands.

IEEE 802.11ah is an amendment to the IEEE 802.11 standard that is focused on IoT applications. The Wi-Fi Alliance has chosen Wi-Fi HaLow as the marketing term to be used for the IEEE 802.11ah amendment. IEEE 802.11ah has some design targets that significantly differ from the high-data rate focused IEEE 802.11 variants. First, IEEE 802.11ah addresses the unlicensed spectrum below 1 GHz, which is in the range 902–928 MHz in the United States and 863–868 MHz in Europe; other regions also have unlicensed spectrum regions somewhere in the range 750–928 MHz [40]. Differences of the sub-1-GHz spectrum versus higher spectrum bands are as follows:

- The propagation conditions sub-1-GHz facilitate longer range. For a wide-area usage and spread of IoT devices, transmission range is a key property to provide sufficient coverage to IoT services with limited amount of access points. At the same time, the IoT devices are expected to transmit only limited amounts of data. For mobile broadband Wi-Fi usage, where devices are expected to transmit a lot of data, the extended range would mean that the channel is blocked for a longer time and the channel access time per device would reduce.
- There is less unlicensed spectrum available than at higher spectrum bands. This also means that the total capacity of data that can be provided within an area is lower than at higher spectrum bands. For mobile broadband focused Wi-Fi usage, this is a disadvantage because one focus is to provide high capacity in combination with high per user data rates. For IoT-focused Wi-Fi this is less of a problem, as the total amount of data transmitted even by a very large group of IoT devices is expected to remain modest.

The IEEE 802.11ah physical layer design is derived from the IEEE 802.11ac [1]. To address the lower spectrum bands, with less available bandwidth, and to enable robust long range transmission, the bandwidth of the IEEE 802.11ah has been scaled down by a factor of 10 compared to 802.11ac. That means that IEEE 802.11ah supports different carrier bandwidths of 2–16 MHz in comparison with the 20–160 MHz carriers of 802.11ac. In addition, an extra robust carrier configuration with 1 MHz bandwidth has been defined. Reference [1] describes a 24.5 dB link budget gain of IEEE 802.11 ah at 900 MHz compared with 802.11n at 2.4 GHz. The gains stem from reduced path loss at low frequency (8.5 dB), reduced noise bandwidth due to narrower carriers (10 dB), further reduced noise bandwidth and repetition coding gains of the new robust 1 MHz carrier configuration

(6 dB). The achievable data rates with IEEE 802.11ah are between 150 kbps and 347 Mbps. Several MAC features have been introduced to reduce power consumption for a client device and support more devices being connected to the same access point. IEEE 802.11 applies LBT in form of CSMA/CA. A larger number of connected devices lead to increased collision probabilities, which can be accentuated with the increased effect of hidden nodes with outdoor deployments [1]. To reduce the collision probability, the *restricted access window* (RAW) has been introduced. It divides the contention period into up to 64 RAW slots. Devices are allocated to particular RAW slots; and the number of devices, which are contending simultaneously for channel access, can be reduced to those devices being allocated to the same RAW slot. Device battery consumption can be significantly reduced, by enabling communication in uplink and downlink direction in new *bidirectional transmission opportunities*, where reverse link traffic can follow closely on forward link traffic. This enables long sleep cycles for devices. In addition, with a new *target wake time* the device and an access point can agree on certain fixed time periods, when data that the access point receives for a device shall be forwarded to the device. This reduces the amount of activity of a device to be able to receive data. Furthermore, the maximum idle period for a device has been extended in IEEE 802.11ah so that devices can be configured with sleep periods of up to around five years, and such devices only need to connect once every maximum idle period to the access point to avoid being automatically disassociated from the access point. IEEE 802.11ah also introduces new frame formats, which reduce the overhead of control information added in messages. This is significant for IoT traffic because the data payloads are often very small (e.g., a few bytes for a meter reading) and control info can quickly introduce significant overhead. For data transmission a short MAC frame format is added, and for control messages a *null data packet* has been introduced.

A more extensive description and evaluation of Wi-Fi IEEE 802.11ah can be found in Refs. [39–43]. The specification was published in early 2017 [52] and as of today no commercial chipsets are on the market.

14.2.1.4 Capillary networks

Short-range radio technologies provide the ability to build out connectivity efficiently to devices within a specific local area. Typically, these local, or capillary, networks need to be connected to the edge of a wide area communication infrastructure so that they have the ability, for example, to reach service functions that are hosted somewhere on the Internet or in a service cloud.

A capillary network needs a backhaul connection, which can be well provided by a cellular network. Their ubiquitous coverage allows backhaul connectivity to be provided practically anywhere, simply and, more significantly, without additional installation of network equipment. Furthermore, a capillary network might be on the move, as is the case for monitoring goods in transit, and therefore cellular networks are a natural solution. To connect a capillary network through a cellular network, a gateway is used between the cellular network and the capillary network, which acts just like any other cellular device toward the cellular network.

Fig. 14.4 illustrates an architecture, which comprises three domains: the capillary connectivity domain, the wide-area connectivity domain, and the data domain. The capillary connectivity domain spans the nodes that provide connectivity in the capillary network,

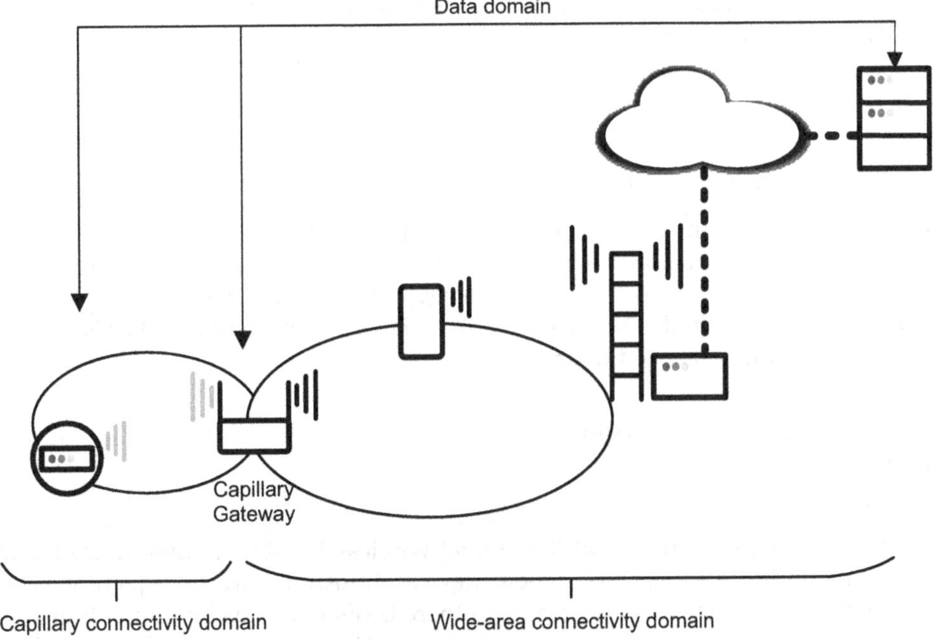

FIG. 14.4 Capillary networks.

and the wide-area connectivity domain spans the nodes of the cellular network. The data domain spans the nodes that provide data processing functionality for a desired service. These nodes are primarily the connected devices themselves as they generate and use service data through an intermediate node, such as a capillary gateway. The capillary gateway would also be included in the data domain if it provides data processing functionality (for example, if it acts as a CoAP mirror server).

All three domains are separate from a security perspective, and end-to-end security can be provided by linking security relationships in the different domains to one another.

The ownership roles and business scenarios for each domain may differ from case to case. For example, to monitor the in-building sensors of a real estate company, a cellular operator might operate a wide-area network and own and manage the capillary network that provides connectivity to the sensors. The same operator may also own and manage the services provided by the data domain and, if so, would be in control of all three domains.

Alternatively, the real estate company might own the capillary network, and partner with an operator for connectivity and provision of the data domain. Or the real estate company might own and manage both the capillary network and the data domain with the operator providing connectivity only. In all these scenarios, different service agreements are needed to cover the interfaces between the domains specifying what functionality will be provided.

In large-scale deployments, some devices will connect through a capillary gateway, while others will connect directly. Regardless of how connectivity is provided, the bootstrapping and management mechanisms used should be homogenic to reduce implementation complexity and improve usability.

A more extensive discussion of IoT connectivity via capillary networks can be found in Ref. [44].

14.2.2 Long-range radio solutions

For unlicensed spectrum usage, short-range radio systems are most common. However, for IoT applications that require very low data rates, it is possible to trade lower data rate for a longer transmission range. Many technology concepts have been developed in recent years for unlicensed LPWAN. Many different variants of unlicensed LPWAN exist; some of which are more often referred to are as follows:

- LoRa
- Sigfox Ultra-Narrow Band (UNB)
- LTE-M-U
- NB-IoT-U

All of those have in common that they target wireless M2M/IoT communication over a long range of multiple kilometers, where devices transmit only infrequently very low amounts of data. Message sizes are small and there is often a focus on uplink transmission. Devices are desired to be simple and battery-powered operation should be possible over extended time periods. All these technologies are proprietary and not standardized in standards developing organizations.

In the following we provide a briefly overview of the LoRa and Sigfox technologies. LTE-M-U and NB-IoT-U are in detailed described in Chapter 15.

14.2.2.1 LoRa

LoRa is a network technology designed to provide long-range connectivity to battery operated devices; it is specified within an industry alliance. The LoRa Alliance claims to provide a Maximum Coupling Loss of 155 dB in the European 867–869 MHz band, and 154 dB in the United States 902–928 MHz band [45]. LoRa has the target to provide secure bidirectional communication. LoRa operates in the sub-GHz unlicensed frequency bands. The physical layer is based on Chirp spread-spectrum modulation, which is a technique using frequency modulation to spread the signal. A radio bearer is modulated with up and down chirps, where an up chirp corresponds to a pulse of finite length with increasing frequency, while a down chirp is a pulse of decreasing frequency. The channel bandwidth is mainly 125 kHz for European spectrum bands, and 125 or 500 kHz for US spectrum bands. Different data rates are supported and are reported to lie in the range of 300 bps–50 kbps. The selection of data rate is a trade-off between transmission duration, i.e., the time during which the message is transmitted over the air and range. The LoRa Alliance claims to provide a Maximum Coupling Loss of 155 dB in the European 867–869 MHz band, and 154 dB in the US 902–928 MHz band [46].

LoRa does not deploy LBT but instead uses the duty cycle restrictions required by regulation and the maximum dwell time (i.e., the maximum time that a device may continuously occupy a channel). The system architecture comprises LoRa end devices, LoRa gateways, and a network server. LoRa gateways correspond to base stations in a cellular network. Communication is between the end device and the network server. The communication between the network server and the gateway is based on IP communication; the communication between the end device and a gateway is based on LoRa specific protocols without IP. IP communication is terminated at the LoRa gateway. When a device transmits in uplink, the message can be received by one or more gateways.

For bidirectional communication a downlink transmission opportunity is provided after an uplink transmission. If downlink data arrives for a device in between uplink transmissions, the data needs to be buffered in the network and can only be transmitted during the devices downlink receive window, which follows on an uplink transmission by the device.

For more information on LoRa see Ref. [46].

14.2.2.2 Sigfox

Sigfox is a proprietary technology and of which until recentently no specifications have been publicly available. As of March 2019 Sigfox published a radio protocol specification describing the device operation. Some indicative properties of the so-called UNB communication scheme of Sigfox was however provided already in the Third Generation Partnership Project (3GPP) contribution [47] and technical reports of an industry specification group in ETSI [48,49].

UNB is targeted for operation in sub-GHz unlicensed spectrum bands. The supported channel bandwidths are 600 Hz and 100 Hz [49]. The uplink physical layer uses differential BPSK, which implies that the data rates are limited in the order of some few hundreds of bits per second. The maximum payload size for uplink data is 12 bytes. UNB does not use LBT but applies duty cycle limitations per transmitter. The channel access scheme is based on ALOHA, which starts to deteriorate at higher loads when the channel utilization exceeds around 15% [47]. Sigfox claims to support a maximum path loss, taking receive and transmit antenna gains into accout, of up to 163 dB [49].

The Sigfox network architecture comprises devices, which communicate with Sigfox servers. The radio communication is between the devices and Sigfox access points or base stations. Devices can transmit at any time without prior synchronization to the network. Typically, messages are transmitted on three different uplink channels, which are randomly selected. The base stations observe the entire system bandwidth to detect and decode uplink data. Messages can be received by different base stations, which provide selection diversity.

Downlink transmission is "piggybacked" onto uplink transmissions. After an uplink transmission a device maintains an open receiver window to receive downlink data for a certain time. The server sends buffered data to the device after receiving an uplink message. If the server has received uplink data via multiple base stations, it selects one of the base stations for downlink transmission.

References

[1] M. Park. IEEE 802.11ah: sub-1-GHz license-exempt operation for the internet of things. Comput. Commun., March 2015, Vol. 58, 53–69.

[2] Federal Communications Commission. Title 47 of the Code of federal regulations, Part 15 on radio frequency devices, 2017. Available from: https://www.fcc.gov/general/rules-regulations-title-47.

[3] European Conference of Postal and Telecommunications Administrations (CEPT). Available from: http://www.cept.org/cept/.

[4] CEPT Electronic Communications Committee (ECC). ERC recommendation 70-03, relating to the use of short-range devices (SRD), 2016. Edition of October 2016.

[5] European Telecom Standards Institute. EN 300 220-1 V3.1.1, short range devices (SRD) operating in the frequency range 25 MHz to 1 000 MHz; Part 1: technical characteristics and methods of measurement, February 2017.

[6] European Telecom Standards Institute. EN 300 220-2 V2.4.1 electromagnetic compatibility and radio spectrum matters (ERM); short range devices (SRD); radio equipment to Be used in the 25 MHz to 1 000 MHz frequency range with power levels ranging up to 500 mW; Part 2: harmonized EN covering essential requirements under article 3.2 of the R&TTE directive, May 2012.

[7] European Telecom Standards Institute. EN 300 440-1 V1.6.1, electromagnetic compatibility and radio spectrum matters (ERM); short range devices; radio equipment to Be used in the 1 GHz to 40 GHz frequency range; Part 1: technical characteristics and test methods, August 2010.

[8] European Telecom Standards Institute. EN 300 440 V2.1.1, short range devices (SRD); radio equipment to Be used in the 1 GHz to 40 GHz frequency range; harmonised standard covering the essential requirements of article 3.2 of directive 2014/53/EU, January 2017.

[9] European Telecom Standards Institute. EN 300 328 V2.1.1, wideband transmission systems; data transmission equipment operating in the 2,4 GHz ISM band and using wide band modulation techniques; harmonised standard covering the essential requirements of article 3.2 of directive 2014/53/EU, November 2011.

[10] J. A. Gutierrez, M. Naeve, E. Callaway, M. Bourgeois, V. Milter, B. Heile. IEEE 802.15.4: a developing standard for low-power low-cost wireless personal area networks. IEEE Network., September/October 2001, Vol. 15, No. 5, 12–9.

[11] E. Callaway, P. Gorday, L. Hester, J. A. Gutierrez, M. Naeve, B. Heile, V. Bahl. Home networking with IEEE 802.15.4: a developing standard for low-rate wireless personal area networks. IEEE Commun. Mag., August 2002, Vol. 40, No. 8, 70–7.

[12] J. Zheng, M. J. Lee. Will IEEE 802.15.4 make ubiquitous networking a reality? a discussion on a potential low power, low bit rate standard. IEEE Commun. Mag., June 2004, Vol. 42, No. 6, 140–6.

[13] I. F. Akyildiz, W. Su, Y. Sankarasubramaniam, E. Cayirci. Wireless sensor networks: a survey. Comput. Netw., 2002, Vol. 38, 393–422.

[14] I. F. Akyildiz, W. Su, Y. Sankarasubramaniam, E. Cayirci. A survey on sensor networks. IEEE Commun. Mag., August 2002, Vol. 40, No. 8, 102–14.

[15] A. Willig. Recent and emerging topics in wireless industrial communications: a selection. IEEE Transactions on Industrial Informatics, May 2008, Vol. 4, No. 2, 102–24.

[16] N. Langhammer, R. Kays. Performance evaluation of wireless home automation networks in indoor scenarios. IEEE Transactions on Smart Grid, December 2012, Vol. 3, No. 4.

[17] M. R. Palattella, N. Accettura, X. Vilajosana, T. Watteyne, L. A. Grieco, G. Boggia, M. Dohler. Standardized protocol stack for the internet of (important) things. IEEE Communications Surveys and Tutorials, 2013, Vol. 15, No. 3.

[18] J.-S. Lee. Performance evaluation of IEEE 802.15.4 for low-rate wireless personal area networks. IEEE Trans. Consum. Electron., August 2006, Vol. 52, No. 3.

[19] Internet Engineering Task Force (IETF). Request for comments 4944, transmission of IPv6 packets over IEEE 802.15.4 networks, September 2007.

[20] J. W. Hui, D. E. Culler. IPv6 in low-power wireless networks. Proc. IEEE, November 2010, Vol. 98, No. 11.

[21] J. W. Hui, D. E. Culler. Extending IP to low-power, wireless personal area networks. IEEE Internet Computing, July/August 2008.

[22] I. Ishaq, D. Carels, G. K. Teklemariam, J. Hoebeke, F. Van den Abeele, E. De Poorter, I. Moerman, P. Demeester. IETF standardization in the field of the internet of things (IoT): a survey. J. Sens. Actuator Netw., April 2013, Vol. 2, No. 2, 235–87.
[23] Internet Engineering Task Force (IETF). Request for comments 6550, RPL: IPv6 routing protocol for low-power and lossy networks, March 2012.
[24] K.-H. Chang, B. Mason. The IEEE 802.15.4g standard for smart metering utility networks. In: Proceedings IEEE international conference on smart grid communications, Tainan city, Taiwan, 5–8 November 2012.
[25] C.-S. Sum, F. Kojima, H. Harada. Coexistence of homogeneous and heterogeneous systems for IEEE 802.15.4g smart utility networks. In: Proc. IEEE international symposium on dynamic spectrum access networks (DySPAN), Aachen, Germany, 3–6 May 2011.
[26] C.-S. Sum, H. Harada, F. Kojima, L. Lu. An interference management protocol for multiple physical layers in IEEE 802.15.4g smart utility networks. IEEE Commun. Mag., April 2013, Vol. 51, No. 4, 84–91.
[27] J. Haartsen. Bluetooth—the universal radio interface for ad hoc, wireless connectivity. Ericsson Rev. Engl. Ed., 1998, Vol. 75, No. 3, 110–7. Available from: http://ericssonhistory.com/Global/Ericsson%20review/Ericsson%20Review.%201998.%20V.75/Ericsson_Review_Vol_75_1998_3.pdf.
[28] J. C. Haartsen. The bluetooth radio system. IEEE Personal Communications, August 2002, Vol. 7, No. 1, 28–36.
[29] Bluetooth Special Interest Group. Bluetooth core specification v4.0, June 2010.
[30] L. F. Del Carpio, P. Di Marco, P. Skillermark, R. Chirikov, K. Lagergren, P. Amin. Comparison of 802.11ah and BLE for a home automation use case. In: Proc. IEEE 27th Annual international symposium on personal, indoor, and mobile radio communications (PIMRC), 4–8 September 2016.
[31] L. F. Del Carpio, P. Di Marco, P. Skillermark, R. Chirikov, K. Lagergren. Comparison of 802.11ah and BLE in a home automation use case, to appear in springer international journal of wireless information networks.
[32] P. Di Marco, R. Chirikov, P. Amin, F. Militano. Coverage analysis of bluetooth low energy and IEEE 802.11ah for office scenario. In: Proc. IEEE int. Symposium on personal, indoor and mobile radio communications. Workshop M2M Communication, August 2015.
[33] Bluetooth Special Interest Group. Internet protocol support profile (IPSP), December 2014.
[34] Internet Engineering Task Force (IETF). Request for comments 7668. IPv6 over BLUETOOTH Low Energy, October 2015.
[35] Bluetooth Special Interest Group. Internet gateways, Bluetooth white paper, February 2016.
[36] Bluetooth Special Interest Group. Bluetooth core specification v5.0, December 2016.
[37] P. Di Marco, P. Skillermark, A. Larmo, P. Arvidson, R. Chirikov. Performance evaluation of the data transfer modes in Bluetooth 5. IEEE Communications Standards Magazine, July 2017, Vol. 1, No. 2, 92–7.
[38] Bluetooth Special Interest Group. Bluetooth specification mesh profile, v1.0.1, January 2019.
[39] W. Sun, O. Lee, Y. Shin, S. Kim, C. Yang, H. Kim, S. Choi. Wi-fi could Be much more. IEEE Communications Magazine, November 2014.
[40] W. Sun, M. Choi, S. Choi. IEEE 802.11ah: a long range 802.11 WLAN at sub 1 GHz. Journal of ICT Standardization, July 2013, Vol. 1, No. 1, 83–108.
[41] T. Adame, A. Bel, B. Bellalta, J. Barcelo, M. Oliver. IEEE 802.11ah: the WiFi approach for M2M communications. IEEE Wireless Communications, December 2014, Vol. 21, No. 6, 144–52.
[42] E. Khorov, A. Lyakhov, A. Krotov, A. Guschin. A survey on IEEE 802.11ah: an enabling networking technology for smart cities. Comput. Commun., March 1, 2015, Vol. 58, 53–69.
[43] B. Badihi, L. F. Del Carpio, P. Amin, A. Larmo, M. Lopez, D. Denteneer. Performance evaluation of IEEE 802.11ah actuators. In: Proceedings 83rd IEEE vehicular technology conference, Nanjing, China, May 2016, pp. 15–8.
[44] J. Sachs, N. Beijar, P. Elmdahl, J. Melen, F. Militano, P. Salmela. Capillary networks – a smart way to get things connected. Ericsson Review, September 2014.
[45] LoRA Alliance Technical Marketing Workgroup 1.0. White paper, LoRaWAN, what is it? A Technical Overview of LoRa® and LoRaWAN, November 2015.
[46] LoRa Alliance. LoRaWAN™ specification, V1.0, January 2015.
[47] Sigfox Wireless. GPC150052, coperative ultra narrow band technology for cellular IoT, 3GPP TSG GERAN1-GERAN2 ad'hoc meeting, sophia antipolis, France, 2–5 February 2015.

[48] ETSI Industry Specification Group on Low Throughput Networks. (ISG LTN), GS LTN 002 V1.1.1, low throughput networks (LTN); functional architecture, September 2014.
[49] ETSI Industry Specification Group on Low Throughput Networks. (ISG LTN), GS LTN 003 V1.1.1, low throughput networks (LTN); protocols and interfaces, September 2014.
[50] European Comission, Comission Decision of 9 November 2006 on harmonisation of the radio spectrum for use by short-range device (2006/771/EC), Edition of 18.8.2017.
[51] S. Andreev, O. Galinina, A. Pyattaev, M. Gerasimenko, T. Tirronen, J. Torsner, J. Sachs, M. Dohler, Y. Koucheryavy. Understanding the IoT connectivity landscape: a contemporary M2M radio technology roadmap. IEEE Commun. Mag., September 2015, Vol. 53, No. 9, 32–40.
[52] IEEE 802.11ah-2016 - IEEE standard for information technology–telecommunications and information exchange between systems - local and metropolitan area networks–specific requirements - Part 11: wireless LAN medium access control (MAC) and physical layer (PHY) specifications amendment 2: sub 1 GHz license exempt operation, May 2017.

CHAPTER

15

MulteFire Alliance IoT technologies

OUTLINE

Cellular Internet of Things, Second Edition
https://doi.org/10.1016/B978-0-08-102902-2.00015-7

© 2020 Elsevier Ltd. All rights reserved.

Abstract

In this chapter, we describe how the MulteFire Alliance have adapted LTE-M and NB-IoT to support operation in unlicensed frequency bands according to FCC and ETSI regulations. The physical layer design as well as the procedures performed in idle and connected mode are presented for both technologies. The MulteFire design closely follows the 3GPP design, and the description is focused on the adaptations introduced to facilitate operation in unlicensed frequency bands. The chapter also presents the technical potential of the technologies in terms of the performance that can be expected when deploying these technologies.

15.1 Background

The MulteFire Alliance (MFA) is a standardization organization that develops wireless technologies for operation in unlicensed and shared spectrum. The MFA specifications are using the 3GPP technologies as baseline and add adaptations needed for operation in unlicensed spectrum. In its first Release 1.0, the MFA developed an unlicensed-spectrum version of 3GPP LTE using the LTE *Licensed-Assisted Access* feature as baseline. The specification is also aligned with the 3GPP Release 14 enhanced Licensed-Assisted Access work, which was developed in parallel by 3GPP. In Release 1.1, the MFA turned its attention toward the 3GPP IoT technologies LTE-M and NB-IoT, which are described in Chapters 5 to 8. These two systems were modified for operation in unlicensed spectrum governed by ETSI and FCC.

Section 15.2 describes how LTE-M was redesigned to enable operation in the global unlicensed 2.4 GHz frequency band. The MFA named this technology *eMTC-U*, after the 3GPP Release 13 work item Further LTE Physical Layer Enhancements for MTC [3], which from time to time is referred to as the eMTC work item. In this book we use the term LTE-M when describing the set of LTE MTC enhancements specified in 3GPP (see Chapter 5), and *LTE-M-U* when describing the work done in MFA.

Section 15.3 focuses on the NB-IoT design for operation in the Short Range Device (SRD) and Industrial, Scientific and Medical (ISM) frequency bands defined in the range 863–870 MHz in Europe and 902–928 MHz in the US, respectively. The MFA has named this technology *NB-IoT-U* and in this book we use the same naming convention.

The descriptions of NB-IoT-U and LTE-M-U provided in the next 3 sections are based on the MulteFire Release 1.1 physical layer specifications [3–6] and higher layer specifications [7–9].

15.2 LTE-M-U

The interest among industries and enterprises to use wireless technologies for operating a private network to serve their needs has grown rapidly during the past few years. To use the 3GPP technologies typically requires a partnership with a spectrum license holder, often a mobile network operator, who grants a permit for sub-leasing the needed spectrum in a limited geographical area. An alternative available in certain regions is to use the 3GPP technologies in shared spectrum such as the *Citizens Broadband Radio Service* bands currently made available in the United States. The concept of sub-leasing is still in 2019 rarely used, while the availability of shared bands that allow operation of 3GPP technologies can at best be said to be emerging.

The MFA developed LTE-M-U to lower the threshold for industries and enterprises to make use of 3GPP technologies. The system is designed to operate in the 2400–2483.5 MHz license exempt frequency band. This is an attractive band due to its significant bandwidth and global availability, which makes the technology deployable in most countries in the world.

The next three sections introduce the design principles followed by MFA when developing LTE-M-U, the LTE-M-U physical layer functionality and the procedures used in RRC idle and connected mode.

15.2.1 Radio access design principles

The LTE-M-U design principles are based on three components: The 3GPP LTE-M design described in Chapter 5, the ETSI regulations for the 2.4 GHz band [1] and the FCC regulations for the same band [2]. In Release 1.1 the MFA fully focused on the ETSI and FCC regulations to prioritize an introduction of LTE-M-U in the European and US markets.

The physical layer of LTE-M-U closely follows the LTE-M design. The downlink is OFDM modulated while the uplink used DFT-precoded OFDM. LTE-M-U is designed as a time-division duplex (TDD) frequency hopping system with a system bandwidth of 6 physical resource blocks (PRB). Each PRB is of 180 kHz bandwidth which results in a useful system bandwidth of 1.08 MHz, usually referred to as a narrowband (NB) in the context of LTE-M and LTE-M-U. LTE-M-U supports CE mode A but not CE mode B (cf. Section 5.1.2.2). The rationale for this is that LTE-M-U is designed for small-cell private network deployments. For this use case, the CE mode A coverage range was deemed fully adequate. This design baseline has been carefully chosen to provide a single solution that complies with both FCC and ETSI regulations.

The next two sections introduce the FCC and ETSI regulations. Detailed descriptions of parts of the regulations are also provided e.g. in Sections 15.2.2.2.3, 15.2.2.2.4, 15.2.3.3 and 15.2.3.4.

15.2.1.1 FCC regulations

In the US, the FCC regulations specify support for three alternative system designs in the 2.4 GHz band: digitally modulated systems, frequency hopping spread spectrum (FHSS) systems or a hybrid system design which is a combination of the first two options.

The LTE-M-U downlink is operating according to the FCC rules for a digitally modulated system while the uplink is designed according to the regulations specified for a hybrid system. A digitally modulated system is required to have a transmit bandwidth of at least 500 kHz and meet a conducted power spectral density (PSD) limitation of 8 dBm/3 kHz, which corresponds to 25.8 dBm/180 kHz. As LTE-M-U base stations are mandated to transmit Cell-Specific Reference Signals (CRS) over the full downlink narrowband bandwidth of 1.08 MHz, and the base station conducted output power is limited to at most 30 dBm, which corresponds to a PSD of 4.4 dBm/3 kHz, both requirements are fulfilled. The uplink transmission bandwidth occupies between 1 and 6 physical resource blocks (PRBs), which implies that the digital modulation bandwidth requirement is not always met. Instead the uplink design complies with the FCC regulations for a hybrid system which is required to frequency hop and at the same time comply with the digital modulation PSD limitation of 8 dBm/3 kHz. As the LTE-M-U narrowband is frequency hopping over 16 or 32 carrier frequencies according to the FHSS regulations and the device PSD never exceeds 20 dBm/180 kHz, both the hybrid system frequency hopping and PSD criterions are met for the uplink design.

The uplink could in principle also be qualified as a frequency hopping system instead of a hybrid system. For a frequency hopping system, the device power class is defined by the maximum peak power, and not the maximum average power which is the case for digitally modulated and hybrid system designs. For LTE-M-U, being an OFDM system with a significant peak to average power ratio (PAPR), it is a benefit to declare the output power according to average, and not peak, power measurement procedures. This explains why the LTE-M-U uplink is declared as a hybrid system, and not a FHSS system when operating under FCC regulations.

As the downlink frequency hops with the uplink one may be led to believe that also the downlink can be declared as a FHSS. The FCC FHSS regulation however mandates an equal visitation of each frequency and a pseudo-random frequency hopping pattern. This condition is not fulfilled for the downlink as the anchor channel, carrying synchronization and broadcast signaling, is mapped on a fixed carrier frequency and is visited more frequently than the data channels. Note that visitation of a channel is here considered to be equivalent to transmission on the channel.

15.2.1.2 ETSI regulations

In Europe, ETSI specifies two categories of wideband data transmission systems in the 2.4 GHz band: FHSS systems and wideband modulation systems which include techniques such as OFDM and direct sequence spread spectrum modulation. The equipment can be adaptive or non-adaptive. Adaptive equipment is not permitted to transmit on an occupied radio channel and is required to sense the channel occupancy before accessing a channel. Non-adaptive equipment is not required to sense the channel, but needs to follow a specified *medium utilization* (MU) cycle which limits the channel occupancy time.

The ETSI wideband modulation regulation defines a PSD limitation of 10 dBm per MHz. Due to this restriction, LTE-M-U is declared as an FHSS system. The uplink is designed according to the non-adaptive FHSS regulations which implies that the LTE-M-U devices are limited to use a 10% MU when operating at 20 dBm *Equivalent Isotropically Radiated Power* (EIRP). While 3GPP specifications commonly refer to conducted power, ETSI and FCC regulations often define limits in terms of radiated power. To avoid a limitation of the downlink resource utilization, the LTE-M-U base stations are designed to follow the adaptive FHSS regulation. This requires the base station to perform channel sensing before accessing the medium.

Table 15.1 summarizes the radio access design principles followed by LTE-M-U. It is important to understand that LTE-M-U has a single design baseline that is defined by the most stringent requirements in the ETSI and FCC regulations. This means for example that the base station is expected to perform channel sensing not only when operating under ETSI regulations, but also when operating in the US under FCC regulations.

TABLE 15.1 LTE-M-U radio access design principles.

Regulation	Link direction	Design basis
FCC	Downlink	Digitally modulated
	Uplink	Hybrid
ETSI	Downlink	Adaptive FHSS
	Uplink	Non-adaptive FHSS

15.2.2 Physical layer

The description of the physical layer is distributed in four parts. The Physical resources section 15.2.2.1 presents the basic time and frequency resources used for transmitting the LTE-M-U signal in the 2.4 GHz band. The Transmission schemes section 15.2.2.2 presents generic aspects and functionality that applies to the physical layer design. The last two sections 15.2.2.3 and 15.2.2.4 present the details of the downlink and uplink physical channels and signals.

15.2.2.1 Physical resources

15.2.2.1.1 Channel raster

In 3GPP the channel bandwidth is defined as the *occupied bandwidth* of a modulated waveform which corresponds to the frequency range containing 99% of the total power of the modulated signal. An LTE-M-U narrowband of 6 PRBs have a 1.4 MHz channel bandwidth according to this definition which is reused by ETSI. The FCC regulations defines the channel bandwidth by its *20-dB bandwidth* requirement. This corresponds to the bandwidth where the signal emissions are attenuated by 20 dB relative the power measured in the center of the carrier. The 20-dB bandwidth definition is more stringent than the 3GPP and ETSI occupied bandwidth requirement. MFA consequently concluded that the 20-dB bandwidth of an LTE-M-U narrowband should be within 1.8 MHz.

LTE-M-U supports the 2.4 GHz license exempt frequency band, defined by the frequency range 2400–2483.5 MHz. Across this band 43 radio frequency (RF) channels are distributed using a channel raster with a granularity of 1.8 MHz, which is aligned with the assumed 20-dB bandwidth. Three RF channels are reserved for the so-called anchor channel, which carries the synchronization signals and the most important broadcast and system information messages.

Each LTE-M-U cell is associated with one out of the three anchor channels. When an LTE-M-U device makes its initial cell selection it needs to scan only the three anchor RF channels in its search for a cell to camp on. The rationale for limiting the number of anchor frequencies to three is to limit the time it takes for a device to perform the initial system acquisition, and to maximize the number of channels available for data transmission.

The remaining 40 channels are dedicated for data transmission. They are grouped into 10 sets, each containing 4 channels. When frequency hopping is configured over 16 data channels, 4 of the 10 sets are configured for carrying physical uplink and downlink control

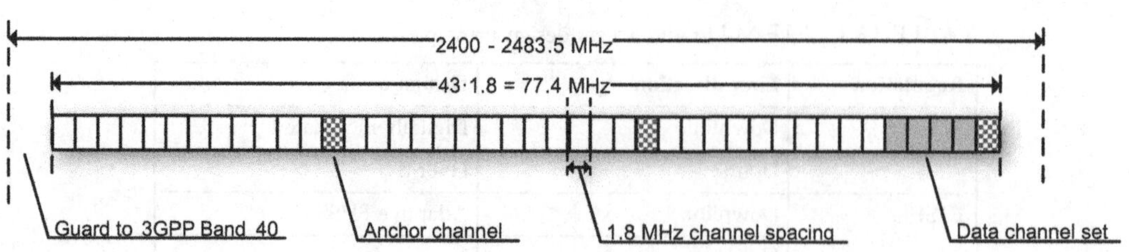

FIG. 15.1 Potential LTE-M-U channel arrangement in the 2.4 GHz band.

channels, shared channels and reference signals. With frequency hopping configured over 32 channels, 8 of the sets are activated. The frequency hopping is further described in Section 15.2.3.6. Fig. 15.1 provides an illustration of a potential arrangement of the 43 LTE-M-U channels distributed over in total 77.4 MHz of the totally available 83.5 MHz in the 2.4 GHz band.

The upper edge of 3GPP band 40 is directly adjacent to the lower edge of the 2.4 GHz band. 3GPP Release 16 is studying a new band located immediately above the 2.4 GHz band. To secure coexistence with 3GPP systems operating in the adjacent bands, the outer LTE-M-U channel locations are expected to be shifted from the 2.4 GHz band edges. This will provide guard-band between the 3GPP bands and LTE-M-U. Coexistence between MFA and 3GPP technologies are for natural reasons important.

15.2.2.1.2 Frame structure

LTE-M-U introduces a new frame structure, called Type 3M. It is a TDD type of frame structure which is a natural choice for the 2.4 GHz unlicensed band since it is defined as a single unpaired band, without any reserved frequency ranges for uplink and downlink transmissions.

The frame structure is illustrated in Fig. 15.2, including the new *mframe* concept. One mframe contains 8 radio frames and spans 80 ms. The mframe numbering run across 8 hyperframes before it wraps around and starts over. Besides the mframe, the LTE-M-U frame numbering follows the same conventions as LTE.

The mframe is shown in Fig. 15.3. The first 5 ms is known as the anchor segment. It is dedicated to downlink transmissions of the LTE-M-U PSS, SSS, PBCH and the SIB1-A message, which are described in detail in Section 15.2.2.3. It is transmitted on an anchor frequency channel associated with the camped-on cell, as discussed in Section 15.2.2.1.1. Grouping these signals and channels together in the beginning of the mframe facilitates a frequency hopping design of the second part of the mframe known as the data segment.

The data segment spans 75 ms and is divided into uplink and downlink portions according to a set of specified uplink-downlink configurations introduced in Section 15.2.2.2. The uplink and downlink parts of the data segment support both control and data channel transmission that are mapped on a set of frequencies according to the configured frequency hopping pattern.

The subframes in an mframe are numbered according to a relative subframe number defined in the range 0–79.

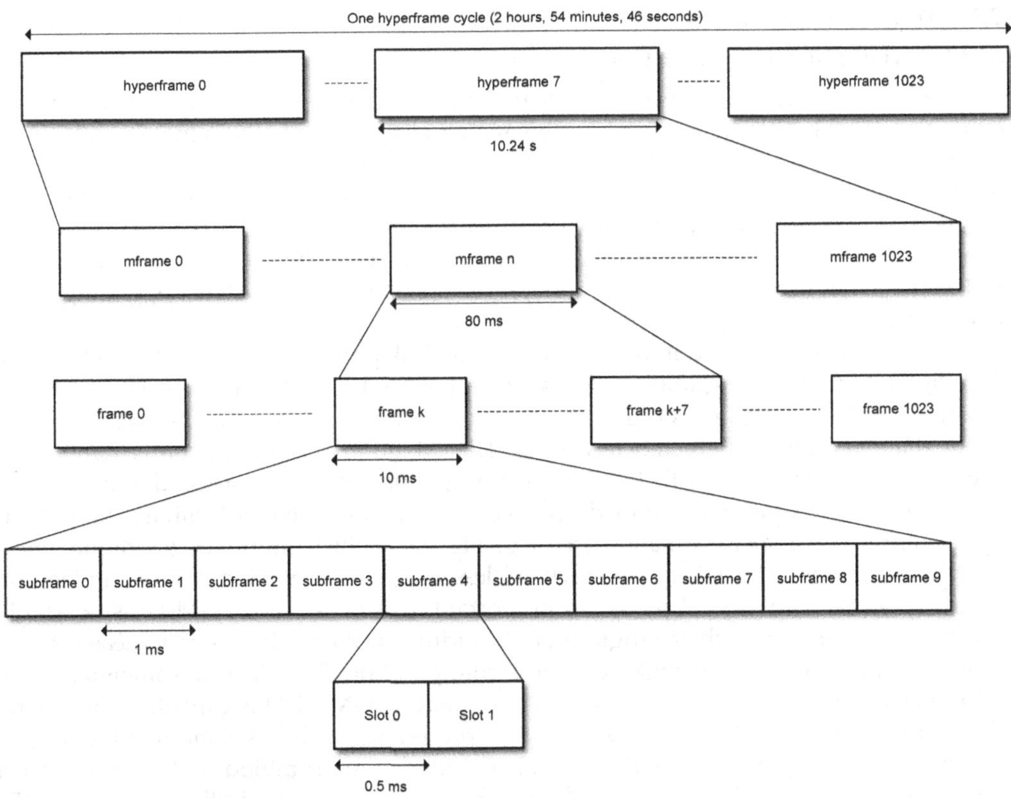

FIG. 15.2 LTE-M-U Frame structure 3M including the mframe concept.

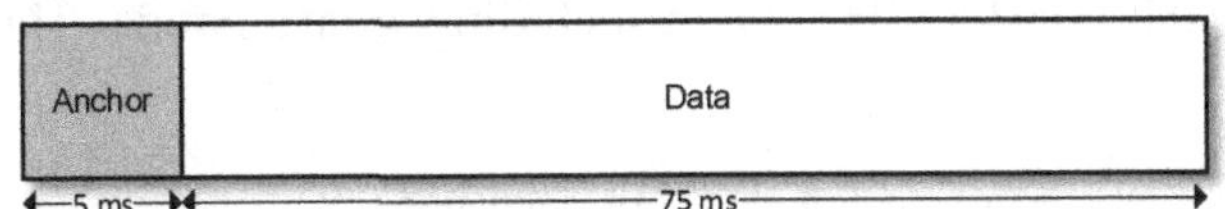

FIG. 15.3 LTE-M-U mframe including anchor and data segments.

15.2.2.1.3 Resource grid

The resource grid is just as for LTE-M defined by a PRB pair transmitted during a subframe of length 1 ms. In the frequency domain the PRB spans 180 kHz. LTE-M-U transmissions are mapped over 1 to 6 PRBs, with a granularity of 1 PRB.

Each PRB pair is mapped onto 2 consecutive time slots. Each slot is defined by a resource grid specified by 7 OFDM symbols (OS) in time and 12 subcarriers in frequency domain. Each subcarrier is 15 kHz wide.

The smallest building block in the resource grid is the resource element, defined by one OFDM symbol transmitted on one subcarrier. Section 5.2.1.3 provides an illustration of a PRB pair.

15.2.2.2 Transmission schemes

15.2.2.2.1 Anchor and data segment transmissions

The operation of LTE-M-U is based on the mframe concept introduced in Section 15.2.2.1.2. The first 5 ms in the mframe are used by the anchor channel while the remaining 75 ms are reserved for data transmissions. The anchor channel segment is dedicated to downlink transmissions. The exclusive use of downlink transmissions avoids near-far uplink-to-downlink interference situations that can otherwise be difficult to avoid in shared spectrum with uncoordinated deployments. The anchor frequencies are hence protected from uplink interference from LTE-M-U devices, but not from interference caused by other systems operating in the shared band.

The data segment is divided into uplink and downlink portions according to a set of specified uplink-downlink configurations. These are presented in Table 15.2. In LTE-M-U, the uplink-to-downlink switching occurs once or twice across 80 subframes, which is less frequent than in LTE-M. Remember that the LTE-M uplink-downlink configurations are defined across 10 subframes with 1 or 2 switching points. The long consecutive intervals of uplink or downlink subframes defined for LTE-M-U are intended to facilitate the concept of cross-subframe channel estimation (see e.g. Fig. 8.2) which improves operation under low SNR conditions. The taken approach of less frequent switching also optimizes the LTE-M-U channel capacity. This is an important aspect since LTE-M-U is a single-narrowband system where the configuration of additional narrowbands to increase system capacity is not supported. Uplink-downlink configuration 7 defines a somewhat more frequent switching to support a latency closer to what LTE-M TDD is capable of achieving.

The last two OS in every downlink subframe preceding an uplink subframe are left unused. This leaves room for RF switching and provides a guard-period that supports a cell radius of roughly 21 km. This explains why no special subframes are indicated in Table 15.2 (cf. Table 5.1 for LTE-M).

The system bandwidth is defined by a single narrowband containing 6 PRBs spanning a bandwidth of 1.08 MHz, or 1.4 MHz when including guard-bands. The anchor and data

TABLE 15.2 LTE-M-U data segment uplink-downlink configurations.

Configuration	Sequence of downlink (DL) and uplink (UL) subframes in the data segment							
0		DL: 55				UL: 20		
1		DL: 45				UL: 30		
2		DL: 35				UL: 40		
3		DL: 20				UL: 55		
4	DL: 25		UL: 15		DL: 20		UL: 15	
5	DL: 15		UL: 25		DL: 10		UL: 25	
6	DL: 30		UL: 10		DL: 20		UL: 15	
7	DL: 15	UL: 10		DL: 15	UL: 10	DL: 10		UL: 15

segments downlink transmissions are always spanning the full narrowband thanks to the CRS transmissions. The uplink transmission bandwidth ranges between 1 and 6 PRBs. Thanks to the frequency hopping the system may operate across the full 2.4 GHz band excluding the guard bands toward adjacent 3GPP bands.

15.2.2.2.2 Transmission modes

The LTE-M downlink transmission modes TM1, TM2, TM6 and TM9 for up to 4 antenna ports are supported for Physical Downlink Shared Channel (PDSCH) transmissions on the data segment. The antenna port is a logical concept for defining the relation between the data and reference signal transmissions. It is also fundamental for supporting the precoding functionality categorized by the transmission modes. For all transmission modes LTE-M-U, just as LTE-M and in contrast to LTE, only supports single layer transmissions.

TM1 is defined for single antenna port transmission. TM1 can typically be associated with a single beamforming radiation pattern intended to provide coverage across an entire served cell. The applied precoder is transparent and common to all devices in a cell. TM2 is specified for 2 or 4 antenna ports and supports transmit diversity by means of Space Frequency Block Coding (SFBC). TM6 is for closed-loop precoding, where the base station applies a precoder, selected from a set of specified precoders, to create a beam optimizing the link quality. The device estimates the channel state based on the CRS transmissions and suggests an optimal precoder, from the available set, to the base station. TM9 is for open-loop precoding where the precoder applied by the network is transparent to the device. This requires the introduction of device-specific reference symbols that are precoded together with the data symbols. Section 5.2.4.7 in some further detail discusses precoding. Section 5.2.4.3 presents the LTE-M-U downlink reference signals and their use in the different transmission modes.

15.2.2.2.3 Listen-before-talk

Adaptive equipment operating according to ETSI regulations, i.e. LTE-M-U base stations, with an EIRP exceeding 10 dBm must perform a *listen-before-talk* based *detect-and-avoid* procedure before transmitting. This means that a base station must perform a *clear channel assessment* (CCA), and possibly an *extended clear channel assessment* (eCCA), in the beginning of each downlink dwell. A dwell is equivalent to a period of downlink or uplink transmission opportunities.

The CCA and eCCA are based on that no signal is detected at a power level exceeding a threshold *TL* defined as:

$$TL = -70\frac{\text{dBm}}{\text{MHz}} + 10\log_{10}\left(\frac{P}{100\text{mW}}\right) \tag{15.1}$$

P corresponds to the output EIRP of the LTE-M-U base station. A base station operating at the maximum output power of 20 dBm EIRP should hence refrain from transmitting on a downlink dwell in the beginning of which an energy exceeding -70 dBm per MHz is measured. At lower detected signal levels, the channel is available for transmissions.

In the context of low-power wide-area networks and compared to the sensitivity offered by a typical 3GPP system -70 dBm can be considered to correspond to a fairly high received signal level. For LTE-M-U being designed for industrial use cases with more moderate requirements on range and receiver sensitivity -70 dBm is expected to be a more representative receive signal level and therefore suitable as CCA threshold.

The CCA should be performed for 0.2% of the intended channel occupancy time, but for no less than 18 μs. In the context of CCA, occupancy is equivalent to transmission. If the CCA fails, the base station must perform an eCCA with a duration of up to 5% of the channel occupancy time. The application of the CCA and the eCCA procedures are in further detail described in Section 15.2.2.3.1 for the anchor segment and in Section 15.2.2.3.2.1 for the data segment.

15.2.2.2.4 Frequency hopping

Non-adaptive FHSS equipment operating according to ETSI regulations, i.e. LTE-M-U devices, shall meet the following conditions:

- Frequency hop over at least N frequency channels defined by 15 MHz divided by the channel separation. With a channel separate of 1.8 MHz, N equals 9.
- Each channel should be visited at least once every four frequency hopping cycles.

A hybrid system operating according to FCC regulations, i.e. LTE-M-U devices, shall in addition:

- Frequency hop over at least 15 channels.
- Follow a pseudo random frequency hopping pattern.
- Visit each channel at most 400 ms during $N \cdot 400\ ms$, with N defined as the number of channels in the hopping set.

Adaptive FHSS equipment operating according to ETSI regulations, i.e. LTE-M-U base stations, shall meet the following conditions:

- The base station should be able of frequency hopping across 70% of the 2.4 GHz frequency band.
- The accumulated transmit time should not exceed 400 ms during $N \cdot 400\ ms$, with N defined as the number of channels in the frequency hopping set.
- Each channel should be visited at least once every four frequency hopping cycles.

These ETSI and FCC requirements are met by LTE-M-U transmissions which hops over 16 or 32 frequencies, located on a 1.8 MHz frequency raster and spread across the band according to the selected frequency allocation. The hopping follows a pseudo random pattern that guarantees an equal visitation of each channel. It takes the physical-cell identity and mframe number as input and generates a cell-specific hopping sequence that repeats itself after every eighth hyperframe.

Fig. 15.4 illustrates the LTE-M-U frequency hopping including the anchor segment periodicity, which equals 80 or 160 ms when hopping over 16 or 32 data channels, respectively. The anchor segment is described in detail in Section 15.2.2.3.1.

15.2.2.3 Downlink physical channels and signals

LTE-M-U supports the basic set of logical and transport channels defined for LTE-M in 3GPP Release 13, which are depicted in Section 5.2.4. In the downlink the Primary and Secondary Synchronization Signals (PSS and SSS), the Cell-Specific and Demodulation

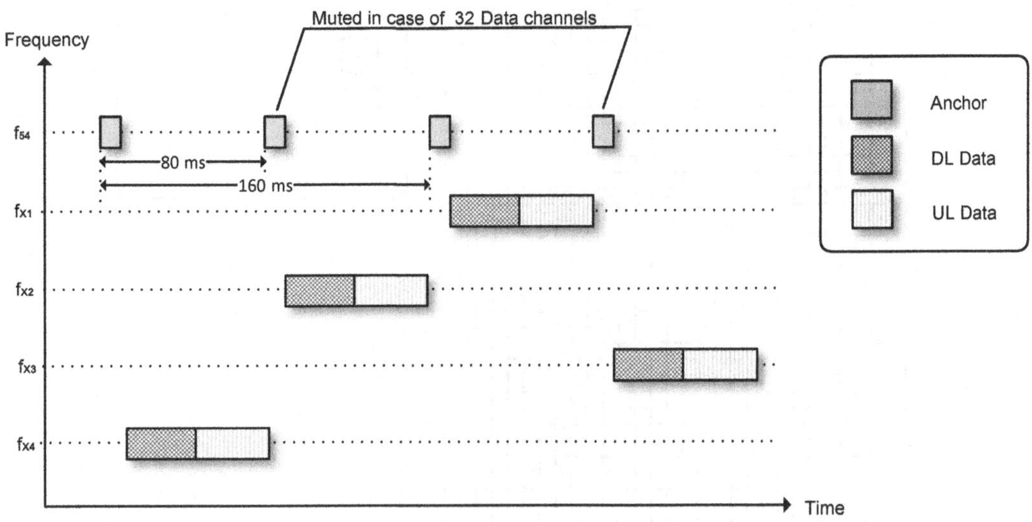

FIG. 15.4 LTE-M-U frequency hopping.

Reference Signals (CRS and DMRS), the MTC Physical Downlink Control Channel (MPDCCH) and the PDSCH are all supported. In addition, LTE-M-U supports the *Presence Detection Reference Signal* (PDRS) for indicating that a basestation has successfully performed a CCA in the beginning of the mframe data segment.

The main change compared to LTE-M design is the introduction of the CCA procedure. It has an influence on all the downlink transmissions. Other noteworthy changes are the introduction of the PDRS and new mappings of the LTE-M-U PSS, SSS and PBCH on the resource grid. The LTE-M-U PDSCH and MPDCCH design is close to identical to the LTE-M design.

In LTE-M a bitmap is broadcasted for indicating which subframes are valid for LTE-M downlink transmissions (see Section 5.2.4.1). In LTE-M-U all downlink subframes being part of the uplink-downlink configuration are available for downlink operation.

To distinguish between the LTE-M physical channels and signals and the corresponding LTE-M-U counterparts, a prefix 'u' is from this point on added to the latter.

15.2.2.3.1 Synchronization signals

The LTE-M-U Primary and Secondary Synchronization Signal (uPSS, uSSS) transmissions start after an RF retuning from the frequency hopping data segment and a CCA have been performed. The CCA, introduced in Section 15.2.2.2.3, is performed to sense that the channel is free and available for transmission. As the anchor channel occupancy time is at most ~5 ms, it is sufficient to perform the CCA over the minimum required time of 18 μs.

A total budget of two OS is reserved for the combination of RF retuning and CCA at the beginning of the anchor channel segment. In case the CCA is successfully performed, the synchronization signals are transmitted starting from OFDM symbol 2 in the first subframe $n = 0$. This is depicted in Fig. 15.5 which shows that each of uPSS and uSSS is repeated 4 times during the first two subframes.

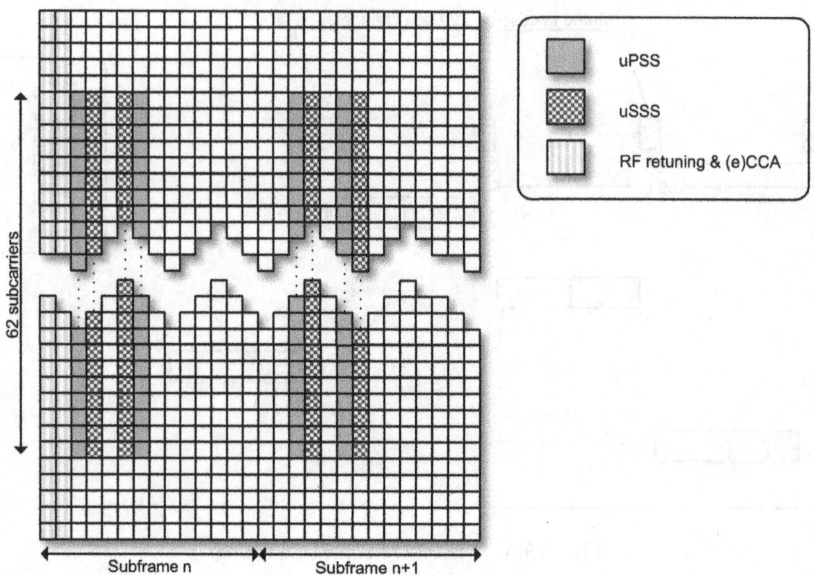

FIG. 15.5 uPSS and uSSS mapping on the center 62 subcarriers, excluding the DC subcarrier, of the anchor channel resource grid in subframe n = 0 or 1 depending on the CCA procedure.

If the CCA fails, the base station must perform an eCCA. For a 5-ms channel occupancy the eCCA should be performed for a random time drawn between 18 and 250 μs. If the eNB is able to perform RF retuning, the CCA and the eCCA within the reserved 2 OS then it can start the synchronization signal transmission according to Fig. 15.5 with the subframe number $n = 0$. If the time needed exceeds 2 OS the transmission is postponed 1 ms and may start first from OFDM symbol 2 in subframe $n = 1$. In this case, the channel occupancy time is ~4 ms which results in a required eCCA period of up to 200 μs.

In case neither the CCA or the eCCA is successful then the anchor channel transmission is canceled.

The uPSS and uSSS transmissions in the anchor segment are densified compared to LTE-M. This is to compensate for the infrequent anchor segment transmission. When the data segment is frequency hopping over 16 data channels the anchor segment is transmitted every 80 ms on the anchor frequency in the beginning of every mframe. With frequency hopping performed over 32 data channels only every other anchor segment is transmitted. This is intended to decrease the anchor channel occupancy and reduce anchor channel interference levels across cells. The periodicity then equals 160 ms as shown in Fig. 15.4. The densification of the uPSS and uSSS increases the likelihood for a device to detect and synchronize to the anchor channel segment. The increased detection probability will to some extent compensate for the long anchor channel periodicity.

15.2.2.3.2 uPSS

Relative subframe	0, 1 or 1, 2 depending on the (e)CCA
Periodicity	80 ms, 160 ms
Subcarrier spacing	15 kHz
Bandwidth	62 subcarriers (not including the DC subcarrier)
Frequency location	At the center of the LTE-M-U narrowband

The uPSS fills the same purpose as the LTE-M PSS, that is to provide time and frequency synchronization and part of the Physical Cell Identity (PCID). LTE-M-U supports, just as LTE-M, 504 PCIDs divided into 167 physical-layer cell-identity groups N_1, each containing 3 cell identities N_2. The uPSS waveform is identical to the PSS and is defined by a 63-element Zadoff-Chu sequence with root indices u equal to 25, 29 or 34.

$$p_u(n) = e^{\frac{-ju\pi n(n+1)}{63}}, \quad n = 0, 1, \ldots, 62 \tag{15.2}$$

The sequence is mapped to the 63 subcarriers located around the center in LTE-M-U narrowband. The center subcarrier itself, which is commonly associated with a DC component in the receiver, is punctured, and not transmitted. This means that the uPSS waveform in practice is defined by the Zadoff-Chu elements 0–30 and 32–62.

Each of the 3 roots is associated with one of the 3 cell identities N_2, defined within a physical-layer cell-identity group. When detecting the uPSS a device determines the root index and the corresponding cell identity.

The Zadoff-Zhu sequence provides both good auto-correlation properties and peak-to-average power ratio (PAPR) characteristics. In the time domain the uPSS sequence is first mapped on OFDM symbol (OS) 2 and repeated on OS 6 in the same subframe, and then on OS 2 and 5 in the next subframe. Fig. 15.5 shows this asymmetric mapping of the uPSS, and the uSSS, onto the resource grid across subframes 0 and 1. This allows a device synchronizing to the downlink frame structure to distinguish between the first and second subframes and obtain radio frame timing.

15.2.2.3.3 uSSS

Relative subframe	0, 1 or 1, 2 depending on (e)CCA
Periodicity	80 ms, 160 ms
Subcarrier spacing	15 kHz
Bandwidth	62 subcarriers (not including the DC subcarrier)
Frequency location	At the center of the LTE-M-U narrowband

The uSSS shares definition with the LTE SSS mapped to subframe 0. It is defined by the 62 symbol sequence *p(n)* which is determined by the interleaving of 2 length-31 m-sequences $s_0^{(m_0)}$ and $s_1^{(m_1)}$, which in turn are scrambled by m-sequences $z_1^{(m_0)}$ and c_0:

$$\begin{aligned} p(2n) &= s_0^{(m_0)}(n)c_0(n) \\ p(2n+1) &= s_1^{(m_1)}(n)c_0(n)z_1^{(m_0)}(n) \\ n &= 0, 1, \ldots, 30 \end{aligned} \tag{15.3}$$

The sequence $s_0^{(m_0)}$, $s_1^{(m_1)}$ and $z_1^{(m_0)}$ are all generated by cyclic shifts m_0 and m_1 of m-sequences *s(n)* and *z(n)*, respectively. 167 unique pairs of shifts (m_0, m_1) are defined, where each pair is associated to one of the 167 physical-layer cell-identity groups N_1.

$c_0(n)$ is generated by a cyclic shift of the m-sequence *c(n)*, with the length of the cyclic shift dependent on the cell identity N_2. As the cell identity N_2 is signaled by the uPSS, this creates a coupling between the uPSS and uSSS transmissions from the same cell.

The m-sequence is a suitable choice for providing physical-layer cell identity signaling by means of sequence detection thanks to its good auto-correlation properties. It is also straightforward to generate the m-sequences by means of shift registers. An interesting property of the m-sequence is that a cyclic shift of the sequence results in a new m-sequence.

The uSSS is just as the uPSS mapped in the frequency domain on the center 62 sub-carriers, excluding the DC subcarrier. In the time domain the signal is mapped on OS 3 adjacent to the first uPSS. It is repeated on OS 5 in the same subframe, and on OS 3 and 6 in the next subframe.

LTE-M concatenates the sequences $s_0^{(m_0)}$ and $s_1^{(m_1)}$ in two alternative fashions to create two versions of the SSS. The two versions are then mapped on different subframes, e.g. 0 and 5 for frequency-division duplex (FDD) LTE, to allow a device to obtain frame timing. LTE-M-U defines a single uSSS version that is repeated on all four OS. An LTE-M-U device therefore needs to use the uPSS and uSSS mapping to determine if it has detected the first or second subframe. It also needs to detect up to two uPSS and uSSS OS pairs to determine the OS timing within the subframe.

15.2.2.3.4 Downlink reference signals

In the downlink LTE-M-U supports the Cell-Specific Reference Signal (CRS), the Device-Specific Reference Signal and the PDRS. The CRS follows the LTE design and supports similar functions as LTE, i.e. channel estimation and frequency tracking for coherent uPDSCH demodulation and radio link monitoring in connected and idle mode. The CRS supports uPDSCH demodulation for transmission modes 1, 2 and 6. For the sake of radio link monitoring a device can assume that the CRS is present in case the channel sensing is successful which is indicated by the transmission of the uPSS, uSSS and uPBCH on the anchor channel and the PDRS on the data channel.

The Device-Specific Reference Signal, also known as the Demodulation Reference Signal (DMRS), supports demodulation of the uMPDCCH, and the uPDSCH in case of transmission mode 9. The DMRS is transmitted on the same antenna ports as the uMPDCCH and uPDSCH to support e.g. device-specific beamforming.

More information on the downlink reference signals can be found in Section 5.2.4.3. Next, we describe the PDRS which is unique to LTE-M-U.

15.2.2.3.4.1 PDRS

Relative subframe	5 for CCA or 6, 7 or 8 for eCCA
Basic TTI	1 ms
Repetitions	1
Subcarrier spacing	15 kHz
Bandwidth	72 subcarriers (not including the DC subcarrier)
Frequency location	Across the LTE-M-U narrowband

A base station is required to perform the CCA procedure (see Section 15.2.2.2.3) before initiating a downlink transmission on the mframe data segment. Use of the downlink parts in the applicable uplink-downlink configuration is only permitted if the channel is clear. The PDRS serves as an indication to the devices in a cell that the channel is clear and that CRS, uPDCCH and uPDSCH transmissions can be expected. Its functionality can be compared to that offered by the Wake-Up Signal specified for LTE-M in Release 15 (see Section 5.2.4.5).

The channel occupancy time for a downlink data transmission is defined from the first downlink subframe to the last downlink subframe in a data segment mapped on the mframe. This results in a channel occupancy time in the range of 20–60 ms depending on the uplink-downlink configuration. The corresponding CCA must be performed for a duration between 40 and 120 μs, i.e. less than 2 OS. The eCCA needs to be performed during in-between 1 and 3 ms.

The PDRS is a defined by a cell-specific pseudo-random bit sequence, generated just as the CRS, that is modulated into QPSK symbols. The QPSK symbols are mapped over a full narrowband on the last 12 OS of a subframe. Fig. 15.6 for simplicity illustrates the mapping on 1 of the 6 PRBs for the case of 2 base station antenna ports.

The first two OS on subframe n are left unused to support the CCA procedure. In case the channel is clear the PDRS transmission may start in subframe 5, which corresponds to the first subframe in the data segment. If the eCCA needs to be performed for $k = 1$, 2 or 3 subframes, the PDRS transmission starts first in subframe $n = 5 + k$. This means that the PDRS may be transmitted in subframes $n = 5$, 6, 7, or 8, and the number of valid subframes for the transmission of downlink physical channels will be reduced accordingly.

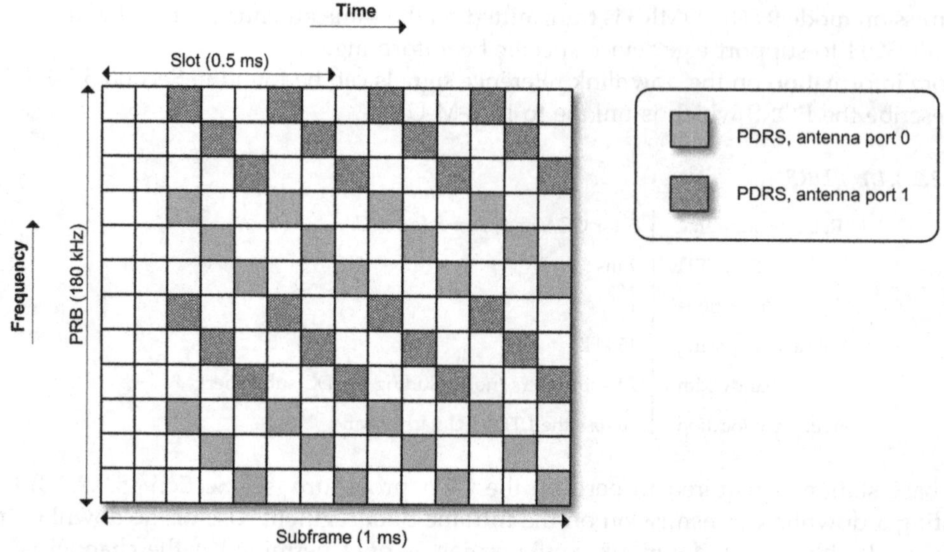

FIG. 15.6 PDRS resource element mapping illustrated for 2 antenna ports.

15.2.2.3.5 uPBCH

Relative subframe	0, 1 or 1, 2 depending on (e)CCA
Basic TTI	2 ms
Periodicity	80 ms, 160 ms
Repetitions	3
Subcarrier spacing	15 kHz
Bandwidth	62 subcarriers (not including the DC subcarrier)
Frequency location	At the center of the LTE-M-U narrowband

The uPBCH design is inspired by the LTE PBCH. It carries the Master Information Block (MIB) that together with the *System Information Block A* (see Section 15.2.3.1) contains the most vital parts of the system configuration information. The 24-bit MIB is convolutionally encoded and rate matched before being repeated three times on the center 72 subcarriers, excluding the DC subcarrier, on the available OS in subframes n and $n+1$ where $n=0$ or 1 depending on the CCA procedure (see Section 15.2.2.2.3). This is depicted in Fig. 15.7, where the CRS mapping for simplicity is not depicted. Each repetition covers six OS:

- The first repetition is mapped on OS 4, 7, 8, 9, 10, 11 in subframe n. This differs from the mapping of the LTE-M core part and instead follows the PBCH mapping specified in MulteFire Release 1.0.
- The second repetition is mapped on OS 12, 13 in subframe n, and on OS 0, 1, 4, 7 in subframe $n+1$.
- The third repetition is mapped on OS 8, 9, 10, 11, 12 and 13 in subframe $n+1$.

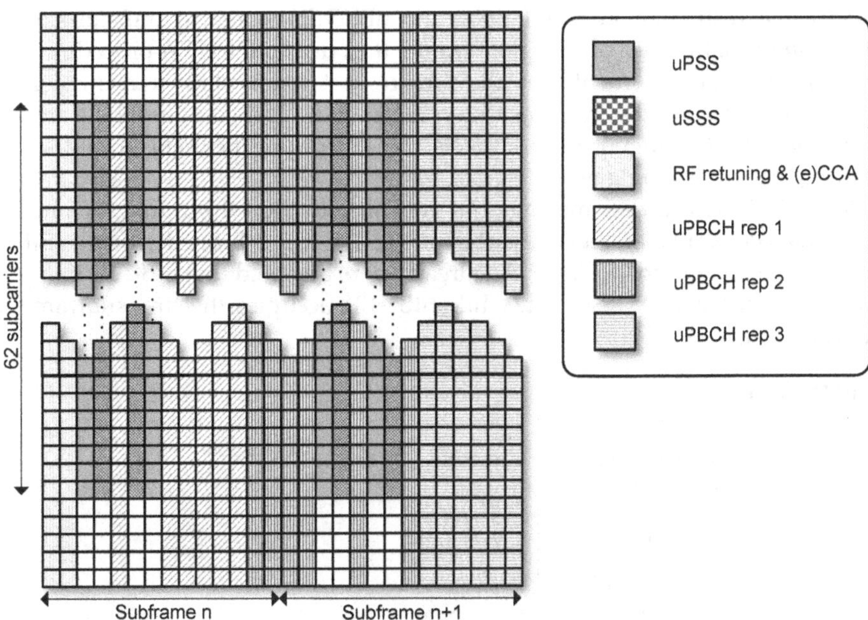

FIG. 15.7 uPBCH mapping on the anchor channel resource grid with subframe n = 0 or 1 depending on the CCA procedure. CRS is not depicted.

The in total 18 uPBCH OS results in a similar code rate as the LTE-M PBCH core part which is mapped over 16 OS in 40 ms. For LTE-M an optional configuration of 4 repetitions of the core part is available for system bandwidths larger than 1.4 MHz. LTE-M-U does not offer any configurability in this aspect and the uPBCH is always mapped over $3 \cdot 6 = 18$ OS.

15.2.2.3.6 uPDSCH

Relative subframe	2 - 4 for SIB-A. Any between 6 and 65 for other transmissions.
Basic TTI	1 ms
Repetitions	1, 2, 4, 8, 16, 32
Subcarrier spacing	15 kHz
Bandwidth	1 - 6 PRBs
Frequency location	Within the narrowband

The uPDSCH supports unicast data and system information transmissions, including SIB-A and SIB1-BR (known as *SIB1-BR-MF* in the MFA specifications). The uPDSCH design closely follows the Release 13 LTE-M PDSCH CE mode A design. It supports up to 32 repetitions using 4 redundancy versions. The maximum transport block size equals 1000 bits. Transmission modes 1, 2, 6 and 9 are supported. More information about the PDSCH physical layer definition is found in Section 5.2.4.7.

One noticeable restriction imposed upon the uPDSCH is that a transmission must be confined within the subframes of a single mframe. It is, due to the CCA procedure performed in the start of each data segment, not allowed to start a downlink transmission in a first mframe *n* and then follow the frequency hopping pattern and continue the transmission in a second mframe *n+1*. This not only impacts the scheduling flexibility but also limits the downlink frequency hopping gain.

The uPDSCH SIB-A is a new message but is similar to the MIB in that it contains critical system information parameters. It is described in Section 15.2.3.1 together with a SIB1-BR and the other system information blocks. It can already here be noted that SIB-A is mapped on the last subframes of the anchor segment, while SIB1-BR occupies the first subframe after the PDRS transmission.

15.2.2.3.7 uMPDCCH

Relative subframe	Any between 7 and 65
Basic TTI	1 ms
Repetitions	1, 2, 4, 8, 16
Subcarrier spacing	15 kHz
Bandwidth	2, 4, 6 PRBs
Frequency location	Within the narrowband

The uMPDCCH supports scheduling of the uPDSCH, uPUSCH and the uPUCCH transmissions. Its physical layer definitions are very similar to the LTE-M MPDCCH described in Section 5.2.4.6. In the frequency domain the MPDCCH can be transmitted over 2, 4 or 6 PRBs using an aggregation level up to 24. The maximum supported number of repetitions have been limited to 16 to support similar coverage as CE mode A, but not CE mode B. This gives a transmission time that spans between 1 and 16 subframes.

The uMPDCCH transmission can occupy several downlink portions of an uplink-downlink configuration, but it must due to the CCA be confined within a single mframe. This is the same restriction as applies to the uPDSCH.

The scheduling of the uPDSCH and uPUSCH is controlled by means of Downlink Control Information (DCI) messages. LTE-M-U supports transmission of DCI formats 6-0A and 6-1A for operation in CE mode A. This is in further detail discussed in Section 5.3.2.1.

The resource used for uPUCCH, providing the HARQ acknowledgment to a uPDSCH transmission, is determined based on the lowest enhanced control channel element (ECCE) that is part of the uMPDCCH transmission scheduling the uPDSCH. The channelization and resource identification of the uPUCCH is discussed in Section 15.2.2.4.4.

It should be noted that the ECCE numbering is not device-specific. The MPDCCH DCI therefore signals a HARQ ACK/NACK resource offset indication that is combined with the lowest ECCE number to make the uPUCCH resource mapping device-specific. This mechanism is intended to support multiplexing of uPUCCH transmissions from multiple devices.

15.2.2.4 *Uplink physical channels and signals*

LTE-M-U supports the same set of physical channels and signals as LTE-M Release 13 with close to identical design as used by LTE-M. The LTE-M-U uplink transmission does not rely on the CCA procedure, so the differences compared to LTE-M are in short limited to the mappings on the physical resources, the information content of the transmitted messages and that the CE operation is limited to CE mode A.

In LTE-M a bitmap is broadcasted for indicating which subframes are valid for uplink transmissions (see Section 5.2.5.1). In LTE-M-U all uplink subframes being part of the uplink-downlink configuration are available for transmissions.

15.2.2.4.1 uPRACH

Relative subframe	According to higher layer configuration
Basic TTI	1 ms
Repetitions	1, 2
Subcarrier spacing	1.25 kHz
Bandwidth	839 subcarriers
Frequency location	Within a narrowband

The uPRACH is identical to the LTE PRACH (see Section 5.2.5.2). Its defined by a length 839 Zadoff-Chu sequence, which is used to generate 64 unique preamble sequences per cell, based one or more configured root indexes complemented by a set of cyclic shifts.

A difference compared to LTE-M is that only PRACH Format 0 is supported for LTE-M-U. Format 0 is designed with a cyclic prefix length of 103 μs while the Zadoff-Chu sequence corresponds to a length of 800 μs after the DFT-precoded OFDM modulation. When mapped to a single subframe a guard period of 97 μs is obtained, which corresponds to a maximum supported cell size of ~15 km if collisions with transmissions in the next subframe are to be avoided.

Another difference compared to LTE-M is that only a repetition factor of 1 or 2 can be configured for the uPRACH. This was deemed sufficient for the coverage range needed to support industrial small cell deployments. The uPRACH transmissions are mapped onto the available uplink subframes according to 1 out of 29 specified configurations. The applicable mapping is indicated to the devices in a cell by means of system information broadcast signaling.

15.2.2.4.2 Uplink reference signals

LTE-M-U supports the uplink Demodulation Reference Signal (DMRS) and the Sounding Reference Signal (SRS). Their purpose and design follow LTE-M described in Section 5.2.5.3 with some small deviations. Since the uPUSCH transmission bandwidth is restricted to one narrowband, so is the DMRS bandwidth for the support of coherent uPUSCH demodulation. DMRS for uPUCCH formats 1, 1a, 2 and 2a are supported, just as for LTE-M operating as a FDD system. uPUCCH format 3 was added to the LTE-M-U design baseline to support

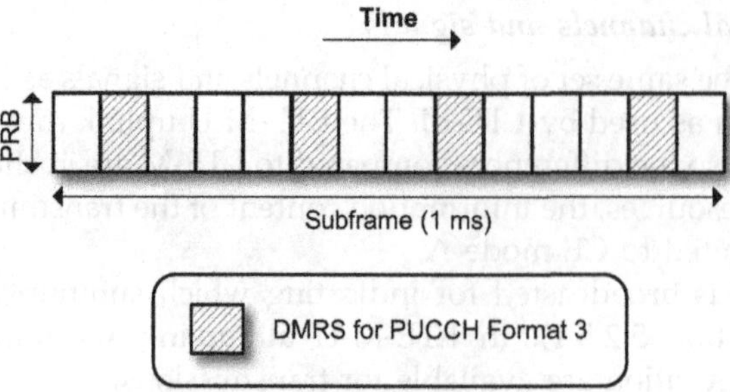

FIG. 15.8 DMRS symbol mapping for uPUCCH Format 3.

ACK/NACK of up to 10 HARQ processes. For uPUCCH Format 3 the DMRS is mapped on symbols 1 and 5 in each slot as depicted in Fig. 15.8.

The SRS is mapped over the center four PRBs in subframes fulfilling $n_{sf}^{rel}\, mod\ T_{SFC} \in \Delta_{SFC}$ in mframes satisfying $n_{mframe}\ mod\ T_{mframe} = 0$. n_{sf}^{rel} is the relative subframe numbering, T_{SFC} is the SRS periodicity within the applicable mframes, Δ_{SFC} defines the subframe for SRS transmissions within the period T_{SFC}. T_{mframe} is finally the SRS mframe periodicity. 32 different SRS configurations are supported. In its most dense configuration the SRS is transmitted once every five uplink subframes. In the least dense configuration two SRS subframes are transmitted every four mframes.

15.2.2.4.3 uPUSCH

Relative subframe	Any between 21 and 65
Basic TTI	1 ms
Repetitions	1, 2, 4, 8, 16, 32
Subcarrier spacing	15 kHz
Bandwidth	1 − 6 PRBs
Frequency location	Within the narrowband

The uPUSCH supports unicast data transmission and aperiodic channel state information reporting. Its design largely follows the LTE-M design described in Section 5.2.5.4. The transmission bandwidth is limited to at most 6 PRBs. The number of repetitions is limited to 32 as only CE mode A is supported.

The LTE-M-U uplink is designed as a non-adaptive frequency hopping system to meet the ETSI regulations. This means that the uplink transmissions need to meet the *MU* requirement described in Section 15.2.3.4. In addition the uPUSCH may not be transmitted for more than 5 consecutive subframes before resting at least 5 ms, and in total not transmit for more than 15 ms during any given data segment. To facilitate this transmission scheme the uplink part of the data segment is divided in 2 *subframe sets* 0 and 1. Fig. 15.9 illustrates the concept for uplink-downlink configuration 1. A device may only make use of one of the subframe sets for its uPUSCH transmissions. A uPUSCH transmission of a length exceeding 5 ms, starting

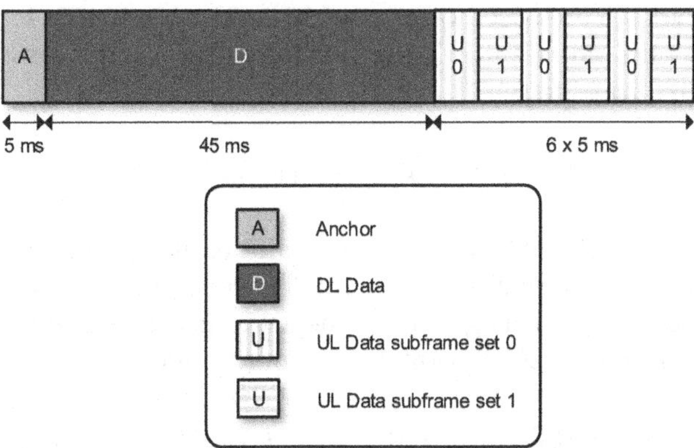

FIG. 15.9 Uplink-downlink configuration 1 with the uplink part portioned in subframe sets 0 and 1.

in the *nth* subframe set 0 will be postponed during the *nth* subframe set 1 before it continues in the next subframe set 0.

While the downlink transmissions need to be confined within a single mframe due to the channel sensing regulation the uplink transmissions may continue across mframes. This since the uplink operates as a non-adaptive frequency hopping system. This may give repeated uplink transmissions a frequency hopping gain, which is not present for the downlink transmissions.

15.2.2.4.4 uPUCCH

Relative subframe	Any between 21 and 65
Basic TTI	1 ms
Repetitions	1, 2, 4, 8
Subcarrier spacing	15 kHz
Bandwidth	1 PRB
Frequency location	Within the narrowband

The uPUCCH follows the LTE-M PUCCH design described in Section 5.2.5.5. LTE-M supports PUCCH Formats 1, 1a, 2 and 2a for FDD and Formats 1b and 2b for TDD. LTE-M-U uses LTE-M FDD from 3GPP Release 13 as its design baseline and implements support for PUCCH Formats 1, 1a, 2 and 2a. It also supports the LTE PUCCH Format 3 to enable HARQ ACK/NACK for up to the 10 supported HARQ processes to be transmitted in a single uPUCCH message.

The uPUCCH is mapped to uplink subframes according to the uplink-downlink configuration and the uplink subframe set concept. In the frequency domain it is mapped to the outermost PRBs of the narrowband. While the LTE PUCCH is frequency hopping across the system bandwidth the uPUCCH is transmitted only on the lower edge of the narrowband. To support efficient multiplexing of the uPUCCH and uPUSCH on a single narrowband it is important to minimize the frequency resources used by the different PUCCH formats.

PUCCH Format 1 and 1a are based on a set of 30 base sequences $\overline{r}_u(k)$. These are the same sequences that defines the single-PRB DMRS transmissions. Each base sequence is uniquely defined by a set of tabulated phases $\varphi_u(k)$:

$$\overline{r}_u(k) = e^{\frac{j\varphi_u(k)\pi}{4}}, \quad k = 0,.,11, \quad u\in\{0,.,29\} \tag{15.4}$$

When a single HARQ ACK/NACK bit d is transmitted it is mapped to a complex-valued symbol $\tilde{d}$ which is modulated by a α_m phase rotated version of $\overline{r}_u(k)$ and multiplied with a block spreading sequence $w_{n_{oc}}$. This generates the set of complex symbols $z(k,l)$ mapped to resource elements (k,l) in a PUCCH Format 1a resource block:

$$z(k,l) = w_{n_{oc}}(l) \cdot e^{j\alpha_m k} \cdot \overline{r}_u(k) \cdot \tilde{d}, \quad k = 0,..,11, \quad l = 0,1,5,6, \quad m\in\{0,..,11\} \tag{15.5}$$

Note that index k identifies the subcarrier, while index l identifies the OFDM symbol associated to a certain resource element. A DMRS sequence, supporting coherent demodulation of the uPUCCH, is in similarity to Eq. (15.5) phase rotated by α_m and spread over symbols $l = 2$, 3, 4. The 12 available phase shifts combined with the length-3 block spreading sequence applied on the DMRS allows in theory up to $3 \cdot 12 = 36$ different PUCCH transmissions to be multiplexed on a single uplink resource block. As a rule of thumb only every second phase shift is useful if orthogonality between users is to be kept under time dispersive channel conditions. This reduces the PUCCH Format 1 multiplexing capacity to $3 \cdot 6 = 18$ users per resource block.

PUCCH Format 2 contains up to 11 bits per subframe. Each of the first 10 bits is modulated by a α_m phase rotated version of $\overline{r}_u(k)$ and is mapped to the resource elements of data symbols 0, 2, 3, 4 and 6 on each of the two slots in the subframe. The use of the same set of base sequences $\overline{r}_u(k)$ allows PUCCH Format 1 and 2 to be multiplexed on the same PRB using unique subsets of the 12 phase shifts α_m, which gives orthogonality between the multiplexed transmissions.

PUCCH Format 3 follows a design different compared to PUCCH Formats 1 and 2. It carries 22 information bits that are block coded into 48 bits mapped to 2 sets of 12 QPSK symbols. Each set of 12 QPSK symbols is multiplied with a length-5 block spreading sequence and mapped to the 12 resource elements of each of the 5 data symbols in each slot. As a result, PUCCH Format 3 cannot be multiplexed on the same resource blocks as the PUCCH Format 1 and 2 transmissions. Fig. 15.10 illustrates an example where PUCCH Formats 1 and 2 are multiplexed on PRB 0, and PUCCH Format 3 is mapped on PRB 1. This leaves at least 4 PRBs for PUSCH transmission. The uPUCCH resource blocks may be used by uPUSCH transmissions when no uPUCCH is scheduled for transmission.

15.2.3 Idle and connected mode procedures

In this section we describe the LTE-M-U idle and connected mode procedures. The design principle of LTE-M-U was to maximize the reuse of the LTE-M procedures described in Section 5.3. Here we therefore focus on the procedures that deviate from the LTE-M baseline, including new procedures required by ETSI and FCC regulations.

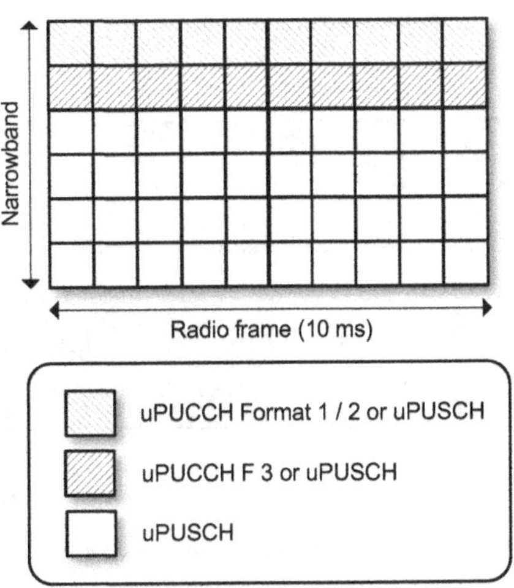

FIG. 15.10 uPUCCH and uPUSCH multiplexing on a narrowband.

15.2.3.1 Cell selection and system information acquisition

The first task a device performs when turned on is to search for, evaluate the suitability of, and select a cell. While an LTE-M device scans all its supported frequency bands in the search for a suitable cell, an LTE-M-U device only scans the 3 anchor channels in the 2.4 GHz frequency band, depicted in Fig. 15.1, in its search for an LTE-M-U cell. The rationale for limiting the set of anchor channels is partly due to the frequency hopping nature of the LTE-M-U system, which leads to a sparse transmission of the anchor channel including its synchronization and MIB blocks. If multiple attempts to synchronize are needed the system acquisition time quickly grows as the uPSS is transmitted only every 80 ms or even only every 160 ms. With only 3 potential anchor channels this is no longer an issue of significance.

After acquiring synchronization to the downlink frame structure and the physical-cell identity via the uPSS and uSSS, the device reads the uPBCH MIB and the uPDSCH SIB-A. The MIB contains the system frame number information, which gives the mframe index within a hyper frame, the SIB-A and the SIB1-BR scheduling information.

15.2.3.1.1 SIB-A

SIB-A is a new system information message, which is similar to the MIB in that it contains critical system information parameters. SIB-A is mapped on the last 2 or 3 subframes on the anchor channel. If the channel is cleared during the first 2 OS the uPBCH transmission takes place during subframes 0 and 1 and the SIB-A is transmitted in subframes 2, 3 and 4. If the enhanced CCA needs to be performed during subframe 0, and is successful, the uPBCH is transmitted on subframes 1, 2 and the SIB-A is only mapped over subframes 3 and 4. Fig. 15.11 illustrates the two possible SIB-A configurations.

In total, SIB-A carries 28 bits including for example the *Channel Set* information element that signals the channel groups that defines the set of data channels used to frequency hop over. For 16 data channels, 4 channel groups will be selected and for 32 data channels, 8 channel groups will be selected. This signaling allows LTE-M-U to configure a set of 16 or 32 data

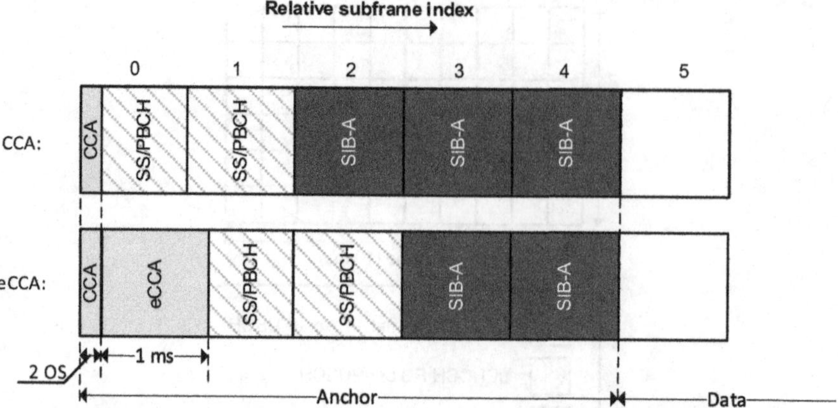

FIG. 15.11 SIB-A mapping on the anchor channel resource grid.

channels that are exposed to as little interference as possible by other systems. The SIB-A information elements (IE) are summarized in Table 15.3. The Paging Indication IE allows devices to skip paging monitoring to save power. The System Information Value Tag IE has been moved to SIB-A to enable early detection of the need to acquire a fresh copy of the other system information messages.

15.2.3.1.2 SIB1-BR

SIB1-BR informs the devices in a cell about the LTE-M-U access mode of operation. A MulteFire cell can operate either using the PLMN access mode or using the *Neutral Host Network* (NHN) access mode. In case of the PLMN access mode, LTE-M-U is connected to an Evolved Packet Core network in standalone mode or as an additional radio access network, e.g. complementing an LTE-M deployment. In the NHN access mode, LTE-M-U operates as a standalone system providing local access for a set of service providers.

TABLE 15.3 SIB-A information elements.

Parameter	Description
Channel set	Identifies the set of 16 or 32 data channels configured for data transmission.
Hyper system frame number	The three least significant bits of the hyper SFN. This gives, combined with the master information frame number information, the complete mframe index.
mframe configuration	Indicates the configured uplink-downlink configuration.
Paging indication	Indicates if there will be a paging during the next data segment.
PDRS window size	Indicates the subframe window in which the PDRS may be transmitted if the channel is cleared.
System information value tag	Informs the devices if a change in the system information has occurred.

In PLMN mode, the PLMN selection is based on the established 3GPP procedures. In case of NHN, the NHN identifier identifies the network, while the set of service providers utilizing the network are identified by a *Participating Service Provider* identity, and the PLMN information element broadcasted in SIB1-BR is set to a single reserved PLMN identity which is common for all MulteFire NHNs.

The LTE-M-U SIB1-BR is transmitted with a periodicity ranging between 80 and 640 ms using 8 or 4 repetitions. The SIB1-BR scheduling is presented in Table 15.4 under the assumption that the PDRS is transmitted in subframe 5. The first subframe in the data segment after the PDRS transmission always contains a SIB1-BR transmission.

The content and scheduling mechanism for the remaining SIB information elements follow the LTE-M specification as described in Section 5.3.1.2.

15.2.3.2 Paging

The LTE-M-U paging is based on the procedure described for LTE-M in Section 5.3.1.4. Both DRX and eDRX are supported. In LTE-M, the start of the common search space for the MPDCCH containing the P-RNTI is determined by a paging frame, and a subframe within the paging frame known as the paging occasion. In the extreme case of high paging loads up to 4 subframes in every radio frame can be configured as paging occasions. In LTE-M-U, a new concept called a *Paging Occasion Window* (POW) is defined. A POW spans 1 to 4 mframes. There is only one paging occasion per mframe, which starts at the first valid downlink subframe within the mframe, i.e. it starts at a subframe not containing system information. If a paging transmission in the first paging occasion fails, then the network may attempt to page the device in the next paging occasion, which is sent in the next mframe within the POW.

To reduce device power consumption from monitoring paging occasions for the case there is no page, LTE-M-U supports a paging indication bit sent in SIB-A. It informs the devices in a cell if they can expect a page or not during the next anchor segment period.

15.2.3.3 Power control

The LTE-M-U uplink power control follows the definitions and procedures specified for LTE-M CE mode A as described in Section 5.3.2.4. An important difference is the device power control accuracy. While 3GPP allows a band dependent maximum output power

TABLE 15.4 LTE-M-U SIB1-BR scheduling.

Period	Repetitions	SIB1-BR subframe index
80 ms	4	6, 7, 8, 9
80 ms	8	6, 7, 8, 9, 10, 11, 12, 13
160 ms	4	6, 7
320 ms	4	6
640 ms	4	6

tolerance of 2–2.5 dB, the ETSI and FCC specifications impose strict limitations on the maximum output power. For ETSI, following the design of a non-adaptive FHSS system implies a maximum EIRP of 20 dBm. In case of FCC, the limit for a hybrid system is 8 dBm/3 kHz which equates to 25.8 dBm over 1 PRB. The first LTE-M-U device category is expected to support power class 5 operating at a max power of 20 dBm, due to the ETSI requirements.

The situation is similar for the downlink. 3GPP allows tolerances in the maximum basestation output power of at most 2.5 dB. ETSI has a strict limit of 20 dBm EIRP for adaptive FHSS systems, while FCC allows up to 30 dBm for a digitally modulated system. There is no acceptance for exceeding these maximum output power levels.

15.2.3.4 Medium utilization

Non-adaptive equipment operating according to the ETSI regulations, i.e. LTE-M-U devices, with an EIRP exceeding 10 dBm must comply with a medium utilization (MU) factor of at most 10%. The MU factor balances the power and time resources used by a certain transmitter and is defined as:

$$MU = \frac{P}{100\ \text{mW}} \cdot DC \qquad (15.6)$$

Where P corresponds to the output EIRP, and the DC is the *duty cycle*. A device operating at the maximum output power of 100 mW (20 dBm) is limited by a duty cycle of 10%. The duty cycle allowance increases with decreasing output power.

15.3 NB-IoT-U

The development of EC-GSM-IoT, LTE-M and NB-IoT was partly triggered by the competition from various low power wide area network (LPWAN) technologies operating in unlicensed spectrum. 3GPP provides through EC-GSM-IoT, LTE-M and NB-IoT a highly competitive offering for close to all types of licensed spectrum arrangements. But for enterprises lacking access to licensed spectrum a LPWAN solution operating in free, unlicensed spectrum is still attractive. The MFA therefore developed NB-IoT-U to complement LTE-M-U and make a 3GPP based LPWAN available in unlicensed spectrum.

The next three sections introduce the design principles followed by MFA when developing NB-IoT-U, the NB-IoT-U physical layer functionality and the procedures used in RRC idle and connected mode.

15.3.1 Radio access design principles

The NB-IoT-U system design is in the first release targeting operation in the FCC ISM band defined by the frequency range 902–928 MHz and the ETSI SRD Band 54 defined by the 250 kHz band located between 869.4 and 869.65 MHz [10]. EU, through the European Commission, has recently decided to make Band 47b containing the frequency ranges 865.6–865.8, 866.2–866.4, 866.8–867 and 867.4–867.6 MHz available for LPWAN [11].

The MFA focused on ETSI band 54 during its work, and this is reflected in the descriptions in this book. Band 54 and 47b are subject to similar regulations, which is expected to make an extension from operation in Band 54 to Band 47b straightforward. These sub-GHz frequency bands are all attractive for low-power wide-area technologies due to their favorable radio propagation characteristics.

NB-IoT-U uses the NB-IoT stand-alone design as its baseline. This is a natural choice, since the system is intended to operate in non-3GPP spectrum. Operation within or in the guard-band of an LTE system is not foreseen in unlicensed spectrum. The intention of NB-IoT-U is to build on the NB-IoT eco-system, to allow a reuse of both device and network implementations. This motivates a reuse of the NB-IoT design to the maximum extent possible. This book distinguishes between the NB-IoT physical channels and signals and the corresponding NB-IoT-U counterparts, by a prefix 'u' that is added to the names of the latter.

NB-IoT-U is designed as a time division duplex type of system. Despite this, it is based on the 3GPP Release 13 NB-IoT FDD specifications, and not the 3GPP Release 15 specifications that supports NB-IoT in TDD operation. The explanation for this is that MFA Release 1.1 specifying NB-IoT-U and 3GPP Release 15 were developed in parallel.

NB-IoT-U shares similarities with LTE-M-U in that it follows a TDD design with a frame structure that is divided into a fixed anchor and frequency hopping data segments. The anchor channel is carrying the synchronization signals and the broadcast channel. The data channels support both uplink and downlink unicast control and data transmissions.

The next two sections introduce the FCC and ETSI regulations. Detailed descriptions of parts of the regulations is also provided later in the text. Section 15.3.2.2.2 presents the frequency hopping rules, while Section 15.3.3.2 describes the power control regulations.

15.3.1.1 FCC regulations

In US the NB-IoT-U anchor segment transmissions are following the FCC rules [2] for a digitally modulated system and is mapped on 3 PRBs. Synchronization and broadcast channels are located at the 2 outer PRBs to secure that the 500 kHz bandwidth requirement for digital modulation is always met. The anchor segment is transmitted on a cell specific frequency which allows devices to sequentially scan the supported frequency bands when performing cell selection. The base stations' digitally modulated transmissions are not permitted to exceed an average conducted power of 30 dBm, and a PSD of 8 dBm/3 kHz. The maximum allowed average radiated power is defined by an EIRP of 36 dBm.

The data segment follows the frequency hopping regulations. It is modulated on a single PRB that frequency hops over 64 frequencies. The NB-IoT-U downlink is a good example of a hybrid system with the anchor part being digitally modulated while the data part is frequency hopping. Due to the required random location of the frequency hopping channels, the anchor segment does not qualify as a frequency hopping system. The frequency hopping PRB may be transmitted using a conducted peak output power of 30 dBm, and a maximum peak EIRP of 36 dBm. The frequency hopping regulations defines the power in terms of peak power, and not average power as in the case of the digital modulation regulations. This may seem like a minor detail but is an important aspect for systems such as NB-IoT-U with a non-zero PAPR. The PAPR of the transmitted OFDM waveform needs to be taken into account when determining the maximum allowed average output power.

The anchor part is reserved for downlink operation, meaning that the devices in the uplink are only permitted to operate over the frequency hopping data segments. The uplink is hence declared as a frequency hopping system.

15.3.1.2 ETSI regulations

In Europe NB-IoT-U is designed according to the regulations defined in *EN 300 220 for Non-specific Short Range Devices (SRD)* [10]. The standard is also applicable to LPWANs although its name may suggest that it does not cover LPWANs. The use of Band 54 allows a peak, not average, output power of 29.2 dBm EIRP, in combination with a 10% duty cycle.

The duty cycle rule is limiting the transmit activity of NB-IoT-U base stations and devices. The impact is most pronounced on the ability of the base stations to provide the needed downlink capacity for serving all devices in a cell. Due to the expected infrequent and small uplink data transmissions the impact on the devices is assumed to be lower.

Most ETSI frequency bands in the range 25 MHz to 1 GHz are associated with a duty cycle budget well below 10%. This makes Band 54 an attractive choice despite its limitations. Table 15.5 summarizes the NB-IoT-U design choices made to comply with the FCC and ETSI regulations.

15.3.2 Physical layer

The description of the physical layer is distributed in four parts. The Physical resources section 15.3.2.1 presents the basic time and frequency resources used transmitting the NB-IoT-U signal in the FCC ISM and ETSI SRD bands. The Transmission schemes section 15.3.2.2 presents generic aspects and functionality that applies across large parts of the physical layer design. The last two sections 15.3.2.3 and 15.3.2.4 present the details of the downlink and uplink physical channels and signals.

15.3.2.1 Physical resources

15.3.2.1.1 Channel raster

NB-IoT-U is in the first release targeting the FCC ISM band 902–928 MHz, and ETSI SRD Band 54 spanning the range 869.4–869.65 MHz. The specified channel raster is dependent on the channel bandwidth definition which differs between ETSI and FCC.

TABLE 15.5 NB-IoT-U radio access design principles.

Regulation	Link direction	Design basis
FCC	Downlink	Hybrid
	Uplink	FHSS
ETSI	Downlink	Non-specific Short Range
	Uplink	Non-specific Short Range

In ETSI the NB-IoT-U channel bandwidth is defined by the *occupied bandwidth* of the modulated waveform. Its definition aligns with 3GPPs occupied bandwidth concept and corresponds to the frequency range containing 99% of the total power of the studied signal. In 3GPP the NB-IoT occupied bandwidth equals 200 kHz which fits in the center of ETSI band 54.

In FCC the frequency range 902–928 MHz have been divided in 3 anchor channels and 64 data channels. The anchor channel baseband signal spans 3 PRBs while the data channel baseband waveform is defined by a single PRB. For the channel bandwidth, the FCC regulations define a *20-dB bandwidth* requirement. This corresponds to the bandwidth where the signal emissions are attenuated by 20 dB relative to the power measured in the center of the carrier. The 20-dB bandwidth definition is more stringent than the 3GPP and ETSI occupied bandwidth requirement which motivates a wider channel bandwidth declaration in FCC compared to ETSI and 3GPP. The single PRB data channel 20 dB bandwidth was agreed to 345 kHz after careful study of the 3GPP spectrum emission mask. The intent of MFA was to make sure that the NB-IoT-U RF requirements are not more stringent than the 3GPP RF requirements. The 345 kHz channel bandwidth for simplicity also defines the data channel frequency raster.

15.3.2.1.2 Frame structure

NB-IoT-U introduces frame structure Type 3N. It is a TDD type of frame structure which is a natural choice since ETSI band 54 and FCC band 902–928 MHz are both unpaired bands. Frame structure Type 3N builds on the 3GPP frame structure with the addition of the new *nframe* concept. A nframe spans $N_{Frame} = 2, 4,$ or 8 radio frames. The nframe index wraps around every 4 hyperframes as illustrated in Fig. 15.12.

The nframe defines the lengths of the TDD uplink-downlink configurations presented in Section 15.3.2.2. The shorter nframes supports more frequent uplink to downlink switching, while the longer nframes are intended to lower the switching frequency to reduce its associated overhead.

Two versions of the frame structure Type 3N are specified. Frame structure Type 3N1 supports operation in the US according to the FCC regulations. In this frame structure the nframe length is fixed to 20 ms. Every fourth nframe contains the 20 ms anchor segment. The anchor segment concept is inherited from LTE-M-U, and supports uNPSS, uNSSS and uNPBCH transmissions. The anchor definition facilitates a frequency hopping design of the data channels. The nframe length determines the dwell time on each RF during the frequency hopping cycle. In Frame structure Type 3N1 the anchor is 3 PRB wide as described in Section 15.3.2.1.3. Besides this exception NB-IoT-U follows a single PRB system design. A device is expected to make use of a single PRB receiver, even on the 3 PRB anchor segment.

Frame structure Type 3N2 supports operation according to the ETSI regulations defined for Band 54. It has a 10% duty cycle restriction which motivates less frequent uplink-downlink switching for optimizing the system capacity. It is also beneficial to group the few available downlink subframes for the support of cross subframe channel estimation, described in Section 8.2.3.2. This further motivates a reduced uplink-downlink switching rate. A reduced overhead from uNPSS, uNSSS and uNPBCH transmissions is also attractive. This was achieved by extending the anchor periodicity to 1280 ms.

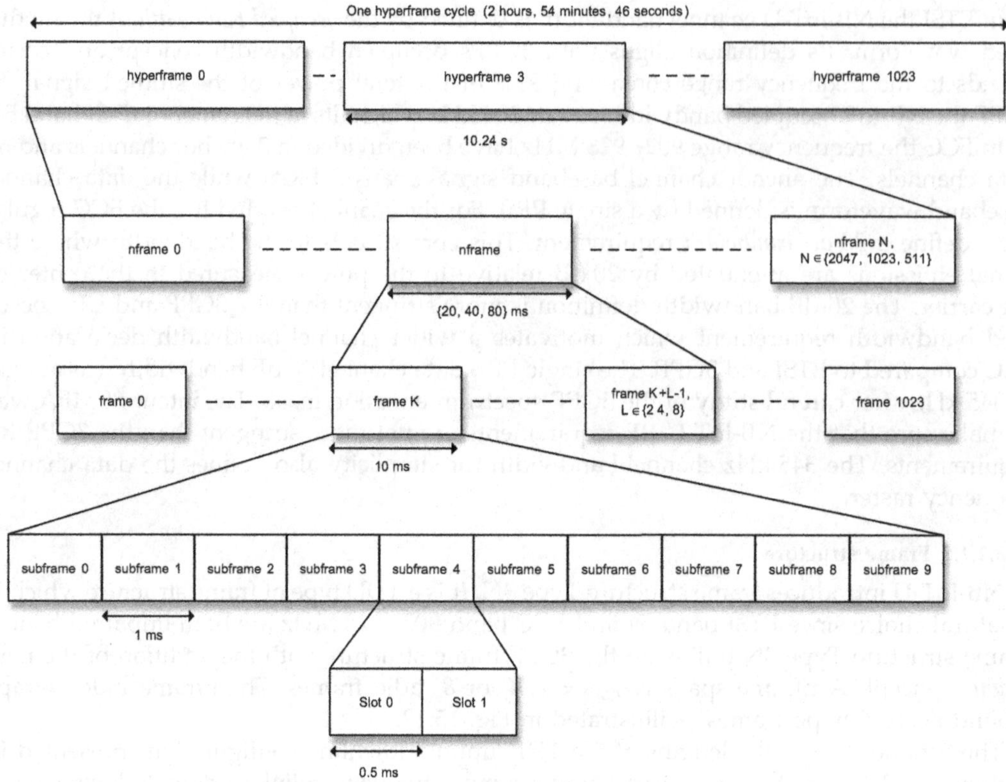

FIG. 15.12 NB-IoT-U frame structure Type 3N including the nframe concept.

Which of the two frame structures that is configured is implicitly indicated by the NB-IoT-U operating band, or more exactly by the frequency location of the anchor carrier configured in a cell.

The subframes in a nframe are numbered between *0* and $10 \cdot N_{frame}-1$ for $N_{frame} = 2, 4$ *or* 8, and are referred to as the relative subframe number. Fig. 15.13 illustrates the two frame structures.

15.3.2.1.3 Resource grid

The basic resource grid follows the same definitions as the LTE-M-U resource grid described in Section 15.2.2.1.2. In ETSI NB-IoT-U follows a single PRB design. In FCC the anchor is digitally modulated and spans 3 PRBs during 20 ms, while the data segment is mapped on a single PRB.

15.3.2.2 Transmission schemes

15.3.2.2.1 Anchor and data segment

NB-IoT-U reuses the NB-IoT transmission schemes presented in Section 7.2.2 with one significant exception: While the NB-IoT TDD uplink-downlink switching period is associated to the 10 ms radio frame, NB-IoT-U uplink-downlink switching periods are defined by the

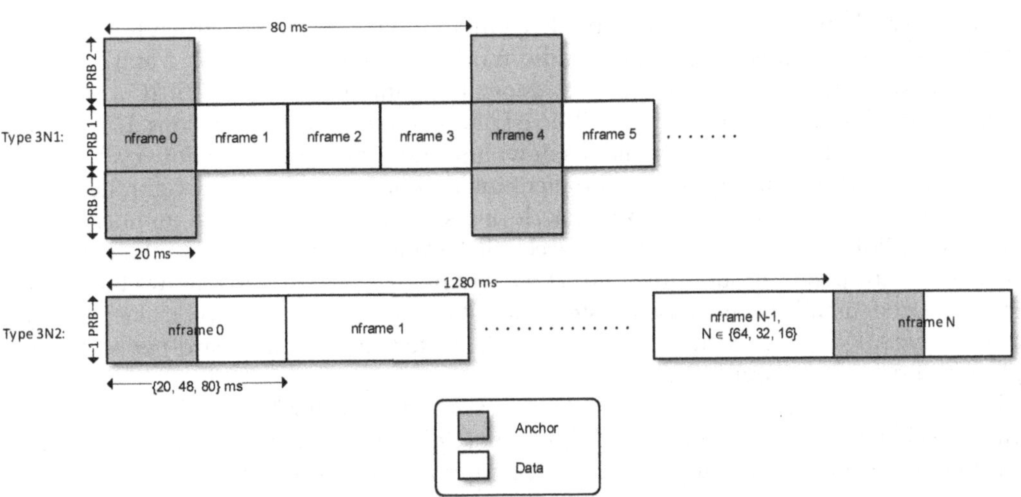

FIG. 15.13 NB-IoT-U nframe and anchor periodicity.

nframe uplink-downlink configurations. Table 15.6 presents the uplink-downlink configurations used for data transmission in frame structure Type 3N1 intended for operation in the US.

In frame structure Type 3N1, 3 types of special subframes are specified. The first and second type supports downlink transmissions followed by a guard-period of either 4 or 7 OS. The third type, indicated as S_G in Table 15.6, is defined as a 1 ms guard-period in case of the first nframe that follows after an anchor segment. In case the nframe follows a nframe used for uplink data transmission then also this special subframe is used for uplink transmissions.

TABLE 15.6 Frame structure Type 3N1 uplink (U) - downlink (D) configurations, including special (S) subframes.

Configuration	Nframe length	Relative subframe number																			
		0	1	2	3	4	5	6	7	8	9	10	11	12	13	14	15	16	17	18	19
0	2	D	D	D	D	D	S	U	U	U	U	U	U	U	U	U	U	U	U	U	U
1	2	D	D	D	D	D	D	D	S	U	U	U	U	U	U	U	U	U	U	U	U
2	2	D	D	D	D	D	D	D	D	D	S	U	U	U	U	U	U	U	U	U	U
3	2	D	D	D	D	D	D	D	D	D	D	D	S	U	U	U	U	U	U	U	U
4	2	S_G	U	U	U	U	U	U	U	U	U	U	U	U	U	U	U	U	U	U	U
5	2	D	D	D	S	U	U	U	U	U	U	D	D	D	S	U	U	U	U	U	U
6	2	D	D	S	U	U	U	U	U	U	U	D	D	S	U	U	U	U	U	U	U
7	2	D	S	U	U	U	U	U	U	U	U	D	S	U	U	U	U	U	U	U	U

Table 15.7 shows the uplink-downlink configurations for frame structure Type 3N2 including nframe lengths of 2, 4 and 8 radio frames. Configurations 0, 2, 4 and 7 reserves 10% of the subframes for downlink transmissions to guarantee that the ETSI 10% duty cycle requirement is never violated by the NB-IoT-U base stations. Configurations 1, 3, 5 and 6 specifies 20, 30 or 40% of the subframes as downlink subframes. Since the duty cycle requirement is measured over a sliding measurement interval of 1 hour, these configurations allow the network to overprovision the downlink during short periods. The total number of transmitted downlink subframes should not exceed 360.000 in 1 h, but it is up to the network to decide when to spend the subframe budget. Downlink overprovisioning allows the network to meet short demands for increased system capacity or improved link level latency. In frame structure type 3N2 the special subframe S_G is dedicated to serve as guard-period between uplink and downlink dwells.

Table 15.8 summarizes the 3 special subframe types defined for NB-IoT-U, including their definitions in terms of downlink (DwPTS), uplink (UpPTS) and guard-period (GP) parts. The guard-period length sets an upper limit on the supported cell size. To avoid that the uplink transmissions of devices near the cell edge interfere the base stations downlink transmissions the cell radius should not exceed ~42.5 km or ~75 km when a guard-period of 4 or 7 OS,

TABLE 15.7 Frame structure Type 3N2 uplink (U) - downlink (D) configurations, including special (S_G) subframes.

		Relative subframe number																
Configuration	**Nframe length**	**0**	**1**	**2**	**3**	**4**	**5**	**6**	**7**	**8**	**9**	**10**	**11**	**12–15**	**16**	**17–19**	**20–39**	**40–79**
0	8	D	D	D	D	D	D	D	D	S_G	U	U	U	U	U	U	U	U
1	8	D	D	D	D	D	D	D	D	D	D	D	D	D	S_G	U	U	U
2	4	D	D	D	D	S_G	U	U	U	U	U	U	U	U	U	U	U	–
3	4	D	D	D	D	D	D	D	D	S_G	U	U	U	U	U	U	U	–
4	2	D	D	S_G	U	U	U	U	U	U	U	U	U	U	U	U	–	–
5	2	D	D	D	D	S_G	U	U	U	U	U	U	U	U	U	U	–	–
6	2	D	D	D	D	D	D	D	D	S_G	U	U	U	U	U	U	–	–
7	2	D	D	S_G	U	U	U	U	U	U	U	D	D	U	U	U	–	–

TABLE 15.8 NB-IoT-U special subframe definitions in terms of OFDM symbols (OS).

Frame structure	Special subframe notation	DwPTS	GP	UpPTS	Cell radius supported by GP
Type 3N1	S	10 OS	4 OS	–	~42.5 km
Type 3N1	S	7 OS	7 OS	–	~75 km
Type 3N1, Type 3N2	S_G	–	14 OS	–	~150 km

respectively, is configured. In the case of the full subframe being used as guard-period it is no longer the special subframe configuration that limits the cell size. Instead other factors, such as the uNPRACH format will limit the cell size.

15.3.2.2.2 Frequency hopping

NB-IoT-U frame structure Type 3N1 is required to follow the FCC frequency hopping regulations. The FCC regulations couples the permitted output power to the number of frequency hopping channels. To be able to transmit the data segments with a peak EIRP of 36 dBm the system is required to frequency hop over at least 50 channels. The hopping is required to be performed in a pseudo random pattern.

NB-IoT-U meets this requirement using a frequency hopping algorithm that is based on the LTE-M-U frequency hopping functionality described in Section 15.2.2.2.4. The frequency hopping algorithm takes the PCID and a counter, that counts nframes carrying data segments, as input and generates a cell specific hopping sequence across the full set of 64 supported frequency channels. This pseudo random hopping pattern repeats itself every 4 hyperframes.

The hopping across the full set of 64 data channels optimizes the frequency diversity. This approach can be compared with that taken for LTE-M-U that ideally makes use of the least interfered subset of specified RF channels at any given moment. In contrast to LTE-M-U, it is not possible for NB-IoT-U to select a subset to hop over as all specified frequency channels (excluding the three anchor channels) are included in a single large frequency hopping set.

Fig. 15.14 illustrates the frequency hopping applied for frame structure Type 3N1. It also illustrates the mapping of the different downlink and uplink physical channels and signals, which in detail is discussed in the next two sections.

15.3.2.3 Downlink physical channels and signals

This section describes the NB-IoT-U downlink physical channels and signals including the Narrowband Primary and Secondary Synchronization Signals (uNPSS, uNSSS), the Narrowband Physical Broadcast Channel (uNPBCH), the Narrowband Physical Downlink Shared

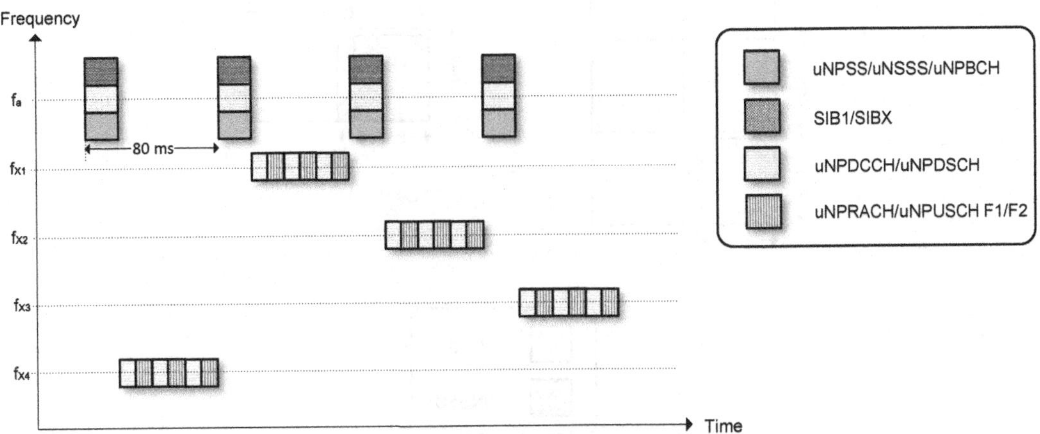

FIG. 15.14 NB-IoT-U frame structure Type 3N1 frequency hopping.

Channel (uNPDSCH) and the Narrowband Physical Downlink Control Channel (uNPDCCH). NB-IoT-U strives to reuse the 3GPP Release 13 NB-IoT downlink design described in Section 7.2.4. The focus of this section is to identify and describe the few differences between the NB-IoT-U and the NB-IoT downlink designs.

In NB-IoT a bitmap signals the downlink subframes that are available for NB-IoT transmissions (see Section 7.2.4.1). For NB-IoT-U all subframes indicated as downlink subframes in the uplink-downlink configuration are available for downlink transmissions.

15.3.2.3.1 Synchronization signals

To support frequency hopping data transmission, it was decided to follow the LTE-M-U design, described in Section 15.2.2.3.1, and group the uNPSS, uNSSS and uNPBCH transmissions together in an anchor segment. The anchor periodicity is specified to 80 ms in FCC and 1280 ms in ETSI.

With such infrequent anchor segment transmissions, it is important that the design of the synchronization signals facilitates a single shot detection of the anchor segment synchronization signals followed by a quick acquisition of the uPBCH MIB. This was accomplished by a densification of the synchronization signals and the uNPBCH transmissions within the anchor segment. In this new design the uNPSS is mapped over 8 consecutive subframes, followed by 2 uNSSS subframes and 10 uPBCH subframes.

Fig. 15.15 illustrates the mapping of the uNPSS and uNSSS for frame structure Type 3N1 and 3N2. In case of frame structure Type 3N1 used in FCC the synchronization signals are mapped on the first of the 3 PRBs.

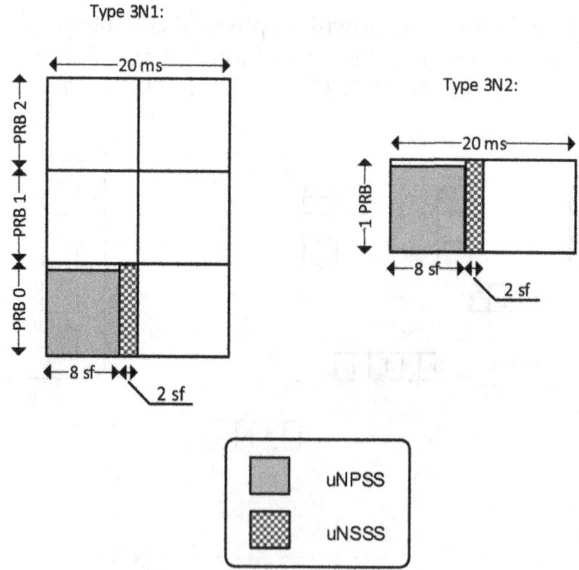

FIG. 15.15 NB-IoT-U synchronization signal mapping on the anchor segment.

15.3.2.3.2 uNPSS

Relative subframe	0–7
Periodicity	80 ms, 1280 ms
Subcarrier spacing	15 kHz
Bandwidth	165 kHz
Frequency location	PRB 0

The uNPSS serves the same purpose as the NPSS and supports cell detection and synchronization to the downlink frame structure in time and frequency. The basic waveform is defined by the same Zadoff-Zhu base sequence that defines the NPSS waveform (see Section 7.2.4.2.1):

$$p(n) = e^{\frac{-j5\pi n(n+1)}{11}}, \quad n = 0,1,\ldots,10 \tag{15.7}$$

In the frequency domain $p(n)$ is mapped over the first 11 sub-carriers of PRB 0, leaving the 12th sub-carrier empty. The NPSS is transmitted in subframe 5 in every radio frame. To leave room for the LTE control region, in case of inband mode of operation, its base sequence $p(n)$ is mapped on the last 11 OS in a subframe. LTE inband operation in unlicensed spectrum is not an expected use case and NB-IoT-U is only designed for stand-alone operation. Consequently, the uNPSS base sequence is mapped over all 14 OS on each of the first 8 subframes in the nframe carrying the anchor segment. A computer-generated cover code of length $8 \cdot 14 = 112$ is applied on top of the 112 repeated copies of the base sequence.

The device can assume that the base station applies the same mapping of the signal to the antenna ports over the first 4 subframes, and over the last 4 subframes. This allows a base station transmitter configured with multiple antenna ports to switch mapping in the middle of the NPSS transmission which introduces spatial diversity in the transmission. This is especially advantageous for devices suffering from limited fading diversity.

The device receiver is expected to detect the uNPSS by performing a sliding auto-correlation across the repeated set of base sequences, while correcting for the cover code. The long cover code supports a high receiver processing gain which facilitates efficient detection of the uNPSS at low SINR levels.

15.3.2.3.3 uNSSS

Relative subframe	8, 9
Periodicity	80 ms, 1280 ms
Subcarrier spacing	15 kHz
Bandwidth	180 kHz
Frequency location	PRB 0

The uNSSS signals the PCID of a cell. 504 PCIDs are, just as for NB-IoT, supported. The uNSSS is as the uNPSS mapped across all 14 symbols of a subframe. The uNSSS spans all 12 sub-carriers of the PRB which means that it in total is mapped over $12 \cdot 14 = 168$ resource elements.

The length 168 uNPSS sequence *s(n)* is generated based on a length 167 Zadoff-Chu sequence, that is multiplied with a length 160 Hadamard sequence $b_q(n'')$ and rotated by a phase θ_l:

$$
\begin{aligned}
s(n) &= b_q(n'') \cdot e^{-j2\pi\theta_l} \cdot e^{\frac{-ju\pi n'(n'+1)}{167}} \\
n' &= n \ mod 167 \\
n'' &= n \ mod 160 \\
n &= 0, 1, \ldots, 167
\end{aligned}
\tag{15.9}
$$

This design is close to identical to the NSSS design. The only differences are that the Zadoff-Chu and Hadamard sequences have been extended in length to support the mapping across all 168 resource elements of a subframe.

The uNSSS is, just as the NSSS, made cell specific thanks to the configured Zadoff-Zhu root u, and through the selection of 1 out of 4 available scrambling sequences $b_q(n)$ determined as:

$$u = (PCID \ mod \ 126) + 3$$

$$q = \left\lfloor \frac{PCID}{126} \right\rfloor$$

For NB-IoT the phase shift θ_l signals the timing of the NSSS transmission.

In case of NB-IoT-U this indication is no longer needed, so the uNSSS phase shift θ_l has for simplicity been fixed to $\frac{42}{168}$.

To support robust performance the uNSSS sequence *s(n)* is repeated twice on relative subframes 8 and 9 in the nframe containing the anchor segment.

15.3.2.3.4 uNRS

Relative subframe	Any DL subframes according to the UL-DL configuration
Periodicity	10 ms
Basic TTI	1 ms
Subcarrier spacing	15 kHz
Bandwidth	180 kHz
Frequency location	Any

The uNRS supports coherent demodulation of the uNPBCH, uNPDCCH and uNPDSCH, and radio resource management measurements. Its design is identical to the NRS, described in Section 7.2.4.3. The only practical difference is where a device can assume to find the uNRS transmissions.

In case of frame structure Type 3N1 and before reading the narrowband version of system information block 1 (SIB1-NB) a device can assume that uNRSs are always present in PRB 0 and PRB 2 in the anchor segment. Relative subframes 10–19 in PRB 0, carries the uNPBCH including NRSs. Relative subframes 0–19 in PRB 2 continuously transmit a uNPDSCH, including uNRSs, carrying system information messages as described in Section 15.3.2.3.5. With the synchronization signals and the uNPBCH mapped to PRB 0 and the system information message always transmitted on PRB2 the Type 3N1 anchor segment can be said to fulfill the FCC digital modulation 500 kHz bandwidth requirement.

In case of frame structure Type 3N2, and before reading SIB1-NB a device can assume that NRSs are always present in the subframes that carry the uNPBCH and in subframes that may contain SIB1-NB transmissions according to the densest SIB1-NB configuration (see Section 15.3.2.3.5 to understand the SIB1-NB mapping).

These default minimal sets of uNRS transmissions are supporting idle mode NRSRP and NRSRQ measurements for the purpose of idle mode radio resource management, including cell selection and reselection.

Besides the default uNRS transmissions the devices can assume that uNRSs are present in any type of scheduled uNPDCCH and uNPDSCH transmissions during idle and connected mode operation.

Compared to the NB-IoT anchor channel the amount of default uNRS transmissions is significantly reduced. This should be seen as an attempt to better cope with the ETSI duty cycle requirement and limit the interference to other systems operating in the same unlicensed spectrum.

15.3.2.3.5 uNPBCH

Relative subframe	10–19
Periodicity	80 ms, 1280 ms
Basic TTI	640 ms, 10.24 s
Subcarrier spacing	15 kHz
Bandwidth	180 kHz
Frequency location	PRB 0

The uNPBCH carries the 34-bit MIB. A 16-bit CRC is appended to the information bits. The resulting 50 bits are convolutional encoded, and rate matched into 2432 bits. The encoded bits are divided into 8 code sub-blocks each of 304 bits. This procedure follows the NPBCH design (see Section 7.2.4.4). It is only the total number of encoded bits that has increased from 1600 for the NPBCH to 2432 for the uNPBCH.

The 304 bits of each code sub-block is modulated into 152 QPSK symbols. These are mapped to the available resource of a subframe where only resource elements for uNRS antenna ports 1 and 2 are reserved. In case of the NPBCH, resources are also reserved for

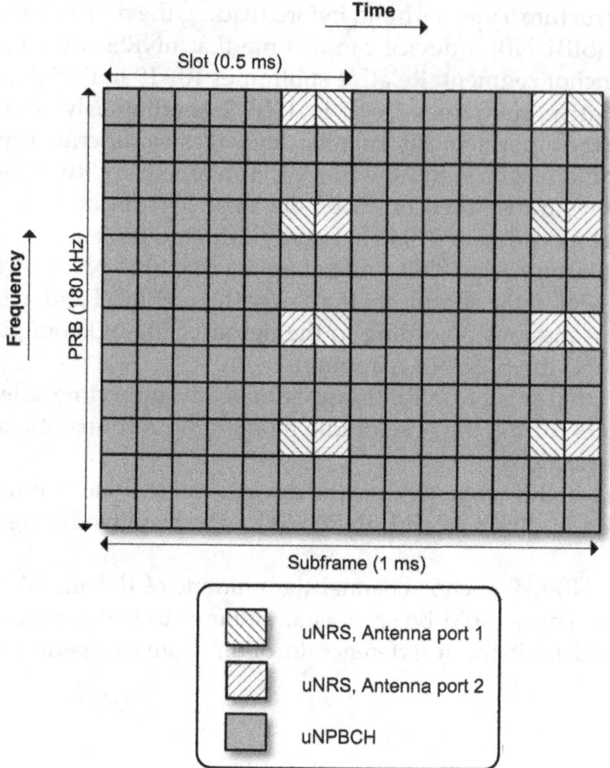

FIG. 15.16 uNPBCH resource element mapping.

the LTE control region and CRS transmissions. Fig. 15.16 shows the code sub-block mapping on a uNPBCH subframe. It can be compared to the NPBCH mapping illustrated in Section 7.2.4.4.

The uNPBCH code sub-block subframe is repeated 10 times on relative subframes 10 to 19. Fig. 15.17 shows the complete mapping of one uNPBCH code sub-block onto the anchor segment for frame structures Type 3N1 and Type 3N2. For Type 3N1 the anchor segment and uNPBCH code sub-block periodicity equals 80 ms. This gives a total transmission time of 640 ms, just as for the NPBCH. For Type 3N2 each anchor segment, including the uNPBCH code sub-block, is transmitted with a periodicity of 1280 ms. This gives a total transmission time of $1.28\ s \cdot 8 = 10.24\ s$.

The transmission scheme use for uNPBCH is the same as for NPBCH, meaning that at most two antenna ports are supported using Space-Frequency Block Code (see Section 7.2.4.4) based transmit diversity.

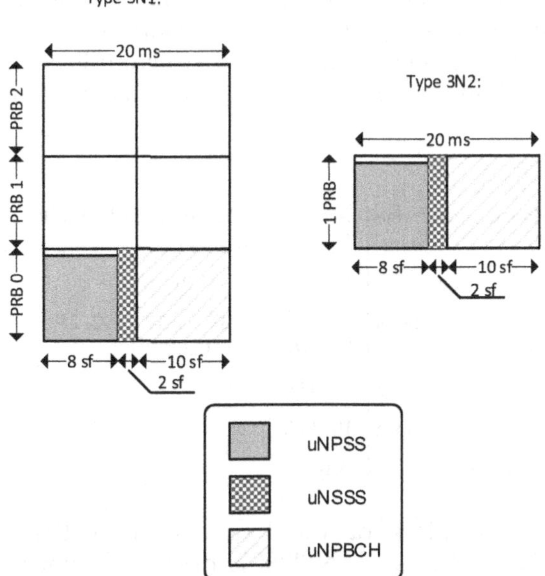

FIG. 15.17 uNPBCH resource element mapping.

15.3.2.3.6 uNPDCCH

Relative subframe	Any DL subframes according to the UL-DL configuration
Basic TTI	1 ms
Repetitions	1, 2, 4, 8, 16, 32, 64, 128, 192, 256, 512, 1024, 2048
Subcarrier spacing	15 kHz
Bandwidth	90 or 180 kHz
Frequency location	Any

The uNPDCCH follows the NPDCCH specification described in Section 7.2.4.5. It carries uplink and downlink scheduling information by means of DCI formats N0 and N1, which are fully reused from NB-IoT (see Section 7.3.2.2). The uNPDCCH transmission opportunities are configured by means of user specific and common search spaces described in detail in Section 7.3.2.1. In frame structure Type 3N1 uNPDCCH transmissions are mapped onto the single PRB nframes, and on the center PRB in the anchor segment. In frame structure Type 3N2 uNPDCCH transmissions can only be mapped on nframe parts not reserved by the anchor segment. The uNPDCCH transmissions must also align with the applicable uplink-downlink configuration. It is, just as the NPDCCH, transmitted from the first valid subframe equal to or after its starting subframe. So in case the starting subframe of a search space is located in an uplink dwell, the actual start of the search space and its uNPDCCH transmission is postponed to the start of the next downlink dwell.

Thanks to the frequency hopping design in frame structure Type 3N1, the uNPDCCH will frequency hop in case its transmission is mapped across one or more nframe borders.

15.3.2.3.7 uNPDSCH

Relative subframe	Any DL subframes according to the UL-DL configuration
Basic TTI	1, 2, 3, 4, 5, 6, 8, 10 ms
Repetitions	1, 2, 4, 8, 16, 32, 64, 128, 192, 256, 384, 512, 768, 1024, 1536, 2048
Subcarrier spacing	15 kHz
Bandwidth	180 kHz
Frequency location	Any

The uNPDSCH supports unicast data and system information message transmissions along the lines of the 3GPP Release 13 NPDSCH design described in Section 7.2.4.6. This means e.g. that the largest supported transport block size is 680 bits. Single Cell Point to Multipoint (SC-PTM) feature specified in 3GPP Release 14 is not supported in the first NB-IoT-U release.

In frame structure type 3N1 uNPDSCH unicast data transmissions are mapped onto the single PRB nframes, and on the center PRB in the anchor segment. In frame structure type 3N2 uNPDSCH transmissions can only be mapped on nframe parts not reserved by the anchor segment. The dynamic scheduling follows that specified for the NPDSCH with the difference that the uNPDSCH transmissions must align with the applicable uplink-downlink configuration, and that transmissions spanning multiple nframes will frequency hop as described in Section 15.2.2.2.

In case of system information transmissions, the uNPDSCH scheduling is described in Section 15.3.3.1.

15.3.2.4 Uplink physical channels and signals

The design for the uplink channels and signals closely follow the NB-IoT Release 13 design described in Section 7.2.5. The main change is that the uplink transmissions follow the NB-IoT-U specific uplink-downlink configurations and, for frame structure type 3N1, is frequency hopping. Uplink transmissions are not supported in the anchor segment. In case of a potential overlap between a scheduled uplink transmission and the anchor segment, the applicable part of the uplink transmission is postponed for the duration of the anchor segment.

15.3.2.4.1 uNPRACH

Relative subframe	According to higher layer configuration
Basic TTI	5.6 or 6.4 ms
Repetitions	1, 2, 4, 8, 16, 32, 64, 128
Subcarrier spacing	3.75 kHz
Bandwidth	3.75 kHz
Frequency location	According to higher layer configuration

The uNPRACH supports the random-access procedure, using a design that follows the NB-IoT Release 13 specification described in Section 7.2.5.1. Both preamble formats 0 and 1 are supported. The use of the longer preamble format, with a basic transmission time of 6.4 ms, is not supported for frame structure Type 3N1 uplink-downlink configuration 5 which contains only 6 consecutive uplink subframes. All other uplink-downlink configurations are designed with at least 7 consecutive uplink subframes to support both preamble formats.

15.3.2.4.2 DMRS

Subframe	Any UL subframes according to the UL-DL configuration
TTI	Same as associated uNPUSCH
Repetitions	Same as associated uNPUSCH
Sub-carrier spacing	Same as associated uNPUSCH
Bandwidth	Same as associated uNPUSCH
Carrier	Any

The DMRS supports coherent demodulation of the uNPUSCH. Its design follows the NB-IoT Release 13 specifications. The only new aspects to take into consideration are the uplink frequency hopping design, and the uplink-downlink configuration.

15.3.2.4.3 uNPUSCH

Subframe	Any UL subframes according to the UL-DL configuration
Basic TTI	1, 2, 4, 8, 32 ms
Repetitions	1, 2, 4, 8, 16, 32, 64, 128, 256
Sub-carrier spacing	3.75, 15 kHz
Bandwidth	3.75, 15, 45, 90, 180 kHz
Carrier	Any

Two uNPUSCH formats are specified. uNPUSCH Format 1 (F1) supports uplink unicast transmissions, while uNPUSCH Format 2 (F2) is used for asynchronous HARQ feedback of uNPDSCH transmissions. The design of both uNPUSCH formats are close to identical to the Release 13 NPUSCH formats. New aspects to take into consideration are the uplink frequency hopping and the uplink-downlink configurations. For long NB-IoT NPUSCH transmissions, transmission gaps are specified to allow a device to perform NRS based frequency and time tracking. For the uNPUSCH such tracking is inherently supported thanks to the uplink-downlink configuration that introduces the needed uplink transmission gaps.

15.3.3 Idle and connected mode procedures

The NB-IoT-U physical layer design closely follows that of Release 13 NB-IoT. The same applies for the higher layer and physical layer procedures. This section presents idle and connected mode procedures and focuses on the few cases where significant differences between the NB-IoT and the NB-IoT-U procedures are identified.

15.3.3.1 Cell selection and system information acquisition

When performing initial selection, a NB-IoT-U device needs to search for anchor segment transmissions in the FCC ISM band 902–928 MHz and in ETSI SRD band 54 covering the frequency range 869.4–869.65 MHz. ETSI SRD band 47b covering the frequency ranges 865.6–865.8, 866.2–866.4, 866.8–867 and 867.4–867.6 MHz may also eventually be available. In the FCC ISM band a sparse anchor raster is specified which supports no more than three anchor carrier frequencies. This design is following the same principles as described for LTE-M-U in Section 15.2.3.1.

After synchronizing to an anchor carrier, a device can determine based on the frequency location of the anchor carrier which of the two frame structures Type 3N1 and Type 3N2 that is configured. Based on this information it may acquire the uNPBCH MIB, SIB1-NB and the other system information blocks.

15.3.3.1.1 System information acquisition for frame structure type 3N1

According to frame structure type 3N1 SIB1-NB is encoded into a codeword that spans over 20 subframes. This can be compared with the 8 subframes used for NB-IoT. The codeword is divided in two code sub-blocks of equal length, i.e. 10 subframes, which are mapped on the first half of PRB 2 on the anchor segment.

In case 16 repetitions are configured the two SIB1-NB code sub-blocks are mapped on the 10 first subframes of PRB 2 in two consecutive anchor segments leading to a code sub-block TTI of 160 ms. In case of 8 repetitions the code sub-block TTI is extended to 320 ms, with the first code sub-block being mapped onto the first anchor segment, and the second code sub-block being mapped on the third anchor segment. For the lowest repetitions count of 4 the code sub-block TTI is further scaled up to 640 ms. The first code sub-block is mapped onto the first anchor segment, and the second code sub-block being mapped on the fifth anchor segment. In all three cases of 4, 8 or 16 repetitions the SIB1-NB total transmission time equals 2560 ms for frame structure type 3N1.

In 3GPP the supported system information blocks 2, 3, 4, 5, 14 and 16 (see Section 7.3.1.2) are transmitted over 2 or 8 subframes. The subframe length is dependent on the transport block size. For NB-IoT-U the transmission time is extended from 2 or 8 to 10 subframes. In case of the 3GPP 2 subframe mapping the NB-IoT-U extension to 10 subframes is done by means of repetitions. In case of the 3GPP 8 subframe mapping the NB-IoT-U extension is achieved by means of rate matching. The 10 subframe system information code blocks are mapped on available subframes in PRB2 not occupied by SIB1-NB transmissions.

The scheduling of the system information blocks is arranged by means of non-overlapping and periodically reoccurring system information windows (SI-windows), like the concept illustrated for NB-IoT in Section 7.3.1.2.3 but each SI message can be configured with its individual SI-window length. In each SI-window only one system information message type is transmitted to fill the resources on PRB2 not consumed by SIB1-NB transmissions.

Fig. 15.18 illustrates the system information scheduling on the anchor segment in frame structure Type 3N1.

15.3.3.1.2 System information acquisition for frame structure type 3N2

In frame structure Type 3N2 the SIB1-NB message is encoded and mapped on 8 subframes. These 8 subframes are spread over the first half of the downlink dwells in a set of consecutive nframes. The second half of the dwell is reserved for control and data channel transmissions. This procedure is repeated 4, 8 or 16 times for sets of nframes that are spread evenly in time across a total SIB1-NB transmission interval of 5120 ms.

Fig. 15.19 shows an example where uplink-downlink configuration 5 (see Table 15.7) and 16 SIB1-NB repetitions are configured. In uplink-downlink configuration 5 the downlink dwell consist of 4 subframes. The SIB1-NB code block is mapped over the 2 first subframes in 4 consecutive nframes, starting from the second subframe after the anchor. The block of 4 nframes is repeated in total 16 times, with the repetitions spread evenly across the SIB1-NB transmission time interval of 5120 ms. This concept shares the principle behind the NB-IoT SIB1-NB mapping described in Section 7.3.1.2.1.

The scheduling of system information blocks 2, 3, 4, 5, 14 and 16 in case of frame structure Type 3N2 reuses the NB-IoT scheduling principles, described and illustrated in Section 7.3.1.2.3, with some minor adaptations to cater for the unlicensed spectrum operation.

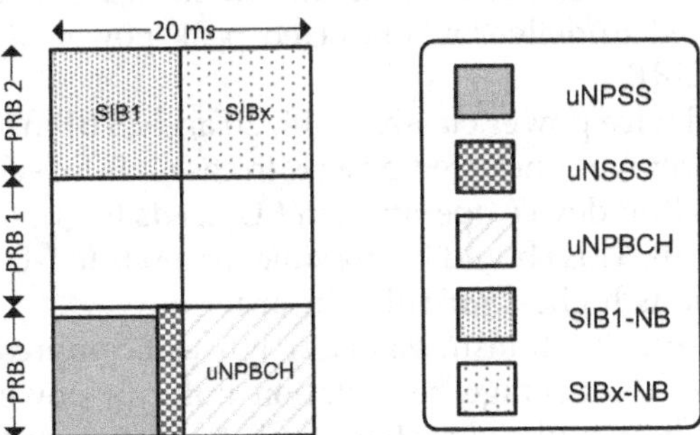

FIG. 15.18 NB-IoT-U frame structure Type 3N1 system information scheduling.

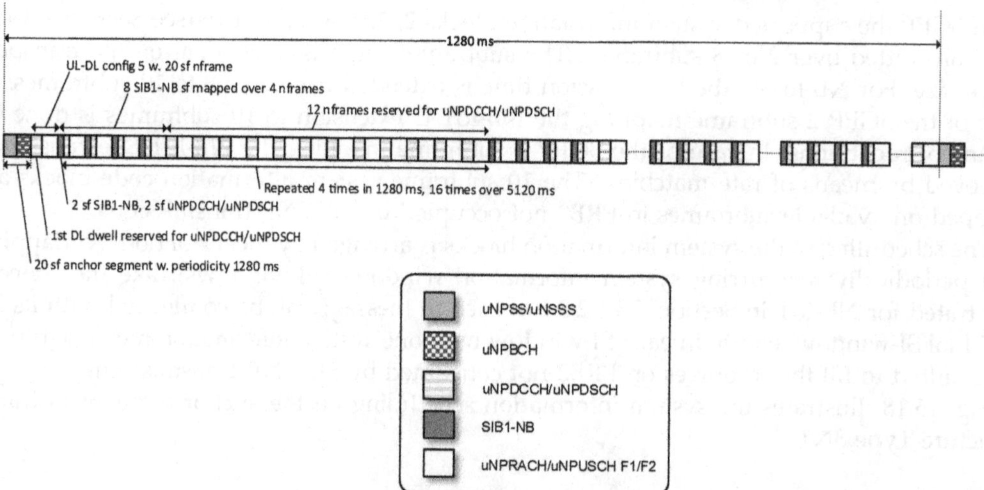

FIG. 15.19 SIB1-NB scheduling example for frame structure Type 3N2.

The content of the MIB, and the system information messages are to large parts aligned with the NB-IoT Release 13 specification. Since operation in inband and guard-band modes are no longer relevant the MIB signaling related to these operations are excluded from the NB-IoT-U signaling. New information elements related to the two new frame Types 3N1 and 3N2, and the uplink-downlink TDD configurations have instead been added. NB-IoT-U supports in similarity to LTE-M-U both PLMN and NHN access modes (see Section 15.2.3.1.1). The needed signaling to support these two options are added to the SIB1-NB information elements.

15.3.3.2 Power control

The NB-IoT-U uplink reuses the NB-IoT open loop power control described in Section 7.3.2.3. The ETSI regulations mandates the use of an adaptive power control, and it was assumed that the existing functionality is sufficient to satisfy this requirement.

Both the ETSI and FCC regulations specify strict power control limits that needs to be met. The radiated peak power in ETSI band 54 equals 29.2 dBm EIRP. EIRP and its part in the path loss calculation performed in Section 15.4.2 is illustrated in Fig. 15.20. In FCC, a frequency hopping system is allowed to deliver a conducted peak power of 30 dBm and a radiated peak power of 36 dBm EIRP.

NB-IoT supports the device power classes of 14, 20 and 23 dBm. The 3GPP power classes are defined in terms of average conducted power. In case NB-IoT-U specifies the same three power classes, then a 23 dBm device operating in EU needs to guarantee that its PAPR does not exceed $29.2 - 23 = 6\ dB$. This should be possible for a NB-IoT-U device using a DFT pre-coded uplink with QPSK as highest modulation order.

In case of downlink single PRB transmissions, an efficient compression of the OFDM PAPR is important to maximize the average base station transmit power. In addition, the FCC requirements for a digitally modulated system needs to be met for the anchor transmissions

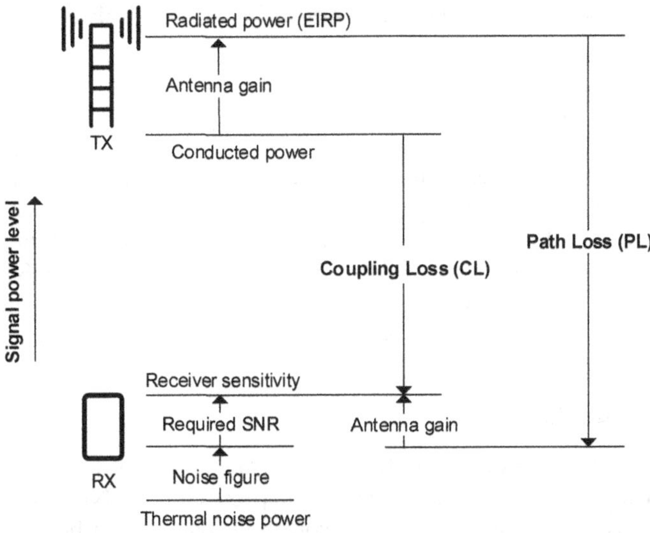

FIG. 15.20 Coupling loss and path loss.

in frame structure Type 3N1. For a digitally modulated system, the average conducted power may not exceed 30 dBm, and the average radiated power may not exceed 36 dBm EIRP. For the 3 PRB anchor this implies a maximum average EIRP of 31.2 dBm per PRB.

15.4 Performance

The performance of systems operating in unlicensed spectrum is not as straightforward to evaluate as the performance of systems operating in controlled licensed spectrum. Uncoordinated systems using the same frequencies may expose each other to high uncontrolled interference levels. Duty cycle limitations do not only put an upper bound on the capacity that can be provided by a system in unlicensed spectrum but may also impact link level key performance indicators such as latency and throughput. Listen-before-talk regulations makes the operation unpredictable as the transmission is dependent on the channels availability.

In this section the NB-IoT-U link level performance is evaluated under the assumption of operation in a controlled radio environment not exposed to uncoordinated interference. This allows a comparison to the link level performance presented for EC-GSM-IoT, LTE-M and NB-IoT in Chapters 4, 6, and 8. When reading this section it should be kept in mind that a controlled radio environment in unlicensed spectrum may be rare in reality.

When it comes to LTE-M-U, limited performance evaluations were available at the time of writing this book. Consequently, performance of LTE-M-U is here only shortly described in general terms. LTE-M-U is intended to serve industry and enterprise networks with relatively small cell deployments. Compared to LTE-M the maximum supported cell size and coverage requirements have been significantly relaxed. Operation in CE mode B is for example not supported. The LPWAN coverage, latency and battery life requirements presented in the next sections are therefore not fully representative for the use cases envisioned for LTE-M-U.

15.4.1 Performance objectives

The NB-IoT-U performance is evaluated in terms of its supported path loss, and the achievable data rates, latency and battery life at the maximum path loss (MPL). The basic framework used in the evaluations is reused from to that presented in Chapter 8. The performance is evaluated both for operation in US using frame structure Type 3N1 and in ETSI using frame structure Type 3N2. The relevant regulations are when motivated taken into consideration.

15.4.2 Coverage and data rates

NB-IoT-U coverage performance is in this section presented for operation according to the ETSI and FCC regulations. Compared to the results presented for NB-IoT in Chapter 8 where coverage is evaluated in terms of maximum coupling loss (MCL) here the results are presented in terms of maximum path loss (MPL). The sum of the transmitting and the receiving nodes antenna gains defines the difference between coupling loss and path loss as can be seen in Fig. 15.20. MPL is a suitable metric for the ETSI and FCC unlicensed regulations since they specify output power limits in terms of EIRP, and not in terms of conducted power levels as traditionally done in 3GPP.

The base station EIRP is for ETSI specified in terms of peak power. The same applies for FCC for the frequency hopping single PRB downlink transmissions. To determine the average EIRP, which is relevant for MPL calculations, the PAPR of the transmissions needs to be assessed. Fig. 15.21 shows PAPR statistics applicable for single PRB uNPBCH,

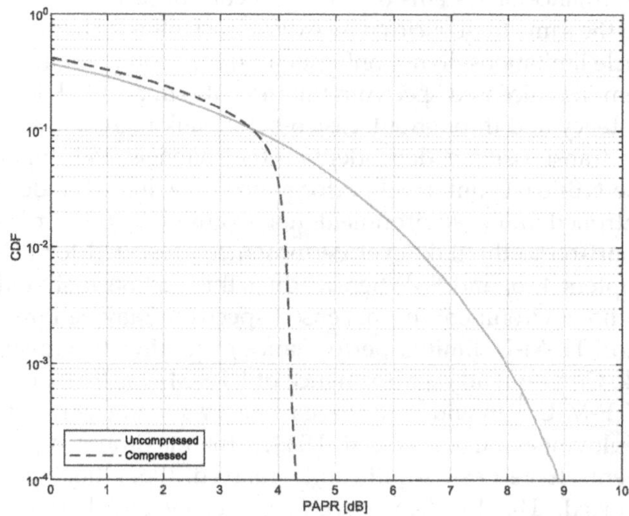

FIG. 15.21 NB-IoT-U uncompressed and compressed PAPR characteristics.

uNPDCCH or uNPDSCH transmission. The maximum PAPR is ~9 dB. The uNPSS has thanks to its Zadoff-Chu base sequence a PAPR around 4 dB. The uNSSS PAPR is dependent on the configured PCID which defines the uNSSS waveform. In ETSI and FCC the allowed downlink peak EIRP equals 29.2 and 36 dBm, respectively. With a PAPR of 9 dB the average EIRP needs to be drastically reduced to 20 and 27 dBm to comply with the ETSI and FCC regulations. To lower the impact, it is possible to compress the uNPDCCH and uNPDSCH waveforms to achieve a reduced PAPR. Fig. 15.21 also presents the PAPR of uNPDCCH and uNPDSCH when the modulated signal amplitude is limited, or compressed, before the base station transmit filter. In this basic compression scheme a PAPR below 5 dB is achieved, which allows a downlink power of 24 and 31 dBm to be configured in ETSI and FCC, respectively.

The FCC digitally modulated anchor segment transmissions may use an average EIRP of 36 dBm. Since the anchor segment is transmitted over 3 PRBs this gives a maximum average EIRP of 31.2 dBm per PRB. Note that signal compression is not motivated by the FCC regulations for the case of digitally modulated waveforms.

The device output power is both in ETSI and FCC defined in terms of peak EIRPs of 29.2 and 36 dBm, respectively. Here we consider a device power class of 20 dBm, and a device antenna gain of 0 dB, which leaves more than sufficient margin to the allowed peak EIRPs to support the full PAPR of the SC-FDMA modulated NB-IoT-U uplink.

Table 15.9 presents the simulation assumptions used in the evaluation of the NB-IoT-U MPL. Besides the EIRP assumptions most parameters can be recognized from the NB-IoT evaluations presented in Chapter 8. In ETSI the uNPDCCH and uNPDSCH performance was evaluated using cross-subframe channel estimation over 4 subframes. This choice matches the number of consecutive downlink subframes in frame structure Type 3N2 uplink-downlink configuration 5 with an nframe duration of 20 ms. This configuration was chosen to reflect that the base station over time needs to respect the 10% duty cycle requirement. In frame structure Type 3N1 uplink-downlink configuration 1 supports uNPDCCH and uNPDSCH cross-subframe channel estimation over 7 subframes (not taking the special subframe into account), which was also the assumption used for the evaluations of the FCC performance. The uNPBCH was in both regulations evaluated using 10 subframes in the cross-subframe channel estimation. For the uNPUSCH F1 and F2 simulations 8 subframes cross-subframe channel estimation was assumed. The device and base station noise figures were reused from the 3GPP Release 13 performance evaluations on NB-IoT presented in Chapter 8.

Table 15.10 presents the performance resulting from link level evaluations performed for NB-IoT-U operation in ETSI. The MCL and MPL calculations follow the illustration in Fig. 15.20. Note that the presented uNPDCCH, uNPDSCH, uNPUSCH F1/F2 and uNPRACH transmission times are not accounting for the spreading in time that the uplink-downlink configurations cause. Some background to the declared BLER targets is given in Section 8.2.1. The choice of a 6 dBi base station antenna gain was inspired by the FCC regulations that explicitly mentions this antenna gain. The uNPBCH MIB was correctly encoded after reading 4 out of the 8 code sub-blocks. Synchronization to the uNPSS and uNSSS was achieved after 2 synchronization attempts.

TABLE 15.9 Assumptions made in the evaluations of NB-IoT-U MPL.

Parameter	ETSI value	FCC value
Frequency band	869.4–869.5 MHz	902–928 MHz
Propagation condition	Typical Urban (TU)	
Fading	Rayleigh, 1 Hz	
Frequency error	uNPSS/uNSSS: ±20 ppm uNPBCH, uNPDSCH, uNPDCCH, uNPUSCH, uNPRACH: ±50 Hz	
Timing error	±2.5 μs	
Device NF	5 dB	
Device antenna configuration	1 transmit antenna, 1 receive antenna	
Device EIRP	20 dBm	
BS NF	3 dB	
BS antenna configuration	2 transmit antennas, 2 receive antennas	
BS EIRP	24 dBm	uNPSS, uNSSS, uNPBCH: 31.2 dBm uNPDSCH, uNPDCCH: 31 dBm
PAPR compression target	uNSSS, uNPBCH, uNPDCCH, uNPDSCH: 5 dB	uNPDCCH, uNPDSCH: 5 dB
Frame structure	Type 3N2	Type 3N1

The achievable MPL in ETSI equals 151 dB. The downlink performance can be said to be the limiting factor. Operation at a lower downlink SINR in ETSI will due to the long anchor segment periodicity result in very long uNPSS, uNSSS and uNPBCH acquisition times. The downlink data rate is also at risk to become unacceptably low.

It can be seen that the MCL corresponding to the declared MPL is limited to 145 dB. This serves as a reference for comparison with the 164 dB MCL achieved by EC-GSM-IoT, NB-IoT and LTE-M (see Chapters 4, 6, 8).

To calculate the uNPUSCH F1 and uNPDSCH MAC-layer data rates at the ETSI MPL the overhead from the uNPDCCH scheduling cycle must be taken into consideration. It depends on the transmission times declared in Table 15.10 weighted by the applied uplink-downlink configuration. For this purpose let's assume that the uplink-downlink configuration 6 (8 DL:1 S: 11 UL subframes) is configured.

In case of NPUSCH F1 scheduling, the UE specific search space can be assumed to be configured using RRC parameters $R_{max} = 256$ and $G = 4$ which determines the search space periodicity to $R_{max} \cdot G$. This gives a uNPDCCH scheduling cycle and uMPUSCH transmission

TABLE 15.10 NB-IoT-U MPL in ETSI for frame structure type 3N2.

Performance/Parameters	Downlink coverage				Uplink coverage		
	NPSS/ NSSS	NPBCH	NPDCCH	NPDSCH	NPRACH	NPUSCH F1	NPUSCH F2
TBS [bits]	–	34	23	680	–	1000	1
Bandwidth [kHz]	180	180	180	180	3.75	15	15
EIRP [dBm]	24	24	24	24	20	20	20
NF [dB]	5	5	5	5	3	3	3
#TX/#RX	2/1	2/1	2/1	2/1	1/2	1/2	1/2
Transmission/acquisition time [ms]	2560	3850	256	1024	25.6	80	2
BLER	10%	10%	1%	10%	1%	10%	1%
SNR [dB]	-11	-11.4	-11	-11	9.3	4.3	-2
MCL [dB]	145.4	145.8	145.4	145.4	146.0	145	151.2
TX antenna gain [dB]	6	6	6	6	0	0	0
RX antenna gain [dB]	0	0	0	0	6	6	6
MPL [dB]	151.5	151.9	151.5	151.5	152.0	151	157.2

opportunity every $4 \cdot 256 = 1024\ ms$ (see Section 7.3.2.1) and an uplink MAC-layer data rate of 879 bps.

$$\text{THP} = \frac{(1 - \text{BLER}) \cdot \text{TBS}}{\text{MPDCCH Period}} = \frac{0.90 \cdot 1000}{1.024} = 879\ \text{bps} \tag{15.10}$$

For the uMPDSCH data rate calculations $G = 1.5$ can be configured to obtain a scheduling cycle of $1.5 \cdot 256 = 384\ ms$. This configuration allows a uMPDSCH transmission every ninth scheduling cycle, i.e. every 3456 ms and gives a downlink MAC-layer data rate of 177 bps.

Based on these configurations downlink duty cycles of 37% and 22% are calculated over the uMPDCCH scheduling cycle for the cases of downlink and uplink data transmissions. Such levels can only be tolerated during short durations to not violate the 10% duty cycle requirement measured over 1 hour. Data rates that can be sustained over a longer period of time while simultaneously meeting the duty cycle requirement will therefore be significantly lower than the just derived data rates.

Table 15.11 presents the performance achievable in FCC for frame structure type 3N1. Compared to ETSI the achievable MPL is increased by 10 dB to 161 dB. In the downlink this is facilitated by a higher permitted EIRP in combination with an 80 ms anchor segment periodicity which allows operation at a lower SINR with a rather limited impact on system acquisition time. In the uplink the SINR operating point is reduced by 10 dB compared to ETSI.

TABLE 15.11 NB-IoT-U MPL in FCC for frame structure type 3N1.

Performance/Parameters	Downlink coverage				Uplink coverage		
	NPSS/ NSSS	NPBCH	NPDCCH	NPDSCH	NPRACH	NPUSCH F1	NPUSCH F2
TBS [bits]	–	34	23	680	–	1000	1
Bandwidth [kHz]	180	180	180	180	3.75	15	15
EIRP [dBm]	31.2	31.2	31	31	20	20	20
NF [dB]	5	5	5	5	3	3	3
#TX/#RX	2/1	2/1	2/1	2/1	1/2	1/2	1/2
Transmission/acquisition time [ms]	340	570	512	2048	102.4	1024	16
BLER	10%	10%	1%	10%	1%	10%	1%
SNR [dB]	−13.3	−13.7	−14.7	−14.3	−0.7	−6.7	−8
MCL [dB]	155	155.3	156.1	155.7	156.0	155.9	157.2
TX antenna gain [dB]	6	6	6	6	0	0	0
RX antenna gain [dB]	0	0	0	0	6	6	6
MPL [dB]	161	161.4	162.2	161.8	162.0	161.9	163.2

The achieved MCL is again clearly inferior to the 164 dB MCL that can be achieved in licensed spectrum by EC-GSM-IoT, NB-IoT and LTE-M (see Chapter 4, 6, 8).

The reduced SINR operating point in FCC results in significantly reduced MAC-layer data rates. Again, a slightly uplink heavy uplink-downlink configuration was used for evaluating the uNPUSCH and uNPDSCH data rates. In this example frame structure type 3N1 uplink-downlink configuration 2 (7 DL: 1 S: 12 UL subframes) is assumed. In the case of the uNPUSCH the UE specific search space was configured by means of $R_{max} = 512$ and $G = 8$, leading to a uNPUSCH transmission every $512 \cdot 8 = 4096\ ms$ and a MAC-layer data rate of 220 bps.

For the uNPDSCH, $G = 1.5$ can be assumed which leads to a scheduling cycle length of $512 \cdot 1.5 = 768\ ms$, and a uNPDSCH transmission every 10th scheduling cycle. This gives a downlink MAC-layer data rate of 80 bps.

Tables 15.12 and 15.13 summarized the uplink and downlink MAC-layer throughput estimates for ETSI and for FCC. It is seen that the uplink data rate for FCC does not meet the 3GPP Release 13 target of 160 bps (see e.g. section 8.1). A more downlink centric uplink-downlink configuration would allow for a slightly improved FCC downlink data rate, but the 160-bps target is still not within reach. In ETSI the estimated data rates only apply for a base station duty cycle significantly exceeding the 10% regulation. If honoring the 10% duty cycle the ETSI downlink data rate would sink below the 160-bps requirement.

TABLE 15.12 NB-IoT-U downlink MAC-layer data rates at the MPL.

Regulation	UL-DL configuration	MPL	DL MAC-layer data rate	BS duty cycle
ETSI	Config. 6	151 dB	177 bps	37%
FCC	Config. 2	161 dB	220 bps	–

TABLE 15.13 NB-IoT-U uplink MAC-layer data rates at the MPL.

Regulation	UL-DL configuration	MPL	UL MAC-layer data rate	BS duty cycle
ETSI	Config. 6	151 dB	879 bps	22%
FCC	Config. 2	161 dB	80 bps	–

15.4.3 Latency

Table 15.14 presents the latency that is achieved when a device's higher layers initiates a transmission of a single uplink packet while the device is in idle mode. The modeled packet size equals 85 bytes on top of the PDCP layer, and 90 bytes when including overhead from PDCP, RLC and MAC as presented in Table 8.23. The overall evaluation framework including the latency definition is aligned with that used for NB-IoT and described in Section 8.4. The RRC Resume procedure is used (see Section 7.3.1.7.1), with the 90-byte packet included in Message 5 carrying the RRC Connection Resume Complete message. The link level performance used for deriving the overall latency corresponds to that presented in Section 15.4.2.

The applied uplink-downlink configurations are for simplicity aligned with those assumed in Section 15.4.2 in which the data rates at the MPL were evaluated. It can be seen in Table 15.13 that the 3GPP target of 10 seconds (see e.g. Section 8.4) is within reach for both regulations. In ETSI this comes at the cost of a base station duty cycle much higher than 10%, which means that latencies around 10 seconds can only be made available in exceptional cases. For frame structure type 3N2 uplink-downlink configuration 2, which

TABLE 15.14 NB-IoT-U latency at the MPL.

Regulation	UL-DL configuration	MPL	Latency	BS duty cycle
ETSI	Config. 6	151 dB	10.2 s	31%
ETSI	Config. 2	151 dB	19.3 s	10%
FCC	Config. 2	161 dB	9.2 s	–

is better aligned with a sustained duty cycle of 10%, the latency equals 19.3 seconds, which is far from the 10 seconds requirement.

Given the higher SINR operating point in ETSI, it may seem contradictive that the latency is slightly better in FCC even when we overprovision the ETSI downlink duty cycle. The actual uplink report transmission is indeed taking longer time in FCC, but this is more than compensated for by the shorter system acquisition time and the fact that the FCC 3 PRB anchor segment supports uNPDCCH and uNPDSCH transmissions.

15.4.4 Battery life

The device power efficiency and ability to operate over many years on limited battery power is crucial for the massive IoT use case. To examine the NB-IoT-U battery life the evaluation performed for NB-IoT in Section 8.5 was repeated for NB-IoT-U. The NB-IoT assumptions and evaluation methods were reused with one exception: The lower device power class of 20 dBm assumed for NB-IoT-U motivates a reduction in the transmit power consumption from 500 mW (see Table 8.26) to 400 mW as shown in Table 15.15. The link level performance corresponds to that presented in Section 15.4.2.

The actual battery life achieved under the assumption of a 200-byte uplink transmission followed by a 65-byte downlink response is presented for FCC and ETSI in Table 15.16. The reporting interval is as seen following a 2-hour or 24-hour periodicity. In case of a diurnal reporting the 10-year target set in 3GPP Release 13 (see Section 8.5) is comfortably met. In case of reporting every 2 hour this is not the case. These results follow the same pattern as presented for NB-IoT in Table 8.27. As the power consumption is dominated by the 200-byte uplink transmission it is no surprise that the battery life in ETSI, thanks to the lower MPL, shows better performance than for FCC.

TABLE 15.15 NB-IoT-U device power consumption.

TX, 20 dBm	RX	Light sleep	Deep sleep
400 mW	80 mW	3 mW	0.015 mW

TABLE 15.16 NB-IoT-U battery life at the maximum path loss.

Regulation	Reporting interval	DL packet size	UL packet size	MPL	Battery life
ETSI	2 h	65 bytes	200 bytes	151 dB	4.9 years
	24 h				24.5 years
FCC	2 h	65 bytes	200 bytes	161 dB	2.2 years
	24 h				16.4 years

References

[1] European Telecom Standards Institute. EN 300 328, V2.0.20. Wideband transmission systems; Data transmission equipment operating in the 2,4 GHz ISM band and using wide band modulation techniques; Harmonised Standard covering the essential requirements of article 3.2 of the Directive 2014/53/EU, March 2016.

[2] U.S. Government Publishing Office. Electronic Code of Federal Regulations, Article 15.247, 7 July 2016.

[3] MulteFire Alliance. Technical specification 36.300, v1.1.1. Evolved Universal Terrestrial Radio Access (E-UTRA) and Evolved Universal Terrestrial Radio Access Network (E-UTRAN); Overall Description; Stage 2, 2019.

[4] MulteFire Alliance. Technical Specification 36.311, v1.1.1. Evolved Universal Terrestrial Radio Access (E-UTRA); Physical Channels and Modulation, 2019.

[5] MulteFire Alliance. Technical Specification 36.312, v1.1.1. Evolved Universal Terrestrial Radio Access (E-UTRA). Multiplexing and Channel Coding, 2019.

[6] MulteFire Alliance. Technical Specification 36.313, v1.1.1. Evolved Universal Terrestrial Radio Access (E-UTRA); Physical layer procedures, 2019.

[7] MulteFire Alliance. Technical Specification 36.304, v1.1.1. Evolved Universal Terrestrial Radio Access (E-UTRA); User Equipment (UE) Procedures in Idle Mode, 2019.

[8] MulteFire Alliance. Technical Specification 36.321, v1.1.1. Evolved Universal Terrestrial Radio Access (E-UTRA); medium access control (MAC) Protocol Specification, 2019.

[9] MulteFire Alliance. Technical Specification 36.331, v1.1.1. Evolved Universal Terrestrial Radio Access (E-UTRA). Radio Resource Control (RRC), 2019.

[10] European Telecom Standards Institute. EN 300 220-1, V3.1.1. Short Range Devices (SRD) Operating in the Frequency Range 25 MHz to 1 000 MHz; Part 1: Technical Characteristics and Methods of Measurement, February 2017.

[11] European Commission. Commission Decision of 9 November 2006 on harmonisation of the radio spectrum for use by short-range device (2006/771/EC), Edition of 18.8.2017.

CHAPTER

16

Choice of IoT technology

OUTLINE

Abstract

This chapter discusses the options of communication technologies for IoT. In a first part, the options of cellular IoT versus non-cellular IoT solutions are assessed.

In a second part, choices of cellular IoT technologies are compared. First, cellular IoT solutions for massive IoT are analyzed, comprising LTE-M and NB-IoT. The characteristics of those technologies are compared with regard to deployment, achievable data rates, latencies, spectral efficiency and battery efficiency. Second, cellular IoT solutions for critical IoT are analyzed, comprising Long-Term Evolution URLLC and NR URLLC.

In a third part, the considerations are described on how to assess the choices of cellular IoT technologies from the perspective of a mobile network operator and an IoT service provider.

Cellular Internet of Things, Second Edition
https://doi.org/10.1016/B978-0-08-102902-2.00016-9

© 2020 Elsevier Ltd. All rights reserved.

16.1 Cellular IoT versus non-cellular IoT

Chapters 3 to 13 present 3GPP cellular IoT solutions. Chapters 14 and 15 provide an overview of interesting unlicensed wireless connectivity solutions for IoT. In this section we discuss how cellular IoT solutions differ from unlicensed connectivity solutions and what benefits they can provide. For a further discussion on IoT connectivity options, see also [1].

One of the differentiators of cellular IoT connectivity is that it decouples connectivity provisioning from the IoT service realization. Cellular IoT is built on the high-level paradigm, that an independent operator provides suitable IoT connectivity essentially everywhere where an IoT service shall be realized. This means that when a new IoT service is established, no dedicated effort needs to be put into installing, managing and operating an IoT connectivity solution. Instead the connectivity is realized via the network of an operator. This is different from unlicensed IoT connectivity solutions. In this case, an installation of a connectivity infrastructure is needed to provide connectivity at the location where the IoT service is to be realized. This comprises installing of base stations or access points, establishing backhaul connectivity, providing *authentication, authorization and accounting* infrastructure, maintaining and updating the connectivity network with security updates, etc. Furthermore, the connectivity needs to be monitored and managed throughout the lifetime of the IoT service. There is the potential that the total cost of ownership for providing and managing a connectivity infrastructure for a wide range of IoT services is lower than the cost of ownership of separate connectivity solutions per IoT service. This is in particular the case when the IoT service and the participating devices are spread over larger areas and are not confined to limited deployments. From the unlicensed technologies, Sigfox provides an operator model for Sigfox-based end-to-end connectivity, where operators build up a dedicated Sigfox infrastructure and connectivity can be purchased by end users.

For critical IoT solutions, providing *critical Machine-Type Communications* (cMTC) services, the connectivity is often tighter integrated into the critical system. Critical systems can be local (e.g. a factory), and often a dedicated deployment of network components is required to guarantee performance in terms of capacity, latency, reliability and availability. Even for critical IoT solutions, the cellular IoT deployment and operation may be decoupled from the operation of the service realization of the end-to-end system (e.g. the automation of the production system). However, due to the tight relationship of the cellular critical IoT solution with the service performance and service assurance, a close coordination between the communication service provider and the end user are needed. This can be done in the form of a stringent *service level agreement* combined with service performance auditing. But it can also be realized in that the communication system is provided by the end user directly with own installation and operation of the cellular IoT system.

One major benefit of cellular IoT solutions is that they provide a reliable long-term and future proof solution. Cellular IoT is based on global standards with very large industry support by a large number of vendors, network and service providers. The technology outlook is independent from the outlook of few individual market players; this is in contrast to proprietary technologies which come with high risk concerning their long-term support.

Cellular IoT solutions are embedded into cellular communication networks, which are, and will be, an essential infrastructure for a society. Deployment plans are made over decades and systems are built to be highly reliable according to standards with high availability. Cellular IoT systems are built for a global market and allow roaming over multiple operator networks. Cellular IoT networks have full support for mobility of devices, which can also be handled over larger areas due to the wide area coverage and high availability. The rollout of cellular IoT capabilities, as well as future updates, takes mainly place as software updates to the installed network infrastructure.

One extremely important benefit of cellular IoT connectivity is that it provides reliable and predictable service performance also for future operation. For wireless critical IoT services, it is only cellular IoT that can guarantee reliable low latency services at scale. Cellular IoT uses dedicated spectrum. Radio resources are managed, interference is coordinated, and full quality of service is supported. Long-term guarantees are challenging to provide for any solution based on unlicensed spectrum. Both mobile broadband services, as well as IoT services are predicted to continue to grow. In particular for IoT devices, an extremely strong growth is predicted leading to hundreds of billions of communicating devices within a decade. Many of those mobile broadband and IoT services will be provided in unlicensed spectrum, which means that a significant increase in utilization of unlicensed spectrum can be expected. This will in particular provide challenges to long-range unlicensed technologies as described in Section 14.1.2 and Chapter 15; but also for critical IoT services where low latency and high reliability are required.

Cellular IoT also follows the continuous evolution of cellular network technologies, where new capabilities and features are continuously added to the networks. This evolution is designed for backwards compatible operation, so that devices that cannot be upgraded to new functionality can continue to operate long-term according to the original capabilities, while new services and devices can simultaneously benefit from newer features.

A disadvantage of cellular IoT is the cost of licensed spectrum resources. This is a cost which solutions for unlicensed bands do not need to bear. Another potential disadvantage of cellular IoT for an IoT service provider can be if the cellular IoT coverage is insufficient for a specific IoT use case. In this case extra connectivity and corresponding network build-out may be needed to cover the entire IoT service area. If the additional coverage buildout is needed in few confined areas, such a buildout may be simpler and more flexibly arranged with a dedicated deployment rather than when an operator needs to be involved. One property that unlicensed long-range radio technologies have benefitted from is their fast time-to-market. Any proprietary technology has a timing benefit over standardized solutions that require harmonization and agreements within an entire industry segment. In case of unlicensed LPWAN versus cellular IoT, a benefit in time-to-market of unlicensed LPWAN has existed in the past while the cellular IoT standards were being developed. Since the first cellular IoT standards were finalized in 3GPP Release 13 and products became widely available, the time-to-market benefit of unlicensed LPWAN has disappeared. Instead the benefit shifts toward cellular IoT deployments, which can reach wide coverage quickly and at low cost due to the reuse of the installed cellular communication network infrastructure.

16.2 Choice of cellular IoT technology

16.2.1 Cellular technologies for massive IoT

Use cases for massive IoT are characterized by low complexity and cost, energy efficient operation, and ubiquitous deployments of massive number of devices. The scale of the deployments requires wireless coverage under the most challenging conditions, and devices that can operate on non-rechargeable batteries for years. The anticipated traffic profile for many massive IoT applications is characterized by small and infrequent data transmission. A good example of a massive IoT deployment is the *Great Britain Smart Metering Implementation Program* where the British government has decided to equip every home and small business in the country with advanced electricity and gas meters. At the end of 2018 12.8 million smart meters had been rolled out [2].

To support a design of the cellular IoT technologies that can meet the demands of massive IoT, 3GPP have agreed the following set of performance objectives:

- Support for a coverage of 164 dB maximum coupling loss (MCL). This corresponds to 20 dB improved coverage compared to the 3GPP non-cellular IoT technologies.
- Support for small and infrequent data transmissions on non-rechargeable batteries for up to 10 years.
- Support for a connection density as high as 1,000,000 connections per square kilometer.
- Support for low to ultra-low device complexity.

Coverage is meaningful first when coupled with a quality of service requirement. 3GPP therefore in addition requires that a cellular IoT system should be able to:

- Provide a sustainable MAC-layer data rate of at least 160 bps at the 164 dB MCL.
- Provide a latency of at most 10 s when transmitting a small data packet from the 164 dB MCL.

The earlier chapters of the book in detail covers the 3GPP Cellular IoT technologies EC-GSM-IoT, LTE-M and Narrowband IoT (NB-IoT). It also presents the work done in MFA to prepare LTE-M and NB-IoT for operation in unlicensed spectrum bands. NB-IoT and LTE-M have, since their introduction in 3GPP Release 13 until March 2019, been commercially deployed in 89 networks across Europe, North and South America, Africa, Asia and Australia [3]. The number of commercial networks is constantly increasing, and many operators are choosing to deploy both technologies. The technologies are supported by the major infrastructure vendors and a significant amount of chipset, module and device providers. EC-GSM-IoT is still waiting for a commercial uptake, while the MFA technologies are still, in March 2019 at the writing of this book, undergoing RF specification work. The following sections therefore naturally focus on the capabilities and performance of LTE-M and NB-IoT.

16.2.1.1 Spectrum aspects

LTE-M and NB-IoT are both LTE technologies and do as such support a long list of frequency bands in the range 450 MHz to just below 3 GHz. Cat-M1 and Cat-M2 devices are capable of operating in E-UTRA FDD bands 1, 2, 3, 4, 5, 7, 8, 11, 12, 13, 14, 18, 19, 20, 21, 25,

26, 27, 28, 31, 66, 71, 72, 73, 74 and 85 in both half-duplex (HD) and full-duplex (FD) FDD mode and in time-division duplex (TDD) bands 39, 40 and 41. Cat-NB1 and Cat-NB2 support E-UTRA bands 1, 2, 3, 4, 5, 8, 11, 12, 13, 14, 17, 18, 19, 20, 21, 25, 26, 28, 31, 41, 65, 66, 70, 71, 72, 73, 74 and 85 in HD-FDD, and band 41 for TDD. The list of bands with their associated frequency ranges is found in 3GPP TS 36.101 [4]. As 3GPP specifies the support for new bands based on market interest also these lists have been growing for every new release of the specification. Both technologies can operate in the NR bands corresponding to the same lists of E-UTRA bands. This is thanks to the forward compatible coexistence functions specified for NR. These are in detail described in Chapters 5 and 7. NB-IoT supports thanks to its configurable mode of operation and small spectrum footprint also operation in GSM spectrum.

LTE-M is natively supported on the LTE system bandwidths of 1.4, 3, 5, 10, 15 and 20 MHz. Cat-M1 devices operate on a channel bandwidth of up to 1.4 MHz, while Cat-M2 can make use of up to 5 MHz for its transmissions. These RF bandwidths are determined based on the spectrum within which 99% of the signal energy falls. In terms of physical resource blocks (PRB) the 1.4 MHz narrowband corresponds to 6 PRBs, while 5 MHz corresponds to 25 PRBs. NB-IoT operates at a minimal RF system bandwidth of 200 kHz, which is equivalent to 1 PRB of 180 kHz. The system capacity can be scaled by increasing the number of carriers. Carrier aggregation (CA) is not supported by NB-IoT, so it is system capacity, and not link capacity that scales with the number of deployed carriers.

LTE-M's limited spectrum footprint makes it an attractive option for deployments in narrow, or fragmented frequency bands, but it can also efficiently and dynamically make use of a larger system bandwidth if available. NB-IoT's even more limited spectrum footprint makes it 3GPP's most flexible system in terms of deployment. Its ability to operate in the guard-band of an LTE carrier is e.g. very attractive as it allows an LTE operator to make better use of its highly valuable spectrum assets. A detailed explanation of the NB-IoT modes of operation, i.e. stand-alone, in-band and guard-band, is provided in Section 7.1.2.5.

LTE-M-U and NB-IoT-U technologies are interesting alternatives for vendors lacking access to licensed spectrum. The current designs are expected to support operation in US and Europe according to the FCC and ETSI regulations, respectively. But it needs to be stressed that the coverage, link quality, reliability and capacity provided by LTE-M-U and NB-IoT-U are not comparable to that provided by LTE-M and NB-IoT.

16.2.1.2 Features and capabilities

LTE-M and NB-IoT have many similarities. This is the result of years of parallel development in 3GPP. But there are important differences. Low complexity and simplicity, both on the device and system side, runs through every aspect of the NB-IoT design. NB-IoT aims at being the low-cost massive IoT carrier with capabilities in terms of reachability, device power efficiency and system capacity. NB-IoT Release 15 does therefore not support features like voice, connected mode mobility, connected mode device measurements and reporting, and closed loop power control which were not deemed relevant for the intended usage of NB-IoT. The Control Plane Cellular Internet of Things EPS optimization feature (see Section 7.3.1.7), which is mandatory for Cat-NB devices, does not support RRC Reconfiguration of the access stratum for a device in RRC connected mode, nor does it support the MAC layer data radio bearer scheduling and prioritization developed to guarantee a targeted quality of service. This is a consequence of routing the data over the control plane on a signaling radio bearer. This functionality is available for Cat-NB devices capable of user plane data transfer, which is an optional capability for NB-IoT.

LTE-M, being an LTE system, is naturally supporting a richer set of features than NB-IoT. LTE-M supports more advanced transmission modes using up to four antenna ports. It supports wideband transmissions, voice, connected mode mobility and full duplex operation. The set of massive IoT use cases to be supported by an operator goes in many cases beyond those characterized by small and infrequent data transmission. This observation has been a driver for keeping LTE-M more capable and advanced compared to NB-IoT. With that said, remember that LTE-M supports two coverage enhancements (CE) modes A and B, with CE mode B providing the most extreme coverage is an optional feature. So, when it comes to supporting operation in extreme coverage NB-IoT can be claimed to be the more capable system.

16.2.1.3 Coverage

The 3GPP 5G objective requires a 5G system for massive IoT to support a coverage of 164 dB MCL [5]. In Section 6.2 and Section 8.9 we provide detailed evaluations of the coverage supported by LTE-M and NB-IoT. The same evaluation assumptions, which are aligned with those agreed by ITU-R for the IMT-2020 evaluation [6], are used in both cases. It is shown that LTE-M and NB-IoT meet the 164 dB requirement. For both technologies the PUSCH can be seen as the limiting channel, i.e. the channel which requires the longest transmission times to reach the 164 dB coverage target. For LTE-M the MPDCCH needs to be configured with 256 repetitions to achieve the 1% BLER target set for the control channel transmission. This is the maximum configurable repetition number for the MPDCCH. So, for LTE-M the MPDCCH coverage is also a limiting factor.

The transmission times used for all channels to reach 164 dB, except for the MPDCCH, are not using the maximum configurable transmission times. We can therefore extend coverage beyond 164 dB if we can accept to reduce the data rates, latencies and battery life presented in in the next sections.

The performance evaluations for EC-GSM-IoT described in Chapter 4 follows slightly different assumptions compared to those used for 5G. The result shows that EC-GSM-IoT can offer similar cell-edge coverage and performance as NB-IoT and LTE-M. When looking at NB-IoT-U and LTE-M-U, described in Chapter 15, we can conclude that NB-IoT-U offers a coverage 10–20 dB below 164 dB. For LTE-M-U which only supports CE mode A a similar conclusion is expected to hold.

16.2.1.4 Data rate

Tables 16.1 and 16.2 summarize the range of data rates achievable for NB-IoT and LTE-M. For simplicity, here we focus on the MAC-layer data rates achievable at the 164 dB MCL, and the MAC-layer data rates observed under error-free conditions. We also present the physical layer data rates. Remember that the MAC-layer data rates correspond to the efficient data rate a device may experience, while the physical layer data rate is defined by the throughput during the actual transmission time interval of the PDSCH or PUSCH. The physical layer data rate, when normalized by the signal bandwidth, should be seen as an indication of the maximum spectral efficiency that can be achieved.

TABLE 16.1 LTE-M and NB-IoT HD-FDD PDSCH data rates.

Technology	MAC-layer at 164 dB MCL	MAC-layer peak	PHY-layer peak
Cat-M1	279 bps	300 kbps	1 Mbps
Cat-M2	> 279 bps	1.2 Mbps	4 Mbps
Cat-NB1	299 bps	26.2 kbps	227 kbps
Cat-NB2	299 bps	127.3 kbps	258 kbps

TABLE 16.2 LTE-M and NB-IoT HD-FDD PUSCH data rates.

Technology	MAC-layer at 164 dB MCL	MAC-layer peak	PHY-layer peak
Cat-M1	363 bps	375 kbps	1 Mbps
Cat-M2	363 bps	2.6 Mbps	7 Mbps
Cat-NB1	293 bps	62.6 kbps	250 kbps
Cat-NB2	293 bps	158.5 kbps	258 kbps

The two tables present Cat-NB1 and Cat-M1 Release 13 performance, and Cat-NB2 and Cat-M2 Release 14 performance. Cat-M1 PUSCH data rate can be further improved compared to the numbers in the table by means of the larger uplink TBS introduced in Release 14, and Cat-M1/M2 PDSCH data rates can be improved by means of the Release 14 feature for HARQ bundling and 10 HARQ processes in downlink. The NB-IoT results are based on the guard-band mode of operation. Chapters 6 and 8 give a richer set of results including data rates for the just mentioned LTE-M features and all three NB-IoT modes of operation.

The result speaks for themselves: The cell-edge MAC-layer data rates for NB-IoT and LTE-M are similar and meet the 5G requirement of 160 bps [5]. LTE-M can offer significantly higher data rates thanks to the larger device bandwidths and the lower processing times. Here we focus on error-free conditions, but this observation holds for a large portion of a cell.

16.2.1.5 Latency

As already mentioned, many massive IoT use cases are characterized by small data transmission. For these the importance of the data rates presented in the previous section is overshadowed by the latency required to set up a connection and transmit a data packet of limited size.

Table 16.3 shows the latency for small data transmission achievable for LTE-M and NB-IoT at the 164 dB MCL based on the two procedures Early Data Transmission (EDT) and RRC Resume. For EDT data is MAC multiplexed with the RRC connection resume request

TABLE 16.3 LTE-M and NB-IoT latency at the 164 dB MCL.

Technology	Method	Latency [s]
LTE-M	EDT	5.0 s
	RRC Resume	7.7 s
NB-IoT	EDT	5.8 s
	RRC Resume	9.0 s

message in Message 3. For RRC Resume the data is multiplexed with the RRC connection resume complete message in Message 5. Figure 6.4 illustrates this MAC layer multiplexing of data and RRC messages. Here we focus on procedures where data is transmitted over the user plane. Message 3 and 5 data transmission is also supported over the control plane. Those methods are expected to provide similar latencies as shown here for the user plane.

LTE-M is seen to perform slightly better than NB-IoT. The reason is that the MPDCCH manages to achieve 164 dB MCL for a transmission time of 256 ms. For the NPDCCH, the same transmission time results in an MCL just below 164 dB. The results are therefore based on an NPDCCH transmission over 512 ms, which supports an MCL of 166 dB. The performance difference presented in Table 16.3 should therefore in practice be negligible.

Both technologies meet the 5G requirement of a 10 s latency with margin [5]. Chapters 6 and 8 present the detailed assumptions behind these results. They also add more results for different radio conditions to give a more complete picture of the massive IoT latency for small data transmission.

16.2.1.6 Battery life

5G massive IoT devices are required to support operation on a non-rechargeable battery power of 5 Wh for over 10 years [5]. Table 16.4 presents the estimated battery lives of Cat-M and Cat-NB devices when operating at the 164 dB MCL under the assumption that a 200-byte uplink report is sent to the network once per day. The RRC Resume procedure was used in these evaluations. The EDT procedure could in principle have been used if it had not been for a limitation of the maximum TBS to 1000 bits for the Message 3 data transmission. The 5G requirement of 10 years battery lifetime is met by both technologies.

The evaluation assumption behind these results are again found in Chapters 6 and 8. Noteworthy is the assumed device transmit power level of 500 mW, which has been deemed realistic by 3GPP [7]. Lab and field measurements performed on commercial NB-IoT and

TABLE 16.4 LTE-M and NB-IoT battery life at the 164 dB MCL.

Technology	Method	Battery life [years]
LTE-M	RRC Resume	11.9 years
NB-IoT	RRC Resume	11.8 years

LTE-M devices suggest that this may be a somewhat optimistic assumption. A higher transmit power level would naturally reduce the supported battery life, unless compensated for by a redimensioning of the battery power supply.

16.2.1.7 Connection density

5G requires that a massive IoT system can serve up to 1,000,000 connections per square kilometer for a traffic model where each device is triggered to send a 32-byte uplink report every second hour [8]. In practice this means that the system should support 1,000,000 connections across 2 h, or on average 139 connection establishments per second.

Fig. 16.1 presents the number of connection establishments LTE-M and NB-IoT can support per cell, second and PRB versus the 99th percentile latency the systems offer [9]. According to the 5G requirement the supported connection density is defined at the load where 99% of the connections are served with a latency of 10 s or less. The results are depicted for four different urban macro scenarios defined by Ref. [10]:

- Base station inter-site distances of 500 and 1732 m.
- Two different channel models named Urban Macro A (UMA A) and Urban Macro B (UMA B).

For LTE-M the minimum system bandwidth is a narrowband defined by 6 consecutive PRBs. To estimate the LTE-M capacity per narrowband the results in Fig. 16.1 should be scaled by a factor 6.

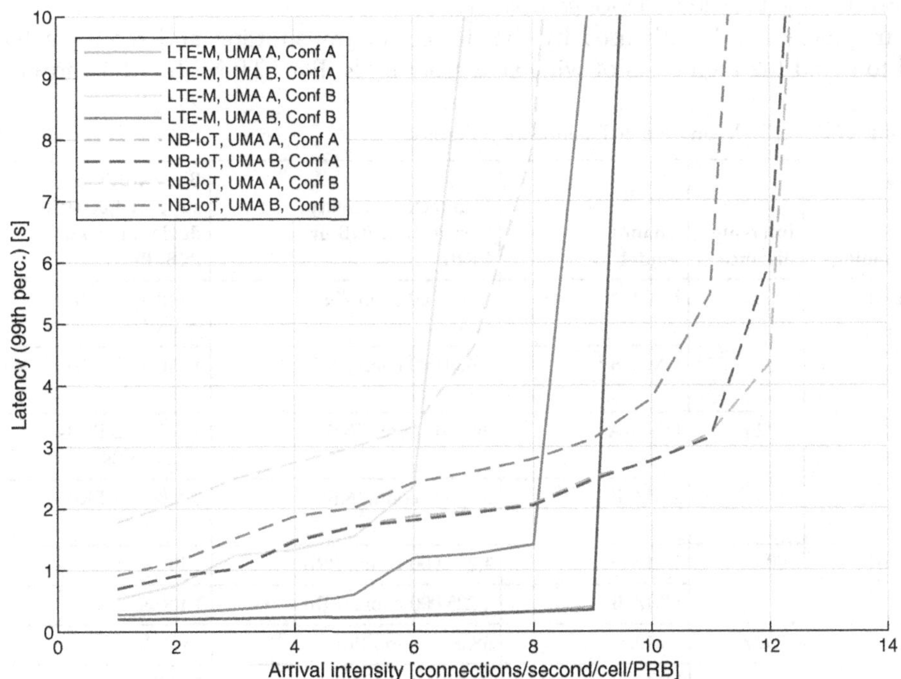

FIG. 16.1 LTE-M and NB-IoT connection establishment rate per PRB.

The knowledge that 1,000,000 connections are equivalent to 139 connection establishments per second allows us to use the simulated results in Fig. 16.1 to calculate the resources needed to support the data traffic generated by the 1,000,000 connections. Table 16.5 shows the results, i.e. the resources needed to support the connection density that 5G is required to support. For LTE-M two PRBs are added to the narrowbands. These are carrying the PUCCH transmissions which are sent outside of the narrowbands close to the cell edges of the LTE system bandwidth.

The general conclusion from the results presented in Fig. 16.1 and Table 16.5 is that both LTE-M and NB-IoT meet the 5G requirement for bandwidths in the range of 1 PRB to 20 PRBs, depending on the scenario. Remember that 1 PRB corresponds to a bandwidth of 180 kHz. The denser deployment making use of a 500 m inter-site distance can carry a significantly higher load that the 1732-m case. The difference in capacity between the two configurations is in parity with the difference in cell area between the two cases. NB-IoT is shown to provide slightly higher capacity, which is explained by the efficient use of subcarrier NPUSCH transmissions in challenging radio conditions. LTE-M is shown to provide better latency, which is due to the ability to use a higher bandwidth in good radio conditions. The LTE-M capacity can be further improved by using the PUSCH sub-PRB feature introduced in Release 15 for users in bad radio conditions.

16.2.1.8 Device complexity

The massive IoT technologies have introduced similar features to reduce device complexity, and thereby enabling low-cost IoT devices. The following design objectives have been pursued for low device complexity:

The frequency bandwidth used by the device for transmitting and receiving has been limited to avoid the high costs of wideband front ends. For LTE-M the RF bandwidth that

TABLE 16.5 LTE-M and NB-IoT connection density.

Technology	Inter-site distance	Channel model	Connection density [connections/NB or PRB]	Bandwidth to support 1,000,000 devices per km^2 [NB, PRB]
LTE-M	500 m	UMA A	5,680,000 conn./NB	1 NB + 2 PRBs = 8 PRBs
		UMA B	5,680,000 conn./NB	1 NB + 2 PRBs = 8 PRBs
	1732 m	UMA A	342,000 conn./NB	3 NBs + 2 PRBs = 20 PRBs
		UMA B	445,000 conn./NB	3 NBs + 2 PRBs = 20 PRBs
NB-IoT	500 m	UMA A	1,233,000 conn./PRB	1 PRBs
		UMA B	1,225,000 conn./PRB	1 PRBs
	1732 m	UMA A	68,000 conn./PRB	15 PRBs
		UMA B	94,000 conn./PRB	11 PRBs

needs to be supported by a device is 1.4 MHz, which is significantly less than the maximum LTE channel bandwidth of 20 MHz. For NB-IoT the RF bandwidth that needs to be supported by the device is 200 kHz.

The physical layer peak data rate has been limited to reduce processing and memory requirements for a device. For Cat-M1 the peak rate has been reduced to 1 Mbps; for Cat-NB1 the peak rate has been limited to below 300 kbps.

Both technologies are specified so that the devices are not required to use more than one antenna to fulfill the performance requirements.

The technologies have been specified to support half-duplex operation. This avoids the needs for a device to integrate one or more costly duplex filters. LTE-M and NB-IoT devices can be implemented with support for half-duplex frequency-division duplex or TDD operation, and LTE-M devices can optionally be implemented with support for full-duplex frequency-division duplex.

The technologies have defined User Equipment categories with lower power classes. This enables a device to use cheaper power amplifiers. It can become an option to implement the power amplifier on the modem chip, and thereby avoiding the costs of a separate component. Three device power classes of 14, 20 and 23 dBm output power are supported.

The features above enable reducing the device costs for the IoT devices. However, it must be noted, that the device cost is not exclusively depending on the communication standard. The cost of the device depends also on what peripherals are added to the device, such as power supply, CPU, or the real-time clock.

In the end, the costs of the device depend on the market success and the market volume of the devices. A large economy of scale will help to reduce the production costs. Due to this many chipset vendors have decided to develop a multi-mode implementation supporting both LTE-M and NB-IoT. For devices targeting the lowest cost, NB-IoT only implementation appears to be the main choice. When it comes to EC-GSM-IoT, commercial chipset and module support remains unavailable on the cellular IoT market.

16.2.2 Cellular technologies for critical IoT

Use cases for critical IoT, or *critical Machine-Type Communications* (cMTC), appear mainly in the field of the *industrial internet of things* (see also Chapter 17) and can generally be grouped into wide-area and local area use cases. 3GPP has characterised several use cases and derived their corresponding requirements in Refs. [11–14]. Wide-area use cases include remote driving of vehicles and automation in intelligent transport systems, automated trains and rail-bound mass transit, or automation of the medium- and high-voltage energy distribution. The upper end of maximum tolerable end-to-end latency latencies for reliable communication over larger distances is in the range of ~100 ms and even ~500 ms for some use cases of train communication, and ~50–100 ms for some of the smart energy grid use cases. On the lower end, 5 ms end-to-end latency is needed for remote driving, 10 ms for critical rail communication over distances up to 2 km, and 5 ms for fast switching and isolation in the power grid over distances up to ~30 km (Fig. 16.2).

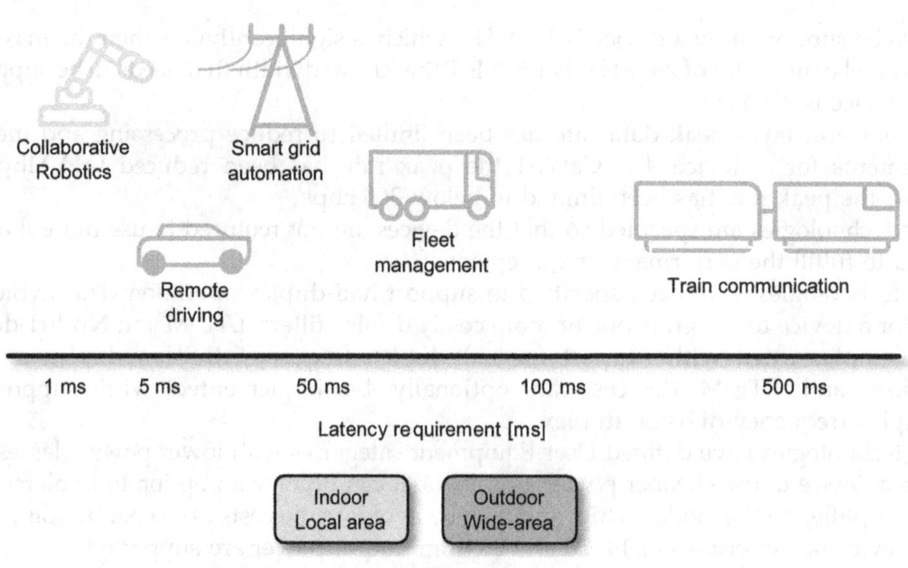

FIG. 16.2 Examples of critical IoT use cases.

Other cMTC use cases are focused on local areas. Most prominent are applications of manufacturing for smart factories: motion control, distributed control and mobile robot applications may require ultra-reliable communication within end-to-end latencies between 0.5 and 10 ms. In these cases, the range of communication is within a confined area, e.g. the factory, and is often indoor. For process control used in the process industries – like for example chemical industry, mining, oil and gas industries – the latency requirements are a bit more relaxed compared to manufacturing. Latency requirements for closed-loop process control can get as low as 10 ms and for process monitoring they are at 100's of ms or even up to seconds. For process control, the confined areas are larger, up to 10's of km^2 including outdoor areas.

What is common in the cMTC use cases above, is that latency plays a critical role and is required to stay within maximum delay bounds. Exceeding a maximum latency in cMTC services can lead to severe consequences, for example a production system comes to a halt, or a short-circuit in the energy grid is not detected and isolated in time.

When using cellular IoT for cMTC services, a suitable spectrum deployment needs to be considered, see Table 16.6. Mobile networks are deployed over different spectrum bands, including low spectrum bands below 1 GHz, and in the higher range 1–6 GHz referred to as mid-band. For 5G NR, also high spectrum bands above 6 GHz are added, which are initially mainly available in the range 24–40 GHz. Mobile network operators have typically licensed a number of spectrum bands and make use of those multiple bands in their deployments. To use them efficiently, mobile networks allow the integration of multiple bands for the transmission with a user equipment, with two defined mechanisms [15]. *Dual connectivity* (DC) allows to set up and aggregate two communication paths that can use radio links of different carriers with separate scheduling entities per carrier on the network side. *CA* allows to pool the radio resource of multiple carriers for the transmission and scheduling

TABLE 16.6 Spectrum bands for critical IoT services.

	Low-band (up to 1 GHz)	**Lower Mid-band (1–2.6 GHz)**	**Higher Mid-band (2.6–6 GHz)**	**High-band (above 6 GHz)**
Duplexing scheme	Mostly FDD	FDD and TDD	mostly TDD	TDD
Carrier bandwidth	Typically small (≤20 MHz)	Mostly small	Medium (20–100 MHz)	Large (>100 MHz)
Candidate critical cMTC technologies	LTE or NR	NR or LTE (for FDD only)	NR	NR
cMTC services	Wide-area (mainly sub-urban or rural)	Wide-area (sub-urban urban, hotspots) Limited local area (urban, hotspots)	Wide-area (sub-urban urban, hotspots) Local area (urban, hotspots)	Local-area (urban, hotspots)

of a user; a joint scheduling entity assigns the resources of the carriers enabling a higher resource efficiency compared to DC. For DC feedback, signaling is needed for each carrier separately, splitting the devices transmission power among those control channels; the control channel can thereby limit the achievable coverage. For CA, a combined feedback is provided on a single control channel which provides better coverage compared to DC.

In sub-urban and rural areas, a mobile network buildout is typically not so dense; low frequency bands up to 1 GHz provide good coverage. Those can be complemented with spectrum carriers in mid-band, but for those bands full coverage is more challenging due to propagation. In denser network deployments, mid-band spectrum bands are used, and in very dense deployments also high band millimeter wave spectrum will be used in future. These spectrum bands differ in their characteristics. In spectrum bands up to 2.6 GHz, FDD allocations dominate, even if some TDD allocations can also be found in the lower mid-band spectrum. In these bands the bandwidth of the carriers available to an operator are typically small, rarely exceeding 20 MHz and often even less. The higher mid-band range of 2.6–6 GHz spectrum is based on TDD, and here larger bandwidth allocations up to e.g. 100 MHz are available. Future spectrum for 5G in the high-band millimeter wave spectrum will also be based on TDD and provide carrier bandwidths that are significantly larger than 100 MHz.

LTE can only be deployed in low-band and mid-band spectrum. NR covers all mentioned spectrum ranges, and new spectrum bands have been identified for 5G in low-band, mid-band and high-band. There are some regional variations among countries about the exact band availability, and also timelines of spectrum auctions vary. In 2019, NR networks are commercially deployed in low, mid and high bands. In addition, NR can be used in spectrum bands already allocated to mobile communications, which allows for a migration of spectrum carriers from 2G, 3G or LTE to NR. A very effective way for fast NR roll-out that reuses the installed network infrastructure and reaches quickly large NR coverage, is to use LTE-NR spectrum sharing [16]. The NR and LTE specifications allow efficient sharing of the radio resources of a carrier among the two radio technologies, so that the same carrier appears as an NR carrier to NR devices, and as an LTE carrier to LTE devices. Radio resources are pooled and can be dynamically distributed between LTE and NR according to instantaneous needs. LTE-NR spectrum sharing is a means of technology migration that is significantly more flexible to traditional spectrum refarming, where the usage of a carrier is first terminated for one radio access technology before a new one is introduced. For NR, spectrum sharing with LTE allows a (slim) NR carrier to be introduced at a low band carrier to provide very quickly wide-area coverage and apply CA to combine low-band coverage with high NR capacity provided by mid-band or high-band NR deployments.

Some of the cMTC use cases are confined to local areas, such as a factory. For such use cases, typically a dedicated network deployment is needed, in order to provide sufficient service coverage and availability for the high QoS requirements. Such a dedicated deployment may be limited to the radio network, extending the outdoor macro network. But most common will be network deployments were larger parts of the cellular IoT network are deployed locally, and possibly even the entire network. A local termination of the user-plane path is motivated for several reasons. By terminating the gateway function locally, the end-to-end latency is significantly reduced compared to routing it via a macro network. This local break-out can be combined with edge computing where application functions can be executed at, or close to, the cellular IoT network gateway, e.g. the *user plane function* of the 5G core network. This has further benefits, that for example, business critical information of the operation of the industrial system is not leaving the premise; also, the integration of the cellular system with existing communication systems and IT infrastructure is simpler. A further discussion on local industrial deployments can be found in Refs. [17,18].

For cMTC services, wider bandwidth allocations are desireable for several reasons. The need for being able to transmit an arriving data packet in short time, implies that instantaneous access to radio resources are needed and cannot be deferred in time. But in several cMTC use cases also a high traffic demand exists, for example, in manufacturing where hundreds of industrial controllers and industrial robots may need to communicate in real-time with ultra-low latency. For those use cases, the upper mid-band spectrum or high-band spectrum are beneficial since they provide larger spectrum allocations, see Table 16.6. In low bands, and largely also in the lower mid bands, only smaller carrier bandwidths are typically available, which limits the capability to support cMTC services, in particular if higher cMTC capacity is required. Methods like carrier aggregation are needed to add additional spectrum bandwidth to the communication system.

As shown in Chapters 10 and 12, both LTE and NR are capable to achieve the IMT-2020 Ultra-Reliable and Low Latency Communications (URLLC) requirement of achieving a 32 B data transmission over the RAN within 1 ms with a reliability of 99.999%. Owing to its design, NR can achieve much lower latencies than LTE, in particular if higher *subcarrier spacing* (SCS) is used. The choice of SCS is coupled to the network deployment and carrier frequency. In high bands, phase noise limits the SCS to higher SCS. At the same time, higher SCS implies shorter cyclic prefix of the OFDM symbols making the communication more sensitive to time-dispersion of the radio signal. Therefore, macro deployments with larger cell sizes, as often used in low band spectrum, require lower SCS. Suitable NR SCS values are listed in Table 16.7.

Table 16.8 shows the latencies that are achievable with high reliability for LTE and NR in different bands; configurations achieving the lowest latencies are selected according to the

TABLE 16.7 NR numerology for different spectrum bands.

	Low-band (up to 1 GHz)	Lower Mid-band (1–2.6 GHz)	Higher Mid-band (2.6–6 GHz)	High-band (above 6 GHz)
Suitable NR SCS	15, 30	30, 60	30, 60	60, 120

TABLE 16.8 Lowest achievable guaranteed latencies.

	Low/mid-band FDD		Mid/high-band TDD	
	Downlink	Uplink	Downlink	Uplink
LTE	0.86 ms	0.86 ms	–	–
NR	0.48 ms	0.48 ms	0.51 ms	0.64 ms

features described in Chapter 9 and 11, and the results are from the evaluations done in Chapter 10 and 12. For LTE, we assume subslot operation; the NR configuration assumes for FDD a 30 kHz SCS and 2-symbol mini-slot, and for TDD a 120 kHz SCS, 4-symbol mini-slot and alternating uplink-downlink TDD slot configuration.

Besides the achievable lower latencies, NR has the benefit that it supports URLLC in both FDD and TDD bands, or combinations of such bands. NR also achieves a higher spectral efficiency and service coverage for URLLC services than LTE, owing to its more flexible configuration, as also presented in Chapter 10 and 12. An additional analysis of both LTE and NR performance for URLLC can also be found in Ref. [19].

Many cMTC services have advanced requirements going beyond only the transmission latency. cMTC for industrial systems often require an interworking with industrial Ethernet or IEEE 802.1 *time-sensitive networking* (TSN), see also Section 17.2. Other requirements are that the communication system must be able to provide high-precision timing information toward devices for being able to synchronize devices with 1 μs precision. LTE provides a basic time synchronization for devices in 3GPP Release-15. For NR and the 5G core network, advanced synchronization features are added in the standardization from 3GPP Release-16, which enables devices to time-synchronize to multiple external time domains, which is needed for industrial automation scenarios. Also, mechanisms for enabling redundant transmission paths for a device throughout the core network are being defined. No activities are currently ongoing for introducing similar features into LTE and the 4G evolved packet core.

In conclusion, both LTE and NR fulfill the 5G requirements from ITU-R (IMT-2020) on URLLC and are able to transmit small packets reliably within 1 ms. NR is more flexible in its design than LTE and has higher performance. NR can not only support URLLC in FDD but also in TDD bands. Further, NR can achieve significantly lower latencies than

LTE, in particular in local deployments, and has also higher spectral efficiency. Finally, NR is more future-proof in its evolution, providing capabilities needed for many cMTC use cases, such as support for redundant transmission paths, time synchronization over the radio interface, and interworking with TSN networks.

16.3 Which cellular IoT technology to select

The choice of a cellular IoT solution is a decision that needs to be taken by different market players. On one hand, it is the mobile network operator that has to decide which cellular IoT technology to add to its existing network. On the other hand, it is the IoT device manufacturer and service provider that have to select for which IoT connectivity options they develop their IoT service. Finally, it is the end customer that may demand what technology to use. The latter is in particular the case for enterprise customers in need of critical IoT services. It can be expected that different options of solutions will coexist.

16.3.1 The mobile network operator's perspective

For a mobile network operator, the decision about which cellular IoT technology to deploy and operate has multiple facets. There are in particular two sides that need to be considered:

- Long-term mobile network strategy and existing assets,
- IoT market segment strategy.

A typical mobile network operator has one or more cellular networks deployed. Increasingly, different radio technologies are provided via a single multi-radio technology network. For example, the same base station can be used for GSM, UMTS/HSPA, LTE or NR transmission. But there are also deployments where the networks for 2G, 3G, 4G and 5G are rather independent in their deployment and operation.

In addition, a mobile network operator has a spectrum license from typically a national regulator, which gives rights to operate a network in the assigned spectrum. Spectrum licenses are long-lasting, e.g. 20 years; this is motivated by providing network operators with an economic safety. A return on investment for an extremely high network installation cost of a new technology can be planned over a long time period. On expiry of a spectrum license, a spectrum licensing contest, like a spectrum auction, is initiated by the regulator for providing a new spectrum license. In general, any network build-out roadmap by an operator is a long-term decision and needs to consider at least the following elements:

- How long are the existing spectrum licenses valid, what technologies are allowed to be operated in the spectrum, and when is a new spectrum re-allocation process planned by the regulator?
- What is the status of the existing network buildout for different radio technologies, in particular GSM and LTE, and what is the number of mobile devices and projected growth for each specific technology?

- What is the network buildout of competing operators and what is their market share?
- What is the strategic intent of an operator concerning IoT?
- What services are planned to be provided, and on what roles does the operator intend to take (e.g. as a connectivity provider or also as a service provider/enabler)?
  - What is the market maturity for IoT services?
  - What IoT segment would the operator like to address?

It shall be noted that the above questions are raised from a perspective of the operation of an operator network in a specific country. However, several operators are active in multiple countries and even on multiple continents. Even if the decision is largely made per country, an operator may want to harmonize decisions over multiple regions in which it operates networks.

When looking at the cellular IoT technology options, the following characteristics can be identified which will influence an operator's decision.

As a baseline, we assume that there is a very large incentive by an operator to reuse existing mobile network infrastructure for deploying any of the cellular IoT technologies. EC-GSM-IoT can be deployed based on a GSM infrastructure and by using GSM spectrum. The GSM network resources and the GSM spectrum would be shared between GSM usage and EC-GSM-IoT usage. LTE-M and NB-IoT can be deployed based on LTE infrastructure and by using the LTE spectrum; LTE network and spectrum resources would be shared between LTE, LTE-M and NB-IoT usage. In most network configurations, it can be expected that the deployment of EC-GSM-IoT, LTE-M and NB-IoT can be realized as a software upgrade to the deployed GSM or LTE networks. This implies that the introduction of the cellular IoT into the market can be realized by operators rather quickly and at a low total cost of ownership. NR has been designed to allow efficient interworking with LTE (including LTE-M) and NB-IoT. This means that LTE-M and NB-IoT can be embedded into an NR carrier in a similar way as they are today integrated into an LTE carrier. In general, it is possible to share a carrier between NR and LTE, where the resources being used for LTE or NR transmission can be dynamically adapted [16]. The LTE-NR coexistence flexibility also enables to migrate an LTE carrier to NR, while continuing LTE-M or NB-IoT devices with long device lifetime to continue operation within the NR carrier after the migration. For IoT, it is expected that many services expect a long lifetime of e.g. a decade. This expectation should be addressed with a cellular IoT network. As a result, the decision of the cellular IoT technology is also coupled to the operator's long-term strategy for mobile networks focusing on telephony and mobile broadband services. If an operator intends to transition GSM deployments to e.g. LTE or NR in the coming future, an introduction of EC-GSM-IoT seems a questionable choice, as any long-term EC-GSM-IoT users would require maintaining the GSM infrastructure operational for a long time. A general trend that is seen globally, is that 2G and 3G spectrum allocations are stepwise migrated to LTE [20]. With the market introduction of NR now and the superior NR capabilities, a refarming of spectrum toward NR is expected for the future. In this step, the compatibility of LTE and NR which allows for e.g. LTE-NR spectrum sharing [16] will make the transition from LTE to NR very smooth and it can be flexibly adapted according to the gradual increase of NR capable devices. Within the light of this migration, it can be noted that

EC-GSM-IoT has as of today not managed to attract significant market interest, while significant deployments of NB-IoT and LTE-M has happened during the last two years [3].

While the reuse of existing network infrastructure and spectrum is an important aspect for an operator, a specific benefit of NB-IoT shall be pointed out in its spectrum flexibility. It is generally expected that existing operator spectrum deployments are extended to also include LTE-M traffic, so the IoT traffic will be on the same spectrum that is already deployed for telephony and mobile broadband services. For NB-IoT, the narrow system bandwidth of NB-IoT makes it suitable to be deployed also in spectrum that is not used for mobile broadband services today. Examples exist where operators have spectrum allocations that do not fit with exact carrier bandwidths provided by LTE. As a result, a remainder of the spectrum allocation remains unused. NB-IoT provides the flexibility to make use of even small portions of idle spectrum resources that an operator may have. Such portions of spectrum resources can even be created by an operator, e.g. by emptying individual GSM carriers from GSM operation and re-use them instead for NB-IoT usage.

For critical IoT services, NR has benefits over LTE, for both wide-area and local (often industrial) use cases. These benefits stem from the higher flexibility, and better performance and spectral efficiency of NR over LTE for URLLC. As a market segment for critical cellular IoT services is at its very beginning, a wide range of innovations and optimizations can be expected for cellular IoT as this segment develops. Already now, several novel features have been identified for future NR standard releases to better address e.g. industrial use cases, by providing time synchronization over the radio interface or interworking with IEEE 802.1 TSN. We foresee that in particular the NR evolution will address these features in future standard releases.

There is a segment of cellular IoT services which has higher demands in performance and capabilities than massive IoT services, but yet without a need for the URLLC required by critical IoT services. This segment has been referred to as *broadband IoT* (see Ref. [21] and Figure 1.5) and it combines requirements on high data rates with massive IoT features like extended coverage and battery saving. There are numerous examples of this category, like advanced wearables, connected vehicles and telematics, video monitoring systems, augmented and virtual reality systems, or connected drones as described in Chapter 13 and [22,23]. Both LTE and NR are well suited to address this segment. Here it should be noted that the LTE-M features for extended coverage and battery saving described in Chapters 5 and 6 can be implemented not only by the massive IoT device categories Cat-M1 and Cat-M2 but also by more high-performing ordinary LTE devices, and that LTE-M can coexist seamlessly with both LTE and NR, so LTE-M can be seen as a suitable candidate solution for use cases requiring extended coverage and battery saving in both the massive IoT and the broadband IoT segment.

The considerations for choosing a cellular IoT technology by a mobile network operator is based on what spectrum and what radio access technology the operator use or plan to use in future. The driving force in this regard is to reuse existing or planned mobile networks in order to achieve a low capital expenditure and operational expenditure for the deployment and operation of cellular IoT connectivity. Another major component in an operator assessment of cellular IoT is the IoT service strategy of the operator. Does the operator target specific IoT market segment? And if so, what are the service requirements in this segment and what connectivity requirements does it imply? In this case the operator decision is largely based on how well a cellular IoT technology fulfills the service requirements, as discussed in Section 16.2.

16.3.2 The IoT service provider's perspective

An IoT service provider targets a set of particular IoT services with its offering. For example, a focus may be on smart city applications, or precision agriculture. The targeted IoT service implies a certain location where the service will be realized, i.e. where IoT devices will be located. For smart city services, this will be in urban areas, for precision agriculture this will be primarily in rural areas, and for industrial IoT solutions it will be at industrial sites. The IoT service characteristics determine what kind of traffic profile needs to be supported. For a smart city this may be regular monitoring of available parking spaces, or notifications when waste containers have reached a certain fill level. For precision agriculture, it can be the monitoring of humidity and fertilization on fields or in green houses, or the tracking of cattle. For industrial IoT solutions, it can be monitoring and control of the industrial precesses and operations. Other IoT service characteristics besides the traffic profile can be the maximum time that a device must operate on a battery.

Based on an analysis of the targeted IoT service, the connectivity requirement of the service becomes clear:

- What data rates need to be supported by the communication?
- Are critical IoT services targeted, and what are the required latency bounds and levels of reliability and availability?
- Do devices need to run on battery for extended time periods?
- What device density is expected?
- Where are the devices located?
- Are devices in particularly hard to reach locations (e.g. in enclosures underground)?
- Are devices mobile over larger areas, possibly even across national borders?

Based on this review a service provider can determine:

- Which cellular IoT technologies provide sufficient performance for the targeted service, see Section 16.2.
- At what locations network coverage is needed.

It can be expected that coverage of multiple cellular IoT technologies is provided by one or more network operators at various locations. In an increasing number of mobile networks, both LTE-M and NB-IoT will be found. For URLLC capabilities, NR is starting to be deployed, but availability will vary between different locations. An IoT service provider will want to select a network operator that provides coverage and connectivity via a suitable cellular IoT technology at the targeted deployment area at a fair price.

Since IoT devices may be deployed and operated over long time spans, flexibility in re-selecting a network provider is desirable. Embedded Subscriber Identity Modules that enable to remotely re-provision devices and re-select network providers will play an increasing role for cellular IoT devices, see Ref. [24].

For critical IoT systems, several IoT use cases require dedicated deployments and installations. The IoT service provider may be a system integrator, potentially a mobile network operator, or the industrial enterprise end user. Even shared responsibilities are imaginable. A specific plan for the system solution is needed, that analyses in detail the requirements and desired capabilities. A specific solution is needed, that considers the local availability of spectrum and is built on standard cellular IoT components.

References

[1] S. Andreev, O. Galinina, A. Pyattaev, M. Gerasimenko, T. Tirronen, J. Torsner, J. Sachs, M. Dohler, Y. Koucheryavy. Understanding the IoT connectivity landscape: a contemporary M2M radio technology roadmap. IEEE Commun. Mag., September 2015, Vol. 53, No. 9, 32–40.

[2] Department for Business. Energy & industrial strategy, smart metering implementation programme, progress report for 2018. London Crown copyright, 2018.

[3] GSMA. Mobile IoT commercial launches, 2019. Available: https://www.gsma.com/iot/mobile-iot-commercial-launches/.

[4] Third Generation Partnership Project, TS 36.101, v16.0.0. E-UTRA UE radio transmission and reception, 2019.

[5] Third Generation Partnership Project, Technical report 38.913, v15.0.0. study on scenarios and requirements for next generation access technologies, 2018.

[6] ITU-R, Report ITU-R M.2412-0. Guidelines for evaluation of radio interface technologies for IMT-2020, October 2017.

[7] Third Generation Partnership Project, Technical report 45.820, v13.0.0. cellular system support for ultra-low complexity and low throughput internet of things, 2016.

[8] ITU-R, Report ITU-R M.2410. Minimum requirements related to technical performance for IMT-2020 radio interfaces(s), 2017.

[9] Ericsson. R1-1903120. IMT-2020 self-evaluation: mMTC non-full buffer connection density. In: 3GPP RAN1 meeting #96, 2019.

[10] ITU-R, Report ITU-R M.2412-0. Guidelines for evaluation of radio interface technologies for IMT-2020, October 2017.

[11] Third Generation Partnership Project, Technical Specifications 22.261, v16.7.1. Service requirements for the 5G system, March 2019.

[12] Third Generation Partnership Project, Technical Specifications 22.104, v16.1.0. Service requirements for cyber-physical control applications in vertical domains, March 2019.

[13] Third Generation Partnership Project, Technical Specifications 22.186, v16.1.0. Enhancement of 3GPP support for V2X scenarios, December 2018.

[14] Third Generation Partnership Project, Technical specifications 22.289, v16.1.0. Mobile communication system for railways, March 2019.

[15] J. Sachs, G. Wikstrom, T. Dudda, R. Baldemair, K. Kittichokechai. 5G radio network design for ultra-reliable low-latency communication. IEEE network, Vol. 32, March-April 2018, 24–31. https://doi.org/10.1109/ MNET. 2018.1700232.

[16] T. Cagenius, A. Ryde, J. Vikberg, P. Willars Simplifying the 5G ecosystem by reducing architecture options, Ericsson Technology Review, November 2018. https://www.ericsson.com/en/ericsson-technology-review/archive/2018/simplifying-the-5g-ecosystem-by-reducing-architecture-options.

[17] J. Sachs, K. Wallstedt, F. Alriksson, G. Eneroth Boosting smart manufacturing with 5G wireless connectivity, Ericsson Technology Review, February 2019. https://www.ericsson.com/en/ericsson-technology-review/archive/2019/boosting-smart-manufacturing-with-5g-wireless-connectivity.

[18] K. Gold, K. Wallstedt, J. Vikberg, J. Sachs Wireless connectivity for industries, book chapter. In: Dastbaz, M. Cochrane, P. editors. Industry 4.0 and engineering for the future. Springer, 2019. ISBN-13: 978-3030129521.

[19] Next Generation Mobile Networks (NGMN). 5G extreme requirements: radio access network solutions, June 2018. https://www.ngmn.org/fileadmin/ngmn/content/downloads/Technical/2018/180605_NGMN_5G_Ext_Req_TF_D2_1_v2.5.pdf.

[20] Ericsson. Ericsson mobility report, 2018. November 2018, [Online]. Available: https://www.ericsson.com/assets/local/mobility-report/documents/2018/ericsson-mobility-report-november-2018.pdf [March 2019].

[21] Ericsson. Cellular IoT evolution for industry digitalization, white paper, 2019. January 2019, [Online]. Available: https://www.ericsson.com/assets/local/trends-and-insights/consumer-insights/reports/wp_evolving-iot-forindustrialdig_jan-312019_revised.pdf [March 2019].

[22] Ericsson. Drones and networks: ensuring safe and secure operations, white paper, 2019. November 2018, [Online]. Available: https://www.ericsson.com/en/white-papers/drones-and-networks-ensuring-safe-and-secure-operations [March 2019].

[23] X. Lin, V. Yajnanarayana, S. D. Muruganathan, S. Gao, H. Asplund, H.-L. Maattanen, M. Bergstrom, S. Euler, Y.-P. E. Wang. The sky is not the limit: LTE for unmanned aerial vehicles. IEEE Commun Mag, July 2017, Vol. 56, No. 4, 204–10.

[24] GSMA. Remote SIM provisioning for machine to machine. Website, 2017 [Online]. Available: http://www.gsma.com/connectedliving/embedded-sim/ [March 2019].

CHAPTER

17

Technical enablers for the IoT

OUTLINE

Abstract

This chapter provides an overview of technical enablers for the Internet of Things. It describes the contribution from various fields on the IoT: from pervasive computing, embedded systems, data analytics and machine learning, cloud computing and cyber-physical systems. It reviews the development of communication technologies that have been developed for IoT and the corresponding activities in the Internet Engineering Task Force. It concludes with an overview of the activities of the Industrial Internet of Things.

The chapter provides an overview of what technical components contribute to the IoT. The technical origins of the IoT fall back onto several disciplines, as shown in Fig. 17.1. The following subsections address the different technology sections in the figure: (1) devices, computing and input/output technologies, (2) communication technologies, (3) internet technologies for IoT, and (4) advanced service capabilities and algorithms.

Cellular Internet of Things, Second Edition
https://doi.org/10.1016/B978-0-08-102902-2.00017-0

© 2020 Elsevier Ltd. All rights reserved.

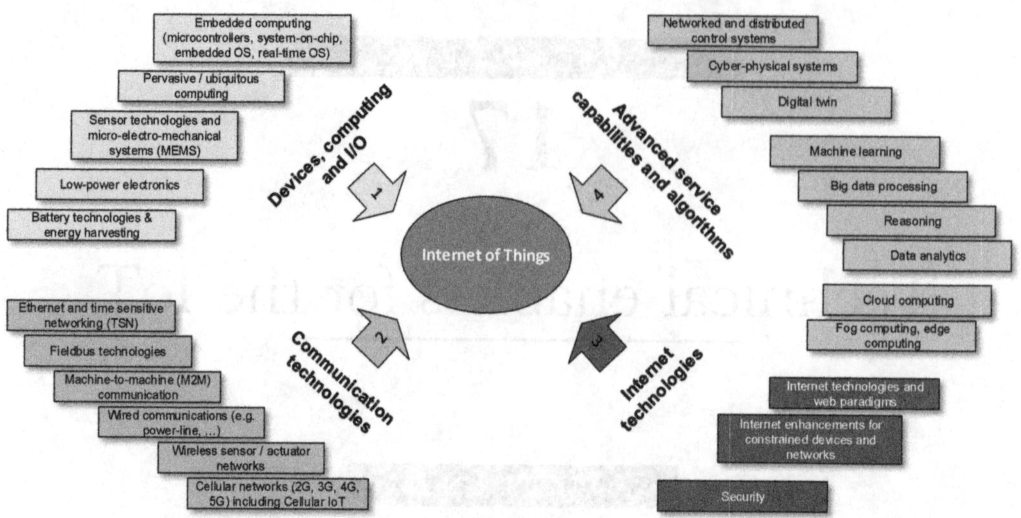

FIG. 17.1 Technology developments enabling the internet of things.

17.1 Devices, computing and input/output technologies

In the 1990's research started on *pervasive computing* or *ubiquitous computing* with the vision that computing platforms could be miniaturized and integrated into physical objects [1–4]. As Mark Weiser stated in his landmark article [1], "the most profound technologies are those that disappear. They weave themselves into the fabric of everyday life until they are indistinguishable from it." The objective was that it should be possible to obtain anytime and anywhere access to information services, where the presence of the sensing and computing devices should become invisible. Ubiquitous computing builds on the evolution and deployments of embedded systems. Embedded systems started to appear in the 1960's when first integrated circuits and microprocessors became integrated into engineering systems. Early application areas where found in the aeronautics and space industries, as well as the automotive industry. Today embedded systems are omnipresent in all areas: consumer electronic devices, entertainment systems, industrial systems, automotive, space and aeronautics, telecommunications systems, and many more. Embedded systems consist of hardware and software that are embedded into larger electronic or mechanical systems [5]. Hardware platforms are based on microcontrollers or system-on-chip solutions. From around the 1990's dedicated operating systems for embedded systems have been introduced, often as real-time operating systems that enable time-critical computing. While the intention with ubiquitous computing has been from the beginning to interconnect and network computing functions, the focus of embedded systems has always been on optimized computing for specific tasks, which are often only localized. Embedded systems are often confronted with several non-functional requirements, like real-time processing capabilities and high reliability, memory size constraints due to e.g. space, size, weight or cost limitations, energy consumption limits and potentially harsh operation conditions. Consequently, embedded systems are often purpose optimized and not based on general-purpose computing platforms. Advances in communication technologies and data center technologies for efficient computing lead to an increasing opportunity of distributing the functionality of embedded systems, and thereby exploiting IoT approaches.

One important component of the internet of things is the interaction with the objects or "things" of the physical world. The key technologies in that regard are *sensors*, which measure some physical properties of the object (e.g. to measure temperature), and *actuators* which can act on the behavior of the object (e.g. to switch on a ventilation). Significant achievements have been made in the development of microscopic sensors and actuators, that can be produced at low costs and can be effectively integrated into physical object. *Micro-electromechanical systems* (MEMS) are a key enabling technology for small scale sensors and actuators. Examples of MEMS-based sensors measure pressure, vibration, acoustic emission, gyroscopes [6–8], acceleration, force, chemical sensors and spectrometers [9]. Microelectro-mechanical systems based actuators are drives in e.g. microgrippers or micromotors, controllable micromirrors, microresonators, microswitches, and others [9,10].

Further technology developments that enable wider spread usage of ubiquitous computing and sensors and actuators, are advancements in low-power electronics and battery technologies with higher energy density and longer battery life. Evolving technologies of energy harvesting can provide further opportunities of embedding devices with reduced dependency on local energy supply [11].

17.2 Communication technologies

A significant number of communication technologies have been developed over the last two decades with significant impact on the IoT. This comprises both wired and wireless communication technologies. In particular, *machine-to-machine* (M2M) communication solutions were developed to connect devices with applications. Most M2M communication solutions are purpose-build and designed to satisfy a very particular application and communication needs. Examples are connectivity for remote-controlled lighting, baby monitors, electric appliances, etc. For many of those systems the entire communication stack has been designed for a single purpose. Even if it enables, in a wider sense, an environment with a wide range of connected devices and objects, it is based on M2M technology silos, usually without end-to-end Internet Protocol (IP) connectivity and instead via proprietary networking protocols. This is depicted on the left-hand side of Fig. 1.3. It is quite different from the vision of the IoT (depicted on the right-hand side in Fig. 1.3), which is based on a common and interoperable IP-based connectivity framework for connecting devices and smart objects, which enables the IoT at full scale.

For wireless communication systems a significant research and development effort has started in the mid-1990's with the development toward *wireless local area networks*, *wireless sensor networks* (WSN) or *wireless sensor and actuator networks* [12], including industrial wireless networks [13]. Those wireless network technologies are further discussed in Chapter 14. In recent years the Third Generation Partnership Project (3GPP) have evolved their cellular technologies to target a wide variety of IoT use cases as described in the earlier chapters of this book.

Wired communication systems targeting machine-to-machine communication include latest standards on *power-line communications*, where the communication is carried over the power conductors used for electricity transfer. M2M use cases for power-line communications include home automation (e.g. HomePlug or LonWorks), smart metering or communication within a smart energy grid. Industrial M2M communications carries often time critical control communication and requires ultra-high reliability and availability. Numerous, typically proprietary Fieldbus technologies have been developed and are used in industrial networks today [14–18]. Ethernet has continuously evolved, and one set of features that are currently being specified for Ethernet networking is the addition of *time-sensitive networking* (TSN) [19]. TSN provides deterministic communication behavior for prioritized traffic. Ethernet TSN is expected to replace over time the Fieldbus technologies for industrial communication [20].

17.3 Internet technologies for IoT

17.3.1 General features

One of the essential components of the Internet of Things, is the advances that have been made in general Internet and web technologies. This includes optimizations that have been introduced to provide connectivity for smart objects and devices, which are light weight to be feasible even for simple devices that are constrained in their processing capabilities and memory size. One driving force behind the Internet of Things is to enable a transition from purpose-build – and often proprietary – machine-to-machine system solutions (as seen on the left-hand side of Fig. 1.3) toward a standardized, open, interoperable communication platform, with in-build and proven security. This is depicted on the right-hand side of Fig. 1.3, and is a basis for making the interconnection of smart objects and applications scale to the 10's of billions of devices that predicted for the Internet of Things in a range of use cases.

The successful technology basis of Internet communication and web protocols has been selected as foundation for the Internet of Things. Communication based on the *IP* stack has proven to provide an open technology basis that scales the Internet to global connectivity. With the introduction of IPv6, the address space of addressable hosts on the internet has increased from the ~4 billion devices in IPv4 to more than 10^{38} addresses; an IPv6 based Internet of Things seems unlimited in number of things that may become connected. Communication between end hosts and servers on the Internet typically takes place via the *world wide web* (WWW), which is a distributed information space of *web resources* via the internet. The original focus of the WWW was to enable human access to information elements distributed over the internet, e.g. via web browsers that present different information objects on a web site. Web services extend the concept by providing communication between different electronic devices via the WWW, where information is encoded in machine-readable file formats. The most common web protocol to access web resources is the *hypertext transfer protocol* (HTTP). HTTP follows a client-server computing models with request-response interactions, where a client sends requests to a server which reacts with response messages. Requests can contain methods – such as GET, PUT, POST, DELETE – in order to access or modify resources, where a resource is identified via a *uniform resource identifier*. The term "resource[1]" refers here to anything that can be addressed via an URI, like for example an information element or a service [21]. In the context of the IoT, a resource can be a measured value of a sensor that can be accessed from client via a GET request, or it can be a control value sent to an actuator via a PUT request.

Significant improvements have been made for the IP suite to support Internet of Things services. An overview is depicted in Fig. 17.2 based on the common Internet model of a layered protocol stack. As a variation to the classical IP stack we have added a *transfer layer* as in Refs. [22–24] to describe protocols that transfer data objects and provide semantics for operations. The center piece is the network layer, where IPv6 provides the internetworking and routing capabilities. Internet packets can be transmitted over a variety of transmission systems, like cellular networks, short-range radio technologies, or fixed transmission technologies like Ethernet or power-line communication, which are referred to as the link and physical layers in the IP model. Above the network layer are Internet transport protocols: the *transmission control protocol* (TCP) for reliable transport and the *user datagram protocol* (UDP) as unreliable transport protocol. *Transport layer security* (TLS) and *datagram transport layer security* (DTLS) provide a secured end-to-end transport connection for TCP and UDP respectively. *Quick UDP internet connections* is a new transport protocol under development that is building on UDP and with integrated security; it is expected to play an increasing role in the future, also for IoT [23–25].

A detailed overview of optimizations of the IP suite can be found in Ref. [22] and also [23,24,26].

17.3.1.1 IoT transfer protocols

HTTP in version 1.1 [27] and version 2 [28] are the most common transfer protocols in the WWW today. They run over TCP transport, can make use of TLS security, and are also common for some IoT applications. The IETF CoRE working group has developed a new transfer protocol: the *constrained application protocol* (CoAP) [23,24,29,30], see Fig. 17.2. The objectives are to have a transfer protocol that is lightweight and with small code size, that can be installed and operated even on constrained devices with limited power, memory and processing capabilities, and has more compact messages compared to e.g. HTTP. CoAP has also been designed for being efficient to communicate over constrained-node networks, where the underlying constrained IP network may be constrained in maximum packet sizes, may experience high packet loss and contain devices that are occasionally in a battery saving sleep mode. Examples of such constrained IP networks are multi-hop sensor networks, e.g. based on IEEE 802.15.4.

[1]It shall be noted that the term "resource" is used with different meanings in the context of IoT. It can be in the meaning of "web resource" referring to the information elements addressed by an URI that can be read or manipulated. Another usage of the term is used in the context of computing, memory or battery resources that are available at a device or required for a transaction. The latter meaning is e.g. applied when talking about resource-constrained IoT devices.

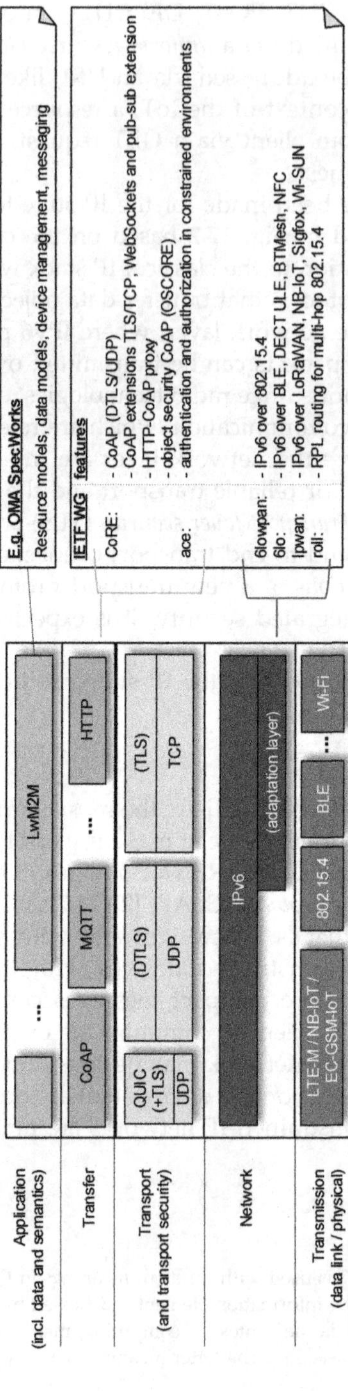

FIG. 17.2 IoT features in the IP stack.

CoAP is based on a *representational state transfer* design, similar to HTTP. This implies an architecture design with [22,29,31,32]:

- Client–server architecture

Data is hosted on servers, and the user interface toward the data is located in a client. The client interacts with the server in order to request access to server resources via a computer network. The only information needed by a client application to interact with a resource on the server is the identifier of the resource and the action to be performed on the resource.

- Statelessness

A server does not maintain any state or context of any client in-between client requests. Any session state is solely maintained within the clients. A service request from the client to the server contains all the information that is required to service the request at the server.

- Uniform interface

A uniform interface exists between components in the system. The typical interface to identify resources is the uniform resource identifier.

- Cacheability

Responses from servers must be specified as cacheable or non-cacheable. This enables a client to reuse information in future request and thereby partially eliminate some client-server interactions. This improves scalability and performance, in particularly for constrained devices.

An IoT device implements a CoAP server. The name "server" may give a feeling that a lot of processing is needed on a high-computing platform. However, CoAP servers can run on simple devices with a very small computing platform, like an 8-bit microcontroller with some kilobytes of memory (see Ref. [22]). Fig. 17.3 depicts a CoAP request-response. A CoAP server is running on an IoT device, which may have one or more sensors or actuators. A CoAP client can request operations on the web resources provided by the sensors/actuators; similar methods as in HTTP are defined: GET/PUT/POST/DELETE. HTTP uses TCP/TLS as a reliable transport protocol, which ensures successful message delivery via

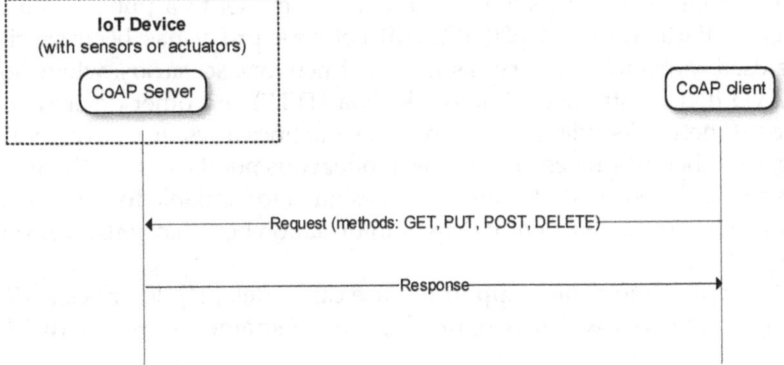

FIG. 17.3 CoAP request-response.

retransmissions. CoAP in contrast has been defined for using the much more lightweight UDP/DTLS protocol, which does not guarantee successful message delivery. For this reason, CoAP implements an own retransmission mechanism. Each CoAP message can be marked as either confirmable or non-confirmable. Confirmable messages are retransmitted after a default timeout with exponential back-off between retransmissions, until an acknowledgment is received from the CoAP receiving instance. This retransmission scheme is more lightweight, compared to using TCP as protocol to provide reliability. An application can select to use unreliable CoAP messaging, e.g. if a series of sensor readings is in progress and the application can tolerate to lose some of the readings.

Another transfer protocol that is frequently used for IoT applications is the *Message Queuing Telemetry Transport* (MQTT), which is specified in OASIS [33]. MQTT follows a publish-subscribe messaging paradigm (see Fig. 17.4) [24,34]. MQTT publishers – as information sources – publish certain information by sending it to a MQTT broker. MQTT subscribers – as information sinks – can subscribe to certain information by subscribing to the MQTT broker. The information the subscriber is interested in, is specified via a topic name. Whenever information is published for a certain topic, the broker forwards this information to all subscribers that have subscribed to that topic. MQTT requires an ordered, lossless transport protocol and is typically used over TCP/TLS.

Besides CoAP, MQTT, HTTP, there are further transfer protocols that are used for IoT services. Those include WebSockets, the *extensible messaging and presence protocol* (XMPP), the *advanced message queuing protocol*, the *data distribution service* [24,34]. A thorough comparison of transfer protocols has been made by Open Mobile Alliance (OMA) SpecWorks [24]. In general, both MQTT and CoAP seem to be the dominating and most flexible protocols for IoT applications. Both protocols have also been evolving with new features, making them more similar.

CoAP has specified an extension for CoAP transfer over TCP/TLS and WebSockets [35], which typically allows better Firewall transversal than transmission over UDP/DTLS. Two CoAP method extensions FETCH and PATCH [36] further enable a client to access parts of a resource, e.g. in case of complex data structures. A new OBSERVE method has been specified [37], which allows a CoAP server to keep a client updated on a resource over a certain time via notifications; this has similarities to a subscription to a resource in a publish-subscribe model. Specification of a publish-subscribe broker is ongoing [22,38], which allows CoAP to extend from the request-response interaction model to a publish-subscribe model similar to MQTT. Furthermore, a HTTP-CoAP network proxy has been specified [22,39], which can be used to realize IoT services in mixed network scenarios, where some devices are connected via non-constrained IP networks (via HTTP) and other devices are connected via constrained IP networks (via CoAP). CoAP also defines a *resource directory* where CoAP servers can register their resources. It serves as rendezvous points for CoAP clients to identify CoAP servers and their resources. Data models designed for scalable IoT, e.g. *Sensor measurement lists* [40], can be used to efficiently deliver information between constrained devices and cloud services [22,23].

MQTT is also evolving to better support IoT use cases. Ref. [24] describes a MQTT variant MQTT for sensor networks which is optimized for constrained sensor networks and uses UDP as transport protocol.

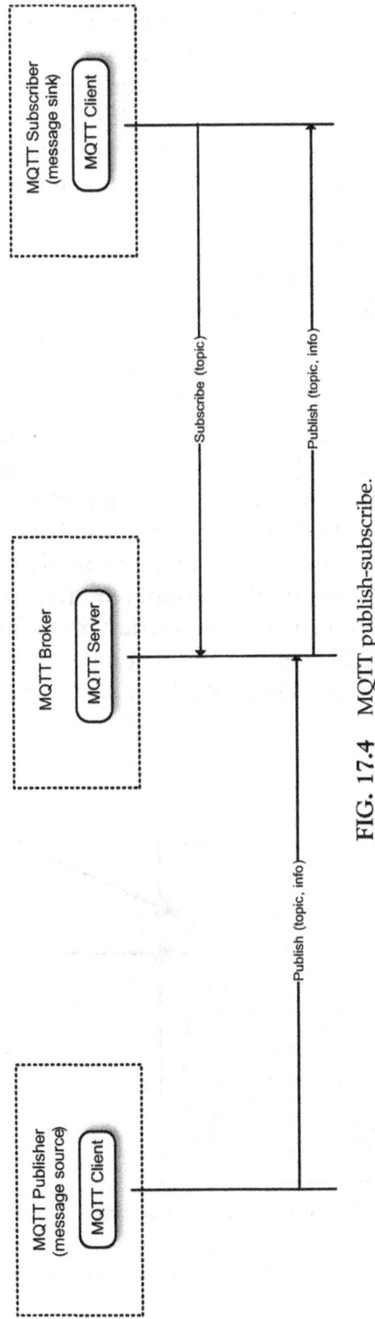

FIG. 17.4 MQTT publish-subscribe.

Security aspects for IoT have received increasing attention during recent years. Security is already provided over a UDP and TCP transport connection via DTLS and TLS. For CoAP also *object security for constrained RESTful environments* has been developed. It provides end-to-end security even in use cases where proxies are used, or the end-to-end path is split into multiple transport layer segments. The IETF working group *authentication and authorization for constrained environments* [41] develops solutions for authentication authorization to enable authorized access to IoT resources that work efficiently in constrained environments [42]. For further discussion on security in IoT environments see e.g. Refs. [23,43,44].

17.3.1.2 IoT application framework

An IoT application can use the IoT protocol stack, comprising the above described transfer, transport and network layers, for realizing IoT services. For the exchange of information and its useful interpretation, a common data structure and semantics are needed between the endpoints. An application framework can be helpful to provide a standardized and reusable specification for IoT applications (see Fig. 17.2). OMA SpecWorks[2] has specified a *lightweight machine-to-machine* (LwM2M) data and device management protocol for IoT applications, that builds on CoAP as transfer protocol [22,23,45–47]. LwM2M defines a model for objects and resources as depicted in Fig. 17.5. A device can have multiple objects, which represent for example sensors, actuators or controllers. Each object can have one or more resources, which describe properties of the object. For example, for a sensor, they could describe the upper and lower measurement range of the sensor, the current reading, and the highest value measured over a certain time period. Objects can also contain device configuration parameters as part of the device management. IPSO Alliance[3] has defined data models for smart objects and their resources [22,23,48]. LwM2M specifies operations for the entire lifecycle of a device.

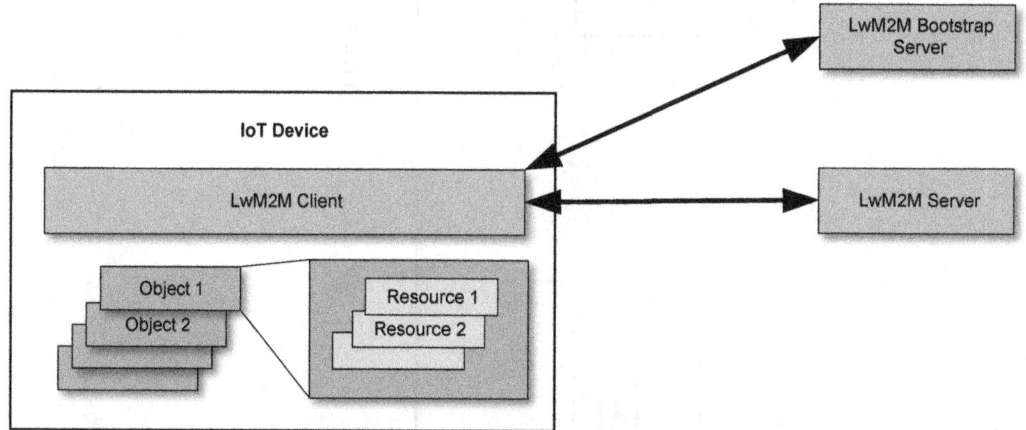

FIG 17.5 LwM2M architecture protocol [22,45,47].

[2]OMA SpecWorks was originally Open Mobile Alliance. It merged with IPSO Alliance in March 2018 to become OMA SpecWorks.

[3]IPSO Alliance has been promoting IP for smart objects and IoT communication down to very simple devices. IPSO alliance has merged with the Open Mobile Alliance (OMA) in March 2018 to form OMA SpecWorks.

Automated bootstrapping is defined with a bootstrapping server. After registration to a LwM2M server, a LwM2M server can interact with the objects and resources of the device, e.g. via read, write, execute operations [22,45]. Interactions are defined for information reporting (e.g. of sensors), and for device management and service enablement. LwM2M also defines objects regarding the firmware running on the device and enables firmware updates to the device from a firmware update server [22,45]. To load large firmware packages onto the device, the firmware package can be partitioned into multiple chunks of e.g. 128 bytes size, which can be transferred to the device via the block-wise transfer of CoAP. OMA LwM2M is one prominent example of an IoT application framework and several implementations exist. Although the usage of LwM2M is not needed for developing an IoT application, it provides a broad set of functions to simplify the development of IoT applications based on a standardized and open framework.

17.3.1.3 IoT link layer adaptations

Often the transmission technology used by Internet of Things devices are limited in their capabilities and performance. This is mostly due to the requirement of IoT devices to be simple, small and battery efficient. An optimized IP protocol stack, e.g. based on IPv6, UDP/DTLS, LwM2M significantly simplifies the software and memory footprint, communication overhead and processing requirements compared to the IP protocol stack used for standard web services. On the lower end an IoT-IP stack may be desired to run on devices with simple 8-bit microcontrollers, where the code size may need to be significantly smaller than 100 kbytes and a RAM memory size smaller than 10 kbytes [22,49,50]. However, even with such optimizations, some transmission technologies – like WSN or some unlicensed low-power wide area radio technologies – may have difficulties to support IoT communication. Reasons can be that the maximum packet size that can be handled by a transmission technology is significantly smaller than the potential size of an IPv6 packet (examples are 127 bytes for IEEE 802.15.4 and 12 bytes or less for SigFox [51]). The IETF has specified features for supporting IPv6 data transfer for IoT services over different transmission technologies within the IETF working groups *6lowpan* [52] (focused on the IEEE 802.15.4 transmission technology), *6lo* [53] (investigating Bluetooth low energy, Bluetooth mesh, and other transmission technologies), *lpwan* [54] (investigating low power wide-area networking technologies, such as SigFox, LoRa, WI-SUN, NB-IoT [55]). Such optimizations are described as an optional *adaptation layer* between IPv6 and the corresponding transmission technology in Fig. 17.2. The following adaptation layer features have been specified [22,23,26,34,56]:

- Static header compression for IPv6/UDP headers [57–59] and CoAP headers [60].
- Fragmentation and reassembly of IPv6 packets to smaller frame sizes that can be handled by an underlying transmission technology [57,59].

WSN like IEEE 802.15.4 transmit over short distance. In order to provide sufficient connectivity multi-hop transmission is applied. The routing protocol *RPL* [61] was specified for this purpose in the IETF *roll* working group [62].

17.3.2 Advanced service capabilities and algorithms

One technology segment that is driving the IoT is the appearance of new service capabilities and algorithms, see Fig. 17.1. If IoT is largely about connecting smart devices and letting them exchange data, then the benefit of IoT does not come from this connectivity per se. Instead some value needs to be created out of the IoT connectivity and the information exchange. Significant enablers for IoT services are data analytics and reasoning. The possibility to provide simple and low-cost connectivity implies that information about physical assets can easily be collected. Many IoT services exploit this capability for asset condition monitoring and tracking, and derive value from that. For example, a rental bicycle fleet can be monitored, the bicycle positions and usage can be tracked. Data can be also used for *health monitoring*, where the collected data can be compared to large data sets of reference measurements to identify anomalies. This could be used in farming, where livestock is monitored, for industrial equipment, or other use cases. For example, Rolls Royce has introduced a new service in 2018 for monitoring its jet engines during operation for the purpose of *predictive maintenance* (estimating when maintenance is needed from health condition data) and optimization of operation [63,64]. The tremendous advances in data analytics, reasoning, and machine learning during the last years, make it possible to develop more meaningful IoT services in an ever-growing field of application areas. Different machine learning techniques exist, within the broader categories of supervised learning, unsupervised learning, or reinforcement learning [65–68]. The overarching principle is that machine learning algorithms build a mathematical model of sample data which can be used to make predictions or decisions without being explicitly programmed to perform the task. While the principles of machine learnings have partly been developed over the last couple of decades, the increasing availability of data together with the tremendous increase in computing capabilities has made machine learning see significant advances in recent years. Application areas are very broad, from the statistical analysis of big data sets, to decision learning for social behavior [69]. In the practical realization, different complexities and dynamics in machine learning can be distinguished. On one side, a homogenous type of data can be collected in large data sets, fed into data models to develop insights, like the detection of unusual patterns or anomalies (as used for predictive maintenance). In other use cases, large-scale systems are being analyzed, where typically some hierarchical decomposition is required. Data can be collected at different locations, data can be of heterogeneous types, and there can be multiple purposes of data analysis. Large scale systems are often distributed, and examples are manufacturing systems and factories, intelligent transport systems or the operation of large networks [66,67,70,71,107,108]. In such systems, machine reasoning and machine learning is often applied to derive decisions to steer or control parts of the system [72]. Such systems are often referred to as cyber-physical systems, as we will describe later. Some of the data processing may happen locally, where data can be consolidated; other data may be centralized; local learning may be blended into more global comprehensive data [71]. Such systems build on distributed computing architectures. Insights can be derived at different levels of the system, and can have different dynamics, from offline analysis to real-time reasoning and control. Some of the challenges and research directions of machine learning are described in Ref. [71]. In real-time decision making, predictions, model updates and inference must be performed on live-streaming data; processing in a centralized cloud is often insufficient and must instead be closer to the data generation. In distributed decentralized intelligence, inferences are made at different levels and places. Local learning may be biased due to small data sets. Via for instance federated learning the scale of data sets on local levels can be increased by aligning multiple local data models [71]. There is a wide range of use cases for new services for machine learning. The Internet of Things is an enabler in collecting and distributing data points, as well as inferred control decisions.

In the analysis, operation and optimization of processes the concept of *cyber-physical systems* has gained significant attention. For a cyber-physical system, a model-based virtual representation is created for the physical components, and it is updated in real-time with sensor measurements from the real system. The digital representation is often referred to as the *digital twin* [73,74]. The control and operation of the physical system can largely be moved into the digital representation, and the steering and control is executed via actuators embedded into the real system. The concept of digital twins and cyber-physical systems has largely been pioneered in the field of manufacturing, i.e. *cyber-physical production systems* [73,75,76]. It can be applied in different lifecycle phases of the production system, where operations are more efficiently performed in the virtual space rather than the physical world. In the production planning, a simulation of various configurations of the production system can be performed in the virtual domain for an intended production process, in order to validate the configuration and optimize resource usage and selecting the best choice to be applied in the physical world. During production, consistency between the virtual model and the physical system is monitored. Disturbances can be identified, but also the models can be refined and calibrated, and the production process can be adapted. After production, a history of the product based on virtual models can be maintained for later maintenance or fault analysis. Cyber-physical systems based on digital twins are an area of active research [], e.g. how the production can be real-time steered and optimized (one example is given in Ref. [77]), or how machine learning can be applied in fault analysis where fault diagnosis models are trained from data collected during the production process [78].

The concept of digital twin is expanding into other fields than production systems. Further examples comprise the management and control of the smart power grid or teleoperations over distance [79–82].

Finally, one technology trend that supports the development and application of IoT is cloud computing. This is basically due to two aspects: the simplicity of implementing IoT services and the scalable and easy access to computing capabilities for advanced services. For a wide range of IoT applications, a solution needs to be deployed, be integrated and – in particular – be operated in a lightweight fashion to be economically viable. This is especially the case for IoT services primarily based on sensing and monitoring of some state, like e.g. monitoring free parking spaces, tracking rental bicycle fleets, etc. If an IoT service would require from an IoT end user, to install and operate a server as computing platform for hosting the backend service, which needs to be non-stop connected to the internet and be maintained with system upgrades, security patches, etc. this would turn many use cases infeasible. This is, even more so the case as often the backend service requirements in terms of computing power and memory storage are rather modest. Cloud computing provides access to computing and storage, including the connectivity to the internet, as a scalable *as-as-service* model. This reduces the service introduction barrier for new IoT services significantly. Service introductions are further simplified, by the usage of IoT application frameworks as described in Section 17.3.1.2. For example, the LwM2M service layer on top an IoT protocol stack and possibly with an integrated data analytics toolbox, are readily available and can be provided as a package together with a cloud service offering.

The other application field of cloud computing for IoT services lies in the capabilities of advanced computing. The capabilities of data analytics are constantly growing, big data analysis with model-based machine learning enable new value services for IoT. However, some of those advances come with the requirement to have high computing and memory demand, which can be well addressed with cloud computing solutions. Advanced computing capabilities are, in particular, of interest for IoT use cases that go beyond data collection, data analysis and reasoning methodology, but those that also provide control and direct interaction with a physical system as in cyber-physical systems. In such systems, a computing infrastructure is essential for the system operation, and it is migrating from a computing paradigm, where most computing is distributed as embedded computing, to centralizing some of the computing on a cloud platform. Cloud computing platforms themselves are also evolving. The original cloud computing paradigm was on having a limited number of highly centralized cloud computing data centers, where computing and storage could be organized in the most effective and cost-efficient way. But a trend toward adding a layer of distributed cloud platforms is ongoing, which is on one hand due to the growth and scaling of cloud computing capacity to meet the growing demand, but also due to new requirements on cloud services. Such requirements are real-time interaction capabilities where the computing platform needs to be close to the physical system, but also availability and survivability aspects, where a dependence on highly-available network connectivity to the data center is prohibitive. The distribution of cloud computing all the way from some few large-scale data centers to many distributed small-scale data centers and even single devices, such as gateways, is often referred to as *edge computing* or *fog computing*. One example for distributed cloud computing is cloud robotics, where the real-time control of a robot is moved to a cloud computing platform [83,84].

17.4 The industrial Internet of Things

A part of the internet of things that receives a significant amount of attention is often labeled as the *industrial internet of things* (IIoT). It can be broadly described as applying IoT approaches in an industrial context, i.e. to support the digital transformation of industry sectors, such as manufacturing, process industries, energy distribution, etc. A landmark white paper was published by Annunziata and Evans of General Electric in November 2012 [85], which defined the *industrial internet* by combining the advances in computing, connectivity and analytics with industrial systems, i.e. "things that spin." They described the "power of 1%," that if the industrial internet could achieve a 1% efficiency gain for industrial systems in the segments of aviation, power, healthcare, rail, and oil and gas, the total global savings over 15 years could accumulate to 276 billion US dollars. A total estimated global benefit of the industrial internet for the time period 2012–30 was identified as 15 trillion US dollars. In 2014 the *industrial internet consortium* (IIC) was founded by AT&T, Cisco, General Electric, IBM and Intel, to promote the industrial internet, develop an ecosystem and describe common reference architecture and framework documents [86]. As of March 2019, IIC had more than 200 founding and contributing members; it addresses a wide range of industrial segments such as energy, healthcare, mining, manufacturing, retail, transport and smart cities. A related activity started in April 2011 when the German initiative *industry 4.0*, or *the fourth industrial revolution* was launched. In 2015 a *platform industry 4.0* was established under the lead of the German government and the industry associations of the information and telecommunications industry (Bitkom), the mechanical engineering industry (VDMA), and of electrical and electronic manufacturers (ZVEI) [87]. The scope of industry 4.0 is similar to what has been described for the industrial internet above, but with a focus on the industry segment of manufacturing. In April 2018, the *5G alliance for connected industries and automation* (5G-ACIA) has been founded as central global forum to discuss how 5G mobile communication can be applied in the industrial IoT [88]. 5G-ACIA brings together members from the operational technologies area and the information and communication technologies area.

One difference of the IIoT from the more consumer-oriented IoT is that in many industrial use cases, the IIoT is part of critical system operation. Consequences of faulty operation may at the worst be production stops, power outages, etc. As a result, ultra-high reliability and availability, resilience, as well as cyber-security are pre-requisites for successful adoption of the IIoT. Many IIoT use cases have also demanding requirements on the real-time capabilities. This is for example the case for real-time control operations or safety alarms, e.g. in manufacturing, or fault protection in a smart grid. Today many industrial systems are isolated, and this will partly continue into the future, leading effectively to an *industrial intranet of things*. But even for isolated systems, the adoption of standardized IIoT protocol stack is beneficial. It is based on open interoperable standards that can be used by many industry players and provides economy of scale without market fragmentation. It is envisioned that the IIoT does also increase the interconnection of originally isolated industrial systems over enterprise boundaries. This can be, for example, to integrate inbound and outbound logistics into a production planning process.

Both the IIC and the German Plattform Industrie 4.0 have developed and published reference architectures for the industrial internet of things [89] and the industry 4.0 respectively, the latter as DIN specification 91345 [90]. The *reference architecture model industry* 4.0 (RAMI4.0) describes system elements according to 3-dimensions, the *architecture* layers, the *lifecycle* phases and the *hierarchy* level, see Fig. 17.6.

- The architecture layers categorize the functionality of assets or combinations of assets and contains the following layers: business (business processes, etc.), functional (functionality provided by the asset), information (the information an asset provides), communication (how information can be accessed), integration (digital representation) and the asset (physical object). The asset layer represents the physical assets, and the integration layer represents the transition from the physical world into the digital world, by mapping between the physical object and a set of information that represents the physical world asset in a meaningful way.
- The lifecycle phases describe the state of an asset at a particular moment of its lifecycle. The lifecycle goes from the development of an asset, via the production of the asset, its usage and maintenance, and finally its end-of-life and decommissioning.

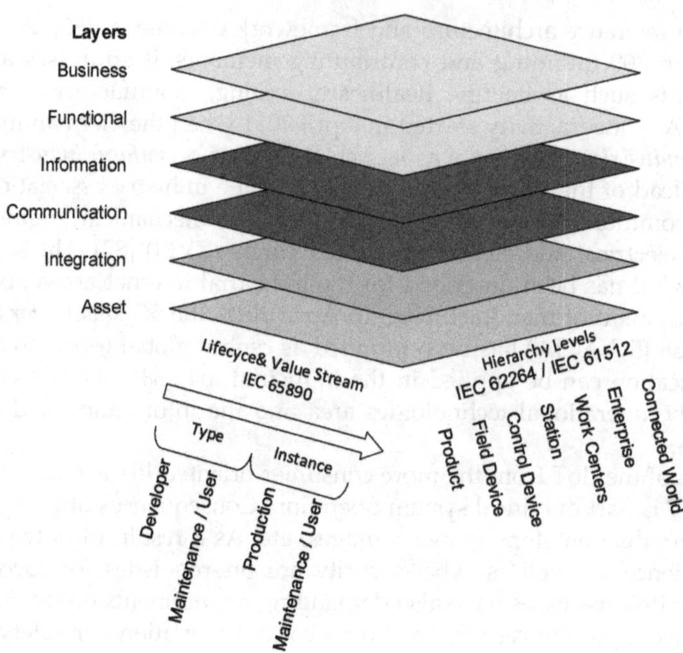

FIG. 17.6 Reference architecture model industrie 4.0 according to Ref. [90].

- The hierarchy is according to and extending existing architecture models of the functionality in a factory, historically structured in an automation hierarchy. It contains the product to be manufactured, the field devices, control devices, stations and work centers in the factory, to the enterprise level where the enterprise control system is integrated, and the connected world crossing enterprise boundaries, where exchange with other installations, like other factories (e.g. a network of factories) takes place.

One important element in RAMI4.0 is the *administration shell* [90], see Fig. 17.7. It is an interface between the physical object and the industry 4.0 (I4.0) system. The administration shell stores data and information about the asset and serves as communication interface to

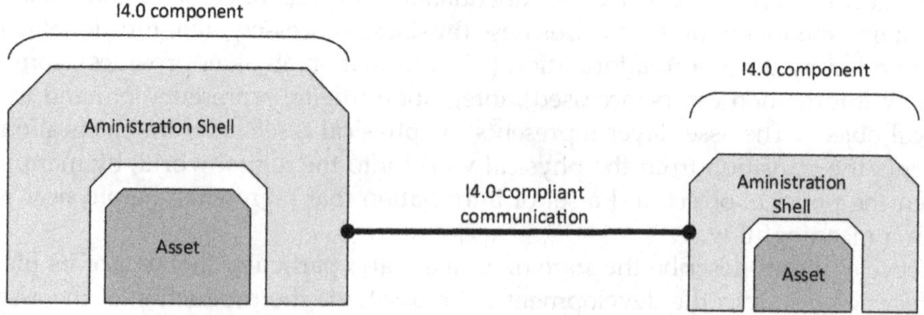

FIG. 17.7 Administration shell according to Ref. [90].

other assets. In other words, the administration shell transfers a component into a I4.0-compatible component. All components with an administration shell can communicate and interact among themselves according to the I4.0-defined methods and semantics.

The industrial internet reference architecture defined by IIC differs from RAMI4.0, as the scope of industrial internet is much broader than in RAMI4.0, which is focused on manufacturing. A comparison and alignment between the two reference architectures has been made as a joint activity between the two associations and is summarized in Ref. [91].

IIC and Plattform Industrie 4.0 describe connectivity and networking for the IIoT. IIC has defined a connectivity framework [92], and summarized industrial networking for the IIoT [93]. Plattform Industrie 4.0 has described IIoT networking in several white papers [94–96]. One difference that can be found between the Industrial IoT and the (more consumer oriented) IoT protocol stack depicted in Fig. 17.2 is that some additional communication technologies play a role. On the networking and transmission level, a number of different industrial fieldbus technologies are used. Those enable reliable communication with very low latencies in industrial real-time control. Different fieldbus technologies are not interoperable. A technology that obtains major attention is the *time-sensitive networking* (TSN) extension of the standard Ethernet technology defined in IEEE 802.1 and 802.3. TSN provides deterministic communication on top of standard Ethernet, with bounded very low latencies, avoidance of congestion related packet loss, and high reliability [97–104]. As a consequence, critical real-time industrial communication can be provided on top of a factory-wide Ethernet network, without a need to deploy specific fieldbus segments in different parts of a factory. At the same time TSN is part of the open and interoperable IEEE Ethernet standard. TSN is expected to replace in future many fieldbus technologies [92,104,105].

Another set of communication technologies that are in particular important for industrial IoT are on the transfer and application layer (see Fig. 17.2). Two such technologies to mention are the *data distribution service* [92] and *open platform communications unified architecture* (OPC UA). In particular OPC UA is considered to become one of the leading communication technologies for industrial communication, in particular in combination with TSN as deterministic real-time transport [104–106]. OPC UA is commonly used in manufacturing for data access in the higher part of the automation functions, like the *enterprise resource planning*, *manufacturing execution system* or *supervisory control and data acquisition* [104]. OPC UA allows data exchange via devices based on a client-server model, and a publish-subscribe option has been added recently. In addition, OPC UA defines a data model and additional services [104].

References

[1] M. Weiser, R. Gold, J. S. Brown. The origins of ubiquitous computing research at PARC in the late 1980s. IBM Syst. J., December 1999, Vol. 38, No. 4.

[2] S. Kurkovsky. Pervasive computing: past, present and future. In: Proceedings 5th international conference on information and communications technology, Cairo, Egypt, 2007.

[3] A. Schmidt. Ubiquitous computing: are we there yet? Computer, February 2010.

[4] M. Weiser. The computer for the twenty-first century. Scientific American, September 1991.

[5] T. A. Henzinger, J. Sifakis. The discipline of embedded systems design. Computer. October 2007. https://doi.org/10.1109/MC.2007.364.

[6] R. Gao, L. Zhang. Micromachined microsensors for manufacturing. IEEE Instrumentation & Measurement Magazine. June 2004. https://doi.org/10.1109/MIM.2004.1304562.

[7] A. A. Barlian, W. Park, J. R. Mallon, A. J. Rastegar, B. L. Pruitt. Semiconductor Piezoresistance for Microsystems. Proceedings of the IEEE 2006, Vol. 97, No. 3, 513–52, March-April 2009. http://ieeexplore.ieee.org/stamp/stamp.jsp?tp=&arnumber=4811093&isnumber=4808261.
[8] D. K. Shaeffer. MEMS inertial sensors: a tutorial overview. IEEE Commun. Mag, April 2013. https://doi.org/10.1109/MCOM.2013.6495768.
[9] J. Bryzek et al. Marvelous MEMS, IEEE circuits and devices magazine, March–April 2006. https://doi.org/10.1109/MCD.2006.1615241.
[10] L. Li. Applications of MEMS actuators in micro/nano robotic manipulators. In: Proceedings 2nd international conference on computer engineering and technology, Chengdu, China, 2010. https://doi.org/10.1109/ICCET.2010.5485670.
[11] M. Ku, W. Li, Y. Chen, K. J. Ray Liu. Advances in energy harvesting communications: past, present, and future challenges. IEEE Communications Surveys & Tutorials, Second quarter 2016. https://doi.org/10.1109/COMST.2015.2497324.
[12] I. F. Akyildiz, W. Su, Y. Sankarasubramaniam, E. Cayirci. A survey on sensor networks. IEEE Commun. Mag., August 2002, Vol. 40, No. 8.
[13] A. Willig. Recent and emerging topics in wireless industrial communications: a selection. IEEE Transactions on Industrial Informatics, May 2008, Vol. 4, No. 2, 102–24.
[14] B. Galloway, G. P. Hancke. Introduction to industrial control networks. IEEE Communications Surveys & Tutorials, Second Quarter 2013. https://doi.org/10.1109/SURV.2012.071812.00124.
[15] P. Gaj, J. Jasperneite, M. Felser. Computer communication within industrial distributed environment—a survey. IEEE Transactions on Industrial Informatics, February 2013. https://doi.org/10.1109/TII.2012.2209668.
[16] P. Danielis et al. Survey on real-time communication via ethernet in industrial automation environments. In: 2014 IEEE emerging technology and factory automation (ETFA), Barcelona, 2014. https://doi.org/10.1109/ETFA.2014.7005074.
[17] T. Sauter, M. Lobashov. How to access factory floor information using internet technologies and gateways. IEEE Transactions on Industrial Informatics, November 2011. https://doi.org/10.1109/TII.2011.2166788.
[18] M. Wollschlaeger, T. Sauter, J. Jasperneite. The future of industrial communication: automation networks in the era of the internet of things and industry 4.0. IEEE Industrial Electronics Magazine, March 2017. https://doi.org/10.1109/MIE.2017.2649104.
[19] J. L. Messenger. Time-sensitive networking: an introduction. IEEE Communications Standards Magazine, June 2018. https://doi.org/10.1109/MCOMSTD.2018.1700047.
[20] D. Bruckner, et al. "An introduction to OPC UA TSN for industrial communication systems," in Proceedings of the IEEE. doi: 10.1109/JPROC.2018.2888703
[21] Internet Engineering Task Force (IETF). Request for comments 3986, uniform resource identifier (URI): generic syntax, January 2005. https://datatracker.ietf.org/doc/rfc3986/.
[22] V. Tsiatsis, S. Karnouskos, J. Höller, D. Boyle, C. Mulligan. Internet of things – technologies and applications for a new age of intelligence. 2nd ed. Academic Press, 2019, ISBN 978-0-12-814435-0.
[23] C. Lundqvist, A. Keränen, B. Smeets, J. Fornehed, C. R. B. Azevedo, P. Von Wrycza. Key technology choices for optimal massive IoT devices. Ericsson Technology Review, January 2019. https://www.ericsson.com/en/ericsson-technology-review/archive/2019/key-technology-choices-for-optimal-massive-iot-devices.
[24] Open Mobile Alliance SpecWorks (OMA SpecWorks). Report, internet of things protocol comparison, October 2018. Available at: https://www.omaspecworks.org/develop-with-oma-specworks/ipso-smart-objects/ip_for_smart_object_publications/.
[25] Internet Engineering Task Force (IETF). Internet-Draft, QUIC: a UDP-based multiplexed and secure transport, draft-ietf-quic-transport-19, March 11, 2019. https://datatracker.ietf.org/doc/draft-ietf-quic-transport/.
[26] I. Ishaq, D. Carels, G. K. Teklemariam, J. Hoebeke, F. Van den Abeele, E. De Poorter, I. Moerman, P. Demeester. IETF standardization in the field of the internet of things (IoT): a survey. J. Sens. Actuator Netw, April 2013, Vol. 2, No. 2, 235–87.
[27] Internet Engineering Task Force (IETF). Request for comments 2616, hypertext transfer protocol - HTTP/1.1, June 1999. https://datatracker.ietf.org/doc/rfc2616/.
[28] Internet Engineering Task Force (IETF). Request for comments 7540, hypertext transfer protocol version 2 (HTTP/2), May 2015. https://datatracker.ietf.org/doc/rfc7540/.
[29] Internet Engineering Task Force (IETF). Request for comments 7252, the constrained application protocol (CoAP), June 2014. https://datatracker.ietf.org/doc/rfc7252/.

[30] C. Bormann, A. P. Castellani, Z. Shelby CoAP: an application protocol for billions of tiny internet nodes. IEEE Internet Computing, March-April. https://doi.org/10.1109/MIC.2012.29.

[31] R. T. Fielding. Architectural styles and the design of network-based software architectures. Doctoral dissertation. Irvine: University of California, 2000. https://www.ics.uci.edu/~fielding/pubs/dissertation/fielding_dissertation.pdf.

[32] Wikipedia, Representational state transfer. https://en.wikipedia.org/w/index.php?title=Representational_state_transfer&oldid=884405762, February 25, 2019.

[33] OASIS standard, message queuing Telemetry transport (MQTT) version 3.1.1 plus errata 01. December 2015. http://docs.oasis-open.org/mqtt/mqtt/v3.1.1/mqtt-v3.1.1.html.

[34] A. Al-Fuqaha, M. Guizani, M. Mohammadi, M. Aledhari, M. Ayyash Internet of things: a survey on enabling technologies, protocols, and applications. IEEE Communications Surveys & Tutorials, Fourthquarter 2015. https://doi.org/10.1109/COMST.2015.2444095.

[35] Internet Engineering Task Force (IETF). Request for comments 8323, CoAP (constrained application protocol) over TCP, TLS, and WebSockets. February 2018. https://datatracker.ietf.org/doc/rfc8323/.

[36] Internet Engineering Task Force (IETF). Request for comments 8132, PATCH and FETCH methods for the constrained application protocol (CoAP), April 2017. https://datatracker.ietf.org/doc/rfc8132/.

[37] Internet Engineering Task Force (IETF). Request for comments 7641, observing resources in the constrained application protocol (CoAP), September 2015. https://datatracker.ietf.org/doc/rfc7641/.

[38] Internet Engineering Task Force (IETF). Internet-Draft, publish-subscribe broker for the constrained application protocol (CoAP), draft-ietf-core-coap-pubsub-08. March 11, 2019. https://tools.ietf.org/html/draft-ietf-core-coap-pubsub-08.

[39] Internet Engineering Task Force (IETF). Request for comments 8075, guidelines for mapping implementations: HTTP to the constrained application protocol (CoAP), February 2017. https://datatracker.ietf.org/doc/rfc8075/.

[40] Internet Engineering Task Force (IETF). Request for comments 8428, sensor measurement lists (SenML). August 2018. https://datatracker.ietf.org/doc/rfc8428/.

[41] Internet Engineering Task Force (IETF), Working Group. Authentication and authorization for constrained environments (ace). March 2019. https://datatracker.ietf.org/wg/ace/about/.

[42] Internet Engineering Task Force (IETF). Internet-Draft, authentication and authorization for constrained environments (ACE) using the OAuth 2.0 framework (ACE-OAuth), draft-ietf-ace-oauth-authz-24. March 27, 2019. https://tools.ietf.org/html/draft-ietf-ace-oauth-authz-24.

[43] K. Mononen, P. Teppo, T. Suihko. End-to-end security management for the IoT, Ericsson Technology Review. November 2017. https://www.ericsson.com/en/ericsson-technology-review/archive/2017/end-to-end-security-management-for-the-iot.

[44] Ericsson. White paper, IoT security – protecting the networked society. June 2017. https://www.ericsson.com/en/white-papers/iot-security-protecting-the-networked-society.

[45] OMA SpecWorks. Technical specification, lightweight machine to machine technical specification: core, approved version: 1.1. July 10, 2018. Available at: http://openmobilealliance.org/wp/index.html.

[46] OMA SpecWorks. Technical specification, lightweight machine to machine technical specification: transport bindings, approved version: 1.1. July 10, 2018. Available at: http://openmobilealliance.org/wp/index.html.

[47] J. Prado. OMA lightweight M2M resource model. In: Position paper, internet architecture board – IoT semantic interoperability workshop, San Jose, USA, March 17–18, 2016. Available at: https://www.iab.org/activities/workshops/iotsi/.

[48] J. Jimenez, M. Kostery, H. Tschofenig. IPSO smart objects. In: Position paper, internet architecture board – IoT semantic interoperability workshop, San Jose, USA, March 17–18, 2016. Available at: https://www.iab.org/activities/workshops/iotsi/.

[49] Internet Engineering Task Force (IETF). Request for comments 7228, terminology for constrained-node networks. May 2014. https://datatracker.ietf.org/doc/rfc7228/.

[50] Internet Engineering Task Force (IETF). Internet draft, terminology for constrained-node networks, version 03. July 2018. https://tools.ietf.org/html/draft-bormann-lwig-7228bis-03.

[51] Internet Engineering Task Force (IETF). Internet-Draft, SCHC over Sigfox LPWAN, version 5, draft-zuniga-lpwan-schc-over-sigfox-05. November 05, 2018. https://datatracker.ietf.org/doc/draft-zuniga-lpwan-schc-over-sigfox/.

[52] Internet Engineering Task Force (IETF). Working Group, IPv6 over Low power WPAN (6lowpan), March 1, 2019. https://datatracker.ietf.org/wg/6lowpan/about/. https://datatracker.ietf.org/wg/lpwan/about/.
[53] Internet Engineering Task Force (IETF). Working Group, IPv6 over networks of resource-constrained nodes (6lo), March 1, 2019. https://datatracker.ietf.org/wg/6lo/about/.
[54] Internet Engineering Task Force (IETF). Working Group, IPv6 over low power wide-area networks (lpwan), March 1, 2019.
[55] Internet Engineering Task Force (IETF). Request for comments 8376, low-power wide area network (LPWAN) overview. May 2018. https://datatracker.ietf.org/doc/rfc8376/.
[56] M. R. Palattella, N. Accettura, X. Vilajosana, T. Watteyne, L. A. Grieco, G. Boggia, M. Dohler, Standardized protocol stack for the internet of (important) things, IEEE Communications Surveys & Tutorials (Third Quarter 2013). https://doi.org/10.1109/SURV.2012.111412.00158.
[57] Internet Engineering Task Force (IETF). Request for comments 4944, transmission of IPv6 packets over IEEE 802.15.4 networks. September 2007. https://datatracker.ietf.org/doc/rfc4944/.
[58] Internet Engineering Task Force (IETF). Request for comments 6282, compression format for IPv6 datagrams over IEEE 802.15.4-based networks. September 2011. https://datatracker.ietf.org/doc/rfc6282/.
[59] Internet Engineering Task Force (IETF). Internet-Draft, LPWAN static context header compression (SCHC) and fragmentation for IPv6 and UDP, version 18, draft-ietf-lpwan-ipv6-static-context-hc-18. December 14, 2018. https://datatracker.ietf.org/doc/draft-ietf-lpwan-ipv6-static-context-hc/.
[60] Internet Engineering Task Force (IETF). Internet-Draft, LPWAN static context header compression (SCHC) for CoAP, version 6, draft-ietf-lpwan-coap-static-context-hc-06. February 05, 2019. https://datatracker.ietf.org/doc/draft-ietf-lpwan-coap-static-context-hc/.
[61] Internet Engineering Task Force (IETF). Request for comments 6550, RPL: IPv6 routing protocol for low-power and lossy networks. March 2012. https://datatracker.ietf.org/doc/rfc6550/.
[62] Internet Engineering Task Force (IETF). Working Group, Routing over Low power and Lossy networks (roll), March 1, 2019. https://datatracker.ietf.org/wg/roll/about/.
[63] https://www.flightglobal.com/news/articles/insight-from-rolls-royce-pioneering-the-intelligent-450103/, March 2019.
[64] https://www.rolls-royce.com/media/press-releases/2018/05-02-2018-rr-launches-intelligentengine.aspx, March 2019.
[65] P. Louridas, C. Ebert. Machine learning. IEEE software, September—October 2016, Vol. 33, No. 5.
[66] Z. M. Fadlullah, et al. State-of-the-Art deep learning: evolving machine intelligence toward tomorrow's intelligent network traffic control systems. IEEE communications surveys & tutorials, Fourthquarter 2017, Vol. 19, No. 4, p. 2432—55. https://doi.org/10.1109/COMST.2017.2707140.
[67] D. Rafique, L. Velasco. Machine learning for network automation: overview, architecture, and applications [Invited Tutorial]. IEEE/OSA journal of optical communications and networking, October 2018, Vol. 10, No. 10, p. D126—43. https://doi.org/10.1364/JOCN.10.00D126.
[68] I. Arel, D. C. Rose, T. P. Karnowski. Deep machine learning - a new frontier in artificial intelligence research [research frontier]. IEEE computational intelligence magazine, November 2010, Vol. 5, No. 4, p. 13—8. https://doi.org/10.1109/MCI.2010.938364.
[69] Y. Chen, C. Jiang, C. Wang, Y. Gao, K. J. R. Liu. Decision Learning: data analytic learning with strategic decision making. IEEE signal processing magazine, January 2016, Vol. 33, No. 1, p. 37—56. https://doi.org/10.1109/MSP.2015.2479895.
[70] A. V. Feljan, A. Karapantelakis, L. Mokrushin, R. Inam, E. Fersman, C. R. B. Azevedo, K. Raizer, R. S. Souza. Tackling IoT complexity with machine intelligence. Ericsson Technology Review April 2017. Available at: https://www.ericsson.com/en/ericsson-technology-review/archive/2017/tackling-iot-complexity-with-machine-intelligence.
[71] Ericsson white paper. Artificial intelligence and machine learning in next-generation systems. available at: https://www.ericsson.com/en/white-papers/machine-intelligence, May 2018.
[72] X. Xu, Q. Hua. Industrial big data analysis in smart factory: current status and research strategies. IEEE Access, 2017, Vol. 5, p. 17543—51. https://doi.org/10.1109/ACCESS.2017.2741105.

[73] F. Tao, M. Zhang. Digital twin shop-floor: a new shop-floor paradigm towards smart manufacturing. IEEE access, 2017, Vol. 5, p. 20418–27. https://doi.org/10.1109/ACCESS.2017.2756069.

[74] K. M. Alam, A. El Saddik. C2PS: a digital twin architecture reference model for the cloud-based cyber-physical systems. IEEE access, 2017, Vol. 5, p. 2050–62. https://doi.org/10.1109/ACCESS.2017.2657006.

[75] M. Grieves. Digital twin: manufacturing excellence through virtual factory replication. White Paper, 2014 [Online]. Available: https://research.fit.edu/media/site-specific/researchfitedu/camid/documents/1411.0_Digital_Twin_White_Paper_Dr_Grieves.pdf.

[76] Q. Qi, F. Tao. Digital twin and big data towards smart manufacturing and industry 4.0: 360 degree comparison. IEEE access, 2018, Vol. 6, p. 3585–93. https://doi.org/10.1109/ACCESS.2018.2793265.

[77] R. Zhao et al. Digital twin-driven cyber-physical system for autonomously controlling of micro punching system. IEEE access, 2019, Vol. 7, p. 9459–69. https://doi.org/10.1109/ACCESS.2019.2891060.

[78] Y. Xu, Y. Sun, X. Liu, Y. Zheng. A digital-twin-assisted fault diagnosis using deep transfer learning. IEEE access, 2019, Vol. 7, p. 19990–9. https://doi.org/10.1109/ACCESS.2018.2890566.

[79] J. Sachs et al. Adaptive 5G low-latency communication for tactile internet services. Proceedings of the IEEE, Februay 2019, Vol. 107, No. 2, p. 325–49. https://doi.org/10.1109/JPROC.2018.2864587.

[80] O. Holland, et al. The IEEE 1918.1 "tactile internet" standards working group and its standards. Proceedings of the IEEE, February 2019, Vol. 107, No. 2, p. 256–79. https://doi.org/10.1109/JPROC.2018.2885541.

[81] H. Laaki, Y. Miche, K. Tammi. Prototyping a digital twin for real time remote control over mobile networks: application of remote surgery. IEEE access, 2019, Vol. 7, p. 20325–36. https://doi.org/10.1109/ACCESS.2019.2897018.

[82] A. El Saddik. Digital twins: the convergence of multimedia technologies. IEEE MultiMedia, April–June 2018, Vol. 25, No. 2, p. 87–92. https://doi.org/10.1109/MMUL.2018.023121167.

[83] M. Puleri, R. Sabella, A. Osseiran. Cloud robotics: 5G paves the way for mass-market automation. Ericsson Technology Review, June 2016. https://www.ericsson.com/en/ericsson-technology-review/archive/2016/cloud-robotics-5g-paves-the-way-for-mass-market-automation.

[84] R. Sabella, A. Thuelig, M. Chiara Carrozza, M. Ippolito. Industrial automation enabled by robotics, machine intelligence and 5G, Ericsson Technology Review. February 2018. https://www.ericsson.com/en/ericsson-technology-review/archive/2018/industrial-automation-enabled-by-robotics-machine-intelligence-and-5g.

[85] M. Annunziata, P. C. Evans. General electric, industrial internet: pushing the boundaries of minds and machines. November 26, 2012. https://www.ge.com/docs/chapters/Industrial_Internet.pdf.

[86] Industrial internet consortium. https://www.iiconsortium.org/, March 2019.

[87] Plattform industrie 4.0. https://www.plattform-i40.de/, March 2019.

[88] 5G alliance for connected and industries and automation (5G-ACIA). https://www.5g-acia.org/, March 2019.

[89] Industrial Internet Consortium. The industrial internet of things, reference architecture. January 2017. Available at: https://www.iiconsortium.org/IIRA.htm.

[90] DIN specification 91345, reference architecture model industrie 4.0 (RAMI4.0), English translation. April 2016. Available at: https://www.din.de/en/about-standards/din-spec-en/current-din-specs/wdc-beuth:din21:250940128.

[91] Industrial internet consortium and industrie Plattform 4.0, white paper, architecture alignment and interoperability - an industrial internet consortium and Plattform industrie 4.0 joint whitepaper. December 2017. Available at: https://www.iiconsortium.org/iic-i40-joint-work.htm.

[92] Industrial Internet Consortium. The industrial internet of things, connectivity framework. February 2018. Available at: https://www.iiconsortium.org/IICF.htm.

[93] Industrial Internet Consortium. White paper, industrial networking enabling IIoT communication. August 2018. Available at: https://www.iiconsortium.org/white-papers.htm.

[94] Plattform industrie 4.0, network-based communication for industrie 4.0. April 2016.

[95] Plattform industrie 4.0, network-based communication for industrie 4.0 – proposal for an administration shell. November 2016.

[96] Plattform industrie 4.0, secure communication for industrie 4.0. November 2016.

[97] N. Finn. Introduction to time-sensitive networking. IEEE communications standards magazine, June 2018, Vol. 2, No. 2, p. 22–8. https://doi.org/10.1109/MCOMSTD.2018.1700076.

[98] J. L. Messenger. Time-sensitive networking: an introduction. IEEE communications standards magazine, June 2018, Vol. 2, No. 2, p. 29–33. https://doi.org/10.1109/MCOMSTD.2018.1700047.

[99] C. Simon, M. Maliosz, M. Mate. Design aspects of low-latency services with time-sensitive networking. IEEE communications standards magazine, June 2018, Vol. 2, No. 2, p. 48–54. https://doi.org/10.1109/MCOMSTD.2018.1700081.

[100] W. Steiner, S. S. Craciunas, R. S. Oliver. Traffic planning for time-sensitive communication. IEEE communications standards magazine, June 2018, Vol. 2, No. 2, p. 42–7. https://doi.org/10.1109/MCOMSTD.2018.1700055.

[101] M. Wollschlaeger, T. Sauter, J. Jasperneite. The future of industrial communication: automation networks in the era of the internet of things and industry 4.0. IEEE industrial electronics magazine, March 2017, Vol. 11, No. 1, p. 17–27. https://doi.org/10.1109/MIE.2017.2649104.

[102] W. Steiner et al. Next generation real-time networks based on IT technologies. In: 2016 IEEE 21st international conference on emerging technologies and factory automation (ETFA), Berlin, 2016, p. 1–8. https://doi.org/10.1109/ETFA.2016.7733580.

[103] S. Kehrer, O. Kleineberg, D. Heffernan. A comparison of fault-tolerance concepts for IEEE 802.1 Time Sensitive Networks (TSN). In: Proceedings of the 2014 IEEE emerging technology and factory automation (ETFA), Barcelona, 2014. p. 1–8. https://doi.org/10.1109/ETFA.2014.7005200.

[104] D. Bruckner et al. An introduction to OPC UA TSN for industrial communication systems. Proceedings of the IEEE, 2019. https://doi.org/10.1109/JPROC.2018.2888703.

[105] P. Drahoš, E. Kučera, O. Haffner, I. Klimo. Trends in industrial communication and OPC UA. In: 2018 cybernetics & informatics (K&I), lazy pod makytou, 2018. p. 1–5. https://doi.org/10.1109/CYBERI.2018.8337560.

[106] A. Eckhardt, S. Müller, L. Leurs. An evaluation of the applicability of OPC UA publish subscribe on factory automation use cases. In: IEEE 23rd international conference on emerging technologies and factory automation (ETFA), Turin, 2018. p. 1071–4. https://doi.org/10.1109/ETFA.2018.8502445.

[107] H. Xu, W. Yu, D. Griffith, N. Golmie. A survey on industrial internet of things: a cyber-physical systems perspective. IEEE access, 2018, p. 78238–59. https://doi.org/10.1109/ACCESS.2018.2884906.

[108] V. K. L. Huang, Z. Pang, C. A. Chen, K. F. Tsang. New trends in the practical deployment of industrial wireless: from noncritical to critical use cases. IEEE industrial electronics magazine, June 2018, Vol. 12, No. 2, p. 50–8. https://doi.org/10.1109/MIE.2018.2825480.

CHAPTER

18

5G and beyond

OUTLINE

Abstract

This chapter provides an outlook into mobile networks beyond 5G. Standardization activities and time plans for the evolution beyond 5G are described. The evolution of cellular IoT is discussed including the areas of massive IoT, critical IoT, broadband IoT and industrial IoT.

The mobile communication networks that have been specified by 3GPP until today, in April 2019, lay a solid foundation for cellular IoT, in addition to enhanced mobile broadband services required by a connected society. Long-Term Evolution (LTE) is, and will remain the dominating radio access technology globally in mobile networks for the many years to come, but a fast introduction of 5G networks based on the 5G New Radio access technology is foreseen reaching 1.5 billion subscriptions by 2024 [1]. These networks address a broad range of cellular Internet of Things (IoT) use cases and requirements, supporting massive IoT applications as well as critical IoT services. LTE-M and Narrowband IoT (NB-IoT) are the technology solutions for massive IoT in cellular networks, as described in Chapters 2, 5–8 and 16. LTE-M and NB-IoT can be well combined and integrated into both LTE and NR networks as shown in Chapters 5 and 7. For critical IoT both LTE and NR provide novel capabilities for *ultra-reliable and low latency communication* (URLLC) as described in Chapters 9–12 and 16. There is a high excitement about the possibilities that critical IoT will bring for new services that may affect many new market segments, like connected energy networks, industrial systems, etc. However, as the related standards have only recently been finished, no commercial deployments for critical IoT exist today; so far real deployments are limited to promising trial systems for technology validation. As LTE URLLC and NR URLLC become ready for market

Cellular Internet of Things, Second Edition
https://doi.org/10.1016/B978-0-08-102902-2.00018-2

© 2020 Elsevier Ltd. All rights reserved.

introduction at the same time, we foresee that NR will become the main technology for critical cellular IoT due to its higher flexibility and even better performance.

Based on this foundation the mobile network evolution will continue. In 3GPP the standardization evolves continuously in phases defined in standard releases. The specification phase for a 3GPP release lasts between 12 and 18 months and brings new features to the standard. Like this, the LTE mobile networks have evolved significantly from the original specification in 3GPP release 8 in 2008 up to the release 15 that was finalized in in mid-2018. In release 15, also the first 5G new radio systems were specified. Currently, the 3GPP standardization is working on its release 16, which will be ready by the end of 2019; and the definition of the next release 17 is in early preparation and is expected to last until the first half of 2021. After that, a continuous evolution in further releases continues.

For massive IoT the following evolution can be anticipated. 3GPP has agreed that an Machine-Type Communications transmission mode in NR that addresses low power wide-area (LPWA) massive IoT use cases is not planned in the first NR releases [2]. This IoT segment is already very well addressed by LTE-M and NB-IoT, and tight integration of LTE-M and NB-IoT into an NR carrier has already been ensured in standardization. Thus, the combination of NR with LTE-M and NB-IoT addresses the 5G needs for LPWA massive IoT. A unique NR-based massive IoT LPWA communication mode could only be motivated if significant improvements over LTE-M/NB-IoT could be achieved and a large market demand could be anticipated. Technical motivations could be, for example, to address new NR frequency bands or define an massive Machine-Type Communications mode which can exploit the full TDD flexibility of NR. We do not foresee that such an approach will become reality for several years to come. One segment, in which an evolution of cellular IoT capabilities is desirable, is the support for demanding sensors. Those can be sensor devices, that do not transmit only infrequently small amounts of data, but higher data volumes. Examples can be industrial sensors for advanced monitoring like machine vision or acoustic sensing. Other examples are sensors which transmit infrequently but require reliable low latency, like alarms. This category can be described by having higher requirements e.g. data rate, latency, reliability than the massive IoT category defined today, while at the same time still striving for cheap devices of low complexity and long battery operation. This category can be placed in the requirements triangle of Figure 1.4 in-between massive MTC, critical Machine-Type Communications and enhanced mobile broadband, and is described as *broadband IoT* in Ref. [3]. Such a category could leverage the design principles of NR, like the flexible numerology, the flexible TDD, the beam management, and address all NR frequency bands [4,5].

For critical cellular IoT services the foundation has been laid in release 15 for LTE and NR with the support of *ultra-reliable and low latency communication* (URLLC), which enables to transmit small messages over the radio access network with a latency not exceeding 1 ms and a reliability of $1\text{-}10^{-5}$. In release 16 further improvements for NR are developed, that increase the reliability level to $1\text{-}10^{-6}$ for even lower latency bounds of 0.5 ms [6], and provide better support for the multiplexing of URLLC traffic flows with different service requirements. Other standard efforts have the objective to make a NR critical IoT solution that better supports communication services of different vertical use cases. Many vertical use cases will introduce 5G as wireless communication solution into an existing communication system. For industrial communication as an example, there are already communication solutions based on fieldbus technologies and Ethernet. 5G would complement such solutions and need to integrate into the existing systems. NR release 16 specifies how the NR system can integrate into such an industrial local area network [7,10–12]. Furthermore, the 5G system shall be able to provide *non-public networks*, which are networks that are intended for non-public use and reserved to a defined set of devices. For example, if the 5G system is used for connecting devices for the monitoring and automation of the energy grid, the 5G system should provide a non-public network access to those devices. A similar situation exists for the devices monitoring and automation of industrial systems, like smart mines, smart harbors or smart factories; also those should be able to make use of a 5G non-public network. A non-public network can be setup in different means. It can either be a *standalone* non-public network, or a non-public network which is integrated into the public 5G network infrastructure. The latter can be realized by a logical separation of public and non-public network services, for example, by means of network slicing or network sharing, where it needs to be ensured that the usage of transmission resources of the non-public network services can be protected from access attempts of non-authorized devices. In order to increase the reliability and availability of the 5G system, also methods for redundant data transmissions are being specified in release 16 [8,9].

One part of the customization of 5G for vertical use cases is to interwork with relevant communication technologies in such areas. The most prominent wired communication technology for critical IoT communication is *time-sensitive networking* as extension of Ethernet. By bringing 5G communication into industrial communication systems, an interworking with time-sensitive networking is a pre-requisite. Work is ongoing in 3GPP to facilitate such interworking [7,9], which comprises time-sensitive communication schemes where the transmission latencies are loss-free, bounded and deterministic. Further, it must be possible to provide time-synchronization to one or more reference clocks, to devices over the communication system.

Extending 3GPP solutions for operation in unlicensed bands has always been a topic of interest. In release 16, 3GPP will introduce NR-based radio access to unlicensed spectrum. This feature is generally referred to as *NR-U*. Although the primary focus of NR-U in release 16 is enhanced mobile broadband use cases, we foresee that NR-U will be further enhanced beyond release 16 to also support IoT use cases.

There are several areas for longer-term evolution of 5G systems [14,15]. From a radio perspective, more spectrum may be addressed in a 5G evolution. This can go beyond the 52.6 GHz defined for 5G in release 16, all the way to 100 GHz and in the long run possibly even to Tera-Hertz spectrum. For the latter a re-evaluation of the fundamental radio access design may be required. Another area is to widen the topology options of a beyond 5G system. *Integrated Access Backhaul* is already investigated in standardization, where the same spectrum band is used for the backhaul connection of a base station and the access connection of the devices. This could be extended to multi-hop and mesh topologies. One motivation may be to provide extreme reliability by managing the inherent redundancy of mesh topologies, but also to provide efficient system coverage in very high spectrum bands with limited reach. Part of the flexible topology and increased reliability and efficiency can motivate the introduction of device-to-device communication with cooperative relaying. By cooperation, multiple devices can for example form a virtual antenna array. This could significantly extend the functionality of device-to-device communication schemes as defined for LTE. A side-effect of such schemes is that the sharp border in mobile networks between network and devices blurs, since devices can become network nodes toward other devices.

In the past few years there has been a growing interest in extending 3GPP's reach from the earth to the sky. LTE has been optimized to support drone communication since Release 15, and NR is expected to become equipped with similar capabilities. It the area of *non-terrestrial networks* 3GPP is currently studying the ability of NR to provide a unified non-terrestrial networks communications solution for satellite constellations in *low earth orbits, medium earth orbits* and *geosynchronous orbits*. This extension of NR has the potential of supporting use cases requiring coverage in areas where it is challenging to provide terrestrial network coverage [16].

Finally, the introduction of machine learning into communication systems is an area for future research [13]. In a first step, machine learning will play a role in optimization and configuration of the network and devices. The mobile network, as well as the devices comprise hundreds or even thousands of configurable parameters. Radio resource management algorithms are used for example for efficient allocation of radio resources, performing handovers and assigning devices to different frequency bands. The usage of machine-learning has potential to improve the network configuration and optimize radio resource management algorithms, in particular when multiple algorithms may interact. As a second step, 5G networks may be evolved to facilitate a broader application in machine learning. Machine learning essentially depends on the availability of data for training and analysis. A beyond 5G system could provide enhanced reporting mechanisms for supporting machine-learning based algorithms.

References

[1] Ericsson. Ericsson mobility report. November 2018. Available at: https://www.ericsson.com/en/mobility-report/reports/november-2018.

[2] Ericsson. Interim conclusions on IoT for rel-16, RP-180581, 3GPP TSG RAN meeting #79, Chennai, India, March 19-22, 2018.

[3] Ericsson. Cellular IoT evolution for industry digitalization. 2018. website, [Online]. Available at: https://www.ericsson.com/en/white-papers/cellular-iot-evolution-for-industry-digitalization.

[4] Ericsson. New SID on NR MTC for industrial sensors, RP-190432, 3GPP TSG RAN meeting #83, Shenzhen, China, March 18–21, 2019.

[5] Ericsson. Motivation for new SID on NR MTC for industrial sensors, RP-190433, 3GPP TSG RAN meeting #83, Shenzhen, China, March 18–21, 2019.

[6] Huawei HiSilicon Nokia and Nokia Shanghai Bell. New SID on physical layer enhancements for NR ultra-reliable and low latency communication (URLLC), RP-182089, 3GPP TSG RAN meeting #81, Gold Coast, Australia, September 10–13, 2018.

[7] Third Generation Partnership Project. New WID: 5GS enhanced support of vertical and LAN services, SP-181120, 3GPP TSG SA meeting #82, Sorrento, Italy, December 12–14, 2018.

[8] Third Generation Partnership Project. New WID on enhancement of ultra-reliable low-latency communication support in the 5G core network, SP-181122, 3GPP TSG SA meeting #82, Sorrento, Italy, December 12–14, 2018.

[9] Nokia and Nokia Shanghai Bell. New WID: support of NR industrial internet of things (IoT), RP-190728, 3GPP TSG RAN meeting #83, Shenzhen, China, March 18-21, 2019.

[10] 5G Alliance for Connected Industries and Automation (5G-ACIA). 5G non-public networks for industrial scenarios, white paper. March 2019. Available at: https://www.5g-acia.org/index.php?id=6958.

[11] 5G Alliance for Connected Industries and Automation (5G-ACIA). 5G for Automation in Industry - primary use cases, functions and service requirements, white paper. March 2019. Available at: https://www.5g-acia.org/index.php?id=6960.

[12] 5G Alliance for Connected Industries and Automation (5G-ACIA). 5G for connected industries and automation - second edition, white paper. February 2019. Available at: https://www.5g-acia.org/index.php?id=5125.

[13] Dahlman, E. Parkvall, S. Peisa, J. Tullberg, H. Murai, H. and Fujioka, M. Artificial intelligence in future evolution of mobile communication. In: 2019 international conference on artificial intelligence in information and communication (ICAIIC), Okinawa, Japan, 2019. Available at: http://doi.org/10.1109/ICAIIC.2019.8669012.

[14] Dahlman, E. Parkvall, S. Peisa, J. Tullberg, H. Murai, H. and Fujioka, M. Future evolution of mobile communication. In: IEICE general conference, Tokyo, Japan, March 19–22, 2019.

[15] Dahlman, E. Parkvall, S. Peisa, J. Torsner, J. and Tullberg, H. Wireless access evolution. In: 6G wireless Summit, Levi, Lapland, Finland, March 24–26, 2019.

[16] Third Generation Partnership Project. New Study item: Study on solutions evaluation for NR to support non-terrestrial network, RP-181370, 3GPP TSG RAN meeting #80, La Jolla, USA, 2018.

Index

Note: 'Page numbers followed by "f" indicate figures and "t" indicate tables.'

D

E

F

G

K

L

M

N

O

S

U

rinted in the United States
Bookmasters